Recent Advances in Plant Stress Physiology

The Editors

Dr. Praduman Yadav is working as Scientist (Biochemistry), at Indian Institute of Oilseeds Research, Hyderabad, India since five years. He did his M.Sc. and Ph.D. from Chaudhary Charan Singh Haryana Agricultural University, Hisar. He has vast experience in the area of purification and characterization of enzymes and plant secondary metabolites. His research areas include oxidative stress and antioxidative defense system during fruit ripening and high temperature stress. Now a days he is working on isolation and characterization of antioxidants and phytochemicals in oilseeds. He has published several research papers in national and international peer- reviewed journals.

Dr. Sunil Kumar is working as Assistant Professor and Head (Botany), at Chhaju Ram Memorial Jat PG College, Hisar - affliated to Kurukshetra University, Kurukshetra, India. He did his M.Sc., M.Phil. and Ph.D. from Kurukshetra University. He has more than eight years of teaching and research experience in field of Plant Biotechnology and Molecular Biology. He has been engaged in the micropropagation of several important crops like *Stevia, Jojoba, Bamboo, Sugarcane, Malkangni* etc. He has conferred Young Researcher Award by Society for Plant Research. He has published several national and international research and review articles in peer-reviewed journals. He is also life member of several national societies.

Dr. Veena Jain, have contributed to the fast growing area of biochemical and molecular mechanism(s) of abiotic stress tolerance in crop plants. She did her M.Sc. and Ph.D. from Chaudhary Charan Singh Haryana Agricultural University, Hisar and PDF from the University of Nottingham, Nottingham (UK) 1996-97. Her research interest includes biochemical studies in relation to improvement of field, vegetable and fruit crops and microbial biocatalysts for food processing industry. A Rafi Ahmed Kidwai Awardee, she is working in CCS HAU, Hisar as Principal Scientist/ Professor for the last fourteen years. She teaches a number of courses *viz.* plant biochemistry, metabolism, enzymology, biochemical techniques, molecular biology and current advances in biochemistry.

Recent Advances in Plant Stress Physiology

— Editors —

Dr. Praduman Yadav

Dr. Sunil Kumar

Dr. Veena Jain

2016

Daya Publishing House®

A Division of

Astral International Pvt. Ltd.

New Delhi – 110 002

Cataloging in Publication Data--DK
Courtesy: D.K. Agencies (P) Ltd. <docinfo@dkagencies.com>

Recent advances in plant stress physiology / editors, Dr. Praduman Yadav, Dr. Sunil Kumar, Dr. Veena Jain.
pages cm
Includes index.

ISBN 978-93-5130-948-2 (International Edition)

1. Plant physiology. 2. Plants--Effect of stress on. I. Yadav, Praduman, editor. II. Kumar, Sunil (Researcher in plant biology), editor. III. Jain, Veena (Principal scientist/professor), editor.

QK711.2.R43 2016 DDC 571.2 23

Published by : **Daya Publishing House®**
A Division of
Astral International Pvt. Ltd.
– ISO 9001:2008 Certified Company –
4760-61/23, Ansari Road, Darya Ganj
New Delhi-110 002
Ph. 011-43549197, 23278134
E-mail: info@astralint.com
Website: www.astralint.com

Laser Typesetting : **Classic Computer Services,** Delhi - 110 035

Printed at : **Thomson Press India Limited**

Kurukshetra University, Kurukshetra, Haryana
(Established by State Legislature Act XII of 1956)
"A" Grade NAAC Accredited

Professor Narender Singh
Chairman (Botany)

Foreword

Plants are very intelligent in their responses towards various external stimuli. There is intrinsic link between the climate change and the crop physiological responses. To ensure the nutritional security with the increasing population load of the country, there is urgent need to identify these physiological responses in plants particularly those involved in mitigating the effect of these stresses. Unless we unrevealed the science behind stress physiology, the aim to develop climate resilient plants could not be achieved.

Scientific and technological advancements during recent years have made it possible to explore the mechanisms involved in stress physiology. The interdisciplinary approach using advanced technologies including Genomics, Transcriptomics, Proteomics, and Metabolomics, with the integrated use of Bioinformatics and Chemo-informatics have led to the identification of genetics, molecular mechanism regulatory points, chemistry of mechanisms underlying stress related pathways etc. The editors have felt the need to compile all these advancements having implications for sustainable growth in agricultural sector.

"Recent Advances in Plant Stress Physiology" covers advances in Physiological, Biochemical, Molecular, Biotechnological ad breeding aspects. The content covered in this book is selective, specific and explained in details with scientific justifications. It will provide researchers the opportunity to use the scientific details provided within the book for designing their experiments. It will also provide the detailed information to the teachers and students to understand the basic mechanism involved in stress physiology.

The editors, Dr. Praduman Yadav, Dr. Sunil Kumar and Dr. Veena Jain have made their best efforts in finalization of the content, selection of individual authors and editing of this book. Authors along with editors of this book deserve appreciation and applause for their sincere efforts for bringing their innovative ideas in the form of this book. I am sure this book will be very useful to all concerned including the researchers, teachers, students and policy makers in public and private sectors.

Narender Singh

Preface

World population is increasing at an alarming rate, generating a concern for progressive and sustainable food production. The task becomes more challenging as plant growth and development is dependent on edaphic and environmental factors. Categorically these are abiotic stresses *viz.,* drought, flooding, salinity, extremes of temperature, mineral deficiency and adverse pH that limit crop production worldwide. Biotic stresses also cause crop yield losses. To cope with the increasing food requirements, understanding the effects of different types of stresses is a rapidly emerging domain to develop better and stress tolerant plants. It is of paramount significance to understand plant responses to stresses that disturb the homeostatic equilibrium at cellular and molecular levels and to identify a common mechanism to multiple stress tolerance.

Salinity, water stress (both drought and flooding), heat and cold are among the major abiotic stresses that adversely affect plant growth and development and cause more than 50 per cent yield losses. All these stresses are often inter-related that lead to oxidative stress and cellular damage. Plant stress tolerance involves changes at cellular, physiological and molecular levels. The recent advances in plant biology and rapid development of physiological, biochemical and molecular techniques have resulted in understanding the mechanisms of plant adaptation to stress at physiological and biochemical levels. The identification of genes/gene determinants involved in plant stress resistance/tolerance using techniques of molecular genetics, transcriptomics, proteomics and metabolomics has further opened the door for new approaches in understanding the mechanisms by which plant adapts to abiotic stresses and ultimately leads to improved and stress tolerant crops. This book provides recent information on all our knowledge of all aspects of plant perception, signaling and transgenic approaches for tolerance and adaptation to various environmental stresses.

Generation of reactive oxygen species (ROS) is responsible for the development of oxidative stress. Higher concentration of ROS has the potential to cause oxidative damage by reacting with biomolecules, whereas, in relatively low levels act as a signal molecule involved in acclimatory signalling, triggering tolerance to different stresses. The chapter 1 delves into oxidative stress and resulting antioxidative defense system in plants.

Being a substantive macronutrient, calcium participates in various structural and physiological phenomena like cell division (formation of microtubules supporting chromosomal movement), cell wall/membrane structure (as calcium pectate), pollen tube formation, nutrient metabolism and cell differentiation. It not only participates in regulation of various developmental processes but also supports plants to sense and respond to abiotic and biotic stimuli. The chapter 2 is focused on calcium signatures, Ca^{2+} sensors and their effect on pathogen defense responses.

Transcription factors (TFs) play vital regulatory roles in abiotic stress responses in plants by interacting with *cis*-elements present in the promoter region of various abiotic stress responsive genes. Now it has been possible to engineer stress tolerance in transgenic plants by manipulating the expression of TFs. This opens an excellent opportunity to develop stress tolerant crops in future. The chapter 3 provides an overview of the role of various transcription factors in crop improvement through transgenic technology.

In plants, hundreds of protein kinases (PKs) are known which are involved in cellular signal transduction services. PKs modify proteins by catalyzing addition of monophosphate groups to the side chains of the most commonly, specific serine, threonine and/or tyrosine residues in the protein backbone. Out of these kinase classes, the most conserved and best characterized protein kinase signaling pathways is mitogen-activated protein kinase (MAPK) cascade. A MAPK cascade consists of a three-tier system where each tier is phosphorylated by upper tier. It is represented as a MAP3K-MAP2K-MAPK module that serves as a link between upstream receptors and downstream targets. The chapter 4 summarizes the role of MAPK cascade in different stress responses in plants. The chapter 5 presents a discussion on recent advances made in understanding the role of plant hormones in modulating plant defense responses against various biotic as well as abiotic stresses.

The adaptation to various stresses has led to the development of common stress transduction pathways and increased synthesis of secondary metabolites, Ca^{2+} fluxes, an oxidative burst and an overlapping set of stress response genes. The main components of stress-induced signalling molecules are the cross-talk between the different signalling pathways allows plants to adjust their responses depending on the combination of stimuli. The chapter 6 focuses on the process of stress regulation in abiotic and biotic stress conditions.

Salinity affects about 20 per cent of the world's cultivated area and nearly half of the world's irrigated lands. As soluble salt levels increases, it becomes more difficult for plants to extract water from soil. The chapter 7 discusses about soil salinity, its effects on plants and tolerance mechanisms which permit the plants to withstand stress, with particular emphasis on ion homeostasis, Na^{+} exclusion and tissue tolerance. *Populus euphratica* is the only tree species naturally distributed at the edge of barren deserts or semi-barren deserts worldwide, and is well known for its high tolerance to salinity and atmospheric drought. It is not belongs in xerophyte but it have some characteristics of xerophyte. It can secrete salt from its body by discharging salty water through portals in its trunks and leaves. Due these all characters, *P. euphratica* considered model tree plant for salt tolerant. The chapter 8 focuses on the adaptability of *P. euphratica* in saline condition and their importance for arid lands.

Low temperature has a huge impact on the survival and geographical distribution of plants. It often affects plant growth and crop productivity, which causes significant crop losses. The changes in expression of hundreds of genes in response to cold temperatures are followed by increases in the levels of hundreds of metabolites, some of which are known to have protective effects against the damaging effects of cold stress. The chapter 9 provides an overview of the physiological and molecular mechanisms of chilling stress tolerance in plants.

High temperatures and drought affect plant growth at all developmental stages; especially, anthesis and grains filling are more susceptible. During these stresses, crops generally coordinate photosynthesis and transpiration, although significant genetic variation in transpiration efficiency has been identified both between and within species. To cope up and survive under such stresses, plants employ various mechanisms both at cellular and molecular level which includes modulation of primary and secondary metabolism, molecules and transcription factors related to cell signaling. Chapter 10 reviews the physiological, biochemical and molecular aspects of heat stress responses and tolerance mechanisms in plants. The chapter 11 highlights the recent development in drought tolerance in Crop plants.

Heavy metal pollution is increasing globally due to mining, industry, road traffic and other natural sources. In the earth crust, Al is the third most abundant element after oxygen and silicon whose naturally occurring forms are stables and does not pose a toxicity hazard in plants. However, under low pH (upon soil acidification) conditions, Al dissolves in various ionic forms [$Al(H_2O)^{3+}$, $Al(OH)^{2+}$ and $Al(OH)_2^{+}$] of which Al^{3+} is potentially phytotoxic to plants. The chapter 12 provides a glance into the potential of mutagenesis and transgenic approach for Al-toxicity improvement in different plants. Among heavy metals, chromium (Cr) is a common contaminant of surface waters and ground waters because of its occurrence in nature, as well as anthropic sources. Cr (VI) is more toxic at lower concentrations than Cr (III), which tend to form stable complexes in soils. Chapter 13 discusses the deleterious effects of chromium on plant health, control measures and ameliorative methods to protect plant from these negative effects of Cr (VI).

Waterlogging is the condition, in which, soil pores are fully saturated with water. Under waterlogged condition, root system of plant is under anaerobic conditions and the shoot system is under normal atmospheric condition. The chapter 14 summarizes the adaptative traits of plants ensuring survival under waterlogging (aerenchyma formation and development of adventitious roots, availability of soluble sugars, antioxidant system, anaerobic metabolism), regulatory mechanisms and signaling events responsible for triggering responses to waterlogging condition in plants.

Anthropogenic stress is the stress resulting from the influence of human beings on nature. Nowadays plant organisms are exposed to a large scale of new stressors related to human activity *viz.* toxic pollutants such as pesticides, noxious gasses (SO_2, NO, NO_2, NOx, O_3 and photochemical smog); photooxidants; soil acidification and mineral deficit due to acid rains; overdoses of fertilizers; heavy metals; intensified UV-B irradiation etc. All these stresses decrease the biosynthetic capacity of plant organisms, alter their normal functions and cause damages which may lead to plant death. The chapter 15 discusses in detail about anthropogenic stress and its impact on plants growth and metabolism.

Now-a-days, agriculture has new huge challenges due to population growth, the pressure on agriculture liability on the environmental conservation, and climate change. To cope with these new challenges, many plant breeding programs have reoriented their breeding scope to stress tolerance in the last years. The chapter 16 and 17 emphasizes on sources of stress tolerance, genetics of stress tolerance, conventional breeding methods and modern molecular biological approaches to develop improved cultivars tolerant to most sorts of abiotic and biotic stresses, respectively.

Transgenic approach is one of the important tools available for modern plant improvement programmes. Transgenic plants are majorly developed by particle bombardment and *Agrobacterium* mediated transformation methods. The chapter 18 highlights the most significant achievements of the transgenic approach to improve tolerance to drought, salinity, chilling and heat stress and the limitations that hinder the commercialization of abiotic stress tolerant transgenic crops. The chapter 19 outlines the major yield-limiting biotic stresses and the plant stress resistance mechanisms and transgenic approaches to overcome biotic stresses.

Submergence has been identified as the third most important constraint for higher rice productivity in eastern India, because it sometimes resulted in near total yield loss. Improvement of germplasm is likely the best option to withstand submergence and stabilize productivity in these environments. The trait for submergence tolerance can be improved if the genetic bases of submergence are well understood. This chapter 20 focuses on physiological understanding of tolerance to submergence in crops and to explore the potential recent technologies like genetic engineering and molecular breeding to enhance the tolerance for submergence in water logged crops.

In order to adapt to stress conditions plants exhibit stress resistance, this involves differential expression of proteins. Recent advancements in the field of proteomics enable the researchers to understand the stress mechanism and deciphering the mechanism of plant protection against the biotic and abiotic stresses. The chapter 21 was intended to describe the various proteomic techniques available and provides information on differentially expressed proteins expressed in plants during biotic and abiotic stresses. This chapter also provides the brief details of different databases and bioinformatics tools available in the field of proteomics.

Molecular marker technology is a modern tool for crop improvement. A substantial progress has been achieved in mapping of genes/QTLs associated with stress tolerance in crops. Marker-assisted selection (MAS) of target genes/QTLs enables plant breeders to select for stress tolerance in breeding populations more precisely and rapidly. The chapter 22 provides an overview on the developments in DNA marker technology and their contributions towards genetics and breeding of stress tolerance in crops.

Dr. Praduman Yadav

Dr. Sunil Kumar

Dr. Veena Jain

Contents

2016, Recent Advances in Plant Stress Physiology *Pages* **1–36**
Editors: ***Praduman Yadav, Sunil Kumar and Veena Jain***
Published by: **DAYA PUBLISHING HOUSE, NEW DELHI**

Chapter 1

Reactive Oxygen Species: Generation, Scavenging and their Role in Cell Signalling in Plants

Praduman Yadav[1*], Sunil Kumar[2], Sandeep Kumar[3], Md Sharif Baba[1] and Iyln Murthy[1]

[1]ICAR – Indian Institute of Oilseeds Research, Rajendranagar, Hyderabad – 500 030, India
[2]CRM Jat PG College, Hisar - 125 001, India
[3]Germplasm Evaluation Division, ICAR – National Bureau of Plant Genetic Resources, New Delhi – 110 012, India

ABSTRACT

Plants are exposed to various environmental stresses throughout their life cycle. Plants possess physical barriers, such as the cuticle, the cell wall and a number of biological and molecular mechanisms to counteract the effect of these stresses, which also includes the synthesis of ROS, namely the oxidative burst. ROS can react with proteins, DNA, and membrane lipids to reduce photosynthesis, increase electrolyte leakage, and accelerate senescence and cell death. These are under strict control through an antioxidative system comprising various enzymes and low molecular weight components. ROS are not only toxic by-products of aerobic metabolism, but are also signalling molecules involved in several developmental processes. Here, the divergent sources of ROS and antioxidative system that is present in plant cells are discussed and also reviewed the role of ROS in the cell signalling.

Keywords: *Antioxidative defence system, Oxidative damage, Reactive oxygen species, Signal transduction.*

**Corresponding author:* E-mail: praduman1311@gmail.com

Introduction

Unlike animals, higher plants are sessile organisms and therefore cannot avoid biotic and abiotic stresses, which adversely affect metabolism, growth and so the yield. One of the consequences of the stressful environment is the activation of paramagnetic oxygen leading to the generation of reactive oxygen species *i.e.* superoxide radicals and hydrogen peroxides, which further react with each other to produce highly reactive hydroxyl radicals. These reactive oxygen species (ROS) are produced in different plant tissues. Generation of ROS is responsible for the development of oxidative stress. Higher concentration of ROS have the potential to cause oxidative damage by reacting with biomolecules, whereas, in relatively low levels act as powerful signalling molecules involved in the regulation of plant growth and development as well as priming acclimatory responses to the stress stimuli (Foyer and Noctor, 2009), and cross-talk (Bowler and Fluhr, 2000), where an encounter with one stress leads to greater response to a second similar exposure, or to other stresses. For example, plants exposure to a sub lethal dose of ozone or UV irradiation, confers increased resistance to infection by virulent pathogens (Sharma *et al.*, 1996). The mode of action of ROS as signalling molecules or causing oxidative damage to the tissues depends on the delicate equilibrium between ROS production and their scavenging. This is determined by the interplay between different ROS-producing and ROS scavenging mechanisms, and can change drastically depending upon the physiological condition of the plant and the integration of different environmental, developmental and biochemical stimuli (Mittler, 2002). Because of the multifunctional roles of ROS, it is necessary for the cells to regulate their level tightly to avoid any oxidative injury and also not to block their generation completely. A complex network of enzymatic and non-enzymatic antioxidants controls the concentration of ROS to prevent it from reaching destructive levels, thus maintaining cellular redox balance. Enzymes such as super-oxide dismutase (SOD), catalase (CAT), peroxidase (POX) and the ascorbate glutathione constitute enzymatic anti-oxidative defense system while the non-enzymatic antioxidants include water soluble (glutathione, ascorbate, phenolic compounds) and lipid soluble (tocopherols, carotenoids) metabolites.

Generation of Free Radicals, Reactive Oxygen Species and Mode of Action

Under anoxic earth atmosphere in the begining, the metabolism was anaerobic. O_2 came into existance ~2.5×10^9 years ago. While O_2 provided advantage to aerobic organisms in terms of high-energy production and better substrate utilization (Figure 1.1), generation of toxic species of O_2, known as ROS, was inevitable. Evolution of antioxidant defence system must have been closely associated with that of photosynthesis and O_2-dependent electron transport mechanisms (Halliwell, 1999).

$$\text{Glucose} + 2ADP + 2P_i \xrightarrow{\text{Anaerobic metabolism}} \text{Lactate} + 2ATP + 2H_2O$$

$$\text{Glucose} + 6O_2 + 36ADP + 36P_i \xrightarrow{\text{Aerobic metabolism}} 6CO_2 + 36ATP + 6H_2O$$

Figure 1.1

A molecule with an unpaired or odd number of electrons is known as a free radical, which is relatively unstable and, therefore, very reactive. Reactive oxygen species (ROS) is a collective term that broadly describes O_2-derived free radicals such as superoxide anion (O_2•–), hydroxyl (HO•), peroxyl (RO_2•), alkoxyl (RO•) radicals and Hydroperoxyl (HO_2•) as well as O_2-derived non-radical species such as hydrogen peroxide (H_2O_2), hypochlorous acid ($HOCl^-$), ozone (O_3), singlet oxygen ($^1O_{2)}$ and peroxynitrite ($ONOO^-$). Hydroxyl, the most toxic radical is generated by H_2O_2 through Haber–Weiss reaction **(Figure 1.2).**

$$Fe^{3+} + \bullet O_2^- \rightarrow Fe^{2+} + O_2$$

$$Fe^{2+} + H_2O_2 \rightarrow Fe^{3+} + OH^- + \bullet OH \text{ (Fenton reaction)}$$

$$\bullet O_2^- + H_2O_2 \rightarrow \bullet OH + OH^- + O_2$$

Figure 1.2

The various roles that different ROS adopt in multiple cellular signalling pathways is not only due to the ability of cells to detoxify or scavenge ROS by many different scavenging enzymes, but also due to the cellular compartmentalization of ROS and multiple receptors or redox-sensitive transcription factors that regulate ROS signalling (Mittler 2002). The regulation of ROS levels is particularly important as ROS also functions as an important signal molecule and can regulate plant growth by modulating gene expression.

ROS signals originating in different organelles have been shown to induce large transcriptional changes and cellular reprogramming that can either protect the plant cell or induce programmed cell death (Foyer and Noctor 2005). ROS signalling is also highly integrated with hormonal signalling networks, processing and transmitting environmental inputs in order to induce plant appropriate responses to environmental constraints (Mittler *et al.*, 2011). Hormonal signals such as abscisic acid (ABA), methyl jasmonate (Me-JA) and auxins; nod factors and phytotoxins, all stimulate generation of ROS that, at least partly, mediate some of the biological effects of these stimuli. Seed germination (Larigueta *et al.*, 2008), stomatal closure (Zhang *et al.*, 2001), root hair initiation (Kwasniewski *et al.*, 2013), UV tolerance (Hidega *et al.*, 2008) root growth and function (Tsukagoshi, 2012) and leaf extension (Rodríguez *et al.*, 2002) are examples of biological processes requiring ROS generation and action.

$\bullet O_2^-$ is a moderately reactive, short-lived ROS with a half-life (Dat *et al.*, 2000) of approximately 2–4 ms. $\bullet O_2^-$ cannot pass through biological membranes as it is readily dismutated to H_2O_2. $^1O_2$ can either transfer its excitation energy to other biological molecules or continue with them, forming endoperoxides or hydroperoxides (Halliwell and Gutteridge, 1989). $^1O_2$ can last for nearly 4 ms in water and 100 ms in a polar solvent (Foyer and Harbinson, 1994). H_2O_2, on the contrary, is moderately reactive, has relatively long half-life (1 ms), and can diffuse some distances from its site of production (Dat *et al.*, 2000). H_2O_2 itself can serve as a signalling molecule for the activation of both local and systemic resistance (Mittler 2002; Mhamdi *et al.*, 2010). Many of the general stress genes are regulated by signalling pathways using H_2O_2 as the secondary messenger (Moller and Sweetlove 2010). Secondary messengers can modulate the intracellular Ca^{2+} level and initiate protein phosphorylation cascades *i.e.* mitogen-activated protein kinase MAPK, calcium dependent protein kinase CDPK, protein phosphatase, SOS3/protein kinase S, etc. The increase in cytosolic Ca^{2+} concentration and the activation of MAPK and CDPK then trigger the release of the reducing agent (RH), the activation of the cell wall peroxidases (CWP), intracellular ROS production, and NADPH oxidases (Figure 1.4). The activation of these mechanisms ultimately results in an H_2O_2 signal (O'Brien, 2012). H_2O_2 may inactivate enzymes by oxidizing their thiol groups. At low concentration it can act as a signal molecule involved in acclimatory signalling, triggering tolerance to different biotic and abiotic stresses. At high concentration it leads to programmed cell death (Quan *et al.*, 2008). ROS also regulate programmed cell death (PCD), such as hypersensitive response (HR) during the defense against biotrophic pathogens (Wrzaczeka *et al.*, 2008), and can drive cell wall reinforcement processes, and also exert toxic effects on the invading pathogen. The first response detected within minutes of an attack by virulent or avirulent pathogen is weak and transient ROS generation, due to a biologically non-specific reaction. After some hours, a second, massive and prolonged ROS production, called oxidative burst, occurs in cells attacked by avirulent pathogens. This two-phase kinetics of ROS production is typical of incompatible plant–pathogen interactions that are characterized by HR (Lamb and Dixon, 1997). Despite the functional importance of ROS signaling, little is known about the molecular mechanisms involved in the regulation of gene expression through ROS.

Sites of ROS Generation

Mitochondria

Mitochondria oxidize imported substrates such as malate and pyruvate to produce ATP. A second important function is generation of precursors for biosynthetic processes, and metabolism of compounds such as glutamate and other amino acids. Plant mitochondria are significantly different from their animal counterparts, with specific electron transport chain ETC) components and functions in processes such as photorespiration and believed to be a major site for ROS production (Figure 1.3) such as H_2O_2 as well as the ROS targets (Rasmusson *et al.*, 2004). The cellular environment of plant mitochondria is also distinctive because of the presence of photosynthesis, which creates an O_2 and carbohydrate (sucrose, glucose and fructose) rich environment (Noctor *et al.*, 2006). The amount of ROS produced by mitochondria and the fraction of total cellular ROS that come from mitochondria are difficult to determine, in part, because ROS levels in general are difficult to measure accurately (Halliwell and Whiteman, 2004). In sunlight rate of production of mtROS is very less as compared to chloroplasts or peroxisomes due to the operation of photosynthesis and photorespiration. However, in the dark or in nongreen tissues, mitochondria appears to be the mainsource of ROS (Puntarulo, 1988). In mitochondria, complex I, ubiquinone and complex III of ETC are the major sites for the generation of $\bullet O_2^-$. It has been estimated that 1-5 per cent of the O_2 consumption of isolated mitochondria results in ROS (H_2O_2) production

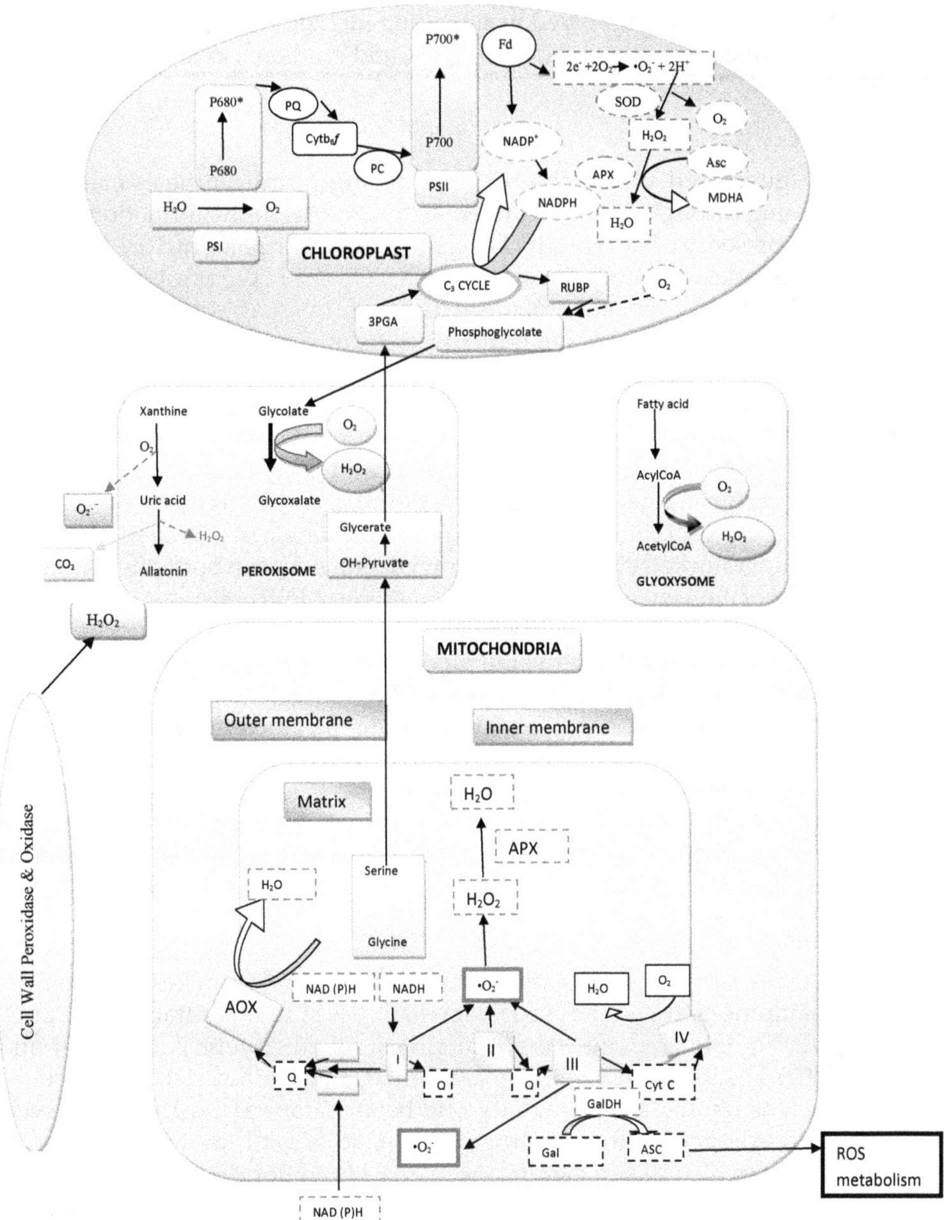

Figure 1.3: Model for Sites ROS Generation in Plant Cell.

(Moller, 2001). This H_2O_2 can react with reduced Fe^{2+} and Cu^{+} to produce highly toxic •OH, and these uncharged •OH can penetrate membranes and leave the mitochondrion (Rhoads *et al.*, 2006). Electrons that reduce ubiquinone can be transferred to complex III and then to complex IV (Cytochrome c Oxidase), or AOX (alternative oxidase). AOX is triggered by ROS accumulation and other signals, although the sites of signal perception are unknown. Mitochondrial redox-linked factors that are potentially involved in signalling include ROS, thiols, ubiquinone, ascorbate, pyridine nucleotides, ADP/O ratios, and gradients in calcium and other ions (Noctor *et al.*, 2006).

Chloroplast

Oxygen generated in the chloroplasts during photosynthesis can accept electrons passing through the photosystems, thus results in the formation of $\bullet O_2^-$. Therefore, the presence of ROS producing centres such as triplet Chl, ETC in PSI and PSII make chloroplasts a major site of ROS ($\bullet O_2^-$,$^1O_2$ and H_2O_2) production (Figure 1.3). In the Mehler reaction, superoxide is mainly formed at photosystem I (PSI), either directly or via ferredoxin, and is rapidly detoxified by superoxide dismutases (SODs), which produce H_2O_2. Ascorbate peroxidases (APXs) then reduce H_2O_2 to water via the oxidation of ascorbate (Asc) to monodehydroascorbate radicals (MDHAz), which are reduced back to ascorbate via glutathione (Pfannschmidt, 2003). Reactive oxygen species are mainly produced by photosystem I; however, under certain circumstances (Extreme temperature, drought, light, excess CO_2 and nutrient deficiency), PSII contributes to the overall formation of ROS in the thylakoid membrane. Under limited electron transport reaction between both photosystems, photoreduction of molecular oxygen by reducing side of PSII generates a superoxide anion radical, its dismutation to hydrogen peroxide and the subsequent formation of a hydroxyl radical terminates the overall process of ROS formation on PSII electron acceptor side (Pospisil, 2012). Furthermore, PSII contains a number of chlorophyll molecules that can act as a photosensitizer for the production of singlet oxygen ($^1O_2$). Studies have revealed that the acceptor side of ETC in PSII also provides sides (QA, QB) with electron leakage to O_2 producing $\bullet O_2^-$ (Cleland and Grace, 1999). It has been demonstrated that electron-hole recombination chemistry can produce a triplet state of chlorophyll that converts triplet molecular oxygen into its reactive singlet form (Vass and Styring, 1993).

Peroxisomes

Peroxisomes can be broadly defined as subcellular organelles bounded by a single membrane that contain basic enzymatic constituents catalase and H_2O_2 producing flavin oxidases and present in almost all eukaryotic cells (Baker and Graham, 2002).The important role of plant peroxisomes in a variety of metabolic reactions such as photorespiration, fatty acid beta-oxidation, the glyoxylate cycle and generation-degradation of hydrogen peroxide is well known (Figure 1.3). Plant peroxisomes also play a significant role in photomorphogenesis (Hu *et al.*, 2002), degradation of branched amino acids, biosynthesis of the plant hormones jasmonic acid and auxin, and the production of the compatible osmosolute Gly betaine (Minorsky, 2002; Reumann *et al.*, 2004). In plants, the cellular population of peroxisomes can proliferate during senescence and under different stress conditions produced by xenobiotics, ozone, cadmium, and H_2O_2 (del R ´o *et al.*, 1998, 2002; Romero-Puertas *et al.*, 1999; Nila *et al.*, 2006). In plants peroxisomes differentiate into at least four different classes; glyoxysomes, leaf peroxisomes, root nodule peroxisomes and unspecialised peroxisomes. Glyoxysomes occur in endosperms and cotyledons of germinating seeds and contain enzymes for the β-oxidation and glyoxylate cycle to convert oil seed reserves into sugars which can be used for germination before the plant is photosynthetically active. On the other hand, leaf peroxisomes house enzymes for the oxidation of glycolate during photorespiration (Figure 1.3). Root nodule peroxisomes contain urate and xanthine oxidase which catalyse the oxidation of xanthine and uric acid produced during nucleotide turnover (Nyathi, Baker, 2006) H_2O_2 is mainly produced by several peroxisomal oxidases (*e.g.*, acyl-CoA oxidase which is involved in the β-oxidation of fatty acids). H_2O_2 is decomposed by catalase and glutathione-peroxidase (GPx) or converted to OH. Hydroxyl radicals can damage the peroxisomal membrane by lipid peroxidation of unsaturated fatty acids. Hydroperoxides formed in this process can be decomposed by catalase and glutathione-peroxidase. Superoxide anions (O_2^-) generated by peroxisomal oxidases (for example, xanthine oxidase (XOx)) are scavenged by manganese superoxide-dismutase (MnSOD) and copper-zinc superoxide-dismutase (CuZnSOD) (Schrader and Fahimi, 2006).

Cell Wall, Plasma Membrane and Apoplast

Cell walls are also regarded as active sites for ROS production. Cell wall peroxidase is able to oxidize NADH and in the process, catalyse the formation of $\bullet O_2^-$. This enzyme utilizes H_2O_2 to catalyse the oxidation of NADH to NAD^+, which in turn reduces O_2 to $\bullet O_2^-$. Superoxide thus generated consequently dismutates to produce H_2O_2 and O_2 (Bolwell and Woftastek, 1997). Diamine oxidases are also involved in production of activated oxygen in the cell wall using diamine or polyamines (putrescine, spermidine, cadaverine, etc.) to reduce a quinone that autooxidizes to form peroxides (Elstner, 1991).

Electron transporting oxidoreductases are ubiquitous at plasma membranes and lead to generation of ROS. Plasma membrane NADPH - dependent oxidase (NADPH oxidase) plays a key role in the production and accumulation of ROS in plants under stress conditions. In addition to NADPH oxidase, pH-dependent cell wall-peroxidases, germin-like oxalate oxidases and amine oxidases have been proposed as a source of H_2O_2 in apoplast of plant cells. Other important sources of ROS in plants that have received little attention are detoxification reactions catalysed by cytochrome P450 in cytoplasm and endoplasmic reticulum (Dybing *et al.*,1976).

Sources of ROS Generation

Heavy Metals

Transition metal complexes, especially of Fe and Cu has also crucial role in lipid peroxidation and ROS generation. They form an activated oxygen complex that can abstract allylic hydrogens or act as a catalyst in the decomposition of existing lipid hydro-peroxides. In the presence of redox active transition metals such as Cu^+ and Fe^{2+}, H_2O_2 can be converted to the highly reactive $\bullet OH$ molecule in a metal-catalyzed reaction known as Fenton reaction (Figure 1.2). Although $\bullet O_2^-$ and H_2O_2 are capable of initiating the reactions but as OH is sufficiently reactive, the initiation of lipid peroxidation is mainly mediated by OH . Loosely bound Fe is also able to catalyse the decomposition of lipid peroxides resulting in the formation of alkoxy and peroxy radicals, which further stimulate the chain reactions of lipid peroxidations.

As Hg^{2+} does not belong to the transition metals, it cannot replace Cu^+ and Fe^{2+} in the Fenton reaction. Hg^{2+} adopts a different mechanism by inhibiting the activities of antioxidative enzymes especially of glutathione reductase, and also raise a transient depletion of GSH (Schutzendubel and Polle, 2002). Thus, as a consequence natural accumulation of ROS takes place.

Cadmium causes oxidative stress probably through indirect mechanisms such as interaction with the antioxidative defense, disruption of the ETC or induction of lipid peroxidation. The activation of lipoxygenase, an enzyme that stimulates lipid peroxidation, has been reported after cadmium exposure (Smeets *et al.*, 2005).

Cr was thought to be a non-redox metal that could not participate in Fenton reaction; however, other studies have shown that Cr can indeed participate in Fenton reaction, proving its redox character (Shi and Dalal, 1989). Chromium can affect antioxidant metabolism in plants. Antioxidant enzymes like SOD, CAT, POX and GR are found to be susceptible to chromium resulting in a decline in their catalytic activities. This decline in antioxidant efficiency is an important factor in generating oxidative stress in plants under chromium stress (Panda, 2003; Panda and Choudhary, 2005). Chromium (Cr) reactivity can be considered from its interaction with glutathione, NADH and H_2O_2, forming OH^- radicals in cell-free systems (Shi and Dalal, 1989; Aiyar *et al.*, 1991).

UV Radiation

Reductions in stratospheric ozone because of anthropogenic activities lead to concomitant increase in UV radiation reaching the earth's surface (Yogamoorthi, 2007). The most damaging part of the UV spectrum reaching earth's atmosphere is UV-B (290–320 nm), constitutes about 7 per cent of the total radiation. UV radiation by itself or in combination with photosynthetically active radiations (PARs) has been shown to impact plants (Jansen, 1998). UV-B radiations induce oxidative stress (Singh *et al.*, 2009); however, the mechanism of formation of reactive oxygen species (ROS) is not well known (Mackerness, 2000). Under elevated UV-B radiation, plant cells produce ROS that induces oxidative damage to DNA, structure of proteins, lipids and other cellular components (Foyer, 1994; Mahdavian, 2008). DNA is a potentially sensitive target molecule for UV-B because it absorbs UV-B efficiently and undergoes transformation that leads to the formation of the cyclobutane pyrimidine dimers (CPD) (Dany *et al.*, 2001) and the pryimidine (6-4) pyrimidinone photoproduct, both of which are formed by covalent bonding of adjacent pyrimidines (Nakajima *et al.*, 1998). The chloroplasts are highly vulnerable to photo oxidation owing to a high content of polyunsaturated fatty acids (PUFA) in their membrane systems, thus signicantly increasing the production of ROS (Nawkar, 2013). Highly energetic photons of the UV-B range are absorbed by the chromophore groups of many biologically important molecules, such as chlorophyll, phycobiliproteins, and quinones. These molecules can act as photosensitizers for the production of ROS (Caldwell *et al.*, 1998). UV-B induces NADPH-oxidase activity which leads to peroxide formation (Rao *et al.*, 1996). Mackerness *et al.* (2001) provided evidence to show that UV-B exposure induced NADPH oxidases and cell wall peroxidases mediated ROS synthesis in the leaves of *Arabidopsis* suggesting that there are multiple sources of H_2O_2 production in response to UV-B radiation. The UV-B radiation-induced inhibition of PSII photochemistry results in excessive excitation energy, which if not dissipated safely may damage PSII due to over reduction of reaction centers (Demmig-Adams, 2003). The alternative way to dissipate this excessive energy is either directly through the Mehler reaction or indirectly through photorespiration which favours the production of $\bullet O_2^-$ and H_2O_2 (Asada, 1999). It may be possible that the plants recognize the UV-B radiation through mechanisms identical to those involved in the pathogen infection.

Pathogen Interaction

ROS play a prominent role in early and later stages of the plant pathogenesis response, putatively acting as both cellular signalling molecules and direct anti-pathogenic agents (Allan and Fluhr, 1997). In plant–pathogen interactions, ROS accumulate in the infected tissue, leading to vigorous oxidative stress (Baker and Orlandi, 1995). The oxidative burst, a rapid production of reactive oxygen species (ROS) released into the apoplast, is described as one of the earliest responses to pathogen infection and is generally associated with hypersensitive response (HR) (Lamb and Dixon, 1997). The principal accumulating ROS is H_2O_2, with the possible participation of $\bullet O_2^-$ as an intermediate (Hammond-Kosack and Jones, 1996). ROS accumulation is closely associated with the induction of plant defence reactions against viral, bacterial, and fungal pathogens, such as the HR, defense gene expression, physical barriers at the large papillae, that are formed at the site of interaction of many pathogens by cross linking of cell wall glycoproteins or via

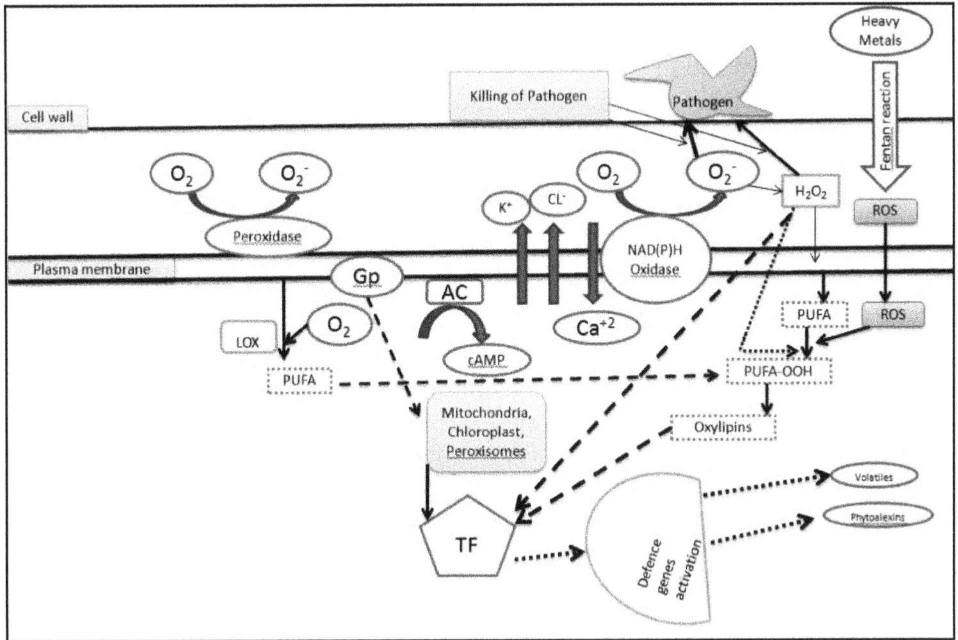

Figure 1.4: Model for Production of ROS in Response to Various Stresses and their Role in Signal Transduction.

oxidative cross linking of precursors during the localized biosynthesis of lignin and suberin polymers (Lamb and Dixon, 1997; Huckelhoven, 2007). They have been also implicated in the destruction of the challenged plant cells, either through lipid peroxidation or through initiation of programmed cell death (Greenberg, 1997) (Figure 1.4). In resistant plants, following specific recognition of a pathogen, an early response occurs immediately, the HR, a plant resistance mechanism leading to cell death, that is notably associated with the generation of ROS in and around the infected cell and afterwards a late response, usually transcription and translation dependent responses that take part in minimizing the long-term effects of the infection and in preventing further infections. The extent to which pathogen susceptibility is dependent upon ROS levels has long been debated, mainly due to the fact that many microbes can still grow in high (mM) concentrations of hydrogen peroxide (H_2O_2). There is also evidence that ROS are necessary for fungal hyphae growth and cellular differentiation (Takemoto *et al.*, 2007).

Pesticide Action

Herbicides and fungicides commonly used in pest management are known to generate ROS either by direct involvement in radical production or by inhibition of biosynthetic pathways (Karuppanapandian, 2011 and Deweza *et al.*, 2005). Xiao *et al.* (2006) studied the phytotoxicities of nine pesticides (paraquat, Xuazifop-*p*-butyl, haloxyfop, Xusilazole, cuproxat, cyazofamid, imidacloprid, chlorpyrifos, and abamectin) in cucumber and found that pesticides induced impairment to photosynthetic apparatus, which results in ROS generation. In plants, paraquat is principally reduced within chloroplasts, where it acts as an alternative electron acceptor taking electron from Fe-S proteins of photosystem I; inhibiting the ferredoxin reduction, the NADPH generation, thengiving rise to other ROS, such as H_2O_2 and OH through a process that can be accelerated by traces of transition metals, such as copper and iron (Kim and Lee, 2005). Paraquat also induces the increase of superoxide radical production in mitochondria, where complexes I and III are the major electron donors. For this reason, paraquat has been widely used to induce mitochondrial oxidative stress in many experimental systems such as isolated mitochondria, cultured cells, and whole organisms including plants, *Saccharomyces cerevisiae, Caenorhabditis elegans, Drosophila melanogaster* and rodents (Cocheme and Murphy, 2008). Herbicides such as triazines (*e.g.*, atrazine) and phenyl ureas (*e.g.*, diuron) have their sites of action in photosystem II (PS II) in the photosynthetic light reaction and inhibit the photosynthesis, thereby resulting in the accumulation of reactive, radical-forming intermediates (Hess, 2000). Diuron and oxyfluorfen induce the formation of ROS by the reaction of triplet chlorophyll or protoporphyrin IX with oxygen (Trebst and Draber, 1986). Fungicides copper sulphate and fludioxonil (*N*-(4-nitrophenyl)-*N*_-propyl-uree) aggravated the production of enzymatic activities of catalase, ascorbate peroxidase, glutathione reductase and glutathione *S*-transferase (Deweza *et al.*, 2005).

Ripening and Senescence

Fruit ripening and senescence are active processes initiated by internal and environmental factors. Harman (1956) in his "free radical theory of aging" postulated a correlation between the formation of free radicals and aging. It is well known that the production of free radicals increases during senescence. Due to their toxic nature, it is supposed that the accumulation of these radicals might cause the senescence processes and the associated degradation events. Increase in ion leakage, decrease in plasma membrane fluidity, loss of membrane integrity and accumulation of lipid peroxides are found in fruits, such as melons (Lester and Stein, 1993), ber (Yadav *et al.*, 2012) apples (Lurie and Ben-Arie, 1983) and tomatoes (Palma *et al.*, 1995). ROS production has also been found to be stimulated during senescence (Paliyath and Droillard 1992), during ripening (Kumar *et al.*, 2011a) and storage of fruits (Kumar *et al.*, 2011b; Rogiers, 1998). Production and accumulation of ROS in all respiring cells is a continuous process and is influenced by physiological stress factors. The H_2O_2 is a relatively stable metabolite that may act as a second messenger, since it could diffuse from the site of production (Cheeseman, 2006) and is also known to induce several other genes and proteins involved in stress defense system like CAT, APX (Morita, 1999). The oxidative burst, leading to generation of superoxide that is rapidly transformed into H_2O_2 was first reported by Doke (1985) in potato tuber.

High Temperature

High temperature (HT) stress causes diverse, and often adverse, alterations in plant growth, physiological processes, and yield. One of the major consequences of HT stress is the excess generation of ROS, which leads to oxidative stress. It has been suggested that, like other abiotic stresses, HT might uncouple enzymes and metabolic pathways which cause the accumulation of unwanted and harmful ROS, most commonly $^1O_2$, O_2•–, H_2O_2 and OH• which are responsible for oxidative stress (Asada, 2006). Thermal stress can induce oxidative stress through peroxidation of membrane lipids and disruption of cell membrane stability by protein denaturation (Huang and Xu, 2008). Exposure to HT increased leaf temperature which reduced the antioxidant enzyme activities resulted into increased malondialdehyde (MDA) content in leaves of rice plant (Hurkman *et al.*, 2009). HTs causes' thermal damage to photosystems, and under such conditions less absorption of photon occurs, if photon intensity is absorbed by PS I and PS II, the excess of which is required for CO_2 assimilation are considered as surplus electrons, those serve as the source of ROS (Halliwell., 2006). Active oxygen species produced under temperature stress brings about heat induced oxidative stress (Larkindale and Knight, 2002), which causes denaturation of functional and structural proteins (Smirnoff, 1998) and activates cell signalling pathway (Zhu, 2002) and cellular responses such as production of stress proteins, upregulation of antioxidants and accumulation of compatible solutes (Cushman and Bohnert, 2000). An increase in H_2O_2 and MDA content, and lipoxygenase activity (LOX) was observed during heat stress in the seedlings of thermo tolerant and thermo susceptible genotypes of *Brassica Juncea* (Rani *et al.*, 2012). Heat stress-response signal transduction pathways and defense mechanisms, involving heat shock transcription factors (HSFs) and heat shock proteins (HSPs), are thought to be intimately associated with ROS (Pnueli *et al.*, 2003). Although the ROS have several negative effects on plant metabolic processes, they have also hypothesized to have signalling role to trigger the heat shock responses towards the development of heat tolerance in plants which are inexplicable and should be divulged (Asada, 2006).

Drought

Drought exacerbates the effect of the other stresses to which plants are exposed (abiotic or biotic) and simultaneous exposure to several different abiotic stresses results in water stress (like salt and cold stresses). One of the inevitable consequences of drought stress is enhanced ROS production in the different cellular compartments, namely in chloroplasts, peroxisomes and the mitochondria. Under drought stress, ROS production is enhanced through multiple ways. For instance, the limitation on CO_2 fixation will reduce $NADP^+$ regeneration through the Calvin cycle, hence provoking an over reduction of the photosynthetic electron transport chain. Under drought stress, the photorespiration is also enhanced, especially when RuBP oxygenation is maximal due to limitation on CO_2 fixation. The involvement of antioxidant enzymes in limitation of ROS production has been examined in potato cultivars differing in the drought tolerance (Boguszewska *et al.*, 2010). Photorespiration is likely to account for over 70 per cent of total H_2O_2 production under drought stress conditions (Noctor, 2002). An increase in thylakoid membrane electron leakage to O_2 under drought stress was also seen in sunflower (Sgherri *et al.*, 1996).

Antioxidative Defense System

Oxygen is essential for energy production for plants but paradoxically, damages key biological sites, threatening their structure and function. Plants employ enzymatic and non- enzymatic antioxidants to protect themselves against ROS. The defense systems of plant cells are not restricted to the intracellular compartments but are also found in the apoplast to a limited extent (Pal *et al.*, 2013). Enzymatic antioxidants include SOD, POX, CAT, APX, MDHAR, DHAR and GR and non-enzymatic antioxidants are GSH, AA and phenolic compounds (water soluble), carotenoids and tocopherols (lipid soluble) (Table 1.1).

Superoxide Dismutase (SOD)

It is well known that superoxide ion (O^{2-}) is the starting point in the chain production of free radicals. The enzyme SOD, a metalloprotein, catalyzing the dismutation of superoxide to H_2O_2 and molecular oxygen (Allen, 1995), is considered to be a key antioxidant in aerobic cells because its activity determines the concentration of the Haber-Weiss reaction substrates, superoxide radical and H_2O_2 (Sabarinath, 2005). The enzyme activity was first described by McCord and Fridovich (1969) with a cupro-zinc protein from bovine erythrocytes. It removes O_2 and hence decreases the risk of HO• formation via the metal catalyzed Haber-Weiss-type reaction. This reaction has a 10,000 fold faster rate than spontaneous dismutation. SODs are classified by their metal cofactors into three known types: the copper/zinc (Cu/Zn-SOD), the manganese (Mn-SOD) and the iron (Fe-SOD). The Cu/Zn-SOD has a central role in scavenging toxic oxygen radicals. It is the major form in the leaves and is responsible for 65-80 per cent of the total activity (Halliwell, 1987). Mn-SOD is localized in mitochondria, whereas Fe-SOD is localized

Table 1.1: Reactive Oxygen Species Localization and Mechanism of Scavenging in Plants

Scavenging System	*Reaction Catalysed*	*Localization*
SOD	$O_2^{\cdot -} + O_2^{\cdot -} + 2H+ \rightarrow 2H_2O_2 + O_2$	Chl, Cyt, Mit, Per, Apo
APX	H_2O_2 + AA → $2H_2O$ + DHA	Chl, Cyt, Mit, Per, Apo
CAT	$2\ H_2O_2 \rightarrow O_2 + 2H_2O$	Per
GR	GSSG + NADPH → 2GSH + NADP+	Cyt, Chl, and Mit
POX	H_2O_2 + DHA → $2H_2O$ + GSSG	CW, Cyt, Vac
MDHAR	2MDHA + NADH → 2AA + NAD+	Chl, Cyt, Mit
DHAR	DHA + 2GSH → AA + GSSG	Chl, Cyt, Mit
AA	Detoxifies H_2O_2, Substrate for APX.	Chl, Cyt, Mit, Per, Apo
GSH	Substrate for various POXs, GSTs and GR. Detoxified H_2O_2, other hydro peroxidases and toxic compounds	Chl, Cyt, Mit, Per, Apo
α-Tocopherol	Protects membrane lipids from peroxidation, detoxifies lipid peroxides, and quenching $^1O_2$	Membranes
Carotenoids	Quench $^1O_2^{\bullet}$ Photosystem assembly, key components of the light harvesting complex, precursors for abscisic acid (ABA)	Chl
Phenolics	Can directly scavenge H_2O_2 and $OH^{\bullet}$	Vac

Chl: Chloroplast; Cyt: Cytosol; Mit: Mitochonderia; Per: Peroxisomes; Apo: Apoplast; CW: Cell wall; Vac: Vacuole.

in chloroplasts. Cu/Zn-SOD is present in three isoforms, which are found in the cytosol, chloroplast, peroxisome and mitochondria. All forms of SOD are nuclear-encoded and targeted to their respective sub-cellular compartments by an amino terminal targeting sequence. A comparison of the effects of drought and water stresses on wheat genotypes suggested that different mechanisms participate in ROS detoxification. For example, water stress did not affect SOD activity, whereas drought resulted in a significant increase in SOD activity (Moussa and Abdel-Aziz, 2008). Significant increase in SOD activity under salt stress in *Lycopersicon esculentum* (Gapinska *et al.*, 2008), during ripening (Kumar *et al.*, 2011a), storage of ber fruits (Kumar *et al.*, 2011b), and under drought stress in *Glycyrrhiza uralensis* Fisch (Pan *et al.*, 2006) has been observed. A minimum of one to a maximum of seven isoforms of SOD have been reported from various sources. Two isoforms of SOD have been reported in pea (Duke and Salin, 1983), sunflower (Fambrini, 1997) and cold acclimated rice plants (Kuk, 2003) while seedlings and needles of Norway spruce (*Picea abies* L.) (Kroniger, 1993), tomato (Ahn, 2002) and ber fruits (Kumar *et al.*, 2014) exhibited three.

Peroxidase (POX)

Peroxidases are another group of non-chloroplastic enzymes that detoxify H_2O_2 in the cytosolic part of cell and are non-specific in utilizing electron donor for oxidation of H_2O_2. A number of peroxidase enzymes play a significant role in response to environmental stresses. Plant peroxidases, heme containing glycoproteins which are usually classified as acidic, neutral, or basic (Yoshida, 2002), are reported to catalyze cell wall softening reactions during fruit ripening (Andrews, 2000). POX decomposes indole-3-acetic acid and has a role in the biosynthesis of lignin and defense against biotic stresses by consuming H_2O_2. POX prefers aromatic electron donors such as guaiacol and pyragallol usually oxidizing ascorbate at the rate of around 1 per cent that of guaiacol. An increase in peroxidase (POX) activity during ripening has been observed in grapes (Calderon, 1993), tobacco leaves (Takahama, 1999) and in chille pepper fruit (Biles, 1997), while it reported to decrease continuously during maturation of orange (Huang, 2007), tomato fruits (Mondal, 2004) and ber fruit (Kumar *et al.*, 2011a). An initial increase and then decrease in POX activity has been reported in potato (Dipierro and Leonardis, 1997), mushroom (Dama, 2007) ber (Kumar *et al.*, 2011b) during storage. However, Mondal (2006) reported continuous decrease in POX activity during storage of tomato. A concomitant increase in POX activity in both the leaf and root tissues of *Vigna radiata* (Panda, 2001), *O. sativa* (Khoji *et al.*, 2009) has also been reported under salinity stress. The GPX activity varies considerably depending upon plant species and stress conditions. It increased in Cd-exposed plants of *Ceratophyllum demersum* L. (Aravind and Prasad, 2003). An enhancement in the activity of POX, suggests that this enzyme serves as an intrinsic defense tool to resist stress-induced oxidative damage in plants (Koji *et al.*, 2009).

Catalase (CAT)

CAT (H_2O_2 oxidoreductase) is a heme-containing enzyme that catalyse the dismutation of H_2O_2 into H_2O and O_2. The enzyme occurs in all aerobic eukaryotes and its function is to remove the H_2O_2 generated in peroxisomes. Most of the catalase activity is associated with peroxisomes where it removes the hydrogen peroxide formed by oxidases involved in β-oxidation of fatty acids, photorespiration, purine catabolism and during oxidative stress (Mittler, 2002; Vellosillo *et al.*, 2010). Therefore, in plants, CAT is often considered to be a peroxisomal marker enzyme because of its presence in these organelles (Corpas, 2006). It has been found to remove the bulk of H_2O_2, whereas POX would be involved in the scavenging of H_2O_2 that is not removed by CAT (Willekens, 1997). CAT is highly sensitive to light and has a rapid turnover rate, similar to that of the D1 protein of PSII. Catalase has one of the highest turnover rates for all enzymes: one molecule of CAT can convert 6 million molecules of H_2O_2 to H_2O and O_2 per minute. This may be a result of light absorption by the heme group or perhaps of H_2O_2 inactivation. Stress conditions that reduce the rate of protein turnover, such as drought, salinity and heavy metals, and senescence, also reduce CAT activity (Karuppanapandian and Manoharan, 2008; Karuppanapandian *et al.*, 2006; Chen *et al.*, 2010; Hojati *et al.*, 2010 Mondal *et al.*, 2003; Pinto, 2001). The CAT activity has been found to increase from mature green stage to fully ripe stage in tomato fruits (Mondal, 2004). Catalase activity was higher in cold acclimated leaves and roots of rice plants than the sensitive one (Kuk, 2003). Increase in CAT activity is supposed to be an adaptive trait possibly helping to overcome the damage to tissue metabolism by reducing toxic levels of H_2O_2 (Thirupathi *et al.*, 2011).

Two CAT variants have been detected in cold acclimated rice plants (Kuk, 2003), ripening and storage of ber fruits (Kumar *et al.*, 2014) and in tissue cultured roots of two species of mountain panax subjected to methyl jasmonate stress (Ali, 2005). Bailly (2004) found that during development and desiccation, increase in CAT activity was due to expression of new CAT isoforms and thus, the second CAT isoform was induced by dehydration. Mittler (2004) characterized three genes for CAT in *Arabidopsis*. Rucinska (1999) detected five isoflorms in lupin roots exposed to Pb^{2+}.

Ascorbate Peroxidase (APX)

Ascorbate peroxidase is a hydrogen peroxide scavenging enzyme that is specific to plant and algae, and is indispensable to protect chloroplast and other cell constituents from damage by hydrogen peroxide and hydroxyl radicals produced from it. Unlike classical plant peroxidases, APX has a remarkable high preference for ascorbate as a reductant. One of the characteristic properties of APX, which distinguishes it from GPOX, C POX and glutathione POX, is its rapid inactivation under conditions where an electron donor is absent. The enzyme has high affinity for both substrates, comparable to physiological concentrations of ascorbate and hydrogen peroxide (Nakano and Asada, 1987). APX has two cytosolic forms, which have defensive roles, and a membrane-bound form, in addition to its role in H_2O_2 scavenging, which modulates quantum efficiency and controls electron transport in conjunction with the ascorbate-glutathione (AsA-GSH) cycle (Foyer and Noctor, 2005).). Increase in APX activity has been observed by Kuk (2003) in leaves and roots of cold acclimatized rice plants compared to non-acclimatized one. Kochhar (2003) reported differential APX activity in superficial scald resistant and susceptible varieties of apple fruit during storage. Hossain *et al.* (2009) noted that during waterlogging, APX activity increased significantly in citrus plants.

Based on amino acid sequences, five isoforms of APX have been reported at different sub-cellular locations in higher plants *i.e.* cytosolic, stromal, thylakoidal, mitochondrial and peroxisomal. APX scavenges H_2O_2 produced within the organelles, whereas cytosolic APX eliminates H_2O_2 produced in the cytosol, apoplast or that diffused from organelles (Mittler and Zilinskas, 1992). In general, plants have been observed to have three isoforms of APX (Amako, 1994). However, two APX isoforms have been reported in tea leaves (Chen and Asada, 1989), lupin roots (Rucinska, 1999), alga *Galdieria partita* (Sano, 2001) and ripening tomato fruit (Ahn, 2002). Gara (2003) reported three distinct isoforms of APX during maturation of *Triticum durum* kernels, of which two were fast migrating anodic forms while one was slow migrating cathodic form. Also, Narendra (2006) reported three major isoforms of APX during growth and development of *Arabidopsis*.

Monodehydro Asorbate Reductase (MDHAR)

MDHA radical produced in APX catalyzed reaction has a short lifetime, and if not rapidly reduced, it disproportionates to full (AsA) and full (DHA). MDHAR is critical in maintaining the proper concentration of ascorbate in cells by reducing the MDHA radical directly to ascorbate at the expense of an NAD(P)H (Asada, 1996). MDHAR is a flavin adenin dinucleotide (FAD) enzyme that is localized in the cytosol, chloroplasts, peroxisomes, mitochondria and plasma membrane. In chloroplasts, MDHAR could have two physiological functions: the regeneration of

AsA from MDHA and the mediation of the photoreduction of dioxygen to O_2•–, when the substrate MDHA is absent (Miyake *et al.*, 1998). MDHAR exhibits a high specificity for monodehydro asorbate (MDHA) as a electron acceptor, preferring NADH rather than NADPH as the electron donor. However, among the antioxidant enzymes much of the research effort has been focused on GR and APX until recently, and only limited information is available concerning the *Mdhar* gene and its expression during oxidative stress. Grantz *et al.* (1995) reported that mRNA levels of Tomato *Mdhar* were increased by wounding, and Nishikawa *et al.* (2003) showed that cytosolic gene expression of broccoli *Mdhar* (*Brassica oleracea*) was stimulated actively after harvest. Overexpression of MDAR in transgenic tobacco increased the tolerance against salt and osmotic stresses (Eltayeb *et al.*, 2007) Schutzendubel *et al.* (2001) have observed enhanced MDHAR activity in Cd-exposed *Pinus sylvestris* and a declined MDHAR activity in Cd-exposed poplar hybrids *Populus Canescens*). Sharma and Dubey (2005) reported that the activities of enzymes involved in regeneration of ASH *i.e.*, MDHAR, DHAR and GR were higher in drought stressed rice seedlings. It has also been reported that the increase in MDAR activity contribute towards chilling tolerance in tomato fruit (Stevens *et al.*, 2008). The isoenzymes of MDHAR have been reported to be present in several cellular compartments such as chloroplasts (Hossain, 1984), cytosol and mitochondria, and peroxisomes (Jimenez, 1997; Dalton, 1993).

Dehydroascorbate Reductase (DHAR)

Dehydroascorbate reductase catalyzes the reduction of DHA to AsA using GSH as the reducing substrate and, thus, regulates the cellular AsA redox state which is crucial for tolerance to various stresses leading to the production of ROS. Despite the possibility of enzymatic and nonenzymatic regeneration of AsA directly from MDHA, some DHA is always produced when AsA is oxidized in leaves and other tissues. DHA, a very short-lived chemical, can either be hydrolyzed irreversibly to 2, 3-diketogulonic acid or recycled to AsA by DHAR. It has been observed that the overexpression of DHAR in tobacco protected the plants against ozone toxicity (Chen and Gallie, 2005). Overexpression of DHAR increased salt tolerance in Arabidopsis (Ushimaru *et al.*, 2006) and drought and ozone stress tolerance in tobacco (Eltayeb *et al.*, 2006). Transgenic potato overexpressing Arabidopsis cytosolic AtDHAR1 showed higher tolerance to herbicide, drought, and salt stresses (Eltayeb *et al.*, 2007). It has been noted that guard cells in DHAR overexpressing plants exhibited a reduced level of H_2O_2, decreased responsiveness to H_2O_2 or abscisic acid signalling, and greater stomatal opening (Chen and Gallie, 2005).

Glutathione Reductase (GR)

Glutathione reductase is another specific and important enzyme of ascorbate-glutathione cycle (Asada and Badger, 1984). It is localized predominantly in chloroplasts, but small amount of this enzyme has also been found in mitochondria and cytosol. In chloroplast, GSH and GR are involved in detoxification of H_2O_2 generated by Mehler reaction. It plays a crucial role in affording protection against oxidative damage in many plants (Foyer, 1994) by maintaining endogenous pool of reduced glutathione (GSH). The mono-dehydroascorbate formed in the above processes can be regenerated to ascorbate by reduced glutathione (GSH), which gets oxidized to form glutathione disulphide (GSSG). For most of the cellular functions, glutathione must be available in its reduced form (GSH). The reduction of oxidized glutathione to reduced glutathione is carried out by GR in a NADPH dependent reaction (Mallick and Mohn, 2000). If the reduced enzyme is not reoxidized by GSSG, it can suffer a reversible inactivation (Mallick and Mohn, 2000).

Salt tolerance in leaves of *Calamus tenuis* has been correlated with elevated level of GR activity (Khan and Patra, 2007). Mandhania (2006) also reported higher GR activity in the salt-tolerant in comparision to salt-sensitive wheat cultivars, indicating the presence of more active ascorbate-glutathione cycle in tolerant cultivars. Tobacco leaves (Kato and Shimizu, 1987) and sunflower seeds (Bailly, 1996), however, exhibited a rapid decline in GR activity during storage. The progressive decrease in GR activity suggests that glutathione-ascorbate cycle, the principal mediator of cellular redox potential is adversely affected (Apel and Hirt, 2004; Ammar, 2008).

Non Enzymatic Defense System

Glutathione (GSH)

Glutathione (γ-glutamyl-cysteinyl-glycine), a low molecular weight water soluble antioxidant, appears to be synthesized both in chloroplast and cytosol. GSH is a product of primary sulphur metabolism and plays major role in transport and storage of reduced sulphur, signal transduction, conjugation of metabolites, detoxication of xenobiotics and also in scavenging of active oxygen species (Dhindsa, 1987; Noctor and Foyer, 1998). Together with cysteine, it forms part of the repertoire of signals that modulate sulphate uptake and assimilation (Kopriva and Rennenberg, 2004). These functions depend on the concentration and/or redox state of the glutathione pool in each cellular compartment. GSH provides a substrate for multiple cellular reactions that yield GSSG (*i.e.*, two glutathione molecules linked by a disulde bond). Thus, maintains the redox equilibrium in the cellular compartments. It has been considered that GSH/GSSG ratio, indicative of the cellular redox balance, may be involved in ROS perception (Li and Jin 2007). It also participates in the regeneration of another powerful water-soluble antioxidant, AA, via the AsA-GSH cycle (Noctor and Foyer, 1998; Halliwell, 2006). Unlike many other redox couples such as ascorbate/dehydroascorbate and NADPH/NADP, glutathione redox potential depends on both GSH/GSSG and absolute glutathione concentration. The concentration-dependent term of the Nernst equation is second-order with respect to GSH but first-order with respect to GSSG; hence, accumulation of GSH can offset the change in redox potential caused by decreases in the GSH/GSSG ratio (Schafer and Buettner, 2001). A central nucleophilic cysteine residue is responsible for the high reductive potential of GSH, which scavenges cytotoxic H_2O_2 and reacts non-enzymatically with other ROS, such as $^1O_2$, $O_2^{\cdot -}$, and $OH^{\cdot}$ (Noctor and Foyer, 1998; Wang *et al.*, 2008).

For overall rejuvenation of the antioxidative system through ascorbate-glutathione cycle, high redox ratio of glutathione system is essential (Apel and Hirt, 2004). Ascorbate-glutathione system is also involved in maintaining a low redox potential and thus, a highly reduced intracellular environment (Tanaka, 1994). The activity of GSH biosynthetic enzymes and thus, the GSH level (Mahan and Wanjura, 2005) has been reported to increase under environmental stresses while ratio of GSH/GSSG decreased during ripening of orange fruits (Huang, 2007). Because of its role in ascorbate reduction, glutathione has been reported to protect membrane by maintaining α-tocopherol and zeaxanthin in the reduced state (Xiang, 2001) (Figure 1.5).

Ascorbic Acid (AsA)

Ascorbic acid is generally the most abundant, water soluble small antioxidant molecule in plants – particularly in leaves, being about 5–10 times more concentrated than glutathione (GSH). Most of AsA, almost more than 90 per cent, is localized in cytoplasm, but unlike other soluble antioxidants a substantial portion is exported to the apoplast, where it is present in millimolar concentration. Apoplastic AsA is believed to represent the first line of defense against potentially damaging external oxidants. It is predominantly present as the ascorbate anion. This readily loses an electron from its *ene*-diol group to produce the MDHA radical. MDHA is a radical with a short lifetime and can spontaneously dismutate into DHA and AsA or is reduced to AsA by NADP (H) dependent enzyme MDHAR (Miyake and Asada, 1996). The ability of ascorbate to donate an electron and the relatively low reactivity of the resulting MDHA radical is the basis of its biologically useful antioxidant and free radical scavenging activity (Buettner and Schafer, 2004). Major function of ascorbate is to protect plant cells against oxidative damage by scavenging oxygen free radicals directly and regenerating α-tocopherol from tocopheroxyl radical. It also acts as a substrate for ascorbate peroxidase enzyme and functions as electron donor for reduction of H_2O_2 both, under normal and stress conditions (Klapheck, 1990; Mittler and Zilinskas, 1991). Ascorbate is oxidized either by direct reaction with oxygen or by serving as a reductant of α-chromanoxyl radical, which in turn disrupts lipid peroxidation reactions by reacting with superoxide radical (Yadav, 2006). AsA acts as a cofactor of violaxantin de-epoxidase thus, sustaining dissipation of excess excitation energy (Smirnoff, 2000).

An initial increase and then decrease in ascorbic acid content has been reported in apple (Joshi, 2004), tomato (Mondal *et al.*, 2004), guava (Mondal *et al.*, 2009) and orange (Huang, 2007) during ripening. The decline in ascorbate content was observed to be linear with increasing salinity level in chickpea (Kukreja, 2006). As under salinity stress, other abiotic stresses like water deficit conditions (Mukherjee and Choudhuri, 1983), UV irradiation (Wise and Naylor, 1987) and light stress (Hernandez, 2004) have also been observed to decrease the ascorbate content.

Tocopherols and Tocotrienols (Toc)

Tocopherols and tocotrienols are synthesized by higher plant plastids and the cyanobacteria. They represent a group of lipophilic antioxidants involved in scavenging of oxygen free radicals, lipid peroxy radicals, and $^1O_2$. Out of four isomers of tocopherols (α-, β-, γ- and δ-) found in plants, α-tocopherol has the highest antioxidant activity due to the presence of three methyl groups in its molecular structure (Kamal-Eldin and Appelqvist, 1996). In seeds, α-tocopherol is associated with triacylglycerols in the oil bodies and possibly in glyoxysomes

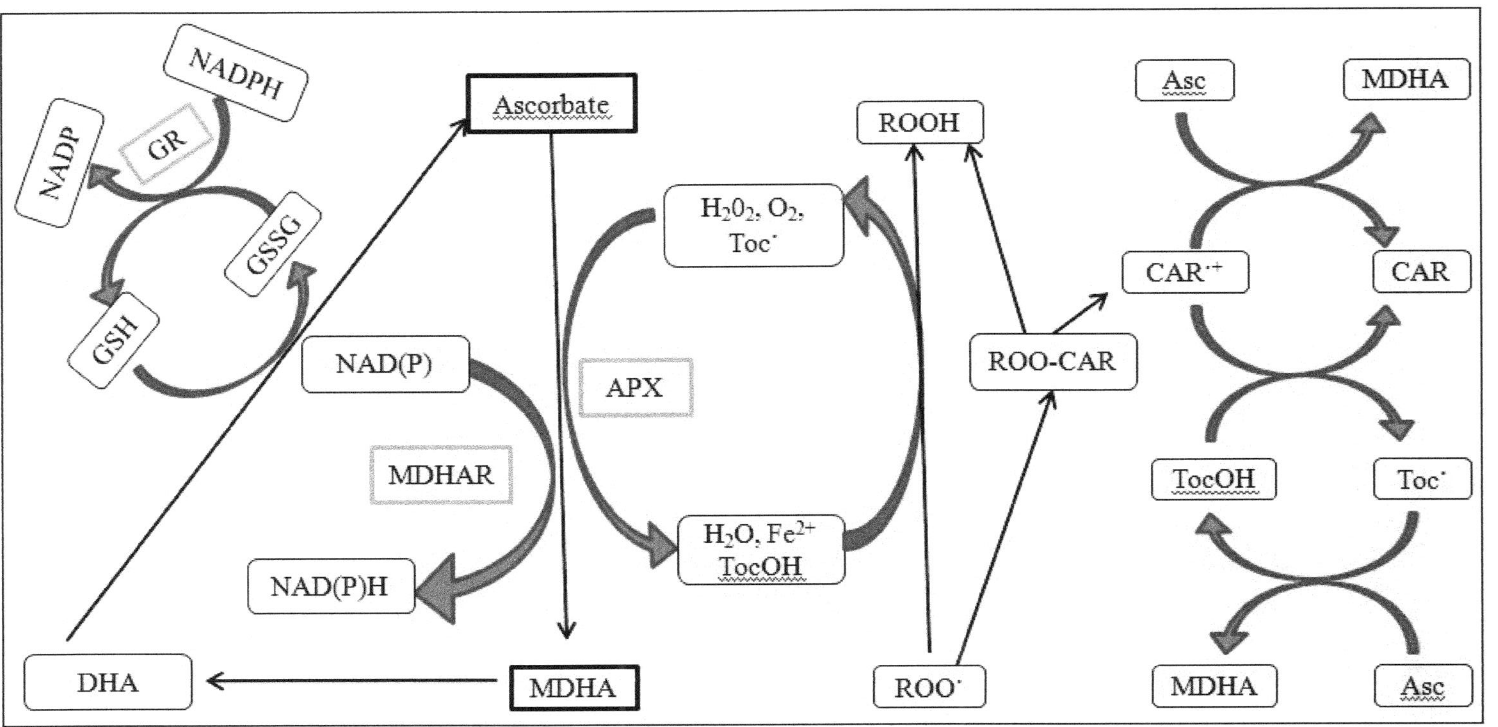

Figure 1.5: Synergistic Interaction between Carotenoids, Tocopherols and Ascorbate in Scavenging Lipid Peroxyl Radicals in Membranes.

where they are oxidised during germination (Sattler, 2004). $^1O_2$ oxygen quenching by tocopherols is highly efficient, and it is estimated that a single α-tocopherol molecule can neutralize up to 220 $^1O_2$ molecules *in vitro* before being degraded (Fukuzawa, 1982). Regeneration of the oxidized tocopherol back to its reduced form can be achieved by AsA, GSH (Fryer, 1992) or coenzyme Q (Kagan *et al.*, 2000) (Figure 1.5). Tocopherols and tocotrienols can quench $^1O_2$ by two ways. Firstly, by absorbing excitation energy, making the singlet oxygen to return to ground state oxygen. Secondly, they can react with singlet oxygen to form an oxidised product, tocopherylquinone. The reaction with superoxide is relatively slow but as expected, very fast with hydroxyl radicals (Halliwell and Gutteridge, 1999). Tocopherol concentration in leaves generally increases on exposure to high light intensity and expression of genes encoding some of the biosynthetic enzymes is also upregulated (Collakova and DellaPenna, 2003). Accumulation of α-tocopherol has been shown to induce tolerance to chilling (Yamaguchi-Shinozaki and Shinozaki, 1994; Bafeel and Ibrahim, 2008), water deficit (Munn´e-Bosch *et al.*, 1999), and salinity (Guo *et al.*, 2006) in different plant species. Both high light and reduced CO_2 assimilation during drought can cause photo-oxidative stress suggesting that tocopherol may be involved in defense. Jain (2004) also proved the oxyradical scavenging capacity of α-tocopherol in chemically generated oxyradical system. The α-tocopherol and related substances like tocols and tocotrienols (water-soluble) are found in high concentration in wheat and maize germ, sunflower and cotton seed oils, and are found to be essential for seed longevity and for preventing lipid peroxidation during germination (Sattler, 2004).

Carotenoids (CAR)

The carotenoids in photosynthetic chloroplast function both as accessory light-harvesting pigments as well as antioxidants (Pandhir and Shekhon, 2006). They absorb light at wavelength between 400 and 550 nm and transfer it to the Chl (Sieferman-Harms, 1987). Carotenoids can quench singlet oxygen in a similar manner to tocopherols and are more effective than tocopherols as chloroplast antioxidants. The carotenoids radical could then be regenerated by tocopherols. Both carotenoid radicals and tocopheryl radicals have the potential to be reduced by ascorbate (Figure 5). Higher carotenoids have been observed in plants, grown under high irradiation (Demmig Adams and Adams, 1992) and subjected to drought stress (Zhang and Kirkham, 1996) in wheat. Mishra (1993) suggested that the increased carotenoid content could be associated with the higher production of singlet oxygen along with oxyradicals, since carotenoids scavenge singlet oxygen and also minimize its formation by absorbing excess energy from excited triplet states of chlorophyll (Pandhir and Shekhon, 2006).

Phenolics

Phenolics, including flavonoids, tannins, hydroxycinnamate esters, and lignin, are diverse secondary metabolites that are excellent antioxidants by virtue of the electron donating activity of the 'acidic' phenolic hydroxyl group. Phenolics show high reactivity as electron donors than those of oxygen radicals such as superoxide O_2^{-}•, ROO•, RO• and HO• radicals, means that these species will readily oxidize phenolics to their respective phenoxyl radicals (Jovanovic, 1994), and phenoxyl radicals are generally less reactive than oxygen radicals (Bors, 1994). Phenolics also have the capability to chelate transition metal ions and to alter peroxidation kinetics by modifying the lipid packing order to decrease the fluidity of the membranes (Arora *et al.*, 2000). These changes could terminating the Fenton reaction, sterically hinder the diffusion of free radicals, scavenge the harmful ROS and inactivate them without promoting further oxidative reactions. They are involved in plant signalling through interaction with plant microbes (Gill and Tuteja, 2010). It has been proved that they are involved in plant responses to stresses such as wounding, drought and metal toxicity (Cle *et al.*, 2008).

Conclusions

ROS – the unwelcome companions of aerobic life were initially thought to be toxic by products of aerobic metabolism, but have now been acknowledged as central players in the complex signalling network of cells. There are several sources for generation of ROS in plants including NADPH oxidase, apoplastic peroxidase, and other oxidases in mitochondria, chloroplasts, and peroxisomes. The use of ROS as environmental indicators and biological signals by plant cells suggests that, during the course of evolution, plants were able to achieve a high degree of control over ROS toxicity and are now using ROS as signalling molecules. They interact both positively and negatively during plants responses to biotic and abiotic stresses. At low concentration ROS can act as a signal molecule involved in acclimatory signalling, triggering tolerance to different stresses. At high concentration it leads to programmed cell death. ROS also communicate with major signalling pathways including those regulated by ABA, JA, SA, ethylene, ion channels to trigger a complex metabolism forming a network that can orchestrate downstream responses.

References

Ahn, T., Schofield, A. and Paliyath, G. 2002. Changes in antioxidant enzyme activities during tomato fruit development. *Physiol. Mol. Biol*, 8 (2): 241-249.

Aiyar, J., Buerkovitis, H.J., Floyd, R.A. and Borges, K. 1991. Reaction of chromium (VI) with glutathione or with hydrogen peroxide: Identification of reactive intermediates and their role in chromium (VI) induced DNA damage. *Environ. Health Persp*, 92:53-62.

Ali, M.B., Thanh, N.T., Yu, K.W., Hahn, E.J., Paek, K.Y. and Lee, H.L. 2005. Induction in the antioxidative systems and lipid peroxidation in suspension culture roots of *Panax ginseng* induced by oxygen in bioreactors. *Plant sci*, 169: 833-841.

Allan A.C. and Fluhr, R. 1997. Two Distinct Sources of Elicited Reactive Oxygen Species in Tobacco Epidermal Cells. *Plant Cell*, 9:1559-1572.

Allen, R.D. 1995. Dissection of oxidative stress tolerance using transgenic plants. *Plant Physiol*, 107: 1049-1054.

Amako, K., Chen, G.X. and Asada, K. 1994. Separate assays specific for ascorbate peroxidase and guaiacol peroxidase and for the chloroplastic and cytosolic isozymes of ascorbate peroxidase in plants. *Plant Cell Physiol*, 35: 497-504.

Ammar, W.B., Mediouni, C., Tray, B., Ghorbell, M.H. and Jemal, F. 2008. Glutathione and phytochelatin contents in tomato plants exposed to cadmium. *Biol. Plant.* 52 (2): 314-320.

Andrews, J., Malone, M., Thompson, D.S., Ho, L.C. and Burton, K.S. 2000. Peroxidase isozyme patterns in the skin of maturing tomato fruit. *Plant Cell Environ,* 23: 415-422.

Apel, K. and Hirt, H. 2004. Reactive oxygen species: Metabolism, oxidative stress, and signal transduction. *Annu. Rev. Plant Biol,* 55: 373-399.

Aravind, P.and Prasad, M.N.V. 2003. Zinc alleviates cadmium-induced oxidative stress in *Ceratophyllum demersum* L. A free-floating freshwater macrophyte. *Plant Physiol Biochem,* 41:391-397.

Arora, A., Byrem, T.M., Nair, M.G. and Strasburg, G.M. 2000. Modulation of liposomal membrane fluidity by flavonoids and isoflavonoids. *Arch. Biochem. Biophys.* 373 (1): 102-109.

Asada, K. 1996. Radical production and scavenging in chloroplasts. Baker N.R. in Photosynthesis and the Environment. *Kluwer. Academic,* Dordrecht. Netherlands, 123-150.

Asada, K. 1999. The water cycle in chloroplasts scavenging of active oxygen and dissipation of excess photons. *Annu. Rev. Plant Physiol.and Plant Mol. Bio.* 50: 601-639.

Asada, K. 2006. Production and scavenging of reactive oxygen species in chloroplasts and their functions. *Plant Physiol,* 141: 391-396.

Bafeel, S.O. and Ibrahim, M.M. 2008. Antioxidants and accumulation of α-tocopherol induce chilling tolerance in *Medicago sativa.Int. J. Agri.Biology,* 10;6: 593-598.

Bailly, C., Benamar, A., Corbineau, F. and Come, D. 1996. Changes in malondialdehyde content and in *superoxide dismutase, catalase and glutathione reductase* activities in sunflower seeds as related to deterioration during accelerated ageing. *Physiol. Plant,* 97: 104-110.

Bailly, C., Leymarie, J., Lehner, A., Rousseau, S., Come, D. and Corbineau, F. 2004. Catalase activity and expression in developing sunflower seeds as related to drying. *J. Exp. Bot,* 55: 475-483.

Baker, A. and Graham, I. 2002. Plant Peroxisomes. Biochemistry, Cell Biology and Biotechnological Applications. *Kluwer. Academic. Publishers.*

Baker, C.J. and Orlandi, E.W. 1995. Active oxygen in plant pathogenesis. *Annu. Rev. Phytopathol,* 33; 299-321.

Biles, C.L., Kuhen, G.D., Wall, M.M., Bruton, B.D. and Wann, E.V. 1997. Characterization of chile pepper fruit *peroxidases* during ripening. *Plant Physiol and Biochem,* 35: 273-280.

Boguszewska, D., Grudkowska,M. and Zagdañska, B. 2010. Drought responsive antioxidant enzymes in potato (*Solanum tuberosum* L). *Potato Res,* 53: 373-382.

Bolwell, G.P. and Woftastek, P. 1997. Mechanism for the generation of reactive oxygen species in plant defense broad perspective. *Physiol Mol and Plant Pathol,* 51:347-366.

Bors, W., Michel, C. and Saran, M. 1994 Flavonoid antioxidants: rate constants for reactions with oxygen radicals. *Methods Enzymol,* 234, 420-429.

Bowler, C. and Fluhr, R. 2000. The role of calcium and activated oxygen as signals for controlling cross-tolerance. *Trends Plant Sci,* 5: 241-245.

Buettner, G.R. and Schafer, F.Q. 2004. Ascorbate as an antioxidant, in Vitamin C. Function and Biochemistry in Animals and Plants. *Bios. Scientific. Publishers. Oxford,*173-188.

Calderon, A.A., Zapata, J., Munoz, R. and Barcelo, A.R. 1993. Localization of *peroxidase* in grapes using nitrocellulose blotting of freezing thawing fruits. Hort. Sci, 28: 38-40.

Caldwell, M.M., Bjorn, L.O., Bornman, J. F., Flint, S. D., Kulandaivelu, G.,Teramura, A. H. and Tevini, M. 1998. Effects of increased solar ultraviolet radiation on terrestrial plants. *Ambio,* 24: 166-173.

Cheeseman, J.M. 2006. Hydrogen peroxide concentrations in leaves under natural conditions. *J. Exp. Bot,* 57: 2435 - 2444.

Chen, G.X. and Asada, K. 1989. Ascorbate peroxidase in tea leaves. Occurrence of two isozymes and the differences in their enzymatic and molecular properties. *Plant Cell Physiol,* 30: 987-998.

Chen, Q., Zhang, M. and Shen, S. 2010. Effect of salt on malondialdehyde and antioxidant enzymes in seedling roots of Jerusalem artichoke (*Helianthus tuberosus* L.). *Acta Physiol. Plant,*.33: 273-278.

Chen, Z.and Gallie, D.R. 2005. Increasing tolerance to ozone by elevating folia ascorbic acid confers greater protection against ozone than increasing avoidance. *Plant Physiol,* 138: 1673-1689.

Cle, D., Hill, L. M., Niggeweg, R., Martin, C. R., Guisez, Y., Prinsen, E. and Jansen. M.A.K. 2008. Modulation of chlorogenic acid biosynthesis in *Solanum lycopersicum* consequences for phenolic accumulation and UV-tolerance, *Phytochemistry,* 69: 2149-2156.

Cleland, R.E., and Grace, S.C. 1999. Voltammetric detection of superoxide production by photosystem II, *FEBS Letters,* 457 (3): 348-352.

Cocheme, H.M. and Murphy, M.P. 2008. Complex I is the major site of mitochondrial superoxide production by paraquat. *J. Biol. Chem,* 283; 4: 1786-1798.

Collakova, E. and DellaPenna, D. 2003. The role of homogentisatephytyl transferase and other tocopherol pathway enzymes in the regulation of tocopherol synthesis during abiotic stress. *Plant Physiology,* 133: 930-940.

Corpas, F.J. Fernandez-Ocana, A., Correras, A., Valderrama, R., Luque, F., Esteban, F.J., Rodriguez, S. M., Chaki, M., Pedrajas, J.R., Sandalio, L.M., Del Rio, L.A. and

Barroso, J.B. 2006. The expression of different superoxide dismutase forms is cell type dependent in olive (*Olea europaea* L.) leaves. *Plant Cell Physiol*, 47: 984-994.

Cushman, J.C.,and Bohnert, H.J., 2000. Genomic approaches to plant stress tolerance. *Curr. Opin. Plant Biol*, 3: 117 - 124.

Dalton, D.A., Baird, L.M.and Langeberg L. 1993. Subcellular localization of oxygen defense enzymes in soybean (*Glycine max* L.). Root nodules. *Plant Physiology*, 102 (2): 481-489.

Dama, C.L. 2007. Effect of temperature on shelf life of mushroom using isozymes as molecular marker.M.Sc. thesis, Department of Molecular Biology and Biotechnology, MPUAT, Udaipur, India.

Dany, A. L., Douki, T., Triantaphylides, C. and Cadet, J. 2001. Repair of the main UV-induced thymine dimeric lesions within *Arabidopsis thaliana* DNA. Evidence for the major involvement of photo reactivation pathways. *J. Photochem and Photobiol*, 65: 127-135.

Dat, J., Van Breusegerm, F., Vandenabeele, S., Vranova, E., Van Montagu, M. and Inze, D. 2000. Active oxygen species and *catalase* during plant stress response. *Cell Mol.and Life Sci*, 57: 779-786.

Del Rio, L.A., Corpas, F.J., Sandalio, L.M., Palma, J.M., Gomez, M. and Barroso, J.B. 2002. Reactive oxygen species, antioxidant systems and nitric oxide in peroxisomes. *J. Exp. Bot*, 53: 1255-1272.

Del Rio, L.A., Pastori, G.M., Palma, J.M., Sandalio, L.M., Sevilla, F., Corpas, F.J., Jimenez, A., Lopez Huertas, E., Hernandez, J.A. 1998. The activated oxygen role of peroxisomes in senescence. *Plant Physio*, 116:1195-1200.

Demmig Adams, B. 2003. Linking the xanthophylls cycle with thermal energy dissipation. *Photosynthetic Research*, 76:73-80.

Demmig Adams, B. and Admas, W.W. 1992. Carotenoids composition in sun and shade leaves of plants with different life forms. *Plant. Cell. Environ*, 15: 411-419.

Deweza, D., Geoffroy, L., Vernet G. and Popovic, R.B. 2005. Determination of photosynthetic and enzymatic biomarkers sensitivity used to evaluate toxic effects of copper and fludioxonil in alga *Scenedesmus obliquus. Aquat Toxicol*, 74:150-159.

Dhindsa, R.S. 1987. Glutathione status and protein synthesis during drought and subsequent rehydration in *Tortula rurali. Plant Physio*,. 83: 816-819.

Dipierro, S. and Leonardis, S.D. 1997. The ascorbate system and lipid peroxidation in stored potato (*Solanum tuberosum* L.) tubers. *J. Exp. Bot*, 48: 779-783.

Doke, N. 1985. NADPH-dependent O_2^- generation in membrane fractions isolated from wounded potato tubers inoculated with *Phytophtora infestans. Physiol. Plant Pathol*, 27: 311 - 322.

Dordrecht, and Minorsky P.V. 2002. Peroxisomes organelles of diverse function. *Plant Physio*, 130: 517-518.

Duke, M.V. and Salin, M.L. 1983. Isoenzymes of cuprozinc superoxide dismutase from *Pisum sativum*. *Phytochem*, 22: 2369-2373.

Dybing, E., Nelson J. R., Mitchell, J. R., Sesame, H. A. and Gillette, J. R. 1976. Oxidation of a methyldopa and other catechols by chytochromes R 450-generated superoxide anion. Possible mechanism of *Methyldopa hepatitis*, 12: 911-920.

Elstner, E.F., Pell, E.J.and Steffen, K.L. 1991. Mechanisms of oxygen activation in different compartments of plant cells, in Active Oxygen/Oxidative Stress and Plant Metabolism. *American Society of Plant Physiologists*.USA, 13-25.

Eltayeb, A.E., Kawano, N., Badawi, G.H., Kaminaka, H., Sanekata, T., Shibahara, T., Inanaga, S. and Tanaka, K. 2007. Overexpression of monodehydro ascorbate reductase in transgenic tobacco confers enhanced tolerance to ozone, salt and polyethylene glycol stresses. *Planta*, 225: 1255-1264.

Eltayeb, A.E., Yamamoto, S., Habora, M.E.E., Yin, L., Tsujimoto, H. and Tanaka, K. 2011. Transgenic potato overexpressing *Arabidopsis* cytosolic AtDHAR1 showed higher tolerance to herbicide, drought and salt stresses. *Breeding Science*. 61 (1):3-10.

Fambrini, M., Sebastiani, L., Rossi, V.D., Cavallini, A. and Pugliesi, C. 1997. Genetic analysis of an electrophoretic variant for the chloroplast-associated form of Cu/Zn *superoxide dismutase* in sunflower (*Helianthus annuus* L.). *J. Exp. Bot*, 48: 1143-1146.

Foyer, C.H. and Mullineauex, P.M. 1994. Photooxidative Stress and Amelioration of Defense Systems in Plants. *CRC Press. Boca Raton. FL.USA*, 1-13.

Foyer, C.H. and Harbinson, J. 1994. Oxygen metabolism and regulation of photosynthetic electron transport.*CRC Press. Boca. Raton. FL.USA*, 1-13.

Foyer, C.H. and Noctor, G. 2009. Redox regulation in photosynthetic organisms. Signaling, acclimation, and practical implications. *Antioxid. Redox. Signal*, 11: 861 - 905.

Foyer, C.H.and Noctor, G. 2005. Oxidant and antioxidant signaling in plants are evaluation of the concept of oxidative stress in a physiological context. *Plant Cell Environ*, 29:1056-107.

Foyer, C.H., Lelandais, M. and Kunert, K. J. 1994. Photooxidative stress in plants. *Physiol. Plant*, 92: 696 - 717.

Fryer, M.J. 1992. The antioxidant effect of thylakoid vitamin-E (α-tocopherol), *Plant. Cell and Enviro*,15 (4): 381-392.

Fukuzawa, K., Tokumura, A., Ouchi S. and Tsukatani, H. 1982. Antioxidant activities of tocopherols on Fe^{2+}-ascorbate-induced lipid peroxidation in lecithin liposomes. *Lipid*, 17 (7): 511 - 514.

Gapinska, M., Sk, M.and Lodowska, B. 2008. Gabara, Effect of short- and long-term salinity on the activities of antioxidative enzymes and lipid peroxidation in tomato roots, *Acta. Physiol. Plant*, 30: 11-18.

Gara, L.D., Depinto, M.C., Moliterni, V.M.C. and Degidio, M.G. 2003. Redox regulation and storage processes during maturation in kernels of *Triticum durum*. *J. Exp. Bot*, 54: 249-258.

Gill, S. S. and Tuteja, N. 2010. Reactive oxygen species and antioxidant machinery in abiotic stress tolerance in crop plants. *Plant Physiol Biochem*, 48(12): 909-30.

Grantz, A.A., Brummell, D.A. and Bennett A.B. 1995. Ascorbate free radical reductase mRNA levels are induced by wounding. *Plant Physiol*, 108: 411-418.

Greenberg, J.T. 1997. Programmed cell death in plant pathogen interactions. *Annu. Rev. Plant Physiol and Plant Mol Biol*, 48: 524-545.

Guo, J., Liu, X., Li, X., Chen, S., Jin, Z. and Liu, G. 2006. Overexpression of VTE1 from *Arabidopsis* resulting in high vitamin E accumulation and salt stress tolerance increase in tobacco plant. *J. Applied. Enviro Biol. Chinese*, 12 (4): 468-471.

Halliwell, B. 1987. Oxidative damage, lipid peroxidation and antioxidant protection in chloroplasts. *Chem. Phys. Lip*, 44: 327 -340.

Halliwell, B. 1999. Antioxidant defence mechanisms: from the beginning to the end. *Free Radic. Res*, 31 (4): 261-272.

Halliwell, B. 2006. Oxidative stress and neuro degeneration where are we now? *J. Neurochem*, 97: 1634-1658.

Halliwell, B. and Gutteridge, J.M.C. 1989. In Free Radicals in Biology Medicine. *Claredon Press, Oxford*.

Halliwell, B. and Whiteman, M. 2004. Measuring reactive oxygen species and oxidative damage in vivo and in cell culture. *J. Pharmacol. Br*, 142: 231-255.

Hammond Kosack, K.E. and Jones, J.D.G. 1996. Resistance gene dependent plant defense responses. *Plant Cell*, 8, 1773-1791.

Harman, D. 1956. Aging a theory based on free radical and radiation chemistry. *J. Gerontol*, 11: 298-300.

Hernandez, J.A., Escobar, C., Creissen, G. and Mullilneaux, P.M. 2004. Role of hydrogen peroxide and the redox state of ascorbate in the induction of antioxidant enzymes in pea leaves under excess light stress. *Func. Plant Biol*, 31: 359-368.

Hess, F.D. 2000. Light dependent herbicides: overview. *Weed Sci*, 48: 160 - 170.

Hidega, E., Manob, J. and Horvátha. G.V. 2008. Adverse effects of gene stacking on improving UV tolerance in *Nicotiana tabacum*. *Physiol. Plant*, 133:135.

Hojati, M., Modarres Sanavy SAM., Karimi, M. and Ghanati, F. 2010. Responses of growth and antioxidant systems in *Carthamus tinctorius* L. Under water deficit stress. *Acta.Physiol. Plant*, 33: 105-112.

Hossain, M.A., Nakano, Y. and Asada, K. 1984. Monodehydro ascorbate reductase in spinach chloroplasts and its participation in regeneration of ascorbate for scavenging hydrogen peroxide. *Plant and Cell Physiology*, 25 (3): 385-395.

Hossain, Z., Lopez climent, M.F., Arbona, V., Perez clemente R.M. and Gomez cadenas, A. 2009. Modulation of the antioxidant system in citrus under water logging and subsequent drainage. *J Plant Physiol,* 166:1391-1404.

Hu, J. P., Aguirre, M., Peto, C., Alonso, J., Ecker, J., and Chory, J. 2002. A role for peroxisomes in photomorphogenesis and development of *Arabidopsis.* Science, 297: 405–409.

Huang, B. and Xu, C. 2008. Identification and characterization of proteins associated with plant tolerance to heat stress. *J. Integr. Plant Biol, 50*:1230-1237.

Huang, R., Xia, R., Hu, L., Lu, Y. and Wang, M. 2007. Antioxidant activity and oxygen scavenging system in orange pulp during fruit ripening and maturation. *Sci. Hort,* 113: 166-172.

Hückelhoven, R. 2007. Cell wall-associated mechanisms of disease resistance and susceptibility. *Annu. Rev. Phytopathol,* 45: 101–127.

Hurkman, W.J., Vensel, W.H., Tanaka, C.K., Whitehand, L.and Altenbach, S.B. 2009. Effect of high temperature on albumin and globulin accumulation in the endosperm proteome of the developing heat grain. *J. Cereal Sci,* 49: 12-23.

Jain, K., Kataria, S. and Guruprasad, K.N. 2004. Oxyradicals under UV-B stress and their quenching by antioxidants. *Indian J. Exp. Biol.* 42: 884-892.

Jansen, M.A.K., Gaba, V. and Greenberg, B.M. 1998. Higher plants and UV-B radiation balancing damage, repair and acclimation.*Trends Plant Sci,* 3:131-135.

Jimenez, A., Hernandez, J.A., Del Rio, L.A. and Sevilla, F. 1997. Evidence for the presence of the ascorbate-glutathione cycle in mitochondria and peroxisomes of pea leaves. *Plant Physiology,* 114: 1:275-284.

Joshi, S.M., Adhikari, K.S. and Bhagat, K.N. 2004. Physico-chemical changes in apple hybrid during fruit development and maturity var. *Chaubattia swarmima. Prog. Hort,* 36 (2): 349-351.

Jovanovic, S.V., Steenken, S., Tosic, M., Marjanovic, B. and Simic, M.G. 1994 Flavonoids as antioxidants. *J. Am. Chem. Soc,* 116: 4846–4851.

Kagan, V.E., Fabisiak, J.P. and Quinn, P.J. 2000. Coenzyme Q and vitamin E need each other as antioxidants. *Protoplasm.*214 (2):11-18.

Kamal-Eldin, A. and Appelqvist L.A. 1996. The chemistry and antioxidant properties of tocopherols and tocotrienols. *Lipids, 31*: 671 - 701.

Karuppanapandian, T. and Manoharan, K. 2008 Uptake and translocation of tri- and hexa-valent chromium and their effects on black gram (*Vigna mungo* L.) roots. *J. Plant Biol,* 51: 192-201.

Karuppanapandian, T., Moon, J.C., Kim, C., Manoharan, K. and Kim, W. 2011. Reactive oxygen species in plants: their generation, signal transduction, and scavenging mechanisms. *AJCS.* 5 (6): 709 - 725.

Karuppanapandian, T., Sinha, P.B., Kamarul Haniya, A. and Manoharan, K. 2006. Differential antioxidative responses of ascorbate-glutathione cycle enzymes and metabolites to chromium stress in green gram (*Vigna radiata* L. Wilczek) leaves. *J.Plant Biol*, 49: 440-447.

Kato, M. and Shimizu, S. 1987. Chlorophyll metabolism in higher plants. VII. Chlorophyll degradation in senescing tobacco leaves phenolic dependent peroxidative degradation. *Can. J. Bot*, 65: 729-735.

Khan, M.H. and Patra, H.K. 2007. Sodium chloride and cadmium induced oxidative stress and antioxidant response in *Calamus tenuis* leaves. *Indian J. Plant Physiol*, 12: 34-40.

Kim, J.H. and Lee, C.H. 2005. In viv*o* deleterious effects specific to reactive oxygen species on photo system I and II after photo-oxidative treatments of rice (*Oryza sativa* L.) leaves. *Plant Sci.* 168: 1115 - 1125.

Klapheck, S., Zimmer, I. and Cosse, H. 1990. Scavenging of hydrogen peroxide in the endosperm of *Ricinus communis* by *ascorbate peroxidase*. *Plant Cell Physiol*, 31: 1005-1013.

Kochhar, S., Watkins, C.B., Conklin, P.L. and Brown, S.K. 2003. A quantitative and qualitative analysis of antioxidant enzymes in relation to susceptibility of apples to superficial scald. *J. Amer. Soc. Hort. Sci*, 128 (6): 910-916.

Koji, Y., Shiro, M., Michio, K., Mitsutaka, T. and Hiroshi, M. 2009 Antioxidant capacity and damages caused by salinity stress in apical and basal regions of rice leaf. *Plant Prod, Sci*, 12: 319-326.

Kopriva, S. and Rennenberg, H. 2004. Control of sulphate assimilation by glutathione synthesis interactions with N and C metabolism, *Journal of Experimental Botany*, 55: 1831-1842.

Kroniger, W., Rennenberg, H. and Polle, A. 1993. Developmental changes of Cu, Zn and Mn-superoxide dismutase isozymes in seedlings and needles of Norway spruce (*Picea abies* L.). *Plant Cell Physiol*, 34: 1145-1149.

Kuk, Y.I., Shin, J.S., Burgos, N.R., Hwang, T.E., Han, O., Cho, B.H., Jung, S. and Guh, J.O. 2003. Antioxidative enzymes offer protection from chilling damage in rice plants. *Crop Sci*, 43: 2109-2117.

Kukreja, S., Nandwal, A. S., Kumar, N., Sharma, S. K., Kundu, B. S., Unvi, V. and Sharma, P. K. 2006. Response of chickpea roots to short-term salinization and desalinization: Plant water status, ethylene evolution, antioxidant activity and membrane integrity. *Physiol. Mol. Biol. Plants*, 12: 67-73.

Kumar S., Yadav, P. Jain, V. and Malhotra,S. 2014. Isozymes of antioxidative enzymes during ripening and storage of ber (*Ziziphus mauritiana* Lamk.). *J. Food. Sci. Technol.* 51 (2): 329-334.

Kumar, S., Yadav, P., Jain V. and Malhotra, S.P. 2011. Evaluation of oxidative stress and antioxidative system in ber (*Zizyphus mauritiana* L). fruits during storage. *J. Food Biochem*, 35 (5): 1434 - 1442.

Kumar, S., Yadav, P., Jain,V. and Malhotra,S.P. 2011.Oxidative Stress and Antioxidative System in Ripening Ber (*Ziziphus mauritiana* Lam). *Fruits Food Technol. Biotechnol.* 49 (4): 453-459.

Kwasniewski, M., Karolina, C., Joianta, K., Julia kusak, Kamil siwiniski, and Iwona szarejko, 2013. Accumulation of peroxidise related reactive oxygen species in trichoblasts correlates with root hair initiation in barley. *J. Plant Physiol.* 170 (2): 185-195.

Lamb, C. and Dixon, R.A. 1997. The oxidative burst in plant disease resistance. *Annu.Rev. Plant Physiol.Plant Mol Biol,* 48: 251-275.

Larigueta, P., Meyerb, M. D., Penelc, C. and Dunandd, C. 2008. Involvement of reactive oxygen species and peroxidases in the multiple pathways leading to *Arabidopsis* seed germination. *Physiol. Plant,* 133: 40.

Larkindale, J. and Knight, M.R. 2002. Protection against heat stress induced oxidative damage in *Arabidopsis* involves calcium, abscisic acid, ethylene, and salicylic acid. *Plant Physiol,* 128: 682 - 695.

Lester, G. and Stein, E. 1993. Plasma membrane physicochemical changes during maturation and postharvest storage of muskmelon fruit. *J. Am. Soc. Hortic. Sci,* 118: 223-227.

Li, J.M., and Jin, H. 2007. Regulation of brassinosteroid signalling, *Trends Plant Science,* 12:37-41.

Lurie, S. and Ben-Arie, R. 1983. Microsomal membrane changes during the ripening of apple fruit. *Plant Physiol,* 73: 636-638.

Mackerness, S.A. 2000. Plant responses to ultraviolet-B (280-320nm) stress. What are the key regulators? *Plant growth Regulation,* 32: 27 - 39.

Mackerness, S.A., John, C. F., Jordan, B. and Thomas, B. 2001. Early signaling components in ultraviolet-B responses. Distinct roles for different reactive oxygen species and nitric oxide. *FEBS Letter,* 489: 237-242.

Mahan, J.R. and Wanjura, D.F. 2005. Seasonal patterns of glutathione and ascorbate metabolism in field grown cotton under water stress. *Crop Sci,* 45: 193-201.

Mahdavian, K.M., Ghorbanli, F. and Kalantari, K.M. 2008. The effects of ultraviolet radiation on the contents of chlorophyll, flavonoid, anthocyanin and proline in *Capsicum annuum* L. *Turk. J. Bot,* 32: 25 - 33.

Mallick, N. and Mohn, F.H. 2000. Reactive oxygen species: response of algal cells, *J. Plant Physiol. 157*: 183 - 193.

Mandhania, S., Madan S. and Sawhney, V. 2006. Antioxidant defense mechanism under salt stress in wheat seedlings. *Biol. Plant,* 50: 227-231.

McCord, J.M. and Fridovich, I. 1969. *Superoxide dismutase-* An enzymic function for erythrocuprein (Hemocuprein). *J. Biol. Chem,* 244: 6049-6055.

Mhamdi, A., Queval, G., Chaouch, S., Vanderauwera, S., Breusegem, F.V. and Noctor, G. 2010. Catalase function in plants a focus on *Arabidopsis* mutants as stress-mimic models. *J. Exp. Bot,* 61:4197-4220.

Mishra, N.P., Mishra, R.K. and Singhal, G.S. 1993. Changes in the activities of antioxidant enzymes during exposure of intact wheat leaves to strong visible light at different temperatures in the presence of protein synthesis inhibitors. *Plant Physiol,* 107: 903-910.

Mittler, R. 2002. Oxidative stress, antioxidants and stress tolerance. *Trends Plant Sci.* 9: 405 - 410.

Mittler, R. and Zilinskas, B.A. 1991. Purification and characterization of pea cytosolic ascrobate peroxidase. *Plant Physiol,* 97: 962-968.

Mittler, R. and Zilinskas, B.A. 1992. Molecular cloning and characterization of a gene encoding pea cytosolic ascorbate Peroxidise. *J.Biological Chemistry,* 267 (30): 21802-21807.

Mittler, R., Vanderauwera, S., Gollery, M. and Van Breusegem, F. 2004. Reactive oxygen gene network of plants. *Trends. Plant. Sci,* 9: 490-498.

Mittler, R., Vanderauwera, S., Suzuki, N., Miller, G., Tognetti, V. B., Vandepoele, K., Gollery, M., Shulaev, V. and Van Breusegem, F. 2011. ROS signaling the new wave? *Trends Plant Sci,* 16 (6): 300-309.

Miyake, C., Schreiber, U., Hormann, H., Sano, S. and Asada, K. 1996. The FAD-enzyme monodehydroascorbate radical reductase mediates photoproduction of superoxide radicals in spinach thylakoid membranes. *Plant and Cell Physiology,* 39 (8): 821-829.

Møller, I. M. and Sweetlove, L. J. 2010. ROS signalling—specificity is required. *Trends Plant Sci.* 15 (7): 370-4.

Moller, M. 2001. Plant mitochondria and oxidative stress,electron transport, NADPH turnover, and metabolism of reactive oxygen species. *Annu. Rev. Plant Physiol. Mol. Biol,* 52:561-591.

Mondal, K., Malhotra, S.P., Jain, V. and Singh, R. 2009. Oxidative stress and antioxidant systems in guava (*Psidium guajava* L.) fruits during ripening *Physiol. Mol. Biol. Plants.* 15;(4): 327-334.

Mondal, K., Sharma, N. S., Malhotra, S. P., Dhawan, K. and Singh, R. 2003. Oxidative stress and antioxidant systems in tomato fruits during storage. *J. Food Biochem.* 27: 515-527.

Mondal, K., Sharma, N.S., Malhotra, S.P., Dhawan, K. and Singh, R. 2006. Oxidative stress and antioxidative systems in tomato fruits stored under normal and hypoxic conditions. *Physiol. Mol. Biol. Plants,* 12 (2): 145-150.

Mondal, K., Sharma, N.S., Malhotra, S.P., Dhawan, K. and Singh, R. 2004. Antioxidant systems in ripening tomato fruits. *Plant. Biol,* 48 (1): 49-53.

Morita, S., Kaminaka, H., Masumura, T. and Tanaka, K. 1999. Induction of rice cytosolic ascorbate peroxidase mRNA by oxidative stress. The involvement of hydrogen peroxide in oxidative stress signaling, *Plant Cell Physiol,* 40: 417 - 422.

Moussa, H.R. and Abdel Aziz, S.M. 2008. Comparative response of drought tolerant and drought sensitive maize genotypes to water stress. *J. Crop. Sci,* 1:31-36.

Mukherjee, S.P. and Choudhuri, M.A. 1983. Implications of water stress induced changes in the levels of endogenous ascorbic acid and hydrogen peroxide in *Vigna* seedlings. *Physiol. Plant,* 58: 166-170.

Munne, S., Bosch, K., Schwarz, and Alegre.1999. Enhanced formation of α-tocopherol and highly oxidized abietane diterpenes in water-stressed rosemary plants. *Plant Physiol,* 121 (3): 1047-1052.

Nakajima, S., Sugiyama, M., Iwai, S., Hitomi, K., Otoshi, E., Kim, S.T., Jiang, C.Z., Todo, T., Britt, A.B. and Yamamoto, K. 1998. Cloning and characterization of a gene (UVR3) required for photo repair of 6–4 photoproducts in *Arabidopsis thaliana. Nucleic Acids Res,* 26: 638-644.

Nakano, Y. and Asada, K. 1987. Purification of ascorbate peroxidase in spinach chloroplasts. Its inactivation in ascorbate depleted medium and reactivation by monodehydroascorbate radical. *Plant Cell Physiol,* 28 (1): 131-140.

Narendra, S., Venkataramani, S., Shen, G., Wang, J., Pasapula, V, Lin, Y., Kornyeyev, D., Holaday, A.S. and Zhang, H. 2006. The *Arabidopsis* ascorbate peroxidise is a peroxisomal membrane-bound antioxidant enzyme and is dispensable for *Arabidopsis* growth and development. *J. Exp. Bot,* 57: 3033-3042.

Nawkar, G.M., Maibam, P., Jung, J.H., Sahi, V.P., Sang S.Y. and Kang H.O. 2013. UV-Induced Cell Death in Plants, *Int. J. Mol. Sci,* 14: 1608 - 1628.

Nila, A.G., Sandalio, L.M., Lopez, M.G., Gomez, M., Del Rio, L.A, and Gomez Lim, M.A. 2006. Expression of a peroxisome proliferator-activated receptor gene (xPPARa) from *Xenopus.*

Nishikawa, F., Kato, M., Hyodo, H., Ikoma, Y., Sugiura, M, and Yano, M. 2003. Ascorbate metabolism in harvested broccoli. *J. Exp. Bot,* 54: 2439-2448.

Noctor, G. and Foyer, C. 1998. Ascorbate and glutathione: Keeping active oxygen under control. *Annu. Rev. Plant Physiol. Plant Mol. Biol,* 49: 249-279.

Noctor, G., Paepe, R.D, and Foyer, C.H. 2006. Mitochondrial redox biology and homeostasis in plants.*Trends Plant Sci,* 12: 125-134.

Noctor, G., Veljovic-Jovanovic, S., Driscoll, S., Novitskaya L. and Foyer, C.H. 2002. Drought and oxidative load in the leaves of C3 plants. A predominant role for photorespiration *Ann. Bot,* 89: 841-850.

Nyathi, Y. and Baker, A. 2006. Plant peroxisomes as a source of signalling molecules. *Biochimica. Biophysica. Acta,* 1763: 1478-1495.

O'Brien, J.A., Daudi, A., Butt V.S. and Bolwell G.P. 2012. Reactive oxygen species and their role in plant defence and cell wall metabolism. *Plant,* DOI 10.1007/s00425-012-1696-9.

Pal, R.S., Agrawal, P.K, and Bhatt, J.C. 2013. Molecular Approach towards the Understanding of Defensive Systems against Oxidative Stress in Plant. A Critical Review *Int. J. Pharm. Sci. Rev. Res,* 22 (2): 131-138.

Paliyath, G, and Droillard, M.J. 1992.The mechanisms of membrane deterioration and disassembly during senescence. *Plant Physiol.Biochem,* 30:789-812.

Palma, T., Marangoni, A.G, and Stanley, D.W. 1995 Environmental stress affect tomato microsomal membrane function differently than natural ripening and senescence. *Postharvest Biol. Technol*, 6: 257-273.

Pan, Y., Wu, L.J, and Yu, Z.L. 2006. Effect of salt and drought stress on antioxidant enzymes activities and SOD isoenzymes of liquorice (*Glycyrrhiza uralensis Fisch*). *Plant Growth Regul*, 49: 157-165.

Panda, S. and Choudhury, S. 2005. Chromium stress in plants. *Brazilian J. Plant Physiol*, 17: 95-102.

Panda, S., Chaudhury, I, and Khan, M, 2003. Heavy metals induce lipid peroxidation and affect antioxidants in wheat leaves. *Biol. Plant*, 46: 289-294.

Panda, S.K. 2001. Response of green gram seeds under salinity stress. *Indian J. Plant Physiol*, 6: 438-440.

Pandhir. and Shekhon. 2006. Reactive oxygen species and antioxidants in plants. An overview. *J. Plant Biochem. Biotech*, 15: 71-78.

Pfannschmidt, T. 2003. Chloroplast redox signals,How photosynthesis controls its own genes. *Trends Plant Sci*,8:1.

Pinto, E., Lentheric, I., Vendrell, M. and Larrigaudiere, C. 2001. Role of fermentative and antioxidant metabolisms in the induction of core browning in controlled atmosphere stored pears. *J. Sci. Food Agric*, 81: 364-370.

Pnueli, L., Liang, H.,Rozenberg M, and Mittler, R. 2003. Growth suppression, altered stomatal responses, and augmented induction of heat shock proteins in cytosolic ascorbate peroxidise. (Apx1) deficient *Arabidopsis* plants. *J. Plant*, 34: 187-203.

Pospíšil, P. 2012. Molecular mechanisms of production and scavenging of reactive oxygen species by photosystem II. *Biochimica. Biophysica. Acta*, 1817: 218-231.

Puntarulo, S., Sanchez, R.A. and Boveris, A. 1988. Hydrogen peroxide metabolism in *soybean* embryonic axes at the onset of germination. *Plant Physiol*, 86: 626-630.

Quan, L.J. Zhang, B., Shi, W.W. and Li, H.Y. 2008. Hydrogen peroxide in plants a versatile molecule of reective oxygen specie network. *Journal Integrat. Plant Biol*, 50: 2-18.

Rani, B., Jain, V., Chhabra, M. L., Dhawan, K., Kumari N. and Yadav P. 2012. Oxidative stress and antioxidative system in *Brassica juncea* under high temperature stress, *Annals Biol.*, 28 (2): 110 - 115.

Rao, M.V., Paliyath, G. and Ormrod, D.P. 1996. Ultraviolet-B and ozone-induced biochemical changes in antioxidant enzymes of *Arabidopsis thaliana. Plant Physiol*, 110:125-136.

Rasmusson, A.G., Soole, K.L, and Elthon, T.E. 2004. Alternative NADPH dehydrogenases of plant mitochondria. *Annu. Rev. Plant Biol*, 55:23-39.

Reumann, S., Ma, C., Lemke, S and Babujee, L. 2004. AraPerox a database of putative *Arabidopsis* proteins from plant peroxisomes, *Plant Physiol, 136*: 2587-2608.

Rhoads, D.M., Umbach, A.L., Subbaiah, C.C, and Siedow J.N. 2006. Mitochondrial reactive oxygen species. Contribution to oxidative stress and inter organellar signaling. *Plant Physiol*, 141: 357-366.

Rodriguez, A.A., Grunberg, K.A. and Taleisnik, E.L. 2002. Reactive oxygen species in the elongation zone of maize leaves are necessary for leaf extension. *Plant Physiol*, 129 (4): 1627-1632.

Rogiers, S.Y., Kumar, G.N.M. and Knowles, N.R. 1998.Maturation and ripening of fruit of *Amelanchier alnifolia* nut are accompanied by increasing oxidative stress, *Ann. Bot*, 81: 203-211.

Romero Puertas, M.C., McCarthy, I., Sandalio, L.M., Palma, J.M., Corpas, F.J., Gomez, M. and Del Rio, L.A. 1999. Cadmium toxicity and oxidative metabolism of *pea* leaf peroxisomes. *Free. Radic. Res*, 31: 25-31.

Rucinska, R., Waplak, S. and Gwozdz, E.A. 1999. Free radicals formation and activity of antioxidant enzymes in lupin roots exposed to lead. *Plant Physiol. Biochem*, 37:187-194.

Sabarinath, S., Bharti, S. and Khanna Chopra. R. 2005. *Superoxide dismutase* and abiotic stress tolerance. *Physiol. Mol. Biol. Plants*, 11 (2): 187-198.

Sano, S., Ueda, M., Kitajima, S., Takeda, T., Shigeoka, S., Kurano, N., Miyachi, S., Miyake, C. and Yokota, A. 2001. Characterization of *ascorbate peroxidases* from unicellular red alga *Galdieria partita*. *Plant Cell Physiol*, 42: 433-440.

Sattler, S.E., Gilliland, L.U., Magallanes L.M., Pollard, M. and Dellapenna, D. 2004. Vitamin E is essential for seed longevity, and for preventing lipid peroxidation during germination. *Plant Cell*, 16: 1419-1432.

Schafer, F.Q. and Buettner, G.H. 2001. Redox environment of the cell as viewed through the redox state of the glutathione disulfide/glutathione couple, *Free Radical Biology and Medicine*, 30: 1191-1212.

Schrader, M. and Fahimi, D. 2006. Peroxisomes and oxidative stress *Biochimica. Biophysica. Acta*, 1763 (12): 1755-1766.

Schutzendubel, A., Schwanz, P., Teichmann, T., Gross, K., Langenfeld-Heyser, R., Godbold, D.L. and Polle, A. 2001. Cadmium induced changes in antioxidative systems, H_2O_2 content and differentiation in pine (*Pinus sylvestris*) roots. *Plant Physiol*, 127: 887- 898.

Schutzendubel. A. and Polle, A. 2002. Plant responses to a biotic stresses heavy metal induced oxidative stress and protection by mycorrhization. *J.Exp.Bot*, 53:1351-1365.

Sgherri, C.L.M., Pinzino, C. and Navari-Izzo, F. 1996. Sunflower seedlings subjected to increasing stress by water deficit Changes in O_2^- production related to the composition of thylakoid membranes. *Plant physiol*, 96:446-52.

Sharma, P. and Dubey R.S. 2005. Modulation of nitrate reductase activity in rice seedlings under aluminium toxicity and water stress: role of osmolytes as enzyme protectant. *J. Plant Physiol*, 162: 854-864.

Sharma, Y.K., Leon, J., Raskin, I. and Davis, K.R. 1996. Ozone-induced responses in *Arabidopsis thaliana* the role of salicylic acid in the accumulation of defense-related transcripts and induced resistance. *Proc. Natl. Acad. Sci. USA*, 93: 5099-6104.

Shi, X. and Dalal, N.S. 1989. Chromium (V) and hydroxyl radical formation during the *Glutathione reductase* catalyzed reduction of chromium (VI). *Biochem. Biophys. Res*, 163:627-634.

Sieferman-Harms, D. 1987. The light harvesting function of carotenoids in photosynthetic membrane, *Plant Physiol*, 69: 561 -568.

Singh, S., Mishra, S., Kumari, R. and Agrawal, S.B. 2009. Response of supplemen-tal ultraviolet-B and nickel on pigments, metabolites, and antioxidants of *Pisum sativum*. L. *J. Environ.biol*, 30.

Smeets, K., Cupers, A., Lambrechts, A., Semane, B., Hoet, P., Vanlaere, A. and Lambrechts, J. 2005. Induction of oxidative stress and antioxidative mechanisms in *Phaseolus vulgaris* after Cd application. *Plant Physiol. Biochem*,.43:437.

Smirnoff, N. 1998. Plant resistance to environmental stress. *Curr. Opin. In Biotech*, 9: 214 - 219.

Smirnoff, N. 2000. Ascorbic acid metabolism and functions of a multifaceted molecule, *Curr. Opin. Plant Biol*, 3: 229 - 235.

Stevens, R., Page, D., Gouble, B., Garchery, C., Zamir, D. and Causse, M. 2008. Tomato fruit ascorbic acid content is linked with monodehydroascorbate reductase activity and tolerance to chilling stress *Plant. Cell. Environ*, 31: 1086-1096.

Takahama, U., Hirotsu, M. and Oniki, T. 1999. Age dependent changes in leaves of ascorbic acid and chlorogenic acid and activities of *peroxidase* and *superoxide dismutase* in the apoplast of tobacco leaves. Mechanism of the oxidation of chlorogenic acid in the apoplast. *Plant Cell Physiol*, 40: 716-724.

Takemoto, D., Tanaka, A. and Scott, B. 2007. NADPH oxidases in fungi diverse roles of reactive oxygen species in fungal cellular diverentiation. *Fungal Genet Biol*, 44:1065-1076.

Tanaka, K. 1994. Tolerance to herbicides and air pollutants. In Causes of Photooxidative Stress and Amelioration of Defense Systems in Plants (eds. Foyer, C.H. and Mullineaux, P.) *CRC Press Inc. Boca Raton*.F.L, 365-378.

Thirupathi, K., Juncheol Moon., Changsoo Kim., Kumariah.,M. and Wookim. 2011. Reactive oxygen species in plants their generation, signal transduction, and scavenging mechanisms *AJCS*, 5 (6): 709-725.

Trebst, A. and Draber, W. 1986. Inhibitors of photo system II and the topology of herbicide and QB binding polypeptide in the thylakoid membrane. *Photosynth. Res*, 10: 381.

Tsukagoshi, H. 2012. Defective root growth triggered by oxidative stress is controlled through the expression of cycle related genes. *Plant Sci*.197: 30-9.

Ushimaru, T., Nakagawa, T., Fujioka,Y., Daicho, K., Naito, M., Yamauchi, Y., Nonaka, H., Amako, K., Yamawaki, K. and Murata, N. 2006. Transgenic *Arabidopsis* plants expressing the rice dehydroascorbate reductase gene are resistant to salt stress. *J. Plant Physiol*, 163: 1179-1184.

Vass, I. and Styring, S. 1993. Characterization of chlorophyll triplet promoting states in photosystem II sequentially induced during photoinhibition. *Biochem*, 32: 3334-3341.

Vellosillo, T., Vicente, J., Kulasekaran, S., Hamberg, M. and Castresana, C. 2010. Emerging complexity in reactive oxygen species production and signaling during the response of plants to pathogens. *Plant Physiol*,154: 444-448.

Wang, J.P, Li, Y.L. and Zhang, J.G. 2008. Effect of high-temperature and excessive-light stress on APX activity in apple peel. *Acta Agri Boreali.Sin*, 23:144-147.

Willekens, H., Chamnongpol, S., Davey, M., Schraudner, M. and Langebartels, C. 1997. Catalase is a sink for H_2O_2 and is indispensible for stress defense in C_3 plants. *EMBO J*, 16: 4806-4816.

Wise, R.R. and Naylor, A.W. 1987. Chilling enhanced photooxidation. Evidence for the role of singlet oxygen and superoxide in the breakdown of pigments and endogenous antioxidants. *Plant Physiol*, 83: 278-282.

Wrzaczeka, M., Broschea, M., Kollistb H. and Kangasjärvia. J. 2008 When proteins go postal a new regulator in ROS induced cell death. *Physiol. Plant*, 13: 3:81.

Xiang, C., Werner, B.L., Christensen, E.M. and Oliver, D.J. 2001. The biological functions of glutathione revisted in *Arabidopsis* transgenic plants with altered glutathione levels. *Plant Physiol*, 126: 564-574.

Xiao Jian Xia., Yue Yuan Huang., Li Wang., Li Feng Huang.,Yun Long Yu., Yan Hong Zhou. and Jing Quan Yu. 2006. Pesticides-induced depression of photosynthesis was alleviated by 24-epibrassinolide pretreatment in *Cucumis sativus* L. Pestic. Biochem and Physiol, 86: 42-48.

Yadav, P., Kumar,S., Jain, V. and Malhotra,S.P. 2012. Cell Wall Metabolism of Two Varieties of Ber (*Ziziphus mauritiana* Lam). Fruit During ripening. *Food Technol. Biotechnol*, 50 (4): 467-472.

Yadav, S., Bhatia, V.S. and Guruprasad, K.N. 2006. Oxyradical accumulation and their detoxifications by ascorbic acid and α-tocopherol in soybean seeds during field weathering. *Indian J. Plant Physiol*, 11 (1): 28-35.

Yamaguchi Shinozaki, K. and Shinozaki, K. 1994. A novel *cis* element in an *Arabidopsis* gene is involved in responsiveness to drought, low-temperature, or high-salt stress. *Plant Cell*, 6 (2): 251-264.

Yang, Y. and Klessig, D.F. 1996. Isolation and characterization of a tobacco mosaic virus-inducible myb oncogene homolog from tobacco. *Proc. Natl. Acad. Sci. USA*, 93: 14972-14977.

Yogamoorthi, A. 2007. Artificial UV-B induced changes in pigmentation of marine diatom *Coscinodiscus gigas*. *J. Environ. Biol*, 28: 327-330.

Yoshida, K., Kaothien, P., Matsui, T., Kawaoka, A. and Shinmyo, A. 2002. Molecular biology and application of plant peroxidase genes. *Appl. Microbiol. Biotechnol,* 60 : 665-670.

Zhang, J. and Kirkham, M.B. 1996. Enzymatic responses of the ascorbate-glutathione cycle to drought in *Sorghum* and sunflower plants. *Plant Sci,* 113: 139-147.

Zhang, X., Zhang, L., Dong, F., Gao, J., David, W., Galbraith and Song, C.P. 2001. Hydrogen Peroxide is involved in Abscisic Acid-Induced Stomatal Closure in *Vicia faba. Plant Physiol,* 126 (4): 1438-1448.

Zhu, J. K. 2002. Salt and drought stress signaling transduction in plants. *Annu. Rev. Plant Biol,* 53: 247 - 273.

2016, Recent Advances in Plant Stress Physiology *Pages* **37–48**
Editors: ***Praduman Yadav, Sunil Kumar and Veena Jain***
Published by: **DAYA PUBLISHING HOUSE, NEW DELHI**

Chapter 2

Calcium Mediated Pathogen Defense System

Manu Pratap Gangola*

Department of Plant Sciences, College of Agriculture and Bioresources University of Saskatchewan, Saskatoon, SK S7N 5A8, Canada

ABSTRACT

Calcium ion (Ca^{2+}) is defined as universal intracellular secondary messenger in plants. It not only participates in regulation of various developmental processes but also supports plants to sense and respond to abiotic and biotic stimuli. Specific spatiotemporal changes in cytosolic Ca^{2+} concentration termed as "calcium signatures" are perceived and decoded by Ca^{2+} sensors like Calmodulins (CaM), Calcium dependent protein kinases (CDPKs), Calcineurin B-like proteins (CBLs) and their interacting kinases (CIPKs). The decoded information stimulates changes at structural, transcriptional, translational and post-translational levels leading to appropriate plant responses to various kinds of stimuli. The present book chapter is focused on calcium signatures, Ca^{2+} sensors and their effect on pathogen defence responses.

Keywords: *Calmodulin, Calcium dependent protein kinase, Plant defense, Signaling.*

Introduction

Plants are sessile organisms, thus have to cope with different kinds ofstresses. Therefore, plants have evolved varioussignaling mechanisms that can sense and respond to different environmental changes (abiotic and biotic factors) leading to their better adaptation. Among such mechanisms, calciumsignaling is one of the fundamental processes. Being a substantive macronutrient, calcium participates in various structural and physiological phenomena like cell division (formation

* *Author*: E-mail: gangola.manu@gmail.com

of microtubules supporting chromosomal movement), cell wall/membrane structure (as calcium pectate), pollen tube formation, nutrient metabolism and cell differentiation (Song *et al.*, 2008; Tuteja, 2009). In addition, recent studies confirmed calcium ion (Ca^{2+}) as an important intracellular secondary messenger in plant cells transforming different types of developmental and environmental stimuli into appropriate physiological responses (Batistic and Kudla, 2012). In *Arabidopsis*, 230 (162+68) calcium responsive (upregulated+downregulated) genes were identified indicating importance of Ca^{2+} signaling in plants (Kaplan *et al.*, 2006). Ca^{2+} signaling regulates gene expression, protein modifications, ionic balance, stomatal aperture, interaction with microbes and light/circadian signaling. Therefore, Ca^{2+} signaling is an essential component of plant defense system against pathogens that cause major loss to agricultural production worldwide (Lecourieux *et al.*, 2006). To further expand the understanding, this chapter briefly summarizes the initiation, perception, decoding and responses of calcium signaling during plant-pathogen interaction.

Signal Genesis: Increase in Free Cellular Ca^{2+} Concentration

Calcium Concentration in Plant Cell

Free Ca^{2+} is used as a secondary messenger in plant cell. Ca^{2+} is translocated from cytoplasm to other cell organelles and extracellular compartments in order to maintain cytosolic Ca^{2+} concentration (Ca^{2+} Cytosol) at 100-200 nM which is about 10^4 to 10^5 times less than that of organelles. Main reservoirs of Ca^{2+} in plant cell are vacuole, endoplasmic reticulum (ER) and apoplast containing 80/0.2-5, 2/0.05-0.5 and 1/0.3 mM of $Ca^{2+}{}_{T}/Ca^{2+}{}_{F}$ (total and free calcium ion concentration), respectively. Other organelles like nucleus, chloroplast, mitochondrion and peroxisome also reserve some $Ca^{2+}{}_{F}$ but in nM concentration range [(Figure 2.1) Reddy, 2001; Stael *et al.*, 2012]. During resting and stress conditions, Ca^{2+} transport is essential for ion homeostasis and signaling events. Ca^{2+} can be transported from extracellular (across plasma membrane) space or intracellular (from different cell organelles) pools into cytosolwhen needed. In plant cells, both intra- and extra- cellular Ca^{2+} transport is modulated by transporters that are also termed as Ca^{2+} signal decoding elements (Tuteja and Mahajan, 2007; Song *et al.*, 2008).

Transport of Ca^{2+}

Ca^{2+} transporters are prerequisite to maintain the cytosolic Ca^{2+} concentration during resting/unstressed phase and after the signaling event (Dodd *et al.*, 2010). Main categories of Ca^{2+}transporters are: (1) calcium-permeable ion channels, (2) Ca^{2+}-ATPases, and (3) Ca^{2+}/H^+ antiporters (Figure 2.1) of which latter two proteins transport Ca^{2+} to secretory system of ER and preclude mineral toxicity by removing other cations like Ni^{2+}, Zn^{2+}, Mg^{2+} and Mn^{2+} from cytosol. Ca^{2+}-ATPases having high Ca^{2+} affinity (1-10 μM) but low translocation capacity, are proposed to maintain $Ca^{2+}{}_{Cytosol}$ during rest phase while on the contrary side Ca^{2+}/H^+ antiporters with low Ca^{2+} affinity (10-15 μM) but high translocation capacity suggested to efflux Ca^{2+}cytosol during signaling process (Hirschi, 2001).

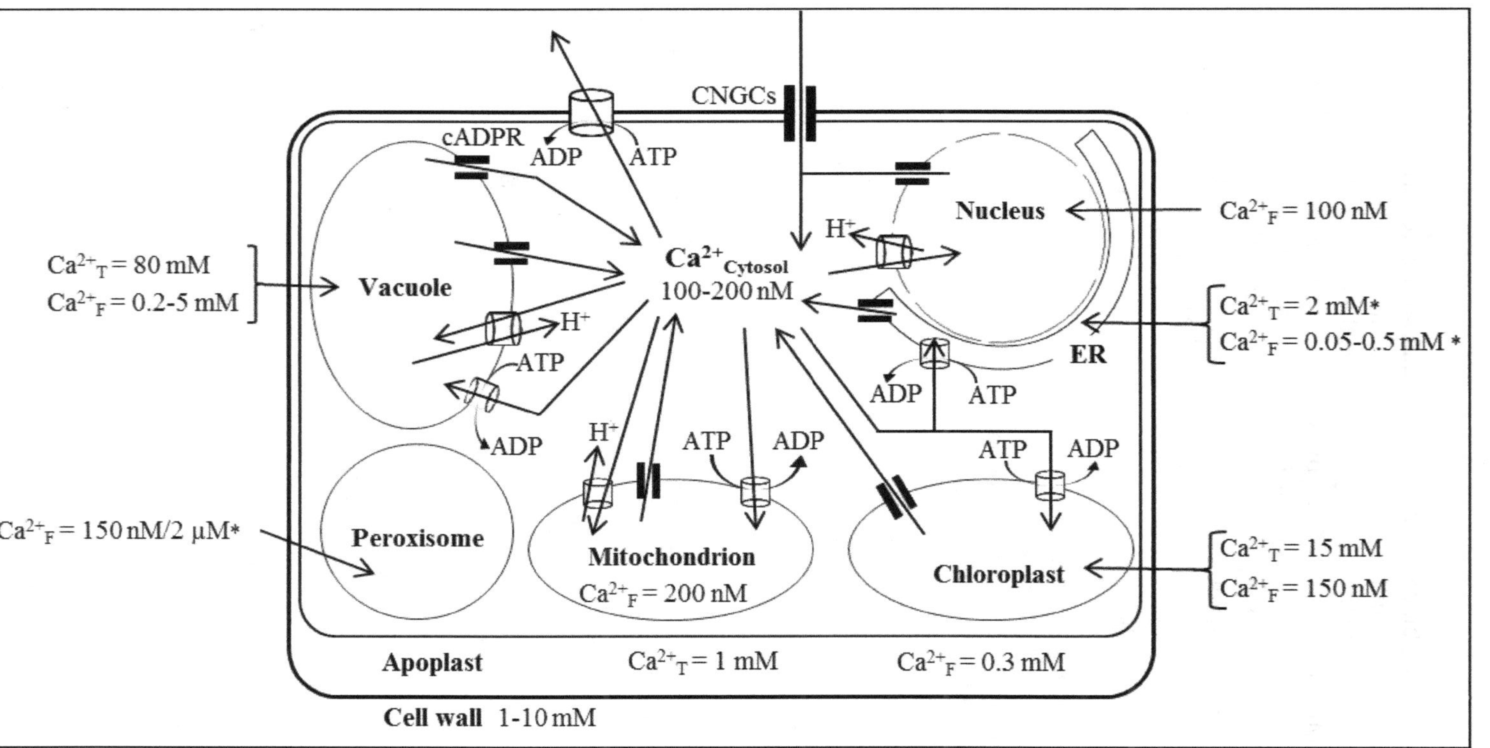

Figure 2.1: Schematic Representation of Ca^{2+} Concentration in different Cell Compartments Together with Ca^{2+} Permeable Channels (▌▌), Antiporters (⌸ with H^+) and Ca^{2+}-ATPases (⌸ with ATP and ADP). ER: Endoplasmic reticulum; Ca^{2+}_T: Total Ca^{2+} concentration; Ca^{2+}_F: Free Ca^{2+} concentration; $Ca^{2+}_{Cytosol}$: Ca^{2+} concentration in cytosol; ATP: Adenosine triphosphate; ADP: Adenosine diphosphate; cADPR: Cyclic ADP ribose; CNGCs: Cyclic nucleotide gated channels. Values taken from animal studies are marked with (*) (Reddy, 2001; Ma and Berkowitz, 2011; Stael *et al.,* 2012).

Calcium Permeable Ion Channels

Electrophysiological, biochemical and molecular studies reported Ca^{2+} channels as regulators of Ca^{2+} influx into cytosol (Song *et al.*, 2008). They are present on all plant membranes. Based on their voltage sensitivity, these channels are categorized into depolarisation-activated (DACC), hyperpolarisation-activated (HACC) and voltage-insensitive Ca^{2+} channels (VICC). Depolarization may be the result of activated anion channels during the perception of different stimuli (environmental, developmental or pathological) leading to the activation of DACC and increase in cytosolic Ca^{2+} concentration. DACC are proposed to be present in plasmamembrane of root, leaf mesophyll and suspension cultured cells. HACC can work either as mechanosensitive channels in guard cell or in cell signaling. They are reported from protoplasts of different cell types performing various physiological functions (White 2000; Sanders *et al.*, 2002). Besides these, other important Ca^{2+} channels are IP3 dependent-, cyclic ADP ribose- (cADPR) and cyclic nucleotide gated- (CNGC) channels. The first two types regulate the influx and efflux of Ca^{2+} across tonoplast (vacuolar membrane), respectively (Reddy, 2001) whereas CNGCs are found at plasma membrane and control the influx of Ca^{2+} into cytosol (Dodd *et al.*, 2010; Jammes*et al.*, 2011). Recent studies suggested the involvement of CNGCs in elevating Ca^{2+} concentration in cytosol during pathogen attack. Receptor proteins responsible for the pathogen signal perception are proposed to activate nucleotidylcyclases thus increasing the concentration of cyclic nucleotides (cAMP/cGMP) and opening CNGCs (Ma and Berkowitz, 2011).

Ca^{2+}-ATPases

Ca^{2+}-ATPases are characterized for ATP based ion translocation and therefore belong to superfamily of P-type ATPases. They represent < 0.1 per cent of the membrane proteins. On the basis of their sequence dissimilarity, Ca^{2+}-ATPases are further divided into two families: (a) IIA/ER (endoplasmic reticulum) type Ca^{2+}-ATPases (ECAs) and (b) IIB/auto-inhibited Ca^{2+}-ATPases (ACAs). ECAs represent broad nucleotide specificity but are devoid of N-terminal autoregulatory domain. Their Ca^{2+} affinity varies from 0.4 to 12 μM. On the other side, ACAs contain N-terminal autoregulatory domain along with a serine residue phosphorylation site. Therefore, ACAs' catalytic activity can be regulated with the binding of CaM (calmodulin; induces the activity) or CDPKs (Ca^{2+} dependent protein kinases; inhibit the activity by phosphorylation). CaM binding sites in these isoforms of ACAs showed diversification indicating different affinities or binding specificity for different CaM isoforms (Tuteja and Mahajan, 2007; Song *et al.*, 2008; Tuteja, 2009). In Arabidopsis, 4 members of ECAs and 10 members of ACAs have been reported by Axelsen and Palmgren (2001).

Ca^{2+}/H^{+} Antiporters

These efflux transporters are operated by pH gradient originated by vacuolar proton pumps thus might be involved in specificity of Ca^{2+} fluctuation in cytosol. Biochemical studies confirmed the presence of these antiporters at cell membrane and tonoplast. Therefore, they are attributed to accumulate Ca^{2+} in vacuole the main pool of Ca^{2+} among cell organelles (Song *et al.*, 2008). The first Ca^{2+}/H^{+} antiporter was identified in a Ca^{2+} hypersensitive mutant of *Saccharomyces cerevisiae* and named as "cation exchangers" (CAX; Hirschi *et al.*, 1996). In comparison to Ca^{2+}-ATPases, antiporters have low affinity but high transport capacity for Ca^{2+}. In addition, they might also transport Mg^{2+}(Tuteja, 2009).

The activity of different kinds of transporters generates specific spatio-temporal changes in cytosolic calcium ion concentration known as "Calcium Signatures". These signatures help a cell to differentiate among various types of stresses (including pathogen attack) leading to a particular biological response. Calcium signatures depend on cell/tissue/organ type, differ in amplitude/frequency and proceed in single transient/oscillation/repeated spikes (Lecourieux *et al.*, 2006). The concept of "Ca^{2+} memory" has also been proposed in plant cells on the basis of decreased elevation of Ca^{2+} concentration in cytosol upon induction of similar kind of stimuli (Tuteja, 2009). To generate a particular biological response against specific stimuli, calcium signatures target specific proteins.

Targets of Calcium Signalling: Ca^{2+} Binding Proteins

Calcium binding proteins associate calcium signatures with appropriate biological responses (Figure 2.2). Hence they are also considered as Ca^{2+} sensors. These proteins can be classified into two groups: (1) sensor relays and (2) sensor responders. Sensor relays includes calmodulin (CaM), CaM related proteins and calcineurin B-like proteins. Ca^{2+} elevation induces conformational change in such proteins that affect their further interaction with target proteins in terms of modulated activity/structure. However, sensor responders like Ca^{2+}-dependent protein kinases (CDPKs) undergo intramolecular conformational change induced by Ca^{2+} modifying their own activity/structure (Lecourieux *et al.*, 2006). These mechanisms provide flexibility and diversity to biological responses activated by Ca^{2+}. Besides this, calcium binding proteins are characterized to have helix-loop-helix(s) known as EF-hand domain attributed to bind with Ca^{2+} with high affinity. However, various proteins have also been reported to bind with Ca^{2+} but do not contain EF-hand domain. So, here we categorized different Ca^{2+} binding proteins on the basis of EF-hand domain presence.

Ca^{2+} Binding Proteins having EF-Hand Domain

EF-hand domain is named after E and F motif in the crystal structure of parvalbumin reported by Kretsinger and Nockolds in 1973. It is a Ca^{2+} binding domain of 29 amino acids composed of an α-helix E (amino acid 1-10), a loop (amino acid 10-21; actually binds to Ca^{2+}) and another α-helix F (amino acid 19-21). The Ca^{2+}affinity of EF-hand depends on the protein sequence especially regarding loop (Tuteja, 2009). To date, the three major classes of EF-hand containing proteins are CaM, CDPKs and calcineurin B-like proteins (CBLs).

Calmodulin (CaM)

CaM is an acidic, conserved and abundantly found Ca^{2+}-sensor in all eukaryotes. Its structure includes two globular domains (each having a pair of EF-hand) connected by an α-helical linker. Binding of Ca^{2+} to EF-hand exposes hydrophobic surfaces leading to the formation of high affinity binding sites for downstream

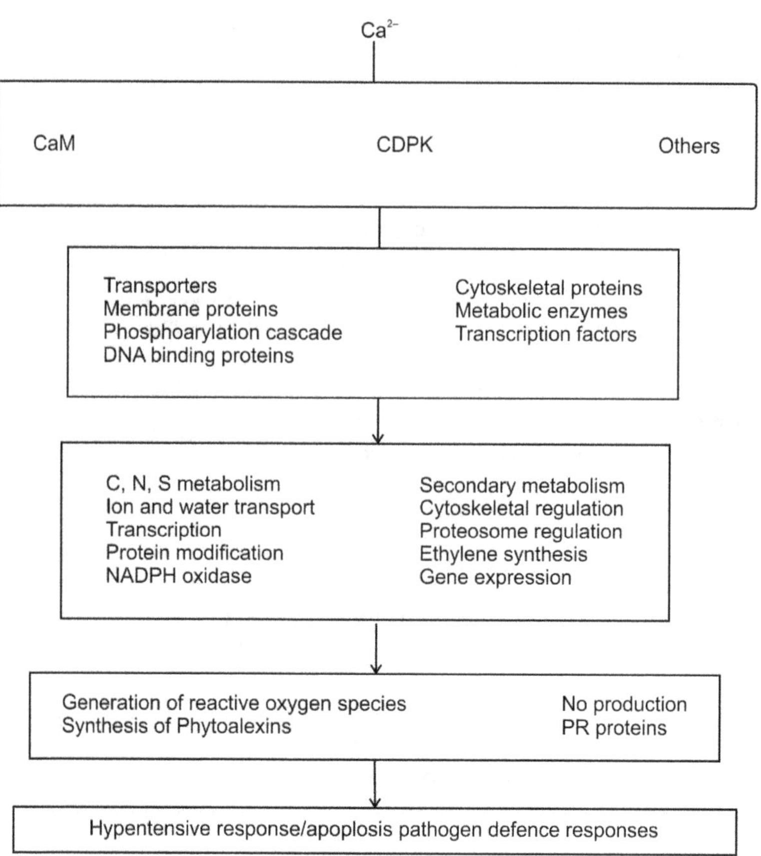

Figure 2.2: Summary of Important Ca^{2+} Sensors/Binding Proteins, their Targets and Biological Responses Specifically in Response to Pathogen Stimuli (Lecourieux *et al.*, 2006; Tuteja and Mahajan, 2007; DeFalco *et al.*, 2010; Dodd *et al.*, 2010).

target proteins (Lecourieux*et al.*, 2006). Some studies also suggest the involvement of electrostatic forces during CaM binding to DNA binding proteins. In contrast to highly conserved sequence of CaM, sequence of CaM binding motif of target proteins varies enormously supporting CaM to modulate the function/activity of diverse array of proteins like phosphorylation cascade, metabolic enzymes, transcription factors, channels pumps and cytoskeletal proteins. In most of the cases, CaM binding domains are Ca^{2+} dependent, however this process can be Ca^{2+} independent and such domains are termed as "IQ motif" (Reddy, 2001; Yang and Poovaiah, 2003; Bouche *et al.*, 2005; Lecourieux *et al.*, 2006; Tuteja, 2009; Cheval *et al.*, 2013).

The targets of CaM binding proteins are keep on increasing day by day.In animals, 3 mechanisms are reported for CaM mediated activation of proteins. The first mechanism called as "relieving autoinhibition" involves the conformational rearrangement of target protein upon CaM binding that results in displacement of pseudosubstrate inhibitory domain and induced enzymatic activity. In such proteins,

CaM binding site falls near/within autoinhibitory domain. The second mechanism is known as "active site remodelling" and explained in relation to anthrax adenyl cyclase (oedema factor). Oedema factor contains four regions able to recognize CaM. Therefore, CaM binding causes a 30 rotational change in the helical domain of the protein away from the catalytic site which in return activates the protein. The third mechanism termed as CaM induced dimerization" is proposed in case of K^+ channel. CaM interact with K^+ motif of Ca^{2+} activated K^+ channels in a Ca^{2+} dependent manner. In this interaction, C-terminal EF-hands support tethering to channel while N-terminal EF-hands cause Ca^{2+} induced dimerization that results in channels gating. Such mechanisms are also being studied in plants (Yang and Poovaiah, 2003).

Calcium Dependent Protein Kinases (CDPKs)

Calcium dependent protein phosphorylation is a thoroughly studied phenomenon in a variety of biological responses during various stresses including pathogen stimuli (Lee and Rudd, 2002). CDPK family includes three distinct classes of protein kinases categorized on the basis of structural (or sequential) differences (Figure 2.3): (1) CPK are conventional CDPKs. In such proteins, C-terminal end of kinase domain associated with regulatory CaM like domain (CaM-LD) that includes four functional Ca^{2+} binding EF-hands. (2) In Ca^{2+} and CaM dependent protein kinases (CCaMK), regulatory domain contains only three functional EF-hands and shows more sequence similarity with visinin (a different EF-hand protein) instead of CaM. (3) CDPK related kinases (CRK) exhibit more structural similarity to CPK rather than to CCaMK. They have the same CaM-LD domain with four EF-hands but are degenerated and non-functional(Reddy, 2001; Harper and Harmon, 2005; Liese and Romeis, 2013).

All the CDPKs consist of N-terminal end that can vary in sequence or length. N-terminal end contains sites for myristoylation which are important for membrane association. In addition, they also possess junction domain having autoinhibitory

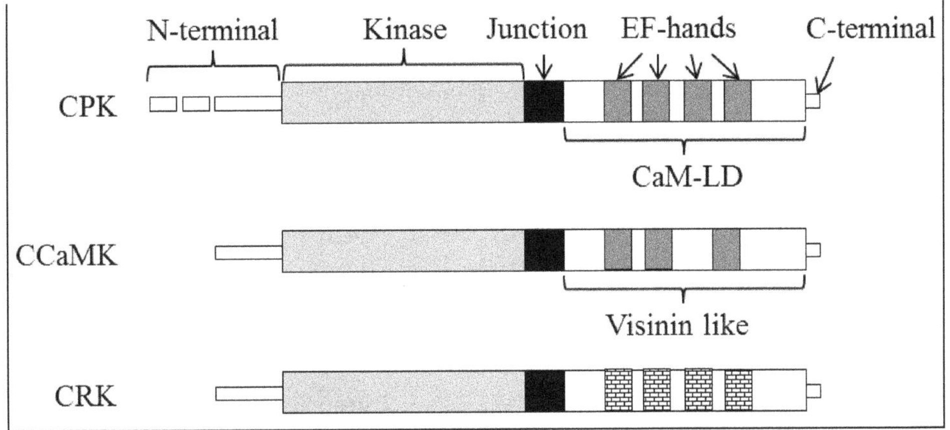

Figure 2.3: Structures of different Types of Calcium Dependent Protein Kinases (Harper and Harmon, 2005; Tuteja and Mahajan, 2007).

pseudosubstrate sequence and binding sites for interaction with other CDPKs. The function of C-terminal end is still unanswered. Functional mechanism of CDPKs is like "relieving autoinhibition" of CaM. Binding of Ca^{2+} with regulatory domain (CaM-LD or visinin like) alters the inatramolecular interaction of regulatory domain with autoinhibitor (junction domain) and activates the enzymatic activity. In plants, CDPKs are present in multiple isoforms at different cellular locations thus are able to induce diverse biological responses against external or internal stimuli (Harper and Harmon, 2005; DeFalco *et al.*, 2010; Hashimoto and Kudla, 2011; Liese and Romeis, 2013). CDPKs are strongly associated with plant-pathogen interaction and are capable of inducing hypersensitive response (HR)/programmed cell death (PCD)/generation of reactive oxygen species (Lee and Rudd, 2002; Hashimoto and Kudla, 2011; Liese and Romeis, 2013).

Calcineurin B-like Proteins (CBL)

It is comparatively new class of Ca^{2+} sensors characterized to have regulatory B-subunit of calcineurin and neuronal Ca^{2+} sensor in animals. For Ca^{2+} signal transduction, CBL interacts in a Ca^{2+} dependent manner to CIPK (CBL interacting protein kinases). Only few CBL-CIPK have been reported to date and only in response to the abiotic stress like cold and salinity stress (Lee and Rudd, 2002; Harper and Harmon, 2005; Lecourieux *et al.*, 2006; Tuteja, 2009; Hashimoto and Kudla, 2011; Kim, 2013).

Other EF-hand Domain Containing Ca^{2+} Binding Proteins

Some other EF-hand containing Ca^{2+} proteins are also studied important for Ca^{2+} mediated pathogen defense system in plants. Rboh (respiratory burst oxidase homologue) are membrane bound proteins found in plants. They regulates pathogen induced oxidative burst [increased generation of reactive oxygen species (ROS)] in a Ca^{2+} dependent mechanism. They are also characterized for having two EF-hand motifs depicting their activation by increased Ca^{2+} concentration. Among different forms of ROS, Rboh regulated the formation of superoxide radicals (Lecourieux *et al.*, 2006).

Centrins/caltractins originally reported from *Chlamydomonas reinhardtii* were also identified in plants. They are described as small acidic protein with four EF-hand motifs like CaM. They are proposed to participate in cell plate formation and Ca^{2+} regulated transport via plasmodesmata. In addition, they are also involved in intracellular arrangements (organization of microtubules and microfilaments) during early stages of pathogen infection (Reddy, 2001; Lecourieux *et al.*, 2006).

Ca^{2+} Binding Proteins without EF-hand Domain

Various Ca^{2+} binding proteins are reported that do not contain EF-hand motif and are summarized in Table 2.1.

Biological Responses

Modulation in the activity of Ca^{2+} sensors/binding proteins in response to various developmental, environmental and biotic stimuli leads to activation of defense signaling cascade that regulates phosphorylation cascade, G-proteins,

NADPH oxidase, ion homeostasis, and production of ROS, salicylic acid (SA), nitric oxide (NO), ethylene and jasmonic acid (JA) (Lecourieux *et al.*, 2006). Phosphorylation/dephosphorylation events modify the activity of defense related proteins. The heterotrimeric guanine-nucleotide-binding (G) proteins are important to transduce extracellular signals to internals targets. G-proteins are proposed to play crucial role during PCD, stomatal closure and activation/suppression of other proteins (Jones and Assmann, 2004; Zhang *et al.*, 2012). NADPH oxidase is related to the production of ROS in cell in response to Ca^{2+} signaling (Lecourieux *et al.*, 2006).

Table 2.1: Important Ca^{2+} Binding Proteins without EF-Hand Motif

Protein	*Description**
Copines	✰ Consists of C2 domain (Ca^{2+} -phospholipid binding site) ✰ Suppressor of hypersensitive response ✰ Determine Ca^{2+} signals' specificity and prevent inappropriate defense response
Phospholipae D (PLD)	✰ Breaks phosphodiester bond of phospholipids ✰ Also contains C2 domain ✰ Play role during plant pathogen defense but function yet to be established
Annexins	✰ 4 to 8 repeats of ~70 amoino acids ✰ Bind to phospholipids ✰ Involved in secretory processes, might have ATPase/peroxidase activity
Calreticulin	✰ Ca^{2+} homeostasis, protein synthesis, molecular chaperone
Pistil expressed Ca^{2+} binding protein (PCP)	✰ High capacity but low affinity Ca^{2+} binding protein help in pollen-pistil interaction and/or pollen development
Calnexin	✰ Chaperone
Florisome	✰ Contractile motor protein regulating of phloem sieve tubes

*Reddy, 2001; Lecourieux *et al.*, 2006; Tuteja and Mahajan, 2007; Tuteja, 2009

SA is a signal molecule involved in HR and systemic acquired resistance (SAR) in response to pathogen attack. SA is proposed to deactivate the activity of catalase (degrades H_2O_2) thus increasing the concentration of hydrogen peroxide (H_2O_2) a potent ROS. The role of SA during SAR is still under investigation (Durner *et al.*, 1997; Shah, 2003). Ethylene and JA are other plant hormones and act as signaling molecules. They activate the synthesis of ROS thus causing oxidative burst and activation of pathogen related proteins and HR (Kachroo and Kachroo, 2007). Most of the pathogen induced biological responses include burst of free radicals like ROS and NO (gaseous). Such free radicals participate in signaling processes, HR, PCD, activation of pathogen related proteins and other defense responses (Apel and Hirt, 2004; Arasimowicz and Floryszak-Wieczorek, 2007).

Conclusion and Future Prospects

The different steps of Ca^{2+} mediated pathogen defense system in plant can be summarized as Figure 2.4. It is now well established that Ca^{2+} signaling is the key

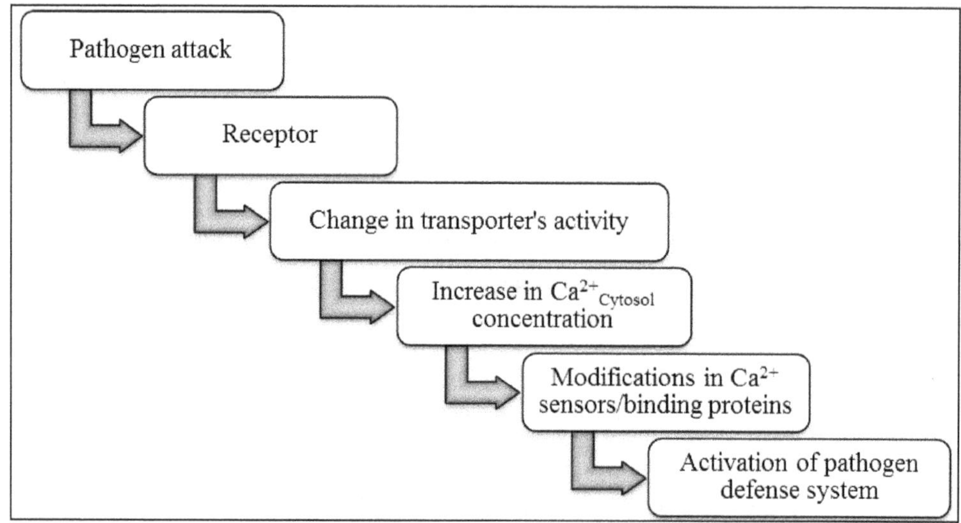

Figure 2.4: A Concise Model of Steps during Ca^{2+} Mediated Pathogen Defense System.

phenomenon in activating biological responses in response to a diverse array of stimuli including plant pathogen interaction.

Concepts in Ca^{2+} signaling that need more attention for better understanding are:

- Mechanisms regarding specific Ca^{2+} channels/transporters and their interaction with other non-specific cation channels.
- How non-specific cation channels affect Ca^{2+} signatures?
- How different Ca^{2+} signatures are recognized and decoded?
- Interconnection among different Ca^{2+} signaling systems.
- Interaction of Ca^{2+} with other signaling pathways (hormonal or secondary messengers).

To date, we are able to answer most of the questions regarding Ca^{2+} signaling in plants but advancement in the above mentioned areas will enhance our knowledge that can be utilized to develop stress resistant/tolerant crop varieties.

References

Apel, K. and Hirt, H., 2004. Reactive oxygen species: metabolism, oxidative stress, and signal transduction. *Annu. Rev. Plant Biol.*, 55: 373-399.

Arasimowicz, M. and Floryszak-Wieczorek, J., 2007. Nitric oxide as a bioactive signaling molecule in plant stress responses. *Plant Sci.*, 172: 876-887.

Axelsen, K.B. and Palmgren, M.G., 2001. Inventory of the superfamily of P-type ion pumps in *Arabidopsis*. *Plant Physiol.*, 126: 696-706.

Batistic, O. and Kudla, J., 2012. Analysis of calcium signaling pathways in plants. *Biochim. Biophys. Acta*, 1820: 1283-1293.

Bouche, N., Yellin, A., Snedden, W.A. and Fromm, H., 2005. Plant-specific calmodulin-binding proteins.*Annu. Rev. Plant Biol.*, 56: 435-466.

Cheval, C., Aldon, D., Galaud, J.P. and Ranty, B., 2013. Calcium/calmodulin-mediated regulation of plant immunity. *Biochim. Biophys. Acta*, 1833: 1766-1771.

DeFalco, T.A., Bender, K.W. and Snedden, W.A., 2010. Breaking the code: Ca^{2+} sensors in plant signaling. *Biochem. J.*, 425: 27-40.

Dodd, A.N., Kudla, J. and Sanders, D., 2010. The language of calcium signaling. *Annu. Rev. Plant Biol.*, 61: 593-620.

Durner, J., Shah, J. and Klessig, D.F., 1997. Salicylic acid and disease resistance in plants. *Trends Plant Sci.*, 2: 266-274.

Harper, J.F. and Harmon, A., 2005. Plants, symbiosis and parasites: a calcium signaling connection. *Nat. Rev. Mol. Cell Biol.*, 6: 555-566.

Hashimoto, K. and Kudla, J., 2011. Calcium decoding mechanisms in plants. *Biochimie*, 93: 2054-2059.

Hirschi, K., 2001. Vacuolar H^+/Ca^{2+} transport: who's directing the traffic? *Trends Plant Sci.*, 6: 100-104.

Hirschi, K.D., Zhen, R.G., Cunningham, K.W., Rea, P.A. and Fink, G.R., 1996. CAX1, an H+/Ca2+ antiporter from *Arabidopsis*. *Proc. Natl. Acad. Sci.*, 93: 8782-8786.

Jammes, F., Hu, H.C., Villiers, F., Bouten, R. and Kwak, J.M., 2011. Calcium-permeable channels in plant cells. *FEBS J.*, 278: 4262-4276.

Jones, A.M. and Assmann, S.M., 2004. Plants: the latest model system for G-protein research. *EMBO Rep.*, 5: 572-578.

Kachroo, A., and Kachroo, P., 2007. Salicylic acid-, jasmonic acid- and ethylene-mediated regulation of plant defense signaling. In: *Genetic Engineering: principles and methods*, (Ed.) J.K.Setlow, Springer Science+Business Media, LLC, 28: 55-83.

Kaplan, B., Davydov, O., Knight, H., Galon, Y., Knight, M.R., Fluhr, R. and Fromm, H., 2006. Rapid transcriptome changes induced by cytosolic Ca^{2+} transients reveal ABRE-related sequences as Ca^{2+}-responsive *cis* elements in Arabidopsis. *Plant Cell*, 18: 2733-2748.

Kim, K.N., 2013. Stress responses mediated by the CBL calcium sensors in plants. *Plant Biotechnol. Rep.*, 7: 1-8.

Kretsinger, R.H. and Nockolds, C.E., 1973. Carp muscle calcium-binding protein. II. Structure determination and general description. *J. Biol. Chem.*, 248: 3313-3326.

Lecourieux, D., Ranjeva, R. and Pugin, A., 2006. Calcium in plant defence-signaling pathways. *New Phytol.*, 171: 249-269.

Lee, J. and Rudd, J.J., 2002. Calcium-dependent protein kinases: versatile plant signaling components necessary for pathogen defence. *Trends Plant Sci.*, 7: 97-98.

Liese, A. and Romeis, T., 2013. Biochemical regulation of *in vivo* function of plant calcium-dependent protein kinases (CDPK). *Biochim. Biophys. Acta*, 1833: 1582-1589.

Ma, W. and Berkowitz, G.A., 2011. Ca^{2+} conduction by plant cyclic nucleotide gated channels and associated signaling components in pathogen defense signal transduction cascades. *New Phytol.*, 190: 566-572.

Reddy, A.S.N., 2001. Calcium: silver bullet in signaling. *Plant Sci.*, 160: 381-404.

Sanders, D., Pelloux, J., Brownlee, C. and Harper, J.F., 2002. Calcium at the crossroads of signaling. *Plant Cell*, 14: S401-S417.

Shah, J., 2003. The salicylic acid loop in plant defense. *Curr. Opin. Plant Biol.*, 6: 365-371.

Song, W.Y., Zhang, Z.B., Shao, H.B., Guo, X.L., Cao, H.X., Zhao, H.B., Fu, Z.Y. and Hu, X.J., 2008. Relationship between calcium decoding elements and plant abiotic-stress resistance. *Int. J. Biol. Sci.*, 4: 116-125.

Stael, S., Wurzinger, B., Mair, A., Mehlmer, N., Vothknecht, U.C. and Teige, M., 2012. Plantorganellar calcium signaling: an emerging field. *J. Exp. Bot.*, 63: 1525-1542.

Tuteja, N., 2009. Integrated calcium signaling in plants. In: *Signaling in plants, signaling and communication in plants*, (Ed.) F. Baluška and S. Mancuso, Springer-Verlag Berlin Heidelberg, 29-49.

Tuteja, N. and Mahajan, S., 2007. Calcium signaling network in plants. *Plant Signaling Behav.*, 2: 79-85.

White, P.J., 2000. Calcium channels in higher plants. *Biochim. Biophys. Acta*, 1465: 171-189.

Yang, T. and Poovaiah, B.W., 2003. Calcium/calmodulin-mediated signal network in plants. *Trends Plant Sci.*, 8: 505-512.

Zhang, H., Gao, Z., Zheng, X. and Zhang, Z., 2012. The role of G-proteins in plant immunity. *Plant Signaling Behav.*, 7: 1284-1288.

2016, Recent Advances in Plant Stress Physiology *Pages* **49–67**
Editors: ***Praduman Yadav, Sunil Kumar and Veena Jain***
Published by: **DAYA PUBLISHING HOUSE, NEW DELHI**

Chapter 3

Transcription Factors in Abiotic Stress Tolerance

Ranjit Singh Gujjar*

Division of Crop Improvement, ICAR – Indian Institute of Vegetable Research Varanasi – 221 305, U.P., India

ABSTRACT

Abiotic stresses such as drought, high salinity, and extreme temperatures are common adverse environmental conditions that significantly reduce the plant productivity. Plant adaptation to these environmental stresses is controlled by cascades of complex molecular networks. Transcription factors play vital regulatory roles in abiotic stress responses in plants by interacting with cis-elements present in the promoter region of various abiotic stress responsive genes. Now it has been possible to engineer stress tolerance in transgenic plants by manipulating the expression of TFs. This opens an excellent opportunity to develop stress tolerant crops in future. This review summarizes the role of various transcription factors in crop improvement through transgenic technology.

Keywords: *Abiotic stress, Drought, Stress tolerance, Transcription factors.*

Introduction

Drought is one of the most devastating abiotic stresses that impacts growth, development and productivity of agricultural crops worldwide. Drought and salinity together affect more than 10 per cent of arable land, resulting in a more than 50 per cent decline in the average yields of major crops. Plants respond and adapt to this stress through a series of molecular, cellular, biochemical as well as physiological reactions and activate a number of defense mechanisms that function to increase tolerance to all kinds of abiotic stresses. The early events of the

**Author*: E-mail: ranjit.gujjar@gmail.com

adaptation of plants to abiotic stress include the perception of stress signals and subsequent signal transduction, leading to the activation of various physiological and metabolic responses. Within the signal transduction networks that are involved in the conversion of stress signal perception to stress-responsive gene expression, various transcription factors (TFs) and *cis*-acting elements contained in stress-responsive promoters function not only as molecular switches for gene expression, but also as terminal points of signal transduction (Figure 3.1). The identification and molecular tailoring of novel TFs have the potential to overcome a number of important limitations encountered in the generation of transgenic crop plants with superior yield under stress conditions. Being sessile, plants have the capability to sense and adjust to abiotic stresses, although the degree of adaptability to specific stresses varies from species to species. The physiological processes involved in the adaptive responses have been the subject of intense molecular and genetic studies during the last decade or so.

Plant engineering strategies for cellular and metabolic reprogramming to increase the efficiency of plant adaptive responses may either concentrate on (1) providing stress tolerance by directly re-programming ion transport processes and primary metabolism or (2) by manipulating signaling and regulatory pathways of the adaptive mechanisms. The second approach seems to be more effective because regulatory factors directly orchestrate the transcriptional and translational signaling components (Golldack *et al.*, 2011). Consequently, many regulatory components have been identified. The advent of microarray technology made it possible to investigate stress responsive transcript changes at the whole genome level (Kreps *et al.*, 2002; Seki *et al.*, 2003).

Novel transcription factors regulating stress-responsive gene expression have been identified by various means (Zou *et al.*, 2007). Elegant genetic works like employing a firefly luciferase fused to a stress-inducible promoter (*RD29A-LUC*) and conventional forward/reverse genetic approaches resulted in the isolation of key stress-signaling components. In many cases, the regulatory components thus identified have been utilized to generate transgenic plants that can tolerate stress to some extent (Zhang *et al.*, 2004). TFs are master regulators that control gene clusters. A single TF (regulatory gene) can control the expression of many target genes (effector genes) through specific binding of the TF to the *cis*-acting element in the promoters of respective target genes. This type of transcriptional regulatory system is called regulon (Shinozaki and Shinozaki, 2005). Several major regulons that are active in response to abiotic stress have been identified in *Arabidopsis*. In general, drought responsive transcription factors fall in two categories: ABA independent transcription factors which do not require stress hormone (Abscisic acid) to operate and ABA-dependent transcription factors requiring stress hormone somewhere in the process of drought responsive signal cascade ending with the final step of structural gene expression (Ali *et al.*, 2011). Dehydration-responsive element binding protein 1 (DREB1)/C repeat binding factor (CBF) and DREB2 regulons function in ABA-independent manner whereas, the ABA-responsive element (ABRE) binding protein (AREB)/ABRE binding factor (ABF) regulon functions in ABA-dependent manner. TFs comprise families of related proteins that share a homologous DNA

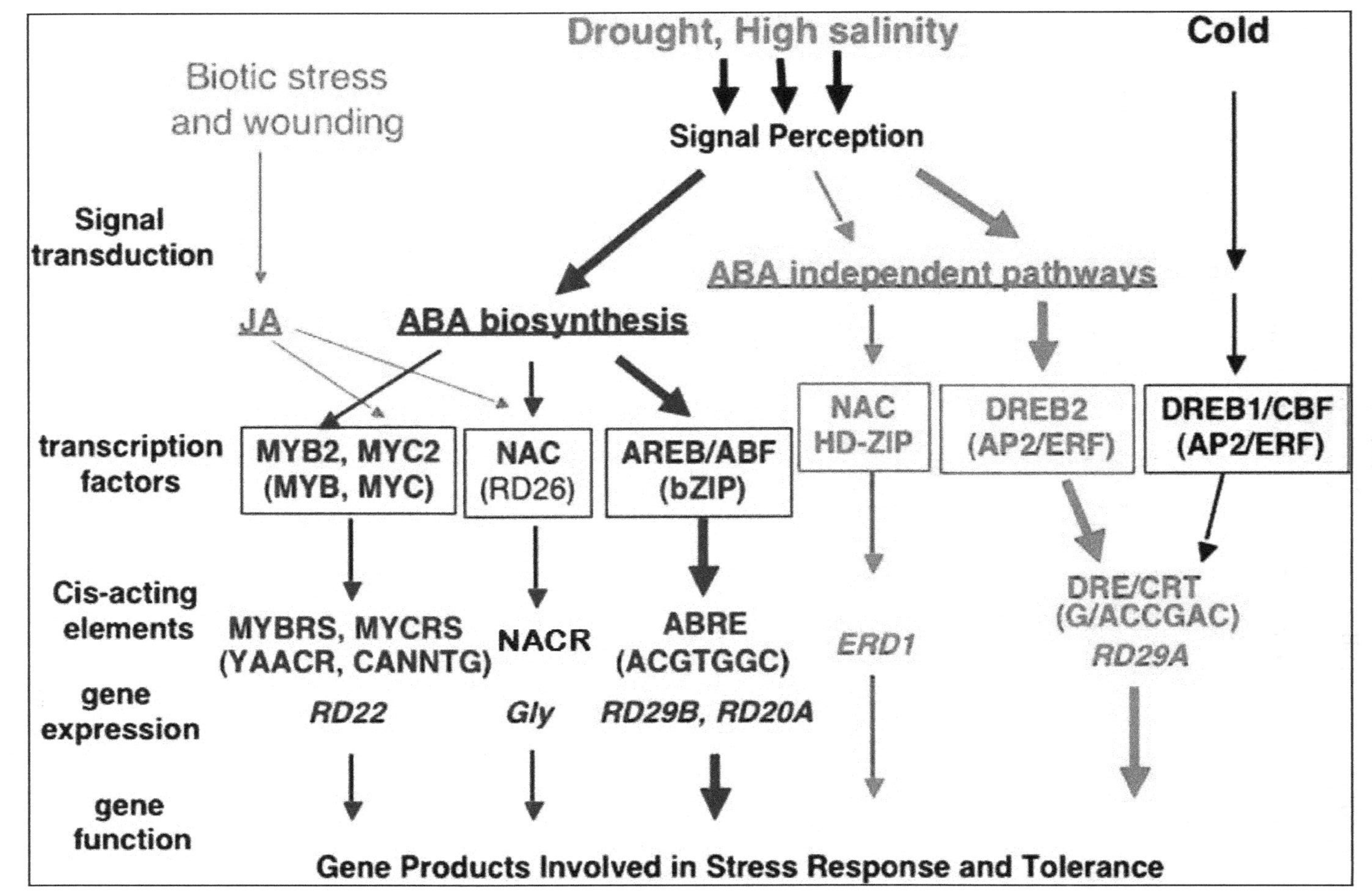

Figure 3.1: Abiotic Stress Signaling Network (adapted from Shinozaki and Shinozaki, 2006).

binding domain. Most common transcription factor families include DREB1/CBF, DREB2, AREB/ABF, MYB/MYC, bHLH, ZFPs, bZIP, WRKY and NAC regulons.

Recent studies demonstrated that DREB1/CBF, DREB2, AREB/ABF, MYB/MYC, bHLH, ZFPs, WRKY and NAC regulons have important roles in abiotic stress responsive gene expression in rice. Expression analysis of these stress inducible genes in *Arabidopsis* indicated the existence of complex regulatory mechanisms between perception of abiotic stress signals and expression of respective stress responsive gene (Zhu, 2002). Over-expression of the genes that regulate the transcription of a number of downstream stress responsive genes seems to be a promising approach in the development of drought, salt and cold resistant/tolerant transgenic plants when compared to engineering individual functional genes. The regulatory genes reported so far not only play a significant role in drought, salinity and cold stresses, but also in submergence tolerance. More recently, an ethylene-response-factor-like gene, named *Sub1A*, has been identified in rice and the overexpression of *Sub1A-1* in a submergence-intolerant variety conferred enhanced submergence tolerance to the plants (Xu *et al.*, 2006), thus confirming the role of this gene in submergence tolerance in rice. Therefore, it is important to enhance regulatory ability of an important transcription factor that activates the expression of many target genes controlling correlated characters. Many key genes have been identified that are involved in the regulatory networks under abiotic stress. Thus transcription factor based transgenic approaches are likely to play a prominent part to develop limited soil moisture, salt and cold tolerant plants. Some of the important transcription factors families involved in abiotic stress tolerance are described below:

DREB Family of Transcription Factors

Probably the best studied group of TFs involved in abiotic stress particularly, in drought and cold tolerance are the DREB1 genes (Reddy and Reddy, 2008). Many osmotic stress inducible genes contain a conserved drought responsive element (DRE) in their promoters. Several cDNAs encoding the DRE binding proteins, DREB1A and DREB2A have been isolated from *Arabidopsis thaliana* shown to specifically bind and activate the transcription of genes containing DRE sequences. In fact, in many studies over-expression of stress inducible DREB transcription factor was found to activate the expression of many target genes having DRE elements in their promoters and the resulting transgenic plants showed improved stress tolerance (Table 3.1). One important way to achieve tolerance to multiple abiotic stresses is to over-express TFs that control multiple genes from various pathways. The best studied and broadly applied example is the over-expression of the *DREB1A* driven either by a constitutive (*CaMV35S*) or the dehydration inducible (*rd29A*) promoter. This approach has activated the expression of many stress tolerant genes under normal growth conditions and resulted in the generation of plants with increased tolerance to drought, freezing and salinity stress (Lata and Prasad, 2011).

CBF/DREB1 transcription factors have been considered to be the important regulators of the cold acclimation response, controlling the level of *COR* (cold-regulated) gene expression, which in turn promotes tolerance to freezing (Gilmour *et al.*, 2000). Therefore, transformation of CBF/DREB1 genes has been found to improve

Table 3.1: Function of DREB Genes in Various Stresses (Lata and Prasad, 2011)

DREB TFs	*Species*	*Accession no.*	*Stress response*
DREB1A	*Arabidopsis thaliana*	AB007787	Cold
DREB2A	*Arabidopsis thaliana*	AB007790	Drought, Salt, ABA
DREB2C	*Arabidopsis thaliana*	At2g40340	Salt, Mannitol, Cold
CBF1	*Arabidopsis thaliana*	U77378	Cold
CBF2	*Arabidopsis thaliana*	AF074601	Cold
CBF3	*Arabidopsis thaliana*	AF074602	Cold
CBF4	*Arabidopsis thaliana*	AB015478	Drought, ABA
OsDREB1A	*Oryza sativa*	AF300970	Cold, Salt, Wounding
OsDREB1B	*Oryza sativa*	AF300972	Cold
OsDREB1C	*Oryza sativa*	AP001168	Drought, Salt, Cold, ABA, Wound
OsDREB1D	*Oryza sativa*	AB023482	None
OsDREB2A	*Oryza sativa*	AF300971	Drought, Salt, faintly to Cold, ABA
OsDREB1F	*Oryza sativa*		Drought, Salt, Cold, ABA
OsDREB2B	*Oryza sativa*		Heat, Cold
OsDREB2C	*Oryza sativa*	AK108143	None
OsDREB2E	*Oryza sativa*		None
OsDREBL	*Oryza sativa*	AF494422	Cold
TaDREB1	*Triticum aestivum*	AAL01124	Cold, Dehydration, ABA
WCBF2	*Triticum aestivum*		Cold, Drought
WDREB2	*Triticum aestivum*	BAD97369	Drought, Salt, Cold, ABA
HvDRF1	*Hordeum vulgare*	AY223807	Drought, Salt, ABA
HvDREB1	*Hordeum vulgare*	DQ012941	Drought, Salt, Cold
ZmDREB2A	*Zea mays*	AB218832	Drought, Salt, Cold, Heat
PgDREB2A	*Pennisetum glaucum*	AAV90624	Drought, Salt, Cold
SbDREB2	*Sorghum bicolor*	ACA79910	Drought
SiDREB2	*Setaria italica*	HQ132744	Drought, Salt
CaDREB-LP1	*Capsicum annum*	AY496155	Drought, Salt, Wounding
AhDREB1	*Artiplex hortensis*		Salt
GmDREBa	*Glycine max*	AY542886	Cold, Drought, Salt
GmDREBb	*Glycine max*	AY296651	Cold, Drought, Salt
GmDREBc	*Glycine max*	AY244760	Drought, Salt, ABA
GmDREB	*Glycine max*	AF514908	Drought, Salt
GmDREB2	*Glycine max*	ABB36645	Drought, Salt
PpDBF1	*Physcomitrella patens*	ABA43697	Drought, Salt, Cold, ABA
PNDREB1	*Arachis hypogea*	FM955398	Drought, Cold
CAP2	*Cicer arietinum*	DQ321719	Drought, Salt, ABA, Auxin
DvDREB2A	*Dendrathema*	EF633987	Drought, Heat, ABA, Cold
DmDREBa	*Dendronthema × moriforlium*	EF490996	Cold, ABA
DmDREBb	*Dendronthema × moriforlium*	EF487535	Cold, ABA
PeDREB2	*Populus euphratica*	EF137176	Drought, Salt, Cold
SbDREB2A	*Salicornia brachiata*	GU592205	Drought, Salt, Heat

environmental stress tolerance to many plants, some of these have the potential to be used as freezing-tolerant verities. Over-expression of the *AtDREB1A* in *Nicotiana tabacum*, wheat, rice and groundnut has been shown to lead to increased drought tolerance and enhanced expression of LEA type genes at least under greenhouse conditions. Recently, transgenic peanut plants overexpressing *Arabidopsis DREB1A* under stress inducible promoter accumulated considerably high levels of some key enzymes compared to wild type plants under drought stress but these enzymes showed no relationship to transpiration efficiency. The orthologue DREB genes from other plants have also been shown to be functional, for example transgenic rice overexpressing *Oryza sativa DREB1A* has demonstrated improved tolerance to drought, high salt and low temperature stresses and also caused accumulation of elevated levels of osmoprotectants such as free proline and various soluble sugars. As a matter of fact, ectopic expression of DREB genes in *Arabidopsis* as well as in heterologous systems such as wheat, barley, soybean, tomato, tobacco, strawberry, rice, oilseed rape, potato, and other grasses has produced enhanced tolerance to one or more types of abiotic stresses.

Transgenic turf grass (*Lolium perenne*) over-expressing *AtDREB1/CBF3* showed enhanced drought and freezing tolerance and the activities of superoxide dismutase (SOD) and peroxidase (POD) were also higher in transgenic plants compared to non-transgenic control plants (Li *et al.*, 2011). Interestingly one of the DREB1 type transcription factor from dwarf apple named *MbDREB1* revealed increase in tolerance to low temperature, drought and salt stresses via both ABA-dependent and ABA-independent pathways (Yang *et al.*, 2011). The *MtDREB1C* gene, isolated from *Medicago truncatula* and driven by the *Arabidopsis rd29A* promoter, enhanced freezing tolerance in transgenic China rose significantly without any obvious morphological or developmental abnormality (Chen *et al.*, 2010). However, little information is available about stress tolerance conferred by different members of the DREB1 subfamily. In case of *Arabidopsis*, DREB1A, DREB1B and DREB1C, were tested through their over-expression under the same freezing conditions, and no significant difference of freezing tolerance and downstream gene expression induction was observed (Gilmour *et al.*, 2004). Similar results were also observed in rice when comparing the freezing or drought tolerance between *OsDREB1A* and *OsDREB1B* over-expressed plants and in *Brassica napus* when comparing freezing tolerance between *BnCBF5* and *BnCBF17* over-expressed plants (Savitch *et al.*, 2005).

bZIP Class Transcription Factors

The bZIP proteins are characterized by a 40 to 80 amino-acid long conserved domain (bZIP domain) that is composed of two motifs: a basic region responsible for specific binding of the TF to its target DNA, and a leucine zipper required for TF dimerization. Genetic, molecular and biochemical analyses indicate that bZIPs are regulators of important plant processes such as organ and tissue differentiation, cell elongation, nitrogen/carbon balance control, pathogen defense, energy metabolism, unfolded protein response, hormone and sugar signaling, light response, osmotic control and seed storage protein gene regulation (Nijhawan *et al.*, 2008). Initially, 50 plant bZIP proteins were classified into five families, taking into account similarities of their bZIP domain. An original investigation of the complete *Arabidopsis thaliana* genome sequence indicated the presence of 81 putative bZIP genes (Correa *et al.*, 2008). Basic leucine zipper proteins contain a DNA binding domain rich in basic residues and a leucine zipper dimerization domain. Several b-ZIP factors that bind to ABA responsive elements (ABREs - PyACGTGGC) are AREB1, AREB2, and AREB3. ABREs were first identified in the wheat *Em* gene and rice *rab16* gene. These genes are expressed in both dehydrated vegetative tissues as well as maturing seeds during late embryogenesis. Region between -294 and +27 in the 5' upstream region of the *rab-16A* gene is sufficient to confer ABA-responsive, transient expression upon the CAT reporter gene in transfected rice protoplasts. Two motifs were found to be conserved in all *rab-16* genes discovered in rice. Motif I has the consensus RTACGTGGR (R is an unspecified purine nucleoside), which is similar to the cAMP-responsive element (TGACGTCA) that binds to the transcription factor CREB. More importantly, motif I has been reported in the 5' upstream regions of five of six *lea* genes and in that of the ABA-responsive wheat *Em* gene. Motif II, which is found in two copies (IIa and IIb) in *rab-16A* and once in *rab-16B-D*, has the consensus CGSCGCGCT, in which S is G or C. It occurs in the *rab-16* genes as part of sequences that are similar to the degenerate deca-nucleotide binding site of SP1, an auxiliary mammalian transcription factor. AREB1 and AREB2 proteins are up-regulated by ABA (Kang *et al.*, 2002).

Till now, 89 bZIP transcription factor-encoding genes have been identified in the rice (*Oryza sativa*) genome. Their chromosomal distribution and sequence analyses suggest that the bZIP transcription factor family has evolved via gene duplication. The phylogenetic relationship among rice bZIP domains as well as bZIP domains from other plants, suggests that homologous bZIP domains exist in plants. Similar intron/exon structural patterns were observed in the basic and hinge regions of their bZIP domains (Nijhawan *et al.*, 2008). *Wlip*19 (a bZIP transcription factor type gene) expressing in transgenic tobacco showed a significant increase in abiotic stress tolerance, especially freezing tolerance by activating *cor/lea* genes (Kobayashi *et al.*, 2008). Over-expressing a bZIP transcription factor *SlAREB*1 cloned from tomato and introduced in tobacco using *Agrobacterium*-mediated transformation showed up-regulation of stress-responsive genes such as *RD29B*, thus mediating the stress response (Yanez *et al.*, 2009). Transgenic *Arabidopsis* plants, overexpressing a bZIP type transcription factor *ABF*3, conferred enhanced tolerance to drought stress using *CaMV 35S* constitutive promoter (Abdeen *et al.*, 2010). The homozygous T-DNA insertional mutants *Osabf*1-1 and *Osabf*1-2 (bZIP transcription factor) were more sensitive in response to drought and salinity treatments compared to wild type plants (Hossain *et al.*, 2010). A novel bZIP gene *ThbZIP*1, cloned from *Tamarix hispida* and introduced in tobacco plant, showed differential regulation in response to treatment with NaCl, polyethylene glycol (PEG) 6000, NaHCO3 and CdCl2, suggesting that *ThbZIP*1 is involved in abiotic stress responses. *ThbZIP*1 transgenic plants conferred stress tolerance by enhancing reactive oxygen species (ROS) scavenging, facilitating the accumulation of compatible osmolytes and inducing the biosynthesis of soluble proteins (Wang *et al.*, 2010).

MYB Family of Transcription Factors

The MYB (myeloblastosis) family of proteins is large, functionally diverse and represented in all eukaryotes. Most MYB proteins function as transcription factors with varying numbers of MYB domain repeats conferring their ability to bind DNA. MYB proteins are key factors in regulatory networks controlling development, metabolism and responses to biotic and abiotic tresses. In plants, the MYB family has selectively expanded, particularly through the large family of *R2R3*-MYB (Dubos *et al.*, 2010). So far, large numbers of MYB genes have been identified in different plant species, comprising 204 members in *Arabidopsis*, 218 members in rice, 279 members in grapevine, 197 members in populus, and 180 members in *Brachypodium* (Zhang *et al.*, 2012). MYB proteins are characterized by a highly conserved DNA-binding domain: the MYB domain. This domain generally consists of four imperfect amino acid sequence repeats (R) of about 52 amino acids, each forming three α helices. The second and third helices of each repeat build a helix–turn–helix (HTH) structure with three regularly spaced tryptophan (or hydrophobic) residues, forming a hydrophobic core in the 3D HTH structure. The third helix of each repeat is the "recognition helix" that makes direct contact with DNA and intercalates in the major groove. During DNA contact, two MYB repeats are closely packed in the major groove, so that the two recognition helices bind cooperatively to the specific DNA sequence motif. MYC and MYB recognition sites present in the promoter region of *RD22* gene (Abe *et al.*, 2003).

The expression patterns of 60 wheat MYB genes under different abiotic stress conditions have been studied till date. Among 60 genes studied, *TaMYB32* encoded an R2R3- MYB-type protein and transgenic *Arabidopsis* over-expressing this gene showed improved tolerance to high salt (Zhang *et al.*, 2012). Several R2R3MYB genes play important role in the stress responses to environmental stimuli. For example, *AtMYB2*, in co-operation with *AtMYC2*, functions as a transcriptional activator in the dehydration and abscisic acid (ABA) inducible gene expression of RD22 (Abe *et al.*, 2003). *AtMyb41* gene from *Arabidopsis* was found to be transcriptionally regulated in response to salinity, limited soil moisture, cold as well as the endogenous plant hormone ABA. Over-expression of *AtMyb41* in transgenic plants resulted in a dwarf phenotype (Lippold *et al.*, 2009). In another study, a rice R2R3-type MYB gene, *OsMYB2*, was up-regulated by salt, cold, and dehydration stress. The OsMYB2-over-expressing plants accumulated greater amounts of soluble sugars and proline and less amounts of H2O2 and malondialdehyde than wild type plants. There was greater up-regulation of stress-related genes, including *OsLEA3*, *OsRab16A*, and *OsDREB2A* and enhanced activities of antioxidant enzymes including peroxidase, superoxide dismutase and catalase (Yang *et al.*, 2012).

Three TaMYB2 TFs were identified from wheat and designated *TaMYB2A*, *TaMYB2B*, and *TaMYB2D*. Among them, *TaMYB2A* was early responsive TF and its over-expression in transgenic *Arabidopsis* confers enhanced tolerance to multiple abiotic stresses as revealed by induction of abiotic stress responsive genes while having no obvious negative effects on phenotype under well-watered and stressed conditions (Mao *et al.*, 2011). Rice MYB transcription factor gene *Osmyb4* expression improved adaptive responses to drought and cold stress in transgenic apples (Pasquali *et al.*, 2008). Transgenic *Arabidopsis* plants over-expressing *MYB96* exhibited enhanced drought resistance with reduced lateral roots (Seo *et al.*, 2009). Transgenic *Arabidopsis* plants over-expressing *R2R3*-MYB transcription factor gene *MdMYB*-10 (a regulatory gene of anthocyanin biosynthesis in apple fruit) showed better tolerance to osmotic stress compared to wild type plants (Gao *et al.*, 2011).

NAC Family of Transcription Factors

In recent years, great achievements have been made in increasing the tolerance of plants to abiotic stresses by identifying potential stress related genes. Among them, NAC (Nacetyl-L-cysteine) family [(NAM (no apical meristem), ATAF1/2 and CUC (cup-shaped cotyledon)] domain proteins comprise of one of the largest plant-specific transcription factors, represented by~117 genes in *Arabidopsis*~151 genes in rice and ~101 genes in soybean genome. NAC TFs are mostly involved in salinity tolerance. They act downstream of auxin and ethylene signaling pathways in addition to ABA pathway (He *et al.*, 2005). Proteins of this family are characterized by a highly conserved DNA binding domain, known as NAC domain in the N-terminal region and a transcriptional activation domain in the C-terminal region which is highly diversified both in length and sequence. NAC domain comprises nearly 160 amino acid residues that are divided into five sub-domains (A-E). A sequence of 60 residues within the NAC domain contains a unique TF fold consisting of a twisted-sheet bounded by a few helical elements. NAC domain has been implicated in nuclear localization, DNA binding and the formation of homo-dimers or hetero-dimers with other NAC domain proteins. The C-terminal regions of NAC proteins confer the regulation diversities of transcriptional activation activity (Hu *et al.*, 2006; Jeong *et al.*, 2010).

NAC TFs are multifunctional proteins and involved in wide range of processes including biotic and abiotic stress responses, lateral root development, flowering, anther dehiscence etc. Evidence for involvement of NAC TFs with the regulation of drought stress response in plants was first reported in *Arabidopsis* where over-expression of *ANAC019, ANAC055* and *ANAC072* altered the expression of many stress-inducible genes in the transgenic plants and conferred a constitutive increase in drought tolerance (Tran *et al.*, 2004). One of the NAC gene named, *CsNAC1* was investigated in citrus and found up-regulated during drought, cold, salt stress and ABA as well (Oliveira *et al.*, 2011). Recently, the role of two rice NAC genes in rice stress adaptation has been characterized. *SNAC*1 was induced mainly in guard cells under drought conditions, and overexpression of this gene in rice resulted in significant increase in drought resistance under field condition at the stage of anthesis. Over-expression of another NAC gene *OsNAC*6/*SNAC*2 in rice resulted in enhanced tolerance to drought, salt, and cold during seedling development (Hu *et al.*, 2006; Zheng *et al.*, 2009).

Over-expression of *OsNAC*10 in rice under the control of the constitutive promoter *GOS*2 and the root-specific promoter *RCc*3 increased the plant tolerance to drought, high salinity and low temperature at the vegetative stage (Jeong *et al.*, 2010). Transgenic rice plants over-expressing *OsNAC*52 were highly sensitive to ABA (Abscisic acid), and over-expressing 35S-*OsNAC*52 transgenic lines activated expression of downstream genes in transgenic *Arabidopsis*, resulting in enhanced tolerance to drought stresses and did not cause growth retardation. This result suggested that *OsNAC52* gene may function as an important transcriptional activator in ABA inducible gene expression and may be useful in improving of plant tolerance to abiotic stresses in plants (Gao *et al.*, 2010). Over-expression of *ZmSNAC1* in *Arabidopsis* led to hypersensitivity to ABA and osmotic stress, and conferred tolerance to dehydration (Lu *et al.*, 2012). In another study, *TaNAC2*, a NAC transcription factor member from common wheat, was cloned and induced under water deficiency, high salinity, low temperature, and ABA. Transgenic experiments indicated that *TaNAC2* increases tolerance to drought, salt, and freezing stresses in *Arabidopsis* without any obvious negative effects on morphology caused by *TaNAC2* overexpression, suggesting a potential for utilization of such NAC TF gene in crop improvement (Mao *et al.*, 2012). A full-length cDNA of *DgNAC1*, containing a typical NAC domain, has been isolated from chrysanthemum and it was observed that 35S:DgNAC1 transgenic tobacco exhibits a markedly increased tolerance to salt with no detectable phenotype defects under normal growth conditions (Liu *et al.*, 2011).

Zn Finger Protein Family of Transcription Factors

C2H2 zinc finger proteins (ZFPs) are characterized by the presence of two cysteine (Cys2) and two histidine (His2) residues in what is called a zinc finger domain, which stabilizes the three-dimensional structure consisting of a two-stranded antiparallel β-sheet and α-helix surrounding a central zinc ion (Kam *et al.*, 2008). ZFPs constitute one of the largest families of transcription factors in eukaryotes. Compared to mammals, the size of the ZFP family in plant genomes is relatively small 176 members in *Arabidopsis* and 189 members in rice (Agarwal *et al.*, 2007). Many plant ZFPs contain a highly conserved QALGGH amino acid motif within the zinc finger domain which forms a plant specific Q-type C2H2 zinc finger subfamily. The QALGGH motif is located at the N-terminus of the α-helix of the zinc finger domain. This α-helix region in animal ZFPs consists of hyper-variable residues that are responsible for recognition of specific DNA bases. The Q-type ZFP family has been implicated in plant response or adaptation to abiotic stresses (Kam *et al.*, 2008; Figueiredo *et al.*, 2012).

Expression analyses have shown that a number of ZFP genes in *Arabidopsis* are responsive to drought and cold stress. Some Q-type ZFP genes, such as *Arabidopsis Zat12*, are involved in oxidative and abiotic stress signaling indicating that the oxidative stress is associated with abiotic stress (Huang *et al.*, 2009). *ZAT12* was described as a negative regulator of the DREB1/CBF regulon (Vogel *et al.*, 2005), while the ZAT10/STZ gene expression was shown to be dependent on DREB1A/CBF3 (Maruyama *et al.*, 2004). C2H2-type TFs are therefore signaling components that can be located either up or downstream of the DREB1/CBF genes. Transgenic *Arabidopsis* plants, constitutively expressing the Cys2/His2 zinc finger protein *Zat7*, had suppressed growth and were more tolerant to salinity stress. Tobacco transgenic plants, over-expressing the *AlSAP* gene (a novel *A20*/*AN*1 zinc-finger transcription factor gene isolated from the halophyte grass *Aeluropus littoralis*) under the control of duplicated *CaMV35S* promoter, exhibited an enhanced tolerance to abiotic stresses such as salinity, drought, heat and freezing (Saad *et al.*, 2010).

Regulation of *OsDREB1B* through zinc finger TFs have also been reported recently (Figueiredo *et al.*, 2012).

Basic Helix-LoopFhelix Family of Transcription Factors

The bHLH proteins, a superfamily of functionally diverse transcription factors, have been intensively studied in plants and animals. This family is defined by the bHLH signature domain, which consists of 60 amino acids with two functionally distinct regions: the basic region at the N-terminal end, involved in DNA binding and the HLH region at the C-terminal end, involved in formation of homo-dimers or hetero-dimers. These proteins have been well characterized and function as important regulating components in transcriptional networks, controlling cell proliferation, determination and differentiation in plants, animals and yeast. Several rice studies investigated genes from this family playing distinct roles in stress response; examples are *OsbHLH*1 in cold response, *RERJ*1 in wound and drought response (Kiribuchi *et al.*, 2005), *OsPTF*1 in tolerance to phosphate starvation and *OsIRO*2 responsible for Fe- deficient conditions (Ogo *et al.*, 2007).

Over-expression of a basic helix-loop-helix transcription factor *OrbHLH*2, isolated from wild rice which encodes a homologue protein of *ICE*1, conferred increased tolerance to salt and osmotic stress in *Arabidopsis* (Zhou *et al.*, 2009). Over-expression of *OrbHLH*2 in *Arabidopsis* conferred increased tolerance to salt and osmotic stress and the stress-responsive genes *i.e. DREB*1*A*/*CBF*3, *RD*29*A*, *COR*15*A* and *KIN*1 were up-regulated in transgenic plants. Abscisic acid (ABA) treatment showed a similar effect on the seed germination or transcriptional expression of stress responsive genes in both wild-type and *OrbHLH*2-over-expressed plants, which implies that *OrbHLH*2 does not depend on ABA in responding to salt stress. *OrbHLH*2 may function as a transcription factor and positively regulate salt-stress signals independent of ABA in *Arabidopsis*, which provides some useful data for improving salt tolerance in crops (Zhou *et al.*, 2009).

WRKY Family of Transcription Factors

WRKY transcription factors, with a conserved WRKYGQK sequence in their DNA binding domain, bind to the W-box (TTGAC) on the promoter region of target gene. They encode for a wide family of transcription factors, characterized by the presence of approximately 60 amino acids containing the amino acid sequence WRKY at its amino terminal end and a putative zinc finger motif at its carboxyl terminal end. WRKY transcription factor proteins, known to mediate the pathogen-induced defense programme, were induced by ABA treatment in rice callus, suggesting the cross talk between ABA response pathway and pathogen-induced defence programme (Shimono *et al.*, 2007).

Till date, approximately 112 members of *WRKY* gene family in rice and 74 members in *Arabidopsis* have been reported. 46 *WRKY* genes have recently been isolated from canola (*Brassica napus* L.) in response to fungal pathogens and hormone treatments (Yang *et al.*, 2009). The role of WRKY transcription factors in abiotic stress response remains obscure and little information is available in limited crops only (Table 3.2). However some reports have determined that WRKY TFs are also involved in abiotic stress responses. For example, *OsWRKY*11 gene, under control of heat shock induced promoter *HSP101*, encodes a transcription factor with *WRKY* domain that is induced by both heat shock and drought stresses in seedlings of rice (Wu *et al.*, 2009). Involvement of an ABA inducible WRKY gene in abiotic stresses has been reported in creosote bush (*Larrea tridentate*) and *Hv-WRKY3* has been found to be involved in cold and drought response in barley (Zou *et al.*, 2007; Mare *et al.*, 2004).

Soybean WRKY-type transcription factor genes, *GmWRKY*13, *GmWRKY*21 and *GmWRKY*54, when introduced into *Arabidopsis*, conferred differential tolerance to abiotic stresses in transgenic *Arabidopsis* plants (Zhou *et al.*, 2008). ABA overly sensitive mutant of *Arabidopsis* (*abo3*) which encodes a WRKY transcription factor *AtWRKY63*, when disrupted by a T-DNA insertion in *At1g*66600, the mutant showed hypersensitivity to ABA and was less tolerant to drought than the wild type, thus uncovering an important role for a WRKY transcription factor in plant responses to ABA and drought stress (Ren *et al.*, 2010). Overexpression of *WRKY25* and *WRKY33* in *Arabidopsis* moderately enhanced the tolerance of transgenic plants to 100mM NaCl solution (Jiang and Deyholos, 2009). *BcWRKY46* gene has been reported to play an important role in regulation of ABA concentration and abiotic stress (Wang *et al.*, 2012). Constitutive expression in transgenic tobacco hosting *BcWRKY46* gene under the control of the *CaMV35S* promoter showed the susceptibility of transgenic tobacco to freezing, ABA, salt and dehydration stresses (Wang *et al.*, 2012). This finding is partly in agreement with previous results obtained with *GmWRKY* genes in soybean where transgenic *Arabidopsis* plants hosting *GmWRKY* genes was shown to increase tolerance to cold, salt, and drought stress. Interestingly, transgenic seedlings over-expressing *VvWRKY11* suggested that its gene product affects the expression of two stress responsive genes namely *AtRD29A* and *AtRD29B*, resulting in drought tolerance. The *VvWRKY11* protein may interact with DREB and/or ABEB, resulting in the increased expression of *AtRD29A* and *AtRD29B* (Wang *et al.*, 2012).

Conclusion

Adaptation to salinity and drought is undoubtedly one of the complex processes, involving numerous changes including attenuated growth the activation/increased expression or induction of genes, transient increases in ABA levels, accumulation of compatible solutes and protective proteins, increased levels of antioxidants and suppression of energy consuming pathways. However, no consensus has been reached in defining the key processes determining tolerance and the secondary follow-up processes. With the advancement of high throughput DNA technologies, several hundred stress-induced or up-regulated genes have been identified. The search for stress-associated genes may have been saturated at least in *Arabidopsis*. However, the functions of only a limited number of gene products are known.

Many plant genes are regulated in response to abiotic stresses such as drought, high salinity, heat and cold, and their gene products function in stress response and tolerance. Understanding the molecular mechanisms of plant stress responses to these abiotic stresses helps in manipulating plants to improve stress tolerance and productivity. The use of transgenes to improve the tolerance of crops to abiotic stresses remains an attractive option and targeting multiple gene regulation through transcription factors appear better than targeting single gene. Temporal control

Table 3.2: Function of WRKY Genes in Various Stresses (Chen *et al.*, 2012)

Gene	*Locus*	*Induced by Abiotic Factors*	*Function in Abiotic Stress*
AtWRKY2	At5g56270	NaCl, mannitol	Negative regulator in ABA signaling
AtWRKY6	At1g62300	H_2O_2, methyl viologen, Pi and B starvation	Negative regulator in low Pi stress and positive regulator in low B stress
AtWRKY18	At4g31800	ABA	ABA signaling, NaCl and mannitao tolerance
AtWRKY22	At4g01250	H_2O_2, dark	Enhanced dark-induced senescence
AtWRKY25	At2g30250	Ethylene, NO, NaCl, mannitol, cold, heat, ABA, cold	Tolerance to heat and NaCl, increased sensitivity to oxidative stress and ABA
AtWRKY26	At5g07100	Heat	Tolerance to heat
AtWRKY33	At2g38470	NaCl, mannitol, cold, H_2O_2, ozone oxidative stress, UV	Tolerance to heat and NaCl, increased sensitivity to oxidative stress and ABA
AtWRKY34	At4g26440	Cold	Negative regulator in pollen specific cold response
AtWRKY39	At3g04670	Heat	Tolerance to heat
AtWRKY40	At1g80840	ABA	ABA signaling
AtWRKY60	At2g25000	Wounding	ABA signaling, NaCl and mannitao tolerance
AtWRKY63	At1g66600	ABA	Negative regulator in ABA signaling while positive regulator in drought tolerance
AtWRKY75	At5g13080	Pi deprivation	Positive regulator in Pi starvation
OsWRKY08	05g50610	Drought, salinity, H_2O_2, ABA, NAA	Tolerance to osmotic stress
OsWRKY11	01g43650	Heat, drought	Tolerance to xerothermic stress
OsWRKY23	01g53260	Salinity, ABA, H_2O_2, Osmotic stress, dark	Enhanced dark-induced senescence
OsWRKY45	05g2577	Cold, ABA	Tolerance to salt and drought stress
OsWRKY72	11g29870	Salinity, heat, ABA, NAA, osmotic stress, sugar starvation	Negative regulator in ABA signaling and sugar starvation
OsWRKY89	11g02520	Salinity, ABA, UV-B, wounding	Tolerance to UV-B radiation
GmWRKY13	DQ322694	Salt, drought	Increased sensitivity to salt and mannitol while decreased sensitivity to ABA
GmWRKY21	DQ322691	Salt, drought, cold	Cold tolerance
GmWRKY54	DQ322698	Salt, drought	Salt and drought tolerance
TcWRKY53	EF053036	Salinity, cold, drought	Negative regulator in osmotic stress
HvWRKY34	DQ863118	Sugar	Sugar signaling
HvWRKY41	DQ863124	Sugar	Sugar signaling
HvWRKY46 (*SUSIBA2*)	AY323206	Sugar	Sugar signaling
NaWRKY3	AY456271	Wounding	JA signaling
NaWRKY6	AY456272	Wounding	JA signaling

of many stress responsive genes is regulated mainly by a combination of TFs and *cis*-acting elements in stress inducible promoters in plants. They play a key role in providing tolerance to multiple stresses in ABA-dependent and ABA independent manner. The TFs can genetically be engineered to produce transgenic plants having higher tolerance to drought, salinity, heat and cold stress using different promoters. Some of the recent studies have shown the importance of DREB TFs as candidate genes in marker-assisted breeding programme and developing proper functional markers, which could be used for allele-mining in crop improvement. Functional markers, based on DREB1 locus on the long arm of chromosome 3B, may be useful in wheat breeding programme for drought tolerance (Lata and Prasad, 2011).

It is reasonable to believe that WRKY genes provide selective advantages for plants to withstand environmental constraints, to cope with increased complexity in developmental and metabolic pathways and to modulate gene expression in a tissue-specific manner. The creation of taxa-specific WRKY proteins is likely to be implicated in controlling defined pathways linked to adaptation to different environmental stimuli or species-specific metabolic pathways. In the near future, the gene can be used for isolating genomic sequence corresponding to this cDNA as this would help in analyzing promoter and other regulatory sequences. Alternatively, gene can be cloned in an expression vector with an objective to characterize its encoded protein product.

References

Abdeen, A., Schnell, J. and Miki, B., 2010. Transcriptome analysis reveals absence of unintended effects in drought-tolerant transgenic plants overexpressing the transcription factor ABF3. *BMC Genomics*, 11: 1-21.

Abe, H., Urao, T., Ito, T., Seki, M., Shinozaki, K. and Yamaguchi-Shinozaki, K., 2003. Arabidopsis AtMYC2 (bHLH) and AtMYB2 (MYB) function as transcriptional activators in abscisic acid signaling. *Plant Cell*, 15: 63–78.

Agarwal, P., Arora, R., Ray, S., Singh, A.K., Singh, V.P., Takatsuj, H., Kapoor, S. and Tyagi, A.K., 2007. Genome-wide identification of C_2H_2 zinc-finger gene family in rice and their phylogeny and expression analysis. *Plant Mol. Biol.*, 65: 467–485.

Ali, K., Gujjar, R.S., Niwas, R., Gopal, M. and Tyagi, A., 2011. A Rapid Method for Estimation of Abscisic Acid and Characterization of ABA Regulated Gene in Response to Water Deficit Stress from Rice. *American Journal of Plant Physiology*, 6: 144-156.

Chen, J.R., Lu, J.J., Liu, R., Xiong, X.Y., Wang, T.X., Chen, S.Y., Guo, L.B. and Wang, H.F., 2010. DREB1C from Medicago truncatula enhances freezing tolerance in transgenic M. truncatula and China Rose (*Rosa chinensis* Jacq.). *Plant Growth Regul.*, 60: 199–211.

Chen, L., Song. Y/, Li. S., Zhang, L., Zou, C. and Yu, D., 2012. The role of WRKY transcription factors in plant abiotic stresses. *Biochimica et Biophysica Acta*, 1819: 120-128.

Correa, L.G.G., Pachon, D.M.R., Schrago, C.G., Santos, R.V., Roeber, B.M. and Vincentz, M., 2008. The Role of bZIP transcription factors in green plant evolution: adaptive features emerging from four founder genes. *PLoS One,* 3: 1-16.

Dubos, C., Stracke, R., Grotewold, E., Weisshaar, B., Martin, C. and Lepiniec, L., 2010 MYB transcription factors in Arabidopsis. *Trends Plant Sci.,* 15(10): 573 – 581.

Figueiredo, D.D., Barros, P.M., Cordeiro, A.M., Serra, T.S., Lourenc, T., Subhash, C.M., Oliveira, M. and Saibo, N.J.M., 2012. Seven zinc-finger transcription factors are novel regulators of the stress responsive gene OsDREB1B. *J. Exp. Bot.,* 63(10): 3643-3656.

Gao, F., Xiong, A., Peng, R., Jin, X., Xu, J., Zhu, B., Chen, J. and Yao, Q., 2010. OsNAC52, a rice NAC transcription factor, potentially responds to ABA and confers drought tolerance in transgenic plants. *Plant Cell Tiss. Organ Cult.,* 100: 255–262.

Gao, J.J., Zhang, Z., Peng, R.H., Xiong, A.S., Xu, J., Zhu, B. and Yao, Q.H., 2011. Forced expression of Mdmyb10, a MYB transcription factor gene from apple, enhances tolerance to osmotic stress in transgenic Arabidopsis. *Mol. Biol. Rep.,* 38: 205-211.

Gilmour, S.J., Fowler, S.G. and Thomashow, M.F., 2004. Arabidopsis transcriptional activators CBF1, CBF2, and CBF3 have matching functional activities. *Plant Mol. Biol.,* 54: 767–781.

Gilmour, S.J., Sebolt, A.M., Salazar, M.P., Everard, J.D. and Thomashow, M.F., 2000. Over-expression of the Arabidopsis CBF3 transcriptional activator mimics multiple biochemical changes associated with cold acclimation. *Plant Physiol.,* 124: 1854–1865.

Golldack, D., Luking, I. and Yang, O., 2011. Plant tolerance to drought and salinity: stress regulating transcription factors and their functional significance in the cellular transcriptional network. *Plant Cell Rep.,* 30: 1383–1391.

He, J.X., Mu, R.L., Cao, W.H., Zhang, Z.G., Zhang, J.S. and Chen, S.Y., 2005. AtNAC2, a transcription factor downstream of ethylene and auxin signaling pathways, is involved in salt stress response and lateral root development. *Plant J.,* 44: 903–916.

Hossain, M.A., Lee, Y., Cho, J., Ahn, C.H., Lee, S.K., Jeon, J.S., Kang, H., Lee, C.H., An, G. and Park, P.B., 2010. The bZIP transcription factor OsABF1 is an ABA responsive element binding factor that enhances abiotic stress signaling in rice. *Plant Mol. Biol.,* 72: 557–566.

Hu, H., Dai, M., Yao, J., Xiao, B., Li, X., Zhang, Q. and Xiong, L., 2006. Over-expressing a NAM, ATAF, and CUC (NAC) transcription factor enhances drought resistance and salt tolerance in rice. *Proc. Natl. Acad. Sci.,* 103: 12987–12992.

Huang, X.Y., Chao, D.Y. and Gao, J.P., 2009. A previously unknown zinc finger protein, DST, regulates drought and salt tolerance in rice via stomatal aperture control. *Gene Dev.,* 23: 1805– 1817.

Jeong, J.S, Kim, Y.S., Baek, K.H., Jung, H., Ha, S.H., Choi, Y.D., Kim, M., Reuzeau, C. and Kim, J.K., 2010. Root-specific expression of OsNAC10 improves drought tolerance and grain yield in rice under field drought conditions. *Plant Physiol.*, 153: 185–197.

Jiang, Y. and Deyholos, M.K., 2009. Functional characterization of Arabidopsis NaCl-inducible WRKY25 and WRKY33 transcription factors in abiotic stresses. *Plant. Mol. Biol.*, 69: 91–105.

Kam, J., Gresshoff, P.M. and Shorter, R., 2008. The Q-type C2H2 zinc finger subfamily of transcription factors in *Triticum aestivum* is predominantly expressed in roots and enriched with members containing an EAR repressor motif and responsive to drought stress. *Plant Mol. Biol.*, 67: 305–322.

Kang, J., Choi, H., Im, M. and Kim, S.Y., 2002. Arabidopsis basic leucine zipper proteins mediate stress-responsive abscisic acid signaling. *Plant Cell,* 14: 343–357.

Kiribuchi, K., Jikumaru, Y., Kaku, H., Minami, E., Hasegawa, M. and Kodama, O., 2005. Involvement of the basic helix-loop- helix transcription factor RERJ1 in wounding and drought stress responses in rice plants. *Biosci. Biotech. Bioch.*, 69: 1042–1044.

Kobayashi, F., Maeta, E., Terashima, A., Kawaura, K., Ogihara, Y. and Takumi, S., 2008. Development of abiotic stress tolerance via bZIP-type transcription factor LIP19 in common wheat. *J. Exp. Bot.*, 59: 891–905.

Kreps, J.A., Wu, Y., Chang, H.S., Zhu, T., Wang, X. and Harper, J.F., 2002. Transcriptome changes for Arabidopsis in response to salt, osmotic, and cold stress. *Plant Physiol.*, 130: 2129–2141.

Lata, C. and Prasad, M., 2011. Role of DREBs in regulation of abiotic stress responses in plants. *J. Exp. Bot.*, 62: 4731–4748.

Li, X., Cheng, X., Liu, J., Zeng, H., Han, L. and Tang, W., 2011. Heterologous expression of the Arabidopsis DREB1A/CBF3 gene enhances drought and freezing tolerance in transgenic *Lolium perenne* plants. *Plant Biotechnol. Rep.*, 5: 61–69.

Lippold, F., Sanchez, D.H., Musialak, M., Schlereth, A., Scheible, W.R., Hincha, D.K. and Udvardi, M.K., 2009. AtMyb41 Regulates transcriptional and metabolic signaling responses to osmotic stress in Arabidopsis. *Plant Physiol.*, 149: 1761–1772.

Liu, Q.L., Xu, K.D., Zhao, L.J., Pan, Y.Z., Jiang, B.B., Zhang, H.Q. and Liu, G.L., 2011. Over-expression of a novel chrysanthemum NAC transcription factor gene enhances salt tolerance in tobacco. *Biotechnol. Lett.*, 33: 2073–2082.

Lu, M., Ying, S., Zhang, D.F., Shi, Y.S, Song, Y.C., Wang, T.Y. and Li, Y., 2012. A maize stress responsive NAC transcription factor, ZmSNAC1, confers enhanced tolerance to dehydration in transgenic Arabidopsis. *Plant Cell Rep.*, 31: 1701–1711.

Mao, X., Jia, D., Li, A., Zhang, H., Tian, S., Zhang, X., Jia, J. and Jing, R., 2011. Transgenic expression of TaMYB2A confers enhanced tolerance to multiple abiotic stresses in Arabidopsis. *Funct. Integr. Genomics,* 11:445–465.

Mao, X., Zhang, H., Qian, X., Li, A., Zhao, G. and Jing, R., 2012. TaNAC2, a NAC-type wheat transcription factor conferring enhanced multiple abiotic stress tolerances in Arabidopsis. *J. Exp. Bot.,* 63(8): 2933–2946.

Mare, C., Mazzucotelli, E., Crosatti, C., Francia, E., Stanca, A.M. and Cattivelli, L., 2004. Hv-WRKY38: a new transcription factor involved in cold- and drought-response in barley. *Plant Mol. Biol.,* 55: 399–416.

Maruyama, K., Sakuma, Y., Kasuga, M., Ito, Y., Seki, M., Goda, H., Shimada, Y., Yoshida, S., Shinozaki, K. and Yamaguchi-Shinozaki, K., 2004. Identification of cold-inducible downstream genes of the Arabidopsis DREB1A/CBF3 transcriptional factor using two microarray systems. *Plant J.,* 38: 982–993.

Nijhawan, A., Jain, M., Tyagi, A.K. and Khurana, J.P., 2008. Genomic survey and gene expression analysis of the basic leucine zipper transcription factor family in rice. *Plant Physiol.,* 146: 333– 350.

Ogo, Y., Itai, R.N., Nakanishi, H., Kobayashi, T., Takahashi, M. and Mori, S., 2007. The rice bHLH protein OsIRO2 is an essential regulator of the genes involved in Fe uptake under Fe-deficient conditions. *Plant J.,* 51: 66–77.

Oliveira, T.M.D., Cidade, L.C., Gesteira, A.S., Filho, M.A.C., Filho, W.S.S. and Costa, M.G.C. 2011. Analysis of the NAC transcription factor gene family in citrus reveals a novel member involved in multiple abiotic stress responses. *Tree Genet Genomes,* 7: 1123–1134.

Pasquali, G., Biricolti, S., Locatelli, F., Baldoni, E. and Mattana, M., 2008. Osmyb4 expression improves adaptive responses to drought and cold stress in transgenic apples. *Plant Cell Rep.,* 27: 1677–1686.

Reddy, G.L. and Reddy, A.R., 2008. Rice DREB1B promoter shows distinct stress-specific responses, and the overexpression of cDNA in tobacco confers improved abiotic and biotic stress tolerance. *Plant Mol. Biol.,* 68: 533–555.

Ren, X., Chen, Z., Liu, Y., Zhang, H., Zhang, M., Liu, Q., Hong, X., Zhu, J.K. and Gong, Z., 2010. ABO3, a WRKY transcription factor, mediates plant responses to abscisic acid and drought tolerance in Arabidopsis. *Plant J.,* 63: 417–429.

Saad, R.B., Zouari, N., Ramdhan, W.B., Azaza, J., Meynard, D., Guiderdoni, E. and Hassairi, A., 2010. Improved drought and salt stress tolerance in transgenic tobacco overexpressing a novel A20/AN1 zinc-finger "AlSAP" gene isolated from the halophyte grass *Aeluropus littoralis. Plant Mol. Biol.,* 72: 171–190.

Savitch, L.V., Allard, G., Seki, M., Robert, L.S., Tinker, N.A., Huner, N.P.A., Shinozaki, K. and Singh, J., 2005. The effect of overexpression of two Brassica CBF/DREB1-like transcription factors on photosynthetic capacity and freezing tolerance in *Brassica napus. Plant Cell Physiol.,* 46: 1525–1539.

Seki, M., Rabbani, M.A., Maruyama, K., Abe, H., Khan, A.M., Katsura, K., Ito, Y., Yoshiwara, K., Shinozaki, K. and Yamaguchi-Shinozaki, K., 2003. Monitoring expression profiles of rice genes under cold, drought, and high-salinity stresses and abscisic acid application using cDNA microarray and RNA gel-blot analyses. *Plant Physiol.*, 133: 1755–1767.

Seo, P.J., Xiang, F., Qiao, M., Park, J.Y., Lee, Y.N., Kim, S.G., Lee, Y.H., Park, W.J., Park, C.M., 2009. The MYB96 transcription factor mediates abscisic acid signaling during drought stress response in Arabidopsis. *Plant Physiol.*, 151: 275–289.

Shimono, M., Sugano, S., Nakayama, A., Jiang, C.J., Ono, K., Toki, S. and Takatsuji, H., 2007. Rice WRKY45 plays a crucial role in benzothiadiazole inducible blast resistance. *Plant Cell*, 19: 2064–2076.

Shinozaki, Y.K. and Shinozaki, K., 2006. Organization of cis-acting regulatory elements in osmotic and cold stress responsive promoters. *Trends Plant Sci.*, 10: 88–94.

Tran, L.S., Nakashima, K., Sakuma, Y., Simpson, S.D., Fujita, Y., Maruyama, K., Fujita, M., Seki, M., Shinozaki, K. and Yamaguchi-Shinozaki, K., 2004. Isolation and functional analysis of Arabidopsis stress-inducible NAC transcription factors that bind to a drought-responsive cis-element in the early responsive to dehydration stress promoter. *Plant Cell*, 16: 2481–2498.

Vogel, J.T., Zarka, D.G., Buskirk, H.A.V., Fowler, S.G. and Thomashow, M.F., 2005. Roles of the CBF2 and ZAT12 transcription factors in configuring the low temperature transcriptome of Arabidopsis. *Plant J.*, 41: 195–211.

Wang, F., Hou, X., Tang, J., Wang, Z., Wang, S., Jiang, F. and Li, Y., 2012. A novel cold-inducible gene from Pak-choi (*Brassica campestris* ssp. *chinensis*), BcWRKY46, enhances the cold, salt and dehydration stress tolerance in transgenic tobacco. *Mol. Biol. Rep.*, 39: 4553–4564.

Wang, Y., Gao, C., Liang, Y., Wang, C., Yang, C and Liu, G., 2010. A novel bZIP gene from *Tamarix hispida* mediates physiological responses to salt stress in tobacco plants. *J. Plant Physiol.*, 167: 222–230.

Wu, X., Yoko, S., Kishitani, S., Ito, Y. and Toriyama, K., 2009. Enhanced heat and drought tolerance in transgenic rice seedlings overexpressing OsWRKY11 under the control of HSP101 promoter. *Plant Cell Rep.*, 28: 21–30.

Xu, K., Xu, X., Fukao, T., Canlas, P. and Maghirang-Rodriguez, R., 2006. Sub1A is an ethylene response- factor–like gene that confers submergence tolerance to rice. *Nature*, 442: 705–708.

Yanez, M., Caceres, S., Orellana, S., Bastias, A., Verdugo, I., Lara, S.R. and Casaretto, J.A., 2009. An abiotic stress-responsive bZIP transcription factor from wild and cultivated tomatoes regulates stress-related genes. *Plant Cell Rep.*, 28: 1497–1507.

Yang, A., Dai, X. and Zhang, W.H., 2012. A R2R3-type MYB gene, OsMYB2, is involved in salt, cold, and dehydration tolerance in rice. *J. Exp. Bot.*, 63(7): 2541–2556.

Yang, B., Jiang, Y., Rahman, M.H., Deyholos, M.K. and Kav, N.V., 2009. Identification and expression analysis of WRKY transcription factor genes in canola (*Brassica napus* L.) in response to fungal pathogens and hormone treatments. *BMC Plant Biol.*, 9: 1-19.

Yang, W., Liu, X.D., Chi, X.J., Wu, C.A., Li, Y.Z., Song, L.L., Liu, X.M., Wang, Y.F., Wang, F.W., Zhang, C., Liu, Y., Zong, J.M. and Li, H.Y., 2011. Dwarf apple MbDREB1 enhances plant tolerance to low temperature, drought, and salt stress via both ABA-dependent and ABA-independent pathways. *Planta*, 233: 219–229.

Zhang, J.Z., Creelman, R.A. and Zhu, J.K., 2004. From laboratory to field, using information from Arabidopsis to engineer salt, cold, and drought tolerance in crops. *Plant Physiol.*, 135: 615–621.

Zhang, L., Zhao, G., Jia, J., Liu, X. and Kong, X., 2012. Molecular characterization of 60 isolated wheat MYB genes and analysis of their expression during abiotic stress. *J. Exp. Bot.*, 63(1): 203–214.

Zheng, X., Chen, B., Lu, G. and Han, B., 2009. Overexpression of a NAC transcription factor enhances rice drought and salt tolerance. *Biochem. Bioph. Res. Co.*, 379: 985–989.

Zhou, J., Li, F., Wang, J., Ma, Y., Chong, K. and Xu, Y., 2009. Basic helix-loop-helix transcription factor from wild rice (OrbHLH2) improves tolerance to salt and osmotic stress in Arabidopsis. *J. Plant Physiol.*, 166: 1296-1306.

Zhou, Q.Y, Tian, A.G., Zou, H.F., Xie, Z.M., Lei, G., Huang, J., Wang, C.M., Wang, H.W., Zhang, J.S. and Chen, S.Y. 2008. Soybean WRKY-type transcription factor genes, GmWRKY13, GmWRKY21, and GmWRKY54, confer differential tolerance to abiotic stresses in transgenic Arabidopsis plants. *Plant Biotechnol. J.*, 6: 486–503.

Zhu, J.K., 2002. Salt and drought stress signal transduction in plants. *Annu. Rev. Plant Biol.*, 53: 247–273.

Zou, X., Shen, Q.J. and Neuman, D., 2007. An ABA inducible WRKY gene integrates responses of creosote bush (*Larrea tridentate*) to elevated Co2 and abiotic stresses. *Plant Sci.*, 172: 997–1004.

2016, Recent Advances in Plant Stress Physiology *Pages* **69–88**
Editors: ***Praduman Yadav, Sunil Kumar and Veena Jain***
Published by: **DAYA PUBLISHING HOUSE, NEW DELHI**

Chapter 4

MAPK Signaling Modules in Stress

K. Prabhakara Rao[1]*, Monika Jaggi[2], K. Sarala[1], T.G.K. Murthy[1]

[1]ICAR – Central Tobacco Research Institute (Indian Council of Agricultural Research
Rajahmundry, Andhra Pradesh
[2]Department of Botany, Mirinda House College,
University of Delhi, New Delhi

ABSTRACT

In recent years MAPK cascade has emerged as one of the most well studied signaling pathways. It plays a crucial role in transmitting extracellular signals to the nucleus in response to various stresses. A MAPK cascade co nsists of a three-tier system where each tier is phosphorylated by upper tier. It is represented as a MAP3K-MAP2K-MAPK module that serves as a link between upstream receptors and downstream targets. MAP2K being the central point of this cascade converge all the signals from upstream MAP3Ks and target genome through downstream MAPKs. Sometimes, MAP4Ks also get involved in linking upstream signaling components to the core MAPK cascade. MAPKs target various genes involved in stress responses as well as cellular and developmental processes. Therefore, in this chapter an effort has been made to summarize the role of MAPK cascade in different stress responses in plants.

Keywords: *MAPK cascade, Plant stress, Signal transduction.*

Introduction

Plant as sessile entities needs to combat with different environmental and endogenous factors during the whole period of ontogenesis. Failure to prevent

* *Corresponding author.* E-mail: prabhakarabt@yahoo.co.in

the negative influence of these factors generally leads to devastating effects on basic metabolism, transport processes, membranes, and cell structure. Plants have developed different response mechanisms at molecular, cellular, organ levels through evolution for effective protection. Depending on the stimuli the cellular responses trigger a cascade of events, consisting perception and molecular recognition of the external stimulus, generation of a signal and its transmittance via specific signaling pathways, which leads to the activation of defined effectors and subsequent molecular reactions (Soropy and Munshi, 1998; Moller and Chua, 1999). Majority of signal transduction mechanisms are modulated by protein phosphorylation and dephosphorylation that is controlled by protein kinases and protein phosphatases. This fast and effective signaling system fulfills all requirements needed to interconnect the sensing system with gene expression processes in coordination with the physiological status of the cell. In plants, hundreds of protein kinases (PKs) are known which are involved in cellular signal transduction services. PKs modify proteins by catalyzing addition of monophosphate groups to the side chains of the most commonly, specific serine, threonine and/or tyrosine residues in the protein backbone. This process is reversible with the help of enzyme phosphatase which leads to the removal of phosphate group. The plant PK superfamily comprise of different classes of kinases that are categorized on the basis of their amino acid sequence similarity. Out of these kinase classes, the most conserved and best characterized protein kinase signaling pathways is mitogen-activated protein kinase (MAPK) cascade. MAP Kinase was first identified by Sturgill and Ray as microtubule associated protein kinase and thus, was christened microtubule associated protein kinases. MAPK comprise a family of serine/threonine protein kinases.The extracellular signals are perceived by the uppermost MAP3Ks and are transferred to substrates via MAPKs (Hardie, 1999). In plants, first MAPK was detected in alfalfa and peasucceeded by cloning of MAPKs from *Arabidopsis thaliana* (Mizoguchi *et al.*, 1993) and *Nicotiana* (Wilson *et al.*, 1996). Upon signal perception, they mediates the transmission of those signals that control numerous cellular processes, including cell division, differentiation, and responses to environmental stresses in eukaryotic systems. Therefore, in this chapter an effort has been made to summarize the role of MAPK cascade in different stress responses in plants

General Architecture of MAPK Cascade and Classification

The MAPK cascade is a major cellular device by which eukaryotic cells adaptively respond to extracellular stimuli. These are conserved signaling modules found in all eukaryotic cells, including plants, fungi and animals. A MAPK cascade minimally consists of three kinases: a MAPKKK (MAPKK kinase), a MAPKK (MAPK kinase) and a MAPK, which phosphorylate and therefore activate each other in a specific way. It is a three-tier system where each tier is phosphorylated by upper tier. MAPKKKs are serine/threonine kinases phosphorylating two amino acids in the S/T-X3–5-S/T motif of the MAPKK activation loop. MAPKKs are dual specificity kinases that activate a MAPK through double phosphorylation of the T-X-Y motif in the activation loop. These act as the central point of the cascade by converging all the signals from upstream MAP3Ks and targeting genome through downstream

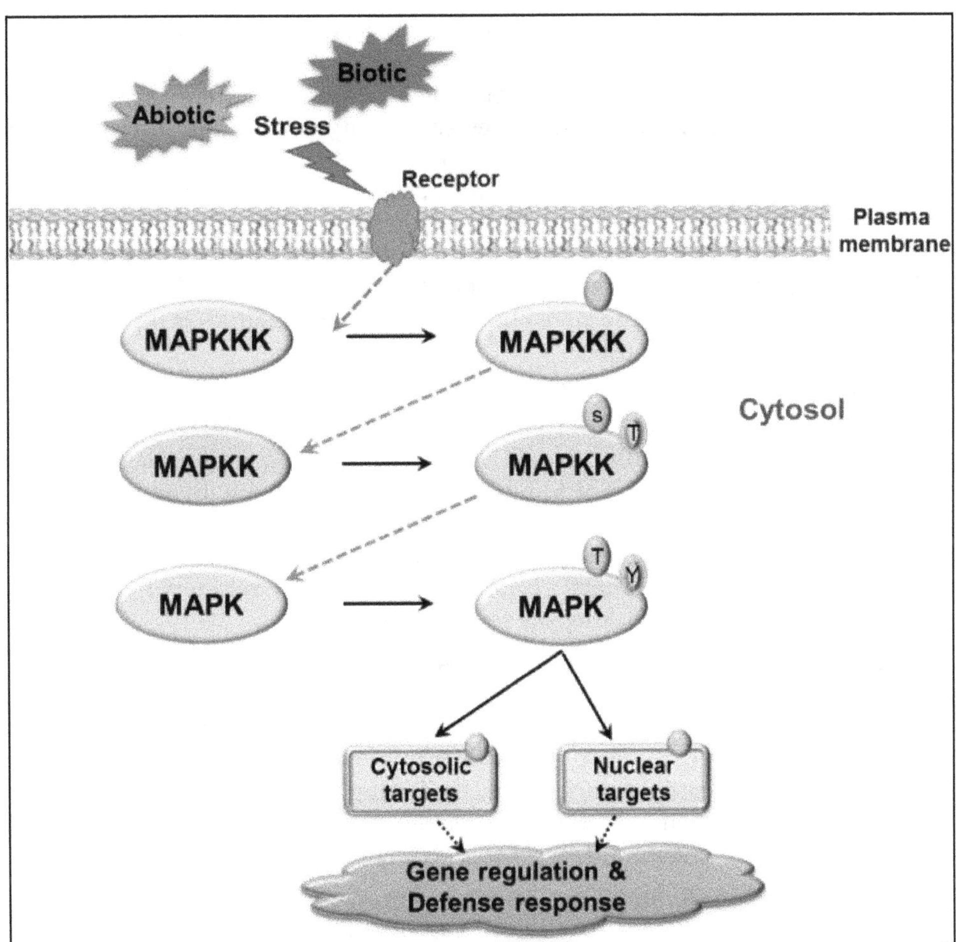

Figure 4.1: A Typical MAP Kinase Cascade Showing Sequential Phosphorylation of MAPKKK, MAPKK and MAPK.

MAPKs (Figure 4.1). MAPKs are serine/threonine kinases able to phosphorylate a wide range of substrates, including other kinases and/or transcription factors. A fourth level of kinases, named MAPKKKKs (MAPKKK kinases), may act as adaptors linking upstream signaling steps to the core MAPK cascades. Interactions between kinases within a MAPK cascade occur through docking sites present in the kinases and/or with the help of external scaffolding proteins.

MAPK proteins have a highly conserved kinase domain containing 11 subdomains which are characteristic for MAPKs. Based on the amino acid motif exist at the phosphorylation site between the sub domains VII and VIII, they are classified as TEY (TEY subtype) and TDY (TDY subtype) types. The TEY subtype can be classified into three Groups, A, B and C, whereas the TDY subtype forms a more distant Group D. The former TEY subtype contains a phosphorylation motif in the activation loop while TDY subtype has motif within its active site buried at its domain surface. The N-terminal domain of MAPK protein comprises of 135 residues which are arranged largely in the form of β-sheets and a glycine-rich loop named as phosphate anchor ribbon which acts as ATP- binding pocket. The C-terminal domain (225 residues) possesses a catalytic base, Mg^{2+} binding sites and the phosphorylation lip (activation loop). The ATP molecule binds between two domains and the protein substrate is believed to be bound on the surface of the C-terminal domain. The N and C terminals are predominantly composed of beta-strand and alpha-helices, respectively. The sequence of phosphorylation and activation loop influences the substrate specificity. The TXY (X-D/E/P) motif is a dual phosphorylation site and phosphorylation of both residues is a prerequisite for the activation of this cascade.

Phylogenetic analysis suggested that MAPKs can be classified into four groups according to their sequences (TEY subtype) and structures. The related A, B and C groups belong to TEY subtype and TDY subtype forms a more distant group D (MAPK group 2002). Group A MAPKs are mostly associated with developmental processes and are activated in response to biotic and abiotic stresses where as group B members are implicated in pathogen defense and abiotic stress responses (Rodriguez, 2010). Group D members possess a C-terminal common docking (CD) domain that may act as a docking site for MAP2Ks.

MAP2Ks could be divided into 4 groups (A, B, C and D). Members of group C and D possess only one exon whereas group A and B contain 8-9 exons. Kinases in group B are characterized by a nuclear transfer factor (NTF) domain (MAPK group 2002) which function to increase the nuclear import of cargo proteins and hence actively involved in cytoplasmic-nuclear trafficking. Classification of MAP2Ks is based upon protein alignment. Different members of same group are activated in the presence of different stimuli (Xu *et al.*, 2008; Zhou *et al.*, 2009).

The MAP3K family forms the largest and most complex group compared to the MAPK and MAP2K classes. Rao *et al.* (2010) categorized MAP3K from rice and Arabidopsis into three sub-groups, *i.e.* Raf, ZIK and MEKK. The latter has a conserved catalytic domain. Most of the proteins belonging to this group participate in MAPK cascades by phosphorylating downstream MAP2Ks. CONSTITUTIVE TRIPLE RESPONSE 1 (CTR1) and ENHANCED DISEASE RESISTANCE1 (EDR1) are two of the well studiedRaf-like MAP3Ks from Arabidopsis which are actively involved in ethylene mediated signaling and defense responses. Most of the Raf-family members possess a long N-terminal regulatory domain and a C-terminal kinase domain. In contrast, ZIK-like members predominantly have N-terminal kinase domain. MEKK family members have comparatively less conserved protein structure with kinase domain located either at N or C terminal or central part of the protein. Ubiquitin-interaction motif and ACT domain which has a regulatory role in a wide range of metabolic enzymes are seen only in the members of Raf family from rice and Arabidopsis (Rao *et al.*, 2010).

MAPK Substrates

MAPKs are characterized by a single pronounced specificity, namely a preference for phosphorylating proteins at a Ser/Thr-Pro motif (Canagarajah *et al.*,

1997). Nevertheless, proteins containing simply a Ser/Thr-Pro motif are often poor MAPK substrates, suggesting that other elements contribute to specific substrate docking. An estimated one-third of all proteins in higher eukaryotes are regulated by phosphorylation by protein kinases (PKs). Although plant genomes encode more than 1000 PKs, the substrates of only a small fraction of these kinases are known. Given these numbers, it is clear that most protein kinases have many different physiological substrates, and that the majority of these substrates remain to be identified. Evidence from other eukaryotes indicates that MAPK phosphorylation regulates transcription factors by altering their activity, localization, and/or stability. A regular MAPK has an active site and a common docking site. Both sites are stationed closely and are involved in recognition and binding of target proteins. MAPK phosphatases and scaffolding proteins play a regulatory role in MAPK signaling specificity, location and duration. Nonetheless, mechanistic evidence for gene regulation in vivo by plant MAPKs via specic transcription factors and their direct transcriptional targets remains sparse.

Since MAPK substrates are evolutionary and structurally very diverse, direct protein-protein interaction and activity screens were used to identify potential MAPK partners in various studies (Bethke *et al.*, 2009). Feilner *et al.* (2005) identified 48 *in vitro* substrates for MPK3 and 39 substrates for MPK6 by employing a protein microarray. Twenty six substrates were found common to both MAPKs. Popescu *et al.* (2009) identified 570 in vitro substrates for 10 different MAPKs by employing Arabidopsis protein microarrays. Several of these substrates are transcription factors (TGA1, WRKY6, WRKY8, WRKY53, WRKY62, and WRKY65) that are known to be involved in the regulation of development, defense and stress responses. There was a 30-40 per cent overlap between the MPK3 and MPK6 substrates in this study, suggesting some functional redundancies of the two MPKs. To date, there are reports of a few MAPK substrates from plants. In Arabidopsis, this includes ACS6, PHOS32, bHLH speechless, VIP1, EIN3, WRKY53, WRKY33, MKS1 and ERF104 (Andreasson and Ellis, 2010). Interactions for a few of them have also been validated in planta. Notably, a MPK4 substrate, MKS1, was reported to interact with MPK4 in a yeast-2-hybrid analysis, and has a putative role in plant defense (Andreasson *et al.*, 2005). The MPK4-MKS1 interaction was validated in planta by co-immuno precipitation and it was found that MKS1 further interacted with two transcription factors, WRKY29 and WRKY33. In another study, specific interaction of ERF104, an ethylene response factor, with MPK6 was identified in Arabidopsis. The interaction of ERF104 was shown by yeast-2-hybrid and was validated by in vivo fluorescence resonance energy transfer (FRET) analysis (Bethke *et al.*, 2009). ERF104 is specifically phosphorylated by MPK6 *in vitro* and is involved in defense responses.

MAPK Modules in Abiotic Stress

Plants for their growth and survival must respond to diverse environmental abiotic stresses including high and low temperature, salinity, wounding, drought, heavy metals and UV irradiation. They developed specific mechanisms to withstand abiotic stress, such as the synthesis of stress hormones like Abscisic acid (ABA) that triggers second responses and the expression of specific sets of genes that result in changes in the composition of the major cell components (Munnik *et al.*, 2001) and other innate immunity mechanisms. MAPKs play central role in transducing the signals during these conditions and it was demonstrated in various plants including *Arabidopsis*, the model plant organism.

MAPKs in Temperature Stress

Extremities in temperatures induce physiological responses in Plant species by changing the properties of plasma membranes: heat increases their fluidity while cold makes them more rigid (Mittler *et al.*, 2012; Wang *et al.*, 2003).

Low temperature (cold stresses) activates MAPK cascade and induce changes in gene expression (Huang *et al.*, 2012; Mishra *et al.*, 2006) and results in varied physiological responses like plasma membrane rigidification (Sangwan *et al.*, 2002), mobilization of Ca^{2+} (Zhu *et al.*, 2013), histidine kinase activation (Nongpiur *et al.*, 2012), and the generation of bioactive phosphatidic acid (Arisz *et al.*, 2013) or phosphatidyl inositol (Delage *et al.*, 2012). The precise mechanisms underlying the cold perception to changes in physiological responses have not been clearly deciphered till now, however the cold stress induced diverse secondary messengers, calcium-dependent protein kinases (CDPKs; Ludwig *et al.*, 2004) and MAPKs (Solanke and Sharma, 2008) were well documented. Many of these modulates the transcriptional regulation of C-repeat binding factors (CBF/DREB), the myc-like transcription factor known as inducer of CBF expression 1 (ICE1), and the myb-like transcription factor SNOW1, which interacts with ICE1 in the regulation of CBF/DREB expression (Lissarre *et al.*, 2010; Zhou *et al.*, 2011). Among the MAPK s MPK4 and MPK6 play important roles in the cold responses of Arabidopsis plants, and become activated within 2 min of cold exposure (Ichimura *et al.*, 2000; Teige *et al.*, 2004). Further protein interaction studies proved the involvement of MEKK1 (MAP3K), MKK2, MKK1 (MAP2Ks) upstream to MPK4 and MPK6, this cascade plays a central role in cold stress responses of Arabidopsis.

During heat stress, when plants experience higher temperature than ambient, complex defense mechanisms will be activated to prevent protein degradation. MAPK pathways mediate the process as it was described in alfalfa and tomato (Link *et al.*, 2002; Sangwan and Dhindsa, 2002). It was reported in Arabidopsis MPK6 phosphorylates a major heat stress transcription factor, HsfA2 within 5 min of the onset of heat stress and relocates HsfA2 from cytoplasm to nucleus (Evrard *et al.*, 2013).

MAPKs in Oxidative Stress

Reactive oxygen species (ROS) commonly induced during periods of abiotic stress, genotoxic irradiation, and drought/water stress (Suzuki *et al.*, 2012), as well as during pathogenic invasions (Obrien *et al.*, 2012) play important roles in cell signaling and homeostasis, and they are intensely involved in MAPK signaling. ROS burust cause the activation of MPK3 and MPK6 (Lumbreras *et al.*, 2010), via MEKK1-MKK1/2 pathway (Pitzschke *et al.*, 2009). OXI1 (oxidative signal-inducible 1) gene transcription was observed along with MPK3/MPK6 activity during oxidative stress in Arabidopsis which maintenance normal root hair growth under stressful conditions. Further it was demonstrated that MKK5 transmit phospho-signals to both MPK6 and MPK3 in responses to ozone-related oxidative stress (Miles *et al.*, 2009; Xing *et al.*, 2013).

MAPKs in Salinity Stress

High concentration of NaCl in soils have strong adverse effects on plant growth and development resulting in salt stress. Plants as a part of tolerance mechanism remove Na^+ from cells via the salinity overly sensitive 1 (SOS1) Na^+/H^+ antiporter, and maintains high intracellular K^+/Na^+ ratios by HKT transporters, and the vacuolar sequestration of Na^+ via the NHX1 tonoplast antiporter (Munns and Tester, 2008). Diverse signaling proteins including various members of the MAPK family will get activated in salt stress through secondary messengers. Activation of a Raf-like MAP3K known as DSM1 (drought hypersensitive mutant 1) in rice (Ning *et al.*, 2010), MKKK20 which lies upstream of MPK6 Arabidopsis (Kim *et al.*, 2012) in high salt concentrations clearly demonstrated the same. MEKK1 in Arabidopsis negatively regulates salt stress whereas MKK2 gets induced in salinity and activatesMPK4 and MPK6 and its null mutant are hypersensitive to salt. Activated of MPK6 rapidly phosphorylates the SOS1 Na^+/H^+ antiporter and thereby induces sodium efflux at the expense of cytosolic acidification (Yu *et al.*, 2010). MPK6 also regulates hyperosmotic and salt induced transcriptional transactivation of responsive genes by directly phosphorylating the ZAT6 zinc finger transcription factor (Liu *et al.*, 2013). The signaling pathway based on MEKK1, MKK2 andMPK4/MPK6 plays a key role in cold stress and also in the transduction of salt stress signals (Teige *et al.*, 2004).

MAPKs in Drought Stress

Drought stress is another significant abiotic stress experienced majorly by land plants due to changes in the soil moisture content. Plants are highly sensitive to drought stress and activates adaptation mechanisms by rapidly inducing the PLD-mediated production of phosphatidic acid (Bargmann *et al.*, 2009; Hong *et al.*, 2010) and the generation of ROS (Choudhury *et al.*, 2013; Yao *et al.*, 2013) which helps in water-deficit induced stomatal closure. MAPK cascade is involved in transducing the low water availability signal and transcriptional regulation of genes for desired plant responses. MPK9 and MPK12, preferentially expressed in stomatal guard cells and are involved in regulation of stomatal movements through changes in ABA levels (Jammes *et al.*, 2011; Salam *et al.*, 2012, 2013). Arabidopsis MAPK MPK6 gets activated by drought induced ROS accumulation and MPK3 downstream to MKK4 together promote drought-induced responses. Activation of diverse MAPKs in relation to drought were reported from different crop plants which includes OsMPK3, 4, 7, 14, 20-4 and 20-5 in rice (Shen *et al.*, 2012), GhMPK2 and GhMPK16 in cotton (Shi *et al.*, 2011; Zhang *et al.*, 2011), and ZmMPK3 in maize (Wang *et al.*, 2010). This illustrates the versatility, redundancy and extensive crosstalk of MAPK cascades in response to various stresses.

MAPKs in Wounding or Mechanical Stress

Wounding stress is very commonly observed in crop plants due to herbivores, cultural operations or other abiotic factors. It rapidly triggers the production of plant hormones like jasmonic acid, ROS and transient Ca^{2+} influxes into the cytoplasm. The elevated ROS and Ca^{2+} levels in Arabidopsis makes MKK3 to activate MPK8, which directly inhibits RbohD (Respiratory burst oxidase homologue an NADPH oxidase) and followed by OXI1 and the transcription factor ZAT12. Further it was demonstrated that MEKK1 interacts with MKK1 during wounding stress and also directly with MPK4, MPK5 and MPK13 as a scaffolding protein. These genes with different upstream factors and secondary messengers were also reported in heavy metal stress of different crop species. Many of the MAPK pathway genes that are involved in transducing the abiotic stress signal are redundant (Figure 4.2) and cross talking was involved at the level of MAPKK and MAPKKK.

MAPK Modules in Biotic Stress

In plants two defensive mechanisms of immune system are known against microbial attack (Jones and Dangl, 2006). The first mechanism involves trans-membrane receptors, named as pattern recognition receptors (PRRs) which recognize conserved pathogen-associated molecular patterns (PAMPs). PAMPs are mostly small molecules derived from different pathogen structures common to a class of pathogens. Bacterial elicitors flagellin (flg22), harpin, and lipopolysaccharides, and the fungal molecules chitin and ergosterol are examples of PAMPs. Under stress, these receptors (PRRs) get activated and induce convergent intracellular signaling pathways which lead to the establishment of PAMP-triggered immunity (PTI). The plants respond to pathogen attack by activating multi-step defense responses, comprising rapid production of reactive oxygen species (ROS), strengthening of cell walls and induction of the HR leading to localized cell death at the sites of infection. Plant defense responses also include synthesis of pathogen-related proteins and phytoalexins (Asai *et al.*, 2002; Lee *et al.*, 2009). The other mechanism is known as effector-triggered immunity (ETI). Effector proteins are secreted by pathogens in to the cells of their host (plants) to inhibit the functions of PTI signaling components. In response to this, plants have also evolved another mechanism to target these effector proteins via immune receptors, called resistance (R) proteins, which trigger the hypersensitive response (HR) (Rodriguez *et al.*, 2010).

MAPKS in PAMP-Triggered Immunity (PTI)

It has been firmly established that MAPKs play a central role in pathogen defense in Arabidopsis, Tobacco, Tomato, Parsley, Brassica and Rice (Frye *et al.*, 2001; Zhang and Klessig, 2001; Cardinale *et al.*, 2002). The PTI requires a signal transmission from receptors to downstream targets and it is mostly achieved via MAPK cascade. The first complete MAPK cascade in regulating plant defense against bacterial pathogen, MAP3K1-MAP2K4/MAP2K5-MAPK3/MAPK6-WRKY22/WRKY29, was proposed to be downstream of one of the best characterized plant PRRs, flagellin receptor kinase (FLS2 LRR), which potentially activates the MAP3K1 by phosphorylating the Ser/Thr residues (Asai *et al.*, 2002) further reported that an elicited FLS2 complex with BRI1-associated kinase (BAK1) induces MAPK signaling cascade. Another MAPK cascade, MAP3K1-MAPKK1/MAPKK2-MAPK4-MKS1 mediate jasmonate- and salicylate dependent defense responses.This cascade negatively regulates plant immune responses as the loss of function of either

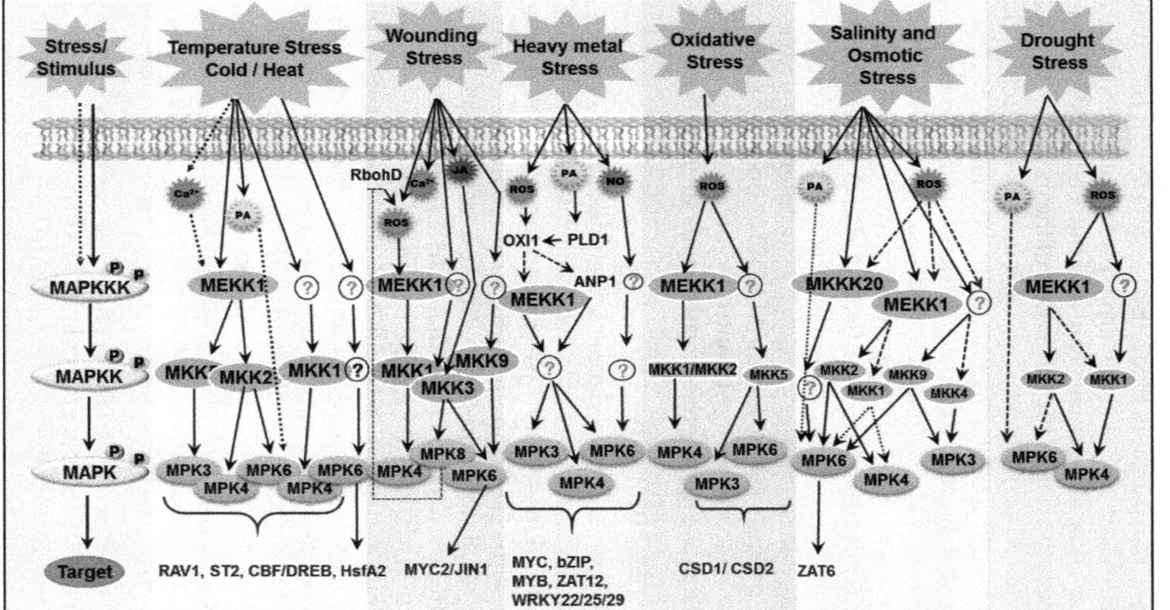

Figure 4.2: MAPK Modules Involved in Various Abiotic Stress Responses from *Arabidopsis thaliana*. A general schematic representation of MAPK cascade are depicted on the left. Solid lines show established signaling pathways, while dashed lines represent putative ones. (Modified from Smekalova *et al.,* 2014).

[The abbreviations denotes as follows: ABA: Abscisic acid; ANP1: Arabidopsis NPK1-like protein kinase (MAP3K); bZIP: Basic region-leucine zipper; CBF/DREB: C-repeat binding factors; CSD1/2: Cu/Zn superoxide dismutase; HsfA2: Heat stress transcription factor; JA: Jasmonic acid; MAPK (MPK): Mitogen activated protein kinase; MAPKK (MKK): Mitogen-activated protein kinase kinase; MAPKKK/MEKK: Mitogen-activated protein kinase kinasekinase; MYC2/JIN1: Member of basic helix–loop–helix (bHLH) transcription factors family; MYB: MYB (myeloblastosis) family of transcription factors; OXI1: Oxidative signal-inducible 1; PA: Phosphatidic acid; PCD: Programmed cell death; PDK1: Phosphoinositide-dependent kinase1; PLD: Phospholipase D; PTI1-4: Pto-interacting1-4; RAV: RAV (Related to ABA-insensitive3/Viviparous1) family of transcription factors; RbohD: Respiratory burst oxidase homologue; ROS: Reactive oxygen species; SOS1: Na/H antiporter Salt-Overly-Sensitive1; STZ: Salt tolerance zinc finger transcription factor; WRKY: Superfamily of transcription factors; ZAT: Zinc finger transcription factor].

MAP3K1 or MAPK4 resulted in the accumulation of salicylic acid and expression of pathogenesis related (PR) proteins. It was proposed that MAPK4 along with its substrate MAP kinase substrate 1 (MKS1) and WRKY33 (transcription factor) regulated the expression of camlexin biosynthetic enzyme Phytoalexin Deficient 3 (PAD3). Upon microbial attack, WRKY 33 is released from this complex and attaches to PAD3 promoter and activates its expression. Flg22 also triggers a swift activation of MPK3, MPK4 and MPK6 (Droillard *et al.*, 2004). Harpin protein (encoded by hrp genes; hypersensitive response and pathogenicity) are also induced by MPK4 and MPK6 which further lead to the activation of PR genes. One of the most well studied plant defense responses has been done in *Botrytis cinerea*. It is known to induce the biosynthesis of a major phytoalexin, camalexin in Arabidopsis. Camalexin provides resistance against *B. cinerea*. The production of camalexin was compromised in *mpk3* and *mpk6* mutants indicating a role of MPK3/6 in defense responses. MKK4/MKK5 as well as MKK9 has also been reported to be apparently involved in camalexin biosynthesis via MPK3/MPK6 but need to be verified. MAP2K3 participates in Jasmonate-mediated developmental signaling and in pathogen defense responses through activation of MAPK6 (Takahashi *et al.*, 2007). The interplay between the positive upregulation of defense responses via MAPK3/6 and the negative regulation via MAPK4 likely enables the tight control of defense responses to promote or restrict growth and make appropriate death versus survival decisions.

MAPKS in Effector-Triggered Immunity

The first evidence that ETI and *R* gene signaling includes MAPK cascades emerged from studies in tobacco and tomato. Wound-induced protein kinase (WIPK) and salicylic acid-induced protein kinase (SIPK) were found to get activated by an infection with tobacco mosaic virus (TMV) in plants that express the N-resistance gene (Zhang and Klessig, 1998). Both of these MAPKs also got induced in the presence of the fungal Avr9 effector in tobacco cells that bears Cf-9 resistance gene. An overexpression of tomato MAP3K, MAPKKKα increased the development of HR lesions whereas its silencing led to the inhibition of HR suggesting the role of MAPKKK in Pto-mediated HR response in tomato (delPozo *et al.*, 2004). Latter, MAP2K1-NTF6 and MAP2K2-SIPK were proposed to act downstream of Pto-resistance gene (Ekengren *et al.*, 2003). AvrPto and AvrPtoB effectors showed interaction with FLS2 and BAK 1. AvrPto can impede the binding of BAK1 to FLS2. AvrPtoB, on the other hand, mediates polyubiquitination and proteasome-dependent degradation of FLS2. In this manner, both effectors negatively regulate MAPK signaling.

Many bacterial pathogens use the HopAI1 effector which possesses a unique phosphor threonine lyaseactivity that dephosphorylates MAPKs (Li *et al.*, 2007). It was observed that HopAI1 inactivated the MAPK3 and MAPK6 which resulted in the suppression of PAMP-induced genes and callose deposition. This effector is also reported to exist in animal/human pathogens interacting with MAPKs ERK1/2 and p38. Similarly another effector molecule HopPtoD$_2$ from *P. syringaepv. tomato* demonstrates phosphatase activity. Its expression in tobacco cells resulted in reduction of cell death induced by expression of the constitutively active MAP2K variant NtMEK2^{DD}. However, HopPtoD$_2$ does not inhibit flg22-mediated

MAPK activation in Arabidopsis. An interesting example of an opportunistic interaction between plants and bacteria where a completely different mode of host MAPK signaling prevails is a hitch hiking mechanism displayed by *Agrobacterium tumefaciens*. It is the causal agent of crown galls and infects plants by integrating a segment of its DNA (transfer DNA) into the host chromosomal DNA. The activation of MPK3 in response to flg22 or Agrobacterium results in the phosphorylation, subsequently followed by nuclear translocation of the host protein VIP1 (virE2 interacting protein 1). Agrobacterium has thus hijacked VIP1 to shuttle their T-DNA into the host nucleus, where it integrates into the host genome (Djamei *et al.*, 2007). Because VIP1 does not only serve as nuclear shuttle for the pathogenic T-DNA complex but can also induce the expression of defense genes nuclear VIP1 would be counteracting Agrobacterium invasion. *Agrobacterium* overcomes this problem through targeting nuclear VIP1 for proteasome degradation by the *Agrobacterium* virulence factor VirF, which encodes an F-box protein.

In plants, stomata are responsible for gas exchange and transpiration but various plant pathogens use them as entrance gate to enter in to their hosts. Pathogen induced stomatal closure restricts the invasion of various pathogens. Therefore, stomata play a pivotal role in protection against both biotic and abiotic stresses. MAPKs have a great impact on stomatal development and function. In drought stress, stomatal closure mediated by ABA involves MKK1, MPK3 and MPK6 (Gudesblat *et al.*, 2007, Hamel *et al.*, 2006). Previous studies have also indicated crosstalk between ABA and defense signaling. The MAPK cascade comprising, YODA-MAP2K4/MAP2K5-MAPK3/MAPK6, plays an important role in regulating stomatal development (Bergmann *et al.*, 2004; Wang *et al.*, 2007; Bayer *et al.*, 2009; Popescu *et al.*, 2009). It was reported that mpk9-1/12-1 double mutants were highly susceptible to *Pseudomonas syringae* DC3000 compared to wild type plants. These results suggested that the regulation of stomatal apertures by MPK9 and MPK12 contributed to the first line of defense against pathogens (Jammes *et al.*, 2011). Whether pathogen- induced and ABA- induced stomatal closures are signaled via a common MAPK cascade remains to be elucidated.

MAPK cascades are also known to get triggered by fungal pathogens (Izumitsu *et al.*, 2009). *Phytophthora infestans* (a plant fungal pathogen) infection led to the swift transcriptional induction of MAP3K19, MAP2K9 and MAP2K4, whereas *Botrytis cinerea* infection resulted in enhanced transcriptional activaton of MAP3K18, MAP3K19 and MAP3K20, Raf43, ZIK2, ZIK8, indicating the distinct signaling pattern in response to bacterial and fungal pathogen attack (Cardinale *et al.*, 2002) described the activation of SIMK and SAMK by various fungal elicitors. Two other alfalfa MAPKs, MMK2 and MMK3 (Medicago MAPK2 and MAPK3, respectively), are also activated by elicitors. Nevertheless, a fungal biocontrol agent *Trichoderma asperellum* has recently been shown to induce systemic resistance in plants through a mechanism that employs jasmonic acid and ethylene signal-transduction pathways resulting in activation of a Trichoderma-induced MAPK (TIPK) gene in cucumber (Shoresh *et al.*, 2006). MAP2K1 is involved in defense responses including flg22-induced activation of MAPK4 (Asai *et al.*, 2002; Meszaros*et al.*, 2006). MAP2K7 is a positive regulator of systemic acquired response (Zhang *et al.*, 2007). MAP2K7/9 is hypothesized to regulate cell death during pathogen defense. Takahashi *et al.* (2007) reported the activation of MAP2K3-MAPK6 module by JA. MAPK6 is also involved in stomatal development, biotic and abiotic responses (Beckers *et al.*, 2009). Several other groups reported proteins *e.g.* NLPs (Necrosis and ethylene-inducing peptide1-like proteins) triggering MAPK activation and inducing defense responses. Modules involving MAPKKs MKK1/4/5/9 and the MAPKs MPK3/6 are shown to be involved in different defense strategies.

Cross-Talk between Biotic and Abiotic Stress

Plants respond to various stresses in a complex manner. It has been observed that they respond to multiple stresses differently from how they do it to individual strategies. For example, in response to a pest or pathogen infection plants become more susceptible to abiotic stress such as drought. Similarly, a long term abiotic stress can led to increased pathogen susceptibility. In contrast described an increased resistance to *B. cinerea* in drought stressed tomato indicating a positive interaction between both stresses. Similarly, virus infection provided protection from drought stress (Xu *et al.*, 2008). In virus infected beet, rice, drought symptoms appeared later and leaves maintained water longer than their uninfected counterparts. Therefore, understanding the mechanism of plant responses to multiple simultaneous stresses is crucial for developing broad spectrum stress tolerant crops. In maize, breeding programs have led to plants which are tolerant to drought and have additional resistance to the parasitic weed *Striga hermonthion*.

MAPKs and CDPKs

MAPKs together with calcium-dependent protein kinases (CDPKs) present two major pathways that are widely used to adapt the cellular metabolism to a changing environment. In many cases, both pathways are activated in parallel by the same stimuli; thus suggesting potential cross-talk between these pathways. Cross-talk between Ca^{2+}-dependent and MAPK signaling pathways has been elaborately studied in animal cells, but is still poorly investigated in plants. Ludwig *et al.* (2005) reported that the expression of a deregulated version of tobacco NtCDPK2 (lacking the C-terminal regulatory and calmodulin-like domains) triggered the production of the phytohormones jasmonic acid and ethylene, and compromised stress-induced MAPK activation in tobacco. Recently, an extensive study of Arabidopsis CDPKs demonstrated a complex network operating in early transcriptional reprogramming mediated by differential activity of CDPKs and MAPK cascades. Analysis of the expression levels of several genes inducible by the bacterial flg22 elicitor peptide in mesophyll protoplasts expressing deregulated AtCDPK5 and AtMKK4 identified MAPK-specific, CDPK-specific, CDPK/MAPK synergistic and predominantly MAPK-regulated target genes; suggesting a possible cross-talk among CDPKs and MAPK cascades during immune response. Thus, different levels of cross-talk between plant Ca^{2+}-dependent and MAPK pathways exist.

Recent Developments

Approximately 74 MAP3Ks, 9 MAP2Ks and 19 MAPKs were identified in maize followed by functional studies. Cai *et al.* (2013) reported the role of ZmMAP2K1 in pathogen defense. It played differential function in necrotrophic versus biotrophic pathogen defense responses. Ectopic expression of ZmMPK5 in tobacco exhibited increased resistance to viral pathogens and detected enhanced expression of PR genes *e.g.* PR1a, PR4, PR5 and EREBP (Zhang *et al.*, 2013). ZmMPK17, a novel maize group D MAP kinase gene, is reported to be involved in multiple stress responses (Pan *et al.*, 2012). In Gossypium, GhMAPK6a over expression resulted in increased sensitivity to the bacterial pathogen *Ralstonia solanacearum* indicating a negative regulatory role of MAPK during bacterial infection. GhMAPK6 was also observed to interact with GhMAP2K4. Another MAP2K from *G. hirsutum*, GhMKK1 showed increased susceptibility to same pathogen by lowering the expression of PR genes but no interaction between MAPK6a and MKK1 has been deciphered so far (Lu *et al.*, 2013).

Polyamines produced in plants are involved in several developmental and physiological processes. It was reported that AtMPK3 and AtMPK6 play a positive role in the regulation of putrescine biosynthesis and latter contributes to bacterial pathogen defense in Arabidopsis (Kim *et al.*, 2013). The *Hordeum vulgare* signaling protein MAP kinase 4 (HvMPK4) showed negative regulation of SA in response to hemi biotrophic fungal pathogen *Magnaporthe grisea.* HvMPK4 antisense lines detected enhanced SA levels and were more resistant to this pathogen (Abass and Morris, 2013). It is known that WIPK and SIPK are induced by biotic and abiotic stresses in tobacco. They positively control the biosynthesis of jasmonic acid (JA) or ethylene (ET) while negatively regulates SA accumulation. Earlier, NtMKP1 (tobacco MAP kinase phosphatase) protein was found to dephosphorylate and inactivate SIPK in vitro, and overexpression of NtMKP1 repressed wound-induced activation of both SIPK and WIPK. Recently, studies involving NtMKP1 silencing showed increased activation of WIPK and SIPK and accumulation of both JA and ET upon wounding. Wound-induced expression of JA-or ET-inducible genes, basic PR-1 and PI-II, was also significantly increased in transgenic lines. NtMKP1 antisense lines exhibited increased resistance against a necrotrophic pathogen, *Botrytis cinerea*, and lepidopteran herbivores, *Mamestra brassicae* and *Spodoptera litura*. It was suggested that NtMKP1 negatively regulates wound response and resistance against both necrotrophic pathogens and herbivorous insects through suppression of JA or ET pathways via inactivation of MAPK.

Conclusion

MAPK signaling has proved its universality among eukaryotes, in this context significant progress has been made in recent years using a combination of physiological, biochemical and genetic approaches. Researchers have identified the transient response of MAPK signaling cascade to multitude of stress elements in various plants and environments. The complete MAPK cascades have been identified for signaling in biotic and abiotic stresses but a few are still missing. So there is a need to elucidate the complete cascade with specific underlying mechanisms of signal transduction using novel approaches and strategies. This will help in engineering the MAPK cascades and their utilization in crop improvement.

References

Abass, M. and Morris, P.C. 2013. The *Hordeum vulgare* signalling protein MAP kinase 4 is a regulator of biotic and abiotic stress responses. *Plant Physiol.*, 170: 1353-9.

Andreasson, E. and Ellis, B. 2010. Convergence and specificity in the Arabidopsis MAPK nexus. *Trends PlantSci.*, 15: 106–13.

Andreasson, E., Jenkins, T., Brodersen, P., Thorgrimsen, S., Nikolaj, H.T., Petersen., Zhu, S.,Qiu, J. L., Micheelsen, P., Rocher, A., Newman, M., Nielsen, H. B., Hirt, H.,Somssich, O., Mattsson., Imre, I. and Mundy, J. 2005. The MAP kinase substrate MKS1 is a regulator of plant defense responses. *EMBO J.*, 24: 2579-89.

Arisz, S.A., VanWijk, R., Roels, W., Zhu J.K., Haring, M.A. and Munnik, T. 2013. Rapid phosphatidic acid accumulation in response to low temperature stress in Arabidopsis is generated through diacyl glycerol kinase. *Front Plant Sci.*, 4: 1.

Asai, T., Tena, G., Plotnikova, J., Willmann, M.R., Chiu, W.L., Gomez-Gomez L, Boller, T., Ausubel, F.M. and Sheen, J.2002 MAP kinase signalling cascade in *Arabidopsis* innate immunity. *Nat.*, 415: 977–83.

Bargmann, B.O., Laxalt, A.M., TerRiet, B., van, Schooten, B., Merquiol, E., Testerink, C., Haring, M.A., Bartels, D. and Munik, T. 2009. Multiple PLDs required for high salinity and water deficit tolerance in plants. *Plant Cell Physiol.*, 50: 78–89.

Bayer, M., Nawy, T., Giglione, C., Galli, M., Meinnel, T. and Lukowitz, W. 2009. Paternal Control of Embryonic Patterning in *Arabidopsis thaliana*. *Science*. 323:1485-8.

Beckers, G.J.M., Jaskiewicz, M., Liu, Y., Underwood, W.R., He, S.Y., Zhang S.Q. and Conrath, U. 2009. Mitogen-Activated Protein Kinases 3 and 6 are required for full priming of stress responses in *Arabidopsis thaliana*. *Plant Cell*, 21: 944-53.

Bergmann, D.C., Lukowitz, W. and Somerville, C.R. 2004. Stomatal development and pattern controlled by a MAPKK kinase. *Science*, 304: 1494–7.

Bethke, G., Unthan, T., Uhrig, J.F., Poschi, Y., Gust, A.A., Scheel, D. and Lee, J. 2009. Flg22 regulates the release of an ethylene response factor substrate from MAP kinase 6 in Arabidopsis via ethylene signaling. *Proc. Natl. Acad. Sci. USA*. 106: 8067-72.

Cai, G.,Wang, G., Wang, L., Pan, J., Liu, Y. and Li, D. 2013. ZmMKK1, a novel group A mitogen-activated protein kinase kinase gene in maize, conferred chilling stress tolerance and was involved in pathogen defense in transgenic tobacco. *Plant Sci.*, 214:57-73.

Canagarajah, B.J., Khokhlatchev, A., Cobb, M.H. and Goldsmith, E.J. 1997. Activation mechanism of MAP kinase ERK2 by dual phosphorylation. *Cell*, 90: 859-69.

Cardinale, F., Meskiene, I., Ouaked, F. and Hirt, H. 2002. Convergence and divergence of stress-induced mitogen- activated protein kinase signaling pathways at the level of two distinct mitogen-activated protein kinase kinases. *Plant Cell*, 14: 703-11.

Choudhury, S., Panda, P., Sahoo, L. and Panda, S.K. 2013. Reactive oxygen species signaling in plants under abiotic stress. *Plant Signal Behav.*, 8: e23681.

Delage, E., Ruelland, E., Guillas, I., Zachowski, A. and Puyaubert, J. 2012. *Arabidopsis* type-III phosphatidylinositol 4-kinases beta1 and beta 2 are upstream of the phospholipase C pathway triggered by cold exposure. *Plant Cell Physiol.*, 53: 565–76.

DelPozo, O., Pedley, K.F. and Martin, G.B. 2004. MAPKKK alpha is a positive regulator of cell death associated with both plant immunity and disease. *EMBO J.*, 23: 3072–82

Djamei, A., Pitzschke, A., Nakagami, H., Rajh, I. and Hirt, H. 2007. Trojan horse strategy in *Agrobacterium* transformation: abusing MAPK defense signaling. *Science*, 318: 453–56

Droillard, M.J, Boudsocq, M., Barbier-Brygoo, H. and Laurière, C. 2004. Involvement of MPK4 in osmotic stress response pathways in cell suspensions and plantlets of *Arabidopsis thaliana*: activation by hypo osmolarity and negative role in hyper osmolarity tolerance. *FEBS Let.*, 574: 42-8.

Ekengren, S.K., Liu, Y., Schiff, M., Dinesh-Kumar, S.P and Martin, G.B. 2003. Two MAPK cascades, NPR1, and TGA transcription factors play a role in Pto-mediated disease resistance in tomato. *Plant J.*, 36: 905–17.

Evrard, A., Kumar, M., Lecourieux, D., Lucks, J., von Koskull-Doring, P. and Hirt, H. 2013. Regulation of the heat stress response in Arabidopsis by MPK6-targeted phosphorylation of the heat stress factor HsfA2. *Peer J.*, 1: e59.

Feilner, T., Hultschig, C., Lee, J., Meyer, S., Immink, R.G., Koenig, A., Possling, A., Seitz, H., Beveridge, A., Scheel, D., Cahill, D.J., Lehrach, H., Kreutzberger, J. andKersten, B. 2005. High throughput identification of potential Arabidopsis mitogen-activated protein kinases substrates. *Mol Cell Proteom.*, 4: 1558-68.

Frye, C.A., Tang, D. and Innes, R.W. 2001. Negative regulation of defense responses in plants by a conserved MAPKK kinase. *Proc. Natl. Acad. Sci. USA*. 98: 373–8.

Gudesblat, G.E., Iusem, N.D. and Morris, P.C. 2007. Guard cell-specific inhibition of Arabidopsis MPK3 expression causes abnormal stomatal responses to abscisic acid and hydrogen peroxide. *New Phytol.*, 173: 713–21.

Hamel, L.P, Nicole, M.C, Sritubtim, S, Morency M.J, Ellis, M, Ehlting J, Beaudoin, N., Barbazuk, B., Klessig, D., Lee, J., Martin, G., Mundy, J., Ohashi, Y., Scheel, D., Sheen, J., Xing, T., Zhang, S., Seguin, A. and Ellis, B.E. 2006. Ancient signals: comparative genomics of plant MAPK and MAPKK gene families. *Trends Plant Sci.*, 11: 192–8.

Hardie, D.G. 1999. Plant protein serine/threonine kinases: classification and functions. *Ann. Rev. Plant Physiol. Plant Mol. Biol.*, 50: 97-131.

Hong, Y., Zhang, W., Wang, X. and Phospholipase, D. 2010. Phosphatidic acid signalling in plant response to drought and salinity. *Plant Cell Environ.*, 33: 627–35.

Huang, G.T., Ma, S.L., Bai, L.P., Zhang, L., Ma, H., Jia, P.,Liu, J., Zhong, M. and Guo, Z. F. 2012. Signal transduction during cold, salt, and drought stresses in plants. *Mol. Biol. Rep.*, 39: 969–87.

Ichimura, K., Mizoguchi, T., Yoshida, R., Yuasa, T. and Shinozaki, K. 2000. Various abiotic stresses rapidly activate *Arabidopsis* MAP kinases ATMPK4 and ATMPK6. *Plant J.*, 24: 655–65.

Izumitsu, K., Yoshimi, A., Kubo, D., Morita, A., Saitoh, Y. and Tanaka, C. 2009. The MAPKK kinase ChSte11 regulates sexual/asexual development, melanization, pathogenicity and adaptation to oxidative stress in *Cochliobolus heterostrophus. Curr. Genet.*, 55: 439-48.

Jammes, F., Yang, X., Xiao, S. and Kwak, J.M. 2011. Two *Arabidopsis* guard cell-preferential MAPK genes, MPK9 and MPK12, function in biotic stress response. *Plant Signal Behav.*,6: 1875–7.

Jones, J.D. and Dangl, J.L. 2006. The plant immune system. *Nat.,* 444: 323–9.

Kim, J.M., Woo, D.H., Kim, S.H., Lee, S.Y., Park, H.Y., Seok, H.Y., Chung, W.S. and Moon, Y,H. 2012. *Arabidopsis* MKKK20 is involved in osmotic stress response via regulation of MPK6 activity. *Plant Cell Rep.*,31: 217–24.

Kim, S. H., Yoo, S. J., Min, K. H., Nam, S. H., Cho, B. H., and Yang, K. Y. 2013. Putrescine regulating by stress-responsive MAPK cascade contributes to bacterial pathogen defense in Arabidopsis. *Biochem. Biophys. Res. Commun.*, 437: 502–508.

Lee, S.C., Lan, W., Buchanan, B.B. and Luan, S. 2009. A protein kinase-phosphatase pair interacts with an ion channel to regulate ABA signaling in plant guard cells. *Proc. Natl. Acad. Sci. USA*. 106: 21419–24.

Li, H., Xu, H., Zhou, Y., Zhang, J., Long C,Li, S., Chen, S., Zhou, J.M. and Shao, F. 2007. The phosphor threonine lyase activity of a bacterial type III effector family. *Science,* 315: 1000–3

Link, V., Sinha, A.K., Vashista, P., Hofmann, M.G., Proels, R.K., Ehness, R. and Roitsch, T. 2002. A heat-activated MAP kinase in tomato: a possible regulator of the heat stress response. *FEBS Lett.*, 531: 179–83.

Lissarre, M., Ohta, M., Sato, A. and Miura, K. 2010. Cold-responsive gene regulation during cold acclimation in plants. *Plant Signal Behav.,* 5: 948–52.

Liu, X.M., Nguyen, X.C., Kim, K.E., Han, H.J., Yoo, J., Lee, K., Kim, M.C., Yun, D. J. and Chung, W.S. 2013. Phosphorylation of the zinc finger transcriptional regulator ZAT6 by MPK6 regulates *Arabidopsis* seed germination under salt and osmotic stress. *Biochem. Biophys. Res. Commun.,* 430: 1054–9.

Lu, W., Chu, X., Li, Y., Wang, C. andGuo, X. 2013. Cotton GhMKK1 induces the tolerance of salt and drought stress, and mediates defence responses to pathogen infection in transgenic *Nicotiana benthamiana. PLoS One*,doi: 10.1371/journal.pone.0068503.

Ludwig, A.A., Romeis, T. and Jones, J.D. 2004. CDPK-mediated signalling pathways: specificity and cross-talk. *J. Exp. Bot.*,55: 181–8.

Lumbreras, V., Vilela, B., Irar,S., Sole, M., Capellades, M., Valls, M., Coca, M. and Pages, M. 2010. MAPK phosphatase MKP2 mediates disease responses in *Arabidopsis* and functionally interacts with MPK3 and MPK6. *Plant J.*, 63: 1017–30.

MAPK group. 2002. Mitogen activated protein kinases cascades in plants: A new nomenclature. *Trends Plants Sci.*, 7: 301-308

Meszaros, T., Helfer, A., Hatzimasoura, E., Magyar, Z., Serazetdinova, L., Rios, G, Bardoczy, V., Teige, M., Koncz, C., Peck, S. and Bogre, L. 2006. The *Arabidopsis* MAP kinase kinase MKK1 participates in defense responses to the bacterial elicitor flagellin. *Plant J.*, 48:485-98.

Miles, G.P., Samuel, M.A. and Ellis, B.E. 2009. Suppression of MKK5 reduces ozone-induced signal transmission to both MPK3 and MPK6 and confers increased ozone sensitivity in *Arabidopsis thaliana. Plant Signal Behav.*, 4:687–92.

Mishra, N.S., Tuteja, R., Tuteja, N. 2006. Signaling through MAP kinase networks in plants. *Arch.Biochem. Biophys.*, 452:55–68.

Mittler, R., Finka, A., Goloubinoff, P. 2012. How do plants feel the heat? *Trends Biochem. Sci.*, 37: 118–25.

Mizoguchi, T., Hayashida, N., Yamaguchi-Shinozaki, K., Matsumoto, K. and Shinozaki, K. 1993. ATMPKs: a gene family of plant MAP kinases in *Arabidopsis thaliana*. FEBS Lett., 336: 440–444.

Mizoguchi, T., Irie, K., Hirayama, T., Hayashida, N., Yamaguchi-Shinozaki, K., Matsumoto, K. and Shinozaki, K. 1996. A gene encoding a mitogen-activated protein kinase kinasekinase is induced simultaneously with genes for a mitogen-activated protein kinase and an S6 ribosomal protein kinase by touch, cold and water stress in *Arabidopsis thaliana. Proc. Natl. Acad. Sci.USA*. 93: 765-9.

Moller, S.G. and Chua, N.H. 1999. Interaction and intersections of plant signaling pathways. *Journal of Molecular Biology*, 293: 219-234.

Munnik, T. and Meijer, H.J. 2001. Osmotic stress activates distinct lipid and MAPK signalling pathways in plants. *FEBS Lett.*, 498: 172–8.

Munns, R. and Tester, M. 2010. Mechanisms of salinity tolerance. *Annu. Rev. Plant Biol.*, 59: 651–81.

Ning, J., Li, X., Hicks, L.M. and Xiong, L. 2010. A Raf-like MAPKKK gene DSM1 mediates drought resistance through reactive oxygen species scavenging in rice. *Plant Physiol.*, 152: 876–90.

Nongpiur, R., Soni, P., Karan, R., Singla-Pareek, S.L. and Pareek, A., 2012. Histidine kinases in plants: cross talk between hormone and stress responses. *Plant Signal Behav.*, 7: 1230–7.

O'Brien, J.A., Daudi, A., Butt, V.S. and Bolwell, G.P. 2012. Reactive oxygen species and their role in plant defence and cell wall metabolism. *Planta*, 236: 765–79.

Pan, J., Zhang, M., Kong, X., Xing, X., Liu, Y., Zhou Y, Liu, Y., Sun, L. and Li, D. 2012. ZmMPK17, a novel maize group D MAP kinase gene, is involved in multiple stress responses. *Planta*, 235: 661–76.

Pitzschke, A., Djamei, A., Bitton, F. and Hirt, H. 2009. A major role of the MEKK1–MKK1/2-MPK4 pathway in ROS signalling. *Mol. Plant*, 2: 120–37.

Popescu, S.C., Popescu, G.V., Snyder, M. and Dinesh S.P. 2009. Integrated analysis of co-expressed MAP kinase substrates in *Arabidopsis thaliana. Plant Signal Behav.* 4:524-7.

Rao, K.P., Richa, T., Kumar, K., Raghuram, B. and Sinha, A.K. 2010. In silico analysis reveals 75 members of mitogen activated protein kinase kinasekinase gene family in rice. *DNA Research*, 17(3): 139-153.

Rao, K.P., Vani, G., Kumar, K., Wankhede, D.P., Misra, M., Gupta, M. and Sinha, A.K. 2011. Arsenic stress activates MAP kinase in rice roots and leaves. *Arch. Biochem. Biophys.*, 506: 73–82.

Rodriguez M.C.S., Petersen M, and Mundy J. 2010. Mitogen-activated protein kinase signaling in plants. *Annu. Rev. Plant Biol.*, 61: 621–49.

Salam, M.A., Jammes, F., Hossain, M.A., Ye, W., Nakamura, Y., Mori, I.C., Kwak, J.M. and Murata, Y. 2012. MAP kinases, MPK9 and MPK12, regulate chitosan-induced stomatal closure. *Biosci. Biotechnol. Biochem.*, 76: 1785–7.

Salam, M.A., Jammes, F., Hossain, M.A., Ye, W., Nakamura, Y., Mori, I.C., Kwak, J.M. and Murata, Y. 2013. Two guard cell-preferential MAPKs, MPK9 and MPK12, regulate YEL signalling in Arabidopsis guard cells. *Plant Biol.*, 15: 436–42.

Sangwan, V., Orvar, B.L., Beyerly, J., Hirt, H., Dhindsa, RS. 2002. Opposite changes in membrane fluidity mimic cold and heat stress activation of distinct plant MAP kinase pathways. *Plant J.*, 31: 629–38.

Sangwan, V. and Dhindsa, R.S. 2002. In vivo and in vitro activation of temperature-responsive plant map kinases. *FEBS Lett.*, 531: 561–4.

Sharrocks, A.D., Yang, S.H. and Galanis, A. 2000 Docking domains and substrate specificity determination for MAP kinases. *Trends in Biochemical Sciences,* 25: 448-453.

Shen, H., Liu, C., Zhang, Y., Meng, X., Zhou, X., Chu, C. and Wang, X. 2012. OsWRKY30 is activated by MAP kinases to confer drought tolerance in rice. *Plant Mol. Biol.*, 80: 241–53.

Shi, J., Zhang., L., An, H., Wu, C. and Guo, X. 2011. GhMPK16, a novel stress-responsive group D MAPK gene from cotton, is involved in disease resistance and drought sensitivity. *BMC Mol. Biol.*, 12: 22.

Shoresh, M., Gal-On A, Leibman, D. and Chet, I. 2006. Characterization of a MAPK gene from cucumber required for *Trichoderma* conferred plant resistance. *Plant Physiol.*, 142: 1169-79.

Smekalova, V., Doskocilova, A., Komis, G. andSamaj, J. 2014. Crosstalk between secondary messengers, hormones and MAPK modules during abiotic stress signalling in plants. *Biotechnol. Adv.*, 32(1): 2-11.

Solanke, A.U. and Sharma, A.K., 2008. Signal transduction during cold stress in plants. *Physiol. Mol. Biol. Plants.,* 14: 69–79.

Soropy, S.K. and Munshi, M. 1998. Protein kinases and phosphatases and their role in cellular signaling in plants. *Critical Review in Plant Science,* 17: 245-318.

Sturgill, T.W. and Ray L.B., 1986. Muscle proteins related to microtubule associated protein-2 are substrates for an insulin stimulatable kinase. *Biochem. Biophys. Res. Commun,* 134: 565-71.

Suzuki, N., Koussevitzky, S., Mittler, R. and Miller, G., 2012. ROS and redox signalling in the response of plants to abiotic stress. *Plant Cell Environ.,* 35: 259–70.

Takahashi, F., Yoshida, R., Ichimura, K., Mizoguchi, T., Seo, S., Yonezawa, M., Maruyama, K.,Yamaguchi-Shinozaki, K. and Shinozaki, K. 2007. The Mitogen-activated protein kinase cascade MKK3-MPK6 is an important part of the jasmonate signal transduction pathway in *Arabidopsis. Plant Cell,* 19: 805-18.

Teige, M., Scheikl, E., Eulgem, T., Doczi, R., Ichimura, K., Shinozaki, K., Danql, J. L. and Hirt, H. 2004. The MKK2 pathway mediates cold and salt stress signaling in *Arabidopsis. Mol. Cell,* 15: 141–52.

Wang, H., Ngwenyama, N., Liu, Y., Walker, J.C. and Zhang, S. 2007. Stomatal development and patterning are regulated by environmentally responsive mitogen-activated protein kinases in *Arabidopsis. Plant Cell,* 19: 63–73.

Wang, J., Ding, H., Zhang, A., Ma, F., Cao, J. and Jiang, M. 2010. A novel mitogen-activated protein kinase gene in maize (*Zea mays*), ZmMPK3, is involved in response to diverse environmental cues. *J. Integr. Plant Biol.,* 52: 442–52.

Wang, W., Vinocur, B. and Altman, A. 2003. Plant responses to drought, salinity and extreme temperatures: Towards genetic engineering for stress tolerance. *Planta,* 218: 1–14.

Wilson, K.P., Fitzgibbon, M.J., Caron, P.R., Griffith, J.P., Chen, W., McCaffrey, P.G., Chambers, S.P. and Su, M.S. 1996. Crystal structure of p38 mitogen-activated protein kinase. *J. Biol. Chem.,* 271: 27696-700.

Xing, Y., Cao, Q., Zhang, Q., Qin, L., Jia, W., and Zhang, J. 2013. MKK5 regulates high light-induced gene expression of Cu/Zn superoxide dismutase 1 and 2 in *Arabidopsis. Plant Cell Physiol.,* 54: 1217–27.

Xu, H., Wang X., Sun X., Shi Q., Yang F. and Du, D. 2008. Molecular cloning and characterization of a cucumber MAP kinase gene in response to excess NO_3– and other abiotic stresses. *Sci. Hortic.,* 117: 1–8.

Yao, Y., Liu, X., Li, Z., Ma, X., Rennenberg, H., Wang, X. and Li, H. 2013. Drought-induced HO accumulation in subsidiary cells is involved in regulatory signaling of stomatal closure in maize leaves. *Planta,* 238: 217–27.

Yu, L., Nie, J., Cao, C., Jin, Y., Yan, M., Wang, F., Liu, J., Xiao, Y., Liang, Y. and Zhang, W. 2010. Phosphatidic acid mediates salt stress response by regulation of MPK6 in *Arabidopsis thaliana. New Phytol.,* 188: 762–73.

Zhang, D., Jiang, S., Pan, J., Kong, X., Zhou, Y., Liu, Y. and Li, D. 2013. The overexpression of a maize mitogen-activated protein kinase gene (ZmMPK5) confers salt stress tolerance and induces defence responses in tobacco. *Plant Biol.*,doi: 10.1111/plb.12084.

Zhang, L., Xi, D., Li, S., Gao, Z., Zhao, S., Shi, J., Wu, C. and Guo, X. 2011. A cotton group C MAP kinase gene, GhMPK2, positively regulates salt and drought tolerance in tobacco. *Plant Mol. Biol.*,77: 17–31.

Zhang, S. and Klessig, D.F. 2001. MAPK cascades in plant defense signaling. *Trends Plant Sci.*, 6: 520.

Zhang, S. and Klessig, D.F. 1998. Resistance gene N-mediated de novo synthesis and activation of a tobacco mitogen-activated protein kinase by tobacco mosaic virus infection. *Proc. Natl. Acad. Sci. USA*. 95: 7433–38.

Zhang, X., Dai, Y., Xiong, Y., Defraia, C., Li, J., Dong, X. and Mou, Z. 2007. Overexpression of Arabidopsis MAP kinase kinase 7 leads to activation of plant basal and systemic acquired resistance. *Plant J.*, 52: 1066-79.

Zhou, C., Cai, Z., Guo, Y. and Gan, S. 2009. An Arabidopsis mitogen-activated protein kinase cascade, MKK9-MPK6, plays a role in leaf senescence. *Plant Physiol.*, 150: 167-77.

Zhou, M.Q., Shen, C., Wu, L.H., Tang, K.X. and Lin, J. 2011. CBF-dependent signaling pathway: a key responder to low temperature stress in plants. *Crit. Rev. Biotechnol.*, 31: 186–92.

Zhu, X., Feng, Y., Liang, G., Liu, N. and Zhu, J.K. 2013. Aequorin-based luminescence imaging reveals stimulus- and tissue-specific Ca^{2+} dynamics in *Arabidopsis* plants. *Mol. Plant,* 6: 444–55.

2016, Recent Advances in Plant Stress Physiology *Pages* ***89–115***
Editors: ***Praduman Yadav, Sunil Kumar and Veena Jain***
Published by: **DAYA PUBLISHING HOUSE, NEW DELHI**

Chapter 5

Plant Hormones and Stress

Vikender Kaur[1]*, Anita Kumari[2] and Sunder Singh[3]

[1]Division of Germplasm Evaluation,
ICAR – National Bureau of Plant Genetic Resources, New Delhi – 110 012
[2]Department of Botany and Plant Physiology,
CCSHAU, Hisar, Haryana
[3]Department of Botany, Maharishi Dayanand University,
Rohtak – 124 001, Haryana

ABSTRACT

Phytohormones are in a prominent position, playing important regulatory roles in plant physiology affecting both developmental processes and responses to a wide range of abiotic and biotic stresses. They play central roles in the ability of plants to adapt to changing environments, by mediating growth, development, nutrient allocation, and source/sink transitions. Although abscisic acid is the most studied stress-responsive hormone, the role of cytokinins, brassinosteroids, and auxins during environmental stress is emerging. A range of abscisic acid-ethylene and abscisic acid-cytokinin relative abundances could represent targets for breeding/managing for yield resilience under a spectrum of stress levels between severe and mild. Three phytohormones-salicylic acid, jasmonic acid and ethylene, are known to play major roles in regulating plant defense responses against various pathogens, pests and abiotic stresses.Recent studies impose brassinosteroids are implicated in plant responses to abiotic environmental stresses and to undergo profound changes in plants interacting with bacterial, fungal and viral pathogens. Crosstalk between the different plant hormones results in synergetic or antagonic interactions that play crucial roles in response of plants to stress adaptation. Rather than being additive, the presence of an abiotic stress can have the effect of reducing or enhancing susceptibility to a biotic pest or pathogen, and vice versa. This interaction between biotic and abiotic stresses is orchestrated by hormone signaling pathways that may induce or antagonize one another. The characterization of

**Corresponding author*: E-mail: vikender@nbpgr.ernet.in, drvikender@yahoo.in

the molecular mechanisms regulating hormone synthesis, signaling, and action are facilitating the modification of hormone biosynthetic pathways for the generation of transgenic crop plants with enhanced stress tolerance. Here, we review recent advances made in understanding the role of these hormones in modulating plant defense responses against various biotic as well as abiotic stresses.

Keywords: *Phytohormones, Plant defense, Stress, Signaling, Pathogen, Tolerance, Yield.*

Introduction

Plants growing under limiting (biotic or abiotic) environments are "stressed". Stress can therefore be defined as any change in the environment that decreases plant growth and reproduction below the genotype's potential until productivity becomes uneconomical and ultimately ceases. Thus, the term *stress* is measurable and meaningful to agriculture. As sessile organisms, plants must regulate their growth and development in order to respond to numerous external stimuli and an ever-changing environment (Wolters and Jurgens, 2009). Plants respond to a stressful environment by a number of morphological and physiological alterations which include the reduced growth of leaves, stems, and hairy roots; stomatal closure; a reduction in the rate of photosynthesis; hormonal imbalance; and a low capacity for nutrient uptake (Chapin, 1991). These adaptive responses are mediated by plant growth regulators. Plant hormones (phytohormones) are organic compounds, synthesized in one part of a plant and translocated to another part that, in very low concentrations, cause a physiological response. This definition sets forth three criteria which separate plant hormones from other nutrients and metabolites:

1. Are endogenously produced,
2. Transported to a target area from the site of synthesis, and
3. Act in low concentrations.

Collectively, plant hormones regulate every aspect of plant growth and development and the responses of plants to biotic and abiotic stresses. Plant growth regulators include the five classical phytohormones: abscisic acid (ABA), ethylene (ET), cytokinin (CK), auxin (IAA), gibberellin (GA). Other hormones such as jasmonate (JA), brassinosteroids (BR), salicylic acid (SA) and peptide hormones also play important roles in regulating developmental processes and signaling networks involved in plant responses to a wide range of biotic and abiotic stresses. Since plants lack a rapid communication system similar to the central nervous system of animals, they adjust their hormonal balance to regulate growth in response to environmental perturbations such as various biotic and abiotic stress responses in plants. Infection of plants with diverse pathogens results in changes in the level of various phytohormones (Adie *et al.*, 2007; Robert-Seilaniantz *et al.*, 2007).

In a large number of instances, it appears that the levels of hormones in a tissue/organ relative to one another are a more important consideration than are their absolute concentrations. They exist in plants at concentrations lower than 10^{-6} M (Weiler *et al.*, 1981), and an endogenous concentration above this is generally considered to be supraoptimal. It is also known that not only does each hormone affect the response of a number of plant parts, but that these responses also depend on the species, the plant organs, its developmental stage, cell and tissue sensitivity, concentration and interaction among the hormones, and the environmental factors. These hormones are detectable in all actively growing plant organs; younger leaves and apical buds are particularly high in auxin, whereas root apices are high in cytokinins, gibberellins, and abscisic acid. Fruits and seeds are generally rich in plant hormones. Therefore, phytohormones are ubiquitous and generally not species specific. In this chapter, we cover the hormonal relations of phytohormones elicited under stress. In recent years, significant research progress contributed to the understanding of processes associated with the biosynthesis of plant hormones, their metabolism, as well as their role in signaling. The identification and characterization of several mutants affected in the biosynthesis, perception and signal transduction of these hormones has been instrumental in understanding the role of individual components of each hormone signaling pathway in plant defense. Substantial progress has been made in understanding individual aspects of phytohormone perception, signal transduction, homeostasis or influence on gene expression. However, the underlying molecular mechanisms by which plants integrate stress induced changes in hormone levels and initiate adaptive responses are poorly understood.

Hormones do not act in isolation but are interrelated by synergistic or antagonistic cross-talk so that they modulate each other's biosynthesis or responses. Reviews on hormone action and signaling of abscisic acid (Cutler *et al.*, 2010; Klingler *et al.*, 2010; Wan *et al.*, 2009), cytokinin (Argueso *et al.*,2010; Perilli *et al.*, 2010; Werner and Schmulling, 2009), ethylene (Stepanova and Alonso, 2009), brassinosteroids (Kim and Wang, 2010; Divi and Krishna, 2009) and jasmonic acid (Wasternack, 2007) and on hormone cross-talk (Jaillais and Chory, 2010; Santner and Estelle, 2009; Wilkinson *et al.*, 2012; Atkinson and Urwin, 2012) have been published recently. Here, we highlight the latest advances in our understanding of the role of hormones and hormone interaction in plant responses to biotic/abiotic stresses. We then discuss the recent progress in the hormone cross talk and engineering of hormone - associated genes aimed at improving crop stress tolerance.

Phytohormones and their Interactions under Stress

The heavy metals action, drought, high salt, changes in temperature and light, wounding, pathogen and pest attack are major abiotic and biotic stresses. In response to these stresses, plant hormones have been heavily involved with "action at a distance" and their ability to move within the plant body has been a paramount consideration. For such stimulus-response coupling, most of the models proposed consist of a sequential four-component system that includes-

1. Perception of the stimulus,
2. Transduction of the signal,
3. Alteration of gene expression leading to a cascade and amplification of the message in the form of a network of biochemical/molecular events,
4. A physiological response in the form of a morphological adjustment/ modification in the growth form.

Under a stressful environment, marked and often rapid changes in the hormonal balance of plants are commonly observed. Since a given stress induces resistance to unrelated stresses, and hormones also influence a stress response, changes in their relative levels may enable the plant to adjust its growth despite suboptimal conditions. Abiotic stress leads to a morphological, physiological, biochemical and molecular changes. Drought, salinity, extreme temperatures and oxidative stress are often interconnected, and may induce similar cellular damage. Oxidative stress, which frequently accompanies high temperature, salinity, or drought stress, may cause disruption of homeostasis and ion distribution in the cells and denaturation of functional and structural proteins. As a consequence, these diverse environmental stresses often activate similar cell signaling pathways and cellular responses, such as the production of stress proteins, up-regulation of antioxidants and accumulation of compatible solutes. Biochemical adaptation in plants involves various changes in the biochemistry of the cell. These changes include the evolution of new metabolic pathways, the accumulation of low molecular weight metabolites, the synthesis of special proteins, detoxification mechanisms and changes in phytohormone levels. Molecular level changes involve alteration in gene expression by inducing or preventing the degradation of transcriptional regulators via the ubiquitin-proteasome system (Santner and Estelle, 2010). Adaptation represents the ability of a living organism to fit into a changing environment, at the same time improving its chances of survival and reproducing itself. Plant responses to stress can be viewed as being orchestrated through a network that integrates signaling pathways characterized by the production of ET, JA and SA. The identified regulatory steps in the network involve transcription, protein-protein interaction and targeted protein destruction. In plants, the mitogen-activated protein (MAP) kinase cascade plays a key role in various abiotic and biotic stress responses and in phytohormone responses that include reactive oxygen species (ROS) signaling (Figure 5.1). The role of individual plant hormone and there interaction with other plant hormone under stress, we are discussing ahead:

Auxins

Auxin is a term derived from the Greek word *auxein*, meaning "to increase." It is a generic name for chemical which typically stimulate cell elongation, but auxins also influence a wide range of other growth and developmental processes. The primary auxin in plants is indole-3-acetic acid (IAA; Figure 5.2). Although other compounds with auxin activity, such as indole-3-butyric acid, phenyl acetic acid, and 4-chloro-IAA, are also present in plants. IAA occurs not only in the free form but also conjugated to amino acids, peptides, or carbohydrates. These IAA conjugates are biologically inactive and appear to serve functions as IAA storage forms in seeds and hormonal homeostasis.

Treatments with auxin have been reported to alleviate some of the adverse effects of stress on germination, seedling growth, and the yield of a number of crop species as well as fresh- and seawater algae. Recently, the role of auxins in drought tolerance was postulated; TLD1/OsGH3.13, encoding indole-3-acetic acid (IAA)-amidosynthetase, was shown to enhance the expression of LEA (late embryogenesis abundant) genes, which correlated with the increased drought tolerance of rice

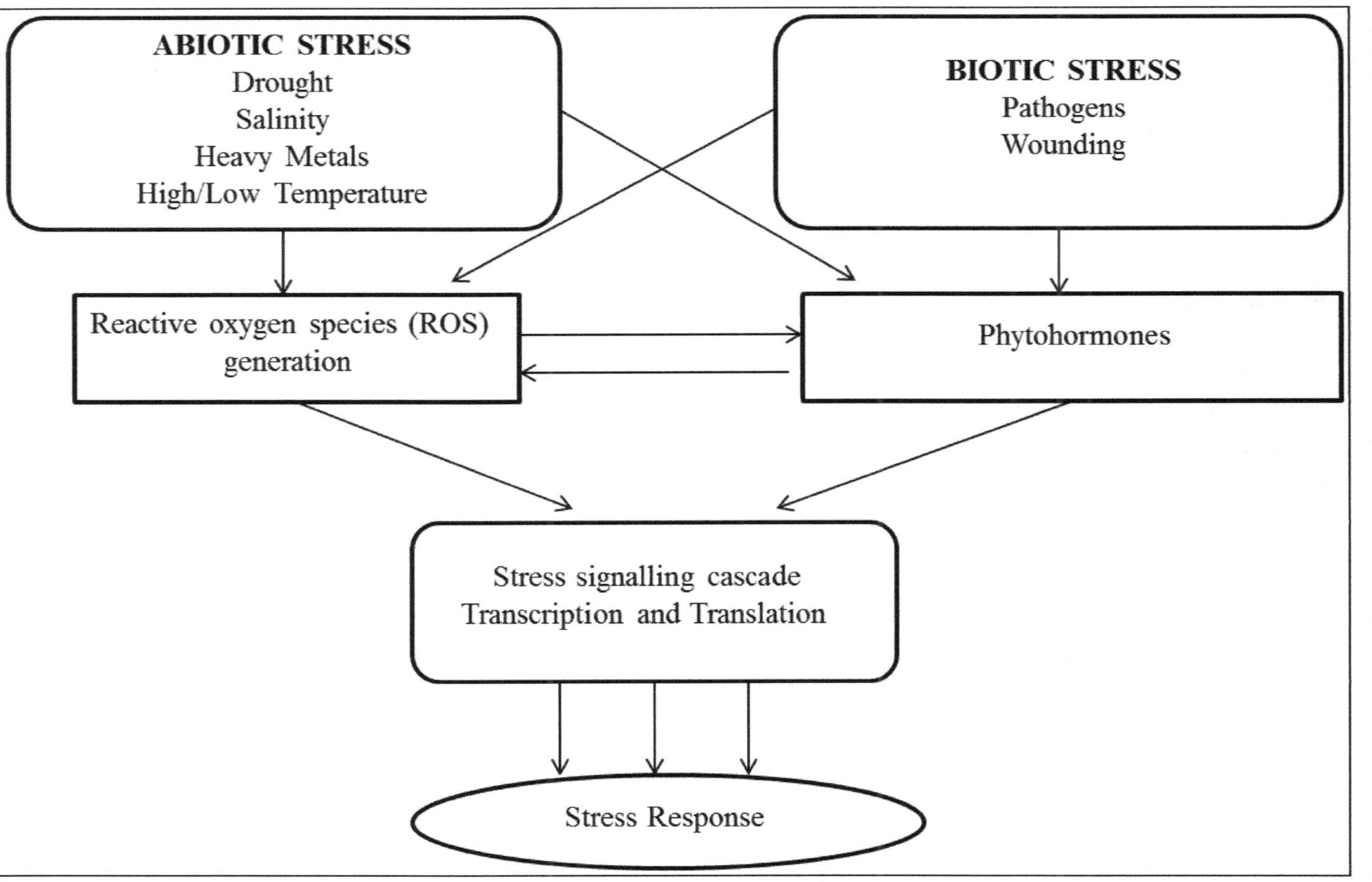

Figure 5.1: Generalized Model of the Plant Stress Signalling Networks.

seedlings (Zhang *et al.*, 2009). The expression of many genes associated with auxin synthesis, perception, and action has been shown to be regulated by ethylene (Stepanova and Alonso, 2009). Among them, are the auxin-responsive factors ARF2 and ARF19, the auxin transporters PIN1, PIN2, PIN4, AUX1, and genes encoding auxin biosynthetic enzymes (ASA1/WEI2/TIR7, ASB1/WEI7, TAA1/SAV3/WEI8) (Stepanova *et al.*, 2005; 2008). Conversely, auxin was found to affect ethylene biosynthesis. Several members of the 1-aminocyclopropane-1-carboxylate synthase (ACS) gene family, encoding rate-limiting enzymes in ethylene biosynthesis, were shown to be regulated by auxin treatment. Recently, CK was also shown to be a positive regulator of auxin biosynthesis, and it was postulated that a homeostatic feedback regulatory loop involving both CK and IAA signaling acts to maintain appropriate CK and IAA concentrations in developing root and shoot tissues (Jones *et al.*, 2010).

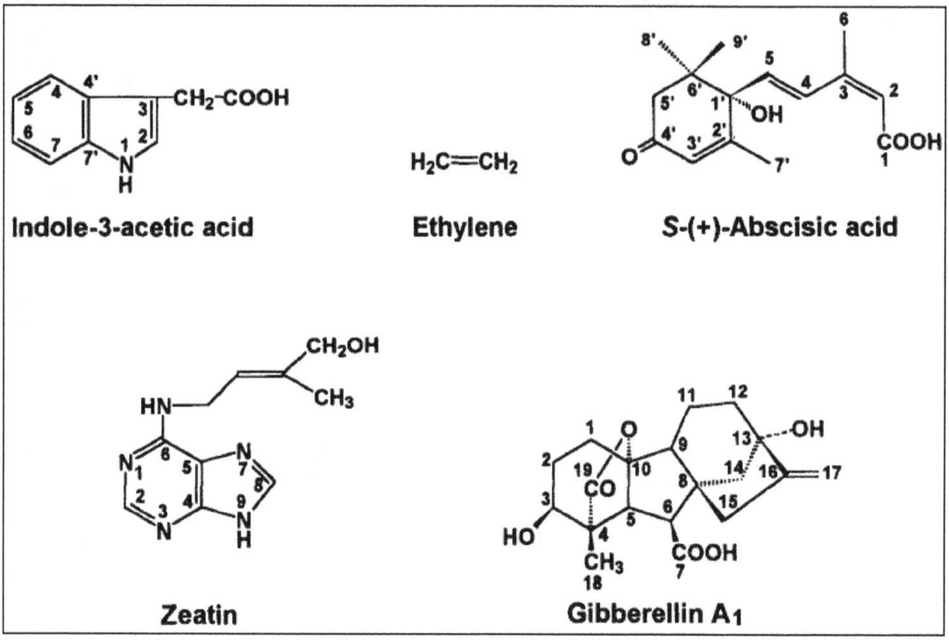

Indole-3-acetic acid

Ethylene

S-(+)-Abscisic acid

Zeatin

Gibberellin A_1

Figure 5.2: Structures of Representatives of the Five Classical Plant Hormones.

Gibberellins

In 1926, Kurosawa discovered that gibberellins increase the growth of plants by greatly elongating the cells. Studying the symptoms of the rice disease *bakanaebyo* ("foolish seedling disease"), the causal pathogen was a soil borne fungus, *Gibberella fujikorai*, the sexual or perfect stage of *Fusarium moniliforme*, which caused infected seedlings to grow abnormally taller and to fall over owing to a spindly stem structure. He further observed that when a pure culture filterate was sprayed onto rice seedlings, it produced the same abnormal growth. This suggested that the filterate contained some soluble substance which caused the growth abnormality. Other Japanese workers demonstrated that the effect was not confined to rice but could be reproduced in many other species. However, in 1938, Yabuta and Sumiki isolated two crystalline active substances from the culture filterates of the fungus and named them gibberellin A and B. Since the first GA from a higher plant, GA_1 (Figure 5.2), was identified many years ago, 112 GAs have been identified to date. Efforts to determine the physiological roles of GA and to elucidate the biosynthetic pathway have been greatly facilitated by the availability of GA- deficient (*i.e.,* dwarf) mutants. Metabolic studies have been conducted with systems that are rich sources of GAs, such as the fungus *Gibberella fujikuroi* and immature seeds of pumpkin, pea, and bean. However, maize is the only higher plant in which the entire biosynthetic pathway has been demonstrated in vegetative tissues by feeding various intermediates (Suzuki *et al.,* 1992; Kobayashi *et al.,* 1996).

The primary action of gibberellins is on stem elongation, which is a consequence of both increased cell division and cell elongation. GA treatment resulted in the improvement in germination and seedling growth of onion (*Allium cepa*), sesame (*Sesamum indicum*), flax (*Linum usitatissimum*), lettuce (*Lactuca sativa*) and the polymorphic seeds and seedlings of atriplex (*Atriplex triangularis*) under stressful environments. Treatment with GA_3 effectively enhanced the α-amylase activity and the coleoptile lengths in wheat (Ansari *et al.,* 1977) and seed germination as well as the seedling growth in barley (Sarin and Narayanan, 1968). In addition to these beneficial effects, GA_3 treatment increased the nutrient uptake and enhanced the yield of field-grown wheat. GA is also associated with SA. The exogenous application of GA (GA_3) induced increased expression levels of isochorismate synthase1 (ICS1) and nonexpressor of pathogenesis related genes 1(NPR1), genes involved in SA biosynthesis and SA action, respectively. Transgenic *Arabidopsis thaliana* plants constitutively overexpressing a GA responsive gene from *Fagus sylvatica* encoding FsGASA4, a member of the GA_3 gene family, showed improved tolerance under abiotic stress and the stress tolerance was correlated with increased endogenous levels of SA (Alonso-Ramirez *et al.,* 2009). GA and BR regulate many common physiological processes. OsGSR1, a member of the GA-stimulated transcript (GAST) gene family, was found to play key roles in both BR and GA signaling pathways, and to mediate the interaction between them (Wang *et al.,* 2005). RNAi transgenic rice plants with reduced OsGSR1 expression displayed phenotypes similar to plants deficient in BR, including short primary roots, erect leaves and reduced fertility.

Cytokinins

It was isolated from immature maize kernel and was named zeatin (Figure 5.2) by Koshimizu and Iwamura (1986). Since then 25 free cytokinins have been isolated and identified from plants. Cytokinins are a group of compounds that stimulate water uptake, increase cell division, promote organ development, and lead to regeneration and proliferation of shoots. Cytokinins have been classified by a substituted base into three groups: zeatin (Z), dihydrozeatin (diHZ), and N6-(Δ-isopentenyl) adenine (2ip). In many species, the zeatin-type cytokinins are the most active and most prevalent forms of cytokinin.

CK content usually declines in water stressed plants (Davies *et al.,* 2005), co-incidentally with ABA accumulation as CK is antagonist to ABA. A causal relationship between these phenomena was confirmed in experiments showing an ABA-induced increase in the expression of the gene coding for cytokinin oxidase (Brugiere *et al.*, 2003) as well as in the activity of enzymes catalysing irreversible CK degradation. Treatment of plants with an inhibitor of ABA synthesis diminished both the activation of cytokinin oxidase in stressed plants, and the decline in CKs (Vysotskaya *et al.*, 2011). Since the decline in CK content in stressed plants is likely to sensitize ABA-induced stomatal closure, it may be beneficial to prevent this CK decline (perhaps via reduced expression of cytokinin oxidase) in plants experiencing mild or erratic soil water deficit, and prevent reductions in carbon gain associated with unnecessarily sensitive stomatal closure. In addition to maintaining carbon gain under mild drought by de-sensitizing stomata to ABA, increasing CK concentrations in droughty plants may also be directly beneficial for grain-filling under stress. This is because, in rice and wheat at least, one yield detriment comes from the strong metabolic dominance of the apical spikelets in grainfilling, and the inhibition of filling to the inferior basal spikelets. CK up-regulation may improve inferior grain filling (Mohapatra *et al.*, 2011), by enhancing the activities of cell-cycle genes to increase cell number, and thus the sink capacity of the developing seeds. Slow filling rates in inferior grains have been linked with low contents of the cytokinins (zeatin and zeatinriboside), auxin, abscisic acid content and high rates of ethylene evolution (Zhao *et al.*, 2007; Zhang *et al.*, 2009). The net development of a plant organ is regulated by a balance between hormones that promote and those that inhibit development.

Some of the work using developmental- or tissue- specific promoters to control increases in isopentenyltransferase (ipt) gene to reduce stress-associated senescence, have coincidentally reduced ABA concentration, alongside the increases in stomatal conductance, photosynthesis, and growth of plants exposed to drought or salt stress (Ghanem *et al.*, 2011). Peleg *et al.* (2011) suggested that robust root development and increased water-absorbing capacity ofthe transgenic peanut plants allowed a higher stomatal conductance even under drought conditions, ensuring a higher CO_2 supply to leaf tissues and therefore a higher photosynthetic rate. Increasing CK concentrations via exogenous applications to crop plants has generally proved ineffective due to in plant regulation of CK homeostasis via cytokinin oxidase. Another way of managing CK increases in plants may be through their inoculation with preparations of CK-producing bacteria, gradually releasing CK in concentrations within the physiological range. Some strains of growth-promoting rhizobacteria have been shown to produce high concentrations of CK. Treatment of plants with one of these strains resulted in an increase in CK content in lettuce and wheat plants accompanied by their faster growth both in well-watered conditions and under soil moisture deficit (Arkhipova *et al.*, 2006).

Ethylene

Ethylene (Figure 5.2) is the simplest organic compound and is biologically active in trace amount. It is biosynthesized and emitted as gas by plant tissues directly into the atmosphere. Besides, the classic *triple response*, characterized by growth retardation, an increase in diameter, and horizontal growth of shoots, is still used as a bioassay to identify and measure ethylene. There is increasing evidence that ethylene influences many plant tissues subjected to injury by a variety of stresses, for example, wounding; chemical; mechanical, and temperature extremes; and pathogens. Since it is a gaseous hormone, its emission has been detected to increase from 2 to 50 times or more depending on the intensity of the stress and the sensitivity of the tissue. However, this surge is short lived, peaking rapidly and returning to normal level within 24 h or less (Tingey, 1980). Kacperska and Kubacka-Zebalska (1989) suggested two prerequisites for such stress induced endogenous ethylene biosynthesis; (a) the promotion of 1-aminocyclopropane-1-carboxylic acid (ACC) growth and developmental processes and interacts with all the other plant hormones (b) activation of a free radical generating system. The latter system was needed for the non-enzymatic conversion of ACC to ethylene and depended on the activation of the membrane-associated lipoxygenase caused by stress-induced alterations in cell membrane properties. Ethylene promotes cell extensibility as well as elongation and has been shown to play an important role in many water plants with aerial parts to adjust to the water level. When plants are submerged, elongation is greatly accelerated until the aerial parts regain the water surface. The organs which elongate may be the stem, petioles, or flower stalks (Metraux and Kende, 1983), but generally the hormone is known to inhibit elongation growth and to accelerate senescence (Penarrubia and Moreno, 1995). Ethylene is also well known to play an important role against pathogenic stress (Raz and Fluhr, 1993).

When up-regulated under stressful conditions such as heat and drought, ethylene can inhibit root growth and development, and, like ABA, it can also reduce shoot/leaf expansion (Sharp, 2002; Pierik *et al.*, 2006). It is probably best known for its involvement in stress-induced leaf senescence and abscission. There is also evidence that stress ethylene may directly reduce photosynthesis (Rajala and Peltonen-Sainio, 2001). Stress ethylene production, for example under heat, flooding, air pollution, soil compaction, and drought, can also induce more direct yield detriments, notably reduced grain-filling rates and/or increased embryo and grain abortion (Hays *et al.*, 2007; Wilkinson and Davies, 2010). As so many of the effects of stress on plants mirror those known to be ethylene-induced, there is now general acceptance that ethylene responses may underlie a portion of the limitation of seed crop yield under abiotic stress, and that the prevention of ethylene production or perception under stress is a desirable genetic or management target. A role for ethylene in stomatal control is also gaining support. It has been shown experimentally to close stomata (Desikan *et al.*, 2006), and instances of stomatal closure via environmentally-induced ethylene accumulation have now been recorded (Vysotskaya *et al.*, 2011). Conversely, ethylene can antagonize drought-and ABA-induced stomatal closure (Tanaka *et al.*, 2005; Wilkinson and Davies, 2009). Under mild drought where effects on leaf senescence and grain abortion are less relevant to yield, the ethylene-induced antagonism of drought/ABA-induced stomatal closure may be beneficial for carbon gain and leaf cooling, both of which are clearly linked to improved yield, at least in hotter environments.

The detrimental effects of ethylene on crop performance and yield are not necessarily confined to a stressful environment, as they may also be linked to genotypic variability in productivity through plant architecture. Slow filling rates in a percentage of the grains lower down on a rice spikelet or wheat ear, called inferior grains, have been linked with low contents of CK and ABA and high rates of ethylene evolution (Zhao *et al.*, 2007; Zhang *et al.*, 2009). Developing grains in the lower part of the panicle/ear that are confined in the flag leaf sheath for a longer period, may be subject to the inhibitory action of ethylene more than those in the upperpart of the ear/panicle, which may be the basis of the fact that a spike or ear always contains at least some inferior, smaller, less plump grains, particularly in some genotypes. Mohapatra *et al.* (2000) used ethylene inhibitors to improve the growth and development of the inferior spikelets, while the application of ethylene promoters inhibited further growth and development. Ethylene has now also been linked to reduced grain-filling in both superior and inferior spikelets under stress (Mohapatra *et al.*, 2011), and the authors envisaged that yield gains in grain crops could be made through screening genotypes for lower retention time in the leaf sheath and/or for shorter leaf sheaths.

Abscisic Acid

Hemberg (1961) is credited with being the pioneer who advanced the idea that plant growth and development are regulated by endogenous levels of both a promotor (auxin) and an inhibitor. Employing an *Avena* bioassay, he observed that potato peels contained a high level of growth inhibitors. Like other planthormones, Abscisic acid also is ubiquitous among vascular plants besides being present in some algae. It is a typical sesquiterpene, consisting of three isoprene units (Figure 5.2), and is known to be synthesized through carotenoid pathways with xanthophylls being immediate precursors. Under stressful conditions, the endogenous level of ABA has been reported to rapidly increase 5- to 100-fold or more; causing stomatal closure and thus reducing transpirational water loss (Davies and Zhang, 1995), it was named the "stress hormone".

ABA influences many physiological processes and its enhancement under stressful environments convinced Quarrie (1991) to remark that, "a plant that cannot make abscisic acid (ABA) is in trouble". Wright (1978) surveyed over 70 plant species and observed an increase in the ABA levels when their excised leaves were subjected to a period of wilting. ABA also has been reported to increase under pathogenic stress (Bosquet *et al.*, 1990), and enhances resistance to pathogens (Amzallag and Lerner, 1994). Under water stress, ABA is actively translocated out of the mature leaves in the phloem and transported to the stem apices, fruits, and seeds. During the past several years, the role of ABA in the control of stomatal closing in the context of the plant water relation has been established. ABA regulates stomatal opening during stress, however, recent studies suggest that other hormones such as CK, ethylene, BR, JA, SA, and NO also affect stomatal function (reviewed by Acharya and Assmann, 2009). While ABA, BR, SA, JA, and NO induce stomatal closure, CK and IAA promote stomatal opening. NO operates as a key intermediate in the ABA-mediated signaling network that regulates stomatal closure. ABA was also shown to inhibit BR-induced responses during the exposure of plants to abiotic stress (Divi *et al.*, 2010).

ABA has been shown to act indirectly by increasing the cytosolic Ca^{2+} (McAinsh *et al.*, 1990). This observation is consistent with the long-known fact that external Ca^{2+} depresses the stomatal aperture. Proline is probably the most widely distributed compatible osmolyte which accumulates in living organisms, including plants, under stressful environments. The enhanced accumulation of proline is induced by ABA but other workers using ABA-deficient plants (Stewart and Voetberg, 1987; Naqvi *et al.*, 1994) or maize cultured cells (Xin and Li, 1993) have demonstrated that stress-enhanced proline was independent of ABA.

ABA synthesis is one of the fastest responses of plants to abiotic stress, triggering ABA-inducible gene expression (Yamaguchi-Shinozaki and Shinozaki, 2006) and causing stomatal closure, thereby reducing water loss via transpiration (Wilkinson and Davies, 2010) and eventually restricting cellular growth. Numerous genes associated with ABA *de novo* biosynthesis and genes encoding ABA receptors and downstream signal relays have been characterized in *Arabidopsis thaliana*. ABA also plays an important role during plant adaptations to cold temperatures. Cold stress induces the synthesis of ABA and the exogenous application of ABA improves the cold tolerance of plants (Xue-Xuan *et al.*, 2010).

Jasmonates

In 1962, jasmonic acid methyl ester (JAME) was isolated for the first time from the essential oil of *Jasminum grandiflorum* (Demole *et al.*, 1962). The first physiological effects of JAME, however, and its free acid (Figure 5.3) were only observed two decades later. In 1980, Ueda's group in Osaka, Japan, described a senescence-promoting effect of JA, and Dathe's group in Halle, Germany, isolated and characterized these compounds as growth inhibitors (Ueda and Kato, 1980; Dathe *et al.*, 1981). JA and its various metabolites are members of the oxylipin family. α-Linolenic acid (α-LeA) released by lipase activity on chloroplast membranes is the substrate for numerous oxygenated compounds collectively called oxylipins including jasmonates, which comprise JA, JAME, JA amino acid conjugates and further JA metabolites. The generation of oxylipins is initiated by lipoxygenases (LOXs), which form hydroperoxides from α-LeA (18:3) or linoleic acid (18:2).

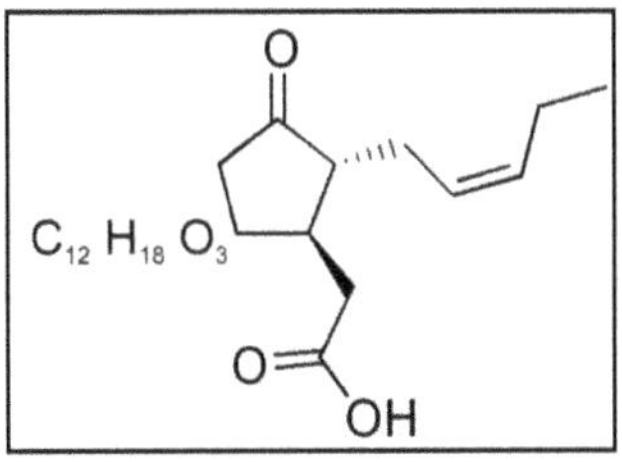

$C_{12}H_{18}O_3$

Figure 5.3: Structure of Jasmonic Acid.

Jasmonates are ubiquitously occurring lipid-derived compounds with signal functions in plant responses to abiotic and biotic stresses, as well as in plant growth and development. Many of them alter gene expression positively or negatively in a regulatory network with synergistic and antagonistic effects in relation to other plant hormones such as salicylate, auxin, ethylene and abscisic acid. One of the first biological activities observed for JA was the senescence-promoting effect. Although it occurs ubiquitously following JA treatment, monocotyledonous plants are more sensitive. Senescence is the last stage in plant development. Consequently, genes expressed during leaf senescence (senescence-associated genes, SAGs) code for proteins with functions in sink/source relationships, photosynthesis, intermediary metabolism, proteolysis and plant defense (Wasternack, 2004; 2006; Wasternack *et al.*, 2006). The role of JA in senescence is linked to (1) down regulation of housekeeping proteins encoded by photosynthetic genes and (2) upregulation of genes active in defense reactions against biotic and abiotic stresses.

One of the best-studied signal-transduction pathways of jasmonates is that of wound-response, which was initially studied mainly in tomato. Briefly, a sequential action of the 18-aa peptide systemin, cleaved from prosystemin upon local wounding, activates JA biosynthetic enzymes such as allene oxide synthase (AOC) and leads to local rise in JA. A proteinase inhibitor (PIN2) was found to accumulate upon wounding (such as that caused by herbivore attack) or by treatment with airborne JAME. Consequently, the plant becomes immunized against a subsequent herbivore attack.

Salicylic Acid

Salicylic acid (Figure 5.4) was identified as a crucial signaling molecule required for the expression of plant defense responses. It is also known as 2-hydroxybenzoic acid. SA is a phenolic phytohormone and is found in plants with roles in plant growth and development, photosynthesis, transpiration, ion uptake and transport. SA also induces specific changes in leaf anatomy and chloroplast structure. SA is involved in endogenous signaling, mediating in plant defense against pathogens. It plays a role in the resistance to pathogens by inducing the production of pathogenesis-related proteins. It is involved in the systemic acquired resistance (SAR) in which a pathogenic attack on one part of the plant induces resistance in other parts. Based on studies in *Arabidopsis*, it is generally assumed that the life-style of a pathogen determines which signal transduction pathway becomes activated, with biotrophic pathogens inducing defense responses via the salicylic acid signaling pathway (McDowell and Dangl, 2000). JA-dependent defense responses, on the other hand, are considered to be activated in *Arabidopsis* in response to infection with necrotrophic pathogens which require host cell death to obtain nutrients (Figure 5.5) (Bari and Jones, 2009).

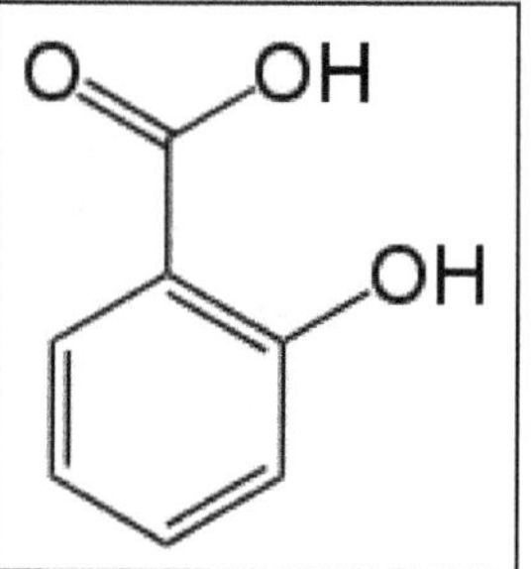

Figure 5.4: Structure of Salicylic Acid.

Brassinosteroids

Brassinosteroids, a class of plant-specific steroid hormones characterized by their poly-hydroxylated sterol structure, were first isolated and purified from *Brassica napus* pollen in 1979 (Grove *et al.*, 1979). The chemical structure of brassinolide (BL), and that of the second steroidal plant hormone, castasterone (CS) (Figure 5.6) discovered in 1982 by Yokota and associates was found to be similar to that of ecdysone, the insect molting steroid hormone (ecdysteroids) and mammalian steroids. Natural BRs have 5a-cholestane skeleton, and their structural variations come from the kind and orientation of oxygenated functions in A- and B-ring. They are divided into free (64) and conjugated (5) compounds (Bajguz and Tretyn, 2003; Bajguz, 2009; Bajguz *et al.*, 2013). The brassinosteroids BL and CS occur ubiquitously in the plant kingdom. The occurrence of BRs has been demonstrated in almost every

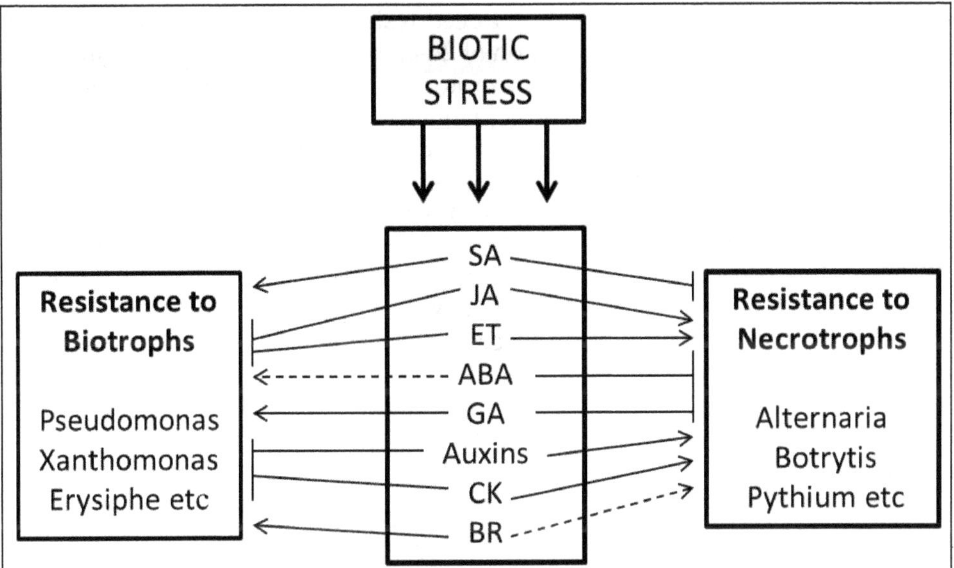

Figure 5.5: A Simplified Model Showing the Inovlvement of Different Hormones in the Positive or Negative Regulation of Plant Resistance to Various Biotrpohic and Necrotrphic Pathogens. The arrows indicate activation or positive interaction and blocked lines indicate repression or negative interaction. See text for further details and abbreviations (Bari and Jones, 2009).

part of plants, such as pollen, flower buds, fruits, seeds, vascular cambium, leaves, shoots and roots. Pollen and immature seeds are the richest sources of BRs. These steroidal compounds occur in free form and conjugated to sugars and fatty acids. To this date, about 70 BRs have been isolated from plants.

Although, BRs are known to influence various developmental processes including seed germination, cell division, cell elongation, flowering, reproductive development, senescence, photo-morphogenesis and abiotic stress responses in plants, very little is known about their role in plant responses to biotic stresses. There are emerging evidences which indicate that BRs are involved in the regulation

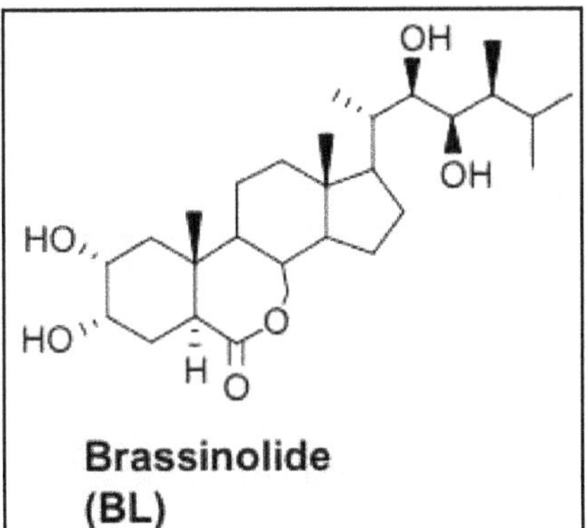

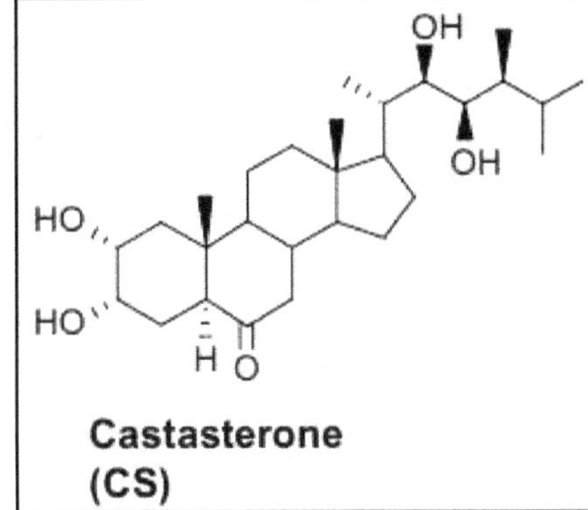

Figure 5.6: The Chemical Structure of BRs-Brassinolide (BL) and Castasterone (CS).

of plant defense responses. This BR induced resistance is not associated with SA biosynthesis *e.g.* exogenous application of 24-epibrassinolide, a BR, was shown to prevent the development of disease symptoms on tomato plants inoculated with *Verticillium dahliae*, whereas untreated plants showed moderate to severe disease symptoms. Similarly, BR sprayed potato plants showed resistance to infection by *Phytophthora infestans* and this resistance was found to be associated with increases in the levels of ABA and ET (Krishna, 2003). This suggests that there is a cross-talk between BRs and other hormone signaling in mediating defense responses in plants. BRs have also been reported to affect the expression of genes involved in defense as well as biosynthesis of other hormones. For example, genes involved in biosynthesis of ET and JA in *Arabidopsis*, were induced by BRs (Muessig *et al.*, 2006). Whether JA or ET is required for BR induced resistance is not known. Collectively, these results indicate that BRs and their signaling components are involved in the modulation of plant defense responses against various pathogens. However, our knowledge on the role of BR on plant defense has started to emerge and the molecular mechanisms involved remains to be understood. More recently, interactions of BRs with other plant hormones, such as auxins, abscisic acid, gibberellins, and ethylene, have also been found to play a major role in plant stress alleviation. Furthermore, ability of BRs to boost antioxidant system of plants is extensively used to confer resistance in plants against a variety of abiotic stresses, such as thermal, drought, heavy metal, pesticides, and salinity. There is ample evidence from both field experiments and greenhouse trials demonstrating the protective effects of exogenous BRs against a fairly broad range of fungal, viral, and bacterial pathogens exhibiting diverse parasitic habits (Bajguz and Hayat, 2009). BR was reported (mainly based on the exogenous application of BR) to induce the expression of stress-related genes, leading to the maintenance of photosynthesis activity, the activation of antioxidant enzymes, the accumulation of osmo-protectants, and the induction of other hormone responses (Divi and Krishna, 2009). The synergistic or antagonistic hormone action and the coordinated regulation of hormone biosynthetic pathways play crucial roles in the adaptation of plants to abiotic stress.

Over the past decade, molecular genetic studies using *Arabidopsis thaliana* and rice (*Oryza sativa*) as model plants have identified numerous genes involved in BR biosynthesis and gene regulation. Coupled with more recent biochemical approaches, these studies have provided fascinating insights into the various aspects of plant steroid signaling, ranging from BR perception at the cell surface to activation of transcription factors (TFs) in the nucleus (Kim and Wang, 2010). BRs are key components in aims to improve the productivity and quality of agricultural products, such as seeds of rice and other cereal species. However, the practical application of BR research in agriculture has not yet been fully explored.

BRs can act efficiently in plants as immune-modulators when applied at the appropriate concentration and at the correct stage of plant development. BRs are implicated in plant responses to abiotic environmental stresses and to undergo profound changes in plants interacting with bacterial, fungal and viral pathogens (Figure 5.7). BR regulated stress response as a result of a complex sequence of biochemical reactions such as activation or suppression of keyenzymatic reactions,

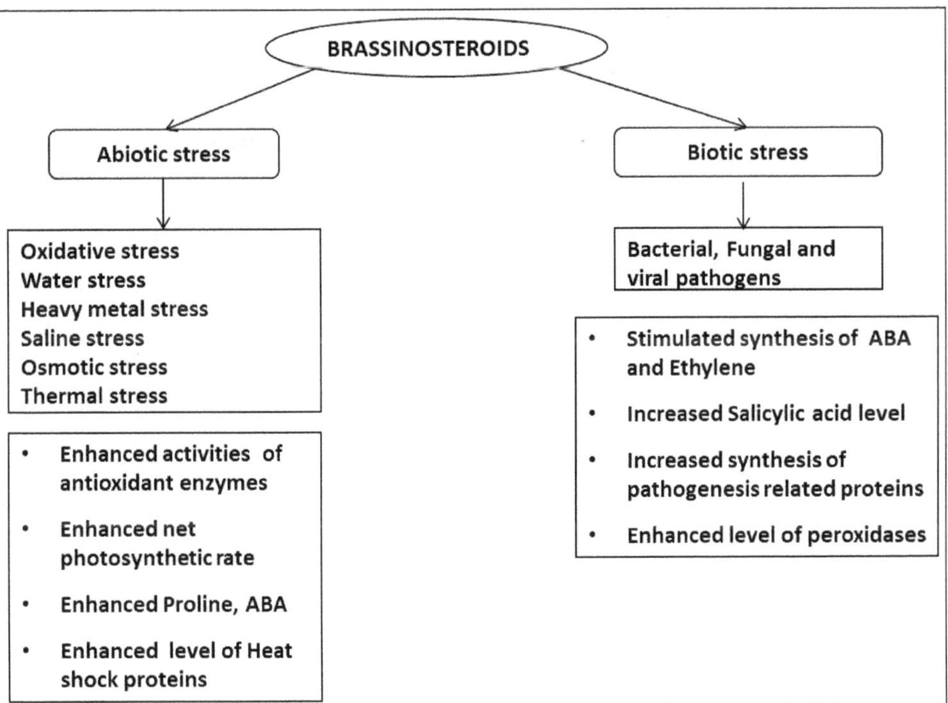

Figure 5.7: Effect of Brassinosteroids on Plants Exposed to Various Stresses.

induction of protein synthesis, and the production of various chemical defense compounds. BRs open up new approaches for plant resistance against hazardous environmental conditions.

Peptide Hormones

Peptide hormones comprise a new class of hormones and are involved in the regulation of various aspects of plant growth and development including defense responses against attacking pathogens and pests (Matsubayashi and Sakagami, 2006; Farrokhi *et al.*, 2008). The peptide hormones are usually processed from larger polypeptide precursors. Examples of plant peptide hormones include systemin, hydroxyproline-rich glycopeptides, AtPep1, CLAVATA3, PHYTOSULFOKINE, POLARIS, ROTUNDIFOLIA4/DEVIL1, inflorescence deficient in abscission gene (IDA), early nodulin 40 (ENOD40), nodule specific cysteine-rich (NCR), and S-locus cysteine-rich protein (SCR).

Defense-related peptide hormones include systemin (Pearce *et al.*, 1991), hydroxyproline-rich glycopeptides systemins (Pearce *et al.*, 2001, 2007; Pearce and Ryan, 2003) from solanaceous plants and AtPep1 peptide from *Arabidopsis* (Huffaker and Ryan, 2007). These peptides are from 18 to 23 amino acids in length, are processed from wound and JAinducible precursor proteins, and play roles in the activation of local and systemic responses against wounding and pest attack. Systemin is synthesized from prosystemin and stored in the cytoplasm. In contrast, hydroxyproline-rich glycopeptides systemins are processed from precursors that are synthesized through the secretory pathway and localized in the cell walls (Narvaez-Vasquez *et al.*, 2005). Systemin acts at or near the site of wounding by inducing and amplifying the JA-derived mobile signal and activates the systemic response (Schilmiller and Howe, 2005). Systemin has been shown to regulate the expression of several genes involved in the octadecanoid pathway and herbivore defense in tomato (Ryan, 2000). Tomato plants overexpressing prosystemin show enhanced resistance whereas plants with reduced systemin levels show more susceptibility to insect herbivory. Both systemin and hydroxyproline-rich glycopeptides can activate the expression of anti-herbivore proteinase inhibitors and polyphenol oxidase in response to wounding and methyl jasmonate. AtPep1, a 23 amino acidpeptide derived from precursor PROPEP1 in Arabidopsis, acts as an elicitor and activates the expression of defense related genes (Huffaker and Ryan, 2007). The gene encoding the precursor can be induced by wounding, methyl jasmonate, ethylene and SA. These results suggest that defense-signaling peptides play important roles in the activation of defense against invaders probably by amplifying the signal initiated by wounding and elicitors. However, the underlying molecular mechanism involved in the activation of these peptide hormones in regulating plant defense remains elusive.

Plant Hormone Interactions: Innovative Targets for Crop Breeding and Management

Abiotic stress responses are largely controlled by the hormone ABA, while defense against different biotic assailants is specified by antagonism between the SA and JA/ET signaling pathways. However, recent findings suggest that ABA acts both synergistically and antagonistically with biotic stress signaling, creating a complex network of interacting pathways with cross-talk at different levels. ABA is therefore likely to be central in the fine-tuning of observed responses to simultaneous biotic and abiotic stresses. Treatment with ABA increases susceptibility in *Arabidopis* to a virulent *P. syringae* strain (Mohr and Cahill, 2003), in tomato to *B. cinerea* and *Erwinia chrysanthemi* (Audenaert *et al.*, 2002; Asselbergh *et al.*, 2008a), in rice to the blast fungus *Magnaporth agrisea* (Koga *et al.*, 2004), and in potato to the pathogens *Phytophthora infestans* and *Cladosporium cucumerinum* (Henfling *et al.*, 1980). In accordance with this, a lack of ABA can induce a higher level of pathogen resistance (Asselbergh *et al.*, 2008a). ABA treatment represses the systemic acquired resistance (SAR) pathway both upstream and downstream of SA induction in *Arabidopsis* and tobacco, as well as inhibiting the accumulation of crucial defense compounds such as lignins and phenyl propanoids (Yasuda *et al.*, 2008; Kusajima *et al.*, 2010). In turn, SA can also interfere with abiotic stress signaling. The exogenous application of SA in maize leads to drought susceptibility (Nemeth *et al.*, 2002), while the artificial induction of SAR in *Arabidopsis* leads to the suppression of abiotic stress responses (Yasuda *et al.*, 2008). In rice, resistance to the rice blast fungus *M. griseais* mediated by a precise balance between ABA and SA levels (Jiang *et al.*, 2010). Ethylene treatment, in turn, activates ABI1 and ABI2, two negative regulators of ABA signaling (Asselbergh *et al.*, 2008b).

A large number of genes associated with *de novo* ABA biosynthesis and genes encoding ABA receptors and downstream signal relays have been characterized in *Arabidopsis*. Overexpression of these genes under the control of constitutive or stress-inducible promoters resulted in a significant increase in transgenic plant yield under stress conditions in the field. The transgenic plants were similar in size to the wild type plants, and under stress they were able to maintain relative growth rates similar to that of the wild-type plants under normal conditions. The modification of genes involved in the regulation of the plant responses to ABA is an alternative approach to enhance of plant stress resistance.

A new model has been proposed regarding the multifaceted role of ABA in pathogen response, whereby the influence of ABA depends on the time scale of infection and also the nature of the attacker (Ton *et al.*, 2009). The model refers to three distinct phases of pathogen infection. In the first, ABA causes stomatal closure, increasing resistance to penetration by pathogens such as bacteria, and thus having a positive effect on the defense response. At this stage, ABA antagonizes SA, JA, and ethylene pathways in order to save resources, as their effects are not yet required. In the second phase, post-invasion defenses centre on callose deposition to strengthen cell walls, a process that is aided by ABA during fungal infection but repressed during bacterial infection. During phase three of infection, PAMPs (pathogen-associated molecular patterns) induce the hormones SA, JA, and ethylene and long-distance signals to regulate a broad spectrum of defensive compounds. The ABA-inducible genes ERD15 and ATAF1 have been identified as switches that may activate ABA-dependent biotic stress responses at the expense of abiotic responses (Ton *et al.*, 2009). However, increased ABA levels arising from abiotic stress conditions may repress the SA, JA, and ethylene responses even during phase three. This hypothesis provides a mechanism for the control of ABA over both biotic and abiotic stress signaling. To complicate matters further, recent research has found that ABA can have either a positive or a negative effect on bacteria-induced callose deposition, depending on other growth conditions such as light and glucose levels, thus switching its effect on the defense response according to the environmental situation (Luna *et al.*, 2011). Such research emphasizes the need for caution when drawing conclusions from pathogen resistance assays. ABA is now considered a global regulator of stress responses that can dominantly suppress pathogen defense pathways, thus controlling the switch in priority between the responses to biotic or abiotic stress and allowing plants to respond to the most severe threat. ABA production may therefore, be the crucial factor in determining how plants respond to multiple stresses.

The ipt-gene, which drives CK biosynthesis, has now been over-expressed in several plant species under different promoters, and the transgenic plants have demonstrated improved stress tolerance (reviewed by Peleg and Blumwald, 2011), including field analyses (Qin *et al.*, 2011).Transgenic cassava plants over-expressing CK show significant drought tolerance as a result of their stay-green capacity under stress (Zhang *et al.*, 2010). The constitutive overexpression of IPT increased endogenous CK concentrations up to 150-fold and resulted in decreased root growth and in water stress. However, the relationship between delayed leaf senescence and improved plant productivity does not always hold (Yang and Zhang, 2010). Constitutive over-expression of ipt results in detrimental pleiotropic effects such as decreased root growth, altered flowering time, and poor tissue water relations. Targeted reductions in CK content in transgenic tobacco and *Arabidopsis* plants over-producing cytokinin oxidase in roots, driven by root-specific promoters, was shown to increase root growth and plant survival under drought (Werner *et al.*, 2010). The results are in conflict with others showing that silencing of cytokinin oxidase genes to increase CK levels in non-stressed barley led to higher plant productivity, partially through improved root biomass (Zalewski *et al.*, 2010). Thus, it is clear that effects of CK on plant productivity are diverse and possibly depend on where and when their content is changed and how the changes are related to other hormones and external conditions.

The use of inducible promoters for the conditional expression of hormone biosynthentic genes makes it possible to control hormone levels without the negative effects on growth and development produced by very large changes in hormone concentrations. The use of maturation- induced and stress-induced promoters (SARK, senescence associated receptor kinase to drive IPT expression in both dicots and monocots provided an alternative approach for the induction of IPT and the concomitant biosynthesis of CK, without the negative effects of constitutively high CK content on plant phenology (Rivero *et al.*, 2007). Antagonism between BR and ABA was recently demonstrated in transgenic P_{SARK}::IPT rice plants, where the increase in CK induced BR-associated genes and repressed ABA-related processes (Peleg *et al.*, 2011). These results further highlight the importance of hormone cross-talk during the response of plants to abiotic stress.

Conclusions and Perspectives

Plant hormones regulate complex signaling networks involving developmental processes and plant responses to environmental stresses including biotic and abiotic stresses. Significant progress has been made in identifying the key components and understanding plant hormone signaling and plant defense responses. Several recent studies provide evidence for the involvement of hormones such as ABA, IAA, GA, CK and BR in plant defense signaling pathways. Treatment of plants with some hormones results in the reprogramming of the host metabolism, gene expression and modulation of plant defense responses against microbial challenge. To understand how plants coordinate multiple hormonal components in response to various developmental and environmental cues is a major challenge for the future. It is important to note that the type of interactions and plant responses to stresses vary depending on the patho-system as well as the time, quantity and the tissue where hormones are produced. Another important question to answer is how different hormone-mediated developmental and defense-related responses are regulated in specific tissues and cell types? Most of the studies in understanding phytohormone signaling has been done using seedlings and there is limited study on mature leaves. More studies using mature leaves are necessary to understand the role of hormone signaling components involved in plant defense against various pathogens. The modification of hormone biosynthetic pathways for the generation of transgenic plants could pave a way for enhanced biotic/abiotic stress tolerance.

We have very limited knowledge on how pathogen effectors confer virulence by modulating hormone signaling components. Recent global expression profiling studies in response to pathogen challenges are providing useful information about different components involved in the complex interactions between hormone-regulated defense signaling pathways. Additional studies involving mature leaves and detailed time course experiment will be necessary to extend our understanding of the complex regulatory mechanisms operating between plant hormone signaling and plant defense responses. A better understanding of phytohormone-mediated plant defense responses is important in designing effective strategies for engineering crops for stress resistance.

References

Acharya, B.R. and Assmann, S.M., 2009. Hormone interactions in stomatalfunction. *Plant Mol. Biol.*, 69: 451–462.

Adie, B.A., Perez-Perez, J., Perez-Perez, M.M., Godoy, M., Sanchez-Serrano, J.J., Schmelz, E.A. and Solano, R., 2007. ABA is an essential signal for plant resistance to pathogens affecting JA biosynthesis and the activation of defenses in *Arabidopsis. Plant Cell,* 19: 1665–1681.

Alonso-Ramirez, A., Rodriguez, D., Reyes, D., Jimenez, J.A., Nicolas, G., Lopez-Climent, M., Gomez-Cadenas, A. and Nicolas, C., 2009.Evidence for a role of gibberellins in salicylic acid-modulated early plant responses to abiotic stress in *Arabidopsis* seeds.*Plant Physiol.*, 150: 1335-1344.

Amzallag, G.N. and Lerner, H.R., 1994. Adaptation of plants to environmental stresses. In:*Handbook of Plant and Crop Physiology*, (Ed.) M. Pessarakli, Marcel Dekker, New York, 557–576.

Ansari, R., Naqvi, S.M., and Azmi, A.R., 1977. Effect of salinity on germination, seedling growth and α-amylase activity in wheat. *Pak. J. Bot.*, 9:163–166.

Argueso, C.T., Raines, T. and Kieber, J.J., 2010.Cytokinin signaling and transcriptional networks. *Curr. Opin. Plant Biol.*, 13: 533-539.

Arkhipova, T.N., Veselov, S.Yu.,Melent'ev, A.I., Martinenko, E.V. and Kudoyarova, G.R., 2006.Comparison of effects of bacterial strains differing in their ability to synthesize cytokinins on growth and cytokinin content in wheat plants.*Russian J. Plant Physiol.*, 53: 507–513.

Asselbergh, B., Achuo, A.E., Hofte, M. and Gijsegem, F.V., 2008a.Abscisic acid deficiency leads to rapid activation of tomato defense responses upon infection with *Erwinia chrysanthemi. Mol. Plant Path.*, 9(1): 11-24.

Asselbergh, B., De Vieesschauwer, D. and Hofte, M., 2008b. Global switches and fine-tuning-ABA modulates plant pathogen defense. *Molecular Plant-Microbe Interactions*, 21: 709-719.

Atkinson, N.J. and Urwin, P.E., 2012. The interaction of plant biotic and abiotic stresses: from genes to the field. *J. Exp. Bot.*, 63(10): 3523-3544.

Audenaert, K., De Meyer, G.B. and Hofte, M.M., 2002. Abscisic acid determines basal susceptibility of tomato to *Botrytis cinerea* and suppresses salicylic acid-dependent signaling mechanisms. *Plant Physiol.*, 128: 491–501.

Bajguz, A., 2009. Isolation and characterization of brassinosteroids from algal cultures of *Chlorella vulgaris*Beijerinck (Trebouxiophyceae). *J. Plant Physiol.*, 166: 1946–1949.

Bajguz, A. and Hayat, S., 2009. Effects of brassinosteroids on the plant responses to environmental stresses. *Plant Physiol. Biochem.*, 47:1–8.

Bajguz, A. and Tretyn, A., 2003. The chemical characteristic and distribution of brassinosteroids in plants. *Phytochemistry*, 62: 1027–1046.

Bajguz, A., Bajguz, J. and Elzbieta, A., 2013. Recent Advances in Medicinal Applications of Brassinosteroids, a Group of Plant Hormones Studies. In: *Natural Products Chemistry*, (Ed.) Atta-ur-Rahman, 40: 33-50.

Bari, R. and Jones, J.D.G., 2009. Role of plant hormones in plant defense responses. *Plant Mol. Biol.*, 69: 473–488.

Bosquet, J.F., Touraud, G., Piollat, M.T., Bosch, U. and Trottet, M., 1990. ABA accumulation in wheat heads inoculated with *Septoria nodorum* in the field conditions. *J. Agron. Crop Sci.* 165: 297–300.

Brugiere, N., Jiao, S.P., Hantke, S., Zinselmeier, C., Roessler, J.A., Niu, X.M., Jones, R.J. and Habben, J.E., 2003. Cytokinin oxidase gene expression in maize is localized to the vasculature, and is induced by cytokinins, abscisic acid, and abiotic stress. *Plant Physiology*, 132: 1228–1240.

Chapin, F.S. III., 1991. Integrated responses of plants to stress: A centralized system of physiological responses. *BioScience*, 41: 29–36.

Cutler, S.R., Rodriguez, P.L., Finkelstein, R.R. and Abrams, S.R., 2010.Abscisic acid: emergence of a core signaling network. *Ann. Rev. Plant Biol.*, 61: 651-679.

Dathe, W., Ronsch, H., Preiss, A., Schade, W., Sembdner, G. and Schreiber, K., 1981. Endogenous plant hormones of the broad bean, *Vicia faba* L.-Jasmonicacid, a plant growth inhibitor in pericarp. *Planta*,155: 530–535.

Davies, W.J. and Zhang, J., 1995. Root signals and the regulation of growth and development of plants in drying soil. *Ann. Rev. Plant Physiol. Plant Mol. Biol.*, 42: 55–76.

Davies, W.J., Kudoyarova, G. and Hartung, W., 2005. Long-distance ABA signaling and its relation to other signaling pathways in the detection of soil drying and the mediation of the plant's response to drought. *Journal of Plant Growth Regulation*, 24: 285–295.

Demole, E., Lederer, E. and Mercier, D., 1962. Isolementet determination de la structure du jasmonate de methyle, constituant odorant characteristique de lessence de jasmin. *Helvetica Chimica Acta*, 45: 675– 685.

Desikan, R., Last, K., Harrett-Williams, R., Tagliavia, C., Harter, K., Hooley, R., Hancock, J.T. and Neill, S.J., 2006. Ethylene-induced stomatal closure in

Arabidopsis occurs via AtrbohF-mediated hydrogen peroxide synthesis. *The Plant Journal*, 47: 907–916.

Divi, U., Rahman, T. and Krishna, P., 2010.Brassinosteroid-mediated stress tolerance in *Arabidopsis* shows interactions with abscisic acid, ethylene and salicylic acid pathways. *BMC Plant Biology*, 10: 151.

Divi, U.K. and Krishna P., 2009.Brassinosteroid: a biotechnological target for enhancing crop yield and stress tolerance. *New Biotechnol.*, 26: 131-136.

Farrokhi, N., Whitelegge, J.P. and Brusslan, J.A., 2008. Plant peptides and peptidomics. *Plant Biotechnol. J.*, 6: 105–134.

Ghanem, M.E., Albacete, A. and Smigocki, A.C., 2011. Root synthesized cytokinins improve shoot growth and fruit yield in salinized tomato (*Solanum lycopersicum* L.) plants. *J. Exp. Bot.*, 62:125–140.

Grove, M.D., Spencer, G.F., Rohwedder, W.K., Mandava, N., Worley, J.F., Warthen Jr., J.D., Steffens, G.L., Flippen-Anderson, J.L. and Cook Jr., J.C., 1979. Brassinolide, a plant growth-promoting steroid isolated from *Brassica napus* pollen. *Nature*, 281: 216–217.

Hays, D.B., Do, J.H., Mason, R.E., Morgan, G. and Finlayson, S.A., 2007. Heat stress induced ethylene production in developing wheat grains induces kernel abortion and increased maturation in a susceptible cultivar. *Plant Sci.*, 172: 1113–1123.

Hemberg, T., 1961.Biogenous inhibitors. In: *Encyclopedia of Plant Physiology*, (Ed.) W. Ruhland, Springer-Verlag, Berlin, 14: 1162–1184.

Henfling, J., Bostock, R. and Kuc, J., 1980. Effect of abscisic-acid on rishitin and lubimin accumulation and resistance to *Phytophthora infestans* and *Cladosporium cucumerinum*in potato-tuber tissue slices. *Phytopathology*, **7:** 1074–1078.

Huffaker, A. and Ryan, C.A., 2007. Endogenous peptide defense signals in *Arabidopsis* differentially amplify signaling for the innate immune response. *Proc. Natl. Acad. Sci.*, 104: 10732–10736.

Jaillais, Y. and Chory, J., 2010.Unraveling the paradoxes of plant hormone signaling integration. *Nature Structural and Molecular Biology*, 17:642-645.

Jiang, C.J., Shimono, M., Sugano, S., Kojima, M., Yazawa, K., Yoshida, R., Inoue, H., Hayashi, N., Sakakibara, H. and Takatsuji, H., 2010.Abscisic acid interacts antagonistically with salicylic acid signaling pathway in rice-*Magnaporth egrisea* interaction. *Molecular Plant Microbe Interactions*,23: 791–798.

Jones, B., Gunneras, S.A., Petersson, S.V., Tarkowski, P., Graham, N., May, S., Dolezal, K., Sandberg, G. and Ljung, K., 2010.Cytokinin regulation of auxin synthesis in *Arabidopsis* involves a homeostatic feedback loop regulated via auxin and cytokinin signal transduction. *Plant Cell*, 22: 2956-2969.

Kacperska, A. and Kubacka-Zebalska, M., 1989. Formation of stress ethylene depends both on ACC synthesis and on the activity of free radical-generating system. *Physiol. Plant.*, 77: 231–237.

Kim, T.W.and Wang, Z.Y., 2010. Brassinosteroid signal transduction from receptor kinases to transcription factors.*Annu. Rev. Plant Biol.,* 61: 681–704.

Klingler, J.P., Batelli, G. and Zhu, J.K., 2010. ABA receptors: the START of a new paradigm in phytohormone signalling. *J. Exp. Bot.,* 61: 3199-3210.

Kobayashi, M., Spray, C.R., Phinney, B.O., Gaskin, P., and MacMillan, J., 1996. Gibberellin metabolism in maize. The step- wise conversion of gibberellin A_{12}-aldehyde to gibberellin A_{20}. *Plant Physiol.,* 110: 413-418.

Koga, H., Dohi, K. and Mori, M., 2004. Abscisic acid and low temperatures suppress the whole plant-specific resistance reaction of rice plants to the infection of *Magnaporth egrisea. Physiol. Mol. Plant Pathol.,* 65: 3–9.

Koshimizu, K., and Iwamura, H. Cytokinins., 1986. In: *Chemistry of Plant Hormones,* (Ed.) N. Takahashi, CRC Press, Boca Raton, Florida, 153–199.

Krishna, P., 2003. Brassinosteroid-mediated stress responses. *J. Plant Growth Regul.,* 22: 289-297.

Kurosawa, E., 1926. Experimental studies on the secretion of *Fusarium heterosporum*on rice plants. *Trans. Natl. Hist. Soc. (Formosa),* 16: 213–227.

Kusajima, M., Yasuda, M., Kawashima, A., Nojiri, H., Yamane, H., Nakajima, M., Akutsu, K. and Nakashita, H., 2010.Suppressive effect of abscisic acid on systemic acquired resistance in tobacco plants. *Journal of General Plant Pathology,* 76: 161–167.

Luna, E., Pastor, V., Robert, J., Flors, V., Mauch-Mani, B. and Ton, J., 2011. Callose deposition: a multifaceted plant defense response. *Molecular Plant-Microbe Interactions,*24:183–193.

Matsubayashi, Y. and Sakagami, Y., 2006. Peptide hormones in plants. *Annu. Rev. Plant Biol.,* 57: 649 674.

McAinsh, M.R.,Brounlee, C. and Hetherington, A.M., 1990. Abscisic acid–induced elevation of guard cell cytosolic Ca^{2+} precedes stomatal closure. *Nature,* 343: 186–188.

McDowell, J.M. and Dangl, J.L., 2000. Signal transduction in the plant immune response. *Trends Biochem. Sci.,* 25: 79 – 82.

Metraux, J.P. and Kende, H., 1983. The role of ethylene in the growth response of submerged deep water rice.*Plant Physiol.,* 72: 441–446.

Mohapatra, P.K., Naik, P.K. and Patel, R., 2000. Ethylene inhibitors improve dry matter partitioning and development of late flowering spikelets on rice panicles. *Australian J. Plant Physiol.,* 27: 311–323.

Mohapatra, P.K., Panigrahi, R. and Turner, N.C., 2011. Physiology of spikelet development on the rice panicle: is manipulation of apical dominance crucial for grain yield improvement? *Advances in Agronomy,* 110: 333–360.

Mohr, P.G. and Cahill, D.M., 2003. Abscisic acid influences the susceptibility of *Arabidopsis thaliana* to *Pseudomonas syringae*pv. Tomato and *Peronospora parasitica. Funct. Plant Biol.,* 30: 461–469.

Muessig, C., Lisso, J., Coll-Garcia, D. and Altmann, T., 2006. Molecular analysis of brassinosteroid action.*Plant Biol.*, 8: 291–296.

Naqvi, S.S.M., Mumtaz, S., Ali, S.A., Shereen, A., Khan, A.H., Ashraf, M.Y. and Khan, M.A.,1994. Proline accumulation under salinity stress. Is abscisic acid involved? *Acta Physiol. Plant*,16: 117–122.

Narvaez-Vasquez, J., Pearce, G. and Ryan, C.A., 2005. The plant cell wall matrix harbors a precursor of defense signaling peptides. *Proc. Natl. Acad. Sci.*, 102: 12974–12977.

Nemeth, M., Janda, T., Horvath, E., Paldi, E. and Szalai, G., 2002. Exogenous salicylic acid increases polyamine content but may decrease drought tolerance in maize. *Plant Sci.*, 162: 569–574.

Pearce, G. and Ryan, C.A., 2003. Systemic signaling in tomato plants for defense against herbivores. Isolation and characterization of three novel defense-signaling glycopeptide hormones coded in a single precursor gene. *J. Biol. Chem.*, 278: 30044–30050.

Pearce, G., Moura, D.S., Stratmann, J. and Ryan, C.A., 2001. Production of multiple plant hormones from a single polyprotein precursor. *Nature*, 411: 817–820.

Pearce, G., Siems, W.F., Bhattacharya, R., Chen, Y.C. and Ryan, C.A., 2007. Three hydroxyproline-rich glycopeptides derived from a single petunia polyprotein precursor activate defensinI, a pathogen defense response gene. *J. Biol. Chem.*, 282: 17777–17784.

Peleg, Z. and Blumwald, E., 2011. Hormone balance and abiotic stress tolerance in crop plants.*Curr.Opin. Plant Biol.*, 14: 290-295.

Peleg, Z., Reguera, M., Tumimbang, E., Walia, H. and Blumwald, E., 2011.Cytokinin-mediated source/sink modifications improve drought tolerance and increases grain yield in rice under water-stress. *Plant Biotech. J.*, 9(7): 747-758.

Penarrubia, L. and Moreno, J., 1995. Senescence in plants and crops. In: *Handbook of Plant and Crop Physiology*, (Ed.) M. Pessarakli, Marcel Dekker, New York, 461–481.

Perilli, S., Moubayidin, L. and Sabatini, S., 2010. The molecular basis of cytokinin function. *Curr. Opin. Plant Biol.*, 13: 21-26.

Pierik, R., Tholen, D., Poorter, H., Visser, E.J.W. and Voesenek, L.A.C.J., 2006. The Janus face of ethylene: growth inhibition and stimulation. *Trends Plant Sci.*, 11: 176-183.

Qin, H., Gu, Q., Zhang, J., Sun, L., Kuppu, S., Zhang, Y., Burow, M., Payton, P., Blumwald, E. and Zhang, H., 2011. Regulated expression of an isopentenyltransferase gene (IPT) in peanut significantly improves drought tolerance and increases yield under field conditions. *Plant Cell Physiol.*, 52: 1904–1914.

Quarrie, S.A., 1991. The role of abscisic acid in regulating water status in plants. *BiolVestn.*, 39(12): 67– 76.

Rajala, A. and Peltonen-Sainio, P., 2001. Plant growth regulator effects on spring cereal root and shoot growth. *Agron. J.*, 93: 936–943.

Raz, V., and Fluhr, R., 1993. Ethylene signal is transduced via protein phosphorylation events in plants. *Plant Cell*, 51: 523-530.

Rivero, R.M., Kojima, M., Gepstein, A., Sakakibara, H., Mittler, R., Gepstein, S. and Blumwald, E., 2007. Delayed leaf senescence induces extreme drought tolerance in a flowering plant. *Proc. Nat. Acad. Sci.*, 104: 19631-19636.

Robert-Seilaniantz, A., Navarro, L., Bari, R. and Jones, J.D., 2007.Pathological hormone imbalances.*Curr.Opin. Plant Biol.*, 10: 372–379.

Ryan, C.A., 2000. The systemin signaling pathway: differential activation of plant defensive genes. *Biochim. Biophys. Acta.*, 1477: 112–121.

Santner, A. and Estelle, M., 2009.Recent advances and emerging trends in plant hormone signaling. *Nature*, 459: 1071-1078.

Santner, A. and Estelle, M., 2010. The ubiquitin–proteasome system regulates plant hormone signaling. *The Plant Journal*,61: 1029-1040.

Sarin, M.N. and Narayanan, A., 1968. Effects of soil salinity and growth regulators on germination and seedling metabolism of wheat. *Physiol. Plant.*, 21: 1201-1209.

Schilmiller, A.L. and Howe, G.A., 2005. Systemic signaling in the wound response. *Curr. Opin. Plant. Biol.*,8: 369–377.

Sharp, R.E., 2002. Interaction with ethylene: changing views on the role of abscisic acid in root and shoot growth responses to water stress. *Plant, Cell Environ.*, 25: 211–222.

Stepanova, A.N. and Alonso, J.M., 2009. Ethylene signaling and response: where different regulatory modules meet. *Curr. Opin. Plant Biol.*, 12: 548-555.

Stepanova, A.N., Hoyt, J.M., Hamilton, A.A. and Alonso, J.M., 2005. A Link between ethylene and auxin uncovered by the characterization of two root-specific ethylene-insensitive mutants in *Arabidopsis. Plant Cell*, 17: 2230-2242.

Stepanova, A.N., Robertson-Hoyt, J., Yun, J., Benavente, L.M., Xie, D-Y., Dolezal, K., Schlereth, A., Jurgens, G., Alonso and J.M., 2008. TAA1-mediated auxin biosynthesis is essential for hormone crosstalk and plant development. *Cell*, 133: 177-191.

Stewart, C.R. and Voetberg, G., 1987. Abscisic acid accumulation is not required for proline accumulation in wilted leaves. *Plant Physiol.*, 66: 230-233.

Suzuki, Y., Yamane, H., Spray, C.R., Gaskin, P., MacMillan, J., and Phinney, B.O.,1992. Metabolism of ent-kaureneto gibberellin A_{12}-aldehyde in young shoots of normal maize. *Plant Physiol.*, 9: 602-610.

Tanaka, Y., Sano, T., Tamaoki, M., Nakajima, N., Kondo, N. and Hasezawa, S., 2005. Ethylene inhibits abscisic acid-induced stomatal closure in *Arabidopsis. Plant Physiol.*, 138: 2337-2343.

Tingey, D.T., 1980. Stress ethylene production-a measure of plant response to stress. Hort. *Science*,15: 630–633.

Ton, J., Flors, V. and Mauch-Mani, B., 2009. The multifaceted role of ABA in disease resistance.*Trends Plant Sci.*, 14: 310–317.

Ueda, J. and Kato, J., 1980. Isolation and identification of a senescence promoting substance from wormwood (*Artemisia absinthium* L.). *Plant Physiol.*, 66: 246–249.

Vysotskaya, L., Wilkinson, S., Davies, W., Arkhipova, T. and Kudoyarova, G., 2011. The effect of competition from neighbours on stomatal conductance in lettuce and tomato plants. *Plant, Cell Environ.*, 34(5): 729-737.

Wan, J., Griffiths, R., Ying, J., McCourt, P. and Huang, Y., 2009.Development of drought-tolerant canola (*Brassica napus* L.) through genetic modulation of ABA-mediated stomatalresponses. *Crop Sci.*, 49: 1539-1554.

Wang, Y., Ying, J., Kuzma, M., Chalifoux, M., Sample, A., McArthur, C., Uchacz, T., Sarvas, C., Wan, J. and Dennis, D.T., 2005. Molecular tailoring of farnesylation for plant drought tolerance and yield protection. *The Plant Journal*, 43: 413-424.

Wasternack, C., 2004. Jasmonates – biosynthesis and role in stress responses and developmental processes. In: *Plant cell death processes*, (Ed.) L.D. Nooden, Elsevier/Academic Press, New York, 143–155.

Wasternack, C., 2006. Oxilipins: biosynthesis, signal transduction and action. In: *Plant hormone signaling: Annual Plant Reviews*,(Ed.) P. Hedden and S. Thomas, Blackwell Publishing Ltd., Oxford, 185–228.

Wasternack, C., 2007. Jasmonates: an update on biosynthesis, signal transduction and action in plant stress response, growth and development. *Ann.Bot.*, 100: 681-697.

Wasternack, C., Stenzel, I., Hause, B., Hause, G., Kutter, C., Maucher, H, Neumerkel, J., Feussner, I. and Miersch, O., 2006. The wound response in tomato–role of jasmonic acid. *J. Plant Physiol.*, 163: 297-306.

Weiler, E.W., Jordan, P.S. and Conrad, W., 1981.Levels of indole-3-acetic acid in intact and decapitated coleoptiles as determined by a specific and highly sensitive solid-phase enzyme immunoassay. *Planta*, 153: 561–571.

Werner, T. and Schmulling, T., 2009.Cytokinin action in plant development.*Curr. Opin. Plant Biol.*,12: 527-538.

Werner, T., Nehnevajova, E., Kollmer, I., Novak, O., Strnad, M., Kramer, U. and Schmulling, T., 2010. Root-specific reduction of cytokinin causes enhanced root growth, drought tolerance, and leaf mineral enrichment in *Arabidopsis* and tobacco. *The Plant Cell*, 22: 3905–3920.

Wilkinson, S. and Davies, W.J., 2009. Ozone suppresses soil drying- and abscisic acid (ABA)-induced stomatal closure via an ethylene dependent mechanism. *Plant, Cell Environ.*, 32: 949–959.

Wilkinson, S. and Davies, W.J., 2010. Drought, ozone, ABA and ethylene: new insights from cell to plant to community. *Plant, Cell Environ.*, 33: 510-525.

Wilkinson, S., Kudoyarova, G.R., Veselov, D.S., Arkhipova, T.N. and Davies, W.J., 2012. Plant hormone interactions: innovative targets for crop breeding and management. *J. Exp. Bot.*, 63(9): 3499–3509.

Wolters, H. and Jurgens, G., 2009. Survival of the flexible: hormonal growth control and adaptation in plant development. *Nature Rev. Genet.*, 10: 305-317.

Wright, S.T.C., 1978. Phytohormones and stress phenomena. In: *Phytohormones and Related Compounds-A Comprehensive Treatise*, (Eds.) D.S. Letham, P.B. Goodwin and T.J.V. Higgins, Elsevier/North Holland, Amsterdam, 495–536.

Xin, Z. and Li, P.H., 1993. Relationship between proline and abscisic acid in the induction of chilling tolerance in maize suspension cultured cells. *Plant Physiol.*, 103: 607–613.

Xue-Xuan, X., Hong-Bo, S., Yuan-Yuan, M., Gang, X., Jun-Na, S., Dong-Gang, G. and Cheng-Jiang, R., 2010. Biotechnological implications from abscisic acid (ABA) roles in cold stress and leaf senescence as an important signal for improving plant sustainable survival under abiotic-stressed conditions. *Crit. Rev. Biotech.*, 30: 222-230.

Yabuta, T. and Sumiki, Y., 1938.On the crystal of gibberellin, a substance to promote plant growth.*J. Agric. Chem. Soc. Japan*, 14: 1526.

Yamaguchi-Shinozaki, K. and Shinozaki, K., 2006.Transcriptional regulatory networks in cellular responses and tolerance to dehydration and cold stresses. *Ann. Rev. Plant Biol.*, 57: 781-803.

Yang, J. and Zhang, J., 2010.Grain-filling problems in 'super' rice. *J. Exp. Bot.*, 61: 1–5.

Yasuda, M., Ishikawa, A. and Jikumaru, Y., 2008.Antagonistic interaction between systemic acquired resistance and the abscisic acid-mediated abiotic stress response in *Arabidopsis. The Plant Cell*, 20: 1678–1692

Yokota, T., Arima, M. and Takahashi, N., 1982.Castasterone, a new phytosterol with plant-hormone potency, from chestnut insect gall. *Tetrahedron Lett.*, 23: 1275–1278.

Zalewski, W., Galuszka, P., Gasparis, S., Orczyk, W. and Nadolska-Orczyk, A., 2010. Silencing of the HvCKX1 gene decreases the cytokinin oxidase/dehydrogenase level in barley and leads to higher plant productivity.*J. Exp. Bot.*, 61: 1839–1851.

Zhang, H., Tan, G., Wang, Z., Yang, J. and Zhang, J., 2009. Ethylene and ACC levels in developing grains are related to the poor appearance and milling quality of rice. *Plant Growth Regulation*, 58: 85–96.

Zhang, P., Wang, W-Q., Zhang, G.L., Kaminek, M., Dobrev, P., Xu, J. and Gruissem, W., 2010. Senescence-inducible expression of isopentenyltransferase extends leaf life, increases drought stress resistance and alters cytokinin metabolism in cassava. *Journal of Integrative Plant Biology*, 52: 653–669.

Zhang, S.W., Li, C.H., Cao, J., Zhang, Y.C., Zhang, S.Q., Xia, Y.F., Sun, D.Y.and Sun, Y., 2009. Altered architecture and enhanced drought tolerance in rice via the down-regulation of indole-3-Acetic Acid by TLD1/OsGH3.13 activation. *Plant Physiol.*, 151: 1889-1901.

Zhao, B., Liu, K., Zhang, H., Zhu, Q. and Yang, J., 2007. Causes of poor grain plumpness of two line hybrids and their relationships to the contents of hormones in rice grain. *Agricultural Sciences in China*, 6: 930-940.

2016, Recent Advances in Plant Stress Physiology *Pages* **117–136**
Editors: ***Praduman Yadav, Sunil Kumar and Veena Jain***
Published by: **DAYA PUBLISHING HOUSE, NEW DELHI**

Chapter 6

Plant Stress Regulation

R.S. Pal[1]*, P.K. Agrawal[1], J.C. Bhatt[1] and Praduman Yadav[2]

[1]ICAR – Vivekananda Institute of Hill Agriculture, Almora, Uttarakhand, India
[2]ICAR – Directarate of Oil Seed Research, Hyderabad, India

ABSTRACT

Plants have a series of fine mechanisms for responding to environmental changes, which has been established during evolution and experiments. These mechanisms are involved in many aspects of physiology, biochemistry, genetics and molecular biology. Alterations in the levels of key signalling metabolites or transcription factors could provide an explanation for how plant metabolism regulation is altered by exposure to various stresses. Plant defence response genes are transcriptionally activated by biotic, as well by different forms of abiotic stress, the induction of specific defence genes in the response against certain pathogens, are dependent on specific environmental conditions, suggesting the existence of a complex signaling network that allows the plant to recognize and protect itself against pathogens and environmental stress. A variety of signal transducers exist in plants, for better understanding let us discuss some of them like Ca^{2+}, reactive oxygen species (ROS), abscisic acid (ABA), salicylic acid (SA), jasmonic acid (JA) and protein kinases involved in phosphorylation cascades. The main components of stress-induced signalling molecules are the cross-talk between the different signalling pathways allows plants to adjust their responses depending on the combination of stimuli. A deep understanding of the mechanisms that respond to stress and sustain stress resistance is required. Here presented an overview of several mechanisms that interact in the stress regulation in plant.

Keywords: *Plant stress, Reactive oxygen species, Signal transduction.*

**Corresponding author*: E-mail: ramesh_bio2006@rediffmail.com

Introduction

Throughout evolution, plants have developed various defence mechanisms manifested through altered physiology to endure environmental abiotic and biotic stress. Abiotic stresses include drought, excess water, salinity, heat, cold, wounding and exposure to UV radiation. Furthermore, plants are also encountered by biotic stresses through microbial pathogens such as fungi, nematodes and bacteria. In their natural environment, plants have to adapt for numerous environmental stresses at the same time and different stresses can occur at different stages of the plant's life cycle. To survive, effective stress regulation and signal transduction mechanisms are needed to initiate a rapid response and are key components to achieve stress tolerance in plants. The adaptation to various stresses has led to the development of common stress transduction pathways and increased synthesis of secondary metabolites, Ca^{2+} fluxes, an oxidative burst and an overlapping set of stress response genes (Cheong *et al.*, 2002, Roberts *et al.*, 2002).

In most stress situations, signals are recognized by receptors, followed by second messenger generation *e.g.* reactive oxygen species (ROS) (Xiong *et al.*, 2002). Secondary messengers can modulate intracellular calcium levels, often initiating a protein phosphorylation cascade, which leads to the activation of proteins directly involved in cellular protection (Figure 6.1).

Tolerance to one environmental stress often affects the tolerance to another stress or condition. This phenomenon is also known as cross-tolerance. The signalling steps that link perception of stresses to the regulation of gene expression involved in biotic defence or abiotic stress adaptation have been studied extensively, yet the overall understanding of different interacting pathways remains poor. In this chapter, we will focus on the process of stress regulation in some of the abiotic and biotic conditions with some examples.

Regulation under Drought Stress

Drought stress will affect vital metabolic functions and maintenance of turgor pressure which control the cell expansion and cell wall formation. In order to minimise water loss, plants respond to lower water availability with the closure of stomata. However, this protective measure will decrease the CO_2 supply within the plant leaves and finally affect photosynthesis. Within the chloroplasts, light is absorbed by the antenna pigments and funnelled to the reaction centres of photosynthesis. When CO_2 limitation causes the decrease of photosynthetic fixation, the absorbed light needs to be diverted to other processes, giving rise to additional plant stress. Even under non-stressed conditions, part of the absorbed energy is used for the reduction of oxygen, creating ROS. ROS including singlet oxygen ($1O_2^-$), superoxide anions (O_2^-), hydrogen peroxide (H_2O_2) and hydroxyl radicals (OH) are highly reactive and damage cells by ROS mediated oxidative processes such as membrane lipid peroxidation, protein oxidation, enzyme inhibition and damage of nucleic acids (Grene, 2002). Consequently, protective mechanisms against this oxidative stress have evolved in all aerobic organisms. Major ROS scavenging enzymes in plants include superoxide peroxidase, ascorbate peroxidase, catalase, glutathione peroxidase and peroxiredoxin. In cooperation with the antioxidants

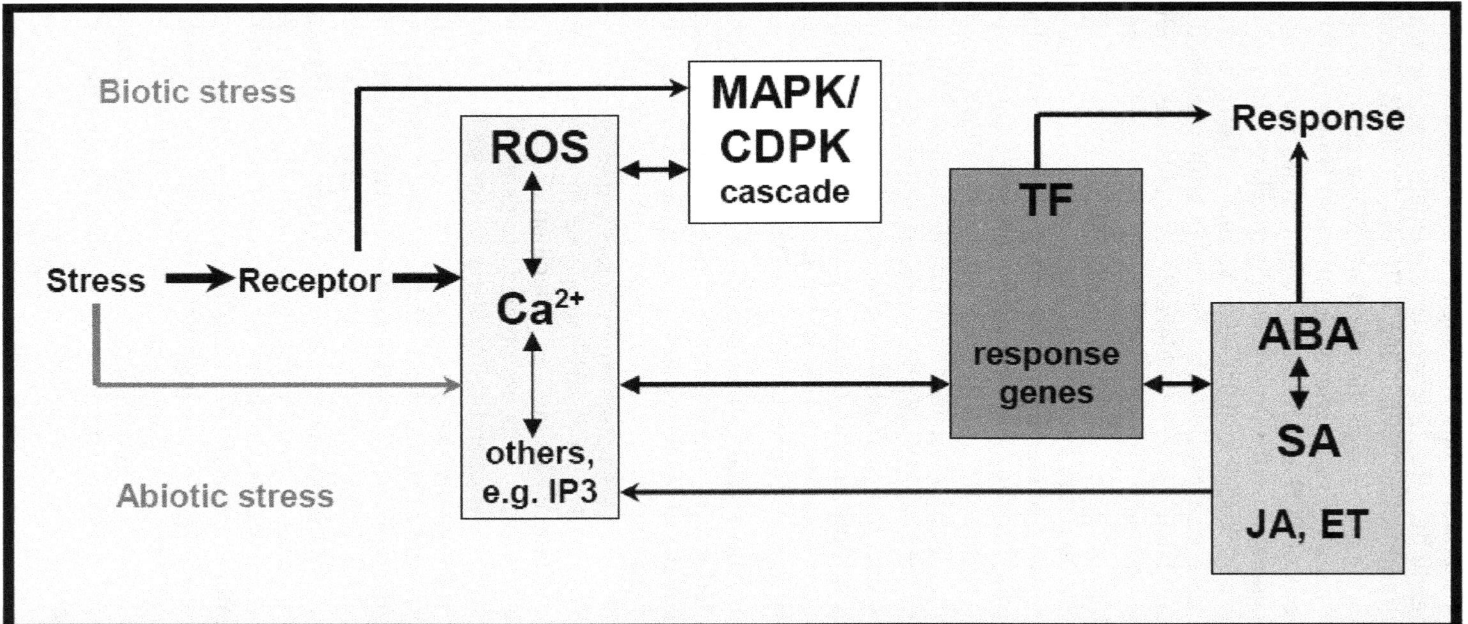

Figure 6.1: Signalling Network in Plants under Biotic and Abiotic Stress. Plants perceive biotic and abiotic stresses via specific receptors and then production of messengers takes place. Production of these messengers can be receptor mediated, however, in many abiotic stresses (*e.g.*, drought, cold, salt stress) membrane over excitation can directly lead to ROS generation. Receptors, ROS or Ca^{2+} dependent signals initiate specific phosphorylation cascades (*e.g.*, mitogen activated protein kinases - MAPK, calcium dependent protein kinases - CDPK) and finally interact with the promotor regions of transcription factors (TF) and response genes. The plant hormones can also influence the ROS and Ca^{2+} levels and initiate a second round of signalling (modified from: Tippmann *et al.,* 2006).

like ascorbic acid and glutathione, these scavenging enzymes detoxify the ROS and prevent serious damage to the cells. Under environmental stress, especially when CO_2 fixation in photosynthesis is limited, the well-tuned balance between ROS producing and scavenging processes is disturbed and ROS accumulate, leading to the above mentioned oxidative stress effects (Scheller and Haldrup, 2005). So the enhanced concentrations of ROS are therefore a common feature of most abiotic stress situations and acclimation to the stress generally correlates with strengthening of the anti-oxidative system. Studies in maize for instance, could show that enhanced activity of anti-oxidative enzymes clearly correlates with increased drought tolerance of the plants (Malan *et al.*, 1990). Nevertheless, enhanced production of ROS during stress poses not only a threat to cells, but it can also act as a messenger for the initiation of stress response. Up to now, only few components of the ROS induced signal transduction pathway of plants have been identified. These include histidine kinase sensors, several members of the mitogen-activated protein kinase (MAPK) cascade, the Ca^{2+} binding protein calmodulin and transcription factors (Apel and Hirt, 2004) indicating overlapping with other signalling pathways (Figure 6.1). Another important messenger during drought stress adaptation is the phytohormone abscisic acid (ABA). ABA biosynthesis is initiated by decreasing water potential, and seems to be essential for the activation of many protective measures towards abiotic stresses. Mutants deficient in ABA show slightly stunted growth in normal and un-stressed conditions, but greatly decreased tolerance to drought, wilting and dying under prolonged stress (Xiong *et al.*, 2001).

In drying soil, an ABA signal can be produced early during stress in the roots and be transported via the phloem to the shoot. There, ABA accumulation represents an important signal for the closure of stomata in response to drought stress which involves many different components such as ABA receptors, G-proteins, protein kinases and phosphatases, transcription factors and secondary messengers (Kuppusamy *et al.*, 2009). Polyamines also participate in ABA-mediated stress responses involved in stomatal closure. The generation of ROS is tightly linked to polyamine catabolic processes, since amino oxidases generate H_2O_2, which is a ROS associated with plant defence and abiotic stress responses (Cona *et al.*, 2006). The occurrence of a still unknown enzyme responsible for direct conversion of polyamines to NO thus cannot be ruled out. In many cases, polyamines appear to regulate stomatal closure by activating the biosynthesis of signalling molecules (H_2O_2 and NO) through different routes (Yamasaki and Cohen, 2006). Altogether, the available data indicate those polyamines, ROS (H_2O_2) and NO act synergistically in promoting ABA responses in guard cells. Biosynthesis of proline for instance increases the concentration of compatible osmoprotectants in the cells (Figure 6.2), while aquaporins (water channels) can facilitate water permeability of cellular membranes and maximise water uptake potential of the plant, and ROS scavenging proteins can limit damage by secondary oxidative stress (Chaves *et al.*, 2003). The specific function of the majority of drought induced genes is, however, still unknown and revelation of their function should give new insights into the plant protective mechanisms against drought.

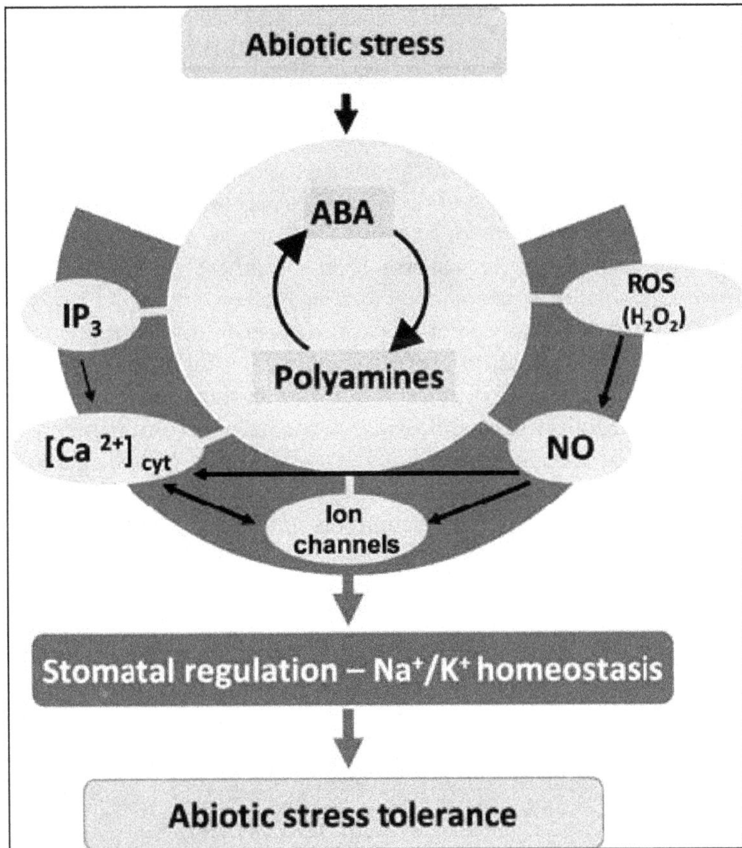

Figure 6.2: Simplified Model for the Integration of Polyamines with ABA, ROS (H_2O_2), NO, Ca^{2+} Homeostasis and Ion Channel Signalling in the Abiotic Stress Response (Modified from: Alcazar *et al.*, 2010).

Regulation under Salt Stress

High concentrations of salt (*i.e.*, ions, mostly Na^+) in the soil solution can impair water and nutrient uptake, reduce growth and photosynthetic activity. Furthermore high salinity can lead to an unfavourable Ca^{2+} or K^+ to Na^+ ratio, toxic intracellular Na^+ concentrations and peroxidation of membrane lipids (Levitt, 1980). The reduced ability of the plant to take up water induces water deficit effects comparable to drought stress. It is therefore not surprising to find similarities in the signalling and response of drought and salt stressed plants. This includes ABA biosynthesis and accumulation, which regulate measures against water loss such as closure of stomata and increased production of compatible osmoprotectants and antioxidants. Limitations in photosynthesis and changes in redox status of the mitochondrion promote increased production of ROS and downstream signalling pathways. More than half of the known drought-responsive genes show also induced gene expression by salt stress. This includes ABA responsive pathways, which are active under both types of stress indicating significant signalling cross-talk (Shinozaki *et al.*, 2003). Tolerancc to thc resulting secondary oxidative stress of salt stress is apparently highly important for stress tolerance: thus, a number of salt tolerant species increase the activity of antioxidant enzymes and accumulate antioxidants in response to salt stress, while salt-sensitive species fail to do so. Additionally to the osmotic stress imposed by high salinity, excessive amounts of salt can enter the plant and accumulate at high concentrations disturbing the nutritional homeostasis in the plant. In the cells, high ion concentration can seriously affect the activity of enzymes and the concentrations of Na^+ and Cl^- but also of salt stress induced Ca^{2+} must be kept under strict cellular control. Increasing Na^+ concentration can further inhibit K^+ uptake, causing K^+ deficiency as a secondary stress. K^+ has a key role in several physiological processes, such as osmotic regulation, protein synthesis, and enzyme activation and substitution of K^+ by Na^+ may lead to an ionic imbalance. Plants have therefore developed protective measures against the accumulation of toxic salt concentration in the cytoplasm, *e.g.*, the active compartmentation of Na^+ into the vacuole, selective import of K^+ or salt excretion by specific glands. Selective ion uptake can be observed in salt tolerant species or plant lines, which show differential activation of K^+ selective channels, enabling these plants to maintain high K^+ concentrations under salt stress (Peng *et al.*, 2004).

Regulation under Temperature Stress

Every plant species performs optimally only in a characteristic temperature range, which depends on the thermo-sensitivity of the plant species. Extremely high or low temperatures affect vital cell functions such as enzyme activity, cell division and membrane integrity. We will consider temperature stresses as follows: heat stress and cold stress including freezing stress.

Heat Stress

Effects of high temperature stress can range from moderate effects such as oxidative stress and enhanced transpiration to fatal consequences for the plant, leading to tissue collapse and plant death. To cope with the high temperature stress, plants usually react with enhanced transpiration rates to achieve evaporative cooling. Since high temperatures are often accompanied with limited water availability, the water potential in the plant cell can decrease, leading to drought conditions and initiate responses similar to drought stress. Under these conditions, different events take place: small heat shock proteins (HSPs), acting as molecular chaperones, are synthesized at high abundance (Vierling, 1997), Ca^{2+} influx is enhanced and ROS accumulate rapidly in different parts of the cell (Foyer *et al.*, 1997). Although the function of many HSPs during heat stress is still unknown, they have been found to accumulate in all prokaryotic and eukaryotic organisms under high temperature stress, and are thought to contribute to the stabilisation of cellular structures and prevent the thermal aggregation of proteins (Lee *et al.*, 1997). Furthermore, the accumulation of HSPs is also seen in oxidative stress and consequently heat tolerance can be achieved by inducing oxidative bursts in plants (Dat *et al.*, 1998). Thus based on the results obtained in different plant species it can be conclude that, ABA, SA, ethylene and Ca^{2+} are involved in heat stress signalling and can also confer protection against heat stress damage (Tuteja, 2007; Mirza *et al.*, 2013).

Cold and Freezing Stress

Chilling stress is a sub-optimal temperature, where the plant faces reduced enzyme activity and maybe water availability, yet the temperature is above the freezing point of water. Cold stress affects mainly metabolic processes, impairing enzyme reactions, substrate diffusion rates and membrane transport properties. There by some reactions are more affected by the cold then others. In particular, the dark reaction of photosynthesis and oxidative phosphorylation seem to be sensitive to chilling. One early response to cold and osmotic stress is the rapid Ca^{2+} influx into the cell. Physical alterations in the cellular structure may cause this Ca^{2+} influx by activation of Ca^{2+} channels and initiate downstream Ca^{2+} dependent signalling pathways (Xiong *et al.*, 2002). When temperature drops below zero degrees, plants can experience freezing stress. Intracellular freezing of water can damage the protoplast membrane structure, mechanically injure and finally kill the cells by the expanding ice crystals. During extracellular freezing, the protoplasm of the plant becomes severely dehydrated when water is transferred to the ice crystals in the intercellular spaces. Freezing acclimation can be achieved by the expression of a class of cold-induced "cold-regulated genes" (COR), which encode hydrophilic polypeptides and stabilize membranes against freeze-induced injury (Artus *et al.*, 1996). The exact role in cold acclimation of these COR genes remain unknown, and several of them are also induced by drought and abscisic acid, as well as by low temperature stress.

Pollutants

Many natural occurring substances, termed xenobiotics, are toxic for plants at high concentrations. Human activity contributed substantially to the accumulation of heavy metals in soils and ground water. Most plants do not possess specific mechanisms to prevent the excessive uptake of heavy metals from the soil. Accumulation of heavy metal ions in the plant can impair membrane integrity, affect enzyme activity and hinder nutrient uptake. Metal ions can generate ROS by auto-oxidation, and disturb the redox status of cells. Metals without redox capacity such as cadmium, mercury and lead can disturb the antioxidative glutathione pool, activate Ca^{2+} dependent systems and iron mediated processes (Pinto *et al.*, 2003). Apart from soil pollution human activity also increased concentration of air pollutants such as O_3, SO_2, NO, NO_2, NH_3, HNO_3 and HF. Through stomata they can enter the leaves and affect the plant metabolism. Entry of O_3 into the cell forces the creation of ROS and induces oxidative stress. It is assumed that the O_3 derived ROS production mimics the signalling pathways following oxidative burst during avirulent pathogen attack (Rao and Davies, 2001). Similar to the pathogen response, O_3 stress also activates the production of SA, JA and ethylene.

UV-B Radiation

UV-B radiation represents a small, but important part of terrestrial solar irradiance. The potentially harmful effects of UV-B radiation on plant life include damage to DNA, the photosynthetic apparatus and membrane lipids. However, during evolution, plants have developed powerful protection and repair mechanisms. Increased UV-B radiation initiates morphological adaptation to highlight environments such as stunted growth, leaf thickening, trichome development, flavanoid biosynthesis and decreased stomatal densities (Gitz and Liu-Gitz, 2003). These photomorphological responses to UV-B are also advantageous for plant exposure to drought stress. UV-B exposure triggers ROS generation from multiple sources (*e.g.* NADPH oxidase, peroxidase) and the source of ROS origin is apparently important for the specific changes in gene regulation. In addition to O_2^- and H_2O_2, nitric oxide (NO) plays a role in UV-B stress response. ROS and NO have also been identified as messengers during pathogen attacks and this might explain the enhanced resistance to pathogens of UV-B treated plants. Furthermore, SA and JA, which are important for gene regulation under pathogen attack, are also up-regulated under UV-B exposure. In particular, UV-radiation has a positive effect on PR protein biosynthesis, similar to oxidative stress. These defence-enhancing effects of UV-B exposure support the idea of cross-talk between the signalling pathways (Mackerness, 2000).

Pathogen Stress

Apart from environmental stresses, plants are permanently exposed to potential enemies including microbial pathogens (viruses, bacteria, and fungi), nematodes and insects. Most microbial pathogens enter the host cell, while others like mildew or rust fungi remain on the outside and grow external structures for nutrient-absorption inside the host. During infection, the integrity of the outer cell wall might be damaged by mechanical wounding when the pathogens enter the cell. Once the infection is established, pathogens can manipulate the host to release water and nutrients necessary for their proliferation, leading to nutrient and water deficiency in the host cells. Infection with pathogens also has a significant impact on the composition of soluble carbohydrates in the infected parts of the plant. Pathogens, especially biotrophs can manipulate the host metabolism in their favour, causing the host plant to accumulate high concentrations of carbohydrates that interfere with the normal metabolism and can cause intracellular osmotic stress (Abood and Losel, 2003). The activation of pathogen stress responses itself poses a stress to the plant, and both disease and defence mechanism might be the cause for reduced yield (Heil *et al.*, 2000).

Signalling Components

The main components of stress-induced signalling pathways (Ca^{2+}, ROS, SA, ABA) and signal cascades seem to have equal importance in responses to environmental changes and pathogen challenge. The cross-talk between the different signalling pathways allows plants to adjust their responses depending on the combination of stimuli. A variety of signal transducers exist in plants, for better understanding let us discuss some of them like ROS, Ca^{2+}, ABA, SA and protein kinases involved in phosphorylation cascades.

Reactive Oxygen Species (ROS)

Two hallmarks of biotic and abiotic stress are changes of Ca^{2+} concentration in the cell and of the intracellular redox state. The production of excess ROS is a common phenomenon in different stress situations. ROS are partially reduced forms of atmospheric oxygen and include H_2O_2, OH and O_2^- (Bowles, 1995; Mittler, 2002). ROS production is induced by a number of stresses, which include hyperosmotic stress, drought, cold, ozone and pathogen attack (Pastori and Foyer, 2002). Plasma membrane localized NADPH oxidases and cell wall peroxidases provide a system, which produces H_2O_2 in response to pathogen and abiotic stress. ROS are capable of unrestricted oxidation in the cell, causing lipid peroxidation and damage to the Photosystem II reaction centre (Smirnoff, 1995). Consequently, in the non-stressed plant, a state of redox homeostasis is kept in a fine tuned balance between ROS production and scavenging by a range of antioxidant systems. The main enzymes of plants include superoxide dismutase (SOD), ascorbate peroxidase (APX), non-specific peroxidase (POX), glutathione reductase (GR) and catalase (CAT). SOD reacts with O_2^- to produce H_2O_2, which is scavenged by POXs and CATs. Among the different peroxidases APX, which uses ascorbate as an electron donor, has an important role in H_2O_2 detoxification and is well described in plants. GR is involved in the reduction of oxidized glutathione to continue H_2O_2 scavenging. CAT has been mainly found in leaves (peroxisomes) to remove H_2O_2 (formed by photorespiration and β-oxidation of fatty acids). The activity of these enzymes is highly differentiated in separate plant organs and under different stresses. In tomato, SOD, CAT and APX are involved in conferring tolerance to salinity mediated oxidative stress (Mittova *et al.*, 2004). In pathogen defence mechanisms, the oxidative burst is one of the earliest responses to pathogen recognition. Plants generate ROS to mediate cell death as part of the host resistance mechanisms (Zhou *et al.*, 2000). Recognition of an invading pathogen results in the activation of plasma membrane-associated NAD(P)H oxidases and peroxidases, which leads to enhanced production of O_2^-. Dismutation of O_2^- by SOD generates H_2O_2, which can diffuse into cells and activate defence mechanisms, ultimately leading to programmed cell death (PCD) (Jabs *et al.*, 1997). At the site of pathogen attack, ROS act as antimicrobial toxic compounds against invading pathogens, reinforce the cell wall by cross-linking proteins and can induce PCD in plant cells. The accumulation of H_2O_2 is thought to provide the signal leading to changes in gene expression and the activation of a localized PCD, the hypersensitive response, or HR (Foyer *et al.*, 1997). Since ROS is also directly active against the pathogen, high concentrations at the point of attempted invasion are needed. To achieve higher local concentrations, the plant suppresses its own detoxification machinery. In tobacco, for instance, tobacco mosaic virus infection leads to suppression of the activity of ROS detoxifying enzymes like cytosolic APX and CAT (Mittler *et al.*, 1998).

There is enough evidence that the ability of a plant to scavenge ROS, which accumulate under different stresses including salinity, irradiation, high ozone or pathogen challenge, can actually determine whether a plant is tolerant to stress. This can eventually result in higher tolerance to other non-related stresses. Hence, mechanisms to control the resulting oxidative stress appear to be one major common component connecting various non-related stresses. On the other hand, since ROS is a signal component, defects in the ROS production pathways impair plant defences significantly. ROS itself can constitute a stress component and higher tolerance to oxidative stress can increase plant stress tolerance. Thus, a picture unfolds where the nature, timing and amount of ROS is important for the effect on the plant: low levels of ROS are involved in stress signalling inducing protective genes, while high concentrations usually lead to the induction of PCD.

Calcium Signalling

Calcium (Ca^{2+}) is a secondary messenger for signalling cross-talk, which is elicited by different stress conditions. Alterations in the concentration mark early cellular responses for a number of processes, for instance stomatal closure or stress adaptation (Hetherington, 2001). Differences in the intracellular Ca^{2+} level as a stress response can be caused by Ca^{2+} influx or release from internal storages and have been observed for a range of stresses. In addition, several stresses induce the biosynthesis of second messengers that require Ca^{2+} for signal transduction, *e.g.*, ROS, ABA and IP3. To describe all signalling components activated by Ca^{2+} is beyond the scope of this chapter. Nevertheless, we will present examples to highlight the importance of Ca^{2+} and the converging nature of different stresses.

Elevation of Ca^{2+} in the cytosol marks early responses of the plant to abiotic stresses such as cold, mechanical, ozone and salt stress (Rao and Davies, 1999) but also to pathogens or pathogen elicitors (Grant, 1999). Rapid increases in cytosolic Ca^{2+} are observed in response to touch or cold shock, involving Ca^{2+} fluxes from internal vacuolar stores and from the plasma membrane (Knight *et al.*, 1996). Transient increases in Ca^{2+} of varying duration are initiated by oxidative stress, pathogen elicitors and hyperosmotic treatments, often in conjunction with an oxidative burst (Rentel and Knight, 2004). During drought and other water limiting stresses, oscillations of Ca^{2+} are important for the regulation of guard cell-mediated stomatal pore mediated by ion channels (Sanders *et al.*, 1999). Such channels are activated by membrane depolarization, reported in response to different stresses such as high light intensity and fungal elicitors. Many different proteins have a role in keeping Ca^{2+} concentrations in the acceptable range, usually by calcium binding. Hence, Ca^{2+} binding proteins (like calmodulin or calcineurin) help to regulate the spatial and temporal release of Ca^{2+} (Sanders *et al.*, 1999) and thereby have multiple roles in stress response: studies on transgenic *Arabidopsis* plants have shown that calcineurin B-like protein functions as a differential regulator of salt drought and cold responses in plants, leading to enhanced tolerance to salt and drought but reduced tolerance to freezing (Cheong *et al.*, 2003). At the cellular level, calmodulins reduce the effective diffusion in the cytosol and restrict calcium to specific locations thereby regulating the activity (Sanders *et al.*, 1999). Consequently, calmodulin gene expression is induced by a variety of stresses like salinity, osmotic stress and wounding in different plants (Phean-O-Pas *et al.*, 2005). Given the redundancy of Ca^{2+} signals in various stresses, how does the cell distinguish between the Ca^{2+}.

Abscisic Acid (ABA)

ABA is involved in many cellular processes like germination, gravitropism and guard cell mediated stomatal opening (Levitt, 1980). Particularly well established is the regulatory function of ABA in water balance and osmotic stress tolerance (Zhu, 2002). Drought and salinity activate *de novo* ABA synthesis to prevent further water loss by evaporation through stomata (Levitt, 1980), mediated by changes in the guard cell turgor pressure. The mechanism involves fast changes in intracellular

Ca^{2+} concentration and stimulates further signalling in the cell. Under osmotic stress, ABA induces the accumulation of proteins involved in the biosynthesis of osmolytes (*e.g.*, proline, trehalose), which increases the stress tolerance of plants (Nayyar *et al.*, 2005). Large-scale gene expression studies in *Arabidopsis* revealed overlapping effects between osmotic, salt stress and ABA application (Takahashi *et al.*, 2000). In guard cells, ABA can activate gene expression of cold regulated genes (Leonhardt *et al.*, 2004). ABA or abiotic stresses, which lead to the accumulation of ABA can significantly impair biotic defence mechanisms. Kim and Triplett (2004) described the over-expression of a single transcription factor, activating ABA responsive genes, which enhances tolerance to various abiotic stresses. A single gene conferring multi-tolerance to different stresses would be highly desirable for breeders to develop plants for dry areas. However, since ABA can negatively affect pathogen responses, it needs to be unravelled whether a highly ABA accumulating plant has fully functional biotic defences. Taken together, one can conclude that in the light of contradicting effects on biotic stress responses, the role of ABA in pathogen defence remains quite complex and is not fully understood. Nevertheless, the role of ABA has to be extended from a major abiotic stress hormone to a multi-stress responsive hormone.

Salicylic Acid (SA)

SA is a phenolic compound, which was identified as a major signal for accumulation of PR proteins in virus infected tobacco (Yang and Klessig, 1996). The infection with avirulent pathogens and the activation of HR often leads to subsequent system wide resistance, SAR (Klessig and Malamy, 1994), which is characterized by a long-lasting defence response against a broad spectrum of pathogens. SA plays a major role in defence signalling and is essential for different defence mechanisms, such as SAR and hypersensitive responses (HRs).

Stress acclimation has been observed for SA mediated stress responses. Sublethal exposure to the stress (*e.g.*, oxidation) can render a plant resistant to a subsequent infection, which is fatal for the non-stressed control plants. One effect of SA is the induction of antioxidant production, which can enhance the tolerance to oxidative stress. This has also effects on temperature stress. Enhanced levels of SA can also be achieved by exogenous application and can affect tolerance to higher temperatures: SA-treated plants show a higher tolerance to heat stress (Larkindale and Knight, 2002). In insensitive plants, SA is not an effective second messenger for induced resistance and activation of defence responses, but rather an important player in the antioxidant response of the plant to oxidative damage caused by biotic or a biotic stress. On the other hand, in SA sensitive plants, it is a direct activator of induced defence responses, but may have a second function as an antioxidant to limit oxidative stress at the infection site (Yang *et al.*, 2004).

Other Signal Components: Protein/MAP Kinases are Key Signal Transmitters

A highly conserved signal mechanism between all eukaryotes is the phosphorylation of proteins by protein kinases. Ca^{2+} dependent protein kinases

(CDPK) and a class of serine/threonine kinases, mitogen activated kinases (MAPK). Specific MAPK cascades are activated by a range of stimuli including cold, drought, salinity, ROS, heat, shaking, wounding pathogens and pathogen elicitors, ethylene, ABA and SA (Tena *et al.*, 2001). MAPKs are central for the transduction of cellular signals by activation and repression of downstream target protein activity (Figure 6.1). Since protein kinases are involved in various stress responses, they are potential integrators of cross-talk between signalling networks, responding to different stresses (Zhang and Liu 2001). On the other hand, MAP kinases can be highly specific and distinct MAPKs are only activated by distinct stimuli. The MAPK activation by fungal and bacterial elicitors and SA is well-documented and MAPKs are involved in regulating a range of defence pathways. Thus, defects can have a significant impact on plant defences and mutations in MAPKs can lead to de-repression of defence mechanisms such as enhanced SA accumulation, constitutive *PR* gene expression and growth inhibition (Xiong and Yang, 2003). In summary, MAP kinase cascades constitute a major mechanism for activation of biotic defence responses and in mediating other signals, such as hormones in response to abiotic stresses like cold and drought, and osmotic stress.

Signalling Networks are Interconnected

Stress sensing may be categorized in separate linear signal chains, but does in fact consist of a network of parallel and interacting signalling chains, which often share common components or have signalling convergence points (Figure 6.1). As a consequence, signalling compounds, which are classical abiotic or biotic stress signals, often induce non-related stress responses. Besides a number of hormones and minor signalling compounds, four or five core elements constitute the core defence machinery of the plant (ROS, Ca^{2+}, ABA and SA) and protein kinases.

Achieving Cross-Tolerance

Cross-Tolerance Emerges for Two Situations

1. An inherent tolerance of a plant species to a stress condition can enable this plant to withstand other stresses, which share an overlapping stress potential. For instance, an effective antioxidative machinery, which allows the plant to cope efficiently with oxidative stress, would lead to enhanced tolerance to other stresses that involve oxidative stress. Additionally, in many species, genes have been identified, which respond to pathogen infection, pathogen defence signal components or elicitors and non-biotic stresses, such as salt or oxidative stress (Hong and Hwang, 2005). A typical example concerns PR proteins, which can confer abiotic stress tolerance, beyond the well known role in pathogen defence. Based on the results from transgenic *Arabidopsis* plants, a scavenging activity of PR proteins or a PR protein-mediated activation of "classical" scavengers such as SOD or CAT has been suggested, thus conferring additional abiotic (oxidative) stress tolerance besides the enhanced pathogen resistance (Figure 6.3). Given the structural similarity (high number of polar residues) to proteins of the dehydrin class, some PR proteins might even act as a late embryogenesis activated/dehydrin proteins to protect other proteins (Pnueli *et al.*, 2002).

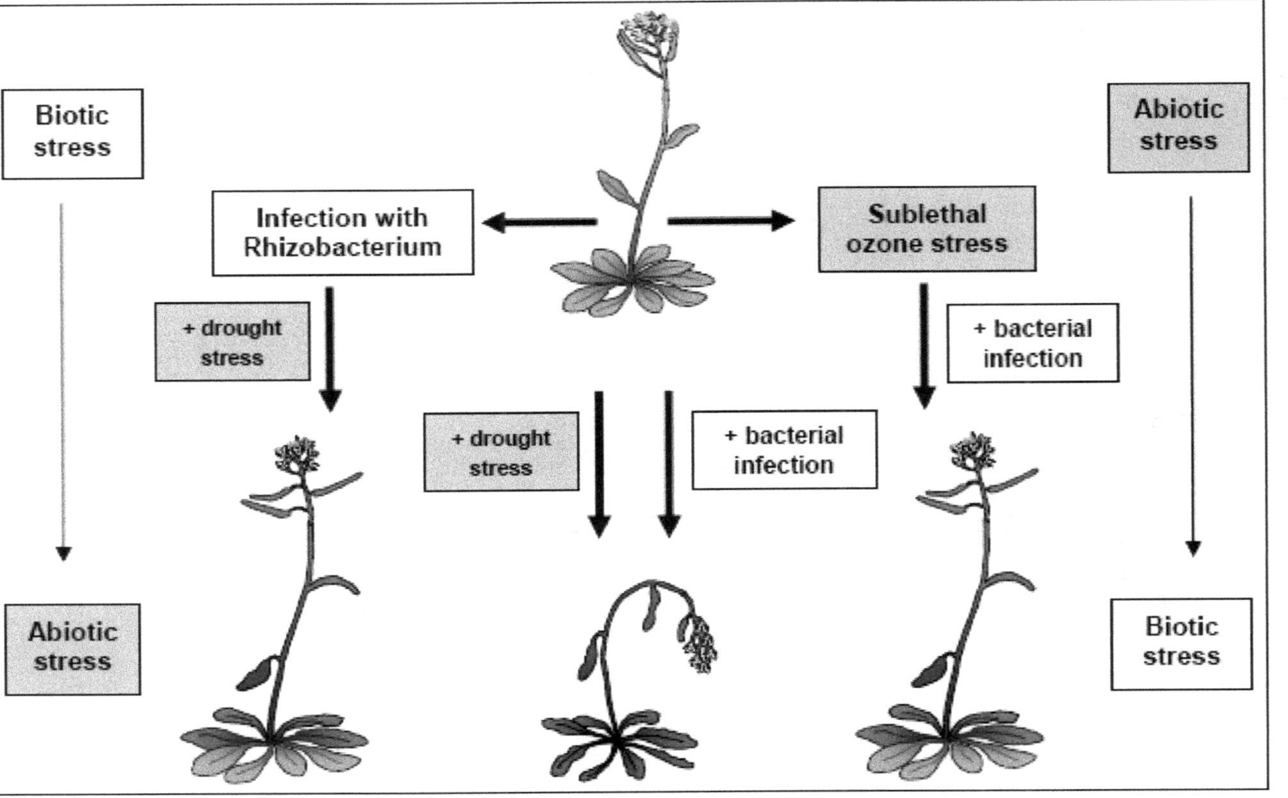

Figure 6.3: Phenomenon of Cross-Tolerance between Biotic and Abiotic Stress in Arabidopsis.

Left: Infection of Arabidopsis with rhizobacteria (*Paenobacillus polymyxa*) can cause mild biotic stress within the plant, but also result in improved tolerance towards subsequent drought stress. Without the pre-treatment, plants are severely damaged by the drought stress (middle). Right: Pre treatment of Arabidopsis with sublethal doses of ozone stress can induce tolerance towards following biotic stress by bacterial infection with *Pseudomonas syringae*. Plants without the pre-treatment will die (middle). (Modified from: Timmusk and Wagner, 1999).

2. The other situation is stress acclimation: usually the exposure of plants to sublethal doses of a stress is referred to as acclimation. A typical example is acquired cold/freezing tolerance of plants following low temperature exposure (Thomashow, 1999). Beyond this well studied phenomenon improving tolerance to the same or related stresses, this can also improve tolerance to non-related stresses (Figure 6.2). Exposure to abiotic stresses can affect the responses of plants to pathogens (Table 6.1), although the effects of the stress and subsequent responses cannot be generalized.

Synergistic and antagonistic effects of biotic and abiotic stresses have been observed and the final outcome depends on the stress signal and the sensitivity of the specific species in question. A general, ecological conclusion is that stresses will often cause plants which are normally resistant to become susceptible to other stresses following stress application (*e.g.*, drought or salinity induced susceptibility to fungal pathogens). This induced susceptibility might even occur after the stress, as observed in barley, where relief from drought stress can lead to the breakdown of powdery mildew resistance conferred by the *mlo* mutation. On the other hand, non-lethal abiotic stresses are known to induce the accumulation of defence transcripts, anti-microbial proteins and compounds, leading to enhanced disease resistance. Even brief periods or non-lethal exposure to stress can induce genes involved in stress "management" and enhance the tolerance to subsequent biotic and abiotic stresses (*e.g.*, low oxidative stress inducing higher drought tolerance or pathogen resistance). The opposite situation, where an infection with beneficial or non-virulent pathogens induces more abiotic stress tolerance, is less well studied. A number of publications showed that rhizo-bacteria infection and colonization with arbuscular mycorrhizal fungi can actually enhance tolerance to saline conditions, mostly by enhanced nutrient uptake and improved water relations (Yano-Melo *et al.*, 2003). A brief overview of different interactions between abiotic and biotic stress tolerance is given in Table 6.1.

Signal Intensity Might Confer Specificity

The understanding of signalling interactions involved in different stresses is complicated by the fact that signalling compounds can act as stress signals and as stress themselves. The effects of these compounds is modulated by their absolute levels and a range of mechanisms is involved in keeping the intracellular concentrations in balance, as small changes result in drastically different responses. High levels of a single second messenger might be beneficial in one plant species, but can be harmful to another. The responses of plants are highly tuned to the absolute levels of these key compounds, distinguishing whether they induce defence gene activity or exhibit secondary stress (*i.e.* oxidation) leading to PCD. The intracellular concentration and localization of these signalling compounds can influence the effect of the stress and activate sensors responding to different stress intensities (Kacperska, 2004). The effects are clearly linked to the sensitivity of the plant species examined, and it might be appropriate to enhance the classification into SA sensitive and insensitive plants to other compounds and define ROS- or ABA sensitive and insensitive plants.

Table 6.1: Cross-Tolerance between Biotic and Abiotic Stresses

Plant Species	*Abiotic Stress*	*Biotic Stress*	*Possible Signal Compound*
	I. positive influence of abiotic stress adaptation towards biotic stress tolerance		
Tobacco	Ozone, UV	Tobacco Mosaic Virus	SA
Arabidopsis	Ozone	*Pseudomonas syringae* (bacterium)	SA, ROS
Bell pepper	Ionic stress, Oxidation	*Pseudomonas syringae* (bacterium)	SA, JA
Medicago	Drought	*Aphanomyces euteiches* (fungus)	ABA
Arabidopsis	Drought	*Peronospora parasitica*	SA
Barley	Salinity	*Botrytis graminis* (fungus)	ABA
Arabidopsis	Salinity, drought	*Botrytis cinerea* (fungus)	JA, ROS
Rye	Cold	*Microdochium nivale* (fungus)	ET, SA
	II. Positive influence of biotic stress adaptation towards abiotic stress tolerance		
Arabidopsis	drought	*Paenibacillus polymyxa* (Rhizobacteria)	SA, ABA?
Banana	salinity	*Glomus mosseae* (Mycorrhizal fungi)	
Tomato	salinity	*Glomus mosseae* (Mycorrhizal fungi)	
Arabidopsis	drought (-) salinity, heat	*Peronospora parasitica* (fungus)	SA
	III. Negative influence of abiotic stress on biotic stress		
Citrus	salinity	*Phytophthora* spp. (fungus), nematodes	?
Red pine	drought	*Sphaeropsis sapinea* (fungus)	?
Oak	drought	*Armillaria* spp. (fungus)	?

(*Source*: Tippmann *et al.*, 2006).

Conclusion

Plants continually encounter stress even under environmental conditions that we think of as normal. The environment changes during the day, day to day and throughout the year, thus plants must respond to stress over the course of each day and often must respond to several stresses at the same time. Study of stress responses show that there is much crosstalk among signaling networks during specific stress responses. Thus, plants may respond to stress perception by an initial global response and follow with specific stress responses. As we discussed in this chapter we can understand stress as a stimulus or influence that is outside the normal range of homeostatic control in a given organism: If a stress tolerance is exceeded, mechanisms are activated at molecular, biochemical, physiological and morphological levels; once stress is controlled, a new physiological state is established, and homeostasis is re-established. When the stress is retired the plant may return to the original state or a new physiological state. Specific factors including various transcription factors and kinases are molecular player, common to multiple networks or involved in crosstalk between stress signaling pathways regulated by abscisic acid, salicylic acid, jasmonic acid and ethylene as well as ROS signaling. Taken together, we have evidence that an increasing number of signalling compounds and components exhibit vivid signal cross-talk and show an increasingly complex network of interacting signal pathways.

References

Abood, J.K. and Losel, M., 2003. Changes in carbohydrate composition of cucumber leaves during the development of powdery mildew infection. *Plant Pathology,* 52: 256-265.

Alcázar, R., Altabella, T., Marco, F., Bortolotti, C., Reymond, M., Koncz, C., Carrasco, P. and Tiburcio, A.F., 2010. Polyamines: molecules with regulatory functions in plant abiotic stress tolerance. *Planta,* 231(6):1237-49.

Apel, K. and Hirt, H., 2004. Reactive oxygen species: metabolism, oxidative stress and signal transduction. *Annu. Rev. Plant Biol.,* 55: 373-399.

Artus, N., Uemura, M., Steponkus, P., Gilmour, S., Lin, C. and Thomashow, M., 1996. Constitutive expression of the cold regulated Arabidopsis thaliana cor15a gene affects both chloroplast and protoplast freezing tolerance. *Proc. Natl. Acad. Sci. USA.,* 93: 13404-13409.

Bowles, D.J., 1995. Signal transduction in plants. *Trends Cell Biol.,* 10: 404-8.

Chaves, M.M., Maroco, J.O. and Pereira, J.S., 2003. Understanding plant responses to drought – from genes to the whole plant. *Funct. Plant Biol.,* 30: 239-264.

Cheong, Y.H., Chang, H.S., Gupta, R., Wang, X., Zhu, T. and Luan S., 2002. Transcriptional profiling reveals novel interactions between wounding, pathogen, abiotic stress, and hormonal responses in Arabidopsis. *Plant Physiol.,* 129: 661-677.

Cheong, Y.H., Kim, K.N., Pandey, G.K., Gupta, R., Grant, J.J. and Luan, S., 2003. Cbl1, a calcium sensor that differentially regulates salt, drought, and cold responses in *Arabidopsis. Plant Cell,* 15: 1833-45.

Cona, A., Rea, G., Angelini, R., Federico, R. and Tavladoraki, P., 2006. Functions of amine oxidases in plant development and defence. *Trends Plant Sci.,* 11: 80–88.

Dat, J., Foyer, C. and Scott, I., 1998. Changes in salicylic acid and antioxidants during induced thermotolerance in mustard seedlings. *Plant Physiol.,* 118: 1455-1461.

Foyer, C., Lopez-Delgado, H., Dat, J. and Scott, I., 1997. Hydrogen peroxide- and glutathione-associated mechanisms of acclimatory stress tolerance and signalling. *Physiol. Plant.,* 100: 241-254.

Gitz, D.C. and Liu-Gitz, 2003. How do UV photomorphogenic responses confer water stress tolerance? *J. Photochem. Photobiol.,* 78: 529-534.

Grant, M.M.J., 1999. Early events in host-pathogen interactions. *Curr. Opin. Plant Biol.,* 2: 312-319.

Grene, R., 2002. Oxidative stress and acclimation mechanisms in plants. In: Somerville CR, Meyerowitz EM (eds) The Arabidopsis Book, The American Society of Plant Biologists, Rockville, USA, pp 1-20.

Heil, M., Hilpert, A., Kaiser, W. and Linsenmair, K.E., 2000. Reduced growth and seed set following chemical induction of pathogen defence: does systemic acquired resistance (SAR) incur allocation costs? *J. Ecol.*, 88: 645-654.

Hetherington, A., 2001. Guard cell signalling. *Cell*, 107: 711-714.

Hong, J.K. and Hwang, B.K., 2005. Induction of enhanced disease resistance and oxidative stress tolerance by overexpression of pepper basic PR-1 gene in Arabidopsis. *Physiol. Plantarum*, 124: 267-277.

Jabs, T., Tschope, M., Colling, C., Hahlbrock, K. and Scheel, D., 1997. Elicitor-stimulated ion fluxes and O_2^- from the oxidative burst are essential components in triggering defense gene activation and phytoalexin synthesis in parsley. *Proc. Natl. Acad. Sci. USA.*, 94: 4800-4805.

Kacperska, A., 2004. Sensor types in signal transduction pathways in plant cells responding to abiotic stressors: do they depend on stress intensity? *Physiol. Plantarum*, 122: 159-168.

Kim, H.J. and Triplett, B.A., 2004. Cotton fiber germin-like protein in molecular cloning and gene expression. *Planta*, 218: 516-524.

Klessig, D.F. and Malamy, J., 1994. The salicylic acid signal in plants. *Plant Mol. Biol.*, 26: 1439-1458.

Knight, H., Trewavas, A.J. and Knight, M.R., 1996. Cold calcium signaling in Arabidopsis involves two cellular pools and a change in calcium signature after acclimation. *Plant Cell*, 8: 489-503.

Kuppusamy, K., Walcher, C. and Nemhauser, J., 2009. Cross-regulatory mechanisms in hormone signaling. *Plant Mol. Biol.*, 69: 375–381.

Larkindale, J. and Knight, M., 2002. Protection against heat stress-induced oxidative damage in Arabidopsis involves calcium, abscisic acid, ethylene, and salicylic acid. *Plant Physiol.*, 128: 682-695.

Lee, G.J., Roseman, A.M., Saibil, H.R. and Vierling, E., 1997. A small heat shock protein stably binds heat-denatured model substrates and can maintain a substrate in a folding-competent state. *EMBO Journal*. 16: 659-671.

Leonhardt, N., Kwak, J.M., Robert, N., Waner, D., Leonhard, G. and Schroeder, J.I., 2004. Microarray expression analyses of Arabidopsis guard cells and isolation of a recessive abscisic acid hypersensitive protein phosphatase 2C mutant. *Plant Cell*, 16: 596-615.

Levitt, J., 1980. Water, radiation, salt and other stresses. In: Kozlowski TT (ed) Responses of Plants to Environmental Stresses (Vol 2), Academic Press NY, pp 365-488.

Mackerness, S.A.H., 2000. Plant responses to ultraviolet-B (UV-B: 280-320nm) stress: What are the key regulators? *Plant Growth Regul.*, 32: 27-39.

Malan, C., Greyling, M. and Gressel, J., 1990 Correlation between cu zn superoxide dismutase and glutathione reductase and environmental and xenobiotic stress tolerance in maize inbreds. *Plant Sci.*, 69: 157-166.

Mirza, H., Kamrun, N., Alam, M., Rajib, R. and Masayuki, F., 2013. Physiological, Biochemical, and Molecular Mechanisms of Heat Stress Tolerance in Plants. *Int. J. Mol. Sci., 14*: 9643-9684; doi:10.3390/ijms14059643

Mittler, R. 2002. Oxidative stress, antioxidants and stress tolerance. *Trends Plant Sci* **7**: 405–410

Mittler, R., Feng, X. and Cohen, M., 1998. Post-transcriptional suppression of cytosolic ascorbate peroxidase expression during pathogen-induced programmed cell death in tobacco. *Plant Cell,* 10: 461-474.

Mittova, V., Guy, M., Tal, M. and Volokita, M., 2004 Salinity up-regulates the antioxidative system in root mitochondria and peroxisomes of the wild salt-tolerant tomato species Lycopersicon pennellii. *J. Exp. Bot.,* 55: 1105-1113.

Nayyar, H., Bains, T. and Kumar, S., 2005. Low temperature induced floral abortion in chickpea: relationship to abscisic acid and cryoprotectants in reproductive organs. *J. Exp. Bot.,* 1: 39-47.

Pastori, G.M. and Foyer, C.H., 2002. Common components, networks, and pathways of cross-tolerance to stress. The central role of "redox" and abscisic acid-mediated controls. *Plant Physiol.,* 129: 460-468.

Peng, Y.H., Zhu, Y.F., Mao, Y.Q., Wang, S.M., Su, W.A. and Tang, Z.C., 2004. Alkali grass resists salt stress through high [K^+] and an endodermis barrier to Na^+. *J. Exp. Bot.,* 55: 939-949.

Phean-O-Pas, S., Punteeranurak, P. and Buaboocha, T., 2005. Calcium signaling-mediated and differential induction of calmodulin gene expression by stress in Oryza sativa L. *J. Biochem. Mol. Biol.,* 4: 432-439.

Pinto, E., Sigand-Kutner, T.C.S., Leitão, M.A.S., Okamoto, O.K., Morse, D. and Colepicolo, P., 2003. Heavy metal induced oxidative stress in algae. *J. Phycol.,* 39: 1008-1018.

Pnueli, L., Hallak-Herr, E., Rozenberg, M., Cohen, M., Goloubinoff, P., Kaplan, A. and Mittler, R., 2002. Molecular and biochemical mechanisms associated with dormancy and drought tolerance in the desert legume Retama raetam. *Plant Journal,* 31: 319-330.

Rao, M.V. and Davis, K.R., 1999. Ozone-induced cell death occurs via two distinct mechanisms in Arabidopsis: the role of salicylic acid. *Plant Journal,* 17: 603-614.

Rao, M.V. and Davis, K.R., 2001. The physiology of ozone induced cell death. *Planta,* 213: 682-690.

Rentel, M.C. and Knight, M.R., 2004. Oxidative stress induced calcium signalling in *Arabidopsis*. *Plant Physiol.,* 135: 1471-1479.

Roberts, M.R., Salinas, J. and Collinge, D.B., 2002. 14-3-3 proteins and the response to abiotic and biotic stress. *Plant Mol. Biol.,* 50: 1031-1039.

Sanders, D., Brownlee, C. and Harper, J.F., 1999. Communicating with calcium. *Plant Cell,* 11: 691-706.

Scheller, H.V. and Haldrup, A., 2005. Photoinhibition of photosystem I. *Planta,* 221: 5-8.

Shinozaki, K., Yamaguchi-Shinozaki, K. and Seki, M., 2003. Regulatory network of gene expression in the drought and cold stress responses. *Curr. Opin. Plant Biol.,* 6: 410-417.

Smirnoff, N., 1995. Antioxidant systems and plant response to the environment. In: Smirnoff N (ed) Environment and Plant Metabolism: Flexibilty and Acclimation, Bios Scientific Publishers, Oxford, UK, pp 217-243.

Takahashi, S., Katagiri, T., Yamaguchi-Shinozaki, K. and Shinozaki, K., 2000. An Arabidopsis gene encoding a Ca2+-binding protein is induced by abscisic acid during dehydration. *Plant Cell Physiol.,* 41: 898-903.

Tena, G., Asai,T., Chiu, W.L. and Sheen, J., 2001. Plant mitogen-activated protein kinase signaling cascades. *Curr. Opin. Plant. Biol.,* 4: 392-400.

Thomashow, M., 1999. Plant cold acclimation: Freezing tolerance genes and regulatory mechanisms. *Annu. Rev. Plant. Physiol. Plant Mol. Biol.,* 50: 571-599

Timmusk, S. and Wagner, E., 1999. The plant growth-promoting rhizobacterium *Paenibacillus polymyxa* induces changes in Arabidopsis thaliana gene expression: a possible connection between biotic and abiotic stress responses. *Mol. Plant Microbe In.,* 12: 951-959.

Tippmann, H.F., Schlüter, U. and Collinge, B., 2006. Common themes in biotic and abiotic stress signalling in plants *In* Floriculture, Ornamental and Plant Biotechnology Volume III Global Science Books. 6: 52-67.

Tuteja, N., 2007. Abscisic acid and abiotic stress signaling. *Plant Signal Beh.,* 13: 135–138. doi: 10.4161/psb.2.3.4156.

Vierling, E., 1997. The small heat shock proteins in plants are members of an ancient family of heat induced proteins. *ACTA Physiologia Planta.,* 19: 539-547.

Xiong, L. and Yang, Y., 2003. Disease resistance and abiotic stress tolerance in rice are inversely modulated by an abscisic acid-inducible mitogen-activated protein kinase. *Plant Cell,* 15: 745-759.

Xiong, L., Ishitani, M., Lee, H. and Zhu, J., 2001. The Arabidopsis LOS5/ABA3 locus encodes a molybdenum cofactor sulfurase and modulates cold and osmotic stress responsive gene expression. *Plant Cell,* 13: 2063-2083.

Xiong, L., Schumaker, K.S. and Zhu, J.,2002. Cell signaling during cold, drought, and salt stress. *Plant Cell,* 14: 165-183.

Yamasaki, H. and Cohen, M.F., 2006. NO signal at the crossroads: polyamine-induced nitric oxide synthesis in plants? *Trends Plant Sci.,* 11: 522–524.

Yang, Y., and Klessig, D.F., 1996. Isolation and characterization of a tobacco mosaic virus-inducible myb oncogene homolog from tobacco. *Proc. Natl. Acad. Sci. USA,* 93: 14972-14977.

Yang, Y., Qi, M. and Mei, C., 2004. Endogenous salicylic acid protects rice plants from oxidative damage caused by aging as well as biotic and abiotic stress. *Plant Journal*, 40: 909-919.

Yano-Melo, A.M., Saggin, O.J. and Maia, L.C. 2003. Tolerance of mycorrhized banana (*Musa* sp. cv. Pacovan) plantlets to saline stress. *Agric. Ecosystems and Environ.* 95 (1): 343-348.

Zhang, S. and Liu, Y., 2001. Activation of salicylic acid-induced protein kinase, a mitogen-activated protein kinase, induces multiple defense responses in tobacco. *Plant Cell*, 13: 1877-1889.

Zhou, F., Andersen, C.H., Burhenne, K., Fischer, P.H., Collinge, D.B. and Thordal-Christensen, H., 2000. Proton extrusion is an essential signalling component in the HR of epidermal single cells in the barley-powdery mildew interaction. *Plant Journal*, 23: 245-254.

Zhu, J.K., 2002. Salt and drought stress signal transduction in plants. *Annu. Rev. Plant Biol.*, 53: 247-273.

2016, Recent Advances in Plant Stress Physiology *Pages* **137–169**
Editors: ***Praduman Yadav, Sunil Kumar and Veena Jain***
Published by: **DAYA PUBLISHING HOUSE, NEW DELHI**

Chapter 7

Salinity Stress in Plants

Reena Devi and Veena Jain*

Department of Biochemistry, College of Basic Sciences and Humanity
CCS Haryana Agricultural University, Hisar – 125 004, Haryana, India

ABSTRACT

Salinity stress is the major factor that limits crop productivity and quality. Soil salinity stresses plants in two main ways; high concentrations of salts in the soil make it harder for roots to extract water leading to osmotic stress, and high concentrations of salts within the plant can be toxic, thereby resulting in nutrient imbalance and inhibition of many physiological and biochemical processes. Metabolic imbalance caused by ion cytotoxicity, osmotic stress and nutritional deficiency under saline conditions leads to oxidative stress, thus affecting all the major processes such as photosynthesis, nitrogen metabolism, carbohydrates metabolism, lipid metabolism, antioxidants system etc. It causes decrease in growth of the plant thereby resulting in decreased plant yield. Plants acclimatize to salinity by osmotic stress tolerance, ion homeostasis and tissue tolerance. Tissue tolerance requires synthesis and accumulation of compatible solutes within the cytoplasm *viz.* glycine-betaine, proline, lea proteins, polyamines, sugar alcohols, trehlose. Survival of plants to salinity stress requires modulation of various genes. The stress is perceived by different receptors (*e.g.* ion channels, serine/threonine kinase or histidine kinase or G-protein-coupled receptors) and the signal is then transduced downstream, which results in the generation of second messengers. These modulate the intracellular calcium level which is sensed by Ca^{2+} sensors. These sensory proteins interact with their respective interacting partners and target the major stress responsive genes or the transcription factors (*e.g.* CBF/DREB, ABF, bZIP, MYC/MYB) controlling these genes. The products of these stress responsive genes ultimately lead to plant adaptation and help the plant to survive under unfavourable conditions.

Keywords: *Compatible solutes, Osmotic stress, Salinity, Tissue tolerance, Transcription factors.*

**Corresponding author*: E-mail: veena.nicejain@gmail.com

Introduction

Salinity is one of the major environmental factors that limits productivity and quality of economically important crops throughout the world (Sharifi *et al.*, 2007) particularly in arid and semi-arid regions, where soil salt content is naturally high (Zhao *et al.*, 2007). According to the FAO Land and Nutrition Management Service (2008), over 6 per cent of the world's land is affected by salinity which accounts for more than 800 million hectare of land. Saline soils are defined by Ponnamperuma (1984) as those contain sufficient salt in the root zone to impose the growth of crop plants. However, since salt injury depends on species, variety, growth stage, environmental factors, and nature of the salts, it is difficult to define saline soil precisely. The most widely accepted definition of a saline soil has been adopted from FAO (1996) as one that has an ECe of $4dSm^{-1}$ or more and soils with ECe's exceeding $15dSm^{-1}$ are considered strongly saline. The hazardous effects of salinization can be appreciated from the fact that it affects about 20 per cent of the world's cultivated area and nearly half of the world's irrigated lands. In addition, the increased salinity of arable land is expected to have devastating global effects, resulting in upto 50 per cent land loss by the middle of the 21st century. Most of this salt affected land has arisen from natural causes, from the accumulation of salts over long periods of time in arid and semi-arid zones (Rengasamy, 2002). The human-induced processes such as use of poor quality water for irrigation, use of fertilizers on a massive scale and inadequate drainage also lead to the accumulation of tons of salts into the soil. The agricultural land has also become saline due to land clearing and deforestation which cause water table to rise and concentrate salt in the root zone.

Soil salinity stresses plants in two main ways; high concentrations of salts in the soil make it harder for roots to extract water leading to osmotic stress (Hussain *et al.*, 2008), and high concentrations of salts within the plant can be toxic, thereby resulting in nutrient imbalance and inhibition of many physiological and biochemical processes. Metabolic imbalance caused by ion cytotoxicity, osmotic stress and nutritional deficiency under saline conditions leads to oxidative stress, thus affecting all the major processes such as photosynthesis, protein synthesis, energy and lipid metabolism (Parida and Das, 2005). During initial exposure to salinity, plants experience water stress, which in turn reduces leaf expansion. The osmotic effects of salinity stress can be observed immediately after salt application and continue for the duration of exposure, resulting in inhibited cell expansion and cell division as well as stomatal closure (Munns, 2002). During long-term exposure to salinity, plants experience ionic stress, which can lead to premature senescence of adult leaves, and thus a reduction in the photosynthesis to support continued growth (Cramer and Nowak, 1992). Sodium ions affect plants by disrupting protein synthesis and interfering with enzyme activity (Hasegawa *et al.*, 2000; Munns, 2002). Many plants have evolved several mechanisms either to exclude salt from their cells or to tolerate its presence within the cells. Hence, complete understanding of molecular mechanisms of salinity tolerance along with plethora of genes involved in the stress signaling net work and subsequent development of salinity-tolerant crops are essential for sustaining food production and achieving future food security. In this chapter, we mainly discuss about soil salinity, its effects on plants and tolerance mechanisms which permit the plants to withstand stress, with particular emphasis on ion homeostasis, Na^+ exclusion and tissue tolerance.

Effects of Salinity Stress on Plants

Alterations in Growth Characteristics of Plants

Salinity has deleterious effects on growth characteristics of crop plants leading to a significant decrease in shoot length, fresh weight and dry weight. The deleterious effects of salinity on growth (Gama *et al.*, 2007; Jamil *et al.*, 2007; Houimli *et al.*, 2008; Rui *et al.*, 2009; Memon *et al.*, 2010) of the plant are associated with osmotic stress, ion-toxicity and nutrient imbalance (Figure 7.1). The soluble salts in high concentration interfere with a balanced absorption of essential nutritional ions by the plants, thus affecting their growth.

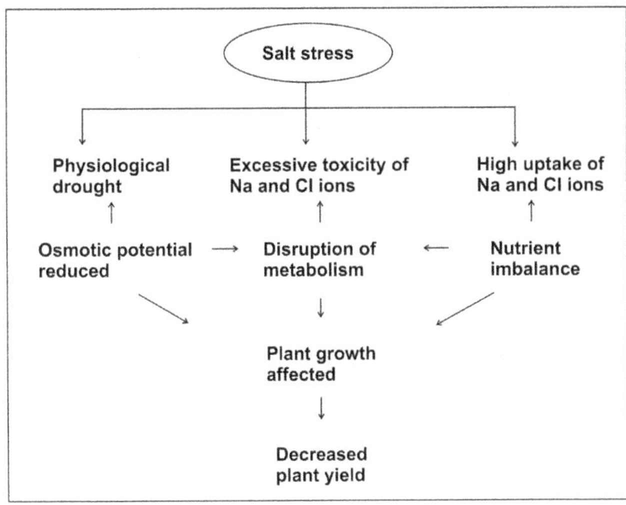

Figure 7.1: Effect of Salt Stress on Plant Growth.
(*Source*: Evelin *et al.*, 2009)

The effects of a saline soil are two-fold: there are effects of the salt outside the roots, and there are effects of the salt taken up by plants. The salt in the soil solution (the "osmotic stress") reduces leaf growth and to a lesser extent root growth, and decreases stomatal conductance and thereby photosynthesis (Munns *et al.*, 1995). The rate at which new leaves are produced depends largely on the water potential of the soil solution, in the same way as for a drought-stressed plant. Salts themselves do not build up in the growing tissues at concentrations that inhibit growth, meristematic tissues are fed largely by the phloem from which salt is effectively excluded, and rapidly elongating cells can accommodate the salt that arrives in the xylem within their expanding vacuoles. The salt within the plant enhances the senescence of old leaves. Continued transport of salt into transpiring leaves over a long period of time eventually results in very high Na^+ and Cl^- concentrations, and they die. The rate of leaf death is crucial for the survival of the plant. If new leaves are continually produced at a rate greater than that at which old leaves die, then there might be enough photosynthesis in leaves for the plant to produce some flowers and seeds. However, if the rate of leaf death exceeds the rate at which new leaves are produced, then the plant may not survive to produce seed.

The effects of salinity other than reduction in leaf area, fresh and dry weight of shoots and roots include slow and insufficient germination of seeds, wilting, desiccation, stunted growth, retarded flowering, sterility and smaller seeds etc. Membrane disorganization, reactive oxygen species, metabolic toxicity, inhibition of photosynthesis and attenuated nutrient acquisition are the starting factors that initiate more catastrophic events. Seed germination is the most critical step in the life cycle of the plant but it is hampered by high salinity stress. A high NaCl concentration causes a reduction in growth parameters (Turhan *et al.*, 2008; Schuch and Kelly, 2008). But decrease in growth characteristics varies species wise (Jamil *et al.*, 2005; Saqib *et al.*, 2006; Turan *et al.*, 2007; Saffan, 2008; Taffouo *et al.*, 2010; Memon *et al.*, 2010). High salinity stress also delays the emergence of nodal roots, leaf and tiller with decrease in relative growth rate (Rui *et al.*, 2009). Salinity affects both vegetative and reproductive development which has profound implications depending on whether the harvested organ is a stem, leaf, root, shoot, fruit, fibre or grain. Salinity generally reduces shoot growth more than root growth (Lauchli and Epstein, 1990) and can reduce the number of florets per ear, increase sterility and affect the time of flowering and maturity in both wheat (Maas and Poss, 1989) and rice (Khatun *et al.*, 1995). It has long been recognized that a crop's sensitivity to salinity varies from one developmental growth stage to the next (Bernstein and Hayward, 1958) and also the definition of salt tolerance is not the same for each growth stage.

Biochemical Changes

Photosynthesis

Photosynthesis, the most fundamental and intricate physiological process in all green plants, is severely affected in all its phases by salt stress. The process of photosynthesis includes light reactions, in which light energy is converted into ATP and NADPH and oxygen is released, and dark reactions, in which CO_2 is fixed into carbohydrates by utilizing the products of light reactions, ATP and NADPH (Dulai *et al.*, 2011). Since the mechanism of photosynthesis involves various components, including photosynthetic pigments and photosystems, the electron transport system, and CO_2 reduction pathways, any damage at any level caused by a stress may reduce the overall photosynthetic capacity of a green plant. The chloroplast is the site for photosynthesis, in which both light and dark reactions of photosynthesis take place and this organelle is highly sensitive to different stressful environments including salinity (Saravanavel *et al.*, 2011). Salinity stress reduces the photosynthetic rate by stress-induced stomatal or nonstomatal limitations (Saibo *et al.*, 2009; Rahnama *et al.*, 2010). Stomatal limitations include closure of stomata and non-stomatal limitations refer to changes in the cholorophyll (Chl) content, chlorophyll fluorescence, damage to the photosynthetic apparatus, chloroplast structure etc. Stomata closure in response to salinity stress generally occurs due to decreased leaf turgor and atmospheric vapor pressure along with root-generated chemical signals (Chaves *et al.*, 2009). Thus, decrease in photosynthetic rate under stressful conditions is normally attributed to suppression in the mesophyll conductance and the stomata closure at moderate and severe stress (Flexas *et al.*, 2004; Chaves *et al.*, 2009), thereby altering photosynthesis and the mesophyll metabolism (Parida *et al.*, 2005; Chaves *et al.*, 2009). Salt stress can also break down chlorophyll (Chl), the effect ascribed to increased level of the toxic cation, Na^+ (Li *et al.*, 2010; Yang *et al.*, 2011) which results in reduction in photosynthetic pigments in crops plants (Ashraf and Sultana, 2000; Akram and Ashraf, 2011; Winicov and Seemann, 1990; Perveen *et al.*, 2010; Pinheiro *et al.*, 2008). However, during the process of Chl degradation, Chl *b* may be converted into Chl *a*, thus resulting in the increased content of Chl *a* (Eckardt, 2009). The decrease in Chl under salt stress (Santos *et al.*, 2001; Santos, 2004; Akram and Ashraf, 2011) has been ascribed to its effect more markedly on Chl biosynthesis than Chl breakdown. Although salt stress reduces the Chl content, the extent of the reduction depends on salt tolerance of plant species. For example, it is generally known that Chl content increases in salt-tolerant species, whereas it decreases in salt-sensitive species under saline regimes (Akram and Ashraf, 2011). In view of this, an accumulation of Chl has been proposed as one of the potential biochemical indicators of salt tolerance in different crops, *e.g.*, in wheat (Sairam *et al.*, 2002; Arfan *et al.*, 2007), pea (Noreen *et al.*, 2010), sunflower (Akram and Ashraf, 2011), alfalfa (Monirifari and Barghi, 2009). Contrarily, in some other studies, Chl accumulation under saline stress is not always associated with salt tolerance (Juan *et al.*, 2005). Therefore, using Chl accumulation as an indicator of salt tolerance depends on the nature of the plant species or cultivar. Carotenoids (Car) are necessary for photoprotection of photosynthesis and they play an important role as a precursor in signaling during the plant development under abiotic/biotic stress. Now a days, enhanced carotenoid contents in plants are of considerable attention for breeding as well as genetic engineering in different plants (Li *et al.*, 2008). Gomathi and Rakkiyapan (2011) found that imposition of salt stress (7–8 dSm^{-1}) in sugarcane at various plant growth stages caused a marked reduction in Chl and Car contents, but salt-tolerant varieties exhibited higher membrane stability and pigment contents.

Photosynthetic pigments present in the photosystems, damaged by salt stress, result in a reduced light-absorbing efficiency of both photosystems (PSI and PSII) and hence a reduced photosynthetic capacity (Akram and Ashraf, 2011; Zhang *et al.*, 2011). Pigment system reduction is attributed to salt induced weakening of protein-pigment-lipid complex or due to increase in destructive enzymes called cholorophyllase (Turan *et al.*, 2007). Photosystem II is a relatively sensitive component of the photosynthetic system with respect to salt stress (Saleem *et al.*, 2011). A reduced activity of the Hill reaction has also been observed in salt-stressed chloroplasts (Zeid, 2009). The inhibitory effect of salt-induced osmotic stress (water deficit) on the rate of photosynthesis has been related to decreased production of ATP due to the impaired electron transport (Moud and Maghsoudi, 2008).

The salt stress-stimulated stomata closure lowers concentration of intercellular CO_2, which in turn causes deactivation of ribulose bisphosphate carboxylase (Rubisco) as well as other enzymes such as sucrose phosphate synthase (SPS) and nitrate reductase (Chaves *et al.*, 2009; Mumm *et al.*, 2011). Furthermore, the salt-induced osmotic effect can adversely affect the activities of a number of stroma enzymes involved in CO_2 reduction (Xue *et al.*, 2008). Flowers *et al.* (1977) reported that Rubisco activity is inhibited *in vitro* by high levels of salt, while Aragao *et al.* (2005) suggested that Rubisco is also affected by salt *in vivo* (Table 7.1). The increase in the amount of Rubisco can be beneficial for the survival of plants under harsh environmental conditions as it enhances the net photosynthetic rate (PN) in most C_3 plants (Taub, 2010; Makino, 2011). However, in contrast, no positive relationship between leaf Rubisco and PN has been reported in a glycophytic species, *Vigna unguiculata,* (Aragao *et al.*, 2005). Thus, in view of such contrasting reports, further research is needed to affirm whether the association between these two traits is positive or negative. It has also been found that not only Rubisco is affected, but also the enzymes involved in regeneration of the Rubisco substrate, ribulose-1,5-bisphophate (RuBP) are regulated by salt stress and they can play a key role in the regulation of the Calvin cycle (Ghosh *et al.*, 2001). Fructose-1,6-bisphosphatase has been considered as one of the potential enzymes that can cause the decline of photosynthetic activity under stressful conditions, because it is involved in the regeneration of RuBP. Phospho*enol*pyruvate carboxylase (PEPC) is another enzyme that is inhibited under saline regimes, however, the enzyme from C_3 plants such as wheat was less sensitive to salt than that from C_4 maize (Abdel-Latif, 2008).

Table 7.1: Stress-induced Increase/Decrease in the Activities of Enzymes Involved in Photosynthesis

Plant Species	*Enzyme*	*Enzyme Activity*	*Reference*
Salt Blackeyed pea (*Vigna unguiculata*)	Rubisco	Decreased	Aragao *et al.*, 2005
Common bean (*Phaseolus vulgaris*)	FBP	Decreased	Seemann and Sharkey, 1982
Sunflower (*Helianthus annuus*)	FBP	Decreased	Gimenez *et al.*, 1992
Potato (*Solanum tuberosum*)	FBP	Increased	Kossman *et al.*, 1994
Rice (*Oryza sativa*)	FBP	Increased	Ghosh *et al.*, 2001
Wheat (*Triticum aestivum*) and maize (*Zea mays*)	PEPC	Decreased	Abdel-Latif, 2008

(Rubisco – ribulose-1,5-bisphosphate carboxylase/oxygenase; FBP – fructose-1,6-bisphosphatase; PEPC – phosphoenol pyruvate carboxylase)

Nitrogen Metabolism

Nitrogen is regarded the single most important growth-limiting factor for crops (Shah, 2008), and most plants can utilize N as NO_3 and NH_4; the available inorganic forms of N absorbed by the roots (Lasa *et al.*, 2002; Wickert *et al.*, 2007; Oh *et al.*, 2008) from the soil, which then undergo complete systems of assimilation, transformation, and mobilization within plants (Oh *et al.*, 2008). Nitrate, taken up by NO_3^- transporters, is reduced to ammonium by nitrate reductase (NR) in the cytosol and nitrite reductase (NiR) in the plastids/chloroplasts. Ammonium derived from the primary nitrate reduction as well as other metabolic pathways is incorporated into glutamate either by glutamine synthetase (GS) and glutamate synthase (GOGAT) pathway or by direct aminating reaction of glutamate dehydrogenase

(GDH). The ammonium assimilation into glutamine and glutamate is vital for plant growth as these two amino acids serve as the precursors for the synthesis of the other amino acids as well as almost all nitrogenous compounds. Alanine aminotransferase (AlaAT) and aspartate aminotransferase (AspAT) are involved in transferring the assimilated nitrogen moieties to organic acids and play an important role in the synthesis of amino acids from glutamate.

Nitrate reduction and ammonium assimilation seem to be highly implicated in the regulation of plant growth in response to salt stress. Changes in the nitrate or ammonium uptake and nitrogen-assimilating enzymes have been observed after salt treatment in several plant species, such as the tomato, *Populus x canescens*, and mulberry (Debouba *et al.*, 2006; Dluzniewska *et al.*, 2007; Surabhi *et al.*, 2008). These investigations have shown that salt stress can affect different plant nitrogen metabolism steps, such as ion uptake, N assimilation, and the synthesis of amino acids and proteins, which may have severe consequences for nitrate assimilation through the inhibition of NR and NiR activities. The key enzyme in nitrogen metabolism, nitrate reductase is very sensitive to NaCl. Salinity stress reduces NO_3^- flux from roots to leaves and impaires the NRA in leaves as well as in roots (Foyer *et al.*, 1998; Flores *et al.*, 2000). The decrease in NO_3^- concentrations by NaCl treatment may result from a disruption of root membrane integrity, an inhibition of NO_3^- uptake and low NO_3^- loading into root xylem (Abd-El Baki *et al.*, 2000). As Cl^- ions inhibit NO_3^- uptake, the decrease in NO_3^- concentrations both in roots and leaves can attribute to competition between Cl^- and NO_3^- for uptake by NO_3^- transporters (Deane-Drummond, 1986), or an inactivation of NO_3^- transporters by toxic effects of salt ions (Lin *et al.*, 1997) or both, thus affecting NRA. Alternatively, Cl^- may have a direct effect on the NRA. The decrease in NRA in plants growing under salt has been reported to be due to the low content of NR protein, and to a limiting nitrate transport to shoots (Carillo *et al.*, 2005; Debouba *et al.*, 2006). The inhibition of NiR activity under saline conditions has also been reported in several plant species (Debouba *et al.*, 2006; Surabhi *et al.*, 2008) and is attributed to a lower availability of nitrate (Debouba *et al.*, 2006). The inhibition of NiRA under saline conditions is considered as a direct consequence of the decrease in NRA and this seems to be necessary to prevent nitrite toxicity and to avoid the waste of electrons by NRA and NiRA. Under salinity, the process of ammonium ion (NH_4^+) incorporation into plant metabolism is also disturbed by causing changes in the activities of GOGAT, GS, and glutamate dehydrogenase. In general, these responses often differ according to the plant species, tissue type, cultivar, and stress conditions (Wang *et al.*, 2007). The changes in these enzyme activities might be associated with a significant decrease in the NO_3^-, NH_4^+, and organic N content when seedlings were treated with NaCl. According to the changes in the enzyme activity under NaCl stress, two types of response have been distinguished (*a*) early responding enzymes, which changed as early as 24 h after adding NaCl, *i.e.* NR and deaminating GDH in leaves, and Fd-GOGAT in the leaves and roots; and (*b*) late responding enzymes, which were affected at least 4d after the NaCl treatment, *i.e.* NADH-GOGAT in leaves, NAD-GDH in roots, and NiR, GS and aminating GDH in both leaves and roots. The increased AlaAT and AspAT activities has been observed in some plant species under salt stress (Ramajulu *et al.*, 1994; Surabhi *et al.*, 2008) which mediate the utilization of ammonia by converting keto acids into amino acids under these stress conditions, which further help coordinate nitrogen (N) metabolism and amino acid synthesis with the availability of carbon skeletons from the Krebs cycle (Hodges, 2002). These changes enable the plant to monitor and adapt to these changes in its N status and supply that result from salt stress, thereby minimizing the harmful consequences of excess salinity.

Plant in saline habitat accumulates lysine, proline, asparagine, glutamine, aspartic acid, glutamic acid, alanine, tyrosine and valine. Ammonia and lysine have toxic effect and their increased quantities cause production of abnormal metabolites (Munns, 2005). However, proline accumulates in large amounts as compared to all other amino acids in salt stressed plants and it acts as an osmoprotectant. Salt stress in legume inhibits nitrogen fixation by reducing nodulation and nitrogenase activity in legumes (Mudgal *et al.*, 2009) and it has also been associated with elevated peroxidase level (Munns *et al.*, 2006; Munns and Tester, 2008). There is a tendency for amides and toxic diamines (putrescine and cudaverine) to accumulate in tissue of stress affected plant due to depression of diamine oxidase activity (Stewart and Lee, 1974). These regulate the entry of NaCl to root and the regulates its entry to shoots.

Carbohydrates Metabolism

Carbohydrates, the main source of carbon and energy, undergo changes under saline conditions and these are of particular importance because of their direct relationship with physiological processes such as photosynthesis, translocation and respiration. Glycolysis, oxidative pentose phosphate (OPP) pathway and the mitochondrial tricarboxylic acid (TCA) cycle are the central respiratory pathways using photosynthesis-derived carbohydrate to supply carbon intermediates for biosynthesis, as well as coupling carbon oxidation with the reduction of NAD(P) to NAD(P)H. These reducing equivalents are then used to support biosynthetic reactions or can be oxidized by the mitochondrial electron transport chain (ETC), localized in the inner mitochondrial membrane. The glucose-6-phosphate dehydrogenase (G6PDH) step in the cytosolic OPP pathway generates NADPH that is critical for the establishment of salinity stress tolerance by providing sufficient substrate for the plasma membrane NADPH oxidase (Scharte *et al.*, 2009). The activation of G6PDH during salt stress is critical in supporting ROS-scavenging systems of the cell (which use NADPH as electron source) to prevent oxidative damage (Dal Santo *et al.*, 2012). In plants exposed to high salt conditions, there is accumulation of hexoses due to lower activity of glucokinase (GK), phosphoglucoisomerase (GPI) and pyruvate kinase (PK). The salinity also induces phosphoenolpyruvate carboxykinase (PEPCK) activity and the increase has been reported to be concentration-dependent. Exposure of kikuyu grass to 150 and 200 mM NaCl caused an increase of $NADP^+$-specific isocitrate dehydrogenase (ICDH) activity in leaves and roots (Muscolo *et al.*, 2003). The sucrose phosphate synthase (SPS) that catalyzes photosynthetic production of sucrose (Krause *et al.*, 1998) found to be increased under salt stress (Chen *et al.*, 2006). Salinity stress alters carbohydrate metabolism differently in salt-tolerant and salt-sensitive cultivars. Salinity stress induced accumulation of total soluble sugars and sucrose in the leaves of salt-sensitive rice cultivars without concomitant increase in the sucrose phosphate synthase. Therefore, it is suggested that accumulation of these sugars may be result of reduction of sink demand due to growth limitation (Walker and Ho, 1977). On the other hand, starch concentration increased markedly in salt-tolerant cultivar when grown under salinity stress. Starch accumulation perhaps results from the increased activity of alkaline invertase activity which hydrolyses sucrose and converts into simple sugars. Although it is unclear whether starch accumulation may play role in salt-tolerant cultivars, it is suggested that partitioning sugars into starch may involve in salinity tolerance by avoiding metabolic alterations (Pattanagul and Walttona, 2008). It is also possible that adjusted carbon partitioning and allocation could have an important implication on the overall plant growth under salinity.

Simple sugars such as fructose, sucrose and glucose act as osmolytes and/ or compatible solutes, are quickly synthesized from polymeric forms in response to salt stress and this increase might be considered to play an important role in osmotic adjustment, which is widely regarded as an adaptive response to water deficit conditions (Kameli and Losel, 1995). The other possible role of sugar may be as a readily available energy source. Raffinose family oligosaccharides (RFOs) that include a number of complex carbohydrates such as stachyose and galactinol have an established role in salt-induced desiccation tolerance of seed and are the second most common storage carbohydrate in seeds. They are also rapidly converted during germination to simple sugars. It has been suggested that RFOs protect seeds against damage during salt induced desiccation and increases storability and longevity. The RFOs possibly interact with cell membranes in the area of the phospholipid head groups, thus increasing their stability during desiccation. Studies with corn have shown that RFOs and monomeric sugars together are able to form highly vitreous states that physically stabilize cellular membrane components (Obendorf, 1997). In transgenic Arabidopsis, two salt and drought induced galactinol synthase genes (*GolS*) when over expressed, resulted in plants to tolerate higher levels of salt and water stress than wild type (Taji *et al.*, 2002). In addition, plants over-expressing the drought response factors DREBA1A/CBF3 have been found more tolerant of dehydration (Kasuga *et al.*, 1999) and contained increased concentrations of sucrose, glucose, fructose, raffinose and other sugars. These results confirm a direct role for raffinose and other sugars as osmoprotectants and osmolytes.

Oxidative Stress and Antioxidants

Oxygen is essential for the existence of aerobic life, but toxic reactive oxygen species (ROS), which include the superoxide anion radical ($\bullet O_2$), hydroxyl radical ($^\bullet OH$) and hydrogen peroxide (H_2O_2), are generated at low levels in all aerobic cells during metabolic processes (Foyer *et al.*, 1994; Asada, 1999; Cai-Hong *et al.*, 2005) and there is a balance between production and scavenging of ROS. This balance may be perturbed by a number of adverse environmental stresses including salinity; thus leading to rapid increase in ROS production which can induce damage to cell. The enhanced production of ROS due to disruption of cellular homeostasis is extremely harmful to the plants. Injury caused by these ROS is known as oxidative stress, which is one of the major damaging factors to plants exposed to environmental stresses. The ROS interact non-specifically with many cellular components and thus, triggering peroxidative reactions which are responsible for causing significant damage to proteins, lipids, and nucleic acids and ultimately leading to cell death. Enhanced production of ROS under saline conditions (Greenway and Munns, 1980; Hasegawa *et al.*, 2000) has been demonstrated to be a major cause of the cellular toxicity by salinity in various plants (Gueta-Dahan *et al.*, 1997; Dionisio-Sese and Tobita, 1998; Mittova *et al.*, 2004; Chawla *et al.*, 2013) thus damaging membrane lipids, proteins and nucleic acids (Hernandez *et al.*, 1993, 1999; 2000; Mansour *et al.*, 2005; Eyidogan and Oz, 2007).

However to detoxify excess ROS and to maintain redox homeostasis, plants havedeveloped the efficient antioxidative scavenging system comprising of enzymatic and non-enzymatic components (Reddy *et al.*, 2004). The enzymatic antioxidant system is one of the protective mechanisms including superoxide dismutase (SOD), that catalyses the disproportion of two O_2 radicals to H_2O_2 and O_2 (Scandalios, 1993). H_2O_2 is eliminated by various antioxidant enzymes such as catalases (CAT) (Kono and Fridovich, 1983; Scandalios, 1993) and peroxidases (POX) (Gara *et al.*, 2003) which convert H_2O_2 to water. The H_2O_2 is also detoxified by ascorbate peroxidase in the ascorbate-glutathione cycle (Asada, 1999; Mittler, 2002). Other enzymes that are very important in the ROS scavenging system and function in the ascorbate-glutathione cycle are glutathione reductase (GR), monodehydro ascorbate reductase (MDHAR) and dehydroascorbate reductase (DHAR) (Candan and Tarhan, 2003; Yoshimura *et al.*, 2000). Ascorbate, glutathione, tocopherols, carotenoids, flavonoids and phenolics serve as the non-enzymatic antioxidants.

The generation of ROS and increased activities of many antioxidant enzymes during salt stress have been reported in various plants such as cotton (Desingh and Kanagaraj, 2007), mulberry (Sudhakar *et al.*, 2001; Harinasut *et al.*, 2003), wheat (Sairam *et al.*, 2002), rice (Chawla *et al.*, 2013) and sugar beet (Bor *et al.*, 2003). In general, the activities of antioxidant enzymes increase in the root and shoot under saline stress. But the increase has been reported to be more significant and consistent in the root (Kim *et al.*, 2005). The quantitative and qualitative aspects of changes are often related to the levels of resistance to salinity. In tomato and citrus, salt-tolerance is attributed to the increased activities of SOD, APX and CAT (Gueta-Dahan *et al.*, 1997; Mittova *et al.*, 2004). Further supporting evidence on the involvement of antioxidant enzymes in salt tolerance has been provided by transgenic plants with a reduced or an increased expression of antioxidant enzymes (Willekens *et al.*, 1997). Increased protection to salt stress has been demonstrated by the over expression of cytosolic APX (Torsethaugen *et al.*, 1997).

In plants, the links between ROS production and photosynthetic metabolism are particularly important (Rossel *et al.*, 2002). Salt stress can lead to stomatal closure, which reduces CO_2 availability in the leaves and inhibits carbon fixation, exposing chloroplasts to excessive excitation energy, which in turn increases the generation of reactive oxygen species (ROS) and induce oxidative stress (Parida and Das, 2005). In chloroplasts, ROS can be generated by the direct transfer of the excitation energy from chlorophyll to produce singlet oxygen or by oxygen reduction in the Mehler reaction (Meloni *et al.*, 2003). H_2O_2 and O_2 may interact in the presence of certain metal ions and chelates to produce highly reactive •OH (Sudhakar *et al.*, 2001). The reduced rate of photosynthesis under salinity increases the formation of ROS.

All the ROS detoxifying mechanisms are present naturally in surfeit, and are woven into the regulatory regimes of the chloroplast to protect the photosystems from photoinhibition that might otherwise occur from the rapidly increasing light loads experienced by leaves under naturally variable situations. If a plant has sufficient capacity to adjust to the instant, large changes in light intensity as the sun emerges from behind a cloud, it has more than enough capacity to adjust to the slower changes in the rate of photosynthesis induced by a saline soil. Therefore, precise understanding the roles of each ROS scavenging enzymes and small molecular antioxidants in stress adaptation and accurate characterization of the complex stress tolerance phenotypes is necessary to develop stress tolerant plants. To maintain the productivity of plants under the salt stress condition, it is important to fortify the antioxidative mechanism of the chloroplasts by manipulating the antioxidant enzymes and small antioxidant molecules in the chloroplast.

Lipids

Lipids, the structural constituents of most of the cellular membranes, protect internal organs and hormones (Singh *et al.*, 2002). They play vital roles in the tolerance to several physiological stressors in a variety of organisms. The mechanism of desiccation tolerance induced by drought or salinity relies on phospholipid bilayers, which are stabilized during water stress by sugars, especially by trehalose. Unsaturation of fatty acids also counteracts water or salt stress. Hydrogen atoms adjacent to olefinic bonds are susceptible to oxidative attack. Lipids are rich in these bonds and are a primary target for oxidative reactions. Lipid oxidation forms the highly reactive species products that modify proteins and DNA (Singh *et al.*, 2002). Wu *et al.* (1998) have analyzed the changes in lipid composition by NaCl stress in root plasma membrane of salt marsh grass (*Spartina patens*) and reported that sterols (including free sterols) and phospholipids decrease with increasing salinity, but the sterol/phospholipid ratio is unaffected by NaCl. However, glycolipid shows a statistically significant increase in the total lipid as salinity in the medium is increased and the content of plasma membrane phosphatidylcholine (PC) and phosphatidyl ethanolamine (PE) decrease by salinity, but the PC/PE ratios are not affected by salinity.

Polyphenols

Polyphenols are a group of chemical substances, characterized by the presence of more than one phenol units or building blocks per molecule. Polyphenols are generally divided into hydrolysable tannins, lignins and flavonoids. Levels of polyphenols increase under increasing levels of salinity, which shows that the induction of secondary metabolism including phenolics is one of the defense mechanisms adapted by the plants to face and survive saline environment (Sadale, 2007).

Changes in Minerals

Plants require different essential minerals for their growth and survival but excessive soluble salts in the soil are harmful to most of the plants. It is well documented that the micronutrients are generally less affected by salt stress than macronutrients. Ions at high concentration in the external solution (*i.e.* Na^+ or Cl^-) are taken up at high rates, which may lead to their excessive accumulation in the tissue. These ions may inhibit the uptake of other ions into the root (*i.e.* K^+ or Ca^{++}) and their transport to the shoot through the xylem, finally leading to a deficiency in the tissue. Vacuolation might provide a means for accumulation of excess ions in plants. Under salinity stress, decrease in the concentrations of root N, K, Ca, and Mg, and increase in Fe and Mn content with no change in P and Cu content occurs (Loupassaki *et al.*, 2002).

Despite the physical and chemical similarity between Na and K, many higher plants have developed a high degree of selectivity for the uptake of K, even in the presence of large amounts of Na. Leaves are more vulnerable than roots to Na^+ simply because Na^+ and Cl^- (chloride) accumulate to higher levels in shoots than in roots. Potassium (K) is needed to sustain plant growth and reproduction as it is essential in nearly all the processes. It plays a vital role in photosynthesis, translocation of photosynthates, protein synthesis, activation of plant enzymes, control of ionic balance and regulation of plant stomata. It is well known that plants deficient in potassium are less resistant to drought, salinity, excess water and varying temperatures. Shoots are selective for K^+ over Na^+ and maintaining K^+ concentrations resulted in relatively high K^+/Na^+ ratio at high salinity. Also excess accumulation of K^+ in roots, might act as a reservoir of K^+ for shoots at high salinity.

Under saline conditions, Na^+ through nonspecific ion channels may cause membrane depolarization that activates Ca^{2+} channels and thus generates Ca^{2+} oscillations and signals salt stress. It is commonly believed that the Ca^{2+} status of the plant are important in salt tolerance. Salinity can inhibit root hair elongation via alterations in Ca^{2+} gradient that regulates root hair growth and by a decrement in cytosolic Ca^{2+}. Low Ca^{2+} is known to impair the selective root permeability to Na^+ (Loupassaki *et al.*, 2002). Ca^{2+} reduces the adverse effect of Na^+ by controlling the intake of toxic ions through the selectivity of the cell membrane (Munns, 2002). Exogenous application of calcium to root growth medium enhances salt tolerance in glycophytic plants (Tattini *et al.*, 1995). However, not much is known about the function of chloride in other metabolic process except its role in photosynthetic light reaction. But it has been implicated that chloride was the main anion used by pepper plants to achieve the osmotic adjustment under saline environment (Martinez-Ballesta *et al.*, 2004).

Plant Responses to Salinity Stress

Plant responses to salinity can occur in two distinct phases through time:

Osmotic Stress

It is a rapid response due to increase in external osmotic pressure and is caused by the salt outside the roots. High concentrations of salts in the soil disturb the capacity of roots to extract water leading to osmotic stress that immediately reduces cell expansion in root tips and young leaves, and causes stomatal closure.

Ionic Stress

It develops over time and is due to a combination of ion accumulation in the shoot and an inability to tolerate the ions that have accumulated (Munns and Tester, 2008).

Na^+ exclusion by roots ensures that Na^+ does not accumulate to toxic concentrations within leaves. A failure in Na^+ exclusion manifests its toxic effect after days or weeks, depending on the species, and causes premature death of older leaves. High concentrations of salts within the plant itself can be toxic, resulting in an inhibition of many physiological and biochemical processes (Munns, 2002; Munns and Tester, 2008). Together, these effects reduce plant growth, development and survival. A two phase model (Figure 7.2) describing the osmotic and ionic effects of salt stress was proposed by Munns *et al.* (1995). It describes that during Phase 1, growth of both salt sensitive and tolerant plants is reduced because of the osmotic effect of the saline solution outside the roots. During Phase 2, old leaves in the sensitive plant die and reduce the photosynthetic capacity of the plant. This exerts an additional effect on growth as shown in Figure 7.2.

The first osmotic phase starts immediately after the salt concentration around the roots increases to a threshold level, making it harder for the roots to extract water. An immediate response to this effect, which also mitigates ion flux to the shoot, is stomatal closure that results in the significant reduction in rate of shoot growth. Shoot growth is more sensitive than root growth to salt-induced osmotic stress probably because a reduction in the leaf area development relative to root growth

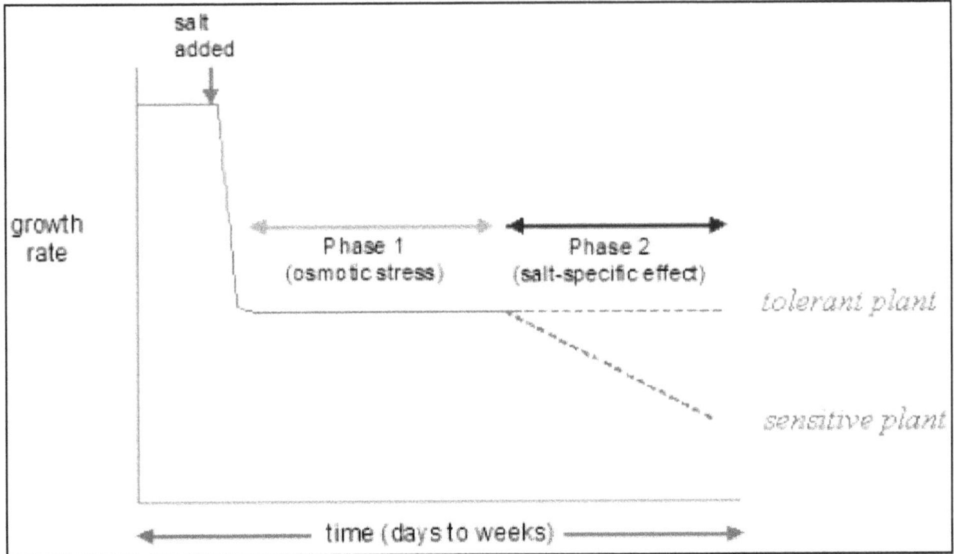

Figure 7.2: Two Phase Growth Response to Salinity.
(*Source*: Munns *et al.*, 1995).

decreases the water use by the plant, thus allowing it to conserve soil moisture and prevent salt concentration in the soil (Munns and Tester, 2008). Reduction in shoot growth due to salinity is commonly expressed by a reduced leaf area and stunted shoots (Lauchli and Epstein, 1990). Final leaf size depends on both cell division and cell elongation. Leaf initiation, which is governed by cell division, was shown to be unaffected by salt stress in sugar beet, but leaf extension was found to be a salt-sensitive process (Papp *et al.*, 1983), depending on Ca^{2+} status. Moreover the salt-induced inhibition of the uptake of important mineral nutrients, such as K^+ and Ca^{2+}, further reduces root cell growth that results in the accumulation of ROS. These ROS can also influence the ROS induced MAPK signal pathway through inhibition of phosphatases or downstream transcription factors (Loredana *et al.*, 2011) (Figure 7.3).

The second ion specific phase corresponds to the accumulation of ions, in particular Na^+, in the leaf blade rather than in the roots (Munns, 2002). Na^+ accumulation is toxic especially in old leaves, which are no longer expanding and so no longer diluting the salt arriving in them as young growing leaves do. In photosynthetic tissues, in fact, Na^+ accumulation affects enzymes and photosynthetic pigments *viz.* chlorophylls and carotenoids (thus affecting rate of photosynthesis (Davenport *et al.*, 2005). The reduction in photosynthetic rate in the salt sensitive plants can also increase the production of reactive oxygen species. Normally, ROS are rapidly removed by antioxidative mechanisms, but this removal can be impaired

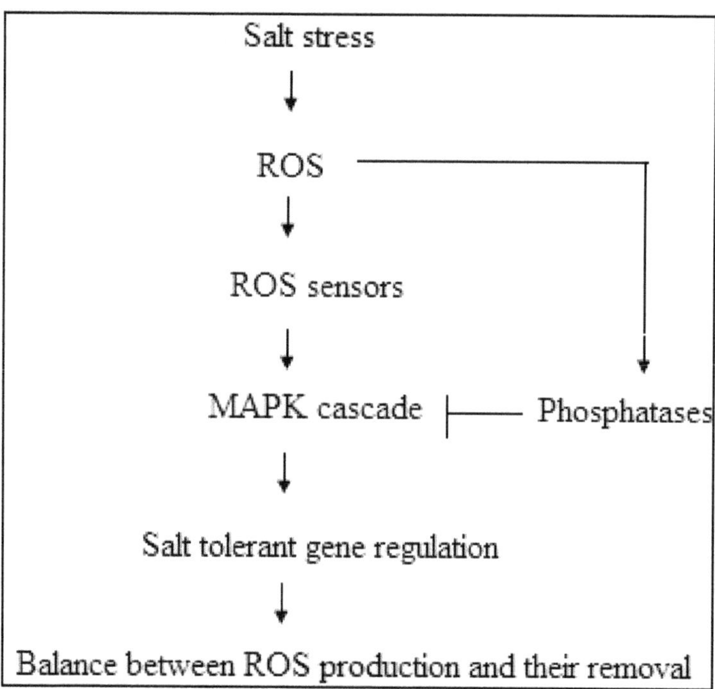

Figure 7.3: ROS Signal Transduction Pathway Under Salt Stress.
(*Source*: Mittler, 2004).

by salt stress. ROS signalling has been shown to be an integral part of acclimation response to salinity (Figure 7.3). ROS, in fact, play a dual role in the response of plants to abiotic stresses including salinity; functioning as toxic by-products of stress metabolism as well as important signal transduction molecules for stress responsive pathway mediated by calcium, hormone and protein phosphorylation (Miller *et al.*, 2010).

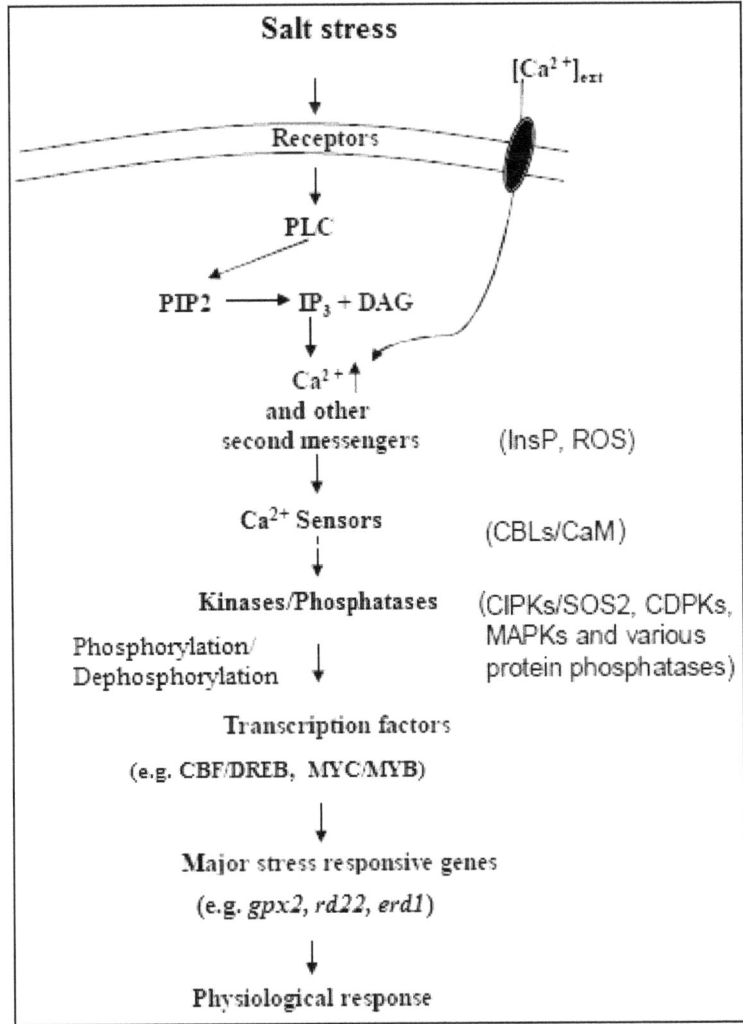

Figure 7.4: Salt Stress Mediated Signal Transduction Pathway.
(*Source*: Tuteja, 2007).

(CBF: C-repeat binding factor; CBL: Calcineurin B-like protein; CDPKs: Cyclin dependent protein kinases; CIPKs: CBL-interacting protein kinases; DAG: Diacylglycerol; DREB: Dehydration-responsive element-binding; *erd1*: Early responsive to hydration1; *gpx2*: Glutathione peroxidase 2; InsP: Inositol phosphate; IP_3: Inositol 1,4,5 trisphosphate; MAPKs: Mitogen activated protein kinases; myc: Myelocytomatosis; myb: Myeloblastosis; PIP2: Phosphatidyl 4,5 bisphosphate).

Stress induces a series of responses both at cellular and systemic level in the plants (Tuteja, 2007). Stress is perceived by different receptors (ion channels, receptors serine/threonine kinase or histidine kinase or G-protein-coupled receptors) located in the plasma membrane of the plant cells (Figure 7.4).

The signal is then transduced downstream, which results in the generation of second messengers including calcium, ROS and inositol phosphates. These second messengers such as inositol phosphates further modulate the intracellular calcium level. This change in cytosolic Ca^{2+} level is sensed by calcium binding proteins, also known as Ca^{2+} sensors. These sensory proteins change their conformation in a calcium dependent manner and then interact with their respective interacting partners often initiating a phosphorylation cascade and target the major stress responsive genes or the transcription factors (*e.g.* CBF/DREB, ABF, bZIP, MYC/MYB) controlling these genes. The products of these stress responsive genes ultimately lead to plant adaptation and help the plant to survive under unfavourable conditions. Thus, plant responds to stresses as individual cells and synergistically as a whole organism.

Mechanism of Salt Tolerance

Plants have evolved several mechanisms to acclimatize to salinity which are categorized into three types (Munns and Tester, 2008)

(1) Tolerance to osmotic stress (2) Ion homeostasis (3) Tissue tolerance

Tolerance to Osmotic Stress

The osmotic tolerance involves the ability of a plant to tolerate the drought aspect of salinity stress and to maintain leaf expansion and stomatal conductance. If the accumulation of salts overcomes the toxic concentrations, the old leaves die (usually old expanded leaves) and the young leaves undergo a reduction in growth and new leaves production due to lack of the export of photosynthates. For this reason increased osmotic tolerance involves an increased ability to continue production and growth of new and greater leaves, and higher stomatal conductance. The resulting increased leaf area would benefit only plants that have sufficient soil water, such as in irrigated food production systems where a supply of water is ensured, but could be undesirable in water-limited systems (Munns and Tester, 2008). However, mechanisms involved in osmotic tolerance related to stomatal conductance and therefore to photosynthetic capacity to sustain carbon skeletons production to meet the cell's energy demands for growth have not been completely unraveled.

Ion Homeostasis

Plants maintain a high concentration of K^+ and a low concentration of Na^+ in the cytosol under salt stress by regulating the expression and activity of K^+ and Na^+ transporters and of K^+ pumps that generate the driving force for transport. Although salt-stress sensors remain elusive, some of the intermediary signaling components have been identified. When Na^+ enters cells and accumulates, its high concentration becomes toxic to enzymes. To prevent growth cessation or cell death, excessive Na^+

has to be extruded or compartmentalized in the vacuole. Unlike animal cells, plant cells do not have Na^+-ATPases or Na^+/K^+-ATPases, and they rely on H^+-ATPases and H^+-pyrophosphatases to create a proton-motive force that drives the transport of all other ions and metabolites.

Calcium plays an important role in providing salt tolerance to plant. Externally supplied Ca^{2+} reduces the toxic effects of NaCl, presumably by facilitating higher K^+/Na^+ selectivity. High salinity results in increased cytosolic Ca^{2+} that is transported from the apoplast as well as from the intracellular compartments (Knight *et al.*, 1997). This transient increase in cytosolic Ca^{2+} initiates the stress induced-signal transduction leading to salt adaptation by SOS pathway. This pathway as depicted in Figure 7.5, involves salt overly sensitive (SOS) genes which are responsible for the exclusion of excess Na^+ ions out of the cell via the plasma membrane Na^+/H^+ anti-porter and helps in reinstating cellular ion homeostasis. Three *SOS* genes (*SOS1*, *SOS2* and *SOS3*) function in a common pathway of salt tolerance (Zhu *et al.*, 1998). The *SOS1* gene encodes a 127-kDa protein with N terminal region composed of 12 trans-membrane domains and a C terminal region with a long hydrophilic cytoplasmic tail (Shi *et al.*, 2000). The trans-membrane region of SOS1 shared substantial sequence homology to the plasma membrane Na^+/H^+ antiporter isolated from bacteria and fungi (Liu *et al.*, 2000). *SOS2* gene was isolated through the genetic screening of mutants oversensitive to salt stress in Arabidopsis. The *SOS2* encodes a novel serine/threonine protein kinase, a CBL interacting protein kinase (CIPK24) that interacts with calcineurin B-like (CBL) protein with an N terminal catalytic and C terminal regulatory domain. Whereas, the N terminal domain shares sequence homology with sucrose non-fermenting kinases (SNF), the C terminal domain is unique to this class of kinases and harbors a 21 amino acid conserved NAF motif (named after prominent conserved amino acids Asn, ala and phe) or FISL motif (Albrecht *et al.*, 2001). FISL motif acts as an auto inhibitory domain and interacts with the catalytic domain thereby keeping the enzyme in an OFF state under normal conditions. The third gene, *SOS3* encodes a Ca^{2+} binding protein also known as calcineurin B-like protein (CBL4) with four helix loop helix structural domain (4 EF hand), Ca^{2+} binding motifs and a myristoylation sequence (MGXXXST/K) at the N-terminus of the protein. Signaling pathway specific to salinity stress involves the Na^+-induced increase in Ca^{2+} that is sensed by SOS3, a calcineurin B-like protein (CBL4). Although the affinity for Ca^{2+} binding of this protein is unknown, physiological increase in cytosolic Ca^{2+} is likely facilitate the dimerization of CBL4/SOS3 and the subsequent interaction with a CBL-interacting protein kinase (CIPK24, originally identified as SOS2) (Halfter *et al.*, 2000). SOS3 interacts with SOS2 via FISL motif and relieves the protein from auto inhibition thereby making the kinase active. SOS3 activates SOS2 protein kinase activity in a calcium dependent manner. The myristoylation motif of SOS3 results in the recruitment of SOS3-SOS2 complex to the plasma membrane, where SOS2 phosphorylates and activates plasma membrane bound Na^+/H^+ antiporter, SOS1 (Qiu *et al.*, 2002; Quintero *et al.*, 2002; Shi *et al.*, 2002). The excess Na^+ ions are expelled out of the cell and cellular ion homeostasis is restored. A low affinity Na^+ transporter, AtHKT1 (*Arabidopsis* low affinity Na^+ transporter/high-affinity K^+ transporter) seems to mediate Na^+ entry into the root cells of *Arabidopsis* during salt stress. SOS3-SOS2

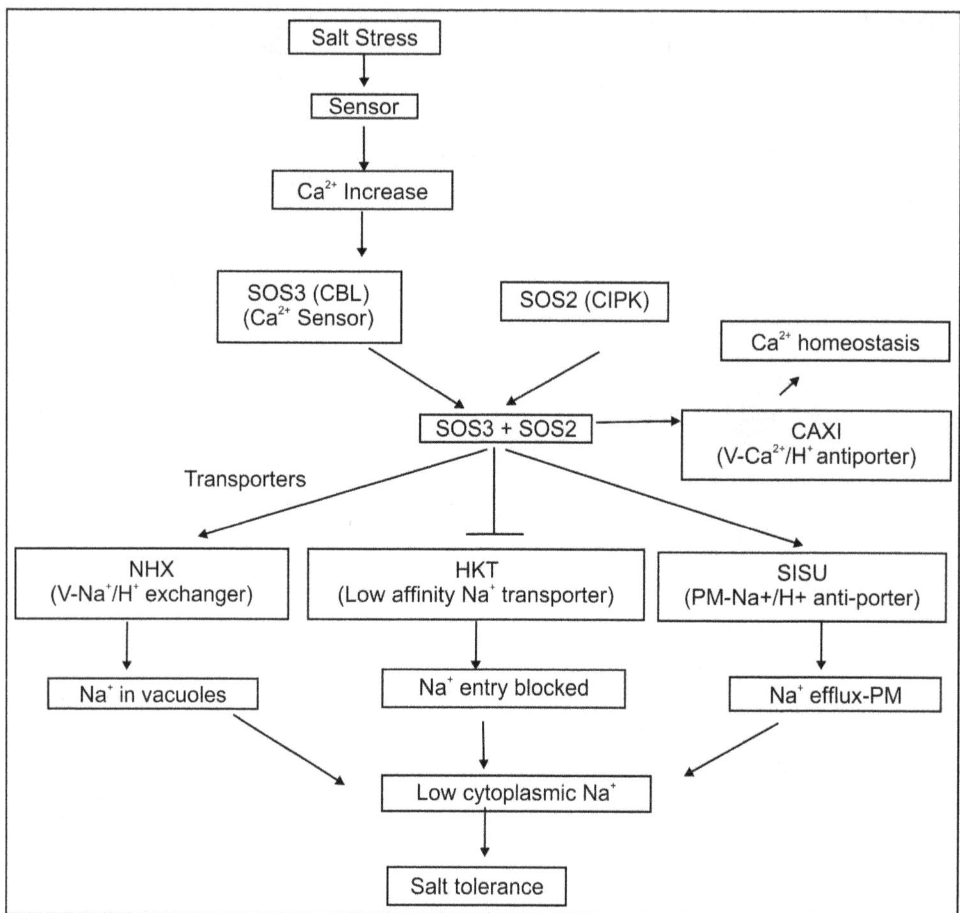

Figure 7.5: Regulation of Ion Homeostasis by SOS and Related Pathways in Relation to Salt Stress Adaptation.
(Source: Tuteja, 2007).

(CAX1: Cation exchanger 1; CBL4: Calcineurin B-like protein 4; CIPKs: CBL-interacting protein kinase; SOS1: Salt overly sensitive 1; SOS2: Salt overly sensitive 2; SOS3: Salt overly sensitive 3; NHX: Na^+/H^+ exchanger; HKT: High-affinity K^+ transporter).

complex functions to down regulate HKT1 gene expression or inactivate the HKT1 protein during salt stress, thereby preventing the Na^+ entry and its build up in the cell. SOS3 and SOS2 seem to negatively regulate the activity of AtHKT1 under salt stress. In addition to controlling SOS1 activity resulting in efflux of excess Na^+ ions, SOS3–SOS2 complex also seems to function in sequestration of excess Na^+ ions in the intracellular compartments. SOS2 is shown to interact with vacuolar Na^+/H^+ antiporter and influence the Na^+/H^+ exchange activity significantly. Recently, it has been shown that SOS2 interacts with the N terminus of CAX1 (H^+/Ca^{2+} antiporter) and regulates its activity (Cheng *et al.*, 2004). This activation of CAX1 via SOS2 is independent of SOS3 and results in maintenance of Ca^{2+} homeostasis.

Tissue Tolerance

The third mechanism, tissue tolerance entails an increase in survival of old leaves. It requires compartmentalization of Na^+ and Cl^- at the cellular and intracellular level to avoid toxic concentrations within the cytoplasm, especially in mesophyll cells in the leaf (Munns and Tester, 2008), and synthesis and accumulation of compatible solutes within the cytoplasm. A major consequence of NaCl stress is the loss of intracellular water. To prevent this water loss from the cell and protect the cellular proteins, plants accumulate many metabolites known as compatible solutes which play a role in plant osmotolerance by various ways, protecting enzymes from denaturation, stabilizing membrane or macromolecules or playing adaptive roles in mediating osmotic adjustment (Ashraf and Foolad, 2007). Water moves from high water potential to low water potential and accumulation of these osmolytes make the water potential low inside the cell and thus prevent the intracellular water loss. The function of the compatible solutes is not limited to osmotic balance but also to protect cellular structures through scavenging ROS (Hasegawa *et al.*, 2000). Compatible solutes are typically hydrophilic, and may be able to replace water at the surface of proteins or membranes, thus acting as low molecular weight chaperones (Hasegawa *et al.*, 2000). Compatible solutes are small molecules, water soluble and uniformly neutral with respect to the perturbation of cellular functions, even when present at high concentrations (Sakamoto and Murata, 2002). They comprise of nitrogen containing compounds such as amino acids, amines and betaines, organic acids, sugars, mainly fructose and sucrose, sugar alcohols and complex sugars like trehalose and fructans.

Glycine-betaine

Glycine betaine (N,N,N-trimethylglycine-betaine; GB) is a major osmolyte and is synthesized by many plants in response to abiotic stresses. Biosynthetic pathway of betaine is a two-step oxidation of choline. Recently, a biosynthetic pathway of betaine from glycine, catalyzed by two N-methyl transferase enzymes, was found. With the inclusion of these genes in various plants and other organisms, it has been shown greater tolerance to salinity. Likewise, direct foliar application of the compound also improves the plant response to a saline environment (Tuteja, 2007). It has been found that the co-expression of N-methyl transferase gene in cyanobacteria caused accumulation of betaine in significant amounts and conferred salt tolerance to a fresh water cyanobacterium to become capable of growth in seawater.

Proline

The synthesis of proline is a frequent response in salt stress. This osmolyte is accumulated in the cytosol and allows proper osmotic adjustment. It is synthesized from glutamic acid by the action of enzymes pyrroline-5-carboxylate synthetase (P5CS) and pyrroline-5-carboxylate reductase (P5CR). The induction of P5CS gene seems be mediated by ABA, as its mRNA accumulates quickly in response to the treatment with this plant hormone (Chinnusamy *et al.*, 2005; Vinocur and Altman, 2005). Proline synthesis, following transcriptional activation of the NADPH-dependent P5C-synthetase (P5CS), could provide a protective valve whereby the regeneration of NADPH could provide the observed protective effect under stress conditions. Furthermore, this amino acid also stabilizes subcellular structures, buffers the redox potential and blocks free radicals.

LEA (Late Embryogenesis–Abundant) Proteins

Osmotic stresses induce late-embryogenesis-abundant (LEA) proteins in vegetative tissues, which impart dehydration tolerance to vegetative tissues of plants. Accumulation levels of these proteins correlate with stress tolerance in various plant species suggesting protective roles under osmotic stress. These proteins constitute a diverse family, but their sequences are enriched in polar amino acid, charged or uncharged, as glycines, glutamic acid or lysines. This hydrophilic character can explain their diverse functions, which include water retention as the protein D-19 from cotton; stabilization of other proteins by hydrophilic and hydrophobic interactions as RAB proteins; formation of a solvation surface around proteins, similar to the sugar solvation surface induced by salinity too; and ion sequestering as HVA1 protein in barley. ABA is inducing LEA protein synthesis in embryogenesis, as has been revealed in some mutants in the ABA signalling, and the same procedure will be controlling the expression of these proteins under salinity (Chourey *et al.*, 2003).

Polyamines

Polyamines have recently gained importance in the escape of seedlings from the adverse effect of salinity. The diamine putrescine and the polyamines spermidine and spermine are present in probably all plants, whereas the diamine cadaverine occurs within the Leguminosae. Putrescine accumulation during environmental stresses is correlated with increased argenine decarboxylase (ADC) activity in oats. Under salinity and drought conditions, polyamines as well as their corresponding enzyme activities are substantially enhanced (Lefevre and Lutts, 2000). Lin and Kao (1995) reported an increase in the level of spermidine under salinity, but a low level of putrescine in the shoot and roots of rice seedlings. Accumulation of spermidine and spermine with the activity of ADC in rice seedlings plays a specific role in salt-tolerance. DNA in plant mitochondria and chloroplasts is regulated and stabilized by putrescine and other polyamines. In addition, many steps of protein biosynthesis are stimulated by polyamines, may be through interaction with nucleic acids. The anti-senescence and anti-stress effects of polyamines are due to their neutralizing and antioxidant properties and for their membrane and cell wall stabilizing abilities.

Sugar Alcohols

Sorbitol, sugar alcohol of glucose accumulates in seeds of many crop plants. In Rosaceae species, it functions as a translocated carbohydrate and is also reported in vegetative parts in the halo-tolerant *Plantago maritime* (Ahmad *et al.*, 1979) which increases several folds in shoot tissues and in root tissues under saline conditions. Accumulation in *P. maritime* serves an osmo-regulatory function and its accumulation in plant seeds suggest that it may contribute to the desiccation tolerance of the mature embryo. The conversion of glucose to its sugar alcohol is catalyzed by aldose reductase. An aldose reductase-like protein accumulates during the period of embryo maturation in barley when desiccation tolerance is obtained (Bartels *et al.*, 1991).

The accumulation of cyclic sugar alcohols, pinnitol and ononitol has been observed in a variety of species which are consistently exposed to saline conditions or accumulate in tolerant species when exposed to saline environments. Facultative halophyte such as *Mesembryanthemum crystallinum* accumulates these compounds only when subjected to water and salinity stresses (Paul and Cockburn, 1989).

Mannitol is correlated with salt-stress tolerance of plants. These solutes are widely believed to function as a protector or stabilizer of enzymes or membrane structures that are sensitive to dehydrations or ionically induced damage. Transgenic tobacco plants that synthesize and accumulate mannitol were engineered by introduction of a bacterial gene that encodes mannitol 1-phosphate dehydrogenase. Growth of plants from control and mannitol-containing lines in the absence and presence of added sodium chloride showed increased ability of mannitol-containing lines to tolerate high salinity (Tarczynski *et al.*, 1993).

Trehalose

Trehalose, a non-reducing sugar, possesses a unique feature of reversible water storage capacity to protect biological molecules from desiccation damages. Recently, there has been growing interest of utilization of trehalose metabolism to ameliorate the effects of abiotic stresses. Garg *et al.* (2002) have demonstrated the expression of trehalose biosynthesis conferred the tolerance to multiple abiotic stresses. The increase in trehalose levels in transgenic rice lines of Pusa Basmati 1 using either tissue specific or stress dependent promoter, resulted into the higher capacity for photosynthesis and concomitant decrease in the extent of photo-oxidative damage during salt, drought and low temperature stresses.

Transcription Factors Involved in Salinity Stress

Transcriptional modulation has been known to play a major role in the control of plant responses to salt stress (Shinozaki and Yamaguchi-Shinozaki, 1996). Transcription factors have been identified based on interaction with promoters of osmotic/salt stress–responsive genes. These factors participate in the activation of stress-inducible genes, and presumably lead to osmotic adaptation. Since the promoters that are controlled by these transcription factors are responsive to several environmental signals, it is not clear which transcription factors, if any, function only in salt stress responses, or if salt-specific transcriptional regulation alone is a requisite component of salt tolerance in plants. ABA-deficient and insensitive mutants have been used to delineate transcription factors as components of osmotic stress signal transduction pathways that either involve or are independent of the growth regulators (Shinozaki and Yamaguchi-Shinozaki, 1997). Promoters of ABA-dependent osmotic stress-responsive genes include regulatory elements that interact with transcription factors *viz.* basic leucine zipper motif (bZIP), MYB, or MYC domains in DNA binding proteins. The bZIP transcription factors interact with the ABA-responsive element (ABRE) (Shinozaki and Yamaguchi-Shinozaki, 1997). The promoter of the osmotic/salt stress-responsive *rd22* gene contains signature recognition motifs for MYC and MYB transcription factors. DNA binding proteins, rd22BP1 (MYC) and Atmyb2 (MYB), interact with their corresponding *cis*-elements and *trans*-activate *rd22*. However, the *rd22* promoter does not contain ABRE motifs, which indicates that ABA dependency is mediated through another process. At least some ABA-dependent transcriptional activation appears to involve Ca^{2+}-dependent protein kinases (CDPKs) (Sheen, 1996). Transcription factors thought to function in osmotic/salt stress gene induction, independent of ABA, include dehydration response element binding (DREB) proteins. Two gene families have been characterized, *DREB1* (which includes the previously identified *CBF1*), and *DREB2* (which *trans*-activates the osmotic/salt-responsive *rd29A*). Both DREB1 and DREB2 family members also have domains that bind ethylene-responsive elements. *DREB1A* overexpressing transgenic plants exhibited constitutive activation of stress-responsive genes and enhanced freezing, dehydration, and salt tolerance (Kasuga *et al.*, 1999). Driving *DREB1A* with the stress-inducible *rd29A* promoter substantially increases salt tolerance, with minimal adverse growth effects in the absence of stress (Kasuga *et al.*, 1999). Ectopic overexpression of *CBF1/DREB1B* in transgenic plants induces cold-responsive genes and enhances freezing tolerance (Thomashow, 1999). Overexpression of the zinc finger transcription factor ALFIN1 activates *MsPRP2* (NaCl-responsive gene) expression and increases salt tolerance of alfalfa.

References

Abd-ElBaki, G.K., Siefritz, F., Man, H.M., Weiner, H., Haldenhoff, R. and Kaiser, W.M., 2000. Nitrate reductase in *Zea mays* L. under salinity. *Plant Cell Environ.*, 23: 515–521.

Abdel-Latif, A., 2008. Phosphoenolpyruvate carboxylase activity of wheat and maize seedlings subjected to salt stress. *Aust. J. Basic Appl. Sci.*, 2: 37-41.

Ahmad, I., Larhar, F. and Stewart, G.R., 1979. Sorbitol: a compatible osmotic solute in *Plantago maritime*. *New Physiol.*, 82: 671–678.

Akram, N.A. and Ashraf, M., 2011. Improvement in growth, chlorophyll pigments and photosynthetic performance in salt-stressed plants of sunflower (*Helianthus annuus* L.) by foliar application of 5-aminolevulinic acid. *Agrochimica*, 55: 94-104.

Albrecht, V., Ritz, O., Linder, S., Harter, K. and Kudla, J., 2001. The NAF domain deWnes a novel protein–protein interaction module conserved in Ca^{2+}-regulated kinases, *EMBO J.*, 20: 1051–1063.

Aragao, M.E.F., Guedes, M.M., Otoch, M.L.O., Guedes, M.I.F., Melo, D.F. and Lima, M.G.S., 2005. Differential responses of ribulose- 1,5-bisphosphate carboxylase/oxygenase activities of two *Vigna unguiculata* cultivars to salt stress. *Braz. J. Plant Physiol.*, 17: 207-212.

Arfan, M., Athar, H.R. and Ashraf, M., 2007. Does exogenous application of salicylic acid through the rooting medium modulate growth and photosynthetic capacity in differently adapted spring wheat cultivars under salt stress? *J. Plant Physiol.*, 6: 685-694.

Asada, K., 1999. The water-water cycle in chloroplasts: scavenging of active oxygens and dissipation of excess photons. *Annu. Rev. Plant Physiol.*, 50: 601-639.

Ashraf, M., and Foolad, M.R., 2007. Roles of glycine betaine and proline in improving plant abiotic stress resistance. *Environ. Exp. Bot.*, 59: 206-216.

Ashraf, M., and Sultana, R., 2000. Combination effect of NaCl salinity and N-form on mineral composition of sunflower plants. *Biol. Plant.*, 43: 615-619.

Bartels, D., Engelhardt, K., Roncarati, R., Schneider, K., Rotter, M. and Salamini, F., 1991. An ABA and GA modulated gene expressed in the barley encodes an aldose reductase related protein. *EMBO J.*, 10: 1037–1043.

Bernstein, L. and Hayward. H.E., 1958. Physiology of salt tolerance. *Ann. Rev. Plant Phyiol.*, 9: 25–46.

Bor, M., Ozdemir, F and Türkan, I., 2003. The effect of salt stress on lipid peroxidation and antioxidants in leaves of sugar beet *Beta vulgaris* L. and wild beet *Beta maritima*. *Plant Sci.*, 164: 77-84.

Cai-Hong, P., Su-Jun, Z., Zhi-Zhong, G. and Wang Bao-Shan, W., 2005. NaCl treatment markedly enhances H_2O_2-scavenging system in leaves of halophyte *Suaeda salsa*. *Physiol. Plant.*, 125: 490-499.

Candan, N. L. and Tarhan, L., 2003. The correlation between antioxidant enzyme activities and lipid peroxidation levels in *Mentha pulegium* organs grown in Ca^{2+}, Mg^{2+}, Cu^{2+}, Zn^{2+} and Mn^{2+} stress conditions. *Plant Sci.*, 163: 769-779.

Carillo, P., Mastrolonardo, G., Nacca, F. and Fuggi, A., 2005. Nitrate reductase in durum wheat seedlings as affected by nitrate nutrition and salinity. *Funct. Plant Biol.*, 32: 209–219.

Chaves, M.M., Flexas, J. and Pinheiro, C., 2009. Photosynthesis under drought and salt stress: regulation mechanisms from whole plant to cell. *Ann. Bot.*, 103: 551–560.

Chawla, S., Jain, S. and Jain, V., 2013. Salinity induced oxidative stress and antioxidant system in salt-tolerant and salt-sensitive cultivars of rice (*Oryza sativa* L.). *J. Plant Biochem. Biotech.*, 22: 27-34 .

Chen, G.L., Li, D., Song, L., Hu, C., Wang, G. and Liu, L., 2006. Effects of salt stress on carbohydrate metabolism in desert soil alga *Microcoleus vaginatus*. *J. Integr. Plant Biol.*, 48: 914-919.

Cheng, N.H., Pittman, J.K., Zhu, J.K. and Hirschi, K.D., 2004. The protein kinase SOS2 activates the *Arabidopsis* H^+/Ca^{2+} antiporter CAX1 to integrate Ca transport and salt tolerance, *J. Biol. Chem.*, 279: 2922–2926.

Chinnusamy, V., Jangerdof, A. and Zhu, J.K., 2005. Understanding and improving salt tolerance in plants. *Crop Sci.*, 45: 437-448.

Chourey, K., Ramani, S. and Apte, S.K., 2003. Accumulation of LEA proteins in salt (NaCl) stressed young seedlings of rice (*Oryza sativa* L.) cultivar Bura Rata and their degradation during recovery from salinity stress. *J. Plant Physiol.*, 160: 1165-1174.

Cramer, G.R. and Nowak, R.S., 1992. Supplemental manganese improves the relative growth, net assimilation and photosynthetic rates of salt-stressed barley. *Physiol. Plant.*, 84: 600-605.

Dal Santo, S., Stampfl, H., Krasensky, J., Kempa, S., Gibon, Y., Petutschnig, E., Rozhon, W., Heuck, A., Clausen, T. and Jonak, C., 2012. Stress-Induced GSK3 regulates the redox stress response by phosphorylating glucose-6-phosphate dehydrogenase in *Arabidopsis*. *Plant Cell*, 24: 3380–3392.

Davenport, R., James, R., Zakrisson-Plogander, A., Tester, M. and Munns, R., 2005. Control of sodium transport in durum wheat. *Plant Physiol.*, 137: 807-818.

Deane-Drummond, C.E., 1986. A comparison of regulatory effects of chloride on nitrate uptake, and of nitrate on chloride uptake into *Pisum sativum* seedlings. *Physiologia Plant.*, 66: 115–126.

Debouba, M., Gouia, H., Suzuki, A. and Ghorbel, M.H., 2006. NaCl stress effects on enzymes involved in nitrogen assimilation pathway in tomato "*Lycopersicon esculentum*" seedlings. *J. Plant Physiol.*, 163: 1247-1258.

Desingh, R. and Kanagaraj, G., 2007. Influence of salinity stress on photosynthesis and antioxidative systems in two cotton varities. *Genet. Appl. Plant Physiol.*, 33: 221-234.

Dionisio-Sese, M.L. and Tobita, S., 1998. Antioxidant responses of rice seedlings to salinity stress. *Plant Sci.*, 135: 1-9.

Dluzniewska, P., Gessler, A., Dietrich, H., Schnitzler, J.P., Teuber, M. and Rennenberg, H., 2007. Nitrogen uptake and metabolism in *Populus x canescens* as affected by salinity. *New Phytol.*, 173: 279-293.

Dulai, S., Molnar, I. and Molnar-Lang, M., 2011. Changes of photosynthetic parameters in wheat/barley introgression lines during salt stress. *Acta Biol.*, 55: 73-75.

Eckardt, N.A., 2009. A new chlorophyll degradation pathway. *Plant Cell*, 21:700

Eyidogan, F. and Oz, M.T., 2007. Effect of salinity on antioxidant responses of chickpea seedlings. *Acta Physiol. Plant.*, 29: 485-493.

Evelin, H., Kapoor, R. and Giri, B., 2009. Arbuscular mycorrhizal fungi in alleviation of salt stress: a review. *Annals Bot.*, 104: 1263-1280.

FAO, 2008. FAO Land and plant nutrition management service. *http://www.fao.org/ag/agl/agll/spush*

FAO, 1996. Fact sheets: World food summit-November 1996. Rome, Italy: FAO

Flexas, J., Bota, J. and Loreto, F., 2004. Diffusive and metabolic limitations to photosynthesis under drought and salinity in C3 plants. *Plant Biol.*, 6: 269-279.

Flores, P., Botella, M.A., Martinez, V. and Cerda, A., 2000. Ionic and osmotic effects of nitrate reductase activity in tomato seedlings. *J. Plant Physiol.*, 156: 552–557.

Flowers, T.J., Troke, P.F. and Yeo, A.R., 1977. Mechanism of salt tolerance in halophytes. *Annu. Rev. Plant Physiol.*, 28: 89-121.

Foyer, C.H., Valadier, M., Migge, A. and Becker, T.W., 1998. Drought–induced effects on nitrate reductase activity and mRNA and on coordination of nitrogen and carbon in maize leaves. *Plant Physiol.*, 117: 283–292.

Foyer, C.H., Lelandais, M. and Kunert, K.J., 1994. Photooxidative stress in plants. *Physiol. Plant.*, 92: 696-717.

Gama, P.B.S., Inanaga, S., Tanaka, K. and Nakazawa, R., 2007. Physiological response of common bean (*Phaseolus vulgaris* L.) seedlings to salinity stress. *African J. Biotech.*, 6: 79–88.

Gara, L.D., Pinto, M.C. and Tommasi, F., 2003. The antioxidant systems vis-a-vis reactive oxygen species during plant-pathogen interaction. *Plant Physiol. Biochem.*, 41: 863-870.

Garg, A.K., Kim, J.K., Owens, T. G., Ranwala, A.P., Choi, Y.D., Kochian, L.V. and Wu, R.J., 2002. Trehalose accumulation in rice plants confers high tolerances levels to different abiotic stresses. *Proc. Natl. Acad. Sci. USA*, 99: 15898-15903.

Ghosh, S., Bagchi, S. and Majumder, A.L., 2001. Chloroplast fructose- 1, 6-bisphosphatase from *Oryza* differs in salt tolerance property from the Porteresia enzyme and is protected by osmolytes. *Plant Sci.*, 160: 1171-1181.

Gimenez, C., Mitchell, V.J. and Lawlor, D.W., 1992. Regulation of photosynthesis rate of two sunflower hybrids under water stress. *Plant Physiol.*, 98: 516-524.

Gomathi, R. and Rakkiyapan, P., 2011. Comparative lipid peroxidation, leaf membrane thermostability, and antioxidant system in four sugarcane genotypes differing in salt tolerance. *Int. J. Plant Physiol. Biochem.*, 3: 67-74.

Greenway, H. and Munns, H., 1980. Mechanisms of salt tolerance in nonhalophytes. *Annu. Rev. Plant Physiol.*, 31: 149-190.

Gueta-Dahan, Y., Yaniv, Z., Zilinskas, B.A. and Ben- Hayyim, G., 1997. Salt and oxidative stress: similar and specific responses and their relation to salt tolerance in citrus. *Planta*, 203: 460- 469.

Halfter U, Ishitani M. and Zhu J.K., 2000. The *Arabidopsis* SOS2 protein kinase physically interacts with and is activated by the calcium-binding protein SOS3. *Proc. Natl. Acad. Sci. USA*, 97: 3735–40.

Harinasut, P., Poonsopa, D., Roengmongkol, K. and Charoensataporn, R., 2003. Salinity effects on antioxidant enzymes in mulberry cultivar. *Sci. Asia*, 29: 109-113.

Hasegawa, P.M., Bressan, R.A., Zhu, J.K., and Bohnert, H.J., 2000. Plant cellular and molecular responses to high salinity. *Annu. Rev. Plant Physiol.*, 51: 463-499.

Hernandez, J.A., Campillo, A., Jimenez, A., Alarcon, J.J., and Sevilla, F., 1999. Response of antioxidant systems and leaf water relations to NaCl stress in pea plants. *New Phytol.*, 141: 241-251.

Hernandez, J.A., Corpas, F.J., Gomez, M., del Rio, L.A. and Sevilla, F., 1993. Salt-induced oxidative stress mediated by activated oxygen species in pea leaf mitochondria. *Physiol. Plant.*, 89: 103-110.

Hernandez, J.A., Jimenez, F., Mullineaux, P. and Sevilla, F., 2000. Tolerance of pea (*Pisum sativum* L.) to long-term salt stress is associated with induction of antioxidant defenses. *Plant Cell Environ.*, 23: 853-862.

Hodges, M., 2002. Enzyme redundancy and the importance of 2-oxoglutarate in plant ammonium assimilation. *J. Exp. Bot.*, 53: 905-916.

Houimli, S.I.M., Denden, M. and Elhadj, S.B., 2008. Induction of salt tolerance in pepper (B0 or Hbt 10 mmtt1) by 24-epibrassinolide. *Eurasia J. Biol. Sci.*, 2: 83–90.

Hussain, T., Chandrasekhar, T., Hazara, M., and Sultan, Z., 2008. Recent advances in salt stress biology - a review. *Mol. Biol. Rep.*, 3: 8-13.

Jamil, M., Lee, C.C., Rehman, S., Lee, D., Ashraf, M. and Rha, E., 2005. Salinity (NaCl) tolerance of brassica species at germination and early seedling growth. *Electron. J. Environ. Agric. Food Chem.*, ISSN: 1579–4377.

Jamil, M., Rehman, S. and Rha, E.S., 2007. Salinity effect on plant growth, ps11 photochemistry and chlorophyll content in sugar beet (*Beta vulgaris* L.) and cabbage (*Brassica oleracea capitata* L.). *Pak. J. Bot.*, 39: 753–760.

Juan, M., Rivero, R.M., Romero, L. and Ruiz, J.M., 2005. Evaluation of some nutritional and biochemical indicators in selecting salt resistant tomato cultivars. *Environ. Exp. Bot.*, 54: 193-201.

Kameli, A. and Losel, D.M., 1995. Contribution of carbohydrates and solutes to osmotic adjustment in wheat leaves under water stress. *J. Plant Physiol.*, 145: 363–366.

Kasuga, M., Liu, Q., Miura, S., Yamaguchi- Shinozaki, K. and Shinozaki, K., 1999. Improving plant drought, salt, and freezing tolerance by gene transfer of a single stress-inducible transcription factor. *Nat. Biotechnol.*, 17: 287-91.

Khatun, S., Rizzo, C. A. and Flowers, T.J., 1995. Genotypic variation in the effect of salinity on fertility in rice. *Plant Soil*, 173: 239–250.

Kim, S., Lim, J., Park, M., Kim, Y., Park, T., Seo, Y. and Yun, S., 2005. Enhanced antioxidant enzymes are associated with reduced hydrogen peroxide in barley roots under saline stress. *J. Biochem. Mol. Biol.*, 38: 218-224.

Knight, H., Trewavas, A.J. and Knight, M.R., 1997. Calcium signalling in *Arabidopsis thaliana* responding to drought and salinity. *Plant J.*, 12: 1067–1078.

Kono, Y. and Fridovich, I., 1983. Inhibition and reactivation of Mn-catalase. *J. Biol. Chem.*, 258: 13646-13468.

Kossman, J., Sonnewald, U. and Willmitzer, L., 1994. Reduction of the chloroplastic fructose-1,6-bisphosphatase in transgenic potato plants impairs photosynthesis and plant growth. *Plant J.*, 6: 637-650.

Krause K.P., 1998. Sucrose metabolism in enid-stored potato tubers with decreased expression of sucrose phosphate synthase. *Plant Cell Environ.*, 21: 285-299.

Lasa, B., Frechilla, S., Aparicio-Tejo, P.M. and Lamsfus, C., 2002. Role of glutamate dehydrogenase and phosphoenolpyruvate carboxylase activity in ammonium nutrition tolerance in roots. *Plant Physiol. Biochem.*, 40: 969-976.

Lauchli, A., and Epstein, E., 1990. Plant responses to saline and sodic conditions. In: *Agricultural salinity assessment and management* (Ed. K.K Tanji), American Society of Civil Engineers, New York, pp. 113-137.

Lefevre, I. and Lutts, S., 2000. Effects of salt and osmotic stress on free polyamine accumulation in moderately salt resistant rice cultivar AIW 4. *Int. Rice Res. Note*, 25: 36–37.

Li, F., Vallabhaneni, R. and Yu, J., 2008. The maize phytoene synthase gene family: overlapping roles for carotenogenesis in endosperm, photomorphogenesis and thermal stress tolerance. *Plant Physiol.*, 147: 1334-1346.

Li, T., Zhang, Y. and Liu, H., 2010. Stable expression of *Arabidopsis* vacuolar Na^+/H^+ antiporter gene *AtNHX1* and salt tolerance in transgenic soybean for over six generations. *Chinese Sci. Bull.*, 55: 1127-1134.

Lin, C.C. and Kao, C.H., 1995. Levels of endogenous polyamines and NaCl inhibited growth of rice seedlings. *Plant Growth Regul.*, 17: 15–20.

Lin, H., Sandra, S.S. and Schumaker, K.S., 1997. Salt sensitivity and the activities of the H-ATPase in cotton seedlings. *Crop Sci.*, 37:190–197.

Liu, J., Ishitani, M., Halfter, U., Kim, C.S. and Zhu, J.K., 2000. The *Arabidopsis thaliana SOS2* gene encodes a protein kinase that is required for salt tolerance, *Proc. Natl. Acad. Sci.* USA, 97: 3730–3734.

Loredana, F.C., Pasqualina, W., Amodio, F., Giovanni, P. and Petronia, C., 2011. Plant genes for abiotic stress. In: *Abiotic stress in plants - Mechanism and Adaptation*, pp. 283-308.

Loupassaki, M.H., Chartzoulakis, K.S., Digalaki, N.B. and Androulakis, I.I., 2002. Effects of salt stress on concentration of nitrogen, phosphorus, potassium, calcium, magnesium and sodium in leaves, shoots and roots of six olive cultivars *J. Plant Nutr.*, 25: 2457-2482.

Maas, E.V. and Poss, J.A., 1989. Salt sensitivity of wheat at different growth stages. *Irrig. Sci.*, 10: 29–40.

Makino, A., 2011. Photosynthesis, grain yield, and nitrogen utilization in rice and wheat. *Plant Physiol.*, 155: 125-129.

Mansour, M.M.F., Salama, K.H.A., Ali, F.Z.M., and Abou Hadid, A.F., 2005. Cell and plant responses to NaCl in *Zea mays* L. cultivars differing in salt tolerance. *Gen. Appl. Plant Physiol.*, 31: 29-41.

Martinez-Ballesta, M.C., Martinez, V. and Carvajal, M., 2004. Osmotic adjustment, water relations and gas exchange in pepper plants grown under NaCl or KCl. *Environ. Exp. Bot.*, 52: 161-174.

Meloni, D.A., Oliva, M.A., Martinez, C.A. and Cambraia, J., 2003. Photosynthesis and activity of superoxide dismutase, peroxidase and glutathione reductase in cotton under salt stress. *Environ. Exp. Bot.*, 49: 69-76.

Memon, S.A., Hou, X. and Wang, L.J., 2010. Morphological analysis of salt stress response of pak Choi. *EJEAF Chem.*, 9: 248–254.

Miller, G.A.D., Suzuki, N., Ciftci-Yilmaz, S., and Mittler, R.O.N., 2010. Reactive oxygen species homeostasis and signalling during drought and salinity stresses. *Plant Cell Environ.*, 33: 453-467.

Mittler, R., 2002. Oxidative stress, antioxidants and stress tolerance. *Trends Plant Sci.*, 7: 405-410.

Mittler, R., Vanderauwera, S., Gollery, M. and Breusegem, F.V., 2004. Reactive oxygen gene network of plants. *Trends Plant Sci.*, 9: 490–498.

Mittova, V., Guy, M., Tal, M. and Volokita, M., 2004. Salinity upregulates the antioxidative system in root mitochondria and peroxisomes of the wild salt-tolerant tomato species *Lycopersicon pennellii*. *J. Exp. Bot.*, 55: 1105-1113.

Monirifari, H., and Barghi, M., 2009. Identification and selection for salt tolerance in alfalfa (*Medicago sativa* L.) ecotypes via physiological traits. *Notulae Sci. Biol.*, 1: 63-66.

Moud, A.M. and Maghsoudi, K., 2008. Salt stress effects on respiration and growth of germinated seeds of different wheat (*Triticum aestivum* L.) cultivars. *World J. Agr. Sci.*, 4: 351-358.

Mudgal, V., Madaan, N., Mudgal, A. and Mishra, S., 2009. Changes in growth and metabolic profile of chickpea under salt stress. *J. Appl. Biosci.*, 23: 1436-1446.

Mumm, P., Wolf, T. and Fromm, J., 2011. Cell type-specific regulation of ion channels within the maize stomatal complex. *Plant Cell Physiol.*, 52: 1365-1375.

Munns, R., 2002. Comparative physiology of salt and water stress. *Plant Cell Environ.*, 25: 239-250.

Munns, R., 2005. Genes and salt tolerance: Bringing them together. *New Phytol.*, 167: 645-663.

Munns, R., James R. A. and Lauchli A., 2006. Approaches to increasing the salt tolerance of wheat and other cereals. *J. Exp. Bot.*, 57: 1025-1043.

Munns, R., Schachtman, D., and Condon, A., 1995. The significance of a two-phase growth response to salinity in wheat and barley. *Funct. Plant Biol.*, 22: 561-569.

Munns, R. and Tester, M., 2008. Mechanisms of salinity tolerance. *Annu. Rev. Plant Biol.*, 59: 651–681.

Muscolo, A., Rosaria, A.P. and Sidari M., 2003. Effects of salinity on growth, carbohydrate metabolism and nutritive properties of kikuyu grass (*Pennisetum clandestinum hochst*). *Plant Sci.*, 164: 1103-1110.

Noreen, Z., Ashraf, M. and Akram, N.A., 2010. Salt-induced modulation in some key gas exchange characteristics and ionic relations in pea (*Pisum sativum* L.) and their use as selection criteria. *Crop Pasture Sci.*, 61: 369-378.

Obendorf, R.L., 1997. Oligosaccharides and galactosyl cyclitols in seed desiccation tolerance. *Seed Sci. Res.*, 7: 63-74.

Oh, K., Kato, T. and Xu, H.L., 2008. Transport of nitrogen assimilation in xylem vessels of green tea plants fed with NH_4-N and NO_3-N. *Pedosphere*, 18: 222-226.

Papp, J.C., Ball, M.C., and Terry, N., 1983. A comparative study of the effects of NaCl salinity on respiration, photosynthesis, and leaf extension growth in *Beta vulgaris* L. (sugar beet). *Plant Cell Environ.*, 6: 675-677.

Parida, A.K., and Das, A.B., 2005. Salt tolerance and salinity effects on plants: A review. *Ecotox. Environ. Safe.*, 60: 324-349.

Pattanagul and Wattana, 2008. Effect of salinity stress on growth and carbohydrate metabolism in three rice (*Oryza sativa* L.) cultivars differing in salinity tolerance. *Indian J. Exptl. Biol.*, 46: 736-742.

Paul, M.J. and Cockburn, W., 1989. Pinnitol, a compatible solute in *Mesembryanthemum crystallinum* L. *J. Exp. Bot.*, 40: 1093–1098.

Perveen, S., Shahbaz, M. and Ashraf, M., 2010. Regulation in gas exchange and quantum yield of photosystem II (PSII) in saltstressed and non-stressed wheat plants raised from seed treated with triacontanol. *Pak. J. Bot.*, 42: 3073-3081.

Pinheiro, H.A., Silva, J.V. and Endres, L., 2008. Leaf gas exchange, chloroplastic pigments and dry matter accumulation in castorbean (*Ricinus communis* L.) seedlings subjected to salt stress conditions. *Ind. Crop. Prod.*, 27: 385-392.

Ponnamperuma, F.N., 1984. Role of cultivar tolerance in increasing rice production in saline lands. In: *Salinity tolerance in plants: Strategies for crop improvement* (Eds.: R.C. Staples and G.H. Toenniessen), John Wiley and Sons, New York. pp. 255-271.

Qiu, Q.S., Guo, Y., Dietrich, M.A., Schumaker K.S. and Zhu, J.K., 2002. Regulation of SOS1, a plasma membrane Na^+/H^+ exchanger in *Arabidopsis thaliana*, by SOS2 and SOS3. *Proc. Natl. Acad. Sci. USA*, 99: 8436–8441.

Quintero, F.J., Ohta, M., Shi, H.Z., Zhu, J.K. and Pardo, J.M., 2002. Reconstitution in yeast of the *Arabidopsis* SOS signaling pathway for Na^+ homeostasis. *Proc. Natl. Acad. Sci. USA*, 99: 9061–66.

Rahnama, A., Poustini, K., Tavakkol-Afshari, R. and Tavakoli, A., 2010. Growth and stomatal responses of bread wheat genotypes in tolerance to salt stress. *Int. J. Biol. Life Sci.*, 6: 216-221.

Ramajulu, S., Vlerajaniyulu, K. and Sudhakar, C., 1994. Short-term shifts in nitrogen metabolism in mulberry *Morus alba* under salt shock. *Phytochem.*, 35: 991-995.

Reddy, A.R., Chaitanya, K.V. and Vivekanandan, M.M., 2004. Drought-induced responses of photosynthesis and antioxidant metabolism in higher plants. *J. Plant Physiol.*, 161: 1189-1202.

Rengasamy, P., 2002. Transient salinity and subsoil constraints to dryland farming in Australian sodic soils: an overview. *Aust. J. Exp. Agric.*, 42: 351–361.

Rossel, J.B., Wilson, I.W., Pogson, B.J., 2002. Global changes in gene expression in response to high light in *Arabidopsis*. *Plant Physiol.*, 130: 1109–1120.

Rui, L., Wei, S., Mu-xiang, C., Cheng-jun, J., Min, W. and Bo-ping, Y., 2009. Leaf anatomical changes of *Burguiera gymnorrhiza* seedlings under salt stress. *J. Trop. Subtrop. Bot.*, 17: 169–175.

Sadale, A.N., 2007. Physiological Studies in *Sesbania grandiflora*. Ph. D. Thesis submitted to Shivaji University, Kolhapur, Maharashtra.

Saffan, S.E., 2008. Effect of salinity and osmotic stresses on some economic plants. *Res. J. Agric. Biol. Sci.*, 4: 159–166.

Saibo, N.J.M., Lourenço, T. and Oliveira, M.M. 2009. Transcription factors and regulation of photosynthetic and related metabolism under environmental stresses. *Ann. Bot.*, 103: 609-623.

Sairam, R.K., Rao, K.V. and Srivastava, G.C., 2002. Differential response of wheat genotypes to long term salinity stress in relation to oxidative stress, antioxidant activity and osmolyte concentration. *Plant Sci.*, 163:1037-1046.

Saqib, M., Zorb, C. and Schubert, S., 2006. Salt resistant and salt-sensitive wheat genotypes show similar biochemical reaction at protein level in the rst phase of salt stress. *J. Plant Nutr. Soil Sci.*, 169: 542–548.

Sakamoto, A. and Murata, N., 2002. The role of glycine betaine in the protection of plants from stress: clues from transgenic plants. *Plant Cell Environ.*, 25: 163-171.

Saleem, A., Ashraf, M. and Akram, N.A., 2011. Salt (NaCl)-induced modulation in some key physio-biochemical attributes in okra (*Abelmoschus esculentus* L.). *J. Agron. Crop Sci.*, 197: 202- 213.

Santos, C., Azevedo H. and Caldeira, G., 2001. *In situ* and *in vitro* senescence induced by KCI stress: nutritional imbalance, lipid peroxidation and antioxidant metabolism. *J. Exp. Bot.*, 52: 351-360.

Santos, C.V., 2004. Regulation of chlorophyll biosynthesis and degradation by salt stress in sunflower leaves. *Sci. Hort.*, 103: 93-99.

Saravanavel, R., Ranganathan, R. and Anantharaman, P., 2011. Effect of sodium chloride on photosynthetic pigments and photosynthetic characteristics of *Avicennia officinalis* seedlings. *Recent Res. Sci. Technol.*, 3: 177-180.

Scandalios, J.G., 1993. Oxygen stress and superoxide dismutases. *Plant Physiol.*, 101: 7-12.

Scharte, J., Schon, H., Weis, E. and von Schaewen, A., 2009. Isoenzyme replacement of glucose-6-phosphate dehydrogenase in the cytosol improves stress tolerance in plants. *Proc. Natl. Acad. Sci. USA*. 106: 8061–8066.

Schuch, U.K. and Kelly, J.J., 2008. Salinity tolerance of cacti and succulents. *Turfgrass, Landscape and Urban IPM Research Summary*, pp. 155.

Seemann, J.R. and Sharkey, T.D., 1982. The effect of abscisic acid and other inhibitors on photosynthetic capacity and the biochemistry of CO_2 assimilation. *Plant Physiol.*, 84: 696- 700.

Shah, S.H., 2008. Effects of nitrogen fertilization on nitrate reductase activity, protein and oil yields of *Nigella sativa* L. as affected by foliar GA_3 application. *Turk. J. Bot.*, 32: 165-170.

Sharif, M., Ghorbanli, M., and Ebrahimzadeh, H., 2007. Improved growth of salinity stressed soybean after inoculation with pretreated mycorrhizal fungi. *J. Plant Physiol.*, 164: 1144–1151.

Sheen, J., 1996. Ca^{2+}-dependent protein kinases and stress signal transduction in plants. *Sci.*, 274: 1900–1902.

Shi, H., Ishitani, M., Kim, C. and Zhu, J.K., 2000. *Arabidopsis thaliana* salt tolerance gene *SOS1* encodes a putative Na^+/H^+ antiporter, *Proc. Natl. Acad. Sci. USA*, 97: 6896–6901.

Shinozaki, K. and Yamaguchi-Shinozaki, K., 1996. Molecular responses to drought and cold stress. *Curr. Opin. Biotechnol.*, 7: 161–67.

Shinozaki, K. and Yamaguchi-Shinozaki, K., 1997. Gene expression and signal transduction in water-stress response. *Plant Physiol.*,115: 327-34.

Singh, S.C., Rajeshwar P.S. and Donat P.H., 2002. Role of lipids and fatty acids in stress tolerance in cyanobacteria. *Acta Protozool.*, 41: 297–308.

Stewart, C.K. and J.A. Lee, 1974. The role of proline accumulation in halophytes. *Planta*, 120: 279-289.

Sudhakar, C., Lakshmi. A. and Giridarakumar, S., 2001. Changes in the antioxidant enzyme efficacy in two high yielding genotypes of mulberry (*Morus alba* L.) under NaCl salinity. *Plant Sci.*, 161: 613-619.

Surabhi, G.K., Reddy, A.M., Jyothsnakumari, G. and Sudhakar, C., 2008. Modulation of key enzymes of nitrogen metabolism in two genotypes of mulberry (*Morus alba* L.) with differential sensitivity to salt stress. *Environ. Exp. Bot.*, 6: 171-179.

Taffouo, V.D., Wamba, O.F., Yombi, E., Nono, G.V. and Akoe, A., 2010. Growth, yield, water status and ionic distribution response of three bambara groundnut (*Vigna subterranean* (L.) *verdc.*) landraces grown under saline conditions. *Int. J. Bot.*, 6: 53–58.

Taji, T., Ohsumi, C., Iuchi, S., Seki, M., Kasuga, M., Kobayashi, M., Yamaguchi-Shinozaki, K., and Shinozaki K., 2002. Important roles of drought- and cold-inducible genes for galactinol synthase in stress tolerance in *Arabidopsis thaliana*. *Plant J.*, 29:417-426.

Tarczynski, M.C., Jensen, R.G. and Bohnert, H.J.,1993. Stress protection of transgenic tobacco by production of the osmolyte mannitol. *Sci.*, 259: 508-510.

Tattini, M., Gucci, R., Coradeschi, M.A., Ponzio, C. and Everard, J.D., 1995. Growth gas exchange and ion content in *Olea europaea* plants during salinity and subsequent relief. *Physiol. Plant.*, 98: 117-124.

Taub, D., 2010. Effects of rising atmospheric concentrations of carbon dioxide on plants. *Nat. Educ. Knowl.*, 1: 21.

Thomashow, M.F., 1999. Plant cold acclimation: freezing tolerance genes and regulatory mechanisms. *Annu. Rev. Plant Physiol.*, 50: 571-99.

Torsethaugen, G., Pitcher, L.H., Zilinskas B.A. and Pell, E.J., 1997. Overproduction of ascorbate peroxidase in the tobacco chloroplast does not provide protection against ozone. *Plant Physiol.*, 114: 529-537.

Turan, M.A., Kalkat, V. and Taban, S., 2007. Salinity-induced stomatal resistance, proline, chlorophyll and ion concentrations of bean. *Int. J. Agric. Res.* 2: 483–488.

Turhan, H., Genc. L., Smith, S.E., Bostanci, Y.B. and Turkmen, O.S., 2008. *African. J. Biot.*, 7: 750-756.

Tuteja, N., 2007. Mechanisms of high salinity tolerance in plants. *Method Enzymol.*, 428: 419-438.

Vinocur, B. and Altman, A., 2005. Recent advances in engineering plant tolerance to abiotic stress: achievements and limitations. *Curr. Opin. Biotech.*, 16: 123-132

Walker, A.J. and Ho. L.C., 1977. Carbon translocation in the tomato: Effects of fruit temperature on carbon metabolism and the rate of translocation. *Ann. Bot.*, 41: 825-832.

Wang, Z.Q., Yuan, Y.Z., Ou, J.Q., Lin, Q.H. and Zhang, C.F., 2007. Glutamine synthetase and glutamate dehydrogenase contribute differentially to proline accumulation in leaves of wheat (*Triticum aestivum*) seedlings exposed to different salinity. *J. Plant Physiol.*, 164: 695-701.

Wickert, S., Marcondes, J., Lemos, M.V. and Lemos, E.G.M., 2007. Nitrogen assimilation in citrus based on CitEST data mining. *Genet. Mol. Biol.*, 30: 810-818.

Willekens, H., Chamnongpol, S., Davey, M., Schraudner, M. and Langebartels, C. 1997. Catalase is a sink for H_2O_2 and is indispensable for stress defense in C3 plants. *EMBO J.*, 16: 4806-4816.

Winicov, I. and Seemann, J.R., 1990. Expression of genes for photosynthesis and the relationship to salt tolerance of alfalfa (*Medicago sativa*) cells. *Plant Cell Physiol.*, 31: 1155-1161.

Wu, J.D., Seliskar, M., and Gallagher, J.L., 1998. Stress tolerance in the salt marsh plant *Spartina patens*: impact of NaCl on growth and root plasma membrane lipid composition. *Physiol. Plant.*, 102: 307-317.

Xue, G.P., McIntyre, C.L., Glassop, D. and Shorter, R., 2008. Use of expression analysis to dissect alterations in carbohydrate metabolism in wheat leaves during drought stress. *Plant Mol. Biol.*, 67: 197-214.

Yang, J.Y., Zheng, W. and Tian, Y., 2011. Effects of various mixed salt-alkaline stresses on growth, photosynthesis and photosynthetic pigment concentrations of *Medicago ruthenica* seedlings. *Photosynthetica*, 49: 275-284.

Yoshimura, K., Yabute, Y., Ishikawa, T. and Shigeoka, S., 2000. Expression of spinach ascorbate peroxidase isoenzymes in response to oxidative stresses, *Plant Physiol.*, 123: 223-233.

Zeid, I.M., 2009. Trehalose as osmoprotectant for maize under salinity-induced stress research. *J. Agr. Biol. Sci.*, 5: 613- 622.

Zhang, L., Zhang, Z. and Gao, H., 2011. Mitochondrial alternative oxidase pathway protects plants against photoinhibition by alleviating inhibition of the repair of photodamaged PSII through preventing formation of reactive oxygen species in *Rumex* K-1 leaves. *Physiol. Plant*. 143: 396-407.

Zhao, G.Q., Ma, B.L. and Ren, C.Z., 2007. Growth, gas exchange, chlorophyll fluorescence and ion content of naked oat in response to salinity. *Crop Sci.*, 47: 123-131.

Zhu, J.K., Liu, J. and Xiong, L., 1998. Genetic analysis of salt tolerance in *Arabidopsis*. Evidence for a critical role of potassium nutrition, *Plant Cell*, 10: 1181–1191.

2016, Recent Advances in Plant Stress Physiology *Pages* **171–184**
Editors: ***Praduman Yadav, Sunil Kumar and Veena Jain***
Published by: **DAYA PUBLISHING HOUSE, NEW DELHI**

Chapter 8

Recent Advances in *Populus euphratica* in Relation to Salinity Stress

Vishnu Dayal Rajput[1,2]* and Chen Yanning[1]

[1]Xinjiang Institute of Ecology and Geography,
University of the Chinese Academy of Sciences, Urumqi, Xinjiang, China
[2]University of the Chinese Academy of Sciences, Beijing 100049, China

ABSTRACT

Plant growth is highly affected by abiotic stress factors like drought or salinity. Populus euphratica is the only tree species naturally distributed at the edge of barren deserts or semi-barren deserts worldwide, and is well known for its high tolerance to salinity and atmospheric drought. Although being a non-halophyte, it grows vigorously in soils containing less than 2.0 per cent salinity and survives up to 5.0 per cent. Populus is long life span woody species, so it evolved a kind of adaptive strategies to survive long time in saline water. In this chapter, we focus on the adaptability of P. euphratica in saline condition and their importance for arid lands.

Keywords: *Desert, Drought, Populus euphratica, Riparian plant, Salinity.*

Introduction

The genus *Populus* L. is a member of the Salicaceae family and dominant tree species of the riparian forest in arid areas. It plays an important role in stabilizing these vulnerable ecosystems. The Euphrates poplar was described by Olivier in 1801 and named after the Euphrates River, where large population of *Populus* existed at that time (Weisgerber, 2000). The total distribution of *Populus euphratica* is much

**Corresponding author*: E-mail: rajput.vishnu@gmail.com

larger and ranges from Morocco in the west over North Africa, West and South Asia to Central Asia as far as Mongolia. It can be found in India, Pakistan, Kazakhstan, Mongolia and Iran as well as in China.

P. euphratica is able to grow in saline environment, but its distribution is restricted to river-banks or areas with deep water tables (Hukin *et al.*, 2005). One of the most remarkable characters of *P. euphratica* is the disparity in shape between young and mature leaves, known as heterophylly. The young plants and twigs on older trees develop narrow lanceolate leaves. *Populus euphratica* also has several specific features for scientific research such as capacity of fast growth, genotype diversity, wide distribution and suitable material for molecular analyses (Rajagopal *et al.*, 2000).

Currently about 6 per cent of the world's land area are salt affected (Munns and Tester, 2008) and increasing due to inappropriate irrigation regimes or use of salt-affected irrigation water. Salinity is the amount of total soluble salts. As soluble salt levels increases, it becomes more difficult for plants to extract water from soil. Soil salinity is expressed as the electrical conductivity of solution extracted from the soil at water saturation. Salinity values are given in units of millimhos per centimeter (mmhos/cm) or decisiemens per meter (dS/m). If the salinity concentration in the soil is high enough, the plant will start wilt and die. Salts generally found in saline soils are NaCl (table salt), $CaCl_2$, gypsum ($CaSO_4$), magnesium sulfate, potassium chloride and sodium sulfate. The calcium and magnesium salts are at a high enough concentration to offset the negative soil effects of the sodium salts. The pH of saline soils is generally below than 8.5. Leaf injury and death is probably due to the high salt load in the leaf that exceeds the capacity of salt compartmentation in the vacuoles, causing salt to build up in the cytoplasm to toxic levels (Munns *et al.*, 2006). Halophytes are salt tolerant plants that can cope with salts concentration in the soil higher than 400 mM. The high salinity responds to modifications in wood anatomy such as changes in tracheary element density, lumen area and alterations of the cell wall, morphology *i.e.* stomatal density, shape and size, plant growth and also effect activities of anti-oxidative enzymes. Water transport in xylem and evaporation by stomata are the foundation for maintaining survival in terrestrial vascular plants under stress conditions. The xylem carries water from the roots to the leaves; it is also the major supporting tissue in woody plants. Long term salt exposure results in excessive accumulation of Na and Cl, which require specific compartmentation to avoid ion induced injury (Chen *et al.*, 2003). Salt stress like drought executes osmotic stress in plants, and it is known that osmotic stress affects auxin transport.

Research on molecular basis of salt resistance in plants has been mainly focused on herbaceous plants (Zhang *et al.*, 2004). The main targets of salt impact are vegetative growth, anatomy, flowering, fruit development and seed germination, accompanied by metabolic dysfunctions, including decreased photosynthetic rates, protein and nucleic acid metabolism and enzymatic activity.

Populus euphratica and Salinity Stress

Currently, salinity is a major problem worldwide. Salinization transforms fertile and productive land to barren land, and often leads to loss of habitat and reduction of biodiversity. *P. euphratica* is not belongs in xerophyte but it have some characteristics of xerophyte. Ma *et al.* (2013) reported the genome sequence of the desert poplar, *P. euphratica*, which exhibits high tolerance to salt stress. They found several genes families likely to be involved in tolerance to salt stress contain significantly more gene copies within the *P. euphratica* lineage. The evolutionary adaptation of *P. euphratica* to saline environments is decreasing transpiration, absorbing ion selectively, reducing ATP consuming, succulent leaf and plasticity (Ottow *et al.*, 2005; Janz *et al.*, 2010; Zhang *et al.*, 2013).

Salinity often alters the morphology and anatomy of woody plants in form of tracheary element density, vessel diameters, lumen area changes, alterations in stomatal aperture, density, shape and size, plant growth and also effect activities of anti-oxidative enzymes. Leaves of plants that grow on saline soils are thicker and more succulent than those of trees growing on salt-free soils (Shannon *et al.*, 1994), salt exposure more than 354 mM NaCl create leaf necrosis in two year old seedlings (Shaoling *et al.*, 2001) whereas it can uptake and transport of salt ions under high levels of external salinity in the longer term. The epidermal cell walls and cuticles of leaves of salinized plants also are thicker. By increasing the internal surface area per unit of leaf surface, leaf succulence may increase CO_2 absorption per unit of leaf area (Shannon *et al.*, 1994). Salinity not only inhibits the rate of cambial growth but also influences the anatomy of plant.

Researcher observed effects of salt and mannitol on *in-vitro* shoot culture of *P. euphratica* with other spp., all *P. euphratica* plantlets were survived at all levels of mannitol and NaCl, while the mortality of other spp. was high (Zhang *et al.*, 2004). *P. euphratica* can secrete salt from its body by discharging salty water through portals in its trunks and leaves (Fu *et al.*, 2012). These all characters, *P. euphratica* considered model tree plant for salt tolerant.

Morphological Characters

P. euphratica is a medium-sized deciduous and anemophilous tree. There are big variation in plant height, 6 M in Spain, 10 M in Turkey and India, 15 M in China and Pakistan; most of them shaped like sparse shrubs, bark light grayish-brown, lower part stripe cracked, sprouting branch slender and round smooth or with villi, bud elliptical smooth brownish and about 7 M long. Leaves of seedlings and sprouting branches are lanceolata or linear lanceolata, entire or irregular sparse sinuate serrate dentate margin branch. The leaves on mature trees are broad-ovate and coarsely serrated. The male catkin grows about two or three, towards the end of the flowering period, up to seven cm in length. The purple red anthers give to the inflorescence a reddish appearance. Female catkins are 2-5 cm long, and grow up to 10 cm in length during ripening period. Female flowers are exists in two different colour morphs, purple and yellowish. When the plant reach is an age of 10–15 years they start to clonally propagate via root suckering.

Reproduction and Ecology

Optimal conditions for germination is bright sunlight, temperature between 25-30 C, water saturated soil and a salt content below 0.2 per cent. Such conditions are increased germination rate of more than 80 per cent (Thevs *et al.*, 2008). Seedling is cohorts form long narrow bands that follow the high water mark. Above-ground development of seedling is limited during the first seasons. The seedling is devoted mainly in the root system and forms prominent tap roots that allow the plant to keep contact to the ground water. Wiehle *et al.* (2009) were measured an average shoot length of 8.1 cm whereas an average root length of 41.7 cm in first year of seedlings. The longest tap root found in one year old seedlings were measured a solid 119 cm. When plants reaches an age of 10–15 years, they started to propagate clonally via root suckers, produced on horizontally running lateral roots, close to the soil surface up to 40 M away from the parent tree. Root suckers also triggered by exposure or damaging of the root system (Wiehle *et al.*, 2009). Since generative regeneration is not possible in established stands (Gries *et al.*, 2005), vegetative reproduction becomes the only means of rejuvenation and expansion, after the first phase of colonization. The distance root suckers have ability to form chains of root suckers (Wiehle *et al.*, 2009). The spatial arrangement and size of old clones is further illustrated the importance of clonal growth in forming adjoining forests (Bruelheide *et al.*, 2004).

P. euphratica is restricted to semi-arid and arid eco-regions; the plants possess surprisingly few adaptations for such as an environment. A waxy cuticle and sunken stomatal aperture are the most prominent concessions to the harsh climatic conditions. The stomata open during the day and remain open even during the hottest midday hours (Wang *et al.*, 1996). These obligate phreatophytes require constant access to the ground water table. Even short term disconnection from their water supply leads to degradation and extermination of whole stands(Gries *et al.*, 2005).

Mechanism of Adaptability under Salinity Stress

P. euphratica leaves have diverse shape, hard, thicker and covered with wax and protect themselves from oxidative damage due to reactive oxygen species (ROS) through production of antioxidants (Ediga *et al.*, 2013). Hydraulic conductivity and embolism are the important factors constraining on plant survival and its productivity. It adjusts their xylem hydraulic traits in various stress conditions through developing more resistant xylem to drought induced cavitation (Awad *et al.*, 2010). *P.euphratica* reproduced photosynthetic rate, evaporation by limiting stomatal aperture, accumulate salts in leaves and roots, deepen tape roots and grow well under dry hot climatic condition. It is growing in arid or semi arid areas but it needs underground water. Normally it founds near the river basin. The Tarim River and the Heihe River is two most important and largest inland rivers basin in Western China. World's 54 per cent area of *P. euphratica* vegetation is spread in these basins (Wang *et al.*, 1996; Zhuang *et al.*, 2010). Its main root penetrate vertically in soil 1 M deep and also a lot of lateral, feeder roots grow on the lower end of the main roots. This special kind of structure of root system is favorable for resisting to the mechanical effect of strong winds in barren desert.

Effect of Salt on *Populus euphratica*

The main targets of salinity on plants are vegetative growth, anatomy, flowering, fruit development and seed germination (Figure 8.1). Salinity is one of the major abiotic stresses which limiting plant growth and productivity. Previous studies have revealed the modifications in *P. euphratica* such as changes in tracheary elemental density, lumen area, vessels diameter, stomatal density, shape and size, plant growth and effect on activities of anti-oxidative enzymes to reduce salinity effects (Meloni *et al.*, 2003; Junghans *et al.*, 2006; Cavusoglu *et al.*, 2008; Chen and Polle, 2010; Ediga *et al.*, 2013). It was recently shown that *P. euphratica* had considerably lower rates of Na^+ net root uptake and Na^+ transport to the shoot under salinity, compared to salinity sensitive *P. tomentosa* (Chen *et al.*, 2003). Gene expression analysis in *P. euphratica* showed that its genome does not contain different genes, compared to the sequenced *P. trichocarpa*, but that the regulation of gene expression is different in response to salinity compared to *P. trichocarpa* (Brosche *et al.*, 2005).

However, little is known about the adjustments in response to salinity in roots of *P. euphratica*, the primary site of NaCl uptake, and the distribution of Na^+ within the plant. Escalante-Perez *et al.* (2009) found xylem differentiation zone was reduced and diameter of lumina has decreased by the exposure of salt for two weeks. In response to soil water deficit less than 1.5 per cent of the genes on the array displayed significant changes in transcription levels. Physiological water deficit are caused by drought, but also by freezing or soil salinity. High salinity may be impede water uptake or even reverse water flux during events of sudden. Long term salt exposure results in excessive accumulation of Na and Cl, which require specific compartmentation to avoid ion induced injury. Because of these attributes, *P. euphratica* is used as a model tree for salt tolerance in trees.

Impact of Salinity on Xylem Anatomy

Xylem forms a specialized conducting channel that carries water and solutes throughout the plant and differentiated from pro-cambium derived from apical meristem and vascular cambium. Xylem develops on the adaxial or internal pole of vascular bundles. Xylem is located at the center of the stem and comprised of conducting tracheary elements called parenchyma cells, which are vessel elements and non-conducting elements called xylary fibers (Dinneny and Yanofsky, 2004). The xylem vessels of halophytic trees are more numerous and narrower than those in mesophytic trees and production of fibers increases (Strogonov, 1964). Salt stress decreased the stem diameter, epidermis cell size, cortex zone thickness, vascular bundle width, cambium thickness, xylem width, trachea diameter (Çavu°oglu *et al.*, 2008). Escalante-Perez *et al.*, 2009 found xylem differentiation zone was reduced and diameter of lumina has decreased by the exposure of salt under hydroponic conditions (Figure 5.2).

Junghans *et al.* (2006) observed salinity reduced vessel diameter in salt sensitive, after exposure to much lower salt concentration than in salt resistant whereas, high salt concentration (150 mM) showed higher reduction in cambial activity in *P. euphratica*, which was reflected by lower number of xylem cells formed under salt stress, and significantly reduced vessel diameters (Figure 5.3).

Figure 8.1: Effect of Saline Water on Seedling of *P. euphratica* (a); Control (b). One month 100 mM Salt treated seedlings.

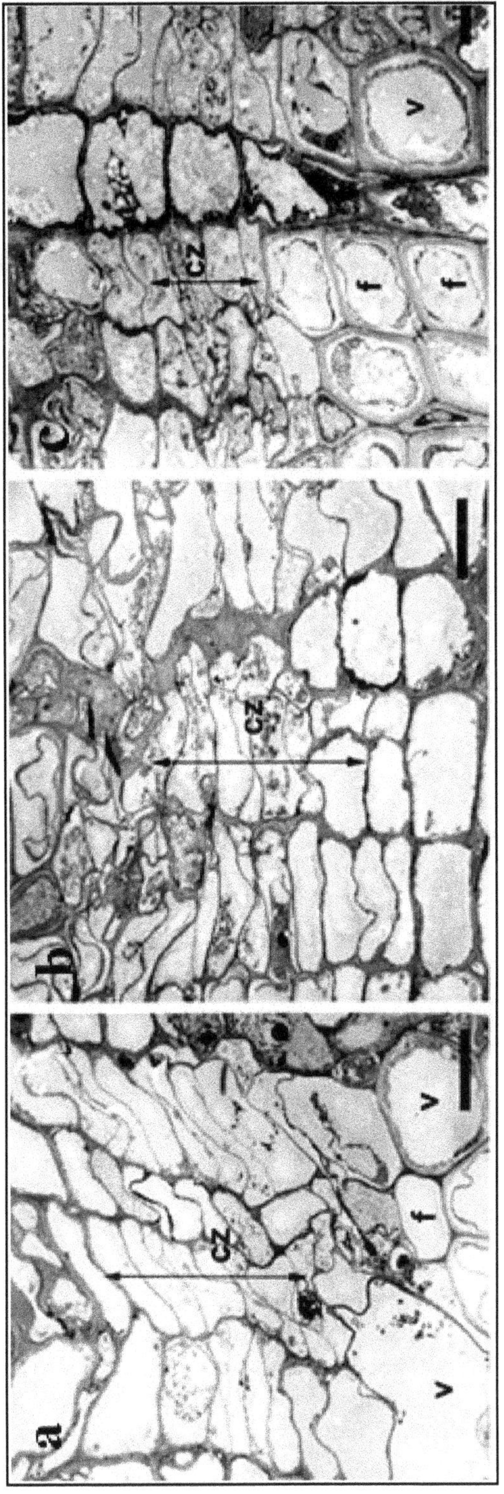

Figure 8.2: TEM Image of Cambial Zone (cz) of Grey Polar, a control, b 1 week salt exposure, c two week salt exposure. After one week salt exposure, the cambial zone consisted 7-9 cell layers and in second week it reduced significantly. F-represent fiber, v-vessel, bar 10 μm (*Source*: Escalante-Perez *et al.*, 2009).

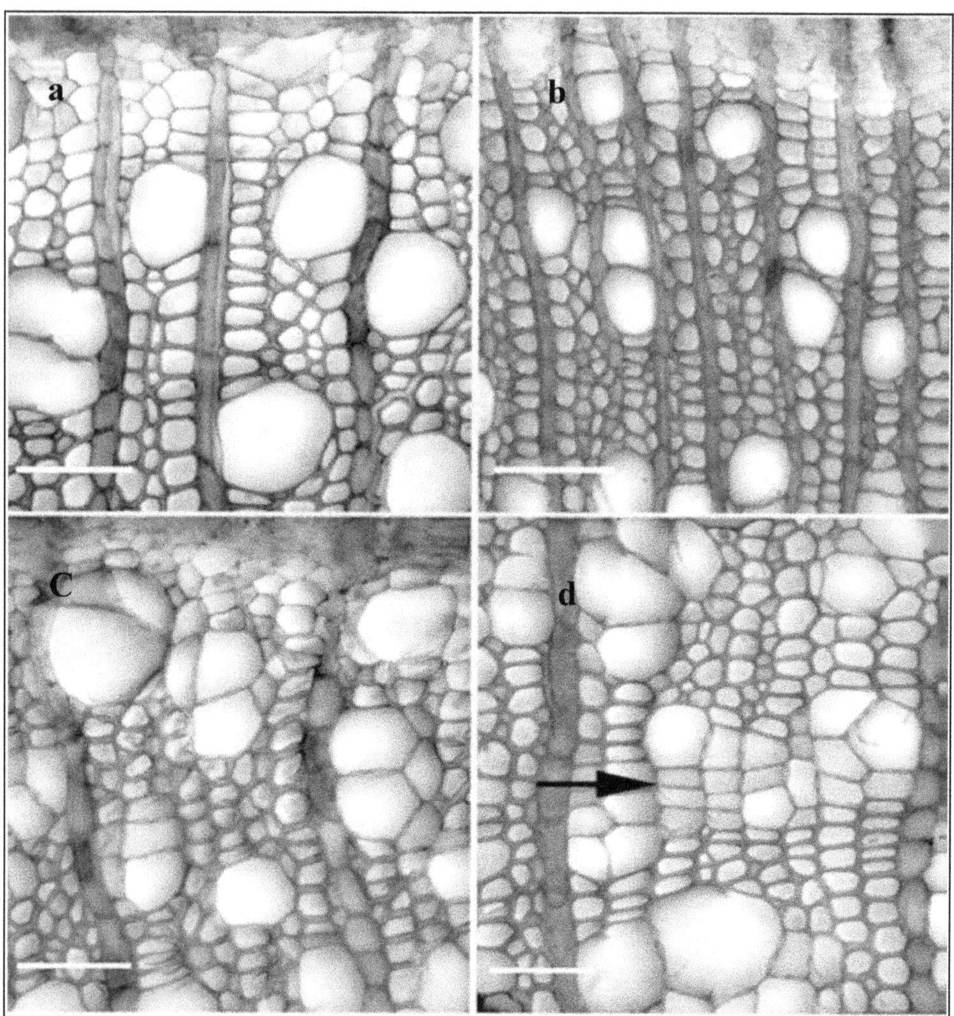

Figure 8.3: Salt-induced Changes in Wood Anatomy of *Populus euphratica* and *Populus* x *canescens*. (a) Control of *P. x canescens* (b) Treatment of *P. x canescens* with 50 mMNaCl for 6 weeks (c) Control of *P. euphratica* (d) Treatment of *P. euphratica* with 50 mMNaCl for 6 weeks. Arrows indicate xylem formed under salt stress. In (b), the whole section shows xylem formed under salt stress. Scale bar (a-d) = 50 μm, (*Source*: Junghans *et al.*, 2006).

Impact of Salinity on Photosynthesis and Stomatal Aperture

P. euphratica response to salinity is a decrease in stomatal aperture and reduced the rate of transpiration. Rate of photosynthesis per unit area in salt treated plants often unchanged, even though stomatal conductance is reduced (James, *et al.*, 2002). Stomata are small structures on plants found on the outer layer of leaf skin, known as epidermis (Figure 8.2). Stomata found on adaxial and abaxial epidermis. They consist of two specialized cells, called guard cells that surround a tiny pore called a stoma. Guard cells regulate stomatal pore size to allow CO_2 uptake while controlling water loss. Stomatal characteristics are size, distribution between leaf surfaces and their densities.

The stomatal aperture limited leaf photosynthetic capacity under salinity stress (Meloni*et al.*, 2003). *P. euphratica* have early mechanism to reduces stomatal size and increase their density to avoid evaporation in stress condition (Abbruzzese *et al.*, 2009). Stomatal density refers to the number of stomata per square millimeter (Figure 8.4).

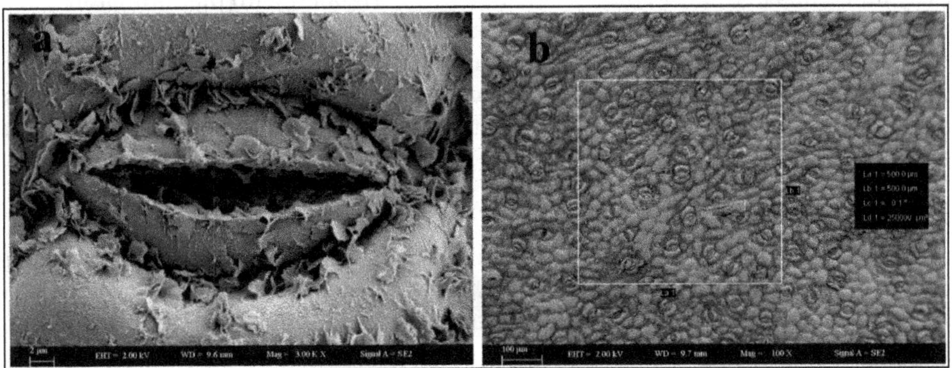

Figure 8.4: (a) Stomata; (b).Stomatal density of two year seedlings of *P. euphratica.*

Antioxidative Response and Effect on Proline under Salinity Stress

Salinity significantly affected the activity of antioxidative enzymes. Plants protect themselves from oxidative damage due to reactive oxygen species (ROS) through both enzymatic and non-enzymatic defense mechanisms (Ardic *et al.*, 2009). Enzymatic ROS-scavenging mechanisms in plants include production of superoxide dismutase (SOD) and peroxidase (POD). The extent of oxidative stress experienced in a cell is determined by the levels of superoxide, H_2O_2 and hydroxyl radicals generated. Additionally, a balance among SOD activities is crucial for suppressing toxic ROS levels within cells (Misra and Gupta, 2005; Ozden*et al.*, 2009). Activities of antioxidative enzymes increased under salinity stress (Ediga *et al.*, 2013). Similar activities have been observed at different level of canopy in *P. euphratica*, the activities of peroxidase (POD) decreased from lower canopy to upper canopy, while the activities of superoxide dismutase (SOD) increased (Pujia *et al.*, 2012). The most studied compound under salinity stress is proline. The amount of proline usually increases under salinity (Khatkar and Kuhad, 2000).

Effect of Salinity on Native Xylem Embolism

Plants can adjust their xylem hydraulic traits in various stress conditions through developing more resistant xylem to drought induced cavitation (Stiller, 2009; Awad *et al.*, 2010), reducing root hydraulic conductance (Triflo *et al.*, 2004), maintaining high hydraulic efficiency in stems and roots (Nardini and Pitt, 1999), and also by increasing whole plant leaf specific conductivity (Cinnirella *et al.*, 2002), changes in xylem sap, salinity can impact hydraulic conductivity in halophytes, plants that tolerate high salt concentrations in the soil. According to Zwieniecki *et al.* (2001) the ionic content of the xylem sap has a significant effect on the hydraulic conductivity of plants, presumably mediated by the swelling and shrinking of hydrogels in pit membrane micro-channels. Availability of nutrients along with factor like abiotic stress is rather variable during plants lifetime, in certain extant plant able to adjust to such changes because its functional traits exhibits phenotypic plasticity. Plant hydraulic traits highly correlated with the xylem structure and their functions (Plavcová and Hacke, 2012). Similar study did by many researchers for hydraulic traits in riparian plants which highly correlated with plants anatomy *viz.* xylem vessel diameter, fibers, pith membrane under abiotic stress (Ayup *et al.*, 2012; Urli *et al.*, 2013; Zhou *et al.*, 2013). The hydraulic stress adaptation of poplar involves cell wall modifications and suppresses xylem development (Janz *et al.*, 2012). This changes occurs due to drought along with salinity, by which xylem development suppress significantly.

Importance

P. euphratica is used for afforestation in arid, semi-arid and in barren lands of India and China, where it is functioning as a sand stabilizer (Fung *et al.*, 1998). In addition, its function for reforestation, wood is used for house construction, as fuel wood and leaves used for fodders.

To Maintain Ecological Equilibrium

Poplar forests are the dominant species of the desert ecosystem in the temperate arid zone. They are fundamental for maintaining the stability of the river-banks forest in arid zones, and have ability to conserve the water resources and provide appropriate habitats for various other plants. They also provide favorable habitats and food resources for micro-organism and animals and the setting in which they can exist and propagate their population. Regions with abundant water and flourishing forests are the migration channels of migratory birds and wild animals. Thus, poplar trees enhance the complicated structure of the food chains of oasis ecosystem and contribute to the stability of oases. The poplar trees are playing an important role in fixing sand dunes, holding back desertification, alleviating dust storms, conserving biodiversity and improving the environment.

Ecological Barrier for Dust Storms

Poplar has strong roots and the main roots reach depths of 6-8 M. horizontally, the roots may extend for tens of meters. These attributes ensure that enough water can be absorbed by the tree. The root can also absorb salt from the soil to increase the osmotic pressure across the cell wall, which enhances the ability to absorb water and prevents salinization. Even if the main trunk is totally withered, the root can catch hold of the soil and sand. Therefore, poplar trees possess a powerful ecological ability to maintain water and soil, as well as withstanding storms and fixing sand dunes.

Conclusion and Future Prospectives

Populus euphratica is morphologically, anatomically and ecologically diverse riparian species, which survives in saline and drought eco-environment. It has numerous salinity and drought tolerant characters such as hard, thicker, waxy leaves; deepen root system, production of antioxidant under stress to avoid damage due to reactive oxygen species (ROS) along-with limiting photosynthetic capacity by reducing stomatal aperture and its density. It is good source for domestic necessities and well known for sand stabilizer in arid environment. Therefore, preserving or planting the poplar forest will play a key role in alleviating dust storms in the world and make possibility to reclaim salinized lands.

All available research information showed that, there is lack of scientific knowledge on the correlation of saline water transport from root to leaf, there effect on transport system, effect on physiology, effect on xylem development and on growth of *P. euphratica*, an important species in riparian vegetations.

References

Abbruzzese, G., Beritognolo, I., Muleo, R., Piazzai, M., Sabatti, M., Mugnozza, G.S. and Kuzminsky, E., 2009. Leaf morphological plasticity and stomatal conductance in three *Populus alba* L. genotypes subjected to salt stress. *Environ. Exp. Bot.*, 66: 381-388.

Ardic, M., Sekmen, A.H. and Tokur, S., 2009. Antioxidant responses of chickpea plants subjected to boron toxicity. *Plant Biol.*,11: 218-228.

Ayup, M., Hao, X., Chen, Y., Li, W. and Su, R., 2012. Changes of xylem hydraulic efficiency and native embolism of *Tamarix ramosissima Ledeb.* Seedlings under different drought stress conditions and after re-watering. *S. Afr. J. Bot.*, 78: 75–82.

Awad, H., Barigah, T., Badel, E., Cochard, H. and Herbette, S., 2010. Poplar vulnerability to xylem cavitation acclimates to drier soil conditions. *Physiol. Plant*, 139: 280–288.

Brosche, M., Vincour, B., Alatalo, E.R., Lamminmäki, A., Teichmann, T., Ottow, E.A., Djilianov, D., Afif, D., Triboulot-Bogeat, M.B., Altman, A., Polle, A., Dreyer, E.Rudd, S., Paulin, L., Auvinen, P. and Kangasjärvim, J., 2005. Gene expression and metabolite profiling of *Populus euphratica* growing in the Negev desert. *Genome Biol.*, 6: R101.

Bruelheide, H., Manegold, M. and Jandt, U., 2004. The genetical structure of *Populus euphratica* and *Alhagi sparsifolia* stands in the Taklimakan Desert. In: Runge M. and Zhang X. (Eds.): Contributions to a Workshop on Sustainable Management of the Shelterbelt Vegetation of River Oases in the Taklimakan Desert, Shaker Verlag, Aachen. 153–160.

Çavuþoglu, K., Kiliç, S. and Kabar, K., 2008. Effects of some plant growth regulators on stem anatomy of radish seedlings grown under saline (NaCl) conditions. *Plant Soil Environ.*, 54: 428–433.

Chen, S. and Polle, A., 2010. Salinity tolerance of Populus. *Plant Biol.*, 12: 317–333.

Chen, S.L., Li, J.K., Wang, S.S., Fritz, E., Huttermann, A. and Altman, A., 2003. Effects of NaCl on shoot growth, transpiration, ion compartmentation, and transport in regenerated plants of *Populus euphratica* and *Populus tomentosa*. *Can. J. Forest. Res.*, 33: 967-975.

Cinnirella, S., Magnani, F., Saracino, A. and Borghetti, M., 2002. Response of mature *Pinuslaricio* plantation to a three-year restriction of water supply: structural and functional acclimation to drought. *Tree Physiol.*, 22: 21–30.

Dinneny, J. R., and Yanofsky, M. F., 2004. Vascular patterning: xylem or phloem. *Curr. Biol.*, 14: 112-114.

Ediga, A., Hemalatha, S. and Balaji, M., 2013. Effect of salinity stress on antioxidant defense system of two finger millet cultivars (*Eleusine coracana* (L.) Gaertn) differing in their sensitivity. *Advan. Biol. Res.*, 7: 180-187.

Escalante-Pérez, M., Lautner, S., Nehls, U., Selle, A., Teuber, M., Schnitzler, J.P., Teichmann, T., Fayyaz, P., Hartung, W., Polle, A., Fromm, J., Hedrich, R. and Ache, P., 2009. Salt stress affects xylem differentiation of grey poplar (*Populus× canescens*). *Planta*, 229: 299-309.

Fung, L.E., Wang, S.S., Altman, A. and Hütterman, A., 1998. Effect of NaCl on growth,photosynthesis, ion and water relations of four poplar genotypes.*For. Ecol. Manage.*,107: 135-146.

Fu, A., Li, W.H. and Chen, Y., 2012. The threshold of soil moisture and salinity influencing the growth of *Populus euphratica* and *Tamarix ramosissima* in the extremely arid region. *Environ. Earth Sci.*, 66: 2519–2529.

Gries, D., Zeng, F., Foetzki, A., Arndt, S.K., Bruelheide, H., Thomas, F.M., Zhang, X. and Runge, M., 2005. Growth and water relations of *Tamarix ramosissima* and *Populus euphratica* on Taklamakan desert dunes in relation to depth to a permanent water table. *Plant Cell Environ.*, 26: 725–736.

Hukin, D., Cochard, H., Dreyer, E., Thiec, D.L. andBogeat-Triboulot, M.B., 2005. Cavitation vulnerability in roots and shoots: does *Populus euphratica* Oliv., a poplar from arid areas of Central Asia, differ from other poplar species?. *J. Exp. Bot.*, 56: 2003-2010.

Janz, D., Behnke, K., Schnitzler, J.P., Kanawati, B., Schmitt-Kopplin, P. and Polle, A., 2010. Pathway analysis of the transcriptome and metabolome of salt sensitive and tolerant poplar species reveals evolutionary adaption of stress tolerance mechanisms. *BMC Plant Biol.*, 10:150.

James, R.A., Rivelli, A.R., Munns, R. and von Caemmerer, S., 2002. Factors affecting CO_2 assimilation, leaf injury and growth in salt stressed durum wheat. *Funct. Plant Biol.*, 29: 1393-403.

Junghans, U., Polle, A., Düchting, P., Weiler, E., Kuhlman, B., Gruber, F. and Teichmann, T., 2006. Adaptation to high salinity in poplar involves changes in xylem anatomy and auxin physiology. *Plant Cell Environ.*, 29: 1519–1531.

Khatkar, D. and Kuhad, M.S., 2000. Short-term salinity induced changes in two wheat cultivars at different growth stages.*Biol. Plant.*, 43: 629-632.

Ma, Tao., Wang, Junyi., Zhou, Gongke., Yue, Zhen., Hu, Quanjun., Chen, Yan., Liu, Bingbing., Qiu, Qiang., Wang, Zhuo., Zhang, Jian., Wang, Kun., Jiang, Dechun., Gou, Caiyun., Yu, Lili., Zhan, Dongliang., Zhou, Ran., Luo, Wenchun., Ma, Hui., Yang, Yongzhi., Pan, Shengkai., Fang, Dongming., Luo, Yadan., Wang, Xia.,

Wang, Gaini., Wang, Juan., Wang, Qian., Lu, Xu., Chen, Zhe., Liu, Jinchao., Lu, Yao., Yin, Ye.,Yang, Huanming., Abbott, Richard., Wu, Yuxia., Wan, Dongshi., Li, Jia., Yin, Tongming., Lascoux, Martin., DiFazio, Stephen., Tuskan, Gerald., Wang, Jun. andQianquan, Liu., 2013. Genomic insights into salt adaptation in a desert poplar. Nature Communications 4, Article number: 2797.

Meloni, Da., Oliva,Ma., Martinez, Ca. andCambraia, J., 2003. Photosynthesis and activity of superoxide dismutase, peroxidase and glutathione reductase in cotton under salt stress.*Environ. Exp. Bot.*, 49: 69-76.

Misra, N. and Gupta, A.K., 2005. Effect of salt stress on proline metabolism in two high yielding genotypes of green gram. *Plant Sci.*, 169: 331-119.

Munns, R. and Tester, M., 2008. Mechanisms of salinity tolerance.*Annu. Rev. Plant Biol.*,59: 651–681.

Munns, R., James, R.A. and Lauchli, A., 2006. Approaches to increasing the salt tolerance of wheat and other cereals. *J. Exp. Bot.*, 57: 1025–1043.

Ottow, E.A., Brinker, M., Teichmann, T., Fritz, E., Kaiser, W., Brosche, M., Kangasjarvi, J., Jiang, X. and Polle, A., 2005. *Populus euphratica* display sapoplastic sodium accumulation, osmotic adjustment by decreases in calcium and soluble carbohydrates, and develops leaf succulence under salt stress. *Plant Physiol.*, 139: 1762–1772.

Ozden, M., Demirel, U. and Kahraman, A., 2009. Effect proline on antioxidant system in leaves of grapevine (*Vitis vinifera* L.) exposed to oxidative stress by H_2O_2. *Sci. Hortic.*, 119: 163-168.

Plavcová, L. and Hacke, U.G., 2012. Phenotypic and developmental plasticity of xylem in hybrid poplar saplings subjected to experimental drought, nitrogen fertilization, and shading. *J. Exp. Bot.*, 63: 6481–6491.

Pujia, Yu.Hailiang, Xu.Wei, Shi.Shiwei, Liu.Qingqing, Zhang.,Xinfeng, Zhao., Wei, Zheng.andPeng, Zhang., 2012. Phsyiological indexes of *Populus euphratica* leaves from different canopy positions in the lower reaches of Tarim River. *Pak. J. Bot.*, 44: 933–938.

Rajagopal, J., Khurana, D.K., Srivastava, P.S. and Lakshmikumaran, M., 2000. Analysis of repetitive DNA elements in Populus species and their use in study of phylogenetic relationships.In: Isebrands J.G., Richardson J. (Eds), IPC- Abstracts of papers and posters presented at the 21st session of the commission. Portland, Oregon, USA: pp. 125–126.

Shannon, M.C., Grieve,C.M. and Francois, L.E., 1994. Whole-plant response to salinity.In Plant-Environment Interactions. Ed. R.E. Wilkinson. Marcel Dekker, New York, pp 199-244.

Strogonov, B.P., 1964. Physiological basis of salt tolerance of plants.Akad.Nauk. USSR. (translation by Israel Progr. Sci. Trans., Jerusalem).

Stiller, V., 2009. Soil salinity and drought alter wood density and vulnerability to xylem cavitation of baldcypress(*Taxodium distichum* (L.) Rich.) Seedlings. *Environ. Exp. Bot.*, 67: 64–171.

Thevs, N., Zerbe, S., Schnittler, M., Abdusalih, N. and Succow, M., 2008. Structure, reproduction and flood-induced dynamics of riparian Tugai forests at the Tarim River in Xinjiang, NW China. *Forestry*, 81: 45–57.

Trifilò, P., Raimondo, F., Nardini, A., Lo Gullo, M.A. andSalleo, S., 2004. Drought resistance of *Ailanthus altissima*: root hydraulics and water relations. *Tree Physiol.*, 24: 107–114.

Urli, M., Porté, AJ., Cochard, H., Guengant, Y., Burlett, R. and Delzon, S., 2013. Xylem embolism threshold for catastrophic hydraulic failure in angiosperm trees. *Tree Physiol.*, 33: 672–683.

Wang, S., Chen, B. and Li, H., 1996. Euphrates Poplar Forest. China Environmental Science Press, Beijing.

Weisgerber, H., 2000. *Populus euphratica* OLIVIER, 1801. In: Schütt, P., Weisgerber, H., Schuck, H. J., Lang, U. M., Roloff, A. (Eds.): Enzyklopädie der Holzgewächse. Handbuchund Atlas der Dendrologie. 22. Erg. Lfg. 12/00, III-2. EcomedVerlagsges., Landsberg/Lech.

Wiehle, M., Eusemann, P., Thevs, N. and Schnittler, M., 2009. Root suckering patterns in *Populus euphratica* (Euphrates poplar, Salicaceae). *Trees*, 23: 991–1001.

Zhang, F., Yang, Y. L., HE, W.L., Zhao, X. and Zhang, L. X., 2004. Effects of salinity on the growth and compatible solutes of callus induced from *Populus euphratica*. *In Vitro Cell. Dev. Biol. Plant*, 40: 491–494.

Zhang, J., Xie, P., Lascoux, M., Meagher, T.R. and Liu, J., 2013. Rapidly evolving genes and stress adaptation of two desert poplars, *Populus euphratica* and *P. pruinosa*. *PLoS One*, 8:66370.

Zhou, H., Chen, Y., Li, W. and Ayup, M., 2013. Xylem hydraulic conductivity and embolism in riparian plants and their responses to drought stress in desert of Northwest China. *Ecohydrol.*, 6: 984–993.

Zhuang, L., Li, W.H., Yuan, F., Gong, W.C. and Tian, Z.P., 2010. Ecological adaptation characteristics of *Populus euphratica* and *Tamarix ramosissima* leaf microstructures in the lower reaches of Tarim River. *Acta Ecol. Sin.*, 30: 62–66.

Zwieniecki, M.A., Melcher, P.J. and Holbrook, N.M., 2001. Hydrogel control of xylem hydraulic resistance in plants.*Science*, 291: 1059–1062.

2016, Recent Advances in Plant Stress Physiology *Pages* **185–199**
Editors: ***Praduman Yadav, Sunil Kumar and Veena Jain***
Published by: **DAYA PUBLISHING HOUSE, NEW DELHI**

Chapter 9

Chilling Stress Tolerance in Plants: Physiology and Mechanisms

Ashok Kumar Yadav*, Somu Kaundal, Akhil Sharma and Sanatsujat Singh

CSIR-Institute of Himalayan Bioresource Technology, Palampur – 176 061, H.P., India

ABSTRACT

Chilling stress is a major environmental factor that limits the agricultural productivity of plants. Low temperature has a huge impact on the survival and geographical distribution of plants. It often affects plant growth and crop productivity, which causes significant crop losses. Many plant species are injured by low temperatures, and exhibit various symptoms of chilling injury such as chlorosis, necrosis, or growth retardation. Defy to these damages, some plants can adapt the chilling tolerance, through mechanisms based on protein synthesis, membrane composition changes, and activation of active oxygen scavenging systems. The changes in expression of hundreds of genes in response to cold temperatures are followed by increases in the levels of hundreds of metabolites, some of which are known to have protective effects against the damaging effects of cold stress. Various low temperature inducible genes have been isolated from plants which appear to be involved in tolerance to cold stress and the expression of some of them is regulated by C-repeat binding factor/dehydration-responsive element binding (CBF/DREB1) transcription factors. Many physiological and molecular changes occur during cold acclimation which divulged that the cold resistance is more complex than perceived and involves more than single pathway.

Keywords: *Cold acclimation, C-repeat binding factor, Transgenesis, Transcription factor.*

**Corresponding author.* E-mail: ashok@ihbt.res.in

Abbreviations

CBF: C-repeat binding factor
COR: Cold responsive gene
LTST: Low temperature induced signal transduction
ROS: Reactive oxygen species
LEA: Late embryogenesis abundant proteins
AFP: Antifreeze proteins.

Introduction

Chilling stress is important factor among the various abiotic stresses that not only limit the economic productivity but also affects the geographical distribution of various plants inhabiting hilly areas. Xin and Browse, (2001) reported that chilling stress causes significant crop losses in hilly regions. Low temperature usually affects the cellular macromolecules, that later on leads to slow metabolism, solidification of cell membranes, and loss of membrane functions. It also cause DNA disruption, reductions in enzymatic activity, destabilization of protein complexes, stabilization of RNA secondary structure, accumulation of reactive oxygen intermediates (ROIs),impairment of photosynthesis and leakage across membranes (Nayyar *et al.*, 2005; Nayyar and Chander, 2004). Chilling has been known to severely inhibit plant reproductive development in many crop plant species such as rice displaying sterility when exposed to chilling temperatures during anthesis (Jiang *et al.*, 2003). Chilling effects include reduced leaf expansion and growth (Sowinski *et al.*, 2005; Rymen *et al.*, 2007), wilting (Bagnall *et al.*, 1983), chlorosis (Yoshida *et al.*, 1996) and may lead to necrosis and impaired development of reproductive components, restricted seed and pod development in sensitive plants species (Kaur *et al.*, 2008; Ohnishi *et al.*, 2010), which ultimately reduces the yield of grain crops (Suzuki *et al.*, 2008). Some plants are able to cope up with stress and acquire chilling tolerance while other (chilling sensitive) plants are unable to cope up properly with chilling exposure. Plant species have developed several adaptations at physiological and molecular levels to maximize cold tolerance through a process known as cold acclimation. Temperature change determines the growth, development and physiology of the plants. Morphological chilling stress symptoms are easily noticeable while biochemical and molecular process involved during the exposure to cold temperatures, needs understanding about the vital processes termed as low temperature induced signal transduction (LTST). The LTST leads to the expression of certain genes of interest in the nucleus, which through central dogma results in the synthesis of some specific proteins. These proteins are either structural or enzymes work for the survival of plant during stress conditions and the plant acquires stress tolerance. In addition to their negative effect on plant cell metabolism, chilling not only induce developmental responses such as flowering (vernalization) or seed germination (stratification) but it is also necessary for the induction of bud dormancy in some plant species such as pear and apple during autumn (Heide and Prestrud, 2005).

Rapid advancement in recombinant DNA technology and development of precise and efficient gene transfer protocols have resulted in efficient transformation and generation of transgenic lines in a number of crop species to develop transgenic crop plants with improved tolerance to cold stress. Number of genes have been isolated and characterized that are responsive to freezing stress. Many studies revealed that cold regulated gene expression is critical in plants for both chilling tolerance and cold acclimation. Advent of molecular tools has made it possible to select directly at the gene level without waiting for the phenotype to show up. Transgenic approach is being pursued actively throughout the world to improve traits including tolerance to biotic and abiotic stresses in a number of crops. Therefore, it is important to use most appropriate tools that help in reaching the goals.

Impact of Chilling Stress on Agriculture Production

Most of the crops of tropical as well as subtropical origin are sensitive to chilling temperatures. Lyons, (1973) reported that the temperature range for chilling injuries varies (0 - 4°C for temperate fruits, 8°C for subtropical fruits and 12°C for tropical fruits) with species and regions of origin. In another study (Lynch, 1990) it was reported that food crops like maize and rice are most sensitive to chilling temperatures (<10°C) while other crops such as soybean, cotton, tomato and banana are injured at temperatures below 10-15°C. Chilling stress is one of the major environmental factors which reduce the crop yield significantly. Sanghera *et al.* (2011) reported that every year, worldwide losses in crop production due to chilling damage amount to approximately $ 2 Billion. Also Christiansen and St. John (1981) estimated annual losses of $60 million to the cotton industry due to chilling temperature. Some of the major losses include the 1995 early fall frosts in the US which caused losses of over $1 billion to corn and soybeans. The occasional freezes in Florida have shifted the citrus belt further south, and California sustained $650M of damage in 1998 to the citrus crop due to a winter freeze (Sanghera *et al.*, 2011).

Plant growth and development are adversely affected by the chilling stress which not only restricts the spatial distribution of plants and agricultural productivity but also prevents the expression of genetic potential of plants due to inhibition of metabolic activity through cold-induced osmotic, oxidative and other stresses directly or indirectly. Cold acclimation or cold hardening is a process by which plants acquire freezing tolerance upon prior exposure to low non-freezing temperatures. Most of temperate plants acquire chilling tolerance to extracellular ice formation in their vegetative tissues. Winter-habit plants (winter wheat, barley, oat, rye, oilseed rape, etc.) have a vernalization requirement, which inhibit the premature transition to reproductive phase before the threat of freezing stress during winter. Thus, vernalization requirement allows plants to survive in low temperature as seedlings. However, after vernalization the cold acclimation ability of winter cereals gradually decreases (Fowler *et al.*, 1996). Many important crops, like rice, maize, soybean, cotton and tomato, are chilling sensitive and incapable of cold acclimation; therefore, they cannot tolerate ice formation in their tissues. Although, in chilling sensitive crops the threshold temperature for chilling damage is lowered by prior exposures to suboptimal low temperatures (Anderson *et al.*, 1994).

Physiological Effect of Chilling Stress

Many plants species exhibit symptoms of injury when exposed to low non-freezing temperatures. These plants including maize (*Zea mays*), soybean (*Glycine max*), cotton (*Gossypium hirsutum*), tomato (*Lycopersicon esculentum*) and banana (*Musa* sp.) are sensitive to temperatures below10-15 C. The symptoms of stress induced injury in these plants appear from 48 to 72 hrs; however, this duration varies from plant to plant and also depends upon the sensitivity of a plant to chilling stress. Various phenotypic symptoms in response to chilling stress include reduced leaf expansion, wilting, chlorosis (yellowing of leaves) and may lead to necrosis (death of tissue). Chilling also inhibit the reproductive development of rice plants when exposed to chilling temperature at the time of anthesis (floral opening) leads to sterility in flowers. During chilling the rate of photosynthesis in the leaves of chilling-sensitive plants is also reduced (Boese *et al.*, 1997; Sonoike, 1998). The physiological reasons for the suppression of photosynthesis are the inhibition of phloem transport of carbohydrates from the leaves, stomatal limitation and destruction of the photosynthetic apparatus (Yordanov, 1992; Nie *et al.*, 1992).

Effects of Chilling Temperature

Chilling stress affects cellular macromolecules, leading to slow metabolism. At low temperature protein content in tissues of chilling-sensitive plants is usually reduced due to a sharp decrease in protein synthesis (Levitt, 1980). Low temperatures also reduce the activity of many enzymes (Guy, 1990). The reasons for this may be the dissociation of multimeric enzymes, protein-lipid and hydrophobic interaction disorders, reversible changes in kinetic properties of enzymes and allosteric regulation (Graham and Patterson, 1982; Matsuo *et al.*, 1994). During chilling stress the concentration of soluble sugars increased and starch content decreased significantly in all the organs of chilling-sensitive plants (Jouve *et al.*, 1993). Changes in the level of carbohydrates caused by the chilling are associated with impaired respiration, photosynthesis, and the activity of enzymes of carbohydrate metabolism (Ebrahim *et al.*, 1998).

Effect of Chilling on Photosynthesis

Photosynthesis is substantially reduced by the low temperatures (chilling stress) as it inhibits the sucrose synthesis in the cytosol which leads to the accumulation of phosphorylated intermediates. This results in the reduction of the available inorganic phosphate and decrease the cycling of inorganic phosphate between the cytosol and the chloroplast (Furbank *et al.*, 1987; Hurry *et al.*, 2000). It reduces the ATP synthesis which is necessary for the regeneration of ribulose-1,5-bisphosphate to support CO_2 fixation. Griffth *et al.* (1984) revealed that low temperatures also inhibit thylakoid electron transport by increasing membrane viscosity and restricting the diffusion of plastoquinone. At low temperatures chlorophyll antenna of photo-system I (PSI) and photo-system II (PSII) trap more energy (Ensminger *et al.*, 2006; Huner *et al.*, 1993) due to which thylakoid membrane becomes over energized. This results in formation of ROS which cause photo damage to PSI. The accumulation of excitation energy in chlorophyll antennae of the photo systems favors the production of triplet excited chlorophyll molecules which interact with O_2 to generate reactive singlet oxygen ($^1O_2$). Over reduction of the photosynthetic electron carrier chain cause reduction of O_2 by PSI, leading to the formation of damaging ROS, such as superoxide (O_2), hydrogen peroxide H_2O_2), and the hydroxyl radical (·OH). The activities of the scavenging enzymes will be lowered by low temperatures, and the scavenging systems will be not able to counter balance the ROS formation that is always associated with mitochondrial and chloroplast electron transfer reactions. Moreover, the chloroplast electron transfer chain will be overreduced during chilling, which will lead to increased ROS formation All these have photo inhibitory effect on both the photo system and thus leading to decrease the rate of photosynthesis.

Effect of Chilling on Membrane Stability

Chilling stress change the physical properties of cell membranes. Low temperature reduce the membrane elasticity, decreasing their compliance and preventing lipid inclusion in their composition, lower lipid fluidity, thereby reducing the activity of several membrane-bound enzymes, including H^+-ATPase, increase the lateral diffusion of phospholipids, sterols and proteins in the plasma membrane (Quinn, 1988; Kasamo *et al.*, 1992; Koster *et al.*, 1994;). The reactive oxygen species produced during chilling exposure lead to lipid peroxidation, resulting in ion leakage. Several types of freezeinduced membrane lesions are formed and all are linked to freezing initiating ice nucleation in the apoplast. The formation of extracellular ice induces cellular water to move to the apoplast in response to change in water potential, thus causing cellular dehydration and osmotic contraction of the cell. Exposure to low temperature also changes the protein components of cell membranes which include: disorders of protein structure and release of non-protein components of enzymes (Graham and Patterson, 1982). Prolonged chilling may also lead to irreversible changes in the state of membrane. These changes will cause loss of membrane integrity and compartmentation, the leakage of solutes, decrease of oxidative activity of mitochondria, the accumulation of toxic substances and the symptoms of chilling injury (Lyons, 1973; Levitt, 1980; Quinn, 1988).

Chilling Stress Responses

Chilling induces plenty of responses in plants which participate in the acquisition of chilling tolerance and freezing tolerance in some species like winter rye (*Secale cereal*), Perennial ryegrass (*Lolium perenne*) and Peach (*Prunus persica*) (Hon, 1995; Pudney, 2003; Wisniewski, 1999). This is the molecular basis of cold acclimation *i.e.* an exposure to chilling temperature results freezing tolerance which is accompanied by the formation of stress related proteins like cold responsive proteins (COR), late embryogenesis abundant proteins (LEA), dehydrins and antifreeze proteins and the initiation of different responsive mechanisms towards the chilling stress tolerance (discussed below):

Formation of Stress Related and Antifreeze Proteins

Accumulation of hydrophilic proteins which form an amphipathic α-helix is one of the best responses to chilling stress. Most of these proteins are known as cold responsive (COR), low temperature induced (LTI), responsive to abscisic acid

(RAB) or early responsive to dehydration (ERD). These also include dehydrins, which define Late Embryogenesis Abundant (LEA) proteins. Dehydrins may have a role in freezing tolerance by preventing the membrane destabilization that occurs during osmotic contraction associated with freezing. Some dehydrins also possess cryoprotective properties, antifreeze activity or as oxygen radical scavengers (Bies-Etheve *et al.*, 2008).

During winter, when the temperature in cold areas drops below the freezing point, the ice forms in intercellular spaces of plant cell and heat energy is released as a result of the heat of fusion of water. Further exposure to low temperatures (-40°C) causes water vapor to move through the plasma membrane from the unfrozen protoplast to the cell wall, causing ice crystals to grow within the intercellular spaces. The formation of ice crystals on cell walls or in the protoplasm requires the presence of ice nucleation points (*i.e.* initial ice formation) on which crystals can be initiated and grow. Steponkus, (1984) reported that the ice formation in plant tissues is considered to be lethal to the cell because the plasma membrane is irreversibly ruptured, resulting in the membrane destabilization and the loss of semipermeable characteristics.

In order to prevent ice nucleation and ice crystal, overwintering plants secrete thermal hysteresis or antifreeze proteins (AFPs) which are secreted into the apoplast where they bind irreversibly to the surface of ice and are incorporated into the ice crystal lattice. However, plants have no constitutive AFP activity, it is induced by cold. Antikainen *et al.* (1996) have reported that in winter rye (*Secale cereal*), AFP accumulated in response to cold in the apoplast of leaves and crowns while in another study of Griffith *et al.* (2005) reported that AFPs appear to directly interact with ice and reduce freezing injury by slowing the growth and re-crystallization of ice.

Chilling Tolerance Mechanisms

The mechanisms involved in cold acclimation are related to a number of processes which include molecular and physiological modifications occurring in plant membranes, accumulation of cytosolic Ca^{2+}, increased level of ROS and the activation of ROS scavenger systems, changes in the expression of cold related genes and transcription factors.

Membrane Modification

Membranes are initial site of cold-induced injury. The studies of Uemura and Steponkus, (1999); Orvar *et al.* (2000) reported that the membrane rigidification are coupled with cytoskeletal rearrangements and calcium influxes. In acclimated plants (cold tolerant plant) the lipid composition of plasma membranes and chloroplasts envelopes changes such that the threshold temperature for membrane damage is lowered relative to that for non-acclimated plants (Uemura and Steponkus, 1999). Chilling resistant plants have a greater abundance of unsaturated fatty acids than chilling-sensitive plants and the proportion of unsaturated fatty acids increases during acclimation to cold temperature (Palta *et al.*, 1993).This modification allows membranes to remain fluid by lowering the temperature at which the membrane lipids experience a gradual phase change from fluid to semi-crystalline. Therefore, desaturation of fatty acids provides protection against damage from chilling temperatures (Khodakovskaya *et al.*, 2006). The acclimation process also activates mechanisms that protect membrane fluidity by ensuring the optimal activity of associated enzymes (Matteucci *et al.*, 2011).

Photosynthesis-Related Pigments

Photosynthesis is strongly affected by chilling stress. The cessation of growth which is resulting from cold stress reduces the capacity for energy utilization and causing feedback inhibition of photosynthesis (Ruelland and Zachowski, 2010). In another study Goulas *et al.* (2006) reported that the photosynthetic activity is maintained by increasing the activity of various Calvin cycle enzymes in cold acclimated plants. As such conifers exhibit long term changes on the organization of the photosynthetic apparatus (PSII reaction centers, a loss of light harvesting chlorophylls and the formation of a large thylakoids/protein aggregate involving light harvesting complex II and PS I and PS II, Savitch *et al.*, 2002).

Quenching pigments like antheraxanthin and zeaxanthin become unable to pass their excitation to chlorophyll-a, resulting in a protective mechanism that lowers the delivery of energy to PSII (part of non-photochemical quenching: NPQ). At chilling temperatures, due to the thermodynamic effect of cold on enzymatic reactions resulting in the photo stasis imbalance which further tends to diminish this imbalance by converting PS II from a state of efficient light harvesting to a state of high thermal dissipation. Winter annual plants acclimated by developing new leaves and again get adapted to new thermal regime, which is effectively associated with increase in thylakoid plastoquinone-A content with a concomitant increase in the apparent size of the intersystem electron donor pool to PS I. Similarly in leaves of *Arabidopsis thaliana* through phenlypropanoid pathways accompanied by the controlled key enzymes like phenylalanine ammonia-lyase and chalcone synthase mRNAs, accumulation of anthocyanin occurs in the leaves and stems in response to the low temperatures and changes in light intensity. The studies reports on its protective function from photo-inhibition during the growth of seedlings at low temperatures (Havaux and Kloppstech, 2001).

Role of Calcium

Exposure to low temperature instantly increases the concentration of calcium ions in the cytosol. Calcium acts as a mediator of stimulus-response coupling in the regulation of plant growth, development, and responses to environmental stimuli (Sanders *et al.*, 2002; Du and Poovaiah, 2005). Cold stress-induced rigidification of plasma membrane micro domains can cause actin cytoskeletal rearrangement. This may be followed by the activation of Ca^{2+} channels and increased cytosolic calcium levels which involved in the cold acclimation process (Sangwan *et al.*, 2001).The Ca^{2+} released from internal cellular reserves, mediated by inositol triphosphate, is upstream of the expression of CBFs (C-repeat binding factors) and COR (cold responsive) genes in the cold-signaling pathway (Chinnusamy *et al.*, 2007). Doherty *et al.* (2009) provided more evidence for a link between calcium signalling and cold induction of the CBF pathway, showing that calmodulin binding transcription activator (CAMTA) factors bind to a regulatory element in the CBF2 gene promoter. As the CAMTA proteins are calmodulin binding transcription factors, they may act directly in the transduction of low temperature-induced cytosolic calcium signals into downstream regulation of gene expression. Similarly, CRLK1, a calcium/ calmodulin-regulated receptor-like kinase was reported to be crucial for cold tolerance in plants (Yang *et al.*, 2010)

Role of Reactive Oxygen Species

Reactive oxygen species (ROS) are toxic oxygen free radicals, which are produced in the plants out of phyto-reactions and cellular oxidation by products under normal conditions (Finkel and Holbrook, 2000). The most primitive responses of plant cells under various environmental stresses is the generation of the oxidative burst, during which large quantities of ROS such as superoxide, hydrogen peroxide, hydroxyl radicals, peroxy radicals, alkoxy radicals, singlet oxygen, etc. are generated (Bhattacharjee, 2005). They are having the potential to cause cellular damage when they accumulate to certain toxic levels. However, these ROS are also having an important role as their accumulation activated defense-signaling pathways thus reduced cellular damage. Since, Suzuki *et al.* (2011) reported that ROS involved in processes leading to plant stress acclimation. This finding indicates that ROS are not simply toxic by-products of metabolism, but act as signalling molecules that modulate the expression of various genes which encodes antioxidant enzymes and modulators of H_2O_2 production.

Role of Transcription Factors

In cold acclimation various regulatory pathways are activated, but the most important pathway involves the C-repeat binding factor (CBF) regulon (Thomashow, 1999). CBF genes represent one of the most significant discoveries in the field of low temperature adaptation and signal transduction. All important crops and few vegetables species have contained this gene. The C-repeat/dehydration-responsive-element binding factor genes (CBF/DREB) are transcriptional activators involved in governing the plants responses to low temperatures (Schwager *et al.*, 2011). These include CBF1, CBF2, and CBF 3 (Gilmour *et al.*, 2004). These cold-induced CBF transcriptional regulatory factors control the cold-responsive expression of a major regulon of COR genes that increases plant freezing tolerance (Van Buskirk and Thomashow, 2006). Hsieh *et al.* (2002) reported that the over expression of CBF1 increased chilling tolerance in tomato by enhancing CATALASE1 gene expression and enzyme activity, and oxidative stress tolerance. Over-expression of AtCBF1/3 enhanced tolerance against cold stress in *Brassica* species (Jaglo *et al.*, 2001), and rice (Oh *et al.*, 2005).

Mantyla *et al.* (1995) reported that within 15 minutes, CBF transcript levels start to increase when plants are exposed to low temperatures and transcripts from the targeted CBF/DRE-regulated COR genes start to accumulate within 2 hours. The actual mechanism through which the CBF genes are activated by low temperature does not involve auto regulation (Gilmour *et al.*, 2004) but is controlled by a set of redundant interacting transcription factors (Vogel *et al.*, 2005; Chinnusamy *et al.*, 2007).

Transgenic Approaches for Cold Stress

Transgenic technology has provided much aspect to improve chilling stress in plants by introduction or removal of gene or genes that regulate a specific trait (Kumar and Bhatt, 2006). It also offers various opportunities to improve the genetic potential of plants in the form of development of specific crop varieties that are more resistant to chilling stresses with enhanced nutritional level. Transgenic plants have been prepared against cold stress to achieve cold tolerance (Table 9.1). These plants have one or more alien genes from stranger or their wild relatives, which over-express and regulate the functioning of metabolic process in a positive manner against stressful temperatures. Tayal *et al.* (2005) also revealed that the analysis of transgenic plants over expressing one or more genes which provide an understanding of basic mechanism and functioning of stress genes during cold stress exposure.

Table 9.1: Transgenic Plants Developed for Cold Tolerance

Gene Engineered	*Transgenic Crop*	*Effect of Gene Engineering*	*References*
sacBLevansucrase	*N. tabacum*	Transformants were more tolerant to freezing and *PEG*-mediated water stress than the wild type	Pilon *et al.*, 1995
cor15a Cold regulated gene	*A. thaliana*	Transformants showed *in vivo* enhanced freezing tolerance of protoplasts and the chloroplasts	Artus *et al.*, 1996
DRE-binding transcription factor gene, *GhDREB* from *Gossypium hirsutum*	*Triticum aestivum*	Improved tolerance to freezing stresses through accumulating higher levels of soluble sugar and chlorophyll in leaves after stress treatments	Gao *et al.*, 2009
CBF3 DRE-binding protein	*Arabidopsis*	Increased freezing tolerance of cold-acclimated plants	Gilmour *et al.*, 2002
SCOF1cold-inducible zinc finger protein	*Glycine max*	activate *COR* gene expression and increase freezing tolerance in non-acclimated transgenic plants	Kim *et al.*, 2001
HOS10 Encodes an R2R3-type protein	*O. sativa*	Enhanced cold tolerance	Zhu *et al.*, 2005

Advancement in plant biotechnology leads to the identification and isolation of a number of transcription factor related to chilling stress tolerance. The genes selected for transformation encoded enzymes for modifying membrane lipids, LEA proteins, and detoxification enzymes. The genes encoding protein factors which regulate gene expression and signal transduction is useful for improving the chilling tolerance of plants by gene transfer as they can regulate many stress-inducible genes involved in chilling stress tolerance.

Conclusion and Future Prospects

Chilling stress affects virtually all aspects of cellular function. Low temperature results in a number of physiological responses, including inactivation of metabolism, accumulation of ROS, reduced photosynthesis and the rigidification of membranes. In some plants species exposure to chilling temperature is perceived as a pre-emptive signal to trigger a response leading to chilling tolerance which includes accumulation of AFPs and dehydrins. The transcriptional factor CBF has an important role in cold acclimation in diverse plant species. The CBF/DREB genes integrate several components of the cold acclimation response by which plants increase their tolerance to low temperatures. By using genetic and molecular approaches, a number of relevant genes have been identified and new information continually emerges to enrich the CBF cold-responsive pathway.The knowledge generated through this study should be utilized in making transgenic plants that would be able to tolerate stress condition without showing any growth and yield penalty.

References

Anderson, M.D., Tottempudi, K P., Barry, A.M. and Cecil, R.S., 1994. Differential gene expression in chilling acclimated maize seedlings and evidence for the involvement of abscisic acid in chilling tolerance. *Plant Physiology*, 105: 331-339.

Antikainen, M., Griffth, M., Zhang, J., Hon, W.C., Yang, D. and Pihakaski-Maunsbach, K., 1996. Immunolocalization of antifreeze proteins in winter rye leaves, crowns, and roots by tissue printing. *Plant Physiology*, 110:845–857.

Artus, N.N., Uemura, M., Steponkus, P.L, Gilmour, S.J, Lin, C. and Thomashow, M.F., 1996. Constitutive expression of the cold-regulated *Arabidopsis thaliana* COR15a gene affects both and protoplast freezing tolerance. *Proc. Natl. Acad. Sci. USA.*, 93:13404–13409.

Bagnall, D., Wolfe, J.O.E. and King, R.W., 1983. Chill-induced wilting and hydraulic recovery in mung bean plants. *Plant Cell Environment*, 6:457–464.

Bhattacharjee, S., 2005. Reactive oxygen species and oxidative burst: roles in stress, senescence and signal transduction in plant. *Curr. Sci.*, 89:1113–1121.

Bies-Etheive, N., Gaubier-Comella, P., Debures, A., Lasserre, E., Jobet, E., Raynal, M., Cooke, R. and Delseny, M. 2008. Inventory, evolution and expression profiling diversity of the LEA (late embryogenesis abundant) protein gene family in *Arabidopsis thaliana. Plant Molecular Biology*, 67: 107–124.

Boese, S.R., Wolfe, D.W. and Melkonian, J.J., 1997. Elevated CO2 mitigates chilling-induced water stress and photosynthetic reduction during chilling. *Plant, Cell and Environment*, 20(5): 625–632.

Chinnusamy, V., Zhu, J. and Zhu, J.K., 2007. Cold stress regulation of gene expression in plants. *Trends Plant Sci.*, 12: 444-451.

Christiansen, M.N. and St. John, J.B., 1981. The nature of chilling injury and its resistance in plants. pp.1-16. *In* C. R. Olien and M. N. Smith (eds.) Analysis and improvement of plant cold hardiness. CRC Press Inc. Boca Raton U. S. A.

Doherty, C.J., Van Buskirk, H.A., Myers, S.J. and Thomashow, M.F., 2009. Roles for *Arabidopsis* CAMTA transcription factors in cold regulated gene expression and freezing tolerance. *Plant Cell*, 21:972–984.

Du, L. and Poovaiah, B.W., 2005. Ca^{2+} calmodulin is critical for brassinosteroid biosynthesis and plant growth. *Nature*, 437:741–745.

Ebrahim, M.K.H., Vogg, G., Osman, M.N.E.H. and Komor, E., 1998. Photosynthetic performance and adaptation of sugarcane at suboptimal temperatures. *Journal of Plant Physiology*, 153(5–6): 587–592.

Ensminger, I., Busch, F. and Huner, N. P. A., 2006.Photostasis and cold acclimation: Sensing low temperature through photosynthesis. *Physiologia Plantarum*, 126: 28–44.

Finkel, T. and Holbrook, N.J., 2000.Oxidants, oxidative stress and the biology of ageing. *Nature*, 408:239–247.

Fowler, D.B., Limin, A.E., Wang, S.Y. and Ward, R.W., 1996. Relationship between low-temperature tolerance and vernalization response in wheat and rye. *Can. J. Plant Sci.*, 76:37–42.

Furbank, R.T., Foyer, C.H. and Walker, D.A. 1987. Regulation of photosynthesis in isolated spinach chloroplasts during orthophosphate limitation. *Biochimica. Biophysica. Acta.*, 894:552–561.

Gao, S.Q., Chen, M., Xia, L.Q., Xiu, H.J. and Xu, Z.S., 2009. A cotton (*Gossypium hirsutum*) Dre-binding transcription factor gene, GHDREB, confers enhanced tolerance to drought, high salt, and freezing stresses in transgenic wheat. *Plant Cell Rep.*, 28:301–311.

Gilmour, S.J., Fowler, S.G. and Thomashow, M.F., 2004. *Arabidopsis* transcriptional activators CBF1, CBF2, and CBF3 have matching functional activities. *Plant Mol. Biol.*, 54:767–781.

Gilmour, S.J., Sebolt, A.M., Salazar, M.P., Everard. J.D. and Thomashow, M.F., 2000. Over expression of the *Arabidopsis* CBF3 transcriptional activator mimics multiple biochemical changes associated with cold acclimation. *Plant Physiol.*, 124:1854–1865.

Goulas, E., Schubert, M., Kieselbach, T., Kleczkowski, L.A., Gardestrom, P., Schroder, W. and Hurry, V., 2006. The chloroplast lumen and stromal proteomes of *Arabidopsis thaliana* show differential sensitivity to short and long-term exposure to low temperature. *Plant J.*, 47:720–734.

Graham, D. and Patterson B.D., 1982. Responses of plants to low, non-freezing temperatures: proteins, metabolism, and acclimation. *Annual Review of Plant Physiology*, 33: 347–372.

Griffth, M., Elfman, B. and Camm, E.L., 1984. Accumulation of plastoquinone-A during low temperature growth of winter rye. *Plant Physiology*, 74: 727–729.

Griffth, M., Lumb, C., Wiseman, S.B., Wisniewski, M., Johnson, R.W. and Marangoni, A.G. 2005. Antifreeze proteins modify the freezing process in plant. *Plant Physiology*, 138: 330–340.

Guy, C.L., 1990. Cold acclimation and freezing stress tolerance: role of protein metabolism. *Annual Review of Plant Physiology* and *Plant Molecular Biology*, 41: 187–223.

Havaux, M. and Kloppstech, K., 2001. The protective functions of carotenoid and flavonoid pigments against excess visible radiation at chilling temperature investigated in *Arabidopsis* npq and tt mutants. *Planta*, 213: 953–966.

Heide, O.M. and Prestrud, A.K., 2005. Low temperature, but not photoperiod, controls growth cessation and dormancy induction and release in apple and pear. *Tree Physiol.*, 250(1): 109-114.

Hon, W.C., 1995. Antifreeze proteins in winter rye are similar to pathogenesis-related proteins. *Plant Physiol.*, 109: 879–889.

Hsieh, T.H., Lee, J.T., Yang, P.T., Chiu, L.H. and Charng, Y.Y., 2002. Heterology expression of the *Arabidopsis* C-repeat/dehydration response element binding factor 1 gene confers elevated tolerance to chilling and oxidative stresses in transgenic tomato. *Plant Physiol.*, 9:1086–1094.

Huner, N.P.A., Oquist, G., Hurry, V.M., Krol, M., Falk, S. and Griffth, M., 1993. Photosynthesis, photoinhibition and low temperature acclimation in cold tolerant plants. *Photosynthesis Research*, 37: 19–39.

Hurry, V., Strand, A., Furbank, R. and Stitt, M., 2000. The role of inorganic phosphate in the development of freezing tolerance and the acclimatization of photosynthesis to low temperature is revealed by the pho mutants of *Arabidopsis thaliana*. *Plant Journal*, 24: 383–396.

Jaglo, K.R., Kleff, S., Amundsen, K.L., Zhang, X., Haake, V; Zhang, J.Z., Deits, T. and Thomashow, W.M.F., 2001. Components of the *Arabidopsis* C-repeat/ dehydration-responsive element binding factor cold-response pathway are conserved in *Brassica napus* and other plant species. *Plant Physiology*, 127(3): 910-917.

Jiang, I.C., Oh, S.J., Seo, J.S., Choi, W.B. and Song, S.I., 2003. Expression of a bifunctional fusion of the *Escherichia coli* genes for trehalose-6-phosphate synthase and trehalose-6-phosphate phosphatise in transgenic rice plants increases trehalose accumulation and abiotic stress tolerance without stunting growth. *Plant Physiol.*, 131:516–52.

Jouve, L., Engelmann, F., Noirot, M. and Charrier, A., 1993. Evaluation of biochemical markers (sugar, proline, malonedialdehyde and ethylene) for cold sensitivity in microcuttings of two coffee species. *Plant Science*, 91(1):109–116.

Kaur, G., Kumar, S., Nayyar, H. and Upadhyaya, H.D., 2008. Cold stress injury during the pod- filling phase in chickpea (*Cicer arietinum* L.): effects on quantitative and qualitative components of seeds. *J.Agron.Crop. Sci.*, 194:457–464.

Kasamo, K., Kagita, F., Yamanishi, H. and Sakaki, T., 1992. Low temperature induced changes in the thermotropic properties and fatty acid composition of the plasma membrane and tonoplast of cultured rice (*Oryza sativa* L.) cells. *Plant and Cell Physiology*, 33(5): 609–616.

Khodakovskaya, M., McAvoy, R., Peters, J., Wu, H. and Li, Y., 2006. Enhanced cold tolerance in transgenic tobacco expressing a chloroplast o-3 fatty acid desaturase gene under the control of a cold inducible promoter. *Planta*, 223:1090–1100.

Kim, J.C., Lee, S.H., Cheong, Y.H., Yoo, C.M., Lee, S.I., Chun, H.J., Yun, D.J., Hong, J.C., Lee, S.Y., Lim, C.O. and Cho, M.J., 2001. A novel cold-inducible zinc finger protein from soybean, SCOF-1, enhances cold tolerance in transgenic plants. *Plant J.*, 25:247–259.

Koster, K.L., Tengbe, M.A., Furtula, V. and Nothnagel, E.A., 1994. Effects of low temperature on lateral diffusion in plasma membranes on maize (*Zea mays* L.) root cortex protoplasts. Relevance to chilling sensitivity. *Plant, Cell and Environment*, 17(12): 1285–1294.

Kumar, N. and Bhatt, R.P. 2006. Transgenic: An emerging approach for cold tolerance to enhance vegetables production in high altitude areas. *Indian J. Crop Sci.*, 1:8–12.

Levitt, J., 1980. Responses of plants to environmental stresses. Chilling, freezing and high temperatures stresses. New York, 1:426.

Lynch, D.V., 1990. Chilling injury in plants: the relevance of membrane lipids. In: Katterman F (ed) Environmental injury to plants. Academic, New York, pp. 17–34.

Lyons, J.M., 1973. Chilling injury in plants. *Annu. Rev. Plant Physiol.*, 24:445–466.

Mantyla, E., Lang, V. and Palva, E.T., 1995. Role of abscisic acid in drought induced freezing tolerance, cold acclimation, and accumulation of LT178 and RAB18 proteins in *Arabidopsis thaliana*. *Plant Physiol.*, 107:141–148.

Matsuo, T., Graham, D., Patterson, B.D. and Hockley, D.B., 1994. An electrophoretic method to detect cold-induced dissociation of proteins in crude extracts of higher plants. *Analytical Biochemistry*, 223(2): 181–184.

Matteucci, M.D., Angeli, S., Errico, S., Lamanna, R., Perrotta, G. and Altamura, M.M., 2011. Cold affects the transcription of fatty acid desaturases and oil quality in the fruit of *Oleaeuropaea* L. Genotypes with different cold hardiness. *J. Exp. Bot.*, 62:3403–3420.

Nayyar, H. and Chander, S., 2004. Protective effects of polyamines against oxidative stress induced by water and cold stress in chickpea. *J.Agron. Crop Sci.*, 190:355–365.

Nayyar, H., Chander, K., Kumar, S. and Bains, T., 2005. Glycine betaine mitigates cold stress damage in Chickpea. *Agron. Sustain Dev.*, 25:381–388.

Nie, G.Y., Long, S.P. and Baker, N.R., 1992. The effects of development at suboptimal growth temperatures on photosynthetic capacity and susceptibility to chilling-dependent photoinhibition in *Zea mays*. *Physiologia Plantarum*, 85(3): 554–560.

Ohnishi, S., Miyoshi, T. and Shirai, S., 2010. Low temperature stress at different flower developmental stages affects pollen development, pollination, and pod set in soybean. *Environ. Exp. Bot.*69:56–62.

Oh, S.J., Song, S.I., Kim, Y.S., Jang, H.J., Kim, S.Y., Kim, Y.K., Nahm, B.H., and Kim J.K., 2005. *Arabidopsis* CBF/DREB1A and ABF3 in transgenic rice increased tolerance to abiotic stress without stunting growth. *Plant Physiology*, 138(1): 341-351

Orvar, B.L., Sangwan, V., Omann, F. and Dhindsa, R.S., 2000. Early steps in cold sensing by plant cells: the role of actin cytoskeleton and membrane fluidity. *Plant J.*, 23:785–794.

Palta, J.P., Whitaker, B.D. and Weiss, L.S., 1993. Plasma membrane lipids associated with genetic variability in freezing tolerance and cold acclimation of *Solanum* species. *Plant Physiol.*, 103:793–803.

Pilon-Smits, E.A.H., Ebskamp, M.J.M., Paul, M.J., Jeuken, M.J.W., Weisbeek, P.J. and Smeekens, S.C.M., 1995. Improved performance of transgenic fructan-accumulating tobacco under drought stress. *Plant Physiol.*, 107:125–130.

Pudney, P.D.A., 2003. The physico-chemical characterization of a boiling stable antifreeze protein from a perennial grass (*Lolium perenne*). *Arch. Biochem. Biophys.*, 410: 238–245.

Quinn, P.J., 1988. Effect of temperature on cell membrane. Plants and Temperature: Symposium of Society for Experimental Biology. – Cambridge, UK, p. 237–258.

Ruelland, E. and Zachowski, A. 2010. How plants sense temperature. *Environ. Exp. Bot.*, 69:225–232.

Rymen, B., Fiorani, F., Kartal, F., Vandepoele, K. and Inze, D., 2007 Cold nights impair leaf growth and cell cycle progression in maize through transcriptional changes of cell cycle genes. *Plant Physiol.*, 143:1429–1438.

Sanders, D., Pelloux, J., Brownlee, C. and Harper, J.F., 2002. Calcium at the crossroads of signalling. *Plant Cell,* 14:S401–S417.

Sanghera, G.S., Wani, S.H., Hussain, W. and Singh, N.B., 2011. Engineering Cold Stress Tolerance in Crop Plants. *Curr Genomics.*, 12(1): 30-43.

Sangwan, V., Foulds, I., Singh, J. and Dhindsa, R.S., 2001. Cold-activation of *Brassica napus* BN115 promoter is mediated by structural changes in membranes and cytoskeleton, and requires Ca^{2+} influx. *Plant J.*, 27: 1–12.

Savitch, L.V., Leonardos, E.D., Krol, M., Jansson, S., Grodzinski, B., Huner, N.P.A. and Oquist, G., 2002. Two different strategies for light utilization in photosynthesis in relation to growth and cold acclimation. *Plant Cell and Environment*, 25: 761–771.

Sonoike, K., 1998. Various aspects of inhibition of photosynthesis under light/ chilling stress. Photoinhibition at chilling temperatures versus chilling damage in the light. *Journal of Plant Research*, 111(1101): 121–129.

Sowinski, P., Rudzinska-Langwald, A., Adamczyk, J., Kubica, I. and Fronk, J. 2005. Recovery of maize seedling growth, development and photosynthetic efficiency after initial growth at low temperature. *J. Plant Physiol.*, 162:67–80.

Steponkus, P.L., 1984. Role of plasma membrane in freezing injury and cold acclimation. *Annual Review Plant Physiology*, 35: 543–584.

Suzuki, K., Nagasuga, K. and Okada, M., 2008. The chilling injury induced by high root temperature in the leaves of rice seedlings. *Plant Cell Physiol.*, 49:433–442.

Suzuki, N., Koussevitzky, S., Mittler, R., and Miller, G., 2011. ROS and redox signaling in the response of plants to abiotic stress. *Plant Cell Environment,* 35:259–270.

Tayal, D., Srivastava, P.S. and Bansal, K.C., 2005. Transgenic crops for abiotic stress tolerance. In: Plant biotechnology and molecular markers. *Springer,* Netherlands, 346–365.

Thomashow, M.F., 1999. Plant cold acclimation: freezing tolerance genes and regulatory mechanisms. *Plant Mol. Biol.,* 50:571–599.

Uemura, M. and Steponkus, P.L., 1999. Cold acclimation in plants: relationship between the lipid composition and the cryostability of the plasma membrane. *J. Plant Res.,* 112:245–254.

Van Buskirk, H.A. and Thomashow, M.F., 2006. *Arabidopsis* transcription factors regulating cold acclimation. *PlantPhysiol.*126: 72-80.

Vogel, J.T., Zarka, D.G., Van Buskirk, H.A., Fowler, S.G. and Thomashow, M.F., 2005. Roles of theCBF2andZAT12 transcription factors in configuring the low temperature transcriptome of *Arabidopsis. Plant J.,* 41:195–211.

Wisniewski, M., 1999. Purification, immunolocalization, cryoprotective, and antifreeze activity of PCA60: a dehydrin from peach (*Prunus persica*). *Physiol. Plant.,* 105: 600–608.

Xin, Z. and Browse, J., 2001. Cold comfort farm: the acclimation of plants to freezing temperatures. *Plant Cell Environment.* 23:893–902.

Yang, T., Chaudhuri, S., Yang, L., Du, L. and Poovaiah, B.W., 2010. A calcium/ calmodulin-regulated member of the receptor-like kinase family confers cold tolerance in plants. *J.Biol.Chem.,* 285:7119–7126.

Yordanov, I., 1992. Response of photosynthetic apparatus to temperature stress and molecular mechanisms of its adaptations. *Photosynthetica,* 26(4): 517–531.

Yoshida, R., Kanno, A. and Kameya, T., 1996. Cool temperature-induced chlorosis in rice plants. II. Effects of cool temperature on the expression of plastid-encoded genes during shoot growth in darkness). *Plant Physiol.,* 112:585–590.

Zhu, J., Verslues, P.E., Zheng, X., Lee, B.H., Zhan, X., Manabe, Y., Sokolchik, I, Zhu, Y., Dong, C.H., Zhu, J.K., Hasegawa, P.H. and Bressan, R.A., 2005. HOS10 encodes an R2R3-type MYB transcription factor essential for cold acclimation in plants. *Proc. Natl. Acad. Sci.USA.,* 102:9966–9971.

2016, Recent Advances in Plant Stress Physiology *Pages* **201–218**
Editors: ***Praduman Yadav, Sunil Kumar and Veena Jain***
Published by: **DAYA PUBLISHING HOUSE, NEW DELHI**

Chapter 10

Heat Stress Response in Plants

Torit Baran Bagchi* and Soham Ray
ICAR – Central Rice Research Institute,
Cuttack – 753 006, Odisha, India

ABSTRACT

By the end of the 21st Century, the earth's climate is predicted to be warmer by an average of 2-4 °C due to both anthropogenic and natural factors. Rise of temperature more than 10-15°C from the ambient is considered as heat stress for the plants. High temperatures affect plant growth at all developmental stages; especially, anthesis and grains filling are more susceptible. Heat stress induces structural changes in tissues and cell organelles, disorganization of cell membranes, disturbance of leaf water relations, and impediment of photosynthesis which adversely affect the whole physiology and metabolism of the plant. At cellular level, high temperature causes reduced photosynthesis via denaturation of proteins and pigments including chlorophyll in photosystems, membrane destabilization via ROS mediated lipid peroxidation and consequent electrolite linkage and purterbation in water relation. Similarly at molecular level, it can cause denaturation of enzymes like Rubisco and chaperons like Rubisco binding protein REP due to the loss of their tertiary structure under high temperature. To cope up and survive under such stress, plants employ various mechanisms both at cellular and molecular level which includes modulation of primary and secondary metabolism, molecules and transcription factors related to cell signaling. Plant hormones are potent signaling molecules which participate in inducing heat tolerance mechanisms of plants along with other signaling molecules like Calmoduline, Ca^{2+}, MAPK. Synthesis of HSPs in the plant cells under heat stress, which is directed by heat stress transcription factors Hsfs, also play a critical role in its survival. Natural osmolytes like soluble sugar and proline act as potent osmoprotectant during plants' acclimatization under heat stress. Ascorbat, glutathione, tocopherol and carotene also protect plants against oxidative stress. Elevated temperature deteriorates grain quality by altering the ratio of amylose to amylopectin by reducing the activity of various grain filling enzymes and the lipid content of the grain.

Keywords: *Enzymes, HSPs, Osmolites, Photosynthesis.*

**Corresponding author.* E-mail: torit.crijaf09@gmail.com

Abbreviations

REP: Rubisco binding protein
ABA: Abscisic acid
MAPK: Mitogen activated Protein Kinase
HSP: Heat Sock Protein
Hsf: Heat stress transcription factor
HS: Heat stress.

Introduction

A report of the Intergovernmental Panel on Climatic Change IPCC, suggests a global mean temperature rise 0.3 C per decade (Jones *et al.*, 1999). The rise of temperature beyond the ambient lead to heat stress which is often defined as the rise in temperature beyond a threshold level for a period of time sufficient to cause irreversible damage to plant growth and development (Hall, 2001). In general, a transient elevation in temperature, usually 10-15°C above ambient, is considered heat shock or heat stress (Wahid *et al.*, 2007). Many researchers found that it is a complex phenomenon as the physiological and biochemical change of plant is dependent on temperature and diurnal fluctuations, moisture content and nutritional status of soil and relative humidity. Besides, temperature threshold of stress varies from crop to crop as well as between different developmental stages of the same plants. Supra-optimal temperatures restrict plant growth by altering various metabolic processes including synthesis and degradation of primary metabolites such as soluble sugars (Rizhsky *et al.*, 2004), amino acids, polyamines (Cvikrova *et al.*, 2012), and organic acids (Lakso and Kliewer, 1975). Decline in the content of fructose and glucose (Bingru, 2013), higher accumulation of compatible solutes *e.g.*, proline, sucrose, *myo*-inositol and putrescine form of polyamines in leaves of *Agrostis scabra* are few of the biochemical response under heat stress. The percentage of seed germination and photosynthetic efficiency declines when plants encounter heat stress. During the reproductive growth period, spikelet sterility occurs due to non swelling of pollen resulting poor dehiscence of anther, altered hormonal balance in the florate, disturbance in the availability and transport of photosynthates to karnel lack of ability of the floral buds to mobilize carbohydrates and changes in the activities of starch and sugars biosynthesis related enzymes. All these phenomena affect crop yield and sometimes survival of the plants. A molecular study shows that heat tolerance of rice during flowering is under polygenic control (Zhang and Sharkey, 2009). Heat stress factor Hsf like HsfA1a, HsfA4a, HsfA8 and Hsps (HSP100, HSP90, HSP70, HSP60, sHSPs, Ubiquitin, HSP8.5) play a central role in the heat stress and acquired thermotolerance in plants. Hsf serves as the terminal component of signal transduction and mediates the expression of Hsp. Reactive oxygen species ROS acts as a signaling molecule to induce HSPs. Heat-stress during greening led to stimulation in foliar superoxide *dis*mutase activity, indicating a potential source for generation of its substrate, superoxide radical (O^{-2}) in light. Ascorbate peroxidase, a key H_2O_2-metabolizing enzyme, was also stimulated at high temperature (HT).

Heat Stress Response to Plants

Molecular Biology of Heat Stress Response

Exposed to higher temperature, plant exhibits a spectrum of changes at cellular and molecular level, few of which are ubiquitous to all the living organisms while the others are typical to plants. Some of these recognizable signatures of heat stress are decreased rate of photosynthesis and synthesis of normal proteins as well as inactivation of proteins due to heat induced conformational changes, increased synthesis of heat stress related proteins, disorganization of cellular structure due to disassembly of cytoskeleton and cell organelles and change in membrane fluidity. Growing body of evidences also beginning to suggest that, changes in super coiling and secondary structure of DNA and RNA, at least in some cases, might serve as the mechanism to sense temperature. Many of these subtle changes are even recognized by the plant system itself to initiate defense response towards heat stress.

Reduction in photosynthesis is one of the most common and initial responses produced by the plants under any sort of stress-biotic or abiotic. Photosynthesis is a highly energy demanding process. Hence, the plant under stress temporarily down-regulate photosynthesis to divert energy towards strengthening immediate defense needs. There are plenty of reports indicating reduction in photosynthesis during heat stress (Sage and Kubien, 2007; Salvucci, 2008, Yamori *et al.*, 2009; Zhang and Sharkey, 2009; Yamori *et al.*, 2010). The major reason behind this is supposed to be the thermal denaturation of Rubisco activase which is highly sensitive to heat (Salvucci, 2001). Denaturation of Rubisco activase leads to inactivation of Rubisco, the enzyme which fix atmospheric CO_2 (RuBP + $CO_2 \rightarrow$ 3-PGA) to initiate photosynthesis process. Beside squeezing the supply starting material for photosynthesis (3-PGA), denaturation of Rubisco activase also leads to imbalance in cellular photostasis (Takahashi and Murata, 2008; Hideg *et al.*, 2008) which induce generation of reactive oxygen species ROS (Hideg *et al.*, 2008, Guy *et al.*, 2008; Zhang *et al.*, 2002; Kristiansen *et al.*, 2009). Since some ROS scavenging enzymes are also thermo-sensitive (Panchuk *et al.*, 2002, Hideg *et al.*, 2008), at higher temperature ROS are free to cause damage to the photosynthetic enzymes, which is another reason of reduced photosynthesis at higher temperature. Heat stress also affects electron transport during photosynthesis by adversely affecting photosystems I and II PSI and PSII and causing oxidation of primary quinone, Q_A (Pshybytko *et al.*, 2008; Zhang and Sharkey, 2009).

Denaturation of cellular proteins and disassembly of cytoskeleton are also well documented features of heat stress. These are also one of the molecular cues identified by the plant itself to initiate defense response (Sugio *et al.*, 2009; Suri and Dhindsa, 2008). Heat induced protein denaturation and microfilament destabilization are the triggers for the synthesis of heat stress transcription factors Hsfs and some of the Heat shock proteins HSPs (Yamada *et al.*, 2007; Suri and Dhindsa, 2008). One of the interesting cases is of HSP90.2 which is a cytoplasmic chaperon of *Arabidopsis* and at normal temperature it remains bound with AtHsfA1d preventing its trimerization (active form) needed for transcription factor activity. During heat stress HSP90.2 gets denatured through an unknown mechanism which frees the AtHsfA1d monomers to trimerize and initiate heat stress induced transcription by binding to the heat stress elements HSEs in DNA (Yamada *et al.*, 2007).

Cell membrane, a moving mosaic of proteins in lipid bi-layers is largely composed of myriad of densely packed phospholipid or galactolipid molecules. These lipid molecules can rotate on its own axis, diffuse within a monolayer or flip-flop between two layers. All these movements are temperature dependent; hence shift in temperature causes rigidification (due to cold) or fluidization (due to heat) of the membrane. Earlier protein denaturation was considered as the only way of heat perception by plants but recently it has been found that increase in membrane fluidity is also equally important mechanism employed by plants for heat sensing. Increased membrane fluidity can trigger the responses like production of Heat Activated MAP Kinases HAMKs, HSP70 and down-regulate Fatty acid desaturase FAD which in turn enhance membrane fluidity (Sangwan *et al.*, 2002, Matsuda *et al.*, 2005; Suri and Dhindsa, 2008).

Recently it has been found that temperature induced change in DNA supercoiling pattern is also capable to induce temperature dependent gene expression. Though it is more frequent in prokaryotes and for cold induced gene expression (Los, 2004) but it opens up a possibility of similar mode of gene regulation, especially in mitochondria and plastids, which are of prokaryotic origin. Temperature induced fluctuation of RNA secondary structure and consequent differential downstream processing *viz.*, alternate splicing, RNA editing or export of RNA from the nucleus to cytoplasm is already reported in the literature (Iida *et al.*, 2004; Filichkin *et al.*, 2010). Ribosome binding to the mRNA can also be regulated by temperature fluctuations. A typical example is the case of *Hsp10* gene expression. 5-UTR of Hsp10 transcript contains an internal ribosomal entry segment IRES, which in ambient temperature is quickly sequestered in a secondary structure defying ribosome binding. But during heat stress the secondary structure is dissolved allowing ribosome binding and subsequent gene expression (Dinkova *et al.*, 2005). Since all these events of transcriptional and translational regulation occur prior to effector step *i.e.*, the gene expression, it opens up a whole new possibility of DNA and RNA working as inbuilt thermometer in plants. Though more experimental data are necessary to establish this fact but obviously the prospect is worth investigating.

Molecular Biology of Heat Stress Tolerance

Plants' ability of to cope up under heat stress is regarded as thermotolerance. Plats are known to be one of the organisms which can achieve thermotolerance rather quickly, within hours, and survive otherwise lethal temperature (Vierling, 1991). Thermotolerance may be further achieved by preconditioning - acute high temperature treatment or chronic low temperature treatment or may be achieved naturally by gradual increase in temperature gradient (Vierling, 1991). Lead from the studies in *Arabidopsis* shows that induction of thermo-tolerance is a function of complex interplays between the several signaling networks which involves expression of HSPs such as HSP101 and HSP32, phytohormone signaling majorly ABA and SA and ROS signaling pathway (Larkindale and Hugan, 2005; Charng *et al.*, 2006). Acquirement of thermo-tolerance has several commonalities with other

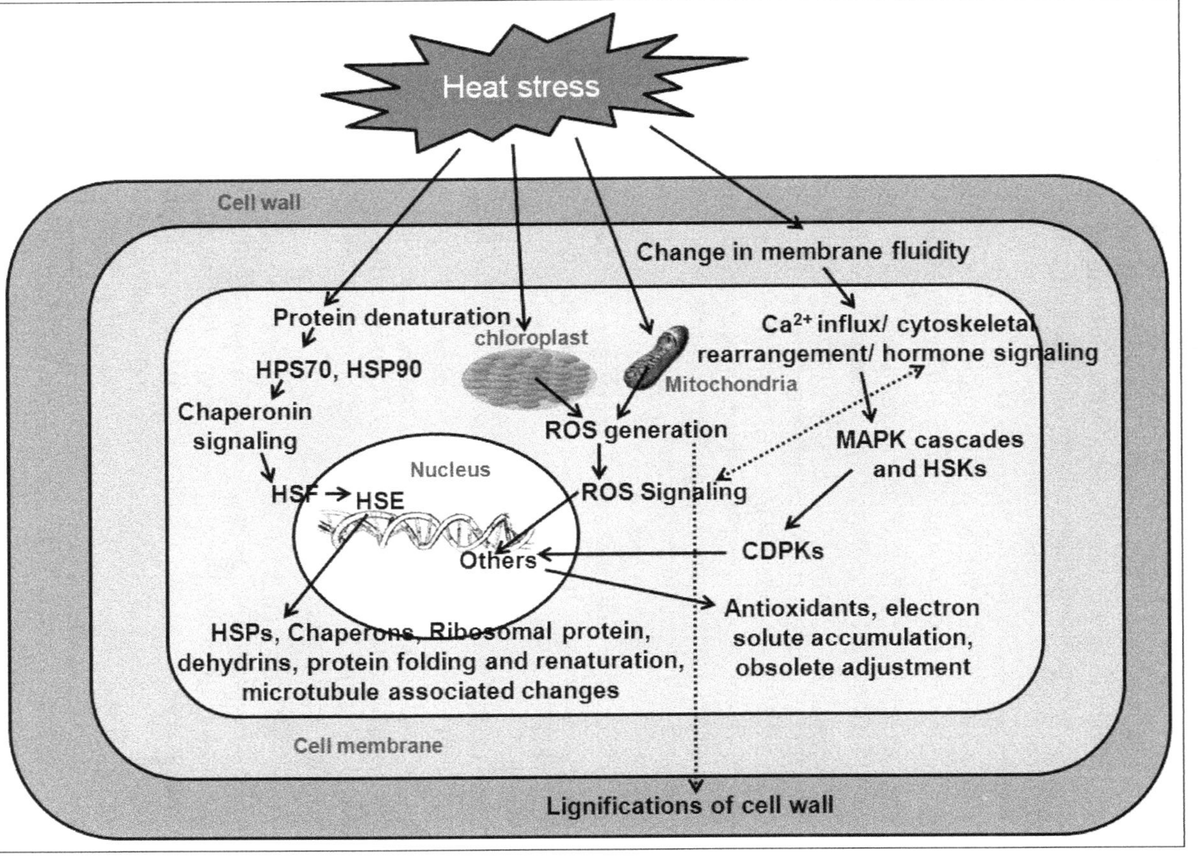

Figure 10.1: Molecular Signaling in a Typical Plant Cell during Heat Stress Response (HSR). Perception of heat stress to producing HSR involves intricate cascades of signaling network and involves plasma membrane, cytosol and other organelles of the plant cells. (*Source*: Modified from Wahid *et al*., 2007).

abоiotic stress responses, especially drought and cold tolerance (Rizhsky *et al.*, 2002). The transient reprogramming in the pattern of gene expression in response of heat stress is known as heat shock response HSR-a fascinating field of study to unzip some of the mysteries regarding thermo tolerance (Schoffl *et al.*, 1999). The most well acknowledged HSR is the transient synthesis of HSPs, though direct support for HSPs being involved in imparting thermo tolerance has been difficult to obtained (Bruke, 2001; Schoffl, 2005). The hypothesis of HSP triggered thermo tolerance stems basically from three facts. Firstly, these are well conserved across genera starting from bacteria to higher organisms (Vierling, 1991). Secondly, these are induced when the organism is under stress. And lastly, their biosynthesis is extremely fast under heat stress and many a times in other stress too. Increased HSP synthesis has been observed in plants under abrupt or gradual increase in temperature (Nakamoto and Hiyama, 1999; Schoffl *et al.*, 1999).

There are several classes of HSPs as differentiated and distinguished by their molecular weights; their initial proportion and level of induction varies depending on the plant species. Among these HSP70, HSP90 and Low molecular weight LMW or small HSPs SmHSP are of greater importance as these are highly induce during heat stress (>200 fold upregulation at transcriptional level). SmHSPs have a special ability to remain as aggregate in granule forms which are termed as heat shock granules HSGs which are presumed to play crucial role in plants' survival under sub-lethal temperature (Miroshnichenko *et al.*, 2005). HSPs prevent missfolding and denaturation of cellular protein and thus aid in stress tolerance. Wheat, rye and maize, when subjected to heat stress 42 C, maize showed better stress tolerance than the other two. Further investigation revealed that five different types of mitochondrial LMW-HSPs were synthesized in maize compared to only one in case of wheat and rye. This might be one of the possible explanations of better ability of maize to withstand heat stress (Korotaeva *et al.*, 2001). Accumulation of HSGs coupled with increased thermo-tolerance has been observed in tomato where HSGs are presumed to play role in protecting the machinery for protein synthesis under heat stress (Miroshnichenko *et al.*, 2005). To sum up, the conformational and aggregation dynamics of HSPs and their function as molecular chaperon play a significant role in heat tolerance.

Apart from HSPs, the effect of heat stress can be noted at different levels such as at plasma membrane, at cytosol, cell organelles like chloroplast and mitochondria and at nucleus. The initial effects of heat stress can be observed on plasma membrane where the lipid layer becomes more fluid. This enhanced fluidity induce Ca^{2+} signaling and cytoskeletal modification by switching on the Ca^{2+} dependent protein kinases CDPKs and Mitogen activated protein kinases MAPKs based signaling cascades (Sangwan and Dhindsa, 2002). Ca^{2+} binding protein, Calmodulin, probably plays the role of mediator during this process. These cascades of events lead to expression of genes involved in the production of secondary metabolites, antioxidants, compatible osmolites etc. the ROS produce in chloroplast and mitochondria also plays a critical role in thermotolerance. These induce lignification making the cell wall rigid as well as play an important role as signaling molecules which ultimately induce antioxidant production (Maestri *et al.*, 2002, Suzuki and Mittler, 2006). Apart from these phytohormones like ABA, SA, and Me-SA, ACC, ethylene and molecules like H_2O_2 can also work as signaling molecule to activate pathways leading to heat stress response HSR.

So in brief, initiating from perception of heat stress to production of HSR involve intricate cascades of signaling network and involve all the layers and organelles of the plant cells. Phytohormones, ROS, Ca^{2+}, ect. etc. as signaling molecules to modulate the cellular response according to need. And finally, HSPs, antioxidants, secondary metabolites, compatible osmolites, etc. cumulatively play very crucial roles to achieve successful thermo tolerance in plants.

Physiological and Biochemical Aspects of Heat Stress Response

Almost all macromolecules in the cells, such as protein complexes, membranes and nucleic acid polymers, 'perceive' the heat simultaneously. The increased kinetic movement of these macromolecules is expected to concomitantly cause reversible physical changes, such as increased membrane fluidity, partial melting of DNA and RNA strands, protein subunit dissociation and exposure of protein hydrophobic cores. Therefore, all macromolecules might, in principle, serve as thermosensors by providing an output in the form, for example, of a transient loss of function. Yet, to properly answer the question 'How do plants feel the heat?' we need to identify which among the many heat-responsive macromolecules in the plant cell acts as a primary heat sensor. This sensor must be able to not only precisely perceive and differentially react to various temperature increments but also differentially trigger a unique signaling pathway that can specifically unregulated hundreds of HSR genes.

Study of Physiology and Biochemical Cases in Respect of Heat Stress

With the recent threat of global warming, heat stress has become a major area of concern in crop production. Extensive research is being conducted worldwide to develop strategies to cope with abiotic stresses through the development of heat- and drought-tolerant varieties, shifting crop calendars and resource management practices, etc. (Venkateswarlu and Shankar, 2009). In rice, a 7-8 per cent yield decrease has been observed for each 1 C increase daytime maximum/nighttime minimum in temperature from 28/21 to 34/27 C (Baker *et al.*, 1992). Harvest Index, spikelet fertility, grain yield, vegetative biomass, total grain weight, grain filling, pollen production, pollen reception, pollen viability are decreased if the rice plant are exposed to HS 5-10 C more than ambient throughout the growing period but the rate of photosynthesis does not vary significantly due to high temp only (Prasad *et al.*, 2006).

Heat-tolerant cultivars have better anther dehiscence and shed more pollen grains on stigma. The tolerant cultivar N-22 had smallest decreases in spikelet fertility, grain yield and harvest index at elevated temperature, while susceptible cultivars had larger decreases in all the above yield parameters. Significant decrease in the photosynthetic rate (P_n), carboxylation efficiency (CE), the maximal photochemical efficiency of photosystem II, and the light-saturated photosynthetic rate, which were related to the reduction of CO_2 assimilation, inactivation of photosystem II and photosynthetic electron transport. PSII is considered to be the

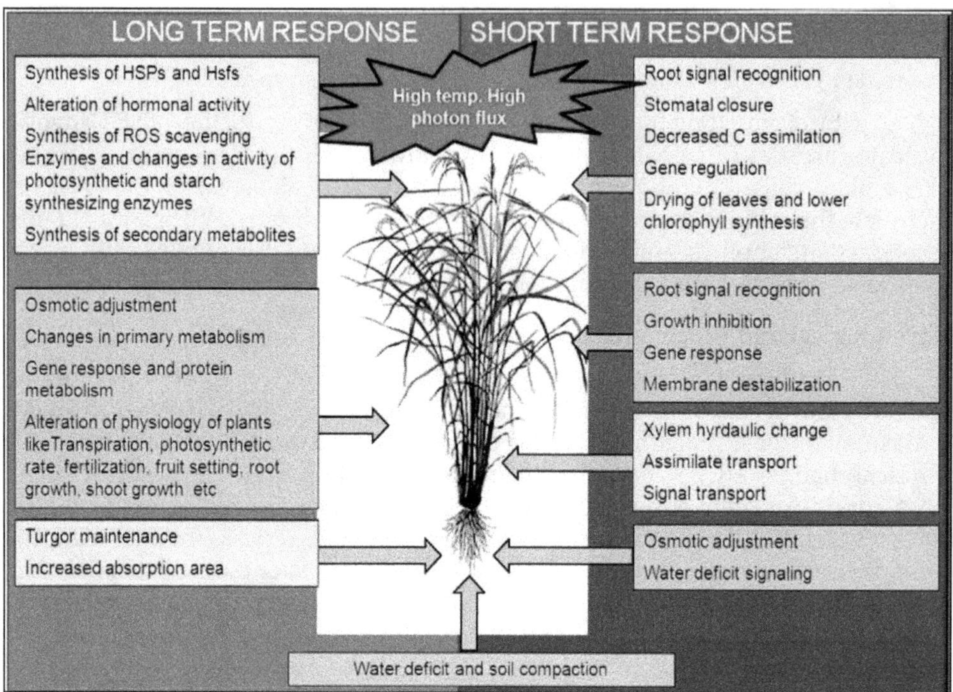

Figure 10.2: Long Term and Short Term Responses of Plant under Heat Stress. (*Source*: Concept taken from the book "Responses of Organisms to Water Stress" by Alexandre Bosco de Oliveira *et al.*, 2013).

most thermo-sensitive component of thylakoid membranes. It has been shown that the water oxidizing complex (WOC) is the most susceptible component of PSII to heat stress, although both the PSII reaction center and the light harvesting complexes can also be disrupted by high temperature (Keren *et al.*, 1997). The thermal inactivation of PSII is due to the extraction of divalent Ca^{2+} and Mn^{2+} cations and Cl^- anion from the PSII pigment-protein complex and the release of extrinsic 18, 24 and 33 kDa polypeptides from thylakoid membranes (Pastenes, 1996). One of the primary responses of plants to water deficit is stomata closure. This response minimizes water loss, but also lowers the intercellular CO_2 concentration (C_i), causing a stomatal or diffusional limitation to photosynthesis (Chaves *et al.*, 2003) because Rubisco, the enzyme that catalyzes CO_2 assimilation in photosynthesis, has a relatively low affinity for CO_2. In addition, reduced transpiration due to stomatal closure under hot, sunny conditions decreases the capacity for leaf cooling, increasing leaf temperature and, consequently, the incidence of heat stress. Photosynthesis is acutely sensitive to inhibition by moderate heat stress *i.e.* <10 C above the thermal optimum, and this inhibition has been linked to various causes, including the thermal instability of Rubisco's molecular chaperone, Rubisco activase (Salvucci and Crafts-Brandner, 2004). Reduction in plant growth is a major consequence of growing under stress conditions. This occurs mainly due to a reduction in net photosynthesis rate and generation of reducing powers as well as interference with mitochondrial functions. It is suggested that during light reactions increased leaf temperature induces

ATP synthesis to balance ATP consumption under heat stress possibly by cyclic electron flow (Bukhov *et al.*, 1999). During dark reactions of photosynthesis, rubisco activation in Calvin cycle has been determined as a critical step, being inhibited at 35-40 C, which results in decreased net CO_2 assimilation and the production of carbohydrates (Crafts-Brandner and Salvucci, 2000; Dubey, 2005). In mitochondria, environmental stress normally causes NAD+ breakdown, ATP over-consumption and higher rate of respiration. This is partially due to a breakdown in the NAD+ pool caused by the enhanced activity of poly ADP ribose polymerase PARP, which uses NAD+ as a substrate to synthesize polymers of ADP-ribose. This poly-ADP ribosylation is a post-translational modification of nuclear proteins that seems to be initiated by oxidative and other types of DNA damage. Stress-induced depletion of NAD+ results in a similar depletion of energy, since ATP molecules are required to resynthesize the depleted NAD+. Collectively, these reactions deplete the energy of the plant and enhance the production of ROS, which eventually lead to cell death (De Block *et al.*, 2005).

A strategy of improving stress tolerance in plants by maintaining the plant's energy homeostasis under stress is the production of transgenic plants with lowered polyADP ribosylation activity; such transgenics appear to be tolerant to multiple stresses by preventing energy overconsumption under stress, thereby allowing normal mitochondrial respiration (De Block *et al.*, 2005). In short, heat tolerance in plants is a cost-intensive process and consumes considerable cellular energy to cope with adversaries of high temperature. In tomato, reproductive processes were adversely affected by high temperature, which included meiosis in both male and female organs, pollen germination and pollen tube growth, ovule viability, stigmatic and style positions, number of pollen grains retained by the stigma, fertilization and post-fertilization processes, growth of the endosperm, proembryo and fertilized embryo (Foolad, 2005). Also, the most noticeable effect of high temperatures on reproductive processes in tomato is the production of an exerted style *i.e.* stigma is elongated beyond the anther cone, which may prevent self-pollination. Poor fruit set at high temperature has also been associated with low levels of carbohydrates and growth regulators released in plant sink tissues (Kinet and Peet, 1997). Considerable attention has been devoted to the induction of heat tolerance in existing high-yielding cultivars. Among the various methods to achieve this goal, foliar application of, or pre-sowing seed treatment with, low concentrations of inorganic salts, osmoprotectants, signaling molecules *e.g.*, growth hormones and oxidants *e.g.*, H_2O_2 as well as preconditioning of plants are common approaches. Preconditioned tomato plants exhibited good osmotic adjustment by maintaining the osmotic potential and stomatal conductance. Heat acclimated, compared to non-acclimated, turfgrass leaves manifested higher thermostability, lower lipid peroxidation product malondialdehyde (MDA) and lower damage to chloroplast upon exposure to heat stress (Xu *et al.*, 2006). Among the growth stages of plant the germination is affected first of all. Heat stress exerts negative impacts on various crops during seed germination though the ranges of temperatures vary largely on crop species (Johkan *et al.*, 2011). Reduced germination percentage, plant emergence, abnormal seedlings, poor seedling vigor, reduced radicle and plumule growth of geminated seedlings are major impacts caused by heat stress

Table 10.1: Physiological and Biochemical Response of Heat Stress Tolerance of different Crops

Crops	*Stress Temp.*	*Physiological and Biochemical Responses*	*References*
Rice	>35°C 39.6°C maximum at flowering and 38°C at anthesis	Sterility at the flowering and set no seed due to poor anther dehiscence low pollen production resulting low germinating pollen grain. Soluble protein and amino acids content of grain are increased. Higher temperature during grain-filling induces Chalkiness of grains; amylose concentration declines in many rice varieties and gelatinization temperature increases. Reduced ATPase activity in grains in heat-sensitive genotypes, increased malondialdehyde content in leaves and reduced root activity and photosynthetic rate of flag leaf.	Zhang *et al.*, 2002; Zhen *et al.*, 1997; Satake and Yoshida, 1978; Matsui *et al.*, 2000 and 2001; Prasad *et al.*, 2006.Yun-Ying *et al.*, 2009; Lisle *et al.*, 2000; Fitzgerald and Resurreccion, 2009; Chen *et al.*, 2008; Cuevas *et al.*, 2010.
Wheat	>40°C	Heat stress significantly reduced the protein and starch yields but the quantity of Glutenin macropolymer GMP significantly decreased and the size of GMP particles increased.	Don *et al.*, 2005.
Maize	30-45°C	Net photosynthesis was inhibited, Transpiration rate increased Progressively, maximum quantum yield of photosystem II *Fv*/*Fm* was relatively insensitive to leaf temperatures up to 45°C. zeatin riboside or zeatin level is much less but ABA maintained steady concentration in kernel.The activation state of phospho-enolpyruvate carboxylase decreased marginally at leaf temperatures above 40°C, and the activity of pyruvate phosphate dikinase was insensitive to temperature up to 45°C.	Nordine and Robert, 1994.
		The activation state of Rubisco decreased at temperatures exceeding 32.5°C, with nearly complete inactivation at 45°C. Levels of 3-phosphoglyceric acid and ribulose-1, 5-bisphosphate decreased and increased, respectively.	Steven and Salvucci, 2002.
Potato	35°C-40°C	Tubers grown at elevated temperatures are also more susceptible to a physiological disorder, internal heat necrosis, breakdown of parenchyma tissue, and dramatically reduce potato quality. So, tuberization is significantly inhibited.	Wannamaker and Collins, 1992.
		smHSP accumulation at above 35°C.	Sterrett *et al.*, 1991.
Oilseeds	22°C-27°C up to 40°C max.	Final grain weight and oil grain percentage was reduced 30-40 per cent, on average through increase in the pericarp:embryo ratio.	Rondanini *et al.*, 2003.
		Modify the oleic/linoleic ratio in standard sunflower cultivars by reducing the linoleic content may be caused in part by the reversible inhibition of oleate desaturase ODS enzyme. Increments in oleic percentage with higher night temperatures applied during early stages of grain development.	Izquierdo *et al.*, 2002.

Contd...

Table 10.1–*Contd...*

Crops	*Stress Temp.*	*Physiological and Biochemical Responses*	*References*
Fiber crops	>40°C	In cotton Heat stress led to decreased 3-phosphoglyceric acid content and increased ribulose-1,5bisphosphate content, which is indicative of inhibited metabolite flow through Rubisco. Heat stress inhibited CER primarily by decreasing the activation state of Rubisco via inhibition of Rubisco activase.	Law and Crafts, 1999.
Sugar crops	40-45°C	Hsp90, SsHsp17.2 and SsHsp17.9 are the most highly expressed Hsp, and class I sHsps in sugarcane and having high chaperon activity.	Viviane *et al.*, 2010.

documented in various cultivated plant species (Kumar *et al.*, 2011). Inhibition of seed germination is also well documented in HT which often occurs through induction of ABA (Essemine *et al.*, 2010).

Physiology and Biochemistry of Heat Stress Tolerance Mechanism

In order to cope with heat stress, plants implement various mechanisms, including maintenance of membrane stability, scavenging of ROS, production of antioxidants, accumulation and adjustment of compatible solutes, induction of mitogen-activated protein kinase MAPK, calcium-dependent protein kinase CDPK cascades and most importantly, chaperone signaling and transcriptional activation. In addition to genetic approaches, crop heat tolerance can be enhanced by preconditioning of plants under different environmental stresses or exogenous application of osmoprotectants such as glycinebetaine and proline. Acquiring thermotolerance is an active process by which considerable amounts of plant resources are diverted to structural and functional maintenance to escape damages caused by heat stress. Moreover, trehalose scavenges H_2O_2 and $O_2^{\bullet -}$ greatly in a concentration-dependent manner, reaching the maximal scavenging rates of 95 per cent for $H_2O_2$and 78 per cent for $O_2^{\bullet -}$ respectively, at 50mM trehalose concentration. These results suggest that trehalose plays a direct role in eliminating H_2O_2 and O_2 –• in wheat under heat stress. The main antioxidant enzymes which eliminate ROS include superoxide dismutases (SODs), peroxidases (PODs), ascorbate catalases (CATs), and ascorbate peroxidases (APX). Most of the secondary metabolites are synthesized from the intermediates of primary carbon metabolism via phenylpropanoid, shikimate, mevalonate or methyl erythritol phosphate (MEP) pathways (Wahid and Ghazanfar, 2006). High-temperature stress induces production of phenolic compounds such as flavonoids and phenylpropanoids. Phenylalanine ammonia-lyase (PAL) is considered to be the principal enzyme of the phenylpropanoid pathway. Increased activity of PAL in response to thermal stress is considered as the main acclamatory response of cells to heat stress. Heat acclimated, compared to non-acclimated, turfgrass leaves manifested higher thermostability, lower lipid peroxidation product malondialdehyde (MDA) and lower damage to chloroplast upon exposure to heat stress (Xu *et al.*, 2006). In pearl millet, pre-sowing hardening of the seed at high temperature 42 C resulted in plants tolerant to overheating and dehydration and showing higher levels of water-soluble proteins and lower amounts of amide-N in leaves compared to non-hardened plants (Tikhomirova, 1985).

Seed Quality and Management Strategies under Heat Stress

Heat affects the rate of carbon and nitrogen deposition in the grain as well as the formation of high molecular starch and protein particles. HT also affects the protein content of grain and glutenin particle size of wheat. Stone and Nicolas (1998) reported that heat stress increased grain protein percentage despite the fact that protein content per grain was reduced by heat. High temperatures during grain filling can significantly elevate protein concentration while lowering the functionality of protein (Corbellini *et al.*, 1997). Genotype was considered the most important determinant of wheat grain quality and end-use quality (Souza *et al.*,

2004). But the influence of management practices to tolerate heat stress in crops was not yet thoroughly studied.

Conclusion and Future Prospectus

It has been suggested that there is a potential for genetic improvement to advance flowering to an earlier time-of-day in present high yielding cultivars. A search for cultivars with earlier time-of-day of flowering among *O.sativa* would be useful for developing cultivars to avoid high temperature stress at critical stages of flower development anthesis, pollen shed, pollen tube growth and fertilization. The effect of high temperature on grain quality and the influence of nutritional and water status of soil should be studied thoroughly. Heat stress tolerance mechanism is a complex trait to investigate but details understanding of it at biochemical, physiological and molecular level will open up novel possibilities to develop designer crops in future, capable of withstanding high temperature, which might be very much important from global worming perspective.

References

Alexandre, B. de O., Nara, L., Mendes, A. and Eneas, G.F., 2013. In "Response of organisms to water stress", ISBN *978-953-51-0933-4*.

Baker, J.T., Allen, Jr.L.H. and Boote, K.J., 1992. Temperature effects on rice at elevated CO_2 concentration. *J. Exp. Bot.*, 43: 959-964.

Bingru, H., Yan, X. and Hongmei, Du., 2013. Identification of metabolites associated with superior heat tolerance in thermal bentgrass through metabolic profiling. *Crop Sci.*, 53: 1626-1635.

Bukhov, N.G., Wiese, C., Neimanis, S. and Heber, U., 1999. Heat sensitivity of chloroplasts and leaves: leakage of protons from thylakoids and reversible activation of cyclic electron transport. *Photosyn. Res.*, 59: 81-93.

Corbellini, M., Carnevar, M.G., Mazza, L., Ciaffi, M., Lafiandra, E. and Borghi, B., 1997. Effect of the duration and intensity of heat shock during grain filling on dry matter and protein accumulation, technological quality and protein composition in bread wheat and durum wheat. *Aust. J. Plant Physiol.*, 24: 245-260.

Charng, Y., Liu, H., Liu, N., Hsu, F. and Ko, S., 2006. Arabidopsis Hsa32, a novel heat shock protein, is essential for acquired thermo-tolerance during long term recovery after acclimation. *Plant Physiol.*, 140: 1297-1305.

Chaves, M.M., Maroco, J.P. and Pereira, J.S., 2003. Understanding plant responses to drought - from genes to the whole plant. *Funct. Plant Biol.*, 30: 239-264.

Chen, M.H., Bergman, C., Pinson, S. and Fjellstrom, R., 2008. Waxy gene haplotypes: associations with apparent amylose content and the effect by the environment in an international rice germplasm collection. *J. Cereal Sci.*, 47: 536-545.

Crafts-Brandner, S.J. and Salvucci, M.E., 2000. Rubisco activase constrains the photosynthetic potential of leaves at high temperature and CO_2. *Proc. Natl. Acad. Sci., U.S.A.* 97: 13430- 13435.

Cuevas, R., Daygon, V., Corpuz, H., Waters, D.L.E., Reinke, R.F. and Fitzgerald, M.A., 2010. Melting the secrets of gelatinization temperature. *Funct. Plant Biol.*, 37: 439-447.

Cvikrova, M., Gemperlova, L., Dobra, J., Martincova, O., Prasil, I.T., Gubis, J. and Vankova, R., 2012. Effect of heat stress on polyamine metabolism in proline over-producing tobacco plants. *Plant Sci.*, 182: 49-58.

De Block, M., Verduyn, C., De Brouwer, D. and Cornelissen, M., 2005. PolyAD Pribose polymerase in plants affects energy homeostasis, cell death and stress tolerance. *Plant J.*, 41: 95-106.

Dinkova, T.D., Zepeda, H., Martinez-Salas, E., Martinez, L.M., Nieto-Sotelo, J. and de Jimenez, E.S., 2005. Cap-independent translation of maize Hsp101. *Plant J.*, 41: 722-731.

Don, C., Lookhart, G., Naeem, H., MacRitchie, F. and Hamer, R.J., 2005. Heat stress and genotype affect the glutenin particles of the glutenin macropolymer-gel fraction. *J. Cereal Sci.*, 42: 69-80.

Dubey, R.S., 2005. Photosynthesis in plants under stressful conditions. In: Pessarakli, M. Ed., *Handbook of Photosynthesis. CRC Press, Boca Raton, Florida*, pp. 717-737.

Essemine, J., Ammar, S. and Bouzid, S., 2010. Impact of heat stress on germination and growth in higher plants: Physiological, biochemical and molecular repercussions and mechanisms of defence. *J. Biol. Sci.*, 10: 565-572.

Filichkin, S.A., Priest, H.D., Givan, S.A., Shen, R., Bryant, D.W., Fox, S.E., Wong, W.K. and Mockler, T.C., 2010. Genome-wide mapping of alternative splicing in *Arabidopsis thaliana*. *Genome Res.*, 20: 45-58.

Fitzgerald, M.A. and Resurreccion, A.P., 2009. Maintaining the yield of edible rice in a warming world. *Funct. Plant Biol.*, 36: 1037-1045.

Foolad, M.R., 2005. Breeding for abiotic stress tolerances in tomato. In: Ashraf, M., Harris, P.J.C. Eds., Abiotic Stresses: Plant Resistance Through Breeding and Molecular Approaches. *The Haworth Press Inc., New York, USA*, pp. 613-684.

Guy, C., Kaplan, F., Kopka, J., Selbig, J. and Hincha, D.K., 2008. Metabolomics of temperature stress. *Physiol. Plant.*, 132: 220-235.

Hall, A.E., 2001. Crop Responses to Environment. *CRC Press LLC*, Boca Raton, Florida.

Hideg, E., Kos, P.B. and Schreiber, U., 2008. Imaging of NPQ and ROS formation in tobacco leaves: heat inactivation of the water-water cycle prevents down-regulation of PS II. *Plant Cell Physiol.*, 49: 1879-1886.

Iida, K., Seki, M., Sakurai, T., Satou, M., Akiyama, K., Toyoda, T., Konagaya, A. and Shinozaki, K., 2004. Genome-wide analysis of alternative pre-mRNA splicing in *Arabidopsis thaliana* based on full-length cDNA sequences. *Nucleic Acids Res.*, 32: 5096-5103.

Izquierdo, N., Aguirrezabal, L., Andrade, F. and Pereyra, V., 2002. Night temperature affects fatty acid composition in sunflower oil depending on the hybrid and the phenological stage. *Field Crops Res.*, 77: 115-126.

Johkan, M., Oda, M., Maruo, T. and Shinohara, Y., 2011. Crop production and global warming. In: Global Warming Impacts-Case Studies on the Economy, Human Health, and on Urban and Natural Environments; Casalegno, S., Ed.; In Tech: *Rijeka, Croatia*, pp. 139-152.

Jones, P.D., New, M., Parker, D.E., Mortin, S. and Rigor, I.G., 1999. Surface area temperature and its change over the past 150 years. *Rev. Geophys*. 37: 173-199.

Kinet, J.M. and Peet, M.M., 1997. Tomato. In: Wien, H.C. Ed., The Physiology of Vegetable Crops. *CAB International*, Wallingford, UK, pp. 207-258.

Korotaeva, N.E., Antipina, A.I., Grabelynch, O.I., Varakina, N.N., Borovskii, G.B. and Voinikov, V.K., 2001. Mitochondrial low-molecular-weight heat shock proteins and tolerance of crop plant's mitochondria to hyperthermia. *Fiziol. Biokhim Kul'turn. Rasten.*, 29: 271-276.

Kristiansen, K.A., Jensen, P.E., Moller, I.M. and Schulz, A. 2009. Monitoring reactive oxygen species formation and localisation in living cells by use of the fluorescent probe CM- H2DCFDA and confocal laser microscopy. *Physiol. Plant.*, 136: 369-383.

Kumar, S., Kaur, R., Kaur, N., Bhandhari, K., Kaushal, N., Gupta, K., Bains, T.S. and Nayyar, H., 2011. Heat-stress induced inhibition in growth and chlorosis in mungbean *Phaseolus aureus* Roxb. is partly mitigated by ascorbic acid application and is related to reduction in oxidative stress. *Acta Physiol. Plant.*, 33: 2091-2101.

Lakso, A.N., and Kliewer, W.M., 1975. The influence of temperature on malic acid metabolism in grape berries. *Plant Physiol.*, 56: 370-372.

Larkindale, J. and Huang, B., 2005. Effects of abscisic acid, salicylic acid, ethylene and hydrogen peroxide in thermotolerance and recovery for creeping bentgrass. *Plant Growth Regul.*, 47: 17-28.

Law, R.D. and Crafts, B.S.J., 1999. Inhibition and acclimation of photosynthesis to heat stress is closely correlated with activation of ribulose-1,5 bisphosphate carboxylase/oxygenase. *Plant Physiol.*, 120: 173-181.

Lisle, A.J, Martin, M. and Fitzgerald, M.A., 2000. Chalky and translucent rice grains differ in starch composition and structure and cooking properties. *Cereal Chem.*, 77: 627-632.

Los, D.A., 2004. The effect of low-temperature-induced DNA supercoiling on the expression of the desaturase genes in Synechocystis. *Cell. Mol. Biol.*, 50: 605-612.

Maestri, E., Klueva, N., Perrotta, C., Gulli, M., Nguyen, H.T. and Marmiroli, N. 2002. Molecular genetics of heat tolerance and heat shock proteins in cereals. *Plant Mol. Biol.*, 48: 667-681.

Matsuda, O., Sakamoto, H., Hashimoto, T. and Iba, K., 2005. A temperature-sensitive mechanism that regulates post-translational stability of a plastidial omega-3 fatty acid desaturase FAD8 in *Arabidopsis* leaf tissues. *J. Biol. Chem.*, 280: 3597-3604.

Matsui, T., Omasa, K. and Horie, T., 2000. High temperatures at flowering inhibit swelling of pollen grains, a driving force for thecae dehiscence in rice *Oryza sativa* L., *Plant Prod. Sci.*, 3: 430-434.

Matsui, T., Omasa. K. and Horie. T., 2001. The difference in sterility due to high temperatures during the flowering period among japonica rice varieties. *Plant Prod. Sci.*, 4: 90-93.

Miroshnichenko, S., Tripp, J., Nieden, U., Neumann, D., Conrad, U. and Manteuffel, R., 2005. Immunomodulation of function of small heat shock proteins prevents their assembly into heat stress granules and results in cell death at sub-lethal temperatures. *Plant J.*, 41: 269- 281.

Keren, N., Berg, A.,Van Kan, P.J.M., Levanon, H. and Ohad, I.,1997.Mechanism of photosystem II photo inactivation and D1 protein degradation at low light: The role of back electron

flow. *Proc. Natl. Acad. Sci., U.S.A.* 944: 1579-1584.

Nakamoto, H. and Hiyama, T., 1999. Heat-shock proteins and temperature stress. In: Pessarakli, M. Ed., *Handbook of Plant and Crop Stress.* Marcel Dekker, New York, pp. 399-416.

Nordine, C. and Robert J.J., 1994. Disruption of maize kernel growth and development by heat stress. *Plant Physiol.*, 106: 45-51.

Panchuk, I.I., Volkov, R.A. and Schoffl, F., 2002. Heat stress- and heat shock transcription factor- dependent expression and activity of ascorbate peroxidase in Arabidopsis. *Plant Physiol.*, 129: 38-853.

Pastenes, P.H., 1996. Effect of high temperature on photosynthesis in beans. I. Oxygen evolution and chlorophyll fluorescence. *Plant Physiol.*, 112: 1245-1251.

Prasad, P.V.V., Boote, K.J., Allen, L.H., Sheehy, J.E. and Thomas, J.M.G., 2006. Species, ecotype and cultivar differences in spikelet fertility and harvest index of rice in response to high temperature stress. *Field Crops Research*, 95: 398-411.

Pshybytko, N.L., Kruk, J., Kabashnikova, L.F. and Strzalka, K., 2008. Function of plastoquinone in heat stress reactions of plants. *Biochim. Biophys. Acta.*, 1777: 1393-1399.

Rizhsky, L., Liang, H.J., Shuman, J., Shulaev, V., Davletova, S. and Mittler, R., 2004. When defense pathways collide. The response of *Arabidopsis* to a combination of drought and heat stress. *Plant Physiol.*, 134: 1683-1696.

Rizhsky, L., Hongjian, L. and Mittler, R., 2002. The combined effect of drought stress and heat shock on gene expression in tobacco. *Plant Physiol.*, 130: 1143-1151.

Rondanini, D., Savin, R. and Hall, A.J., 2003. Dynamics of fruit growth and oil quality of sunflower Helianthus annuus L. exposed to brief intervals of high temperature during grain filling. *Field Crop Res.*, 83: 79-90

Sage, R.F. and Kubien, D.S., 2007. The temperature response of C_3 and C_4 photosynthesis. *Plant Cell Environ.*, 309: 1086-106.

Salvucci, M.E., 2008. Association of Rubisco activase with chaperonin-60beta: a possible mechanism for protecting photosynthesis during heat stress. *J. Exp. Bot.*, 59: 1923-1933.

Salvucci, M.E., Osteryoung, K.W., Crafts-Brandner, S.J. and Vierling, E., 2001. Exceptional sensitivity of Rubisco activase to thermal denaturation *in vitro* and *in vivo*. *Plant Physiol.*, 127: 1053-1064.

Salvucci, M.E. and Crafts-Brandner, S.J., 2004. Relationship between the heat tolerance of photosynthesis and the thermal stability of Rubisco activase in plants from contrasting thermal environments. *Plant Physiology*, 134: 1460-1470.

Sangwan, V. and Dhindsa, R.S., 2002. *In vivo* and *in vitro* activation of temperature responsive plant map kinases. *FEBS Lett.*, 531: 561-564.

Satake, T. and Yoshida, S., 1978. High temperature induced sterility in indica rices at flowering. *Jpn. J. Crop Sci.*, 47: 6-17.

Schoffl, F., 2005. The role of heat shock proteins in abiotic stress response and the development of plants. *Universitat Tubingen, ZMBP, Allgemeine Genetik.*

Schoffl, F., Prandl, R. and Reindl, A., 1999. Molecular responses to heat stress. In: Shinozaki, K., Yamaguchi-Shinozaki, K. Eds., Molecular Responses to Cold, Drought, Heat and Salt Stress in Higher Plants. *R.G. Landes Co., Austin, Texas*, pp. 81-98.

Souza, E.J., Martin, J.M., Guttieri, M.J., O'Brien, K.M., Habernicht, D.K. and Lanning, S.P., 2004. Rice yield decline with higher night temperature from global warming. *Proc. Natl. Acad. Sci., U.S.A.* 101: 9971-9975.

Sterrett, S.B., Henninger, M.R. and Lee, G.S., 1991. Relationship of internal heat necrosis potato to time and temperature after planting. *J. Am. Soc. Hort. Sci.*, 116: 697-700.

Steven, J.C.B. and Salvucci, M.E., 2002. Sensitivity of photosynthesis in a C_4 plant, maize, to heat stress. *Plant Physiol.*, 129: 1773-1780.

Stone, P.J. and Nicolas, M.E., 1998. The effect of duration of heat stress during grain filling on wheat varieties differing in heat tolerance: grain growth and fractional protein accumulation. *Aust. J. Plant Physiol.*, 25: 13-20.

Sugio, A., Dreos, R., Aparicio, F. and Maule, A.J., 2009. The cytosolic protein response as a subcomponent of the wider heat shock response in *Arabidopsis*. *Plant Cell*, 21: 642- 654.

Suri, S.S. and Dhindsa, R.S., 2008. A heat-activated MAP kinase HAMK as a mediator of heat shock response in tobacco cells. *Plant Cell Environ.*, 31: 218-226.

Suzuki, N. and Mittler, R., 2006. Reactive oxygen species and temperature stresses: a delicate balance between signaling and destruction. *Physiol. Plant*, 126: 45-51.

Takahashi, S. and Murata, N., 2008. How do environmental stresses accelerate photoinhibition? *Trends Plant Sci.*, 134: 178-82.

Tikhomirova, E.V., 1985. Changes of nitrogen metabolism in millet at elevated temperatures. *Field Crops Res.*, 11: 259-264.

Venkateswarlu, B. and Shanker, A.K., 2009. Climate change and agriculture: daptation and mitigation strategies. *Indian J. Agron.*, 54: 226-230.

Vierling, E., 1991. The role of heat shock proteins in plants. *Annu. Rev. Plant Physiol. Plant Mol. Biol.*, 42: 579-620.

Viviane, C.H. da Silva., Thiago, C., Cagliari, T.B.L., Fabio C.G.C. and Ramos, H.I., 2010. Conformational and functional studies of a cytosolic 90 kDa heat shock protein Hsp90 from sugarcane. *Plant Physiol. Bioch.*, 68: 16-22.

Wannamaker, M.J. and Collins,W.W., 1992. Effect of year, location, and harvest on susceptibility of cultivars to internal heat necrosis in North Carolina. *Am. Potato J.*, 69: 221-228.

Wahid, A., Gelani, S., Ashraf, M. and Foolad, M., 2007. Heat tolerance: An overview. *Environ. Exp. Bot.*, 61: 199-223.

Xu, S., Li, J., Zhang, X., Wei, H. and Cui, L., 2006. Effects of heat acclimation pretreatment on changes of membrane lipid peroxidation, antioxidant metabolites, and ultrastructure of chloroplasts in two cool-season turfgrass species under heat stress. *Environ. Exp. Bot.*, 56: 274-285.

Yun-Ying, C.A.O., Duan, H., Yang, L.N., Wang, Z.Q., Liu, L.J. and Yang, J.C., 2009. Effect of high temperature during heading and earlyfilling on grain yield and physiological characteristics in *Indica* rice. *Acta. Agron. Sin.*, 353: 512-521.

Yamada, K., Fukao, Y., Hayashi, M., Fukazawa, M., Suzuki, I. and Nishimura, M., 2007. Cytosolic HSP90 regulated the heat shock response that is responsible for heat acclimation in *Arabidopsis thaliana*. *J. Biol. Chem.*, 282: 37794-37804.

Yamori, W., Ko, N., Kouki, H. and Ichiro, T., 2010. Phenotypic plasticity in photosynthetic temperature acclimation among crop species with different cold tolerance. *Plant Physiol.*, 152: 388-399.

Yamori, W. and Von Caemmerer, S., 2009. Effect of Rubisco activase deficiency on the temperature response of CO_2 assimilation rate and Rubisco activation state: insights from transgenic tobacco with reduced amounts of Rubisco activase. *Plant Physiol.*, 151: 2073-2082.

Zhang, L., Wu, D.Y., Zhu, B.Y., Zhang, M.G. and Li, J.Q., 2002. Effects of temperature and light on the dynamic change in soluble protein and sugar in rice leaves and grains during the milk-filling stage. *J. South China Norm Univ. Nat. Sci.*, 52: 98-101.

Zhang, R. and Sharkey, T.D., 2009. Photosynthetic electron transport and proton flux under moderate heat stress. *Photosynth. Res.*, 100: 29-43.

Zhen, H., Huang, C.L., Chen, Y., Li, M.L., Liu, X.Z., Pan, D.J. and Zhang, H.Q., 1997. Study on the protein content of wild rice resource. *J. South China Agri. Univ.*, 184: 16-20.

2016, Recent Advances in Plant Stress Physiology *Pages* **219–236**
Editors: ***Praduman Yadav, Sunil Kumar and Veena Jain***
Published by: **DAYA PUBLISHING HOUSE, NEW DELHI**

Chapter 11

Drought Tolerance in Crops

Anup Singh*

Jawahar Navodaya Vidyalaya,
Devrala (a unit of MHRD), Govt. of India

ABSTRACT

Drought tolerance refers to the degree to which a plant is adapted to arid or drought conditions. A drought tolerant plant is one that will survive in the typical or somewhat less than typical amount of rainfall in any region. The plant hormone abscisic acid (ABA) is behind crops responses to stressful situations such as drought, but how it does so has been a mystery for years. Abscisic acid triggers an array of plant drought-tolerance mechanisms. Pyrabactin, a new synthetic chemical, mimics a naturally produced stress hormone in plants to help them cope with drought conditions and it also turns on the ABA signaling pathway in Arabidopsis, a small flowering plant used widely in plant biology laboratories as a model organism. Plants naturally produce ABA in modest amount to help them survive drought by inhibiting growth. Pyrabactin is relatively inexpensive, easy to make, and not sensitive to light.

Molecular markers can be used to explore germplasm through segregation and association mapping to identify useful alleles in both cultivated varieties and wild relatives. Many stress-related genes have been isolated and characterized in the last two decades in a variety of crop species. The transgenic plants showed increased stress tolerance as well as the over induction of downstream stress related genes and/or higher levels of soluble sugars and proline. An emerging strategy to broaden stress tolerance in crops is to maintain energy homeostasis under stress conditions. The transgenic crops show improved water stress tolerance and a higher photosynthesis rate due to modified stomatal conductance under water deficit conditions. A well developed root system enables plants to take up more water during water deficit stress, resulting in a more favorable plant water status and less injury. Recent studies have found the way to increase root size through single gene transformation.

Breeders now have new perspectives for plant improvement for drought. First, they will soon have new tools that will provide an opportunity to move the selection from phenotype to genotype. Second, they will have a new role, to blend together all knowledge of the traits sustaining yield under drought and to accumulate randomly dispersed trans-genes into elite genotypes. This so called

**Author*: E-mail: yadav.anupsingh@gmail.com

"breeding by design" strategy will lead to new highly yielding cultivars able to improve performance in both high and low yielding environments. A significant example showing how the integration of stress physiology and genomics can lead to an integrated view of plant breeding is represented by the studies on transpiration efficiency. During drought stress, crops generally coordinate photosynthesis and transpiration, although significant genetic variation in transpiration efficiency has been identified both between and within species.

Keywords: *Abscisic acid, Drought, Pyrabactin, PYR1 protein, Quantitative Trait Locus* (*QTL*).

Introduction

Drought is when an area gets less than its normal rainfall over months or even years. Crops and other plants need water to grow, and land animals need it to live. It can become dangerous to people and other animals; causing famine and even creating deserts. Drought is a natural event, caused by other weather events like El Nino and high pressure systems. Drought can also be triggered by deforestation, global warming and diverting rivers or emptying lakes.

Drought is defined as:

1. Deficiency of available soil moisture which produces water deficits in the plant sufficient to cause a reduction in plant growth. Or
2. Drought is a period of inadequate or no rainfall over extended time creating soil moisture deficit and hydrological imbalances.

Classification of Drought

On the Basis of Water Availability

Meteorological Drought

Ramdas (1960) defined as the actual rainfall which is deficient by more than twice the mean deviation. Indian Metrological Department (IMD) has defined meteorological drought as the situation when actual rainfall is less than 75 per cent of the normal rainfall over an area. This is accepted principally because of its simplicity.

Hydrological Drought

Linsley *et al.* (1975) considered hydrological drought as "a period during which stream flows are inadequate to supply established used under given water management system". The frequency and severity of hydrological droughts often defined on the basis of water depletion or shortage in reserve basins, reservoirs lakes wells etc. This drought affects industry and power generation.

Agricultural Drought

Heathcote (1974) defined agricultural drought as the shortage of water for agriculture. It is soil moisture deficit that leads to reduction in growth, development and yield of crops. This is a situation resulted from inadequate rainfall when soil moisture falls short to meet the water demands of the crop during the growing period. This affects the crop growth or crop may wilt due to moisture stress resulting in yield reduction.

Socio-Economic Drought

The socio-economic effects of drought can also incorporate the features of meteorological hydrological and agricultural droughts. They are usually associated with the supply and demand of some economic goods. This drought should be linked to precipitation as well as fluctuations in demand.

On the Basis of Time of Occurrence

Droughts differ in time and period of their occurrence. The various types of droughts are:

Permanent Drought

This is the drought area of permanent dry arid or desert regions. Crop production due to inadequate rainfall is not possible without irrigation in these areas. Vegetation like cactus thorny shrubs, xerophytes are generally observed.

Seasonal Drought

In the regions with clearly defined rainy (wet) and dry climate, seasonal droughts may result due to large scale seasonal circulation. This generally happens in monsoon area.

Contingent Drought

This results due to irregular and variability in rainfall especially in humid and sub humid regions. The occurrence of drought may coincide with critical crop growth stages resulting in severe yield reduction.

Drought is a disaster which usually takes place slowly. It is difficult to decide the time of its start and its end. Its effects often build up slowly over a long period of time and may continue from months to years after the end of the event. Drought is a continuous period of dry weather due to lack of rain. Therefore drought is a climatic anomaly, characterized by deficient supply of moisture resulting either from sub-normal rainfall, erratic rainfall distribution, higher water need or a combination of all the factors. The escalating impacts of droughts have increasingly drawn the attention of scientists, planners and society. The vulnerability to drought in relation to the increasing needs of the growing population has become a point of great concern, especially on the food front. In spite of the technological developments in providing improved crop varieties and better management practices, in India, agriculture has been considered a gamble as the agricultural productivity is strongly influenced by the vagaries of the monsoon.

Droughts are the resultant of acute water shortage due to lack of rains over extended periods of time affecting various human activities and lead to crop failure, un-replenished ground water resources, depletion in lakes/reservoirs, shortage of drinking water and, reduced fodder availability etc. Often a region adopts itself to a certain level of water shortage based on the long-term climatic conditions experienced by it. Any negative departure from these levels creates conditions of drought, depending on the intensity and duration of this deficit. Thus drought conditions differ from region to region. Because drought affects many economic and social sectors, scores of definitions have been developed by a variety of disciplines and the approaches taken to define it also reflect regional and ideological variations.

In general, drought means different things to different people. To a meteorologist it is the absence of rain while to the agriculturist it is the deficiency of soil moisture in the crop root zone to support crop growth and productivity. To the hydrologist it is the lowering of water levels in lakes, reservoirs, etc., while for the city management it may mean the shortage of drinking water availability. Thus, it is unrealistic to expect a universal definition of drought for all fields of activity.

Drought is by far the most important environmental stress in agriculture and many efforts have been made to improve crop productivity under water-limiting conditions. While natural selection has favoured mechanisms for adaptation and survival, breeding activity has directed selection towards increasing the economic yield of cultivated species. More than 80 years of breeding activities have led to some yield increase in drought environments for many crop plants. Meanwhile, fundamental research has provided significant gains in the understanding of the physiological and molecular responses of plants to water deficits, but there is still a large gap between yields in optimal and stress conditions. Minimizing the 'yield gap' and increasing yield stability under different stress conditions are of strategic importance in guaranteeing food for the future.

Drought Tolerance

Drought tolerance refers to the degree to which a plant is adapted to arid or drought conditions. A drought tolerant plant is one that will survive in the typical or somewhat less than typical amount of rainfall in any region. There is no one-size-fits-all drought tolerant plant. *Desiccation tolerance* is an extreme degree of drought tolerance. Plants naturally adapted to dry conditions are called *xerophytes*. Frequently the term "drought tolerant" is thought of as being "dry or desert-like," but this is an inaccurate description. They need not be limited to cactus varieties or other dry climate plants; instead they include a wide selection of lush, green plants that are attractive in any landscape. Once established, these plants are able to withstand long periods of dryness without deterioration, going several weeks, or in some cases an entire season, between deep watering. Such plants reduce the impact on limited water supplies.

Water Requirements

When planting drought tolerant species of trees and shrubs, it is necessary to water frequently and deeply for one or two seasons. Once the plant has become established, it can thrive on far less water than we provide. If plants are watered frequently, such as during lawn watering, they become shallow rooted and therefore dependent upon frequent irrigation. On the other hand, less frequent watering promotes deep rooting up to some extent. Trees, shrubs and plants also require less water when proper gardening practices are followed. These include proper soil preparation, selecting the plant for the site and planting correctly. Proper irrigation is also important along with the use of mulch and weed control.

Figure 11.1: Drought Tolerant Crops.

Cultural Factors

A large majority of landscape plants, once established, are relatively drought tolerant. Following gardening practices can be employed for the plants which are not drought tolerant:

1. Using organic materials and/or water holding polymers at time of planting.
2. Installing drip or soaker hose irrigation systems.
3. Using organic mulch on top of ground where plants are growing.
4. Controlling water-robbing weeds adjacent to landscape plants.
5. Using anti-transpirants to reduce excessive water loss from leaves and stems.

Adaptations to Dry Conditions

Drought tolerant crops typically make use of either C_4 carbon fixation or crassulacean acid metabolism (CAM) to fix carbon during photosynthesis. Both are improvements over the more common but more basal C_3 pathway in that they are more energy efficient. CAM is particularly good for arid conditions because carbon dioxide can be taken up at night, allowing the stomata to stay closed during the heat of day and thus reducing water loss.

Many adaptations for dry conditions are structural, including the following:

- Adaptations of the stomata to reduce water loss, such as reduced numbers or waxy surfaces.
- Water storage in succulent above-ground parts or water-filled tubers.
- Adaptations in the root system to increase water absorption.
- Trichomes on the leaves to absorb atmospheric water.

Recent Development in Drought Tolerance in Crops

A Mysterious Hormone: ABA

When drought tolerant plants detect dry conditions, they synthesize abscisic acid, which causes changes from root tips to leaves and flowers. Plants under the influence of this hormone begin to conserve water. Their seeds lie dormant in the ground. Their leaves close microscopic pores to stop water loss. They slow their own growth, and they signal numerous genetic changes, reprogramming themselves to accomplish their single most pressing goal – survival.

ABA triggers an array of drought tolerance mechanisms in plants. It was discovered in the early 1960s, and plant biologists have known for decades that it plays this crucial role in keeping plants alive during drought. Crops and other plants are constantly confronted with adverse environmental conditions which lower the yield. Plants use specialized signals to sense difficult times and adapt to stressful conditions to enhance survival. Of the various stress hormones, Abscisic acid, produced naturally by plants, has emerged over the last 30 years as the key hormone that helps plants cope with drought conditions. Under such stress, plants increase their ABA level, which helps them to survive drought.

This plant hormone triggers crops' responses to stressful situations such as drought. A group of scientists headed by José Antonio Marquez at the European Molecular Biology Laboratory (EMBL) in Grenoble, France, and Pedro Luis Rodriguez at Consejo Superior de Investigaciones Cientificas (CSIC) in Valencia, Spain discovered that the key lies in the structure of a protein called PYR1 and how it interacts with the hormone. Their study, published in *Nature* in Nov 2009, could open up new approaches to increasing crops' tolerant to water shortage. Under normal conditions, protein phosphatase 2C (PP2C) inhibits the ABA pathway, but when a plant is subjected to drought, the concentration of ABA increases. This removes the brake from the pathway, allowing the signal for drought response which turns specific genes on or off, triggering mechanisms for increasing water uptake and storage, and decreasing water loss.

Crops, treated with ABA before a drought occurs, take all their water-saving measures before the drought actually hits, so they are more prepared, and more likely to survive that water shortage *i.e.* they become more tolerant to drought.

PYR1 Protein: Structure and Role

Collaborating closely with Schroeder and his lab, Elizabeth Getzoff, Professor, Department of Molecular Biology and her group have investigated the involvement of PYR1 in drought resistance by observing PYR1 and ABA interaction. Getzoff's lab enlisted the use of a technique called x-ray crystallography. X-ray crystallography is a method that can determine three-dimensional positions for the individual atoms of a protein's structure. The research showed that two copies of PYR1 fit snugly together in plant cells. There, they are targeted by abscisic acid. Each copy of the PYR1 molecule has an internal open space like the inside of a tin can, and when a hormone molecule comes along, it fits neatly into one of the two spaces. This induces part of the PYR1 protein that the team calls the "lid" to close. Further structural changes to other parts of the PYR1 molecule initiate interactions with other proteins, thus triggering plant processes for resisting drought.

To determine the structure of PYR1, the scientists made use of the infrastructure of the Partnership for Structural Biology, including EMBL Grenoble's high-throughput crystallization facilities and the beam lines at the European Synchrotron Radiation Facility, located in the same campus as EMBL Grenoble.

Abscisic Acid versus Pyrabactin

ABA is naturally produced by plants in modest amount to help them survive drought by inhibiting growth. ABA has already been commercialized for agricultural use. But it has at least two disadvantages: it is light-sensitive and costly to synthesize artificially for commercial use. Pyrabactin, on the other hand, is relatively inexpensive, easy to synthesize, and not sensitive to light. But its drawback is that, unlike ABA, it does not turn on all plant receptors that are needed to be activated for drought tolerance.

A receptor is a protein molecule in a cell to which mobile signaling molecules – such as ABA or pyrabactin, each of which turns on stress-signaling pathways in plants – may attach. Usually at the top of a signaling pathway, the receptor functions like a boss relaying orders to the team below that then proceeds to execute particular decisions in the cell.

Each receptor is equipped with a pocket, akin to a padlock, in which a chemical, like pyrabactin, can dock into, operating like a key. Even though the receptor pockets appear to be fairly similar in structure, subtle differences distinguish a pocket from its peers. The result is that while ABA, a product of evolution, can fit neatly in any of these pockets, pyrabactin is less successful. Still, pyrabactin, by being partially effective (it works better on seeds than on plant parts), serves as a leading molecule for devising new chemicals for controlling stress tolerance in plants. Each receptor is equipped with a lid that operates like a gate. For the receptor to be activated, the lid must remain closed. Pyrabactin is effective at closing the gate on some receptors, turning them on, but cannot close the gate on others. The researchers have now cracked the molecular basis of this behavior. In a receptor where the gate closes, pyrabactin fits in snugly to allow the gate to close. In a receptor not activated by pyrabactin, the chemical binds in a way that prevents the gate from closing and activating the receptor.

Impact of Pyrabactin

Pyrabactin has paved the way for manufacturing new molecules that activate or turn on receptors. For it to be a good agriculture chemical, however, it needs to turn on more receptors by fitting into their pockets. A derivative of pyrabactin, capable of turning on all the receptors for drought tolerance, will open new spheres in the field of agriculture. The discovery of pyrabactin by the Cutler lab was heralded as a breakthrough research of 2009 by *Science* magazine.

Protein that Helps Crops Tolerate Drought

Plant roots use their endodermis, or inner skin, as a cellular gatekeeper to control the efficient use and movement of water and nutrients from the soil to the above-ground parts of the plant. A key part of that cellular barrier is the casparian strip, which also helps plants to tolerate stresses such as salinity, drought and flooding. Until recently, little was known about the genes that lead to the formation of the Casparian strip, which is composed of a fine band of lignin, the polymer that gives wood its strength. A protein was discovered that plays a vital role in how plant roots use water and nutrients, a key step in improving the production and quality of crops and bio-fuels. A team including researchers from Dartmouth, the University of Aberdeen and the University of Lausanne, identified a protein, ESB1, involved in the deposition of lignin patches early in the development of the Casparian strip and the fusion of these patches into a continuous band of lignin as the Casparian strip matures. Plants use lignin deposition in many different cell types and in response to various environmental stresses. A better understanding of lignin deposition may eventually help scientists to manipulate lignin content in plants and boost crop production under unfavorable conditions.

Developing Drought-Tolerant Crop

Using a method called chemical genomics, Sean Cutler at University of California, Riverside, identified pyrabactin, a new synthetic chemical that mimics a naturally produced stress hormone in plants to help them cope with drought conditions and it also turns on the ABA signaling pathway in *Arabidopsis*, a small flowering plant used widely in plant biology laboratories as a model organism.

Drought is the most significant environmental stress in agriculture worldwide and improving yield under drought is a major goal of plant breeding. A review of breeding progress pointed out that selection for high yield in stress-free conditions has, to a certain extent, indirectly improved yield in many water-limiting conditions. Conventional breeding requires the identification of genetic variability to drought among crop varieties, or among sexually compatible species, and introducing this tolerance into lines with suitable agronomic characteristics. Although conventional breeding for drought tolerance has and continues to have some success, it is a slow process that is limited by the availability of suitable genes for breeding. Some examples of conventional breeding programs for drought tolerance are the development of rice, wheat and Indian mustard varieties tolerant to salt and to alkali soils by the Central Soil Salinity Research Institute, Karnal, Haryana; the development of maize hybrids with increased drought tolerance; efforts to incorporate salt tolerance to wheat from wild related species; and the incorporation of drought tolerance as a selection trait in the generation of new maize and wheat germplasm by the International Maize and Wheat Improvement Center.

The development of tolerant crops by genetic engineering requires the identification of key genetic determinants underlying stress tolerance in plants, and introducing these genes into crops. Drought triggers a wide array of physiological responses in plants, and modifies the activity of a large number of genes: gene expression experiments have identified several hundred genes which are either induced or repressed during drought.

A number of physiological studies have identified some traits that are linked with plant's adaptability to drought-prone environment. Among them, traits, such as small plant size, reduced leaf area, early maturity and prolonged stomatal closure lead to a reduced total seasonal evapo-transpiration, and yield potential (Fischer and Wood, 1979, Karamanos and Papatheohari, 1999). Depending on the stress factors, some adaptive traits can be considered for yield improvement under drought if they enable plants to cope with a stress event that tends to occur every year at the same growth stage. For instance, a good level of earliness is an effective breeding strategy for enhancing yield stability in Mediterranean environments where wheat and barley are exposed to terminal drought stress. Late heading and flowering, followed by a short grain-filling period can be associated with higher yield when drought stress is experienced early in the season, during the vegetative phase (Van Ginkel *et al.*, 1998). A more general "xerophytic" breeding strategy to improve plant survival through the limitation of evapo-transpiration can be applied in extremely harsh environments. Nevertheless, every breeding strategy for drought-prone environments also has to consider that the timing and intensity of the stress events vary significantly from year to year and plants designed to cope with a specific type of drought may under-perform when the stress conditions are different or absent.

In mild to moderate drought conditions characterized by a wheat/barley grain yield between 2 and 5 mg ha^{-1}, selection for high yield potential has frequently led to some yield improvements under drought conditions (Araus *et al.*, 2002). In these cases the breeders have selected plants characterized by high yield potential and high yield stability, with the latter being attributed to a minimal G E interaction. This implies that traits maximizing productivity normally expressed in the absence of stress, can still sustain a significant yield improvement under mild to moderate drought (Tambussi *et al.*, 2005). An example is the success of wheat and rice varieties bred at CIMMYT and IRRI where selection under stress-free environments identified genotypes with high yield in a wide range of conditions including regions with a low yield potential.

Further progress will depend on the high yielding genotypes that are able to improve drought tolerance without detrimental effects on yield potential, thus reducing the gap between yield potential and yield in drought-prone environments. This goal can be achieved through identification of drought tolerance-related traits and the subsequent manipulation of the corresponding genes using marker assisted selection (MAS) and/or gene transformation.

Physiological Bases for Yield Under Drought

The physiologically relevant integrators of drought effects are the water content and the water potential of plant tissues. They in turn depend on the relative fluxes of water through the plant within the soil-plant-atmosphere continuum.

Thus, apart from the resistances and water storage capacities of the plant, it is the gradient of water vapour pressure from leaf to air, and the soil water content and potential that imposes conditions of drought on the plant. Once a drop in water potential develops, responses of a wide range of physiological processes are induced. Some of these responses are directly triggered by the changing water status of the tissues while others are brought about by plant hormones that are signaling changes in water status (Chaves, 2003). Physiological traits relevant for the responses to water deficits and/or modified by water deficits span a wide range of vital processes (Table 11.1). As a consequence, it can be expected that there is no single response pattern that is highly correlated with yield under all drought environments.

Table 11.1: Physiological Traits Relevant for Response to Drought Conditions

Plant Traits	*Effects Relevant for Yield*	*Modulation under Stress*	*References*
Stomatal conductance/ leaf temperature	More/less rapid water consumption. Leaf temperature reflects the evaporation and hence is a function of stomatal conductance	Stomatal resistance increases under stress	Jones (1999), Lawlor and Cornic (2002)
Photosynthetic capacity	Modulation of concentration of Calvin cycle enzymes and elements of the light reactions	Reduction under stress	Lawlor and Cornic (2002)
Single plant leaf area	Plant size and related productivity	Reduced under stress (wilting, senescence, abscission)	Walter and Schurr (2005)
Rooting depth	Higher/lower tapping of soil water resources	Reduced total mass but increased root/ shoot ratio, growth into wet soil layers, growth on stress release	Hoad *et al.* (2001), Sharp (2002)
Photosynthetic pathway	C_3/C_4/CAM, greater heat tolerance of C_4 and CAM		Cushman (2001)
Accumulation of stress-related proteins	Involved in the protection of cellular structure and protein activities	Accumulated under stress	Ramanjulu and Bartels (2002), Cattivelli *et al.* (2008)

Source: Luigi Cattivelli (Genomic Research Centre, Italy), Anna M. Mastrangelo (Cereal Research Centre, Italy) and Franz-W. Badeck (Potsdam Institute for Climate Impact Research, Germany).

Molecular Markers to Dissect Drought Tolerance-Related Traits

Molecular markers can be used to explore germplasm through segregation and association mapping to identify useful alleles in both cultivated varieties and wild relatives. Association mapping is intrinsically more powerful than 'classical' genetic linkage mapping because it scrutinizes the results of thousands of generations of recombination and selection. Most of the data available up-to-date on drought tolerance are based on segregation mapping and QTL analysis. Many efforts have been dedicated to understanding the genetic basis of physiological traits conferring advantages in dry environment while less attention has been given to understanding the high yield stability in dry and wet environment. Drought tolerance is a typical quantitative trait, however single genes, such as those controlling flowering time, plant height and ear type may have important roles in adaptation to drought-prone environments (Forster *et al.*, 2000). The association between variation in drought-related quantitative traits and, ultimately, the effects of these traits on yield in drought and favourable environments is the main goal for present and future research.

Genes and Metabolites Conferring Drought Tolerance

New chances to further improve yield and/or yield stability under limiting conditions come from the last 10 years' progress in the identification of the genetic determinants of the physiological responses related to stress tolerance. Adaptation of plants to drought and to the consequent cellular dehydration induces an active plant molecular response. This response significantly improves the tolerance to negative constraints and it is to a great extent under transcriptional control. Many stress-related genes have been isolated and characterized in the last two decades in a variety of crop species (Ramanjulu and Bartels, 2002). From model plants, genetic information is being moved to crops exploiting genome synteny, taking advantages of conserved molecular pathways, including that controlling stress tolerance.

The transgenic plants showed increased stress tolerance as well as the over induction of downstream stress related genes and/or higher levels of soluble sugars and proline. A recent report has shown that rice plants over expressing the SNAC1 (stress-responsive NAC1) transcription factor showed improved drought tolerance and yield potential under field conditions. The leaves of SNAC1-overexpressing plants lost water more slowly showing an increased stomatal closure and ABA sensitivity. Metabolic engineering for increasing osmolytic contents was successful in several crops. The first example of metabolic engineering for drought tolerance was the overproduction of proline in transgenic tobacco and rice (Zhu *et al.*, 1998), resulting in an enhanced biomass under stress conditions. Garg *et al.* (2002) developed drought tolerant transgenic rice lines showing tissue- or stress-inducible accumulation of trehalose, which accounted for higher soluble carbohydrate levels, a higher capacity of photosynthesis and more favourable mineral balance under both stress and non-stress conditions, without negative effects. A significant improvement of wheat tolerance to water deficit was achieved by Abebe *et al.* (2003), through the ectopic expression of the mannitol-1-phosphate dehydrogenase (*mtlD*) gene that caused a small increase in the level of mannitol.

An emerging strategy to broaden stress tolerance in crops is to maintain energy homeostasis under stress conditions. In fact, it is known that the high energy consumption of the crop's stress response increases respiration rate with a linked production of reactive oxygen species (Tiwari *et al.*, 2002). De Block *et al.* (2005) obtained *Brassica napus* plants tolerant to multiple stresses by preventing over activation of mitochondrial respiration and high energy consumption.

The transgenic crops show improved water stress tolerance and a higher photosynthesis rate due to an altered stomatal conductance under water deficit conditions (Yan *et al.*, 2004). Since an earlier-than-normal stomatal closure in a crop is considered a positive trait to improve water use efficiency in drought environments, developing transgenic plants with a drought-avoidance phenotype represents a possible strategy for crop improvement. A phenotype with decreased conductance and higher water use efficiency was obtained in tobacco plants (Laporte *et al.*, 2002). The implication of abscisic acid hormone as a molecular signal in drought activated pathways and in the control of stomatal closure makes ABA synthesis and response a possible target for improving drought tolerance. When a farnesyl-transferase acting as a negative-regulator of ABA sensing was down-regulated in a drought inducible manner in *Brassica napus* the transgenic plants showed enhanced ABA sensitivity, as well as a significant reduction in stomatal conductance and transpiration under drought stress conditions (Wang *et al.*, 2003).

A more robust root system enables plants to take up greater amounts of water during water deficit stress, resulting in a more favourable plant water status and less injury. Recent studies have found the way to increase root size through single gene transformation. The gene coding for the vacuolar H^+-pyrophosphatase (H^+-PPase)-*AVP1*-plays an important role in root development through the facilitation of auxin fluxes. Over expression of *AVP1* in tomato resulted in more pyrophosphate-driven cation transport into root vacuolar fractions, an increased root biomass, and an enhanced recovery of plants from an episode of soil water deficit stress (Gaxiola *et al.*, 2001; Park *et al.*, 2005).

Even if the use of biotechnology in agriculture offers great potential benefits to farmers, the effectiveness of a transgenic-based breeding strategy is nowadays hampered by non-biological constraints, related to the commercialization of transgenic crops, particularly in Europe. The commercial success of these products will depend upon the development of regulatory frameworks that are clearly defined and scientifically based, and upon public acceptance of transgenic plant products.

Future Implications and Directions

Research in the last three decades has opened up three main approaches:

1. Plant physiology provided new tools to understand the complex network of drought-related traits and several drought-related traits useful to improve selection efficiency have been proposed.
2. Molecular genetics has led to the discovery of a large number of loci affecting yield under drought or the expression of drought tolerance-related traits; and

3. Molecular biology has provided genes that are useful for transgenic approaches. The integration of molecular genetics with physiology is leading to the identification of the most relevant loci controlling drought tolerance and drought-related traits.

The success of any selection process relies on the availability of superior alleles for the target trait. A chance to find new useful alleles is represented by the exploitation of wild germplasm. During the domestication process, wild plants carrying promising traits were cultivated, leading to locally adapted landraces. Many undesirable alleles have lost and useful alleles become enriched in the cultivated gene pool (Tanksley and McCouch, 1997). Many studies have demonstrated the value of alleles originating from non-cultivar germplasm showing that centuries of selective breeding have thrown away useful alleles in addition to many useless ones. Notably, when four genomic segments for root length were transferred from the rice cultivar "Azucena" into "Kalinga III", only one of them significantly increased root length in the new genetic background (Steele *et al.*, 2006). Transgenic breeding will also have a role in the future and the possibility of cloning stress-related quantitative trait loci (QTL) will enable the simultaneous engineering of multiple genes governing quantitative traits. However, the scarcity of field trials for drought tolerant transgenic plants does not allow for final conclusions. New transgenic plants where the gene is introduced into elite genotypes have to be tested under optimal as well as drought conditions to evaluate the impact of the transgenes on yield potential and stress tolerance.

As against ten years ago, breeders now have new perspectives for plant improvement for drought tolerance. First, new tools provided an opportunity to move the selection from phenotype to genotype. Second, new approach applied to blend together all knowledge of the traits. This so called "breeding by design" strategy will lead to new high yielding varieties to improve performance in both high and low yielding environments. A significant example showing how the integration of stress physiology and genomics can lead to an integrated view of plant breeding is represented by the studies on transpiration efficiency. During drought stress, crops generally coordinate photosynthesis and transpiration, although significant genetic variation in transpiration efficiency has been identified both between and within species. The isolation of a gene, *ERECTA* that regulates transpiration efficiency in *Arabidopsis* (Masle *et al.*, 2005) and the transcriptional analysis of wheat genotypes with contrasting transpiration efficiency are providing the molecular bases of the isotopic discrimination parameter (Xue *et al.*, 2006). Every day, it becomes more evident that successful breeding for stable high yield under drought conditions will only be possible when a true integration of traditional breeding with physiology and genomics is achieved. Thus, to face drought stress and to achieve sufficient grain yield in the future requires a multidisciplinary approach based on plant genomics, physiology and modeling.

Conclusion

Drought is a natural event, caused by other weather events like El Nino and high pressure systems. Drought is defined as- "Deficiency of available soil moisture which produces water deficits in the plant sufficient to cause a reduction in plant growth". In general, drought means different things to different people. To a meteorologist it is the absence of rain while to the agriculturist it is the deficiency of soil moisture in the crop root zone to support crop growth and productivity. To the hydrologist it is the lowering of water levels in lakes, reservoirs, etc., while for the city management it may mean the shortage of drinking water availability. Thus, it is unrealistic to expect a universal definition of drought for all fields of activity.

Drought tolerance refers to the degree to which a plant is adapted to arid or drought conditions. A drought tolerant plant is one that will survive in the typical or somewhat less than typical amount of rainfall in any region. There is no one-size-fits-all drought tolerant plant. Drought tolerant crops typically make use of either C_4 carbon fixation or crassulacean acid metabolism (CAM) to fix carbon during photosynthesis. Both are improvements over the more common but more basal C_3 pathway in that they are more energy efficient. CAM is particularly good for arid conditions because carbon dioxide can be taken up at night, allowing the stomata to stay closed during the heat of day and thus reducing water loss.

When drought-tolerant plants detect dry conditions, they synthesize abscisic acid, which causes changes from root tips to leaves and flowers. Plants under the influence of this hormone begin to conserve water. Abscisic acid triggers an array of plant drought-tolerance mechanisms. The hormone was discovered in the early 1960s, and plant biologists have known for decades that it plays crucial role in keeping plants alive during drought.

Getzoff and Schroeder discovered how PYR1 was involved in drought resistance. They observed PYR1 and abscisic acid molecules on micro scale and nano-scale. The research showed that two copies of PYR1 fit snugly together in plant cells. There, they are targeted by abscisic acid. Each copy of the PYR1 molecule has an internal open space like the inside of a tin can, and when a hormone molecule comes along, it fits neatly into one of the two spaces. This induces part of the PYR1 protein. Sean Cutler identified pyrabactin, a new synthetic chemical that mimics a naturally produced stress hormone in plants to help them cope with drought conditions and it also turns on the ABA signaling pathway in *Arabidopsis,* a small flowering plant used widely in plant biology laboratories as a model organism. Sean Cutler used pyrabactin to fish out a receptor for ABA. A receptor is a protein molecule in a cell to which mobile signaling molecules may attach.

Adaptation of plants to drought and to the consequent cellular dehydration induces an active plant molecular response. This response significantly improves the tolerance to negative constraints and it is to a great extent under transcriptional control. Many stress-related genes have been isolated and characterized in the last two decades in a variety of crop species. The transgenic plants showed increased stress tolerance as well as the over induction of downstream stress related genes and/or higher levels of soluble sugars and proline. An emerging strategy to broaden stress tolerance in crops is to maintain energy homeostasis under stress conditions. In fact, it is known that the high energy consumption during crop's stress response increases respiration rate with a linked production of reactive oxygen species. The transgenic crops show improved water stress tolerance and a higher photosynthesis rate due to an altered stomatal conductance under water deficit conditions.

References

Abebe, T., Guenzi, A.C., Martin, B. and Cushman, J.C., 2003. Tolerance of mannitol-accumulating transgenic wheat to water stress and salinity. *Plant Physiol.*, 131: 1748–1755.

Araus, J. L., Slafer, G. A., Reynolds, M. P., and Royo, C. 2002. Plant breeding and water relations in C3 cereals: what should we breed for? *Ann. Bot.*, London, 89: 925–940.

Araus, J. L., Slafer, G. A., Reynolds, M. P., and Royo, C. 2003b. Physiology of yield and adaptation in wheat and barley breeding. In : *Physiology and Biotechnology Integration for Plant Breeding*. Part I. Physiological basis of yield and environmental adaptation. 1–49. Nguyen, H. T., Blum, A. Eds Marcel Dekker, Inc. New York, Basel.

Araus, J. L. 2004. The problem of sustainable water use in the Mediterranean and research requirements for agriculture. *Ann. Appl. Biol.*, 144: 229–272.

Araus J. L., Ferrio J. P., Buxo R., Voltas J. 2007a. The historical perspective of dryland agriculture: Lessons learned from 10,000 years of wheat cultivation. *J. Exp. Bot.* 58: 131–145.

Cattivelli, L., Rizza, F., Badeck, F.W., Mazzucotelli, E., Mastrangelo, A.M., Francia, E., Mare, C., Tondelli, A. and Stanca, A.M., 2008. Drought tolerance improvement in crop plants: An integrated view from breeding to genomics. *Field Crops Research*, 105: 1–14.

Cattivelli, L., Delogu, G., Terzi, V. and Stanca, A.M., 1994. Progress in barley breeding. In: Slafer, G.A. (Ed.), *Genetic Improvement of Field Crops*. Marcel Dekker, Inc., New York, 95–181.

Chaves, M.M., Maroco, J.P. and Pereira, J.S., 2003. Understanding plant responses to drought-from genes to the whole plant. Funct. *Plant Biol.*, 30: 239–264.

Cushman, J.C., 2001. Crasulacean acid metabolism. A plastic photosynthetic adaptation to arid environment. *Plant Physiol.*, 127: 1439–1448.

Cutler, S.R., Okamoto, M., Peterson, F.C., Defries, A., Park, S.Y., Endo, A., Nambara, E. and Volkman, B.F., 2013. Activation of dimeric ABA receptors elicits guard cell closure, ABA-regulated gene expression, and drought tolerance. *Proc. Natl. Acad. Sci.*, U.S.A. 16: 12132-12137.

Cutler, S.R., Santiago, J., Rodrigues, A., Saez, A., Rubio, S., Antoni, R., Dupeux, F., Park, S.Y., Márquez, J.A. and Rodriguez, P.L., 2009. Modulation of drought resistance by the abscisic acid receptor PYL5 through inhibition of clade A PP2Cs. *Plant J.*, 60: 575-88.

De Block, M., Verduyn, C., De Brouwer, D., Cornelissen, M., 2005. Poly (ADP-ribose) polymerase in plants affects energy homeostasis, cell death and stress tolerance. *Plant Journal*. 41: 95–106.

Fischer, R.A. and Maurer, R., 1978. Drought resistance in spring wheat cultivars. I. Grain yield response. *Aust. J. Agric. Res.*, 29: 897–912.

Fischer, R.A. and Wood. J.T., 1979. Drought resistance in spring wheat cultivars. Ill. Yield associations with morpho-physiological traits. *Aust. J. Agric. Res.*, 30: 1001-1020.

Forster B. P., Ellis R. P., Thomas W. T. B., Newton A. C., Tuberosa R., This D., El Enein R. A., Bahri M. H., Ben Salem M., 2000. The development and application of molecular markers for abiotic stress tolerance in barley. *J. Exp. Bot.*, 51: 19–27.

Forster, B. P., Thomas, W. T. B., Chloupek, O., 2005. Genetic controls of barley root systems and their associations with plant performance. *Aspects of Applied Biology*, 73: 199–204.

Garg, A., Kim, J., Owens, T., Ranwala, A., Choi, Y., Kochian, L. and Wu, R., 2002. Trehalose accumulation in rice plants confers high tolerance levels to different abiotic stresses. *Proc. Natl. Acad. Sci.*, U.S.A. 99: 15898–15903.

Gaxiola, R.A., Li, J., Undurraga, S., Dang, L.M., Allen, G.J., Alper, S.L., Fink,G.R., 2001. Drought- and salt-tolerant plants result from over-expression of the AVP1 H+-pump. Proc. *Natl. Acad. Sci., U.S.A.*, 98: 11444–11449.

Getzoff, E.D., 2002. Passing a signal: low carbs, less protein. *Chem. Biol.*, 9: 1165-1166.

Getzoff, E.D., Hubbard, K.E., Nishimura, N., Hitomi, K. and Schroeder, J.I., 2010. Early abscisic acid signal transduction mechanisms: newly discovered components and newly emerging questions. *Genes. Dev.*, 24: 1695-1708.

Getzoff, E.D., Gutwin, K.N. and Genick, U.K., 2003. Anticipatory active-site motions and chromophore distortion prime photoreceptor PYP for light activation. *Nat. Struct. Biol.*, 10: 663-668.

Heathcote, R.L., 1974. Drought in South Australia. In *Natural Hazards: Local, National, Global*, edited by White, G.F. New York: Oxford University Press.

Hoad, S.P., Russell, G., Lucas, M.E., Bingham, I.J., 2001. The management of wheat, barley, and oat root systems. *Advances in Agronomy*, 74: 193–254.

Jones, H.G., 1980. Interaction and integration of adaptive responses to water stress: the implications of an unpredictable environment. In: Turner, N.C., Kramer, P.J. (Eds.), Adaptation of Plants to Water and High Temperature Stress. Wiley, New York, 353–365.

Jones, H.G., 2007. Monitoring plant and soil water status: established and novel methods revisited and their relevance to studies of drought tolerance. *J. Exp. Bot.*, 58: 119–130.

Karamanos, A. J. and Papatheohari, A.Y. 1987. Understanding the mechanisms of drought resistance of some crop plants. In: *Drought Resistance in plants. Physiological and Genetic Aspects.* (eds. L. Monti and E. Porceddu), 95-109, Commission of the European Communities, Luxembourg.

Karamanos, A. J. and Papatheohari, A.Y. 1999. The assessment of drought resistance of crop genotypes by means of the later potential index. *Crop Science*, 39: 1792-1797.

Laporte, M.M., Shen, B. and Tarczynski, M.C., 2002. Engineering for drought avoidance: expression of maize NADP-malic enzyme in tobacco results in altered stomatal function. *J. Exp. Bot.*, 53: 699–705.

Lawlor, D.W. and Cornic, G., 2002. Photosynthetic carbon assimilation and associated metabolism in relation to water deficits in higher plants. *Plant Cell Environ.*, 25: 275–294.

Linsley, R.K., Jr, Kohler, M.A. and Paulhus, J.L.H., 1975. *Hydrology for Engineers.* McGraw-Hill, New York.

Masle, J., Gilmore, S.R. and Farquhar, G.D., 2005. The ERECTA gene regulates plant transpiration efficiency in Arabidopsis. *Nature*, 436: 866–870.

Park, S., Li, J., Pittman, J.K., Berkowitz,G.A., Yang, H., Undurraga,S., Morris,J., Hirschi, K.D., Gaxiola, R.A., 2005. Up-regulation of a H+-pyropho-sphatase (H+-PPase) as a strategy to engineer drought-resistant crop plants.Proc. *Natl. Acad. Sci.* U.S.A. 102: 18830–18835.

Ramanjulu, S. and Bartels, D., 2002. Drought- and desiccation-induced modulation of gene expression in plants. *Plant Cell Environ.*, 25: 141–151.

Ramdas, L. A., 1968. *WMO Tech. Note*, 104: 399-404.

Ramdas, L. A. and Atmanathan, S., 1932. The Vertical distribution of air temperature near the ground at night, *Beit. Geophys.*, 37: 116–117.

Ramdas, L. A., 1953. *Proc. Indian Acad. Sci.*, 304-306.

Ramdas, L. A., 1951. Microclimatological Investigations in India. *Archiv. für. Met. Geophys. Biockli.*, 3: 149–167.

Ramdas, L. A. and Malurkar, S. L., 1932. *Indian J. Phys.*, 495-508.

Sharp, R.E., 2002. Interaction with ethylene: changing views on the role of abscisic acid in root and shoot growth responses to water stress. *Plant, Cell and Environment*, 25: 211–222.

Steele, K.A., Price, A.H., Shashidhar, H.E., Witcombe, J.R., 2006. Marker-assisted selection to introgress rice QTLs controlling root traits into an Indian upland rice variety. *Theor. Appl. Genet.*, 112: 208–221.

Tambussi, E.A., Nogués, S., Araus, J.L. 2005. Ear of durum wheat under water stress: water relations and photosynthetic metabolism. *Planta.*, 221: 446-458.

Tambussi, E. A., Bort, J., Guiamet, J.J., Nogués, S., Araus, J.L. 2007. The photosynthetic role of ears in C3 cereals: metabolism, water use efficiency and contribution to grain yield. *Crit. Rev. Plant Sci.*, 26 (1): 1–16.

Tanksley, S.D., McCouch, S.R., 1997. Seed banks and molecular maps: unlocking genetic potential from the wild. *Science*, 277: 1063–1066.

Tiwari, B.S., Belenghi, B. and Levine, A., 2002. Oxidative stress increased respiration and generation of reactive oxygen species, resulting in ATP depletion, opening of mitochondrial permeability transition, and programmed cell death. *Proc. Natl. Acad. Sci.*, U.S.A. 128: 1271–1281.

Van Ginkel, J.H., Gorissen, A. and van Veen, J.A., 1996. Long-term decomposition of grass roots as affected by elevated atmospheric carbon dioxide. *Journal of Environmental Quality*, 25: 1122-1128.

Van Ginkel, J.H. and Gorissen, A. 1998. *In situ* decomposition of grass roots as affected by elevated atmospheric carbon dioxide. *Soil Science Society of America Journal*, 62: 951-958.

Van Ginkel, M., Calhoun,D.S., Gebeyehu,G., Miranda,A., Tian-you,C., PargasLara, R., Trethowan, R.M., Sayre, K., Crossa, J., Rajaram, S., 1998. Plant traits related to yield of wheat in early, late, or continuous drought conditions. *Euphytica.*, 100: 109–121.

Van Ginkel, J.H., Whitmore, A.P. and Gorissen, A. 1999. Lolium perenne grasslands may function as a sink for atmospheric carbon dioxide. *Journal of Environmental Quality*, 28: 1580-1584.

Walter, A. and Schurr, U., 2005. Dynamics of leaf and root growth–endogenous control versus environmental impact. Annals of Botany, 95: 891-900.

Wang, W., Vinocur, B. and Altman, A., 2003. Plant responses to drought, salinity and extreme temperatures: towards genetic engineering for stress tolerance. *Planta*, 218: 1–14.

Xue, G.P., McIntyre, C.L., Chapman, S., Bower, N.I., Way, H., Reverter, A., Clarke, B. and Shorter, R., 2006. Differential gene expression of wheat progeny with contrasting levels of transpiration efficiency. *Plant Mol. Biol.*, 61: 863–881.

Yan, J., He, C., Wang, J., Mao, Z., Holaday, S.A., Allen, R.D., Zhang, H., 2004. Overexpression of the Arabidopsis 14-3-3 protein GF14 lambda in cottonleads to a "stay-green" phenotype and improves stress tolerance undermoderate drought conditions. *Plant Cell Physiol.*, 45: 1007–1014.

Zhu, J.K., 2001. Cell signaling under salt, water and cold stresses. *Curr. Opin. Plant Biol.*, 4: 401–406.

2016, Recent Advances in Plant Stress Physiology *Pages* **237–263**
Editors: ***Praduman Yadav, Sunil Kumar and Veena Jain***
Published by: **DAYA PUBLISHING HOUSE, NEW DELHI**

Chapter 12

Aluminium Stress in Crop Plants

Ana Luisa Garcia-Oliveira[1]*, Subhash Chander[2], Juan Barcelo[3] and Charlotte Poschenrieder[3]*

[1]International Institute of Tropical Agriculture (IITA), C/O ILRI, P.O. Box 30709, Nairobi 00100, Kenya
[2]Oilseeds Section, Department of Genetics and Plant Breeding, CCS Haryana Agricultural University, Hisar, India
[3]Lab Fisiología Vegetal, Facultad de Biociencias, Universidad Autonoma de Barcelona, E-08193 Bellatera, Spain

ABSTRACT

Aluminium (Al) toxicity has a strong negative impact on crop productivity on acid mineral soils. Liming and fertilisation frequently are insufficient to overcome the problem, especially in tropical areas with subsoil acidity. The focus on breeding for acid soil tolerance as a more sustainable tool to achieve better crop performance has strongly stimulated research into the mechanisms of Al toxicity and tolerance. Root tips have been identified as the primary target of Al toxicity in most crops, wherefore primary Al tolerance mechanisms based on exudation of strong Al-ligands also are operating there. Recent studies have already identified several genes responsible for the Al tolerance based on Al-induced root tip exudation of organic acids. In the post genomic era, genomic technologies tied together with physiology and biochemistry have made considerable further advances in our understanding of Al-toxicity and tolerance mechanisms in an integrated way. However, in order to fill the still existing gap between geneticists and breeders a deeper knowledge on the Al-toxicity and tolerance mechanisms at the genetic, molecular and physiological levels is required, not only in crop plants, but also considering the large ecological diversity of naturally adapted species. This will provide a closer relation between genotyping and phenotyping tools allowing faster breeding progress. In addition to these aspects, this chapter will also provide a glance into the potential of mutagenesis and transgenic approach for Al-toxicity improvement in different plants.

Keywords: *Aluminium, Al-toxicity, Abiotic stress, Tolerance, Plants.*

**Corresponding author:* E-mail: a.oliveira@cgiar.org; charlotte.poschenrieder@uab.cat

Introduction

A large portion of the earth's land is acidic, occurring mainly in the northern (cold and temperate) and in the southern (tropical) belts (Figure 12.1). Only 4.5 per cent of the area belonging to these acidic soils is used for arable crops (von Uexkull and Mutert, 1995). Soil acidity is a natural occurrence in tropical and subtropical zones whereas it is an increasing problem in the temperate zones particularly industrial regions of the USA, Canada and Europe due to acidic nature of rainfall (Vitorello *et al.*, 2005). Furthermore, over use of ammonia and amide containing fertilisers and organic matter decay also cause soil acidification and aggravates aluminium toxicity. In these regions, high acidity accentuates the combination of abiotic stresses, such as, aluminium (Al), proton (H^+) and manganese (Mn) toxicities, and phosphorus (P) deficiency which result in sub-optimal plant performance. In the earth crust, Al is the third most abundant element after oxygen and silicon whose naturally occurring forms are stables and does not pose a toxicity hazard in plants. However, under low pH (upon soil acidification) conditions, Al dissolves in various ionic forms [$Al(H_2O)^{3+}$, $Al(OH)^{2+}$ and $Al(OH)_2^+$] of which Al^{3+} is potentially phytotoxic to plants (Kinraide, 1997). In this chapter, the abbreviation for toxic forms of aluminium will be considered Al.

Agronomically, lime (calcium carbonate) application is used to change the soils' pH and is considered to be the best way to ameliorate the soils acidity. However, this is not always economically or physically feasible because the amount of the lime required greatly depends upon the soil's pH and texture. In addition, liming of subsoil layers is also ineffective; ultimately, the consequence is a relatively shallow root system vulnerable to drought. In contrast to lime, application of magnesium (Mg) has been reported to be more efficient in alleviating Al toxicity, but, when Mg is present in excess, it becomes toxic (Bose *et al.*, 2011). Therefore, the understanding of Al-toxicity symptoms and effects, natural variation for Al-tolerance in crop plants, mechanism of Al-tolerance, genes conferring Al-tolerance, efficient screening techniques and the available tools in the breeders kit are highly desirable for the improvement in the Al tolerance of the existing crop species which seems to be most durable and eco-friendly approach in order to enhance food production in acid soils.

Al-toxicity Symptoms and Effects in Plants

Al is highly toxic to plants under acidic conditions and is detrimental to their growth and development. The primary symptom of Al-toxicity in higher plants is a rapid inhibition of root growth which can usually be detected within few minutes to 2 hours (Barcelo and Poschenrieder, 2002). Subsequently, it not only induces the other root stress responses such as reduced cell division, lignin deposition and development of lesions in epidermal and cortical tissues near the elongation zone but also severely changes the root morphology such as curved, cracked, brownish, stubby and stiffed root apices (Kochian *et al.*, 2004; Poschenrieder *et al.*, 2008), which leads to reduced and damaged root system, making it increasingly difficult for the plant to acquire both water and nutrients (Barceló and Poschenrieder, 2002). The symptoms of prolonged Al-toxicity in plants are not easily identifiable and frequently are mixed with symptoms of P-deficiency and in some cases calcium (Ca) deficiencies. Toxicity symptoms in shoots mainly are manifested as overall

Figure 12.1: Global Distribution of Acid Soils; Highlighted in Red Colour. (*Source*: FAO-UNESCO, 2007).

stunting, small and dark leaves with chlorotic patches and marginal necrosis, late maturity, cellular modifications in leaves, reduced stomatal opening and decreased photosynthetic activity. These symptoms are usually considered as a result of Al-induced root damage (Miyasaka *et al.*, 2007).

The root apex is the most sensitive site of Al-toxicity. Within this region, the distal transition zone (localised between the meristematic and elongation zones) appears to be the most susceptible site which can accumulate Al easily (Sivaguru and Horst, 1998). Aluminium in its ionic form of Al^{3+}, is highly reactive, and potential targets of its injury include the cell wall, the plasma membrane surface, the cytoskeleton and the nucleus (Kochian *et al.*, 2004). In cell walls, Al can bind to pectin and modify the synthesis or deposition of polysaccharides, stiffening cell walls causing disruption of normal cell wall growth and inhibition of cell elongation (Gunsé *et al.*, 1997). Furthermore, Al can bind to phospholipids in the plasma membrane and can interfere with protons involved in transport and signal transduction processes. Inside cells, Al can bind to membranes of cellular organelles and can interfere with many metabolic processes and cellular functions resulting in blockage of cell division in root tips, increases the rigidity of DNA double helix by reducing DNA replication, reduces root respiration and alters enzyme activity related to sugar phosphorylation. In addition, Al is reported to cause oxidative stress generating reactive oxygen species (ROS).

Natural Variation for Al Tolerance in Crop Plants

Al-toxicity has been recognised as a major problem in acid soils for the past 100 year (Hartwell and Pember, 1918), but the first conclusive evidence of genetic variation for Al tolerance in crop plants was reported in wheat during the first quarter of the previous century. Scientists observed poor plant growth and a burning effect in wheat on acid soils and it was termed as "crestamento" which in Portuguese literally means burning or toasting. By 1942, "crestamento" was found to be caused by Al-toxicity present in acid soils in Brazil where largest area of acid soil in the world is prevalent (Rajaram *et al.*, 1988). Subsequently due to the recognition of the importance of Al-toxicity in plants, numerous studies were performed which exhibited the existence of considerable genetic variation for Al tolerance at inter- and intra-specific levels in different agriculturally important crop plants (Table 12.1). It has been observed that wild relatives can also be a novel source of Al tolerance. As an example, the wild relative of wheat, known as goatgrass (*Aegilops uniaristata*), has been identified as a useful source of Al tolerance genes (Mohammed *et al.*, 2013).

Table 12.1: Al-tolerance Level of Selected Crop Plant Species

Level of Al Tolerance		*Crop Plant Species*
Highly Sensitive	:	Barley, Durum Wheat, Lettuce, Pea
Sensitive	:	Wheat, Oat
Moderately sensitive	:	Triticale, Maize, Sorghum, Cabbage
Moderately Tolerant	:	Rice, Rye
Tolerant	:	Soybean, Pigeon pea
Highly Tolerant	:	Tea, Buckwheat, *Brachiaria*

Usually, most of the Al-tolerant species or genotypes within a species that have high levels of Al tolerance are endemic to regions with acidic soils. This suggests that in these acidic soils Al-toxicity imposes a strong selection pressure on the plants to evolve strategies to cope with Al stress (Garcia-Oliveira *et al.*, 2014).

Physiological Mechanisms of Al-Tolerance in Plants

Al-toxicity tolerance mechanisms have been extensively studied in numerous plant species, but due to the complex chemistry of Al, these mechanisms are still not well understood. There has been considerable speculation that multiple Al tolerance mechanisms are employed by different plant species. Although these mechanisms can be broadly differentiated into two major categories: A) Avoidance mechanisms that promote external detoxification of Al and exclusion of Al, and B) Tolerance mechanisms or internal detoxification of Al (Liu *et al.*, 2014).

Avoidance Mechanisms

These mechanisms rely on all those factors or strategies adopted by plant species which prevent or restrict the Al-uptake by roots and thereby avoiding the direct interactions of toxic forms of Al with sensitive sites. The evidence of this mechanism implicates cell wall chemistry like formation of mucilage (Miyasaka and Hawes, 2001), efflux of organic anions (Ma *et al.*, 2001), secretion of phosphate (Pellet *et al.*, 1996), and secondary metabolites such as flavonoid type phenolics catechin and quercetin compounds from root apices (Ofei-Manu *et al.*, 2001; Tolrà *et al.*, 2009), and alkalinisation of rhizosphere pH (Degenhardt *et al.*, 1998).

Among the Al exclusion strategies, efflux of organic anions from root apices is one of the most demonstrated by strong genetic and molecular studies in numerous monocot and eudicot species, and seems to be conserved Al tolerance mechanism that plays a central role in exclusion of Al in different plant species (Ma *et al.*, 2001). Among organic acid anions, malate, citrate and oxalate are the most commonly released anions from the roots of Al-tolerant genotypes of different plant species (Delhaize *et al.*, 1993; Zheng *et al.*, 1998; Magalhães *et al.*, 2007; Ryan *et al.*, 2009). These anions form harmless complexes with Al in the apoplast that protect the sensitive root apex and reduce uptake of Al into the roots. It is interesting to note that two specific temporal patterns of secretion of these organic acids have been identified in different plant species (Ma *et al.*, 2001). In case of Pattern I, no discernible delay is observed between the addition of Al and the onset of organic acid secretion like roots of wheat and buckwheat secretes malate and oxalate, respectively, within few minutes after exposure to Al. Contrarily, in case of Pattern II, organic acid secretion is delayed for several hours after exposures to Al, for example, maximal citrate efflux in *Cassio taro* onsets after 4 h exposure to Al.

Besides organic acid exudation, there is also indication that exclusion mechanism also relies on other alternative or additional factors in some plant species. Root mucilage is a gelatinous polysaccharide which is exuded from the outer layers of the root cap. Horst and collaborators (1982) found a higher Al sensitivity of cowpea roots without mucilage and suggested the protective function of the mucilage against Al injury because a positive correlation was observed between the levels of Al content in the root tip tissues and the amount of root mucilage. Similarly, Henderson and Ownby (1991) also detected the direct association between the amounts of mucilage and the level of Al tolerance in wheat roots, and suggested that the mucilage may acts as a diffusion barrier to Al into the root by bounding it in rhizosphere or through the concentration of organic acids that chelated toxic Al ions. However, the mucilage mechanism of Al exclusion in the plant species is not unequivocal, but it cannot be ruled out. Besides these factors, there are also indications that phosphate (Pi) exudation could be an important Al exclusion mechanism through the formation of Al-phosphate complexes, and protonation that might increase pH in the acidic rhizosphere. Pellet and co-workers (1997) observed a constitutive phosphate exudation from the root apex of Al tolerant wheat genotype Atlas66 and suggested that Al tolerance in wheat can be mediated by multiple exclusion mechanisms. So far, no conclusive data associating Pi exudation with Al tolerance have been presented in different plant species. In general, it is usually believed that binding of Al to charged sites on the cell surface is a prerequisite for its uptake and toxicity. Although some reports exhibited the negative correlation between root cation exchange capacity (CEC) and Al tolerance in genotypes of some plant species (Blamey *et al.*, 1993), but such mechanisms do not play significant role in differential Al-tolerance in most of the crop plants (Tang *et al.*, 2002).

Tolerance Mechanisms

Symplastic or internal detoxification mechanism allow plants to cope with Al once it enters the root and shoot symplasm. Al tolerance may rely on multiple cooperating mechanisms: the formation of harmless complexes with organic ligands in the cytosol, compartmentalisation into the vacuoles, involvement of Al binding proteins or enzymes that may elevate tolerance of the enzymatic activity in the cells, by a rapid repair of any lesions incurred, including those resulting from oxidative stress (Kochian *et al.*, 2004; Delhaize *et al.*, 2012). Although, symplastic evidences have yet to be supported with more experimental data. It is noteworthy that some plant species such as tea [*Camelia sinensis* (L.) Kuntze], buckwheat (*Fagopyrum esculentum* Moench) and Hydrangea (*Hydrangea macrophylla* Thunb.) have remarkable ability to accumulate Al in their roots and aerial parts (leaves and shoots). It has been observed that these species detoxify internal Al by forming Al-organic acid complexes and it seems that Al undergoes a ligand exchange with oxalate and citrate depending on whether it is transported into xylem or being sequestered in the leaves (Ma *et al.*, 2001). These species are good examples of internal detoxification of Al mechanism in which Al moves across a range of membranes (from roots to shoots) without revealing the adverse symptoms and even their growth can be stimulated by the presence of Al (Ma *et al.*, 1997; Zheng *et al.*, 1998; Tolrà *et al.*, 2011). Usually most of plant species rely on either Al avoidance or tolerance mechanisms but some plant species particularly forest trees such as *Pinus taeda* adapted these both mechanisms to protect them from Al-toxicity (Nguyen *et al.*, 2003; Nowak and Friend, 2006).

Genetic Mechanisms for Al Tolerance in Plants

Wheat is probably the first crop species in which inheritance of Al tolerance has been studied. The systematic studies in wheat led to the development of Al-tolerant cultivars such as BH 1146 in Brazil and Atlas66 in USA. Most studies suggested that Al tolerance in bread wheat is mainly governed by one to two dominant genes with different alleles conferring varying degrees of Al tolerance (Choudhry 1978; Campbell and Lafever, 1981). Similar results were also obtained in barley, maize, sorghum, pea, chickpea and oat which exhibited monogenic inheritance with multiple alleles (Minnella and Sorrells, 1992; Sibov *et al.*, 1999; Singh and Choudhary, 2010; Singh and Raje, 2011; Castilhos *et al.*, 2011). In the past, Chinese Spring aneuploid stocks lacking a specific chromosome or chromosome arms helped the geneticist to locate numerous Al tolerance loci more precisely on individual chromosomes which were located on chromosome arms 2DL, 3DL, 4BL, 4DL, 6AL, 7AS and chromosome 7D in bread wheat (Aniol and Gustafson, 1984).

In the last quarter of the 20th century, the wider recognition of Al-toxicity as the predominant growth-limiting factor in acid soils has swung the pendulum of attention to Al tolerance and its genetic control (Carver and Ownby, 1995). Furthermore, recent findings in DNA technology and with the development of robust statistical analysis methods facilitated a more precise identification of loci and their relative contributions to governing Al tolerance in different plant species. QTL mapping studies based on the segregation analysis of large F_2 or RILs (recombinant inbred lines) populations clearly demonstrated that Al tolerance is a quantitative trait which is governed by multiple genes. In addition of major consensus loci like Alt_{BH} and *Xcec* loci in wheat on 4DL and 4BL, respectively, *Alp* locus in barley on chromosome 4H, and Alt_{SB} locus in sorghum on chromosome 3, numerous minor loci and their relative contribution to Al tolerance have also been demonstrated by several researchers. On the other hand, inheritance of Al tolerance in rye and rice appear to be more complex. Al tolerance in rye is controlled by at least two major dominant and independent loci (Gallego and Benito, 1997). Similarly, besides the consensus loci such as four loci namely *Alt1, Alt2, Alt3* and *Alt4* in rye on chromosomes 6RS, 3RS, 4RL and 7RS, respectively, and three loci on chromosomes 1, 8 and 9 in rice, several minor loci were also consistently identified in numerous studies. However, the number of minor loci detected in different crop plants depends on both the diversity among the parents for Al tolerance and the size of the mapping population, as well as on the number of polymorphic markers used in the study.

Molecular Mechanisms of Al Tolerance in Plants

Since 1990s, with the routine application of molecular markers in most crop improvement programmes, so far several QTL for Al tolerance have been identified in different crop plants including the model plant *Arabidopsis*. However, only major QTL have been validated at the molecular level with the functional characterisation of candidate genes for Al tolerance. Over the decades, research on physiology and genetics of Al tolerance in wheat has provided an excellent starting position for molecular studies. The first breakthrough in uncovering the molecular basis of Al tolerance in wheat was achieved by the identification of a higher expression of a cDNA in the root apices of Al tolerant plants than in the sensitive plants of a pair of NILs (near-isogenic lines) (Sasaki *et al.*, 2004). This was the first Al tolerance gene identified in any plant species which encodes for a membrane-localised protein

ALMT (Al-Activated Malate Transporter) and controls the malate efflux from bread wheat roots apices. Subsequently, promoter analyses in diverse bread wheat genotypes revealed that transcript expression of *TaALMT1* in wheat relies, in part, on a series of *cis* mutations in the upstream region of *ALMT1* gene and these specific variations could be classified into seven patterns, Type I to Type VII (Sasaki *et al.*, 2006, Garcia-Oliveira *et al.*, 2014). Type I pattern had shown the simplest structure, while the others had duplicated and triplicated blocks of sequence in different arrangements which are responsible for the higher expression of *TaALMT1* in Al-tolerant genotypes of wheat.

Subsequently, fine mapping of Alt_{SB} locus on chromosome 3 and *Alp* locus on chromosome 4H identified the *SbMATE* in sorghum and the *HvAACT1* gene in barley, respectively, both of which belong to the multidrug and toxic compound exudation (MATE) family that encodes citrate efflux (Magalhães *et al.*, 2007; Furukawa *et al.*, 2007). Similar to *ALMT1*, variations in the upstream regions of *SbMATE* in sorghum and *HvAACT1* in barley were also identified which are capable of altering the transcript expression of these genes. In sorghum, MITEs (tourist-like miniature inverted repeat transposable elements) occur upstream of the *SbMATE* gene and the number of these repeats is broadly correlated with the level of *SbMATE* transcript expression (Magalhães *et al.*, 2007). On the other hand, a 1023-bp insertion in the 5 UTR of *HvAACT1* of barley enhanced transcript expression of *HvAACT1* and also altered the location of expression to root apices in some Asian barley cultivars (Fujii *et al.*, 2012). More recently, a previously reported potential locus, Xce_c associated with citrate efflux that confers Al tolerance in bread wheat, at chromosome 4BL accounts for 50 per cent of the phenotypic variation in citrate efflux from roots of Brazilian bread wheat cultivar Carazinho (Ryan *et al.*, 2009), was characterised which also encodes a citrate transporter *TaMATE1* (Tovkach *et al.*, 2013). Interestingly, a SLT (Sukkula-like transposable element) inserted 25 bp upstream of the ATG start site of *TaMATE1B* homoeologue extends *TaMATE1B* transcript expression to the root and shoot tissues which confers citrate efflux and enhances Al tolerance (Tovkach *et al.*, 2013; Garcia-Oliveira *et al.*, 2014).

Contrarily to the role of upstream variations of *ALMT1* for Al tolerance in wheat, copy number and coding sequences variation between Al-tolerant and Al-sensitive genotypes appear to play an important role in Al tolerance of rye. *Alt4* locus on chromosome 7RS contains a cluster of tandemly repeated *ALMT1* homologs in rye. At *Alt4* locus, five *ScALMT1* genes were found in the Al-tolerant genotype whereas only two copies were observed in the Al-sensitive genotype (Collins *et al.*, 2008). In line with this, tandem duplication of MATE paralogues in the maize genome has also been demonstrated for Al toxicity. More recently, Maron *et al.* (2013) revealed that copy number variation in the *ZmMATE1* gene (three copies of *ZmMATE1* gene in Al-tolerant plants compared with only one copy in normal plants) is the basis of the phenotypic variation for Al tolerance in maize. Besides cereals, additional resistance genes were also isolated based on their homology to these *ALMT1* and *MATE* genes and by exploiting the available information on the physiology and genetics of Al resistance in numerous plant species (Table 12.2).

Table 12.2: Details of known Genes Involved in Al-tolerance in different Plants

Gene		*Protein Function*	*Plant Species (References)*
ALMT1	:	*Malate transport efflux*	Wheat (Sasaki *et al.*, 2004); *Arabidopsis* (Hoekenga *et al.*, 2006); *Brassica napus* (Ligaba *et al.*, 2006); *Arabidopsis* (Kobayashi *et al.*, 2007); Rye (Collins *et al.*, 2008)
ALMT2	:	*Malate transport efflux*	*Brassica napus* (Ligaba *et al.*, 2006)
MATE1/ AACT1/ Frdl4	:	*Citrate transport efflux*	Sorghum (Magalhães *et al.*, 2007); Barley (Furukawa *et al.*, 2007); *Arabidopsis* (Liu *et al.*, 2009); Maize (Maron *et al.*, 2010); Rye (Yokosho *et al.*, 2010); Rice (Yokosho *et al.*, 2011); Wheat (Tovkach *et al.*, 2013; Garcia-Oliveira *et al.*, 2014)
MATE2	:	*Citrate transport efflux*	Rye (Yokosho *et al.*, 2010); Maize (Maron *et al.*, 2010)
STOP1	:	*C2H2- type Zn finger TF*	*Arabidopsis* (Iuchi *et al.*, 2007); Wheat (Garcia-Oliveira *et al.*, 2013)
ART1	:	*C2H2- type Zn finger TF*	Rice (Yamaji *et al.*, 2009))
STAR1	:	*UDP-glucose transport Partial ABC protein*	Rice (Huang *et al.*, 2009); *Arabidopsis* (Huang *et al.*, 2010)
STAR2	:	*UDP-glucose transport Partial ABC protein*	Rice (Huang *et al.*, 2009)
ALS1	:	*Partial ABC protein-function unclear*	*Arabidopsis* (Larsen *et al.*, 2007), Rice (Huang *et al.*, 2012)
ALS3	:	*Partial ABC protein-function unclear*	*Arabidopsis* (Larsen *et al.*, 2005)
Nrat1	:	*Transporter specific for trivalent Al*	Rice (Xia *et al.*, 2010)
MGT1	:	*Mg uptake Transporter*	Rice (Chen *et al.*, 2012)

ALMT: Al-activated malate transporter; MATE/AACT/FRDL: Multidrug and toxic compound exudation/ Al-activated citrate transporter/Feric reductase defective 3-like4; STOP: Sensitive to proton rhizotoxicity; ART: Al resistance transcription factor; STAR: Sensitive to Al rhizotoxicity; ALS: Al sensitive; Nrat: Nramp aluminium transporter; MGT: Mg uptake transporter.

Regulation of temporal and spatial expression patterns of specific stress genes is an important part of the plant stress response. Research carried out in past years has demonstrated that many transcription factors, belonging to different families such as AP2/ERF, bZIP, Zn-finger, NAC, MYB, and WRKY are playing a central role in abiotic stress tolerance in different plants species. The role of transcription factors in Al tolerance of plants came only after the identification of a Zn-finger transcription factor, *STOP1* in *Arabidopsis* (Iuchi *et al.*, 2007) and was further substantiated by *ART1* in rice (Yamaji *et al.*, 2009). Transcriptome analysis under Al stress revealed that 101 and 31 genes were down-regulated in the *Arabidopsis STOP1* (Sawaki *et al.*, 2009) and rice *ART1* mutants (Yamaji *et al.*, 2009), respectively. Interestingly, the previously reported genes related to Al tolerance, particularly *ALMT1* and *MATE1* were regulated by these transcription factors. Recently, the contribution of *ART1* locus to the variation for Al tolerance in rice has also been identified in QTL analysis (Famoso *et al.*, 2011).

Besides the major genes which explained most genotypic variation for Al tolerance in different crop plant species, genes encoding transporters have been identified through mutational analyses, especially ABC transporters type (*ALS1* and *ALS3*) in model plants *Arabidopsis* (Larsen *et al.*, 2005 and 2007) and bacterial-type ABC transporters (*STAR1* and *STAR2*) in rice (Huang *et al.*, 2009). These genes are essential for Al tolerance, and seem to be a part of the symplastic or internal detoxification mechanism through uptake and redistribution of Al to less sensitive tissues in plants, although their actual function in Al tolerance mechanism in plant is still unclear. Recently, a plasma membrane-localised transporter gene *Nrat1* for trivalent Al and a tonoplast localised half-size ABC transporter gene *ALS1* regulated by the transcription factor *ART1* were identified in rice. It is plausible that *Nrat1* is required for the transport of absorbed Al from plasma membrane to vacuoles (Xia *et al.*, 2010) whereas *ALS1* appears to be specifically required for internal compartmentation of Al. It was suggested that functional *ALS1* is a prerequisite for *Nrat1* playing a role in Al tolerance of rice (Huang *et al.*, 2012).

Phenotyping for Al-Toxicity Tolerance in Plants

Rapid and precise phenotyping is not only the essential component in any crop improvement programme, but also has paramount importance to dissect the genetic architecture of quantitatively inherited complex traits. From the breeder's point of view, phenotyping technique should be flexible, simple and rapid which also favours to maintain the costs within reasonable limits. In the second half of the last century, the science of Al-resistance phenotyping in plants has evolved a number of techniques that vary from the hydroponic assays to field evaluation.

Hydroponics Assay

For better understanding the dynamics of Al in the solution-root interfaces, so far hydroponic culture is a most useful phenotyping methodology which provides not only an easy access to the root systems for Al analyses, but also enables the application of accurate concentrations of minerals coupled with tight control over nutrient availability and pH. Moreover, hydroponics allows non-destructive measurements of Al tolerance based on root growth (Carver and Ownby, 1995). Contrastingly, in soil screening approaches, it is difficult to control and modify the Al concentration in soil solution under field conditions since the dynamic equilibrium between solid and aqueous phases imply modifications of soluble and exchangeable Al concentrations, and then modifications of pH and availability of other nutrients. Therefore, the interpretation of the specific Al effect on the crop yields is impossible.

In the past, a number of histo-chemical assays such as hematoxylin, morin, Schiff's reagent, eriochrome cyanine R and Evans blue staining have been performed to investigate the Al-toxicity in different plants. Hematoxylin and morin assays are used to localise Al in the plant tissues. The hematoxylin assay is based on the formation of coloured complex between hematoxylin and the root-bounded Al (Polle *et al.*, 1978); whereas morin is a flurochrome which forms a uorescent complex with Al (Ezaki *et al.*, 2000). Eriochrome cyanine R staining has been extensively used for the measurement of root re-growth under Al stress and shows that if root apical meristem is irreversibly damaged, the root tips remained intensively stained with

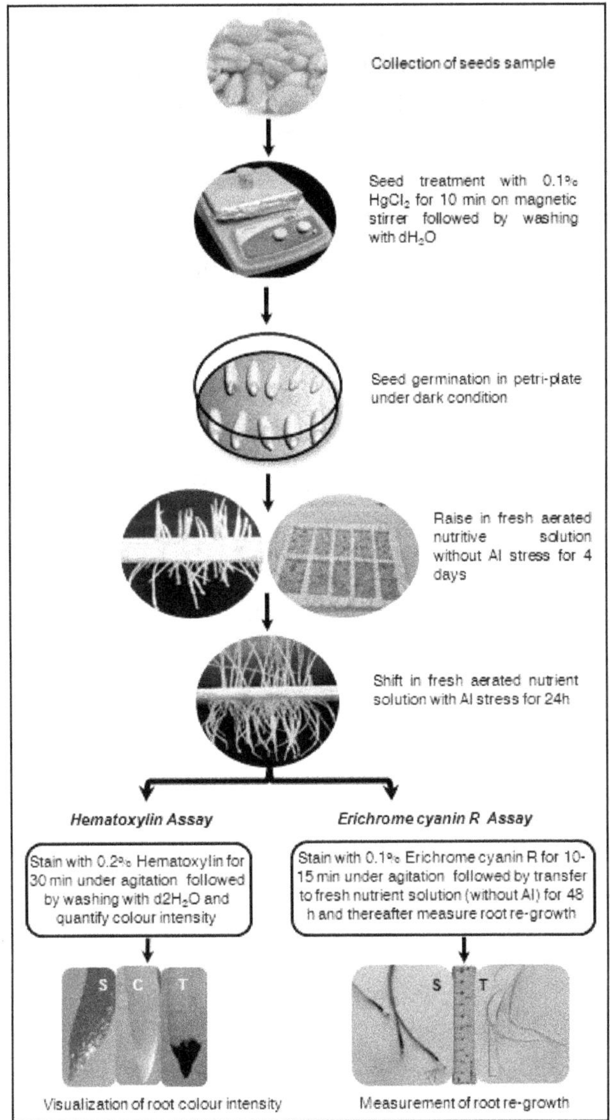

Figure 12.2: Al-tolerance Phenotyping using Hematoxylin and Eriochrome Cyanin R Assay in Hydroponic Conditions.

purple colour, whereas the part of the root which grows after exposure to Al stress remains unstained (Aniol, 1995). In general, Schiff's reagent staining technique is widely used for the assessment of Al-induced lipid peroxidation in the plants' root system detecting aldehyde functions that are originated from the peroxidation of the membrane's lipids and are bound to the membrane's proteins. Schiff's reagent procedure was originally performed to investigate the lipid peroxidation in liver tissue of bromobenzene-intoxicated mice and is based on the intensity of the pink colour development. In addition, the loss of plasma membrane integrity, induced by the toxic Al concentration, could be visualised in the roots by Evans blue staining technique (Yamamoto *et al.*, 2001). Among these histo-chemical assays, hematoxylin and eriochrome cyanine R staining are widely used to screen for Al tolerance in economically important plant species because of their consistent reliability in the staining patterns (Figure 12.2).

Pot Assay with Sand or Acid Soil

Free Al is directly toxic to plant roots and, in most cases, is little absorbed or translocated to the aerial plant parts. Therefore, small pots either may be filled with pre-washed and air dried sands impregnated with nutrient solution or may be filled with acid soils collected from the target regions. This system is suitable for the analysis of aerial and sub-aerial plant part characteristics.

Field Evaluation

Under field condition, the soil matrix may manifest plant diseases and other stresses which can complicate or affect the output. Although the evaluation of genotypes under field condition is imperative, because the improved genotypes for Al tolerance will ultimately grown in those areas which have major problem of Al-toxicity. Moreover, the other advantage of field evaluation is that large population can be screened with relatively low cost and less efforts.

Tandem Phenotyping Assay

Keeping in mind that the ultimate goal is to enhance the economic yield under acid soils, a combination of hydroponic and soil assays seems to be the best approach. Thus, the Al tolerance mechanisms identified in the hydroponic cultures using histo-chemical assays could then be confirmed in the soil bioassay and such tandem phenotyping approach can be helpful for the better understanding of both seedling as well as adult plant tolerance.

Breeding for Al Tolerance

Conventional plant breeding is primarily based on the recognition of natural variation within a crop followed by phenotypic selection of superior plants among segregating populations derived by hybridisation. So far, traditional breeders successfully utilised natural variation (intra- and inter-specific) to develop improved cultivars for numerous agronomic traits in various crop plants using either simple or modified forms of different breeding methodologies such as backcross, pedigree, single seed descent, doubled haploids and recurrent selection. In any crop improvement programme, the choice of breeding methodology relies on a number of factors such as the inheritance of the trait, the availability of adequate variability to allow the target trait to be improved and the genetic background of the trait. Besides breeding methodology, rapid, easy and reliable phenotyping techniques for trait of interest are pivotal for early generation selection. Thus, simplest breeding method for improving Al tolerance is to backcross parents having genes with relatively large effects from locally unadapted Al-tolerant genotypes to well-adapted Al-susceptible parents, because the large effect genes could be easily tracked through consecutive backcross generations by histo-chemical markers (hematoxylin or eriochrome R staining) at the early seedling stage (Carver and Ownby, 1995). Subsequently, the selected BCF_1 seedlings could be easily backcrossed to the re-current parent. The transfer of major gene for Al tolerance using backcross method has been performed in wheat, for example from Carazinho to Egret and Atlas66 to Chisholm and Century using backcross method but the genetic gain was limited which clearly indicated the complex nature of inheritance of Al tolerance in wheat (Carver and Ownby, 1995). Gupta (1997) suggested that pedigree and recurrent selection methods should be employed for the improvement of plants for Al tolerance, but such approach may not be economically justified where soil acidity is less severe.

With the advent of molecular marker technology, now gene pyramiding became more feasible which has long been advocated to improve complex traits, such as Al resistance. The success of marker-assisted selection/marker-assisted backcrossing (MAS/MAB) for pyramiding loci could be improved by developing functional markers rather than random DNA markers linked to loci, so that linkage drag could be reduced. In wheat, functional markers for ALMT1 enabled to identify different alleles correlating with levels of Al tolerance (Sasaki *et al.*, 2006). In near future, the availability of new candidate genes and already available gene sequences related to Al tolerance with similar functions in different plant species will help to develop more polymorphic functional markers and apply them for breeding.

Potential of Mutagenesis and Transgenic Approach for Al-toxicity Improvement

Over the past 70 years, the use of mutagenesis approach has involved forward genetic screens which enable the identification of more than 2500 improved varieties and numerous novel phenotypes in different crops that were exploited in conventional breeding programmes (Parry *et al.*, 2009). The role of mutagenesis has also been reported for the enhancement of Al tolerance in barley (Nawrot *et al.*, 2001). Recent development in reverse-genetic tools and high throughput techniques for mutation screening not only create novel genetic materials for breeding, but also allowed significant progress in functional genomics of Al resistance.

During last 10 years, a number of genes such as *ALS3, STOP1, ALS1* in *Arabidopsis*, and *ART1, STAR1* and *STAR2, Nrat1* and *ALS1* in rice have been identified through mutagenesis which greatly improved our understanding of Al resistance mechanism in plants at the molecular level. In addition, these studies also provided the additional kit to molecular breeders for the development of functional markers. In this context, the contribution of *ART1* and *Nrat1* to the natural variation for Al tolerance in rice has been identified which also could be used for the development of Al-resistant cultivars in rice (Famoso *et al.*, 2011). In the near future, the genome sequencing of major crops can also facilitate the identification of such genes with similar functions which will expedite the breeding through MAS/MAB for Al resistance in crop plants.

Burgeoning world population emphasised the need for genetic modification of crop plants through transgenic approach for improving their performance on acid soils. Since the first transgenic tobacco plant with enhanced Al tolerance was developed by using the bacterial citrate synthase gene (Fuente *et al.*, 1997), so far, a range of genes originating from bacteria, nematode and plants have been used to generate different transgenic plant species with enhanced level of Al tolerance (Table 12.3). These genes could be grouped into following categories:

I. Genes encoding enzymes associated with organic anion metabolism such as Citrate synthase (key enzyme involved in the tricarboxylic acid and glyoxylate cycles), Malate dehydrogenase (oxidises oxaloacetate to form malate) and Pyruvate phosphate dikinase (key enzyme associated in gluconeogenesis and photosynthesis which catalyses the conversion of pyruvate to phosphoenolpyruvate).

II. Al responsive genes such as Glutathione S-transferase, Peroxidase, GDP-dissociation inhibitor 1, Blue copper binding protein, Dehydroascorbate reductase and mitochondrial Manganese superoxide dismutase.

III. Genes encoding transporter proteins that mediate organic anion movement across the plasma membrane to the external medium such as Al-activated malate transporter and Multidrug and Toxic compound Exudation/Al-Activated Citrate Transporter/Ferric reductase defective3.

IV. Other stress responsive genes like Cell wall- associated receptor kinase (*WAK1*), Auxin-like protein, ΔD8 sphingolipid desaturase and *Ced*-9.

The overexpression of these genes *in planta* based on homologous and heterologous systems exhibited 1.2- to 20-fold increase in root re-growth phenotype in the transgenic lines compared to their wild type (Table 12.3). In most of these studies, transgenic plants exhibited modest gains in Al tolerance excepting the overexpression of wheat Al-activated malate transporter gene (*TaALMT1*) in barley, wheat and *Arabidopsis* which demonstrated better relative root re-growth by 20, 8 and 4-fold, respectively. Surprisingly, *TaALMT1* overexpression did not enhance Al resistance in rice (Sasaki *et al.*, 2004) which may be attributed to the very high basal level of Al resistance of rice. Although, as expected, it conferred Al-activated malate efflux in transgenic rice. These results indicate that overexpression of wheat *ALMT1* could provide another avenue to ameliorate acid soil toxicity in highly Al-sensitive crops like barley. Besides improved Al tolerance, it was also observed that *TaALMT1* overexpressing transgenic barley had enhanced Phosphorus-use efficiency (PUE) and improved grain yield when grown on an acid soil (Delhaize *et al.*, 2009). Recently, such dual effects have also been reported in transgenic canola overexpressing a bacterial citrate synthase, which exhibited higher tolerances to Al-toxicity and P-deficiency (Wang *et al.*, 2013). Contrarily to single genes, the multi-gene overexpression approach seems to be more lucrative for the enhancement of Al tolerance in plants as it may provide more effective protection against Al-toxicity. Recently, genes encoding two key enzymes in organic acid metabolism namely phosphoenolpyruvate carboxylase (PEPC) and citrate synthase (CS), were overexpressed individually as well as simultaneously using light-inducible promoter in tobacco leaves (Wang *et al.*, 2012). The measurement for the relative root elongation rates in nutrient solution and plant growth in soil bioassay of the transgenic tobacco plants exposed to Al-stress indicated that Al tolerance in the

Table 12.3: Details of Studies for Al-Tolerance Enhancement in Plants by Genetic Engineering

Name of the Gene	*Transgenic Strategy*			*Relative Root Growth*	*Proposed Mechanism*	*Reference*
	Source of Gene	*Recipient*	*Promoter*			
I. Organic anion metabolism						
CS	*P. aeruginosa*	Tobacco and Papaya	35S	2.0-fold	OA efflux (↑)	Fuente *et al.*, 1997
	P. aeruginosa	Alfalfa	RB7P and Act2P	2.5-fold	–	Barone *et al.*, 2008
	P. aeruginosa	Canola	35S	1.8-fold	Malate and citrate efflux (↑)	Wang *et al.*, 2013
	Carrot[+]	*Arabidopsis*	35S	1.3-fold	Citrate efflux (↑)	Koyama *et al.*, 2000
	Rice	Tobacco	35S	4.5-fold	Citrate efflux (↑)	Han *et al.*, 2009
	Arabidopsis[+]	Canola	35S	2.0-fold	Citrate efflux (↑)	Anoop *et al.*, 2003
MDH	Alfalfa	Alfalfa	35S	2.0-fold	OA efflux (↑)	Tesfaye *et al.*, 2001
	Arabidopsis and *E.coli*	Tobacco	PrbcS	2.4-fold	Malate efflux (↑)	Wang *et al.*, 2010
PPDK	*M.crystallinum*	Tobacco	Patatin B33	1.2-fold	Malate and citrate efflux (↑)	Trejo-Tellez *et al.*, 2010
II. Stress responsive						
GST	Tobacco	*Arabidopsis*	35S	1.7-fold	Oxidative stress tolerance (↑)	Ezaki *et al.*, 2000
Pox	Tobacco	*Arabidopsis*	35S	1.7-fold	Protection from oxidative stress	Ezaki *et al.*, 2000
GDI	Tobacco	*Arabidopsis*	35S	1.7-fold	Stress tolerance (↑)	Ezaki *et al.*, 2000
BCB	*Arabidopsis*	*Arabidopsis*	35S	1.7-fold	Protection from oxidative stress	Ezaki *et al.*, 2000
DHAR	*Arabidopsis*	Tobacco	35S	1.5-fold	Higher ascorbic acid protects from oxidative stress	Yin *et al.*, 2010
MnSOD	Wheat[+]	Canola	35S	2.5-fold	Protection from oxidative stress	Basu *et al.*, 2001

Contd...

Table 12.3–*Contd...*

Name of the Gene	*Transgenic Strategy*			*Relative Root Growth*	*Proposed Mechanism*	*Reference*
	Source of Gene	*Recipient*	*Promoter*			
III. Transporter						
ALMT1	Wheat	Barley	Ubiquitin	20-fold	Malate efflux (↑)	Delhaize *et al.*, 2004
	Wheat	Wheat	Ubiquitin	8.0-fold	Malate efflux (↑)	Pereira *et al.*, 2010
	Wheat	*Arabidopsis*	35S	4.0-fold	Malate efflux (↑)	Ryan *et al.*, 2011
	Barley	Barley	Ubiquitin	3.4-fold	Malate efflux (↑)	Gruber *et al.*, 2011
	Arabidopsis	*Arabidopsis*	$AtMATE_P$	*negligible*	Malate efflux(=)	Liu *et al.*, 2012
	Soybean	*Arabidopsis*	35S	2.0-fold	Malate efflux (↑)	Liang *et al.*, 2013
	Alfalfa	Tobacco	35S	2.0-fold	Malate efflux (↑)	Chen *et al.*, 2013
*AACT1/	Barley	Tobacco	35S	2.3-fold	Citrate efflux (↑)	Furukawa *et al.*, 2007
MATE1/Frd3	Barley	Barley	Ubiquitin	2.0-fold	Citrate efflux (↑)	Zhou *et al.*, 2013
	Barley	Wheat	Ubiquitin	1.3-fold	Citrate efflux (↑)	Zhou *et al.*, 2013
	Maize	Barley	Ubiquitin	2.0-fold	Citrate efflux (↑)	Zhou, 2012
	Sorghum	*Arabidopsis*	35S	2.5-fold	Citrate efflux (↑)	Magalhães *et al.*, 2007
	Sorghum	Barley	Ubiquitin	2-3.0-fold	Citrate efflux (↑)	Zhou, 2012
	Arabidopsis	*Arabidopsis*	35S	2.0-fold	Citrate efflux (↑)	Durrett *et al.*, 2007
	Arabidopsis	Barley	Ubiquitin	2-3.0-fold	Citrate efflux (↑)	Zhou, 2012
	Maize	*Arabidopsis*	35S	3.0-fold	Citrate efflux (↑)	Maron *et al.*, 2010
	Rice bean	Tomato	35S	2.5-fold	Citrate efflux (↑)	Yang *et al.*, 2011
	Arabidopsis	*Arabidopsis*	$AtALMT1_P$	3.0-fold	Citrate efflux (↑)	Liu *et al.*, 2012
H^+-PPase	Endogenous genes	*Arabidopsis*, Tomato and Rice	35S	1.8-fold	OA efflux (!)	Yang *et al.*, 2007

Contd...

Table 12.3–*Contd...*

Name of the Gene	*Transgenic Strategy*			*Relative Root Growth*	*Proposed Mechanism*	*Reference*
	Source of Gene	*Recipient*	*Promoter*			
IV. Other						
WAK1	*Arabidopsis*	*Arabidopsis*	35S	3.0-fold	Stress responsiveness (↑)	Sivaguru *et al.*, 2003
AuLP	*Arabidopsis*	*Arabidopsis*	35S	3.0-fold	Endocytosis (↓)	Ezaki *et al.*, 2007
SLD	*S. hamata*	*Arabidopsis*	35S	2.0-fold	Changes to membrane lipid composition	Ryan *et al.*, 2007
Ced-9	*C. elegans*	Tobacco	35S	2.0-fold	Al^{3+}-induced PCD (↓)	Wang *et al.*, 2009

CS: Citrate synthase; **MDH**: Malate dehydrogenase; **PPDK**: Pyruvate phosphate dikinase; **GST**: Glutathione S: transferase; **Pox**: Peroxidase; **GDI**: GDP: dissociation inhibitor; **BCB**: Blue copper binding protein; **DHAR**: Dehydroascorbate reductase; **MnSOD1**: Manganese superoxide dismutase 1; **ALMT**: Aluminium activated malate transporter; **MATE1/AACT1/Frd3**: Multidrug and Toxic compound Exudation1/Al: Activated Citrate Transporter1/Ferric reductase defective3; **H^+: PPase**: H^+: pyrophosphatase AVP1; **AuLP**: Auxilin: like protein; **WAK1**: Cell wall: associated receptor kinase; **SLD**: D8 sphingolipid desaturase.

Note: * mean homologs; + Mitochondrial cDNA.

double gene overexpression lines was stronger than that of the transgenic *CS* or *PEPC* lines and wild type plants.

Notably, in most of these transgenic studies, transgenes were driven by constitutive promoters either 35S or Ubiquitin. Thus, it is reasonable to speculate that the mechanism of efflux of organic acid anions constitutes a significant carbon cost to plants and may affect plant growth and that transcription of genes consumes additional energy. The fact that the expression of Al tolerance genes requires induction by Al, *i.e.* the expression only occurs under Al-toxicity conditions, constitutes an efficient resource-saving strategy (Liu *et al.*, 2013). Traditionally, optimisation of metabolic output relied on classical mutagenesis and genetic selection to create and identify variants with increased output of secondary metabolites. Thus, dissection of *cis*-acting elements that regulate Al-inducible gene expression could be useful for genetic improvement of Al tolerance through promoter reconstruction. In contrast to the overexpression approach for maximising metabolic throughput, it may be sensible to fine-tune the expression of key enzymes in the pathway, for instance, in case of Al tolerance mechanism, citrate is a much more effective Al-chelator than malate. Recently, Liu *et al.* (2012) utilised promoter swap strategy to test if the expression of the *AtMATE* gene driven by the strong AtALMT1 promoter will result in increased Al tolerance in *Arabidopsis*. Their results indicated that the AtALMT1: AtMATE promoter swap not only increased Al resistance of the transgenic plants, but also enhanced carbon usage efficiency for Al tolerance. Thus, utilisation of transgenic approach by numerous ways could be a useful strategy for genetic improvement of plant performance under acidic soils.

Conclusion

The recognition of Al toxicity in crop plants goes back to the beginning of the 20th century. Since then much effort has been made to both better crop management for amelioration of Al toxicity and development of improved cultivars through breeding for Al resistance. Plant physiologists made remarkable contribution by identifying primary toxicity targets in root tips and in the development of fast screening methods using different histo-chemical markers for the identification of Al resistant genotypes at early seedling stages. Although Al resistant genotypes identified through histo-chemical assays need to be confirmed in the soil bioassay. Tandem phenotyping seems to the best approach to efficient screening for Al resistance both at the seedling and the adult stage; moreover, keeping in mind that the ultimate goal is to enhance the economic yield under acid soil conditions.

Our understanding of Al toxicity and resistance mechanisms in plants has been progressing with the fast technological development of plant sciences. However, now mechanisms of Al resistance seem to be more complex than what was thought before. In fact, it is getting more and more evident that different plant species and/or genotypes within the same species have different mechanisms of Al resistance. Among all implied mechanisms, root tip exudation of organic acids (OAs) is one of the best studied ways for Al exclusion and seems to play an important role to cope with Al-toxicity under acidic conditions. Organic acids regulate the cell pH and chelate Al, therefore, protecting the roots from Al-toxicity. Generally, the kinds of organic acids secreted from the roots under Al stress differ among plant species. Among the cereal crops, mainly malate is secreted in bread wheat, whereas citrate is released in sorghum and barley. Noticeably, in addition to the genotypic differences within a species in the Al-induced rate of exudation of a particular organic acid, also differences in the kind of exudated organic acids have been observed. For example, some Al resistant bread wheat genotypes exudate either malate or citrate.

At the molecular level, mechanisms of Al resistance are extremely complex and still not fully understood. In cereals, numerous genetic studies clearly revealed that Al resistance is governed by multiple genes that cause minor to moderate level of phenotypic effects, making the identification of these genes difficult. A thorough understanding of both the genetics and physiology of Al resistance played a pivotal role in the identification of the genes implicated in Al tolerance, particularly in cereals like *ALMT1* in wheat, and *MATE1* in sorghum and barley. More recently, there is also strong evidence supporting the important role of regulatory genes in plant resistance to Al. Al may trigger resistance mechanisms by either directly binding to transcription factors or indirectly by activating different signaling pathways (Liu *et al.*, 2014). Fast induction by Al of changes in cytosolic Ca^{2+} levels (Rengel, 1992), reactive oxygen species (Ramirez-Benitez *et al.*, 2011), and phytohormone-mediated signal transduction in plants (Massot *et al.*, 2002) could operate here. To connect these Al-induced signalling pathways with the expression of Al resistance mechanisms is the challenge for the immediate future in this field of research.

References

Aniol, A. and Gustafson, J.P., 1984. Chromosome location of genes controlling aluminum tolerance in wheat, rye, and triticale. *Can. J. Genet. Cytol.*, 26: 701-705.

Aniol, A.M., 1995. Physiological aspects of aluminium tolerance associated with the long arm of chromosome 2D of the wheat (*Triticum aestivum* L.) genome. *Theor. Appl. Genet.*, 91(3): 510-516.

Anoop, V.M., Basu, U., McCammon, M.T., McAlister-Henn, L. and Taylor, G.J., 2003. Modulation of citrate metabolism alters aluminium tolerance in yeast and transgenic canola overexpressing a mitochondrial citrate synthase. *Plant Physiol.*, 132: 2205-2217.

Barceló, J. and Poschenrieder, C., 2002. Fast root growth responses, root exudates and internal detoxification as clues to the mechanisms of aluminum toxicity and resistance: a review. *Environ. Exp. Bot.*, 48: 75-92.

Barone, P., Rosellini, D., Fayette, P.L., Bouton, J., Veronesi, F. and Parrott, W., 2008. Bacterial citrate synthase expression and soil aluminum tolerance in transgenic alfalfa. *Plant Cell Rep.*, 27: 893- 901.

Basu, U., Good, A.G. and Taylor, G.J., 2001. Transgenic Brassica napus plants overexpressing aluminium-induced mitochondrial manganese superoxide dismutase cDNA are resistant to aluminium. *Plant Cell Environ.*, 24: 1269-1278.

Blamey, F.P.C., Robinson, N.J. and Asher, C.J., 1993. Interspecific differences in aluminium tolerance in relation to root cation-exchange capacity. In Genetic Aspects of Plant Mineral Nutrition. *Develop. Plant Soil Sci.*, 50: 91-96.

Bose, J., Babourina, O. and Rengel, Z. 2011. Role of magnesium in alleviation of aluminium toxicity in plant. *J. Exp. Bot.*, 62: 2251-2264.

Campbell, L.G. and Lafever, H.N., 1981. Heritability of aluminum tolerance in wheat. *Cereal Res. Comm.*, 9: 281-287.

Carver, B.F. and Ownby, J.D., 1995. Acid soil tolerance in wheat. *Adv. Agron.*, 54:117-173.

Castilhos, G., Farias, J.G., Schneider, A.B., Oliveira, P.H., Nicoloso, F.T., Schetinger, M.R.C. and Delatorre, C.A., 2011. Aluminum-stress response in oat genotypes with monogenic tolerance. *Environ. Exp. Bot.*, 74: 114-121.

Chen, Q., Wu, K.H., Wang, P., Yi, J., Li, K.Z., Yu, Y.X. and Chen, L.M., 2013. Overexpression of *MsALMT1*, from the aluminum-sensitive *Medicago sativa*, enhances malate exudation and aluminum resistance in tobacco. *Plant Mol. Biol. Rep.*, 31:769-774.

Chen, Z.C., Yamaji, N., Motoyama, R., Nagamura, Y. and Ma, J.F., 2012. Up-regulation of a magnesium transporter gene *OSMGT1* is required for conferring aluminum tolerance in rice. *Plant Physiol.*, 159: 1624-1633.

Choudhry, M.A., 1978. Genetic differences for aluminum tolerance in five wheat crosses. Agr Abstr Madison, *Am. Soc. Agr.*, 151.

Collins, N.C., Shirley, N., Saeed, M., Pallotta, M. and Gustafson, J.P., 2008. An *ALMT1* gene cluster controlling aluminium tolerance at the *Alt4* locus of rye (*Secale cereal* L). *Genetics*, 179: 669-682.

Degenhardt, J., Larsen, P.B., Howell, S.H. and Kochian, L.V., 1998. Aluminum resistance in the *Arabidopsis* mutant alr-104 is caused by an aluminum-induced increase in rhizosphere pH. *Plant Physiol.*, 117: 19-27.

Delhaize, E., Ma, J.F. and Ryan, P.R., 2012. Transcriptional regulation of aluminium tolerance genes. *Trends Plant Sci.*, 17(6): 341-348.

Delhaize, E., Ryan, P.R., Hebb, D.M., Yamamoto, Y., Sasaki, T. and Matsumoto, H., 2004. Engineering high-level aluminum tolerance in barley with the *ALMT1* gene. *Proc. Nat. Acad. Sci. USA*, 101: 15249-15254.

Delhaize, E., Taylor, P., Hocking, P.J., Simpson, R.J., Ryan, P.R. and Richardson, A.E., 2009. Transgenic barley (*Hordeum vulgare* L.) expressing the wheat aluminium resistance gene (*TaALMT1*) shows enhanced phosphorus nutrition and grain production when grown on an acid soil. *Plant Biotechnol. J.*, 7: 391-400.

Delhaize, E., Ryan, P.R. and Randall, P.J., 1993. Aluminum tolerance in wheat (*Triticum aestivum* L.) (II. Aluminum-stimulated excretion of malic acid from root apices). *Plant Physiol.*, 103(3): 695-702.

Durrett, T.P., Gassmann, W. and Rogers, E.E., 2007. The *FRD3*-mediated efflux of citrate into the root vasculature is necessary for efficient iron translocation. *Plant Physiol.*, 144: 197-205.

Ezaki, B., Gardner, R.C., Ezaki, Y. and Matsumoto, H., 2000. Expression of aluminum-induced genes in transgenic *Arabidopsis* plants can ameliorate aluminum stress and/or oxidative stress. *Plant Physiol.*, 122: 657-665.

Ezaki, B., Kiyohara, H., Matsumoto, H. and Nakashima, S., 2007. Overexpression of an auxilin-like gene (F9E10.5) can suppress Al uptake in roots of Arabidopsis. *J. Exp. Bot.*, 58: 497-506.

Famoso, A.N., Zhao, K., Clark, R.T., Tung, C.W., Wright, M.H., Bustamante, C., Kochian, L.V. and McCouch, S.R., 2011. Genetic architecture of aluminum tolerance in rice (*oryza sativa*) determined through genome-wide association analysis and QTL mapping. *PLoS Genet*, 7(8): e1002221.

Fuente, J.M.dl., Ramirez-Rodriguez, V., Cabrera-Ponce, J.L. and Herrera-Estrella, L., 1997. Aluminum tolerance in transgenic plants by alteration of citrate synthesis. Science (Washington) 276: 5466-1568.

Fujii, M., Yokosho, K., Yamaji, N., Saisho, D., Yamane, M., Takahashi, H., Sato, K., Nakazono, M. and Ma, J.F., 2012. Acquisition of aluminium tolerance by modification of a single gene in barley. *Nat. Commun.*, 3: 713.

Furukawa, J., Yamaji, N., Wang, H., Mitani, N., Murata, Y., Sato, K., Katsuhara, M., Takeda, K. and Ma, J.F., 2007. An aluminum-activated citrate transporter in barley. *Plant Cell Physiol.*, 48: 1081-1091.

Gallego, F.G. and Benito, C., 1997. Genetic control of aluminum tolerance in rye (*Secale cereale* L.). *Theor. Appl. Genet.*, 95: 393-399.

Garcia-Oliveira, A.L., Benito, C., Prieto, P., Menezes, R.A., Rodrigues-Pousada, C., Guedes-Pinto, H. and Martins-Lopes, P., 2013. Molecular characterization of *TaSTOP1* homoeologues and their response to aluminium and proton (H^+) toxicity in bread wheat (*Triticum aestivum* L.). *BMC Plant Biol.*, 13: 134.

Garcia-Oliveira, A.L., Martins-Lopes, P., Tolrá, R., Poschenrieder, C., Tarquis, M., Guedes-Pinto, H. and Benito, C., 2014. Molecular characterization of the citrate transporter gene *TaMATE1* and expression analysis of upstream genes involved in organic acid transport under Al stress in bread wheat (*Triticum aestivum* L.). *Physiol. Plant*, 152(3):441-452.

Gruber, B.D., Delhaize, E., Richardson, A.E., Roessner, U., James, R.A., Howitt, S.M. and Ryan, P.R., 2011. Characterisation of *HvALMT1* function in transgenic barley plants. *Func. Plant Biol.*, 38: 163-175.

Gunsé, B., Poschenrieder, C. and Barceló, J., 1997. Water transport properties of roots and root cortical cells in proton- and AI-stressed maize varieties. *Plant Physiol.*, 113: 595-602.

Gupta, U.S., 1997. Stress tolerance. Low pH tolerance. *Crop improve.*, 2:33-59.

Han, Y.Y., Zhang, W.Z., Zhang, B.L., Zhang, S.S., Wang, W. and Ming, F., 2009. One novel mitochondrial citrate synthase from *Oryza sativa* L. can enhance aluminum tolerance in transgenic tobacco. *Mol. Biotechnol.*, 42: 299-305.

Hartwell, B.L. and Pember, F.R., 1918. The presence of aluminium as a reason for the differene in the effect of so-called acid soils on barley and rye. *Soil Sci.*, 6: 259-269.

Henderson, M. and Ownby, J.D., 1991. The role of root cap mucilage secretion in aluminum tolerance in wheat. *Biochem. Curr. Topics Plant Physiol.*, 10: 134-141.

Hoekenga, O.A., Maron, L.G., Piñeros, M.A., Cançado, G.M.A., Shaff, J., Kobayashi, Y., Ryan, P.R., Dong, B., Delhaize, E., Sasaki, T., Matsumoto, H., Yamamoto, Y., Koyama, H. and Kochian, L.V., 2006. *AtALMT1*, which encodes a malate transporter, is identified as one of several genes critical for aluminum tolerance in *Arabidopsis*. *Proc. Natl. Acad. Sci. USA*, 103(25): 9738-9743.

Horst, W.J., Wagner, A. and Marschner, H., 1982. Mucilage protects root meristems from aluminium injury. *Z. Pflanzenphysiol*, 105: 435-444.

Huang, C.F., Yamaji, N. and Ma, J.F., 2010. Knockout of a bacterial-type ABC transporter gene, *AtSTAR1*, results in increased Al sensitivity in *Arabidopsis*. *Plant Physiol.*, 153: 1669-1677.

Huang, C.F., Yamaji. N., Mitani, N., Yano, M., Nagamura, Y. and Ma, J.F., 2009. A bacterial-type ABC transporter is involved in aluminium tolerance in rice. *Plant Cell*, 21: 655-667.

Huang, C.F., Yamaji, N., Chen, Z. and Ma, J.F., 2012. A tonoplast-localized half-size ABC transporter is required for internal detoxification of aluminum in rice. *Plant J.*, 69(5):857-867.

Iuchi, S., Koyama, H., Iuchi, A., Kobayashi, Y., Kitabayashi, S., Ikka, T., Hirayama, T., Shinozaki, K. and Kobayashi, M., 2007. Zinc finger protein STOP1 is critical for proton tolerance in Arabidopsis and coregulates a key gene in aluminum tolerance. *Proc. Nat. Acad. Sci. USA*, 104: 9900-9905.

Kinraide, T.B., 1997. Reconsidering the rhizotoxicity of hydroxyl, sulphate, and fluoride complexes of aluminium. *J. Exp. Bot.*, 48: 1115-1124.

Kobayashi, Y., Hoekenga, O.A., Itoh, H., Nakashima, M., Saito, S., Shaff, J.E., Maron, L.G., Pineros, M.A., Kochian, L.V. and Koyama, H., 2007. Characterization of *AtALMT1* expression in aluminum-inducible malate release and its role for rhizotoxic stress tolerance in *Arabidopsis*. *Plant Physiol.*, 145: 843-852.

Kochian, L.V., Hoekenga, O.A. and Piñeros, M.A., 2004. How do crop plants tolerate acid soils? Mechanisms of aluminum tolerance and phosphorus efficiency. *Annu. Rev. Plant Biol.*, 55: 459-93.

Koyama, H., Kawamura, A., Kihara, T., Hara, T., Takita, E. and Shibata, D., 2000. Overexpression of mitochondrial citrate synthase in *Arabidopsis thaliana* improved growth on a phosphorus-limited soil. *Plant Cell Physiol.*, 41: 1030-1037.

Larsen, P.B., Cancel, J., Rounds, M. and Ochoa, V., 2007. *Arabidopsis ALS1* encodes a root tip and stele localized half type ABC transporter required for root growth in an aluminum toxic environment. *Planta*, 225: 1447-1458.

Larsen, P.B., Geisler, M.J.B., Jones, C.A., Williams, K.M. and Cancel, J.D., 2005. *ALS3* encodes a phloem-localized ABC transporter-like protein that is required for aluminum tolerance in *Arabidopsis*. *Plant J.*, 41: 353-363.

Liang, C.Y., Piñeros, M.A., Tian, J., Yao, Z.F., Sun, L.P., Liu, J.P., Shaff, J., Coluccio, A., Kochian, L.V. and Liao, H., 2013. Low pH, aluminum, phosphorus, coordinately regulate malate exudation through *GmALMT1* to improve soybean adaptation to acid soils. *Plant Physiol.*, 161 (3): 1347-1361.

Ligaba, A., Katsuhara, M., Ryan, P.R., Shibasaka, M. and Matsumoto, H., 2006. The *BnALMT1* and *BnALMT2* genes from rape encode aluminum-activated malate transporters that enhance the aluminium resistance of plant cells. *Plant Physiol.*, 142: 1294-1303.

Liu, J.P., Luo XP, Shaff, J., Liang CY, Jia XM, Li ZY, Magalhães, J.V. and Kochian, L.V., 2012. A promoter-swap strategy between the *AtALMT1* and *AtMATE* genes increased Arabidopsis aluminium resistance and improved carbon use efficiency for Al resistance. *Plant J.*, 71(2): 327-337.

Liu, J.P., Magalhães, J.V., Shaff, J. and Kochian, L.V., 2009. Aluminum activated citrate and malate transporters from the MATE and ALMT families function independently to confer *Arabidopsis* aluminium tolerance. *Plant J.*, 57: 389-399.

Liu, J.P., Piñeros, M.A. and Kochian, L.V., 2014. The role of aluminium sensing and signalling in plant aluminium resistance. *J. Integr. Plant Biol.*, 56: 221-230.

Liu, M.Y., Chen, W.W., Xu, J.M., Fan, W., Yang, J.L. and Zheng, S.J., 2013. The role of *VuMATE1* expression in aluminium-inducible citrate secretion in rice bean (*Vigna umbellata*) roots. *J. Exp. Bot.*, 64(7): 1795-1804.

Ma, J.F., Ryan, P.R. and Delhaize, E., 2001. Aluminium tolerance in plants and the complexing role of organic acids. *Trends Plant Sci.*, 6(6):273-278.

Ma, J.F., Hiradate, S., Nomoto, K., Iwashita, T. and Matsumoto, H., 1997. Internal detoxification mechanism of Al in Hydrangea (Identification of Al form in the leaves). *Plant Physiol.*, 113(4): 1033-1039.

Magalhães, J.V., Liu, J., Guimarães, C.T., Lana, U.G., Alves, V.M., Wang, Y.H., Schaffert, R.E., Hoekenga, O.A., Piñeros, M.A., Shaff, J.E., Klein, P.E., Carneiro, N.P., Coelho, C.M., Trick, H.N. and Kochian, L.V., 2007. A gene in the multidrug and toxic compound extrusion (*MATE*) family confers aluminum tolerance in sorghum. *Nat. Genet.*, 39: 1156-1161.

Maron, L.G., Guimarães, C.T., Kirst, M., Albert, P.S., Birchler, J.A., Bradbury, P.J., Buckler, E.S., Coluccio, A.E., Danilova, T.V., Kudrna, D., Magalhães, J.V., Piñeros, M.A., Schatz, M.C., Wing, R.A. and Kochian, L.V., 2013. Aluminum tolerance in maize is associated with higher MATE1 gene copy number. *Proc. Nat. Acad. Sci. USA*, 110: 5241-5246.

Maron, L.G., Pineros, M.A., Guimaraes, C.T., Magalhães, J.V., Pleiman, J.K., Mao, C.Z., Shaff, J., Belicuas, S.N.J. and Kochian, L.V., 2010. Two functionally distinct members of the *MATE* (multi-drug and toxic compound extrusion) family of transporters potentially underlie two major aluminum tolerance QTLs in maize. *Plant J.*, 61: 728-740.

Massot N., Nicander, B., Barceló, J., Poschenrieder, C. and Tilberg, E., 2002. A rapid increase in cytokinin levels and enhanced ethylene evolution precede Al^{3+}-induced inhibition of root growth in bean seedlings (*Phaseolus vulgaris* L.). *Plant Growth Regul.*, 37: 105-112.

Minella, E. and Sorrells, M., 1992. Aluminum tolerance in barley genetic relationships among genotypes of diverse origin. *Crop Sci.*, 32: 593-598.

Miyasaka, S.C. and Hawes, M.C., 2001. Possible role of root border cells in detection and avoidance of aluminum toxicity. *Plant Physiol.*, 125: 1978-1987.

Miyasaka, S.C., Hue, N.V. and Dunn, M.A., 2007. Aluminum - Handbook of plant nutrition, in Handbook of Plant Nutrition, D.J.P. Allen, V Barker, Editor, CRC Press. p. 439-498.

Mohammed, Y.S.A., Eltayeb, A.E. and Tsujimoto, H., 2013. Enhancement of aluminum tolerance in wheat by addition of chromosomes from the wild relative *Leymus racemosus*. *Breeding Sci.*, 63: 407-416.

Nawrot, M., Szarejko, I. and Maluszynski, M., 2001. Barley mutants with increased tolerance to aluminium toxicity. *Euphytica*, 120: 345-356.

Nguyen, N.T., Nakabayashi, K., Thompson, J. and Fujita, K., 2003. Role of exudation of organic acids and phosphate in aluminum tolerance of four tropical woody species. *Tree Physiol.*, 23: 1041-1050.

Nowak J., Friend, A.L., 2006. Loblolly pine and slash pine responses to acute aluminum and acid exposures. *Tree Physiol.*, 26: 1207-1215.

Ofei-Manu,P., Wagatsuma, W., Ishikawa, S., Tawaraya, K., 2001. The plasma membrane strength of the root-tip cells and root phenolic compounds are correiated with Al tolerance in several common woody plants. Soil Sci. *Plant Nutr.*, 47(2):359-375

Parry, M.A.J., Madgwick, P.J., Bayo, C., Tearall, K., Hernandez-Lopez, A., Baudo, M., Rakszegi, M., Hamada, W., Al-Yassin, A., Ouabbou, H., Labhilili, M. and Phillips, A.L., 2009. Mutation discovery for crop improvement. *J. Exp. Bot.*, 60(10): 2817-2825.

Pellet, D.M., Papernick, L.A., Jones, D.L., Darrah, P.R., Grunes, D.L. and Kochian, L.V., 1997. Involvement of multiple aluminium exclusion mechanisms in aluminium tolerance in wheat. *Plant Soil*, 192: 63-68.

Pellet, D.M., Papernik, L.A. and Kochian, L.V., 1996. Multiple aluminum-resistance mechanisms in wheat: roles of root apical phosphate and malate exudation. *Plant Physiol.*, 112: 591-597.

Pereira, J.F., Zhou, G., Delhaize, E., Richardson, T. and Ryan, P.R., 2010. Engineering greater aluminium resistance in wheat by over-expressing *TaALMT1*. *Ann. Bot-London*, 106: 205-214.

Polle, E., Konzak, C.F. and Kittrick, J.A., 1978. Visual detection of aluminium tolerance levels in wheat by hematoxylin staining of seedling roots. *Crop Sci.*, 18: 823-827.

Poschenrieder, C., Gunse, B., Corrales. I. and Barceló, J., 2008. A glance into aluminum toxicity and resistance in plants. *Sci. Total Environ.*, 400: 356-368.

Rajaram, S., Pfeiffer, W. and Singh, R., 1988. Developing bread wheat for acid soils through shuttle breeding. In Kohli and Rajaram eds. Wheat breeding in acid soils. Review of Brazilian/CIMMYT Collaboration, 1974-1986 pp: 51-58.

Ramírez-Benítez, E., Muñoz-Sánchez, J.A., Becerril-Ch, K.M., Miranda-Ham, M.L., Casro-Concha, L.A. and Hernández-Sotomayor, S.M.T., 2011. Aluminum induces changes in oxidative burst scavenging enzymes in *Coffea Arabica* L. Suspension cells with different Al tolerance. *J. Inorg. Biochem*, 105: 1523-1528.

Rengel, Z., 1992. Role of calcium in aluminum tolerance. *New Phytol.*, 121: 499-513.

Ryan, P.R., Liu, Q., Sperling, P., Dong, B., Franke, S. and Delhaize, E., 2007. A higher plant *Delta 8 sphingolipid desaturase* with a preference for (Z)-isomer formation confers aluminum tolerance to yeast and plants. *Plant Physiol.*, 144: 1968-1977.

Ryan, P.R., Raman, H., Gupta, S., Horst, W.J. and Delhaize, E., 2009. A second mechanism for aluminum resistance in wheat relies on the constitutive efflux of citrate from roots. *Plant Physiol.*, 149: 340-351.

Ryan, P.R.,Tyerman, S.D., Sasaki, T., Furuichi, T., Yamamoto, Y., Zhang, W.H and Delhaize, E., 2011. The identification of aluminium-resistance genes provides opportunities for enhancing crop production on acid soils. *J. Exp. Bot.*, 62(1): 9-20.

Sasaki, T., Ryan, P.R., Delhaize, E., Hebb, D.M., Ogihara, Y., Kawaura, K., Noda, K., Kojima, T., Toyoda, A., Matsumoto, H. and Yamamoto, Y., 2006. Sequence upstream of the wheat (*Triticum aestivum* L.) ALMT1 gene and its relationship to aluminum resistance. *Plant Cell Physiol.*, 47: 1343-1354.

Sasaki, T., Yamamoto, Y., Ezaki, B., Katsuhara, M., Ahn, S.J., Ryan, P.R., Delhaize, E. and Matsumoto, H., 2004. A wheat gene encoding an aluminum-activated malate transporter. *Plant J.*, 37: 645-653.

Sawaki. Y., Iuchi, S., Kobayashi, Y., Ikka, T., Sakurai, N., Fujita, M., Shinozaki, K., Shibata, D., Kobayashi, M. and Koyama, H., 2009. STOP1 regulates multiple genes that protect *Arabidopsis* from proton and aluminum toxicities. *Plant Physiol.*, 150(1): 281-294.

Sibov, S., Gaspar, M., Silva, M., Ottoboni, L., Arruda, P. and Souza, A., 1999. Two genes control aluminum tolerance in maize: Genetic and molecular mapping analyses. *Genome*, 42: 475-482.

Singh, D. and Raje, R.S., 2011. Genetics of aluminium tolerance in chickpea (*Cicer arietinum*). *Plant Breed.*, 130: 563-568.

Singh, S. and Choudhary, A.K., 2010 Inheritance pattern of aluminium tolerance in pea. *Plant Breed.*, 129: 688-692.

Sivaguru, M., Ezaki, B., He, Z.H., Tong, H.Y., Osawa, H., Baluska, F., Volkmann, D. and Matsumoto, H., 2003. Aluminum-induced gene expression and protein localization of a cell wall-associated receptor kinase in *Arabidopsis*. *Plant Physiol.*, 132: 2256-2266.

Sivaguru, M. and Horst, W., 1998. The distal part of the transition zone is the most aluminum-sensitive apical root zone of maize. *Plant Physiol.*, 116: 155-63.

Tang, Y., Garvin, D.F., Kochian, L.V., Sorrells, M.E. and Carver, B.F., 2002. Physiological genetics of aluminum tolerance in the wheat cultivar Atlas 66. *Crop Sci.*, 42(5): 1541-1546.

Tesfaye, M., Temple, S.J., Allan, D.L., Vance, C.P. and Samac, D.A., 2001. Overexpression of *malate dehydrogenase* in transgenic alfalfa enhances organic acid synthesis and confers tolerance to aluminum. *Plant Physiol.*, 127: 1836-1844.

Tolrà, R., Barceló, J. and Poschenrieder, M. 2009. Constitutive and aluminium-induced patterns of phenolic compounds in two maize varieties differing in aluminium tolerance. *J. Inorg. Biochem.*, 103(11): 1486-1490.

Tolrà, R., Vogel-Mikuš, K., Hajiboland, R., Kump, P., Pongrac, P., Kaulich, B., Gianoncelli, A., Babin, V., Barceló, J., Regvar, M. and Poschenrieder, M., 2011. Localization of aluminium in tea (*Camellia sinensis*) leaves using low energy X-ray fluorescence spectro-microscopy. *J. Plant Res.*, 124(1): 165-172.

Tovkach, A., Ryan, P.R., Richardson, A.E., Lewis, D.C., Rathjen, T.M., Ramesh, S., Tyerman, S.D. and Delhaize, E., 2013. Transposon-mediated alteration of *TaMATE1B* expression in wheat confers constitutive citrate efflux from root apices. *Plant Physiol.*, 161(2): 880-892.

Trejo-Tellez, L.I., Stenzel, R., Gomez-Merino, F.C. and Schmitt, J.M., 2010. Transgenic tobacco plants overexpressing *pyruvate phosphate dikinase* increase exudation of organic acids and decrease accumulation of aluminum in the roots. *Plant Soil*, 326: 187-198.

Vitorello, V.A., Capaldi, F.R.C. and Stefanuto, V.A., 2005. Recent advances in aluminum toxicity and resistance in higher plants. *Braz. J. Plant Physiol.*, 17: 129-143.

von Uexkull, H.R. and Mutert, E., 1995. Global extent, development and economic impact of acid soils. In *Plant-Soil Interactions at Low pH: Principles and Management*, ed. RA Date, NJ Grundon, GE Raymet, ME Probert, pp. 5–19. Dordrecht, The Neth:Kluwer Academic.

Wang, Q., Yi, Q., Hu, Q., Zhao, Y., Nian, H., Li, K., Yu, Y., Izui, K.and Chen, L., 2012. Simultaneous overexpression of *Citrate Synthase* and *Phosphoenolpyruvate Carboxylase* in leaves augments citrate exclusion and Al resistance in transgenic tobacco. *Plant. Mol. Biol. Rep.*, 30: 992-1005.

Wang, Q.F., Zhao, Y., Yi, Q., Li, K.Z., Yu, Y.X. and Chen, L.M. 2010. Overexpression of *malate dehydrogenase* in transgenic tobacco leaves: enhanced malate synthesis and augmented Al-resistance. *Acta Physiol. Plant*, 32: 1209-1220.

Wang, W.Z., Pan, J.W., Zheng, K., Chen, H., Shao, H.H., Guo, Y.J., Bian, H.W., Han, N., Wang, J.H. and Zhu, M.Y., 2009. *Ced-9* inhibits Al-induced programmed cell death and promotes Al tolerance in tobacco. *Biochem. Bioph. Res. Co.*, 383: 141-145.

Wang, Y., Xu, H., Kou, J., Shi, L., Zhang, C. and Xu, F., 2013. Dual effects of transgenic *Brassica napus* overexpressing *CS* gene on tolerances to aluminum toxicity and phosphorus deficiency. *Plant Soil*, 362: 231-246.

Xia, J., Yamaji, N., Kasai, T. and Ma, J.F., 2010. Plasma membrane-localized transporter for aluminum in rice. *Proc. Nat. Acad. Sci. USA*, 107: 18381-18385.

Yamaji, N., Huang, C.F., Nagao, S., Yano, M., Sato, Y., Nagamura, Y. and Ma, J.F., 2009. A zinc finger transcription factor *ART1* regulates multiple genes implicated in aluminum tolerance in rice. *Plant Cell*, 21: 3339-3349.

Yamamoto, Y., Kobayashi, Y. and Matsumoto, H., 2001. Lipid peroxidation is an early symptom triggered by aluminum, but not the primary cause of elongation inhibition in pea roots. *Plant Physiol.*, 125: 199-208.

Yang, H., Knapp, J., Koirala, P., Rajagopal, D., Peer, W.A., Silbart, L.K., Murphy, A. and Gaxiola, R.A., 2007. Enhanced phosphorus nutrition in monocots and dicots over-expressing a phosphorus-responsive type I H+-pyrophosphatase. *Plant Biotechnol. J.*, 5: 735-745.

Yang, X.Y., Yang, J.L., Zhou, Y., Pineros, M.A., Kochian, L.V., Li, G.X. and Zheng, S.J., 2011. A *de novo* synthesis citrate transporter, *Vigna umbellata* multidrug and toxic compound extrusion, implicates in Al-activated citrate efflux in rice bean *Vigna umbellata*) root apex. *Plant Cell Environ.*, 12: 2138- 2148.

Yin, L.N., Wang, S.W., Eltayeb, A.E., Uddin, M.I., Yamamoto, Y., Tsuji, W., Takeuchi, Y. and Tanaka, K., 2010. Overexpression of *dehydroascorbate reductase*, but not *monodehydroascorbate reductase*, confers tolerance to aluminum stress in transgenic tobacco. *Planta*, 231: 609- 621.

Yokosho, K., Yamaji, N. and Ma, J.F., 2010. Isolation and characterisation of two MATE genes in rye. *Funct. Plant Biol.*, 37: 296-303.

Yokosho, K., Yamaji, N. and Ma, J.F., 2011. An Al-inducible MATE gene is involved in external detoxification of Al in rice. *Plant J.*, 68: 1061-1069.

Zheng, S.J., Ma, J.F. and Matsumoto, H., 1998. High aluminum resistance in buckwheat: I. Al-induced specific secretion of oxalic acid from root tips. *Plant Physiol.*, 117: 745-51.

Zhou, G., 2012. Enhancing aluminium resistance in barley through over-expression of MATE genes. PhD thesis. Univ Tasmania, pp. 37-118.

Zhou, G., Delhaize, E., Zhou, M. and Ryan, P.R., 2013. The barley MATE gene, *HvAACT1*, increases citrate efflux and Al^{3+} tolerance when expressed in wheat and barley. *Ann. Bot.*, 112(3): 603-612.

2016, Recent Advances in Plant Stress Physiology *Pages* **265–282**
Editors: ***Praduman Yadav, Sunil Kumar and Veena Jain***
Published by: **DAYA PUBLISHING HOUSE, NEW DELHI**

Chapter 13

An Insight into Plant Growth and Metabolism in Relation to Hexavalent Chromium

Punesh Sangwan[1] and Vinod Kumar[2]*

[1]Department of Biochemistry, C.C.S. Haryana Agricultural University, Hisar, India
[2]Akal School of Biotechnology, Eternal University, Baru Sahib, H.P., India

ABSTRACT

Heavy metals are the intrinsic component of the environment with both essential and non essential types. Heavy metal accumulation in soils, due to industrial and fertilizer applications, is of concern in agricultural production. Their excessive levels pose a threat to plant growth and yield. Some heavy metals are toxic to plants even at very low concentrations. Among heavy metals, chromium (Cr) is a common contaminant of soils, as Cr (III) and Cr (VI). Cr (VI) is highly toxic to plants even at lower concentrations, due to its higher solubility and bioavailability. Cr accumulated preferentially in the roots and its absorption and translocation in plants depends on multiple factors. This hexavalent Cr is shown to be adversely affecting seed germination, early plant growth, photosynthesis, metabolism, micronutrient content and nutritional status of crop plants. Plants growing in Cr-polluted sites exhibit altered metabolism, growth reduction, lower biomass production and metal accumulation. In this chapter, we have discussed some of these very important aspects of plant growth and metabolism under influence of hexavalent Cr followed by an introduction to some of the ameliorative measures to remove these phytotoxic effects. Understanding of these plant responses will enable us to design proper ameliorative model to protect plants in presence of this heavy metal.

Keywords: *Amelioration, Hexavalent chromium, Phytotoxicity, Plant growth.*

**Corresponding author.* E-mail: sangwan.vinod@yahoo.com

Introduction

Heavy metal pollution is increasing globally due to mining, industry, road traffic and other natural sources. Besides endangering health, elevated levels of metal pollutants further jeopardize human welfare by lowering the productivity potential of agricultural land due to their deleterious impact on plant growth. Trace metal toxicity reduces growth and vigor; interfere with photosynthesis, respiration, water-relations and reproduction. A high concentration of trace metals is one of the most important environmental stress factors. The extent to which a plant can survive is determined by its sensitivity to metal toxicity. The effect of heavy metals *i.e.* Cd, Cr, Cu, Zn, Pb, Al on crops and vegetables has been well established. Among heavy metals, chromium (Cr) is a common contaminant of surface waters and ground waters because of its occurrence in nature, as well as anthropic sources (Babula *et al.*, 2008). In soils, Cr exists primarily in the water-insoluble form. In this form, it firmly attached to soil and displayed little mobility within the soil strata. A small proportion of Cr in soils is water soluble, which may be mobile and contaminate ground water. Most soil conditions favors thermodynamically stable Cr (III) over Cr (VI). Cr (III) and Cr (VI), being the most stable forms in nature, are important in terms of environmental contamination. The most important sources of Cr (III) are fugitive emissions from road dust and industrial cooling towers. Cr (VI) compounds are still used in the manufacture of pigments, in metal-finishing and Cr-plating, in stainless steel production, hide tanning, as corrosion inhibitors and in wood preservation (Shtiza *et al.*, 2008). Of what is known due to its higher solubility and thus, bioavailability, Cr (VI) is more toxic at lower concentrations than Cr (III), which tend to form stable complexes in soils (Lopez-Luna *et al.*, 2009). The toxic effects of Cr (VI) on plant health includes several growth parameters, nutritive value (Sangwan *et al.*, 2014a), enzymes of nitrogen metabolism (Sangwan *et al.*, 2014b), antioxidant enzymes (Kumar and Joshi, 2008), carbohydrate content (Sangwan *et al.*, 2013) and others as reflected in Figure 13.1. In the following section of this chapter, these effects leading to deleterious plant health and productivity have been discussed in detail individually. Thereafter, a brief outline of control measures and ameliorative methods to protect plant from these negative effects of Cr (VI) are suggested.

Transport and Accumulation of Chromium

The inter-conversion of different forms of Cr in soil and hydroponic media is quite frequent and there are conflicting results regarding which form of Cr is taken up and accumulated by plants. Huffman *et al.* (1973) detected no differences in the uptake of either Cr (III) or Cr (VI). In studies with bean and wheat. Very few studies have attempted to elucidate the transport mechanisms of Cr in plants, but factors like oxidative Cr state or its concentration in substrate play important roles (Babula *et al.*, 2008). Cr absorption by plants is depends on experimental conditions, plant species, presence of other metals, minerals in soil, modified by soil pH, organic matter content and chelating agents (Han *et al.*, 2004). Cr accumulated preferentially in the roots and low transport was detected to the aerial parts (Kumar *et al.*, 2010). Cr in the roots makes them a metal excluder that restricts metal from being translocated to aerial parts of the plants or maintains lower levels over a wide range of its concentrations

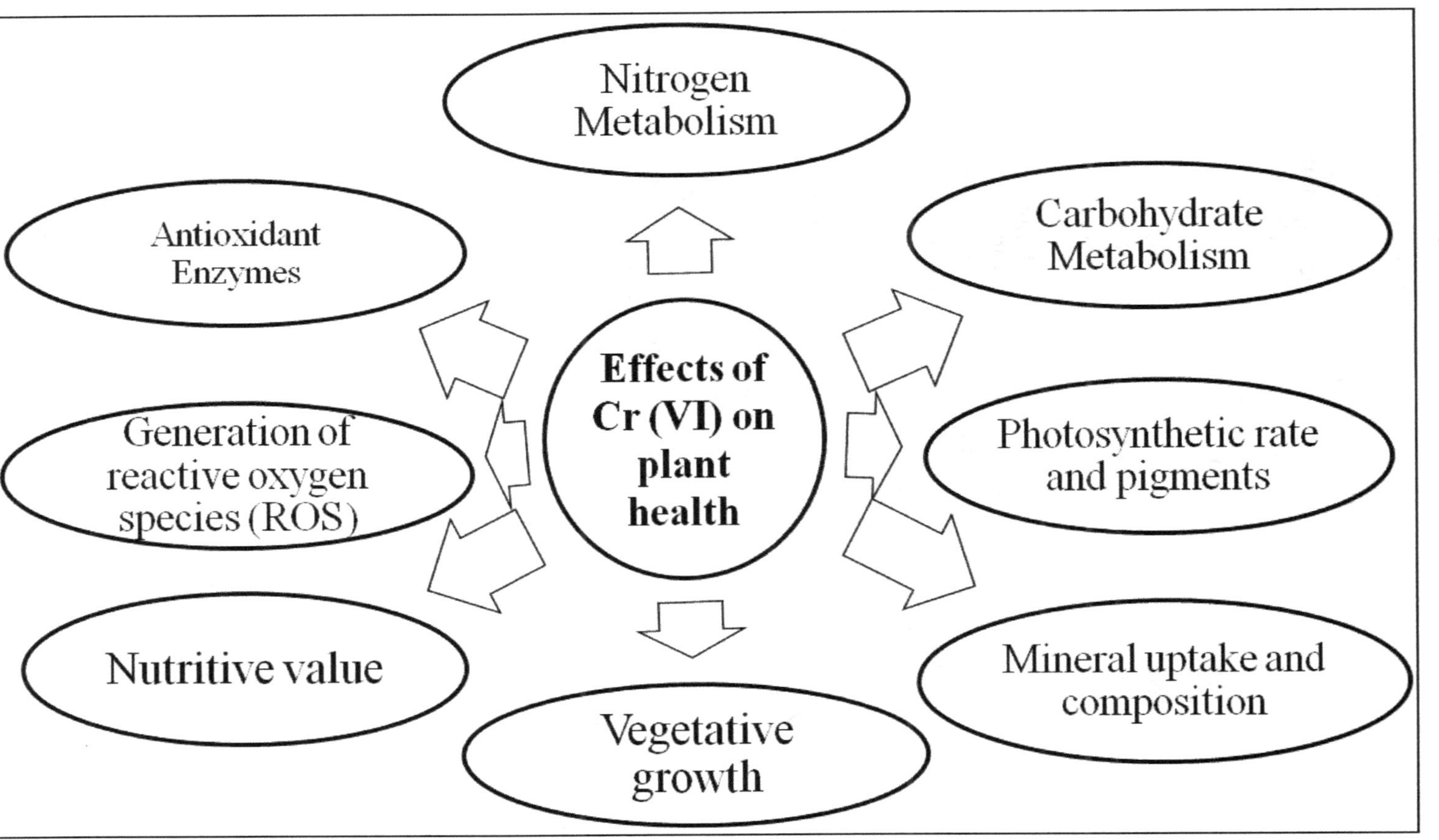

Figure 13.1: Different Plant Growth and Development Parameters Affected by Cr(VI) Phytotoxicity.

in the soil (Raskin *et al.*, 1997). The high metal concentration in the root tissues may be due to immobilization of metal by cell wall and extracellular carbohydrates, which may be an important defence strategy, adopted by plants. Further, efficient binding and sequestration to the vacuoles by GSH (reduced glutathione) and PCs (phytochelatins) contributes to this high accumulation (Mishra *et al.*, 2008). This might also be attributed to Cr compartmentalization in the root vacuoles, as a defence mechanism against its toxicity and a potential tolerance mechanism operating in plants under metal stress, or its retention in the cation exchange sites of the vessel walls of xylem parenchyma cells in roots (Hayat *et al.*, 2012; Diwan *et al.*, 2012). Reduction of Cr (VI) to Cr (III), which reduces its mobility from roots to stem might also a probable reason of comparatively lower accumulation of chromium in leaves than root (Rai *et al.*, 2004). With increase in the age of the plant, however, the ability to transfer the metal to aerial parts increased, leading to a higher Cr accumulation in the aboveground plant parts, especially in the stem (Diwan *et al.*, 2012). There are reports of metal accumulation in different parts of plants like *Brassica napus* (Najafian *et al.*, 2012), eucalyptus (Cipriani *et al.*, 2012) and barley (Ali *et al.*, 2012). Brassica spp. showed a higher ability to take up and accumulate Cr and other heavy metals than other plant species (Kumar *et al.*, 1995). The sulfate transport system may be responsible for the high Cr uptake of these plants as Cr uptake was reported to be increased by *Brassica* species, in presence of high content of sulpher (S) (Gardea-Torresdey *et al.*, 2004). Under such conditions, plants produced higher concentrations of amino acids and other organic compounds derived from S metabolism (Lee and Leustek, 1999). According to Moreno *et al.* (2005), direct cover over plants also enhance Cr uptake in *Brassica* plants and may be used as a part of Cr remediation. Arbuscular micorrhizal fungi are known to play important in mineral exchange in plant root cortical cells and alter Cr uptake potential of plants. Davies *et al.* (2011) observed an increased Cr uptake and survival of sunflower plants in presence of arbuscular micorrhizal fungi, *Glomus intraradices*.

Effect of Chromium on Plants

Effect on Growth and Development

Plant growth and development are essential processes of life and propagation of species. They mainly depend on external resources present in soil and/or air (Shanker *et al.*, 2005). Cr is one of the heavy metal that makes high toxicity in the plant growth and development. The effects of Cr on seed germination and plants growth have been studied thoroughly. Seed germination is the first physiological process affected by Cr. The germination ability of a seed in a medium containing Cr would be an expression of its level of tolerance to this metal (Peralta *et al.*, 2001). Reduced seed germination or early seedling development due to Cr toxicity have been reported by Sharma *et al.* (1995).

Plant yield is dependent on leaf growth, leaf area, and its number. As Cr affects most of the biochemical and physiological process in plants, productivity and yield are also affected. Grain weight and yield (kg ha^{-1}) of paddy (*Oriza sativa*) was decreased up to 80 per cent, when irrigated with water containing 200 mg L^{-1}

of Cr (Sundaramoorthy *et al.*, 2010). Cipriani *et al.* (2012) evaluated Cr (VI) toxicity in *Eucalyptus urophylla* and revealed that Cr at the rates of 0.08 mmol L^{-1} or higher significantly decreased ($P<0.01$) dry matter yield of shoot and root. Watermelon plants growing in the presence of Cr (VI) showed reduced number and size of leaves, which turned yellow, wilted, due to loss of turgor and hung down from petioles. With continued Cr supply the lamina of affected old leaves became necrotic, permanently wilted, dry, and shed down (Dube *et al.*, 2003). According to Purohit *et al.* (2003) and Gautam *et al.* (2012), Cr (VI) caused reduction in root length, shoot length and decreased number of branches in tomato and brinjal, respectively. Number of stomata and epidermal cells decreased with increasing level of Cr in both crops. Deformed and less differentiated stomata were frequently present in the treated plants. There was reduction in leaf area, width of primary vein, secondary vein and areoles. A gradual decrease was observed for various morphological parameters like fresh and dry weight of shoot and root and plant height of sunflower (*Helianthus annuus*) with increase in Cr levels (Fozia *et al.*, 2008). There are multiple reports of adverse effect of Cr toxicity on plant morphology *viz.* fresh and dry weight (Subrahmanyam, 2008), shoot length (Tiwari *et al.*, 2009), roots length (Chen *et al.*, 2001; Mallick *et al.*, 2010), wilting and plasmolysis in root cells (Vernay *et al.*, 2007). Cr provoked growth inhibition of roots has also been reported in species like wheat (Chen *et al.*, 2001), *Zea mays* (Mallick *et al.*, 2010) and mung bean (Samantaray *et al.*, 2001).

Although the growth was impaired in all plant parts under Cr toxicity, the effect was more pronounced in roots as compared to shoots (Shah *et al.*, 2008). General response of decreased root growth due to toxicity could be due to inhibition of root cell division/root elongation or inhibition of cell cycle in root. Under high concentration, the reduction in root growth could be due to the direct contact of seedling root with Cr causing a collapse and subsequent inability of the roots to absorb water from the medium (Shanker *et al.*, 2005). Aerial growth of plants has also been proven to be negatively affected by Cr in rice (Singh *et al.*, 2006), wheat, oat and sorghum (Lopez-Luna *et al.*, 2009). Justifications to these facts were proposed by Shanker *et al.* (2005) by stating that root growth inhibition, as well as its consequent and/or causal nutrient imbalance, could be behind low shoot development. In fact, Cr, due to its structural similarity with some essential elements, can affect mineral nutrition of plants in a complex way and there has been innumerous considerations regarding this issue, especially in crop species. Hewitt (1953) has demonstrated that very low concentrations of Cr (0.05-1.0 mg/L) promoted growth and increased nitrogen fixation and yield in *legumionosae,* while higher concentrations of this metal reduced the uptake of the essential elements Fe, K, Mg, Mn, P and Ca in *Salsola kali* (Gardea-Torresdey *et al.*, 2005), and K, Mg, P, Fe and Mn in roots of soybean (Turner and Rust, 1971). The justification of nutrient imbalance has been pointed to competitive binding to common carriers by Cr (VI), to inhibition of the activity of plasma membrane's H^+-ATPase and to reduced root growth and impaired penetration of the roots into the soil due to Cr toxicity (Shanker *et al.*, 2005).

Effect of Chromium on Plant Physiology and Metabolism

Photosynthesis

Like other metals, Cr affects photosynthesis, which ultimately translates in loss of productivity and death. The amount of photosynthetic pigments was noticed as a common parameter, which tends to decrease, when plants or algae are exposed to high doses Cr (Rodriguez *et al.*, 2011, Subrahmanyam, 2008, Vernay *et al.*, 2007, Shanker *et al.*, 2005). Also, accumulation of Cr by plants can reduce growth, induce chlorosis in young plants, decrease pigment content, mutate enzymatic functions and cause ultrastructural modifications of the chloroplast and cell membrane (Choudhury and Panda, 2004). Reduced chlorophyll content was observed in leaves of *Brassica napus* (Najafian *et al.*, 2012) and pea plants grown in sand under different concentrations of Cr (VI) (Tiwari *et al.*, 2009). A concentration dependent reduction in chlorophyll content over control was also observed in the leaves of *Eichhornia crassipes* (Mishra *et al.*, 2009) and *Brassica juncea* (Ghani, 2011; Hamid *et al.*, 2012). Inhibition of chlorophyll biosynthesis has been reported in different plant species exposed to Cr (VI). According to Vajpayee *et al.* (2001), impaired chlorophyll biosynthesis resulted in reduction in total chlorophyll content as Cr-induced primary phytotoxic effects. Cr reduced the foliar contents of total chlorophyll, chl *a*, and chl *b* in *Ocimum tenuiflorum*. The content of carotenoid, which served as accessory pigments for photosynthesis and also protected the plants from photo-oxidation, was also decreased in *O. tenuiflorum* under Cr stress. Similar results were also observed in lettuce and soybean (Rai *et al.*, 2004; Ganesh *et al.*, 2008). This was attributed to the Cr toxicity degradation of anino-levulinic acid dehydratase, which reduces the availability of prophobilinogen required for chlorophyll biosynthesis, thereby affecting the amino levulinic acid (ALA) utilization. It causes ALA build up and finally reduces the chlorophyll level (Diwan *et al.*, 2012). The impaired δ-aminolaevulinic acid dehydratase activity leading to reduced photosynthetic pigments has also been observed in Cr-treated *Nymphaea alba* (Vajpayee *et al.*, 2000). Besides these effects, both Cr (III) and Cr (VI) can alter chloroplast and membrane ultra structure in plants (Choudhury and Panda, 2004; Panda *et al.*, 2003). Such alterations in chloroplast have also been observed in case of plants like *Lemna minor*, Pistia species, *Taxithelium neplense* and are accompanied by changes in thylakoid arrangement. Moreover, at high concentration (1 mM), complete distortion of chloroplastic membrane was observed together with severe disarrangement of thylakoids indicating that Cr in hexavalent form can replace many Mg^{2+} ions from active enzymes and cause severe phytotoxic effects (Hamid *et al.*, 2012). Cr (VI) inhibits photosynthesis at several targeted site in algae and higher plants where sensitive sites includes PS II (photosystem II) oxygen-evolving complex (Susplugas *et al.*, 2000) and reducing side (Prasad *et al.*, 1991), electron transport between QA and QB electron carriers (Pan *et al.*, 2009), and it decrease the efficiency of energy transfer from light-harvesting complex (LHC) to PS II reaction center (Mallick and Mohn, 2003). Pan *et al.* (2009) demonstrated the inhibitory effect of Cr (VI) on PS II activity of *Synechocystis* sp. Cr (VI) reduced O_2 evolution, and the water-splitting system was a primary target site for Cr (VI) toxicity and it resulted in increased proportion of PSIIX centers and accumulation of PSIIb.

Nitrogen Metabolism

Nitrogen is one of the major constituents of biological systems. A significant decrease in N_2 fixation in response to chromium was observed in *Pisum sativum* (Bishnoi *et al.*, 1993). According to Kumar and Joshi (2008), higher doses of Cr adversely affected the key enzymes of nitrogen metabolism in forage sorghum (*Sorghum bicolor*). Total as well as specific enzyme activity of nitrate reductase, nitrite reductase, glutamine synthetase, glutamate dehydrogenase (GDH) and urease in various plant organs at different growth stages decreased with an increase in Cr (VI) levels. In general, the enzyme activity increased with advancement of growth to reach maximum at 50 per cent flowering stage (70 DAS) and thereafter decreased at grain filling stage (90 DAS). However, nitrite reductase (NR) activity in shoot increased continuously till grain filling stage. Similar result have been observed by Rai *et al.* (2004), where chromium toxicity resulted in the reduction of nitrate reductase activity through impaired substrate utilization in *Ocimum tenuiflorum*. Chromium-induced toxicity resulted in reduced activity of NR was also reported in *Nymphaea alba* (Vajpayee *et al.*, 2000), *Vallisneria spiralis* (Vajpayee *et al.*, 2001), sorghum (Kumar and Joshi, 2008) and Indian mustard (Diwan *et al.*, 2012). Study on wheat (Panda and Patra, 2000), *Eichhornia crassipes* (Mishra *et al.*, 2009) and Indian mustard (Diwan *et al.*, 2012) revealed that lower doses of chromium had stimulatory, while higher doses had inhibitory effect on NR activity. A significant increase in NR activity was observed corresponding to Cr concentration in *Brassica juncea* (Hamid *et al.*, 2012). Effects of Cr (III and VI) on spring barley and maize biomass yield and content of nitrogenous compounds was studied by Wyszkowski and Radziemska (2010). Spring barley yield was markedly decreased by Cr compounds, particularly Cr (VI). In contrast, maize yield was significantly increased by Cr (VI). Cr (VI) exerted a greater effect than the Cr (III) form, on nitrogen levels in spring barley. A significant increase in ammonia nitrogen content in maize was detected under Cr stress. In this regard, it was suggested that Cr may induce premature senescence of plants through increased level of NH_4^+ and its accumulation within cell in higher concentrations, led nutrient deficiency and chlorosis, altered intracellular pH and osmotic balance, inhibited ATP synthesis and eventually secondary growth (Gerendas *et al.*, 1997).

Effect on Nutritional Quality of Crops

Heavy metal soil pollution takes place, when the metal concentration of soil exceeds natural background level and causes ecological destruction and deterioration of the environment. Nutritive value in terms of various quality parameters like protein, hydrocyanic acid (HCN) content, structural carbohydrates and *in vitro* dry matter digestibility (IVDMD) of forage sorghum (*Sorghum bicolor*) was found to be influenced by Cr (VI) in soil (Kumar *et al.*, 2010). Fibre components and HCN content increased but protein content and IVDMD decreased with increasing Cr levels. The Cr ions also affect the mobilization of important seed storage materials such as starch, proteins and sugars and decreased the vigor index (Sharma *et al.*, 2011a). It is undoubtedly true that Cr stress retards the plant growth through its influence on the accumulation of lignin, which is considered as one of the constituents of the secondary cell wall. The accumulation of insoluble phenols, such as lignin, in the secondary cell wall was reported in plants exposed to heavy metals and could be associated with an increase in the activity of lignifying peroxidases (Schutzendubel *et al.*, 2001). Lignification decreases IVDMD (Luthra *et al.*, 1989), cell-wall plasticity and therefore, reduces the cell growth (Schutzendubel and Polle, 2002). Further, Cr (VI) treatment might have decreased the activity of cellulase enzyme due to distortion of its quaternary structure, resulting in an increase in NDF content which is positively correlated with ADF (Kumar *et al.*, 2010). Tiwari *et al.* (2009) also found that exposure to increasing concentration of Cr caused a decrease in the amount of non-reducing sugars while the reverse was observed for reducing sugars. Such decline in sugar content is correlated with suppression in pigment concentration.

Effect on Mineral Content

Cr, being structurally similar to other essential elements, may also affect plant mineral nutrition. Mallick *et al.* (2010) found that Cr exposure decreased Cu absorption in Zea mays roots. Uptake of both macronutrients (*e.g.*, N, P, K) and micronutrients decreased with increase of Cr (VI) in paddy (Sundaramoorthy *et al.*, 2010). Cr also, decreased uptake of the micronutrients Mn, Fe, Cu, and Zn (Liu *et al.*, 2008). The reduction in nitrogen compounds, K and P could be due to reduced root growth and impaired penetration of the roots in the soil due to Cr toxicity (Azmat and Khanum, 2005). Ali *et al.* (2012) also reported that chromium stress decreased K, Fe and Zn concentration and accumulation in barley plants. Increasing concentrations of chromium also caused marked decrease in K contents in *Brassica* (Ghani, 2011). Reduced uptake of these elements may be due to breakdown of membrane function and inhibition of H^+-ATPase, which hampers nutrient and water uptake (Najafian *et al.*, 2012), and plays an important role in the transport of multiple ions through plasma membrane (Shi and Zhu, 2008). It has been reported that inhibition in activity of H^+-ATPase membrane pump by Cr might be due to changes in structure and destruction of membrane by free radicals formation. Lower Fe concentration attributed to competition between Cr and Fe for transport binding or higher Fe efflux from the plants. Cr influence uptake of Fe by inhibiting Fe reductase activity and photosynthesis (Liu *et al.*, 2008). Moral *et al.* (1995) studied the effects of Cr (0, 50, and 100 mg Cr/L in a nutrient solution) on the nutrient content and morphology of tomato and reported the effects of Cr (III) on the distribution of N, P, K, Na, Ca, Mg and Cr in the plant and the growth and yield of a tomato plant. In general, the nutrient concentration in stem and branches was significantly affected by the Cr treatments. Cr accumulated preferentially in the roots and low transport was detected to the aerial parts. Growth was diminished due to presence of Cr in the nutrient solution. Total yield was not affected, however the number of fruits was diminished and the mean fresh weight of fruit increased with each increment of Cr in the nutrient solution. Cr was not detected in the edible part (fruit) of the plant.

Chromium Stress Generates Reactive Oxygen species

Reactive oxygen species (ROS) are toxic by-products, generated at low levels in normal plant cells in chloroplasts and mitochondria, and also by cytoplasmic, membrane-bound or exocellular enzymes involved in redox reactions. High amounts of ROS are generated under stressful conditions including pathogen attacks, wounding, herbivore feeding, UV light and heavy metals (Michalak, 2006). Cr toxicity led generation of ROSs can damage cellular membranes by inducing lipid peroxidation (Panda *et al.*, 2003). They also can damage DNA, proteins, lipids and chlorophyll. The most popular ROS which are generated by Cr are $^{\bullet}O_2^-$ (superoxide radical), H_2O_2 (hydrogen peroxide), and $^{\bullet}OH$ (hydroxyl radical) originating from one, two or three electron transfers to dioxygen (O_2) (Choudhary and Panda, 2003; Panda *et al.*, 2003; Panda and Choudhary, 2005).

Effect of Chromium Toxicity on Plant Antioxidant Enzymes

In general, induction and activation of antioxidative defense system is one of the major detoxification mechanisms in higher plants under stress conditions. In order to protect plant from the toxic action of various ROS generated during heavy metal stress, plants have evolved various enzymatic antioxidants, such as peroxidase (PRX), ascorbate peroxidase (APX), catalase (CAT), glutathione reductase (GR), superoxide dismutase (SOD) and guaiacol peroxidase (POX). This antioxidant potential usually varied with type of stress, plant species and growth conditions. *Brassica* plants are considered to have high tolerance to Cr toxicity due to high activity of antioxidant enzymes and consequent detoxification of ROS. Zaimoglu *et al.* (2011) conducted study to assess antioxidative responses (CAT, APX and GR) of *Brassica juncea* and *Brassica oleracea* plants considering their phytoremediation potential. They observed a similar trend by both species but the total enzyme activities were higher in *Brassica oleracea* than in *Brassica juncea,* in response to Cr (VI) treatments. Cr (VI) treatments significantly decreased the activities of CAT and after the significant increase, a sharp decrease was observed in the activity of APX and GR in both species. The coordinated increase in APX and GR activities in both *Brassica* species under Cr stress played a role as signals to protect the plants from Cr-induced stress. Cr mediated changes in antioxidant system was also indicated by decreased SOD and CAT activities in roots and shoots of *T. aestivum* plants grown in the presence of Cr (VI) while the activity of POX was reportedly decreased in roots and increased in shoots in this case (Dey *et al.*, 2009). An increase in activity of antioxidant enzymes including PRX, CAT, glucose-6-phosphate dehydrogenase and SOD was observed in mungbean plants exposed to different Cr concentrations (Samantary, 2002). Decreased CAT activity and increased SOD and POX was reported in *A. viridis* exposed to increased concentration of Cr (VI) (Liu *et al.*, 2008).

Amelioration of Chromium Toxicity

As evident from existing information, hexavalent Cr proved to be highly toxic for plant growth and development. Therefore, in order to overcome or protect plant from adverse effects of Cr or other heavy metals, methods are needed to ameliorate Cr toxicity and also to decrease Cr content in crops. These include production of metal tolerant genotypes, use of improved irrigation and proper management of disposal of waste containing Cr compounds. Further, use of plant hormone and chemical messengers produced in one cell or tissue may also regulate plant growth and development (Gangwar *et al.*, 2011). There are several evidences in support of Cr ameliorative potential of various strategies. Among several possibilities, few strategies which may be promising in alleviating toxic effects of Cr (VI) from plants includes phytoremediation using Cr tolerance and hyper-accumulating plants species, bioremediation using microbial inoculants tolerant to high Cr concentrations, use of metal chelators, application of plant growth regulators and other effectors like silicon and sulphur during plant growth conditions. Some of these are discussed in this section of chapter and these can be used alone or in combination to protect plants from such toxicity.

One way of protecting plants from Cr phytotoxicity may be application of Iron (Fe), which is a cofactor for approximately 140 enzymes that catalyze unique biochemical reactions (Brittenham, 1994). According to Sinha *et al.* (2005), supplemental Fe on roots act as: a) a shield that protects roots by coprecipitation of other heavy metals, b) a nutrient reservoir, and c) a reservoir for active ferrous (Fe^{2+})-Fe inside cells that could compete with heavy metals for metabolically sensitive sites inside plants. Adverse effect of Cr was found to be nullified by the supply of suitable amount of Fe and Zn in moong, gram and pea plants possibly due to importance of these two essential nutrients in growth and metabolism of plants (Vazquez *et al.*, 1987). Yadav *et al.* (2007) observed that the fresh weight, shoot length and chlorophyll in leaves of 2 mM Cr^{+} 0.2 mM Fe treated maize plants, was higher than in 2 mM Cr treated plants, both after 45 and 90 days. To overcome the toxic effects of chromium in *Raphanus sativus*, Nath *et al.* (2009) used iron in the recovery treatments and reported a significant recovery in most of the studied plant growth parameters. Hasegawa *et al.* (2012) also observed that foliar spray of Fe or increased Fe supply to roots also ameliorated the chlorosis in rice plants under exposure to high Ni concentrations. Sinha *et al.* (2005) conducted an experiment to determine, whether the ill effects of excess Cr can be ameliorated by iron application in spinach by withdrawal of Cr by iron application through different modes and recovery from ill effects of Cr was observed as reflected by increase in biomass, concentrations of chlorophylls, total and ferrous iron, Hill reaction activity, relative water content and recovery in activity of catalase, peroxidase, ribonuclease and starch phosphorylase along with lowered Cr concentration. In other approach, Hayat *et al.* (2010) reported that pre-soaking of pea seeds in salicylic acid (SA) had a beneficial effect on growth and photo-synthesis with decreased oxidative injuries caused by heavy metal stress. Sharma *et al.* (2011b) evaluated effect of a brassinosteroid, 28-homobrassinolide (28-HBL), pre-treatment on the seeds of *Raphanus sativus* raised under various concentrations of Cr (VI) and found that the toxic effects of Cr in terms of reduced growth, lowered contents of chlorophyll (Chl), protein, proline; increased malondialdehyde (MDA) content and elevated metal uptake were ameliorated by applications of 28-HBL. Gangwar *et al.* (2011) studied of exogenous gibberellic acid (GA) application on growth, protein and nitrogen contents, ammonium (NH_4^+) content, enzymes of nitrogen assimilation and antioxidant system in pea seedlings under Cr (VI) phytotoxicity. They showed that exogenous application of GA (10 μM) together with Cr was able to alleviate Cr phytotoxicity appreciably. Recently, role of reduced GSH in maintaining cellular oxidation balance and protection against drought, salinity and heavy metals has also been proved (Zeng *et al.*, 2012). In one such study, a hydroponic experiment was conducted to determine the possible effect of exogenous glutathione (GSH) in alleviating Cr stress through examining plant growth, chlorophyll contents, antioxidant enzyme activity, and lipid peroxidation in rice seedlings exposed to Cr toxicity (Zeng *et al.*, 2012). In addition to above, chelating agents such as low molecular weight organic acids and synthetic chelators are commonly applied for chemically assisted phytoextraction of metals from soils (do Nascimento *et al.*, 2006). Such substances are capable of forming complexes with metal ions, thereby increasing the bioavailability of heavy metals in soils. Synthetic chelators like EDTA form chemically and microbiologically stable complexes with heavy metals which otherwise contaminate ground water (Satroutdinov *et al.*, 2000). In a supporting evidence, Mohanty and Patra (2011) observed that total chlorophyll content in the rice (*Oryza sativa* L.) seedlings treated with Cr (VI)-EDTA (10μM) solution was more as compared to the untreated. In a study using rice plants, Zeng *et al.* (2011) investigated the alleviatory effect of Si on Cr toxicity using a hydroponic experiment with two Cr levels. The results showed that toxic effects of Cr treatments on antioxidant enzymes (superoxide dismutase and ascorbate peroxidase in leaves; catalase and ascorbate peroxidase in roots) and other parameters were greatly alleviated due to Si addition to the culture solution. It was concluded that Si alleviated Cr toxicity mainly through inhibiting the uptake and translocation of Cr and enhancing the capacity of defence against oxidative stress induced by Cr toxicity.

Future Prospects

A great deal of studies is undertaken on Cr toxicity under different environmental conditions and on different plant species. The response of plants of different species had variation and lethality of Cr concentration also depends on plant types and experimental conditions. Although, this information have great implication in selecting plants depending on their resistance and suitability in these kind of toxic environments, the exact mechanism of Cr action on different metabolic pathways in specific manner is still needs to be explored. The work is also need to be done on how Cr 'cross talk' exactly with other mineral elements and other metabolites. Further, in addition, amelioration of heavy metal toxicity has emerged as a very interesting area of research in recent years and act as powerful tool in enhancing the growth productivity and combating ill effects generated by various heavy metals stress. Different strategies are being adapted by different workers and it provided a base for further strengthening of better strategies. These strategies are not generalized and have their own limitations due to variation in results depending on experimental conditions of plant growth, different metals and concentrations, plant species and soil types. A comprehensive review of these approaches will lead to better understanding and designing of an approach for management of good plant health even under influence of heavy metal toxicity. The future applications of this study lies in enhancing efforts for better understanding and removal of toxic effects of Cr toxicity in plants as mentioned above.

References

Ali, S., Cai, S., Zeng, F., Qiu, B. and Zhang, G., 2012. Effect of salinity and hexavalent chromium stresses on uptake and accumulation of mineral elements in barley genotypes differing in salt tolerance. *J. Plant Nutr.*, 35: 827-839.

Azmat, R. and Khanum, R., 2005. Effect of chromium metal on the uptakes of mineral atoms in seedlings of bean plant *Vigna radiata* (L.) wilckzek. *Pak. J. Biol. Sci.,* 8: 281-283.

Babula, P., Adam, V., Opatrilova, R., Zehnalek, J., Havel, L. and Kizek, R., 2008. Uncommon heavy metals, metalloids and their plant toxicity: A review. *Environ. Chem. Lett.,* 6: 189-213.

Bishnoi, N., Dua, A., Gupta, V. and Sawhney, S., 1993. Effect of chromium on seed germination, seedling growth and yield of peas. *Agri. Ecosys. Environ.,* 47: 47-57.

Brittenham, G., 1994. New advances in iron metabolism, iron deficiency and iron overload. *Curr. Opin. Hematol.,* 1: 101-105.

Chen, N., Kanazawa, S. and Horiguchi, T., 2001. Effect of chromium on some enzyme activities in the wheat rhizosphere. *Soil Microorganisms.,* 55: 3-10.

Choudhury, S. and Panda, S. K., 2004. Induction of oxidative stress and ultrastructural changes in moss *Taxithelium nepalense* (Schwaegr.) broth under lead (Pb) and arsenic (As) phytotoxicity. *Curr. Sci.,* 87: 342-348.

Cipriani, H. N., Bastos, A. R. R., de Carvalho, J. G., da Costa, A. L. and Oliveira, N.P., 2012. Chromium toxicity in hybrid eucalyptus (*Eucalyptus urophylla* st blake x grandis w. Hill ex. Maiden) cuttings. *J. Nutr.,* 35: 1618-1638.

Davies Jr, F.T., Puryear, J.D., Newton, R.J., Egilla, J.N. and Saraiva Grossi, J.A., 2011. Mycorrhizal fungi enhance accumulation and tolerance of chromium in sunflower (*Helianthus annuus*). *J. Plant Physiol.,* 158: 777-786.

Dey, S.K., Jena, P.P. and Kundu, S., 2009. Antioxidative efficiency of Triticum aestivum L. exposed to chromium stress. *J. Environ. Biol.,* 30: 539-544

Diwan, H., Ahmad, A. and Iqbal, M., 2012. Chromium-induced alterations in photosynthesis and associated attributes in Indian mustard. *J. Environ. Biol.,* 33: 239-243.

do Nascimento, C.W.A., Amarasiriwardena, A. and Xing, B., 2006. Comparison of natural organic acids and synthetic chelates at enhancing phytoextraction of metals from a multi-metal contaminated soil. *Environ. Poll.,* 140: 114–123.

Dube, B., Tewari, K., Chatterjee, J. and Chatterjee, C., 2003. Excess chromium alters uptake and translocation of certain nutrients in citrullus. *Chemosphere.,* 53: 1147-1153.

Fozia, A., Muhammad, A. Z., Muhammad, A. and Zafar, M.K., 2008. Effect of chromium on growth attributes in sunflower (*Helianthus annuus* L.). *J. Environ. Sci.,* 20: 1475-1480.

Ganesh, K.S., Baskaran, L., Rajasekaran, S., Sumathi, K., Chidambaram, A.L.A. and Sundaramoorthy, P., 2008. Chromium stress induced alterations in biochemical and enzyme metabolism in aquatic and terrestrial plants. *Coll. Surf. B: Biointer.,* 63: 159-163.

Gangwar, S., Singh, V.P., Garg, S.K., Prasad, S.M. and Maurya, J.N., 2011. Kinetin supplementation modifies chromium (VI) induced alterations in growth and ammonium assimilation in pea seedlings. *Biol. Trace Elem. Res.*, 144: 1327–1343.

Gardea-Torresdey, J. L., Peralta-Videa, J. R., De La Rosa, G. and Parsons, J., 2005. Phytoremediation of heavy metals and study of the metal coordination by X-ray absorption spectroscopy. *Coordination Chem. Rev.*, 249: 1797-1810.

Gardea-Torresdey, J.L., Peralta-Videa, J.R., Montes, M., de la Rosa, G. and Corral-Diaz, B., 2004. Bioaccumulation of cadmium, chromium, and copper by Convolvulus arvensis l.: impact on plant growth and uptake of nutritional elements. *Biores. Technol.*, 92: 229–235.

Gautam, M., Singh, A. K. and Johri, R. M., 2012. Root morphology of tomato and brinjal as influenced by Cr toxicity. *Indian J. Hort.*, 69: 205-208.

Gerendas, J., Zhu, Z., Bendixen, R., Ratcliffe, R. G. and Sattelmacher, B., 1997. Physiological and biochemical processes related to ammonium toxicity in higher plants. *J. Plant Nutr. Soil Sci.*, 160: 239–251.

Ghani, A., 2011. Effect of chromium toxicity on growth, chlorophyll and some mineral nutrients of *Brassica juncea* L. *Sci. Technol.*, 4: 197-202.

Hamid, R., Parray, J. A., Kamili, A. N. and Mahmooduzzafar., 2012. Chromium stress in *Brassica juncea* L. cv. 'Pusa Jai Kissan' under hydroponic culture. *Afr. J. Biotechnol.*, 11: 15658- 15663.

Han, F. X., Sridhar, B., Monts, D. L. and Su, Y., 2004. Phytoavailability and toxicity of trivalent and hexavalent chromium to *Brassica juncea*. *New Phytologist.*, 162: 489-499.

Hasegawa, H., Rahman, M. M., Kadohashi, K., Rahman, M. A., Takasugi, Y., Tate, Y. and Maki, T., 2012. Significance of the concentration of chelating ligands on Fe^{3+}-solubility, bioavailability and uptake in rice plant. *Plant Physiol. Biochem.*, 58: 205-211.

Hayat, Q., Hayat, S., Irfan, M. and Ahmad, A., 2010. Effect of exogenous salicylic acid under changing environment: A review. *Environ. Exp. Bot.*, 68: 14-25.

Hayat, S., Khalique, G., Irfan, M., Wani, A. S., Tripathi, B. N. and Ahmad, A. 2012. Physiological changes induced by chromium stress in plants: An overview. *Protoplasma.*, 249: 599-611.

Hewitt, E.J., 1953. Metal interrelationships in plant nutrition. I. Effects of some metal toxicities on sugar beet, tomato, oat, potato and marrow stem kale grown in sand culture. *J. Exp. Bot.*, 4: 59–64.

Huffman Jr., E.W.D. and Allaway, W.H., 1973. Chromium in plants: distribution in tissues, organelles and extracts and availability of bean leaf Cr to animals. *J. Agric. Food Chem.*, 21: 982-986.

Kumar, P., Dushenkov, V., Motto, H. and Raskin, I., 1995. Phytoextraction: the use of plants to remove heavy metals from soils. *Environ. Sci. Technol.*, 29: 1232-1238.

Kumar, S. and Joshi, U.N., 2008. Nitrogen metabolism as affected by hexavalent chromium in sorghum (*Sorghum bicolor* L.). *Environ. Exp. Bot.*, 64:135-144.

Kumar, S., Joshi, U. and Sangwan, S., 2010. Chromium (VI) influenced nutritive value of forage sorghum (*Sorghum bicolor* L.). *Anim. Feed Sci. Technol.*, 160: 121-127.

Lee, M. and Leustek, T., 1999. Identification of the gene encoding homoserine kinase from Arabidopsis thaliana and characterization of the recombinant enzyme derived from the gene. *Arch. Biochem. Biophys.*, 372: 135–142.

Liu, D., Zou, J., Wang, M. and Jiang, W., 2008. Hexavalent chromium uptake and its effects on mineral uptake, antioxidant defence system and photosynthesis in *Amaranthus viridis*. *Biores. Technol.*, 99: 2628-2636.

Lopez-Luna, J., González-Chávez, M., Esparza-García, F. and Rodríguez-Vázquez, R., 2009. Toxicity assessment of soil amended with tannery sludge, trivalent chromium and hexavalent chromium, using wheat, oat and sorghum plants. *J. Hazard. Mater.*, 163: 829-834.

Luthra, Y. P., Joshi, U. N., Gandhi, S. K. and Arora, S. K., 1989. Phenolics, carbohydrates and mineral elements in guar leaves in relation to bacterial blight. *Ann. Biol.*, 5: 1–7.

Mallick, N. and Mohn, F.H., 2003. Use of chlorophyll fluorescence in metal-stress research: a case study with the green microalga Scenedesmus. *Ecotoxicol. Environ. Saf.*, 55:64–69.

Mallick, S., Sinam, G., Kumar Mishra, R. and Sinha, S., 2010. Interactive effects of Cr and Fe treatments on plants growth, nutrition and oxidative status in *Zea mays* L. *Ecotoxicol. Environ. Saf.*, 73: 987-995.

Michalak, A., 2006. Phenolic Compounds and Their Antioxidant Activity in Plants Growing under Heavy Metal Stress. *Polish J. Environm. Stud.*, 15: 523-530.

Mishra, K., Gupta, K. and Rai, U. N., 2009. Bioconcentration and phytotoxicity of chromium in *Eichhornia crassipes*. *J. Environ. Biol.*, 30: 521-526.

Mishra, V. K., Upadhyay, A. R., Pandey, S. K. and Tripathi, B., 2008. Concentrations of heavy metals and aquatic macrophytes of Govind Ballabh Pant Sagar an anthropogenic lake affected by coal mining effluent. *Environ. Monitoring Assessment.*, 141: 49-58.

Mohanty, M. and Patra, H.K., 2011. Effect of Cr^{+6} and chelating agents on growth, pigment status, proline content and chromium bioavailability in rice seedlings. *Internat. J. Biotech. Appl.*, 3: 91-96.

Moral, R., Pedreno, J. N., Gomez, I. and Mataix, J., 1995. Effects of chromium on the nutrient element content and morphology of tomato. *J. Plant Nutr.*, 18: 815-822.

Moreno, D.A., Víllora, G., Soriano, M.T., Castilla, N. and Romero, L., 2005. Sulfur, chromium, and selenium accumulated in Chinese cabbage under direct covers. *J. Environ. Managem.*, 74: 89-96.

Najafian, M., Kafilzadeh, F., Azad, H. N. and Tahery, Y., 2012. Toxicity of chromium (Cr) on growth, ions and some biochemical parameters of *Brassica napus* L. *World Appl. Sci. J.*, 16: 1104-1109.

Nath, K., Singh, D., Verma, A. and Sharma, Y., 2009. Amelioration of tannery effluent toxicity in radish (*Raphanus sativus*) based on nutrient application. *Res. Environ. Life Sci.*, 2: 41-48.

Pan, X., Chen, X., Zhang, D., Wang, J., Deng, C., Mu, G. and Zhu, H., 2009. Effect of chromium (vi) on photosystem ii activity and heterogeneity of synechocystis sp. (cyanophyta): studied with in vivo chlorophyll fluorescence tests. *J. Phycol.*, 45: 386-394.

Panda, S. and Patra, H., 2000. Does chromium (III) produce oxidative damage in excised wheat leaves? *J. Plant Biol,*. 27: 105-110.

Panda, S. K. and Choudhury, S., 2005. Chromium stress in plants. *Brazilian J. Plant Physiol.*, 17: 95-102.

Panda, S., Chaudhury, I. and Khan, M., 2003. Heavy metals induce lipid peroxidation and affect antioxidants in wheat leaves. *Biol. Plant.*, 46: 289-294

Peralta, J., Gardea-Torresdey, J., Tiemann, K., Gomez, E., Arteaga, S., Rascon, E. and Parsons, J., 2001. Uptake and effects of five heavy metals on seed germination and plant growth in alfalfa (*Medicago sativa* L.). *Bull. Environ. Contam. Toxicol.*, 66: 727-734.

Prasad, S.M., Singh, J.B., Rai, L.C. and Kumar, H.D., 1991. Metalinduced inhibition of photosynthetic electron transport chain of the cyanobacterium *Nostoc muscorum. FEMS Microbiol. Lett.*, 82:95–100.

Purohit, S., Varghese, T. and Kumari, M., 2003. Effect of chromium on morphological features of tomato and brinjal. *Indian J. Plant Physiol.*, 8: 17-22.

Rai, V., Vajpayee, P., Singh, S. N. and Mehrotra, S., 2004. Effect of chromium accumulation on photosynthetic pigments, oxidative stress defense system, nitrate reduction, proline level and eugenol content of *Ocimum tenuiflorum* L. *Plant Sci.*, 167: 1159-1169.

Raskin, I., Smith, R. D. and Salt, D. E., 1997. Phytoremediation of metals: Using plants to remove pollutants from the environment. *Curr. Opin. Biotechnol.*, 8: 221-226.

Rodriguez, E., Santos, C., Lucas, E. and Pereira, M., 2011. Evalution of chromium(VI) toxicity to *Chlorella vulgaris* Beijerinck cultures. *Fresenius Environ. Bull.*, 20: 334-339.

Samantaray, S., Rout, G. R. and Das, P., 2001. Induction, selection and characterization of Cr and Ni-tolerant cell lines of *Echinochloa colona* (L.) Link *in vitro. J. Plant Physiol.*, 158: 1281-1290.

Samantary, S., 2002. Biochemical responses of Cr-tolerant and Cr-sensitive mung bean cultivars grown on varying levels of chromium. *Chemosphere* 47: 1065-1072.

Sangwan, P., Kumar, V. and Joshi U.N., 2014a. Chromium (VI) affected nutritive value of forage clusterbean (*Cyamopsis tetragonoloba* L.). *Internat. J. Agri. Environ. Biotech.*, 6: 217-223.

Sangwan, P., Kumar, V. and Joshi, U.N., 2014b. Effect of Chromium (VI) toxicity on enzymes of nitrogen metabolism in clusterbean (*Cyamopsis tetragonoloba* L.). *Enzyme Res.*, 2014:1-8, 784036.

Sangwan, P., Kumar, V., Khatri, R.S., Joshi, U.N., 2013. Chromium (VI) induced biochemical changes and gum content in cluster bean (*Cyamopsis tetragonoloba* L.) at different developmental stages. *J. Botany.*, 2013:1-8, 578627.

Satroutdinov, A.D., Deedyukhina, *E.G.*, Chistyakova, T.I., Witschel, M., Minkevich, I.G., Eroshin, V.K. and Egli, T., 2000. Degradation of metal- EDTA complexes by resting cells of the bacterial strain DSM 9103. *Environ. Sci. Technol.*, 34: 1715–1720.

Schutzendubel, A. and Polle, A., 2002. Plant responses to abiotic stresses: Heavy metalinduced oxidative stress and protection by mycorrhization. *J. Exp. Bot.*, 53: 1351-1365.

Schutzendubel, A., Schwanz, P., Teichmann, T., Gross, K., Langenfeld-Heyser, R., Godbold, D. L. and Polle, A., 2001. Cadmium-induced changes in antioxidative systems, hydrogen peroxide content, and differentiation in scots pine roots. *Plant Physiol.*, 127: 887-898.

Shah, F., Ahmad, N., Masood, K. and Zahid, D., 2008. The influence of cadmium and chromium on the biomass production of shisham (*Dalbergia sissoo* Roxb.) seedlings. *Pak. J. Bot.*, 40: 1341-1348.

Shanker, A. K., Cervantes, C., Loza-Tavera, H. and Avudainayagam, S., 2005. Chromium toxicity in plants. *Environ. Int.*, 31: 739-753.

Sharma, D., Chatterjee, C. and Sharma, C., 1995. Chromium accumulation and its effects on wheat (*Triticum aestivum* L. cv. HD 2204) metabolism. *Plant Sci.*, 111: 145-151.

Sharma, I., Pati, P.K. and Bhardwaj, R., 2011b. Effect of 28-homobrassinolide on antioxidant defence system in *Raphanus sativus* L. under chromium toxicity. *Ecotoxicol.*, 20: 862-74.

Sharma, P., Mohan, M. and Goyal, S. P., 2011a. Effect of chromium hexavalent on seed germination, early seedling growth and biochemical constituents of pea. *Int. J. Chemical Anal. Sci.*, 2: 25-26.

Shi, Q. and Zhu, Z., 2008. Effects of exogenous salicylic acid on manganese toxicity, element contents and antioxidative system in cucumber. *Environ. Exp. Bot.*, 63: 317-326.

Shtiza, A., Swennen, R. and Tashko, A., 2008. Chromium speciation and existing natural attenuation conditions in lagoonal and pond sediments in the former chemical plant of Porto-Romano (Albania). *Environ. Geol.*, 53: 1107-1128.

Singh, A. K., Misra, P. and Tandon, P., 2006. Phytotoxicity of chromium in paddy (*Oryza sativa* L.) plants. *J. Environ. Biol.*, 27: 283-285.

Sinha, P., Dube, B. and Chatterjee, C., 2005. Amelioration of chromium phytotoxicity in spinach by withdrawal of chromium or iron application through different modes. *Plant Sci.*, 169: 641-646.

Subrahmanyam, D., 2008. Effects of chromium toxicity on leaf photosynthetic characteristics and oxidative changes in wheat (*Triticum aestivum* L.). *Photosynthetica.*, 46: 339-345.

Sundaramoorthy, P., Chidambaram, A., Ganesh, K. S., Unnikannan, P. and Baskaran, L., 2010. Chromium stress in paddy: (i) Nutrient status of paddy under chromium stress; (ii) Phytoremediation of chromium by aquatic and terrestrial weeds. *Comptes Rendus Biologies.*, 333: 597-607.

Susplugas, S., Srivastava, A. and Strasser, R.J., 2000. Changes in the photosynthetic activities during several stages of vegetative growth of *Spirodela polyrhiza*: effect of chromate. *J. Plant Physiol.*, 157:503–12.

Tiwari, K., Dwivedi, S., Singh, N., Rai, U. and Tripathi, R., 2009. Chromium (VI) induced phytotoxicity and oxidative stress in pea (*Pisum sativum* L.): Biochemical changes and translocation of essential nutrients. *J. Environ. Biol.*, 30: 389-394.

Turner, M. and Rust, R., 1971. Effects of chromium on growth and mineral nutrition of soybeans. *Soil Sci. Society Am. J.*, 35: 755-758.

Vajpayee, P., Rai, U., Ali, M., Tripathi, R., Yadav, V., Sinha, S. and Singh, S., 2001. Chromium-induced physiologic changes in *Vallisneria spiralis* L. and its role in phytoremediation of tannery effluent. *Bull. Environ. Contam. Toxicol.*, 67: 246-256.

Vajpayee, P., Tripathi, R., Rai, U., Ali, M. and Singh, S., 2000. Chromium (VI) accumulation reduces chlorophyll biosynthesis, nitrate reductase activity and protein content in *Nymphaea alba* L. *Chemosphere.*, 41: 1075-1082.

Vazquez, M., Poschenrieder, C. and Barcelo, J., 1987. Chromium VI induced structural and ultrastructural changes in bush bean plants (*Phaseolus vulgaris* L.). *Ann. Bot.*, 59: 427-438.

Vernay, P., Gauthier-Moussard, C. and Hitmi, A., 2007. Interaction of bioaccumulation of heavy metal chromium with water relation, mineral nutition and photosynthesis in developed leaves of *Lolium perenne* L. *Chemosphere.*, 68: 1563-1575.

Wyszkowski, M. and Radziemska, M., 2010. Effects of chromium (III and VI) on spring barley and maize biomass yield and content of nitrogenous compounds. *J. Toxicol. Environ. Health, Part A.*, 73: 1274-1282.

Yadav, S. S., Verma, A., Sani, S. and Sharma, Y., 2007. Likely amelioration of chromium toxicity by Fe and Zn in maize (*Zea mays*). *J. Ecophysiol. Occupational Health.*, 7: 111-117.

Zaimoglu, Z., Koksal, N., Basci, N., Kesici, M., Gulen, H. and Budak, F., 2011. Antioxidative enzyme activities in *Brassica juncea* L. and *Brassica oleracea* L. plants under chromium stress. *J. Food Agric. Environ.*, 9: 676-679.

Zeng, F., Qiu, B., Wu, X., Niu, S., Wu, F. and Zhang, G., 2012. Glutathione-mediated alleviation of chromium toxicity in rice plants. *Biol. Trace Elem. Res.*, 148:255–263.

Zeng, F.R., Zhao, F.S., Qiu, B.Y., Ouyang, Y.N., Wu, F.B. and Zhang, G.P., 2011. Alleviation of chromium toxicity by silicon addition in rice plants. *Agri. Sci. China*, 10: 1188-1196.

2016, Recent Advances in Plant Stress Physiology *Pages* **283–296**
Editors: ***Praduman Yadav, Sunil Kumar and Veena Jain***
Published by: **DAYA PUBLISHING HOUSE, NEW DELHI**

Chapter 14

Waterlogging Stress: An Overview

Ruchi Bansal* and Ram Swaroop Jat

ICAR – Directorate of Medicinal and Aromatic Plants Research
Anand – 387 310, Gujarat, India

ABSTRACT

Waterlogging constitutes a major abiotic constraint on growth, species distribution and agricultural productivity. Waterlogging is frequently accompanied with reduction in O_2 concentration, which limits ATP production and plants shifts its metabolism to anaerobic mode. Ethylene mediated responses such as aerenchyma and adventitious root formation enable plants to survive under waterlogging. Availability of soluble sugars and antioxidant properties of plants also play an important role in stress tolerance. In response to waterlogging, plants induce a wide range of morphological, physiological, biochemical and molecular responses that lead to sustainability under stress condition. Genomic and proteomic approaches have generated a vast knowledge of basic adaptive mechanisms. Understanding of these mechanisms is crucial for developing an effective strategy to cope with the stress conditions.

***Keywords**: Aerenchyma, Flooding, Hypoxia, Waterlogging.*

Introduction

Waterlogging is a major abiotic stress affecting crop productivity worldwide due to its detrimental effects. It is a common problem in the areas, where land is not properly leveled. Waterlogged conditions may also result due to faulty irrigation, poor drainage and unpredicted rainfall. Intergovernmental panel on climate change reported that man induced world climate change will increase the frequency of precipitations of higher magnitude as well as cyclone activity. As a consequence, occurrence of waterlogging events is expected to be higher in future. With the

* *Corresponding author.* E-mail: ruchibansal06@gmail.com

pressure of increasing population, marginal lands are being incorporated into cultivation to cope with the rising food demand. These issues lead to the necessity to increase the productivity in the lands frequently exposed to waterlogged conditions. Therefore, understanding of waterlogging tolerant mechanisms is crucial.

Generally the term waterlogging/flooding is used interchangeably. But, waterlogging is the condition, in which, soil pores are fully saturated with water. Under waterlogged condition, root system of plant is under anaerobic conditions and the shoot system is under normal atmospheric condition, while, during flooding, partial/complete submergence of plants take place (Figure 14.1). Waterlogging has also been defined as prolonged soil saturation with water at least 20 per cent higher than the field capacity (Agarwal *et al.*, 2005). Response of a plant to waterlogged condition depends on the developmental stage (*e.g.* seedling *vs.* adult plant), plant growth habit (*e.g.* creeping plant growth *vs.* erect plant growth), and the other traits influencing the plant height. Another important factor affecting the response is the stress period to which the plant is exposed and it has been recognized as a major factor in determining plant's survival following oxygen deprivation. A single species of a same developmental stage that is capable of surviving a short stress period may perish if exposed to longer duration.

Detrimental Effects

Waterlogging stress leads to hypoxia in soil, *i.e.* reduced oxygen concentration, but if it prolongs, it gradually leads to anoxia (absence of oxygen). O_2 concentration is 20.95 per cent in air, but in plants, O_2 concentration varies from tissue to tissue. It ranges from 1-7 per cent in well aerated roots, stem, tubers and developing seeds (Bailey-Seres and Voesenek, 2008). According to Drew (1997), tissues or cells are under hypoxic state, when O_2 partial pressure limits the production of ATP by mitochondria. Waterlogging leads to a severe decrease in the oxygen diffusion rate into the soil because of the 10^4 times lower diffusion of gases into water with respect to air. Soon after, the soil is flooded, the respiration of roots and micro-organisms deplete the remnant oxygen and the environment becomes hypoxic and later anoxic (Bailey-Serresand Voesenek, 2008). Depletion of oxygen is more rapid as the temperature increases. So, the first constraint for plant growth under waterlogging/flooding is the immediate lack of oxygen necessary to sustain aerobic respiration of submerged tissues (Voesenek *et al.*, 2004).

One of the most important consequences of waterlogging stress is energy limitation and altered redox state of the cell. Due to unavailability of oxygen, the intermediate electron carriers in electron transport chain become reduced, affecting redox active metabolic reactions. Therefore, to sustain, the redox characteristics ($NADH/NAD^+$ ratio) of the cells should remain unaffected under waterlogging. Saturated electron transport components, the highly reduced intracellular environment and low energy supply are the factors favorable for reactive oxygen species (ROS) generation. The consequences of ROS formation depend on the intensity of the stress as well as on the physicochemical conditions in the cell (*i.e.*, antioxidant status, redox state and pH). Elevated cytoplasmic Ca^{2+} concentration, cytoplasmic acidification and change in redox state of the cell were observed under

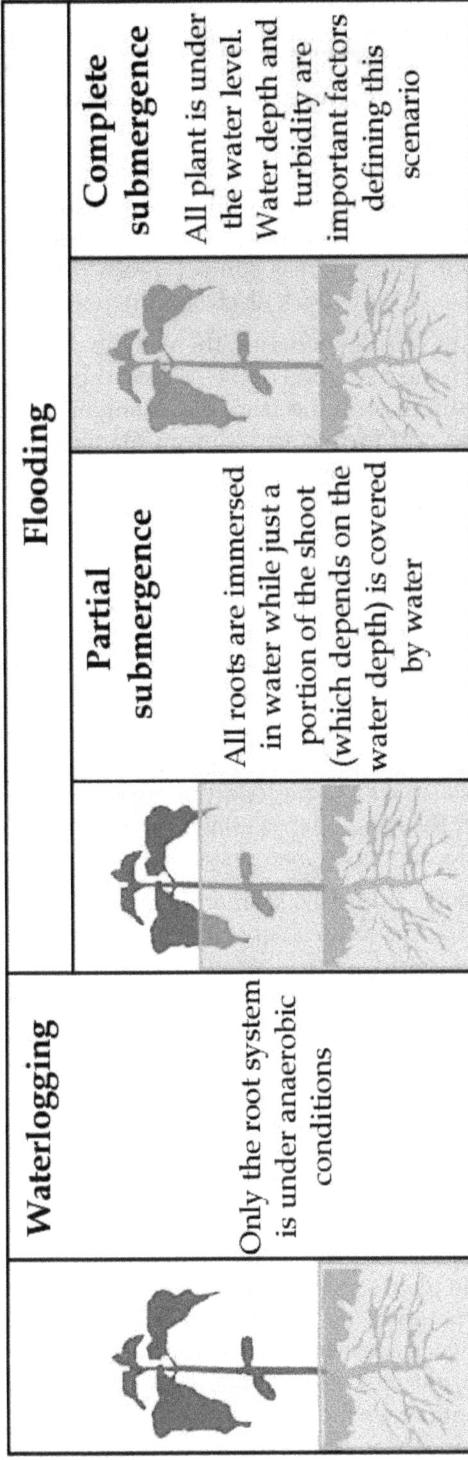

Figure 14.1: Scheme of the Different Scenarios Encountered by Plants being Subjected to Increasing Levels of Water Excess, Ranging from Waterlogging to Flooding (Partial/complete submergence) (Striker, 2012).

waterlogged conditions (Crawford and Braendle, 1996). At low O_2 concentrations, non enzymatic sources of ROS generation include plant terminal oxidases and mitochondrial electron transport chain, while enzymatic sources involve xanthine oxidase, which can use xanthine, hypoxanthine or acetaldehyde as electron donor and NADPH oxidase (Sairam *et al.*, 2008).

Oxidative injury due to waterlogging results in reduction of membrane integrity leading to increase in membrane permeability as a result of lipid peroxidation (Bansal and Srivastava, 2012). Free fatty acids, released as a consequence of membrane disintegration under stress can act as uncouplers in mitochondrial electron transport chain and may block electron transport.

As the duration of the stress increases, there is the progressive decline in the soil redox potential. With the reduction of the soil redox potential, potentially toxic compounds such as sulfides, soluble iron and manganese, ethanol, lactic acid, acetaldehyde, acetic acid and formic acid appear. Manganese oxides are the next electron acceptors followed by iron and sulfates. It leads to build of manganese and iron ions in the soil which may increase up to toxic level. But the solubility of these cations depends on soil pH and varies with the soil type. Manganese tolerance was developed in waterlogging sensitive wheat cultivars to enable their survival under stress conditions (Khabaz-Saberi *et al.*, 2010). Another concern is the impact of waterlogging on nitrogen metabolism in plants. As concentration of free oxygen declines under waterlogging, nitrate is used as alternate electron acceptor by the soil microorganisms. In waterlogged soil, nitrogen is available in the form of NH_4^+ instead of NO_3^-. Therefore, the plants more adaptive to NH_4^+ as compared to NO_3^- are more efficient under prolonged waterlogging. Impeded gas exchange during waterlogged conditions also lead to high partial pressure of CO_2 affecting the root growth and metabolism. Rice and other wetland plants can tolerate high pCO_2, but other species suffer rapid damage (Greenway *et al.*, 2006).

Leaf wilting, epinasty, chlorosis, stomatal closure, reduced photosynthesis and carbohydrate partitioning, reduced growth rate, disruption of cell membranes, reduced mineral uptake and altered growth regulator relationship are the consequences of waterlogging stress in plants. These parameters are affected by a number of factors, *viz.*, availability of water in soil, air temperature and relative humidity, extent of stomatal opening, and plant resistance to water movement. Yield loss is also closely associated with reduced nitrogen uptake as a consequence of oxygen deprivation due to waterlogging (Bennett and Albrecht, 1984).

Adaptation to Waterlogging

Physiological and molecular studies of the mechanisms of hypoxia/ anoxia tolerance in wild plants provide evidences of physiological mechanisms. Waterlogging tolerance in plant species varies from *Oryza glaberrima, Rumex palustris, Trifolium repens, Cyperus rotendus, Hordeum marinum*. Mechanisms exist from waterlogging escape to waterlogging tolerance. Escape from prolonged submergence requires the ability to enhance leaf elongation under water, which,further, increase leaf area and photosynthesis. In *Cyperus rotandus*, a weed in low land rice, submergence tolerance was found to be associated with increased tuber size with carbohydrates mainly stored as soluble sugars and ability to maintain high amylase activity and to maintain anaerobic respiration under stress conditions. Ethylene induced petiole elongation and leaf emergence in *Rumex palustris* ensures the survival under submergence. *Trifolium repens* with more petiole length and aerenchyma was found to be more tolerant towards waterlogging. *Oryza glabberima* escape waterlogging by leaf elongation and growing the shoot biomass.Maturity duration was also found to be important in waterlogging tolerance. Winter wheat is less sensitive to waterlogging as compared to early maturing spring wheat. Plants develop different strategies to mitigate/tolerate waterlogging stress (Table 14.1).

Table 14.1: Adaptations in Response to Waterlogging Stress

Type	*Trait*	*Reference*
Phenology	Seed/seedling vigour	Gardner and Flood, 1993
	Long season	Gardner and Flood, 1993
	Slow growth	McDonald *et al.*, 2001
Morphology/Anatomy	Adventitious root formation	Visser *et al.*, 1996; Mergemann and Sauter, 2000
	Petiole elongation	Sakagami *et al.*, 2009
	Aerenchyma development	Vander Sman *et al.*, 1991; Watkin *et al.*, 1998; Shimamura *et al.*, 2007
Nutrition	Root membrane integrity	Yan *et al.*, 1996; Bansal and Srivastava, 2012
	Root length and depth	McDonald *et al.*, 2001
	Fe tolerance	Huang *et al.*, 1995
	Mn tolerance	Huang *et al.*, 1995; Khabaz-Saberi *et al.*, 2010
Root metabolism	Anoxia tolerance	Johnson *et al.*, 1994; Peng *et al.*, 2001
	High carbohydrate concentration	Huang and Johnson, 1995; Sairam*et al.*, 2008
	Phytotoxin tolerance	Khabaz-Saberi*et al.*, 2010
Post anoxic damage	Antioxidaxive system	Lin *et al.*, 2004; Tan *et al.*, 2008; Hossain *et al.*, 2009; Bansal and Srivastava, 2012

Ethylene Production

During hypoxia, ethylene is responsible for most of the adaptive growth responses in roots and shoots. Ethylene concentration in waterlogged soils and in submerged plant parts has been reported up to 10 $cm^3\ dm^{-3}$. The rate of diffusion of ethylene from the root into the water is 10 times slower than its diffusion into air. Therefore, ethylene is released into aerenchymatous tissue and diffuses from the root to the shoot. During hypoxia, ethylene production is also enhanced in the root and in the aerobic shoot. In ethylene biosynthetic pathway, 1-amino cyclopropane 1-carboxylic acid (ACC) serves as immediate precursor of ethylene and is mainly synthesized in the root. Oxygen is an obligate requirement for conversion of ACC to ethylene. In the absence of oxygen, the reaction is inhibited. Therefore, under hypoxic conditions, ACC moves towards aerenchyma channels or to the shoot and more and more ACC is transported to the shoot. Molecular studies in rice and rumex have shown that ACC synthase and ACC oxidase (regulatory enzymes involved in ethylene biosynthesis) play a substantial role in short term and long term submergence tolerance (Rieu *et al.*, 2005). Increase in ACC oxidase concentration counter balances the reduced enzyme activity caused by low oxygen concentration during submergence, thus sustaining ethylene production under these conditions.

Aerenchyma Formation

An important adaptive response to waterlogging is the development of aerenchyma. Constitutive aerenchyma formation in the stems and roots of aquatic and flooding tolerant species provides long distance oxygenation to hypoxic tissues. The formation of interconnected and gas-filled spaces takes place either by cell separation during development or by programmed cell death *i.e.* by lysigeny or schizogeny. Ethylene plays an important role in inducing aerenchyma formation under waterlogging. Hypoxic roots possess higher concentration of ethylene as well as its precursor, and also show increased activity of enzymes responsible for ethylene biosynthesis (He *et al.*, 1994). Exogenous application of ethylene at rates of 0.1 $cm^3 dm^{-3}$ triggered aerenchyma formation in maize roots, while the application of Ag^+ ions, an inhibitor of ethylene action, blocked the aerenchyma development (Drew *et al.*, 1981).

Under waterlogged condition, aerenchyma formation was observed in number of crop species such as rumex (Vander Sman *et al.*, 1991), rice (Justin and Armstrong, 1991), wheat (Watkin *et al.*, 1998), and luffa (Shimamura *et al.*, 2007) etc. Cellulases along with other enzymes such as xyloglucan endotransglycosylase-1, expansin and pectinases are involved in lysigenous aerenchyma formation (Sairam *et al.*, 2008).

Adventitious Roots Formation

During waterlogging, adventitious roots emerge from the submerged part of the stem. These new roots replace the function of the original root system as an adaptation to the stress. These new roots are closer to the water surface and are connected to the stem at the site of aerenchyma formation. Therefore, they have more access to oxygen than old root system. The flood tolerant rumex species developed new flood-resistant roots in the upper 10 cm of the waterlogged soil, whereas in susceptible species, no change in the root distribution was recorded (Voesenek *et al.*, 2004).

Increase in ethylene concentration along with the maintenance of endogenous auxin level was found to be necessary for adventitious root formation in response to waterlogging. In deep water rice, it was reported that only under waterlogging, adventitious root developed while under normal condition, only initiation of adventitious root primordia took place. The adventitious root developed as a result of ethylene triggered cell death of epidermal cells of primordial tip. Application of ACC and waterlogging both triggered this cell death. It revealed that during waterlogging, adventitious root formation was associated with ethylene mediated cell death process (Mergemann and Sauter, 2000).

Availability of Soluble Sugars

Due to shifting of energy metabolism from aerobic to anaerobic mode under hypoxia/anoxia the energy requirements of the tissue is greatly restricted as only few ATPs are generated per molecule of glucose. A high level of anaerobic metabolism in hypoxic or anoxic roots is therefore urgent requirement to supply the energy in order to sustain metabolism in roots for the survival of plants. Thus, maintaining adequate levels of readily available sugars in hypoxic or anoxic roots is one of the adaptive mechanisms to water logging. The amount of root sugar reserve and activity of sucrose hydrolyzing enzymes play important role in water logging tolerance. In pigeon pea, it was observed that the roots of water logging tolerant genotypes had more sugar content (total, reducing and non-reducing sugar) as compared to susceptible genotype. Increase in the content of reducing sugar was due to more activity of sucrose synthase in tolerant genotype under water logging. The expression of sucrose synthase mRNA was also more in tolerant genotype, while, susceptible genotype show very little expression under waterlogged condition (Sairam *et al.*, 2008). Water logging tolerant wheat genotypes also maintained higher sugar content in the roots as during stress condition (Huang and Johnson, 1995).

Therefore, the availability of sugar reserves in the root as a consequence of more sucrose synthase to provide reducing sugars for metabolic activities is one of the important adaptation imparting waterlogging tolerance.

Anaerobiosis Induced Proteins (ANP)

ANP are the proteins, which are induced during hypoxia/anoxia. Two-dimensional electrophoresis revealed the induction of many proteins under anaerobic condition. Most of the anaerobic proteins induced were found to be enzymes either involved in glycolysis or in sugar phosphate metabolism. The identified ANP were enolase, glucose-6-phosphate isomerase, fructose-1,6-diphosphate aldolase, glyceraldehyde-3-phosphate dehydrogenase, pyruvatede carboxylase (PDC), lactate dehydrogenase (LDH), and alcohol dehydrogenase (ADH) (Chung and Ferl, 1999).

Waterlogged condition leads to hypoxia/anoxia. Since, oxygen is theterminal electron acceptor of mitochondrial electrontransport, in the absence of oxygen, ATP synthesis is inhibited and oxidation of NADH is blocked. As a consequence, the adenylate energy charge of the cell (ratio of ATP to ADP and AMP) declines. In order to maintain the adenylate charge, continuous supply of ATP is required. Therefore to supply the ATP, glycolytic and fermentation pathway are triggered. This process is known as the Pasteur Effect. As compared to aerobic respiration, glycolysis is relatively inefficient for ATP production. Further, pyruvate, the end product of glycolysis should be converted to other product to produce NAD. This pyruvate is converted to lactic acid by LDH. Accumulation of lactic acid in the cell leads to acidification of cytoplasm known as cytoplasmic acidosis. LDH, has a pH optimum above 7.0. As a result of cytoplasmic acidosis, enzyme gets inactivated. PDC and ADH are the enzymes with pH optima below pH 7.0, at lower pH these enzymes are activated. In a pH dependant manner, pyruvate is either converted to ethanol or lactic acid. ADH activity was found to be critical for the recycling of NADH as well as for the continuation of glycolytic pathway (Johnson *et al.*, 1994). Ethylene signal was reported to be essential for induction of *ADH* gene in arabidopsis. Presence of amino oxyactetic acid (inhibitor of ethylene biosynthesis) partially inhibited the induction of *ADH* gene, while, in the presence of ACC, the effect was reversed. It was concluded that ethylene is required for *ADH* induction during hypoxia (Peng *et al.*, 2001).

Under waterlogging, upregulated expression of the enzymes of sugar metabolism also imparts tolerance to stress conditions. During waterlogging, the energy requirement is greatly restricted because only few ATPs are generated per glucose molecule. To meet the energy requirements, readily available sugars are the most important. Recard *et al.* (1998) showed the induction of sucrose synthase gene in maize and sucrose synthase was the main enzyme active in sucrose breakdown in maize roots under anaerobic condition. In pigeon pea also, expression of sucrose synthase mRNA was upregulated under waterlogging (Sairam *et al.*, 2008). Hansch *et al.* (2003) reported the anaerobic induction of glyceraldehyde-3-phosphate dehydrogenase 4(*GapC4*) promotor in arabidopsis (*Arabidopsis thaliana*). The promotor was anaerobically induced in roots, leaves, stems and flowers.

In *T. subterraneum*, sucrose synthase, fructose kinase, lactate dehydrogenase and alcohol dehydrogenase activities were up regulated after 15 days of hypoxic treatment (Aschi-Smiti *et al.*, 2004).

Class 1 Haemoglobins, Nitric Oxide and Waterlogging

Different types of haemoglobins are widely distributed in plants. Symbiotic haemoglobins are well known and are mainly found in the nodules of leguminous plants. Non-symbiotic haemoglobins are not involved in symbiotic nitrogen fixation. On the evolutionary basis, there are three classes of haemoglobins. Class 1 haemoglobins have very high affinity to oxygen and are involved in NO scavenging. Class 2 haemoglobins with low affinity to oxygen facilitate oxygen supply to developing tissues. Class 3 haemoglobins are truncated version and have very less similarity to class 1 and class 2 haemoglobins. On the basis of co-ordination of heme iron in the haemoglobin structure, they have been classified into symbiotic and non-symbiotic haemoglobins (Gupta *et al.*, 2011).

Class 1 haemoglobins were first reported in barley aleurone layers under anaerobic conditions (Taylor *et al.*, 1994). Nie and Hill, 1997 reported that haemoglobin gene expression was induced in the presence of respiratory inhibitors. It was concluded that enhanced expression of haemoglobin gene under hypoxia was due to reduced availability/unavailability of ATP in the tissue.Expression of recombinant barley haemoglobin showed that it had a very slow oxygen dissociation rate (Duff *et al.*, 1997).

Nitric oxide is produced in higher concentration during hypoxia/anoxia. Dordas *et al.*, 2003 reported that in alfalfa root cultures, nitric oxide concentration increased significantly under anaerobic condition. Class 1 nonsymbiotic haemoglobins are involved in nitric oxide scavenging and maintain redox balance by increasing NADH recycling under flooding. Class I haemoglobins are also induced under higher concentrations of nitrite/nitrate and more nitric oxide is produced under such conditions. Perazzolli *et al.*, 2004 reported that in arabidopsis, over expression of these haemoglobins resulted in reduced nitric oxide production. As a consequence, high energy conditions prevailed during the stress period. The reactions involved in NO scavenging under hypoxic conditions constitute the Hb/NO cycle. Nitrate reduction to nitrite, nitrite reduction to NO and oxygenation of NO to nitrite; all these reactions involve oxidation of NAD(P)H, which helps in maintenance of redox potential under waterlogging. ATP synthesis may also occur due to mitochondrial nitrite: NO reductase activity associated with the electron transport chain. Class-1 nsHbs from Arabidopsis detoxify NO to nitrate in an NAD(P)H-dependent manner (Perazzolli *et al.*, 2004).

Antioxidant Activities

Waterlogging stress provides suitable conditions for generation of ROS. Redox potential is very high under anaerobic conditions and metal ions such as Fe^{2+} are present in highly reduced condition in rhizosphere. They reduce atmospheric oxygen to superoxide. Therefore oxygenation conditions are ideal for activation of oxygen. These tissues exhibit enhanced mitochondria-dependent ROS generation, acetaldehyde dependent superoxide formation via xanthine oxidase, lipoxygenase action on membrane lipids and finally lipolytic acyl hydrolase-catalyzed liberation of FFA, this leads to a burst in lipid peroxidation on return to normoxia. The main cellular components susceptible to damage by free radicals are lipids (peroxidation of unsaturated fatty acids in membranes), proteins (denaturation), sugars and nucleic acids. Consequences of hypoxia-induced oxidative stress depend on tissue and/ or species (*i.e.* their tolerance to anoxia), on membrane properties, on endogenous antioxidant content and on the ability to induce the response in the antioxidant system.

A balance between ROS production and ROS scavenging activity plays an important role in deciding the fate of cells under waterlogging stress. Many studies on the role of antioxidant enzymes in hypoxia/waterlogging stress have been conducted; however, contradictory results are available. When plant roots are subjected to waterlogged conditions, superoxide dismutase (SOD) activity increased in barley (Kalashnikov *et al.*, 1994) and citrus (Hossain *et al.*, 2009), decreased in winter wheat (Tan *et al.*, 2008) and remained unaffected in tomato (Lin *et al.*, 2004). Yan *et al.* (1996) reported that short-term flooding enhanced SOD and ascorbate peroxidase (APX) activity in maize (*Zea mays* L.) leaves, and the extended periods of flooding decreased the activities of these enzymes. In mungbean, reduction in catalase (CAT), APX and SOD activities was observed under flooding conditions (Ahmed *et al.*, 2002). In pigeonpea, waterlogging tolerance was found to be associated with increase in antioxidant enzyme activities (Bansal and Srivastava, 2012).

Hypoxic Signaling and Gene Regulation

Plants sense the signals in the environment, which are usually the first step in transduction cascades leading to downstream changes responsible for plant's response to the stress. Two gaseous molecules *i.e.* ethylene and oxygen are very important, because diffusion of these gases is dramatically reduced under waterlogged condition. Ethylene levels increase significantly in submerged tissues because of continuous production and hampered diffusion. These enhanced ethylene levels induce different morphological and anatomical traits such as aerenchyma and adventitious root formation (Ricard *et al.*, 2006). In plants, ethylene is perceived by receptor molecules located in endoplasmic reticulum. Ethylene receptors form a complex with a protein called constitutive triple response. These proteins are activated on binding with the receptors on endoplasmic reticulum and repress downstream responses induced by ethylene (Visser and Vosenek, 2004). In the presence of ethylene, receptors undergo conformational changes and gets inactivated. Further, downstream signals are transduced by several positive regulators (ethylene insensitive 2 (EIN2), EIN5 and EIN6) ending with the transcription factors EIN3 and EIN3-like (EIL). The components of this signal transduction cascade range from the receptor molecules to the transcription factor family EIN3/EIL(Guo and Ecker, 2004).

In roots, oxygen levels decline because oxygen is consumed continuously in respiration and delivery of aerial oxygen is very slow from shoots to roots. Reduced oxygen level leads to inhibition of respiration, lowering of adenylate status, down regulation of TCA cycle and glycolysis. These changes are consistent with the down-regulation of the genes that encode enzymes involved in the biosynthesis of cell walls, lipids and flavanoids, defense responses and protein degradation in Arabidopsis roots exposed to low oxygen (Klok *et al.*, 2002). This inhibition of biosynthetic flux and metabolic rates occurs even at oxygen levels that are much higher than Km of cytochrome oxidase and alternate oxidase. The observed metabolic shift is an adaptive response, because it reduces the oxygen consumption, saves ATP and delays the onset of anoxia.

Conclusions and Perspectives

Waterlogging is one of the major constraints effecting sustained agriculture production in the changing climate scenario. Adaptative traits of plants ensuring survival under waterlogging (aerenchyma formation and development of adventitious roots, availability of soluble sugars, antioxidant system, anaerobic metabolism) are well studied. But, study on regulatory mechanisms and signalling events responsible for triggering responses to waterlogging/flooding conditions in plants is an interesting area of research. Recent development of high throughput technologies provides excellent tools for the exploration of signaling events in plant cells. A comprehensive knowledge of plant's responses under waterlogging may assist in breeding programs as well as in development of management practices for cultivating the crops in flood prone areas.

References

Agarwal, S., Sairam, R.K., Srivastava, G.C., Tyagi, A. and Meena, R.C., 2005. Role of ABA, salicylic acid, calcium and hydrogenperoxide on antioxidant enzymes induction in wheat seedlings. *Plant Sci.*, 169:559-570.

Ahmed, S., Nawata, E. and Sakuratani, T., 2002. Effects of waterlogging at vegetative and reproductive growth stages on photosynthesis, leaf water potential and yield in mungbean. *Plant Prod. Sci.*, 5(2):117-123.

Aschi-Smiti, S., Chaibi, W., Brouquisse, R., Berenice-Ricard, B. and Saglio, P., 2004. Assessment of enzyme induction andaerenchyma formation as mechanisms for floodingtolerance in *Trifolium subterraneum* 'Park'. *Ann. Bot.*, 91:195-204.

Bailey-Serres, J. and Voesenek L.A.C.J., 2008. Flooding Stress: Acclimations and Genetic Diversity. *Annu. Rev. Plant Biol.*, 59:313–39.

Bansal, R. and Srivastava, J.P.,2012. Antioxidative defense system in pigeonpea roots under waterlogging stress. *Acta. Physiol. Plant.*, 34:515-522.

Bennett, J.M. and Albrecht, S.L., 1984. Drought and flooding effects on nitrogen fixation, water relations and diffusive resistance of soybean, *J. Agron.*, 76:735-740.

Chung, H.J. and Ferl, R.J., 1999. *Arabidopsis* alcohol dehydrogenase expression in both shoots and roots is conditioned by root growth environment. *Plant Physiol.*, 121:429-436.

Crawford, R.M.M. and Braendle, R., 1996. Oxygen deprivation stress in a changing environment. *J. Exp. Bot.*47:145-159.

Dordas, C., Hasinoff, B.B., Igamberdiev, A.U., Manach, N.,Rivoal, J. and Hill, R.D., 2003. Expression of a stress-induced hemoglobin affects NO levels produced by alfalfa root cultures under hypoxic stress. *Plant J.*, 35:763-770.

Drew, M.C., 1997. Oxygen deficiency and root metabolism: injury and acclimation under hypoxia. *Annu. Rev. Plant Physiol. Plant mol. Biol.*, 48:223-250.

Drew, M.C., Jackson, M.B., Gifford, S.C. and Campbel, R., 1981. Inhibition by silver ions of gas space (aerenchyma) formation in adventitious roots of *Zea mays* L. subjected to exogenous ethylene or to oxygen deficiency. *Planta*, 153:217-224.

Duff, S.M.G., Wittenberg, J.B. and Hill, R.D., 1997. Expression, purification, and properties of recombinant barley (*Hordeum* sp.) hemoglobin: optical spectra and reactions with gaseous ligands. *J. Biol. Chem.*, 272:16746-16752.

Gardner, W.K., and Flood, R.G., 1993. Less waterlogging damage with long season wheats. *Cereal Res. Commun.*, 21(4): 337-343.

Greenway, H., Armstrong, W. and Colmer, T.D., 2006. Conditions leading to high CO_2 (> 5 kPa) in waterlogged-flooded soils and possible effects on root growth and metabolism. *Ann. Bot.*, 98: 9–32.

Guo, H., and Ecker, J.R., 2004. The ethylene signaling pathway: New insights. *Curr. Opin. Plant Biol.* 7: 40–49.

Gupta, K.J., Igamberdiev, A.U., Manjunatha, G., Segu, S., Moran, J.F., Neelawarne, B., Bauwe, H., and Kaiser, W.M., 2011. The emerging roles of nitric oxide (NO) in plant mitochondria. *Plant Sci.*, 18: 520–526.

Hänsch, R., Mendel, R.R., Cerff, R. and Hehl, R., 2003. Light-dependent anaerobic induction of the maize glyceraldehyde-3-phosphate dehydrogenase 4 (*GapC4*) promoter in *Arabidopsis thaliana* and *Nicotiana tobacum*. *Ann. Bot.*, 91: 149-154.

He, C.J., Drew, M.C. and Morgan, P.W., 1994. Induction of enzymes associated with lysigenous aerenchyma formation in roots of *Zea mays* during hypoxia or nitrogen starvation. *Plant Physiol.*, 105:861–865.

Hossain, Z., Lopez-Climent, M.F., Arbona, V., Perez-Clemente, R.M. and Gomez-Cadenas, A., 2009. Modulation of the antioxidant system in citrus under waterlogging and subsequent drainage. *J. Plant Physiol.*, 166:1391-1404.

Huang, B. and Johnson, J.W., 1995. Root respiration and carbohydrate status of two wheat genotypes in response to hypoxia. *Ann. Bot.*, 75: 427- 432.

Huang, B., Johnson, J.W., Nesmith, D.S. and Bridges, D.C., 1995. Nutrient accumulation and distribution of wheat genotypes in response to waterlogging and nutrient supply.*Plant Soil*, 173:47-54.

Johnson, J.R., Cobb, B.G. and Drew, M.C., 1994. Hypoxic induction of anoxia tolerance in roots of Adh null *Zea mays*. *PlantPhysiol.*, 105: 61-67.

Justin, S.H.F.W. and Armstrong, W., 1991. Evidence for the involvement of ethane in aerenchyma formation in adventitious roots of rice (*Oryza sativa* L.). *New Phytol.*, 118:49-62.

Kalashnikov,Yu.E., Balakhnina, T.I. and Zakrzhevsky, D.A., 1994. Effect of soil hypoxia on activation of oxygen and the system of protection from oxidative destruction in roots and leaves of *Hordeum vulgare*. *Russ. J. Plant Physiol.*, 41:583-588.

Khabaz-Saberi, H., Rengel, Z., Wilson, R. and Setter, T.L., 2010. Variation of tolerance to manganese toxicity in Australian hexaploid wheat. *J. Plant Nutri. Soil Sc.*, 173: 103–112.

Klok, E.J., Wilson, I.W., Wilson, D., Chapman, S.C., Ewing, R.M., Somerville, S.C., Peacock, W.J., Dolferus, R. and Dennis, E.S.,2002.Expression profile analysis of the low-oxygen response in *Arabidopsis* root cultures. *Plant Cell*, 14: 2481-2494.

Lin, K.H.R., Weng, C.C., Lo, H.F. and Chen, J.T., 2004.Study of the root antioxidative system of tomatoes and eggplants under waterlogged conditions.*Plant Sci.*, 167:355-365.

McDonald, M.P., Galwey, N.W. and Colmer, T.D., 2001. Waterlogging tolerance in the tribe Triticeae: The adventitious roots of *Critesionm arinum* have a relatively high porosity and a barrier to radial oxygen loss. *Plant Cell Environ.*, 24:585- 596.

Mergemann, H. and Sauter, M., 2000.Ethylene induces epidermal cell death at the site of adventitious root emergence in rice. *Plant Physiol.*, 124:609-614.

Nie, X.Z. and Hill, R.D., 1997. Mitochondrial respiration and hemoglobin gene expression in barley aleurone tissue. *Plant Physiol.*, 114:835-840.

Peng, H.P., Chan, C.S., Shih, M.C. and Yang, S.F., 2001. Signaling events in the hypoxic induction of alcohol dehydrogenase gene in *Arabidopsis*. *Plant Physiol.*,126:742-749.

Perazzolli, M., Dominici, P., Romero-Puertas, M.C., Zago, E.D.,Zeier, J., Sonoda, M., Lamb, C. and Delledonne, M., 2004. Arabidopsis non-symbiotic haemoglobin AHb1 modulates nitric oxide bioactivity. *Plant Cell*, 16:1–10.

Recard, B., Van Toi, T., Chourey, P. and Saglio, P., 1998. Evidence for the critical role of sucrose synthase for anoxic tolerance o fmaize roots using double mutant. *Plant Physiol.*, 116:1323-1331.

Ricard, B., Aschi-Smiti, S., Gharbi, I. and Brouquisse, R., 2006. Cellular and Molecular Mechanisms of Plant Tolerance to Waterlogging. In: Plant-Environment Interactions, Huang, B. (Ed.). Taylor and Francis, New York, London, 177-208.

Rieu, I., Cristescu, S.M., Harren, F.J.M., Huibers, W., Voesenek, L., Mariani, C. and Vriezen, W.H., 2005. RPACS1,a flooding-induced 1-aminocyclopropane-1-carboxylate synthase gene of *Rumex palustris*, is involved in rhythmic ethylene production. *J. Exp. Bot.*, 56: 841-849.

Sairam, R.K., Kumutha, D., Ezhilmathi, K., Deshmukh, P.S. and Srivastava, G.C., 2008.Physiology and biochemistry of waterlogging tolerance in plants. *Biol. Plantarum*, 52(3):401-412.

Sakagami, J.I., Joho, Y. and Ito, O., 2009. Contrasting physiological responses by cultivars of *Oryza sativa* and *O. glaberrima* to prolonged submergence. *Ann. Bot.*, 103:171–180.

Shimamura, S., Yoshida, S. and Mochizuki, T., 2007. Cortical aerenchyma formation in hypocotyl and adventitious roots of *Luffacyl indrica* subjected to soil flooding. *Ann. Bot.*, 100:1431–1439.

Striker, G.G., 2012.Flooding stress on plants: anatomical, morphological and physiological responses. In: *Botany*, (Ed.) Dr. John Mworia, Intech, 3-28.

Tan, W., Liu, J., Dai, T., Jing, Q., Cao, W. and Jiang, D., 2008.Alterations in photosynthesis and antioxidant enzyme activity in winter wheat subjected to post-anthesis water-logging, *Photosynthetica*, 46(1):21-27.

Taylor, E.R., Nie, X.Z., MacGregor, A.W. and Hill, R.D., 1994. A cereal haemoglobin gene is expressed in seed and root tissues under anaerobic conditions. *Plant Mol. Biol.*, 24:853-862.

Van der Sman, A.J.M., Voesenek, L.A.C.J., Blom, C.W.P.M., Harren, F.J.M. andReuss J. 1991. The role of ethylene in shoot elongation with respect to survival and seed output of flooded *Rumexmaritimus* L. plants. *Funct. Ecol.*, 5:304-313.

Visser, E.J.W., Bogemann, G., Blom C.W.P.M. and Voesenek, L.A.C.J., 1996. Ethylene accumulation in waterlogged Rumex plants promotes formation of adventitious roots. *J. Exp. Bot.*, 47:403-410.

Visser, E.J.W. and Voesenek, L.A.C.J., 2004. Acclimation to soil flooding - sensing and signal - transduction. *Plant Soil*, 254:197- 214.

Voesenek, L.A.C.J., Rijnders, J., Peeters, A.J.M.,Van de Steeg, H.M.V. and De Kroon, H., 2004. Plant hormones regulate fast shoot elongation under water: from genes to communities. *Ecology*, 85:16–27.

Watkin, E.L.J., Thomson, C.J. and Greenway, H. 1998. Root development and aerenchyma formation in two wheat cultivars and one Triticale cultivar grown in stagnant agar and aerated nutrient solution. *Ann. Bot.*, 81:349–354.

Yan, B., Dai, Q., Liu, X., Huang, S. and Wang, Z., 1996. Flooding-induced membrane damage, lipid oxidation and activated oxygen generation in corn leaves. *Plant Soil*, 179:261–268.

2016, Recent Advances in Plant Stress Physiology *Pages* **297–328**
Editors: ***Praduman Yadav, Sunil Kumar and Veena Jain***
Published by: **DAYA PUBLISHING HOUSE, NEW DELHI**

Chapter 15

Anthropogenic Induced Stress and its Impact on Plants

Subodh Kr. Bishnoi[1]* and Naresh Kr. Yadav[2]**

[1]Assistant Professor, [2]Senior Research Fellow,
Agricultural Research Station, (SK Rajasthan Agricultural University, Bikaner),
Sriganganagar – 335 001, Rajasthan, India

ABSTRACT

Study of anthropogenic stress can be an important tool in environmental research and management. Aerial overview of stress can be a good basis of environmental decision making at various levels. To upgrade cooperation on stress through common institutions remains one of the most difficult tasks in environmental science. Stress measures need to integrate evaluation of accumulative impacts across multiple stressors because environmental problems such as acid deposition, climate change etc. lap over in time and space. Atmospheric trace gases such as SO_2 and H_2S containing sulphur compounds are considered to be the major source of anthropogenic air pollution. Their toxic effects on plants such as on plant glutathione metabolism is a serious cause of concern. Glutathione, sulphur-containing tripeptide, is an important anti-oxidant in plants preventing damage to important cellular components caused by reactive oxygen species. Glutathione is linked to sequestration of xenobiotics and heavy metals and is extensively used as an oxidative stress marker in plants.

Keywords: *Global warming, Anthropogenic, Abiotic stress, Biotic stress, Heavy metal.*

Introduction

Anthropogenic stress is the stress that occurs as a result of damage done to ecological components by the activities of individual people or human groups. Anthropogenic stress existed many thousand years ago too when the agricultural

**Author.* E-mail: bishnoisk@gmail.com; yadavnaresh1285@gmail.com

technique involved the cutting and burning of plants in forests to create fields. However, as a result of recent major increases in human population, technological capabilities, and standard of living globally, the amount of anthropogenic stress has increased greatly (Cairns John Jr., 2013). Anthropogenic global warming has been caused by human activity in addition to natural global warming. The science around this is difficult to prove, as identifying a clear indicator of the human contribution which is independent of multiple natural causes is very hard to do in reality.

Since the start of the industrial era (about 1750), the overall effect of human activities on climate has been a warm-ing influence. The human impact on climate during this era greatly exceeds that due to known changes in natural processes, such as solar changes and volcanic eruptions (Solomon *et al.*, 2007). Human activities contribute to climate change by causing changes in Earth's atmosphere in the amounts of greenhouse gas-es, aerosols (small particles), and cloudiness. The largest known contribution comes from the burning of fossil fuels, which releases carbon dioxide gas to the atmosphere. Greenhouse gases and aero-sols affect climate by altering incoming solar radiation and out-going infrared (thermal) radiation that are part of Earth's energy balance. Changing the atmospheric abundance or properties of these gases and particles can lead to a warming or cooling of the climate system.

Nitrogen (78 per cent) and oxygen (21 per cent) constitute 99 per cent of the dry atmosphere. The rest of the gases, including GHGs, are collectively classified as "trace" gases due to their low concentrations.

The recent attention given to the greenhouse effect and global warming is based on the recorded increases in concentrations of some of the greenhouse gases due to human activity. Of particular interest are water vapor, carbon dioxide, methane, nitrous oxide, chlorofluorocarbons, and ozone. With the exception of chlorofluorocarbons, all of these gases occur naturally and are also produced by human activity (Henderson *et al.*, 2002)

Water vapor is the most important GHG on the planet. Unlike most of the other atmospheric gases, water vapor is considered to be a 'variable' gas; that is, the percentage of water vapor in the atmosphere can vary greatly depending on the location and source of the air. For example, over the tropical oceans, water vapor may account for 4 per cent of the total volume of gases, while over deserts or at high altitudes, it may be nearly absent. Water vapor absorbs heat readily. When discussing global warming, however, people often don't consider water vapor. Why not? The main reason is that human activity is not directly changing water vapor content. However, we do directly influence other GHGs. Although other GHGs are individually less important than water vapor, increasing their concentrations may affect global climate in significant and measurable ways.

Carbon dioxide (CO_2) is considered the most important human-influenced GHG. Scientific measurements reveal an unmistakable global increase that is rapid and escalating. This increase arises primarily from the burning of fossil fuels (motorized vehicles, electric power plants, and homes heated with gas or oil) and the burning and clearing of forested land for agricultural purposes.

Methane (CH_4) is largely a product of natural biologic processes, but its output can be accelerated by human activities. CH_4 is emitted from the decay of organic matter in waterlogged soils (for example, wetlands and rice paddies) and from the digestive tracts of grazing animals (for example, ruminants). The additions from human activities include the expansion of rice agriculture, the increased number of livestock, the increased number of landfills, and leakage from natural gas pipelines.

Chlorofluorocarbons (CFCs) have no natural source; they are produced entirely by human activity. CFCs have historically been used widely as refrigerants in air conditioners, refrigerators, freezers, and heat pumps. They are found in some foam plastics and used in some electronics manufacturing. Even though CFC production has been vastly reduced, these compounds remain in the atmosphere for a long time; we shall see their effects as GHGs for many years.

Nitrous oxide (N_2O) is a naturally occurring GHG, which has increased significantly in recent years due to human activity. N_2O is emitted from coal-burning power plants and can be released from the breakdown of chemical fertilizers in the soil.

Ozone (O_3) is also a greenhouse gas. It is important not to confuse the presence of the ozone in the stratosphere (a good thing) with the presence of ozone in the troposphere (a bad thing). In the troposphere, ozone can be a major component of urban smog damaging crops and aggravating respiratory problems as well as enhancing the greenhouse effect.

The concentrations of these GHG are increasing (although, thanks to recent global agreements, CFCs are being largely eliminated and their concentrations have begun to drop in the lower atmosphere). The emissions are not uniformly distributed globally. Most of the emissions come from the more developed countries, where power generation, power consumption, and living standards are highest (Henderson *et al.*, 2002).

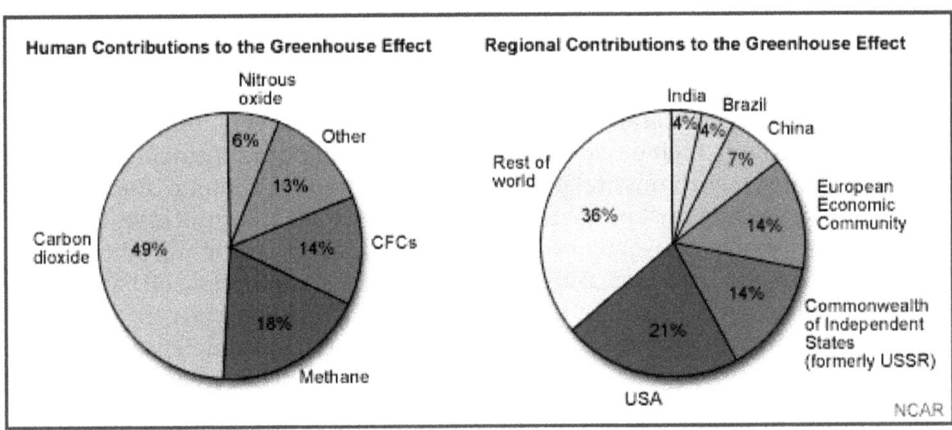

Figure 15.1

Ozone-Induced Changes in Plant Secondary Metabolism

In the global climate change scenario, the relation between the anthropogenic emissions of greenhouse gases, responsible for climate warming and rainfall alteration, and tropospheric ozone is quite complex, especially in the agro-ecosystem context. Crop potential yield depends on defining factors (CO_2, radiation, temperature and crop traits), limiting factors (water, nitrogen and phosphorous) and reducing factors (pests, pathogens, weeds and pollutants) (Goudrian and Zadoks 1995). A deep knowledge of the role played by these factors under altered climatic conditions is necessary to evaluate the responses at crop and agro-ecosystem level (Fuhrer, 2003). It is well known that increasing tropospheric ozone concentration at ambient CO_2, causes a decline in the yield of many crop species, and this negative effect is reduced in a CO_2-enriched atmosphere, probably due to the decrease of stomatal conductance and ozone flux, or to the increase in the activity of anti-oxidant enzymes (Heagle *et al.*, 1998, 1999, 2000). Thus, the detrimental effect of enhanced CO_2 concentration on temperature is in some way, compensate by its protective role in plants under ozone stress, as assessed by biomass and yield stimulation studies in these conditions (McKee *et al.*, 1995, 1997).

The influence of global climate change in plant-pathogen interactions is quite complex too (Violini 1995; Manning and von Tiedemann 1995). On one hand, ozone may induce the same sequel of events involved in plant immunity (Iriti and Faoro, 2007) *i.e.* oxidative burst and hypersensitive response at the onset of systemic acquired resistance (SAR) (Langebartels *et al.*, 2002), on the other, plants weakened by ozone stress may be particularly susceptible to infections (Manning and von Tiedemann 1995). Interestingly, wheat rust (*Puccinia recondite* f. sp. *tritici*) is strongly inhibited by ozone, but unaffected by elevated CO_2, both in presence or absence of ozone stress (von Tiedemann and Firsching 2000). *Vice versa,* a protective effect of rust (*Uromyces fabae*) infection in broad bean (*Vicia faba*) was reported against ozone, sulphur dioxide either alone or combined (Lorenzini *et al.*, 1994).

Finally, regardless of the specific ecophysiological meaning, plants cope with the plethora of stressful abiotic and biotic conditions by modifying their secondary metabolic pathways. In this view, during the evolution, the chemical diversity improved the fitness of plant organisms, thus ensuring their evolutionary radiation. Phytochemicals, with their broad spectrum activities, are primarily involved in plant tolerance against environmental pollutants and worsening climatic conditions, as well as in resistance against pests and pathogens. The metabolic processes activated in these defence responses may be tightly separated or overlapping, according to the stress factor and the plant cultivar, as a result of negative (trade off) or positive (cross resistance) cross talk, that is to say the communication between molecular signals and transduction pathways involved in different plant defence responses. The most important consequence of this cross talk is the alert of defence mechanisms against an abiotic stress in consequence of priming with a biotic elicitor and *vice versa*.

Metabolism under Stress

Drought, salinity and low temperature affect uptake and conductance of water. Environmental factors that affect water supply lead to changes in stomatal opening which can, if stress persists, set in motion a chain of events originating from changes in the concentration of leaf-internal carbon dioxide, consecutively affecting the carbon reduction cycle, light reactions, energy charge, and proton pumping (Nelson *et al.*, 1998, Jain and Selvaraj 1997). Other pathways are affected as a result of increased shuttling of carbon through the photo-respiratory cycle (Bohnert and Jensen 1996). Eventually, carbon and nitrogen allocation and storage require readjustment; reactions that lead to the consumption of reducing power become favored, and development and growth may become altered (Jain and Selvaraj 1997). During the past few years, the complex interrelationship of biochemical pathways that change during stress has become appreciated, although we are far from understanding this complexity; several review articles are available (Bray1997, Nelson *et al.*, 1998, Jain and Selvaraj 1997). Mechanisms for which experimental evidence indicates an important contribution to metabolic adjustments under stress at the cell level are illustrated with the names of proteins, enzymes and metabolites. The significance of these mechanisms is supported by gene discovery, with stress-dependent regulation of the corresponding transcripts, or by biochemical analyses. Further support comes from experiments with transgenic plants (which might be termed 'transgenovars', *e.g.*, *N. tabacum*) expressing proteins encoded by such transcripts.

Plant Growth, Metabolism and Adaptation in Relation to Stress Conditions. Endogenous Levels of Hormones, Minerals and Organic Solutes in *Pisum sativum* Plants as Affected by Salinity

In the present study, the observed decreases in auxins, gibberellins and cytokinins that occurred concurrently with an increase in ABA, at all stages of plant development, in response to the higher levels of salinity used, are in accord with the results reported by other workers (Aharoni and Richmond 1978, Davenport *et al.*, 1980, Zeevaart 1983). Endogenous growth regulators have been shown to be critical determinators of growth and differentiation (Street and Cockburn 1970) and Younis and El-Tigani 1970 stated that growth may depend on the ratios rather than on the absolute levels of these growth substances in the plant. Thus, the promotion of growth of pea plants observed by El-Shahaby *et al.*, 1990 under low salinity levels seems to be correlated with increased levels of auxins, gibberellins and cytokinins that were associated with decreased ABA content. Furthermore, the inhibition of growth of plants treated with high salinity levels appears to coincide with the decline in growth promotery substances and accumulation of ABA. In support of these correlations, El-Shahaby *et al.*, 1990 observed that seed pre-soaking in GA_3 or IAA was in most cases capable of partially or completely counteracting the varied harmful effects maintained in pea plants by the different levels of salinity used. This indicates some tolerance through the inductive changes in the endogenous levels of growth regulators (Younis *et al.*, 1989). Berger and Ges, Horn, Austria, research that salinity induced an accumulation of Na^+ and Ca^{2+} in both shoots and roots at all stages of plant development and this was associated with decreases in K^+ content. In connection with these results, several workers have demonstrated an increase in Na^+ levels and a decline in that of K^+ with a concomitant decline in $KVNa^+$ ratio in plants treated with NaCl (Imamul-Huq and Larher 1983, Hasaneen *et al.*, 1987, 1989). Thus the well known effect of Na^+ on K^+ content of plant tissues was further substantiated by our results. Furthermore, the increase in calcium content with rise in salinity is in agreement with the findings of Ashraf *et al.*, 1986 and Epstein 1961 stated that the maintenance of appreciable levels of Ca^+ clearly indicates some kind of salt resistance of the plant. The low levels of salinity (-0.3 and -0.6 MPa) which increased Mn^{2+} content in shoots caused a decline of that content in roots. With —0.9 and — 1.2 MPa, Mn^{2+} content was either significantly increased in roots or significantly decreased in shoots. These differential changes in Mn^{2+} content under salinity stress are in conformity with the results of Mehrotra 1971 and Heikal *et al.*, 1980. A tentative explanation to the present changes could be based on ion compartmentation which is considered as a component of salt adaptation of glycophyte cells (Binzel *et al.*, 1988). Furthermore, phosphorus uptake and translocation from root to shoot was increased rather than inhibited by high salinity levels. In this respect, Wilson *et al.*, 1970 and El-Shahaby 1981 reported that, whatever may be the salinizing agent and the duration of salinization, the increase in phosphorus content in the plant seems to be a dominant effect of salinity.

A process in soil that is very sensitive to the presence of certain pesticides is nitrification. This seems to be due to the narrow range of microbial species that carry out this transformation.

Several pathogenic microorganisms are also sensitive to certain pesticides. Addition of fungicides to soils, for example, may effectively control plant disease due to the destruction of the pathogenic fungi. In addition, however, other beneficial fungal species may also be killed. For most soil processes, addition of pesticides to soil causes only a small, temporary drop in activity. The addition of a pesticide to soil may, in some cases, bring about selective pressure for those microorganisms capable of degrading the pesticide. Or, the pesticide may stimulate the production of extracellular enzymes that are capable of detoxifying the pesticide molecule. The phenomenon in which a pesticide degrades so rapidly in soil that it no longer performs its intended function of pest control is called enhanced biodegradation or accelerated degradation. An understanding of the soil, biochemical, and microbial factors and the interactions involved that bring about enhanced biodegradation is lacking. Application to an agricultural field of a pesticide that fails to perform as expected can result in large financial losses. On the other hand, if the process of enhanced biodegradation were clearly understood, it could be applied to cases where compounds must be rapidly reduced in their detrimental effect. The persistence of a pesticide in soil causes different types of problems. If the pesticide remains in soil for a long period of time it may (i) be assimilated by plants and accumulate in edible portions, (ii) be transported with eroding soil particles or with surface and subsurface waters to locations where its presence may cause harm or (iii) it may accumulate in the animal food chain and change the ecological balance of nature. Compounds that are short-lived in the environment are less likely to achieve these fates.

Xenobiotics in Soil

Metabolism of Pesticides in Soil: Metabolism of pesticides in soil may be classified into two main categories. The first category is defined by the ability of the chemical to support microbial growth by supplying nutrients such as carbon and nitrogen, as well as supplying a source of energy. The presence of the pesticide will cause a rise in numbers of the microbial subpopulation that can utilize the chemical while at the same time bringing about the disappearance of the pesticide from the soil. Microorganisms of this type can be isolated from soil through enrichment techniques. The second category of metabolic processes is called cometabolism. The pesticide is degraded but does not serve as a source of energy or nutrients. Heterotrophs of this sort cannot be isolated by enrichment culture techniques. Martin Alexander has outlined six different classes of reactions that result in partial or total transformation of a pesticide molecule. Detoxification is the conversion of an inhibitory molecule to a nontoxic product. Complete mineralization is the extreme case of detoxification. However, more common is the removal of a functional group from the pesticide molecule. Degradation is the transformation of a complex substrate molecule to simple products and is often considered synonymous with mineralization. A reaction that results in making the pesticide more complex by combining with itself or other soil compounds is called conjugation, complex formation, or addition reactions. Oxidative coupling is an example of this class of reactions. Activation is the conversion of a nontoxic substrate or a pesticide precursor to a toxic molecule. For example the insecticide phorate is transformed in soil to give metabolites that are also toxic to insects. A changing of the spectrum of toxicity is another reaction in which pesticides may participate. Some pesticides may be toxic to one group of organisms but may be metabolized to become toxic to completely different group of, and possibly nontarget, organisms. Metabolic pathways of pesticide degradation have been established in the laboratory in model systems involving pure cultures for a large number of pesticides. These "type reactions" are very useful in making educated guesses about the types of compounds that may be formed in soil during the degradation of a specific pesticide molecule. However, the type reactions do not allow us to make predictions about which step is the rate-limiting step in the overall degradation process, and thus, which metabolic products may accumulate in the soil. Yet in view of the large number of chemicals that eventually reach the soil and the inability to test each one rapidly and thoroughly, the ability to predict the metabolic fates and products by applying type reactions to the chemical of interest has utility. The rate at which a pesticide degrades in soil is subject to a number of variables. Five conditions have been outlined which must be met before a pesticide molecule may be degraded. These conditions include:

1. An organism effective in metabolizing a pesticide molecule must live or be capable of living in the soil,
2. The compound must be in a chemical and physical form suitable for degradation,
3. The chemical must be located in the same environment as the active agent, *e.g.* amicrobial cell or an extracellular enzyme, of degradation,

4. The compound must be capable of inducing formation of the enzymes or enzymes appropriate for detoxification - low solubility or low concentration of the pesticide and low uptake into the microbial cell may result in lack of induction, and
5. Environmental conditions such as pH, temperature, and organic matter must be suitable for the degrading microorganisms to proliferate and the enzyme to operate.

If any of these conditions are not met, the pesticide will not readily be decomposed and may persist in the soil environment for long periods of time.

The factors that control the rate of degradation of pesticides in soil are characteristic of the soil, microorganism, and the pesticide molecule. An understanding of the direct and indirect effect of all three factors must exist before accurate assessments of the ecological impact of pesticides in the soil can be made.

Global Warming and its Impact on Agriculture

Climate change: it is a statistically significant variation in either the mean state of the climate or in its variability, persisting for an extended period (typically decades or longer).

Global warming: Global warming is when the earth heats up (the temperature rises). It happens when greenhouse gases (carbon dioxide, water vapor, nitrous oxide, and methane) trap heat and light from the sun in the earth's atmosphere, which increases the temperature as shown in the Figure 15.2. This is equally harmful for all people, animals and plants.

Agriculture sector in India contributes 28 per cent of the total GHG emissions (Figure 15.3, NATCOM, 2004a). The global average from agriculture is only 13.5 per cent (Figure 15.3, IPCC, 2007a). In future, the percentage emissions from agriculture in India are likely to be smaller due to relatively much higher growth in emissions in energy-use transport and industrial sectors. The emissions from agriculture are primarily due to methane emissions from rice fields, enteric fermentation in ruminant animals and nitrous oxides from application of manures and fertilizers to agricultural soils (Figure 15.3, NATCOM, 2004b). At world level if we consider on percapita basis India releases 1.02 tonnes/year whereas, developed countries like USA release 20.01 tonnes/year (Figure 15.3, www.indiabudget.nic.in).

During the last ice age (Figure 15.4), which ended only 14 000 years ago, global average surface temperature was 5 °C lower than it is today. At that time, less energy was being received from the sun as a consequence of the position of the earth's orbit. The CO_2 levels in the atmosphere were lower, and heat redistribution by ocean circulation was weaker. Over a period of about 5 000 years, the global surface temperature gradually rose to an average of about 15 °C, where it remained until about 100 years ago. Then, as a result of human activity, what UNFCCC calls a "thickening" of the blanket of greenhouse gases occurred and the earth's average surface temperature started to increase rapidly. Today it has risen to over 15.5 °C, with most of the increase occurring since the 1980. It is projected to rise by a further 2 to 3 °C before the end of the century. This means that, over a period of only 100

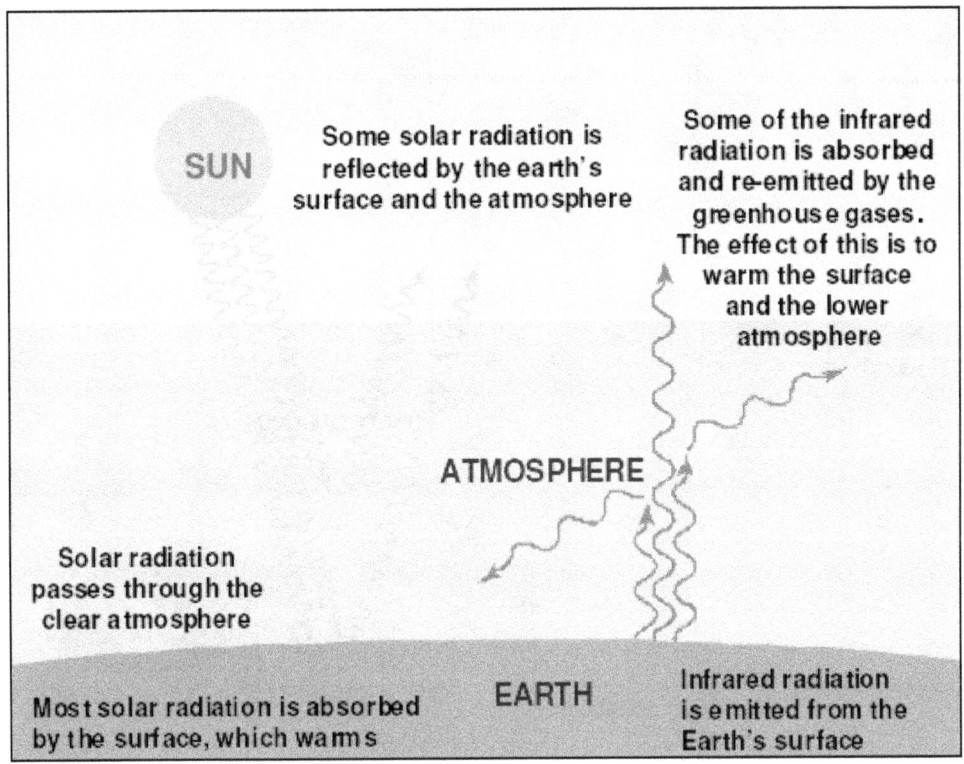

Figure 15.2: Four Major Greenhouse Gases, different Sources and per Capita Emissions (India).

years, the earth will have experienced an increase in global mean temperature comparable to the one that took 2500 years to occur ten millennia ago. Moreover, the increase that is now occurring is pushing global mean temperatures towards what may be an upper limit for human survival.

Impact of Climate Change on Agriculture

- ☆ Soil: Drier, reduced productivity
- ☆ Irrigation: Increased demand, reduced supply
- ☆ Pests: Increased ranges and populations
- ☆ Production: Reduced crop yield, particularly in south Asia
- ☆ Livestock: Increased diseases and heat stress
- ☆ Fishery: Affected abundance and spawning
- ☆ Economic impact: Reduced agricultural output
- ☆ Agricultural productivity in India was estimated to decrease by 2.5 to 10 per cent by 2020(FAO) to 5 to 30 per cent by 2050 (IPCC assessment)

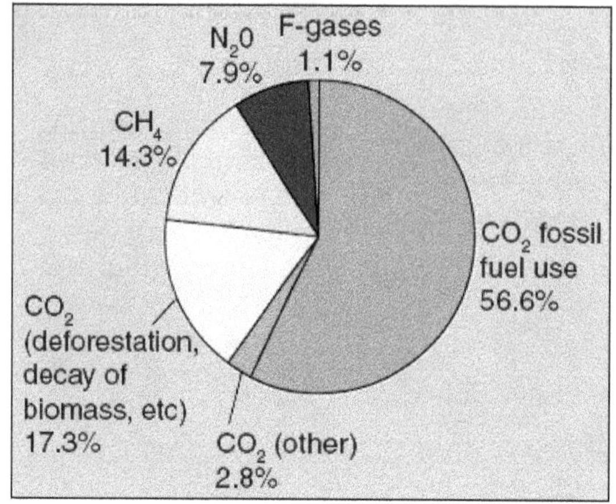

(NATCOM, 2004a) A

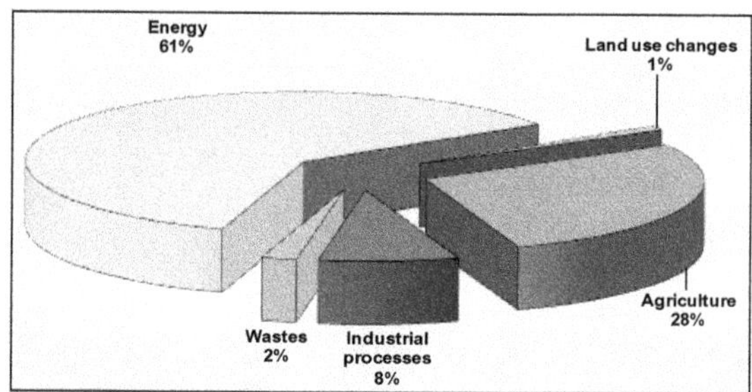

(NATCOM, 2004B) B

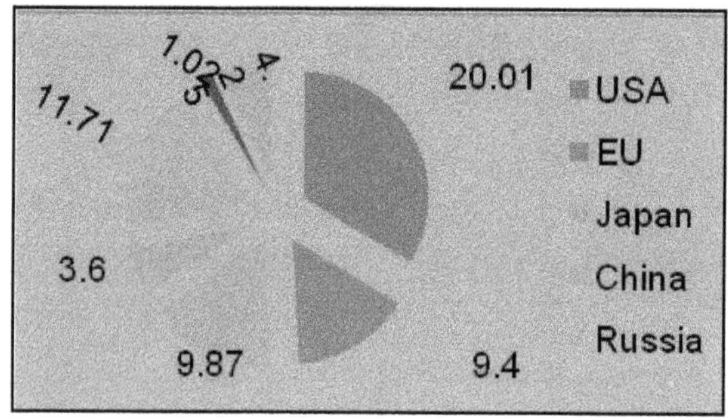

(WWW.INDIABUDGET.NIC.IN) C

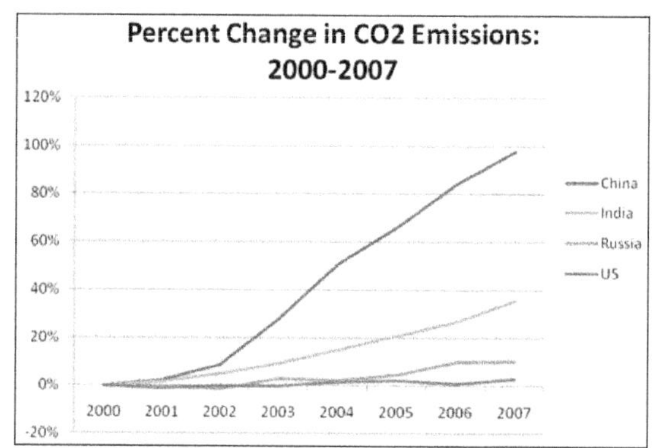

(IPCC, 2007A) D

Figure 15.2 A, B, C and D: Types of Greenhouse Gases Emissions by different Countries.

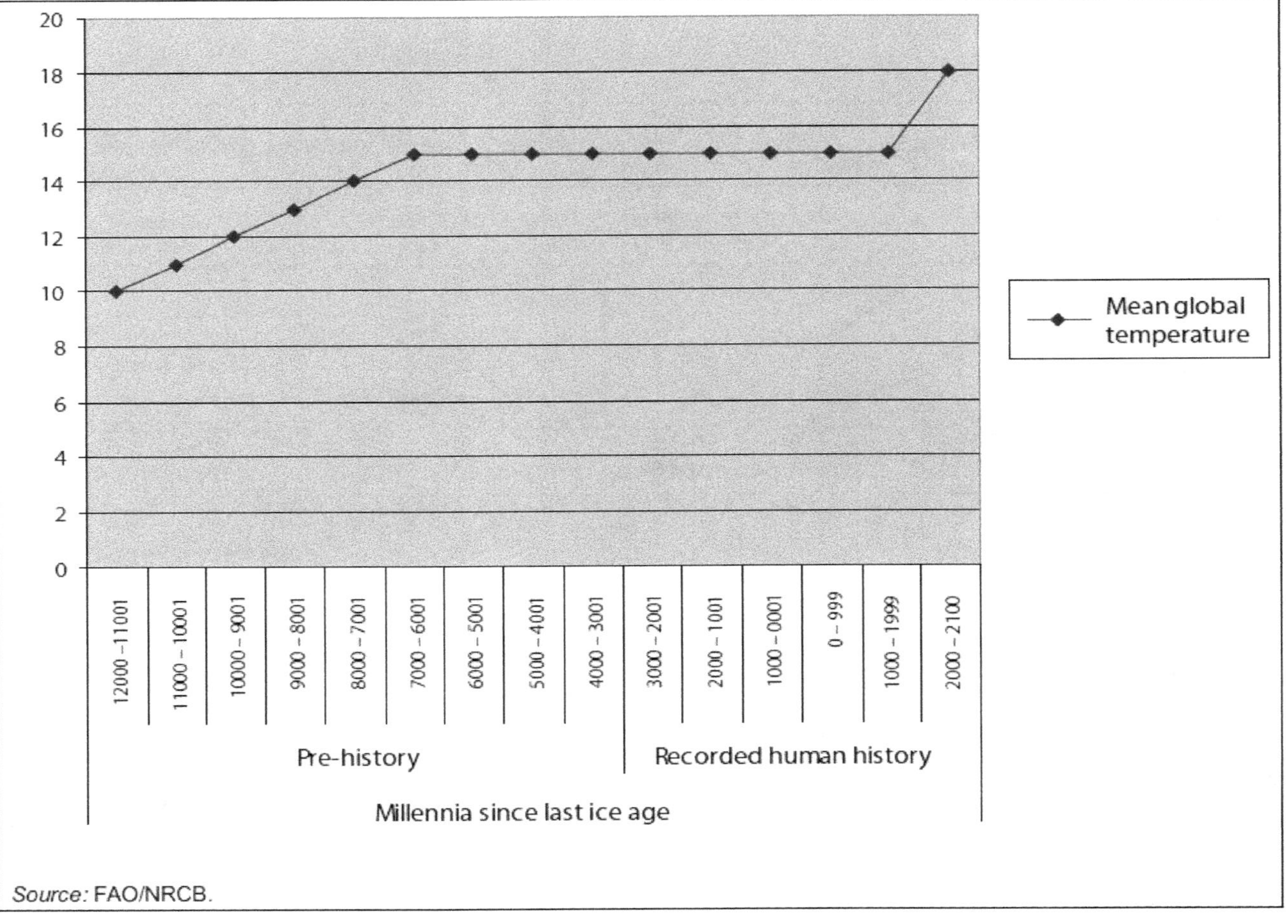

Source: FAO/NRCB.

Figure 15.4: Global Average Surface Temperature of Last more than 3000 Years.

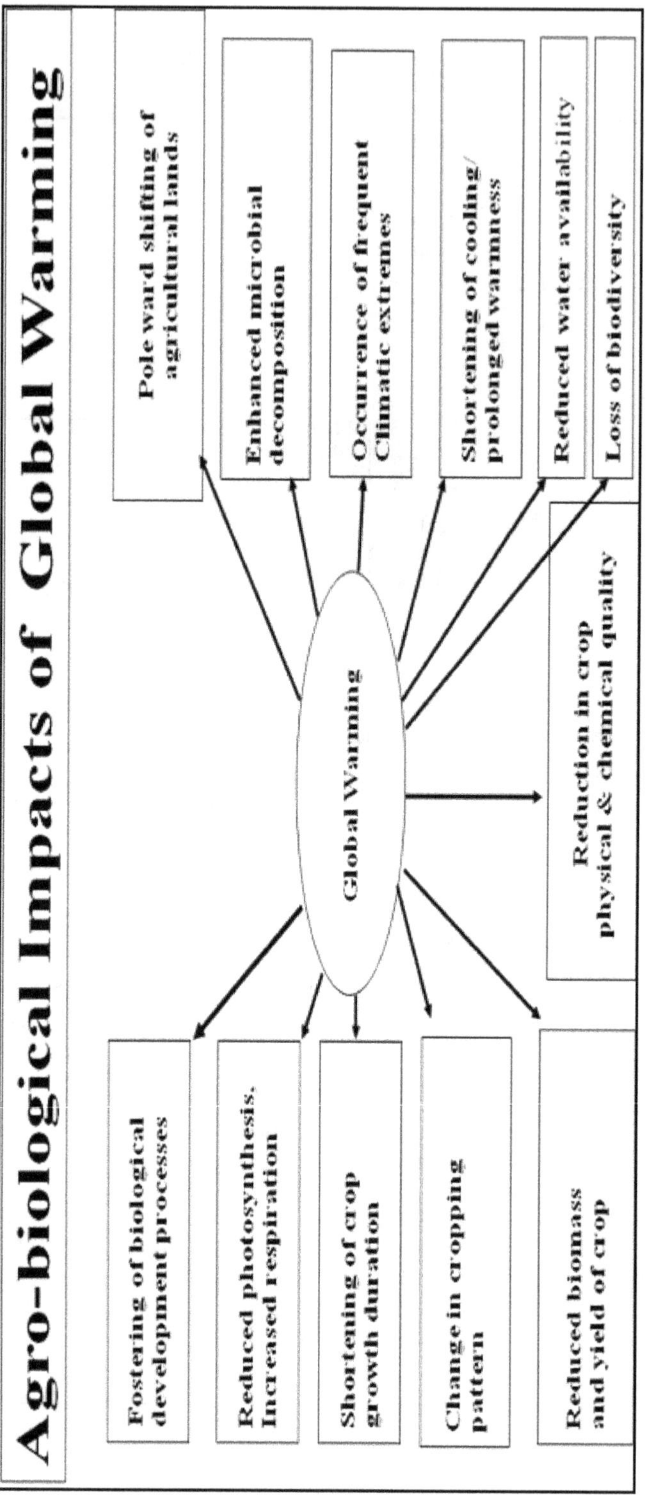

Figure 15.4: The Agro-Biological Impacts of the Global Warming.

Simulated average potential grain yield of sorghum, maize, groundnut and pigeonpea at Bulawayo are presented in Table 15.1, along with the average main effects of the climate change scenarios. APSIM output shows that increasing CO_2 concentrations will have a positive effect on crop yields, on average 8 per cent and 6 per cent for the two legume test crops. Similarly, a reduction in rainfall amount had the expected negative impact on grain yield, although with 6–8 per cent yield reduction across species, it is a lower reduction than the 10 per cent reduction in rainfall. This suggests some improvement in water use efficiency with reduced rainfall in this environment, probably as a result of less runoff.

Table 15.1: Impact of Climate Change on Average Potential Grain Yield of Sorghum, Maize, Groundnut and Pigeon Pea

Crop	*Potential** Grain Yield kgha^{-1}*	*CO_2 Effect on Yield*	*Rainfall Effect on Yield*	*Temperature Effect on Yield*	*CC* Effect on Yield*
Sorghum	2753	n/a	6 per cent	16 per cent	22 per cent
Maize	2125	n/a	8 per cent	16 per cent	25 per cent
Groundnut	1979	+8 per cent	7 per cent	31 per cent	30 per cent
Pigeonpea	1230	+6 per cent	7 per cent	3 per cent	8 per cent

* Climate change – combined effects of increased temperature and reduced rainfall, and increased CO_2 in the case of groundnut and pigeonpea, and of increased temperature and rainfall in the case of sorghum and maize

** Potential yield of the current rainfall, CO_2, temperature and radiation environment averaged over 50 seasons, with no nutrient, pest or disease constraints (Dimes *et al.*, 2008)

Clearly, the scenario of increasing temperature has the most dramatic impact on crop grain yields, at least for the two cereals, which had a reduction of 16 per cent, and particularly for groundnut, which had a 31 per cent reduction. Interestingly, pigeonpea yield was little affected by the temperature increase – its 3 per cent reduction in yield was even less than the reduction due to rainfall. Hence, for the combined effects of climate change, it appears that pigeonpea will be the least affected crop, incurring an 8 per cent reduction in potential grain yield. In contrast, groundnut can be expected to incur a 30 per cent reduction compared to current potential, sorghum a 22 per cent reduction and maize a 25 per cent reduction. However, it must be noted that of the four test crops, shortduration pigeonpea has by far the lowest current yield potential at Bulawayo.

Under the combined effects of climate change, all crops have large reductions in days to maturity (13–18 per cent), and all have a large reduction in total biomass (18–27 per cent) except pigeonpea, which shows a 4 per cent increase (Table 15.2). For maize and groundnut, there is also a decrease in average harvest index (HI) and wateruse efficiency (WUE) because the grain filling period for these two crops suffers a larger reduction (19 and 14 per cent) compared to the vegetative phase (17 and 11 per cent). Sorghum, on the other hand, experiences greater shortening of the vegetative phase (18 per cent) relative to the grain filling phase (14 per cent), resulting in increased HI while retaining a WUE of 6.7 kg ha^{-1} mm^{-1}. The higher HI

Table 15.2: Impact of Climate Change on Total Biomass, Crop Duration, Rain, Harvest Index (HI), Water Use Efficiency (WUE) of Sorghum, Maize, Groundnut and Pigeon Pea

Crop	*Baseline*					*Climate Change*				
	Total Biomass (kg ha^{-1})	*Duration (d)*	*Incrop Rain (mm)*	*HI*	*WUE (kg ha^{-1} mm^{-1})*	*Total Biomass (kg ha^{-1})*	*Duration (d)*	*Incrop Rain (mm)*	*HI*	*WUE (kg ha^{1} mm^{-1})*
Sorghum	6398	107	396	0.41	6.7	4663	88	320	0.44	6.7
Maize	6403	129	433	0.29	4.3	4747	107	352	0.28	3.9
Groundnut	4628	122	416	0.42	4.5	3782	106	345	0.37	3.8
Pigeonpea	4288	165	463	0.27	2.3	4445	136	397	0.24	2.4

(Dimes *et al.*, 2008).

for sorghum thus results from much lower total biomass, resulting in the 22 per cent reduction in grain yield shown in Table 15.1.

In contrast, pigeonpea has the largest reduction in crop duration (18 per cent), yet shows an increase in total biomass and an increase in WUE. A more favourable soil water balance for the pigeonpea explains this result. Under the current unimodal rainfall conditions (and latitude), the crop has a very long duration such that grain filling takes place under declining rainfall and increasing water stress. Higher temperatures under climate change shorten the crop duration so that it matures when the wet season is still active. This is particularly so for the duration of the grain filling period, which is reduced by 31 per cent on average.

In terms of relative change, sorghum and pigeonpea grain yield demonstrated the most resilience of the four test crops under climate change (Table 15.1) as a consequence of increased HI and WUE. However, it can be seen in Table 15.2 that WUE itself is dependent on crop duration and incrop rainfall, both of which are reduced by climate change. An important associated result is an increase in residual soil water under climate change, from 8–22mm (current climate) to 11–34mm across species. If, under climate change, the length of the rainy season is assumed to be unchanged (as in this analysis), then the predicted shortening of crop duration and higher residual soil water are indicative of opportunities for increased cropping intensity, especially on soils of higher water holding capacity.

Growth Attributing Characters of Blackgram under Elevated CO_2 and Ambient Conditions

Shoot length recorded significant increase with enhanced CO_2 levels and the values at 700 ppm were higher 550 ppm at different growth stages when compared with the chamber control and the maximum response was recorded at 45 DAS for both elevated CO_2 concentrations (Figure 15.6).

The stem dry weight increased and the maximum response was at 28 DAS at 700 ppm, 550 ppm and also in case of the increment in CO_2 level from 550 ppm to 700 ppm (Figure 15.7).

The response of root length, root dry weight and root volume was positive and significant at 550 ppm and 700 ppm levels of CO_2, as well (Figure 15.6). The improved response due to CO_2 concentration increased from 550 to 700 ppm was more significant at initial growth stages (7 DAS). The root volume was measured at 30 and 60 DAS and the increase in root volume under 700 ppm was 39.5 and 23 per cent, respectively; however, with 550 ppm the higher response was observed only at 60 DAS (10.8 per cent) and with the increase in CO_2 from 550 to 700 ppm the root volume increased by 37.5 and 13 per cent during these days of observation. The root dry weight at 700 ppm was higher than the chamber control at different growth stages (Figure 15.7).The maximum response was at 28 DAS for root dry weight in all the treatments.

The biggest leaf area was recorded at 45 DAS in all the treatments and the response was strongest at 700 ppm followed by 550 ppm. The response of leaf area to the increase in CO_2 from 550 to 700 ppm was high at early stage of germination

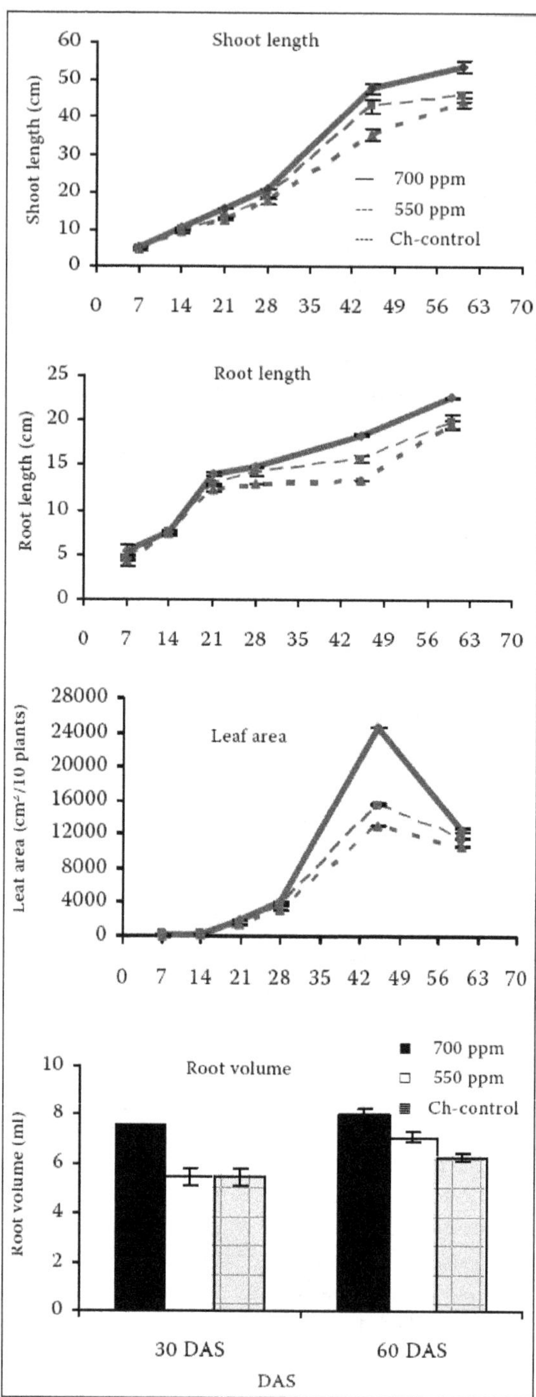

Figure 15.6: The Stem and Root Length, Leaf Area, Root Volume and Root : Shoot Ratio of Blackgram at different Growth Stages under Elevated CO_2 (700 ppm and 550 ppm) and Ambient (365 ppm) Conditions.

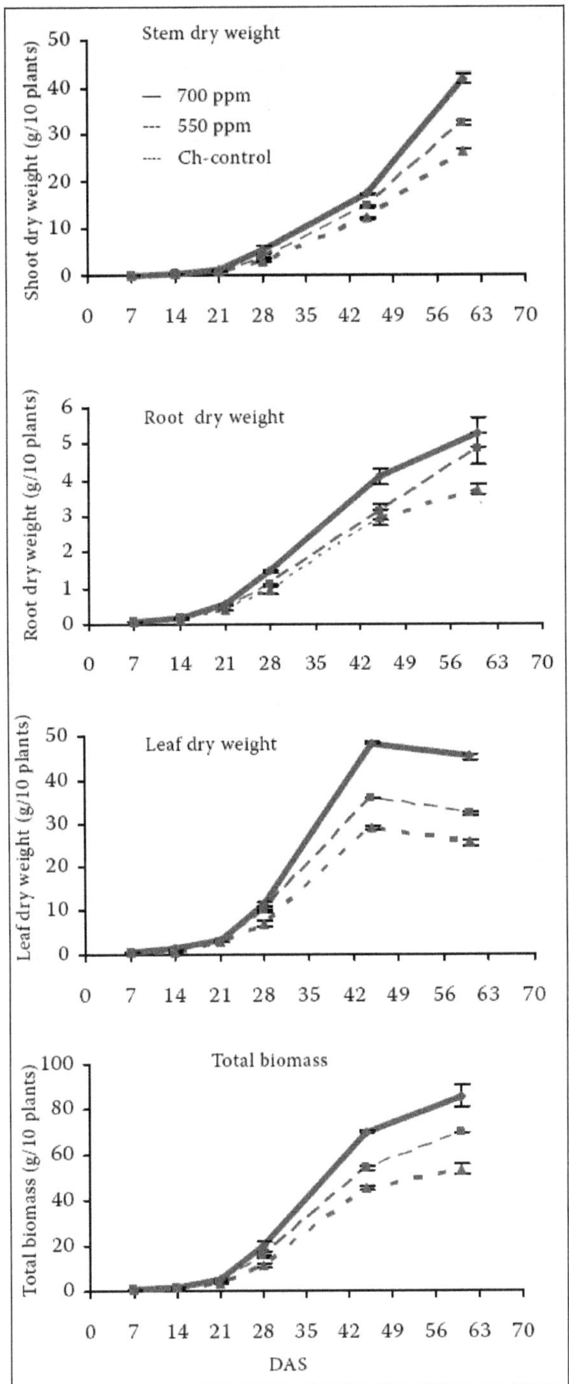

Figure 15.7: The Stem, Root, Leaf Dry Weight and Total Biomass of Blackgram at different Growth Stages under Elevated CO_2 (700 ppm and 550 ppm) and Ambient (365 ppm) Conditions (Vanaja *et al.*, 2007).

(59.4 per cent) and also at 45 DAS (58.7 per cent). The leaf dry weight response followed the trend of leaf area and at 45 DAS it reached the maximum in all the treatments. The response of leaf dry weight to different levels of enhanced CO_2 over chamber control showed a different trend. At initial growth stages (up to 14 DAS) the response was very high to both enhanced CO_2 levels hence the difference between 700 and 550 ppm CO_2 level was not so significant.

The root: shoot ratio showed a different tendency. Under Ch-control condition the ratios were significantly higher at initial growth stages (up to 14 DAS), however the difference between treatments narrowed at later stages. The total biomass showed a significant and positive response at all the growth stages under enhanced levels of CO_2, with much higher response under 700ppm (Figure 15.7). The highest response was recorded at 28 DAS for both levels of CO_2, whereas the response continued to be high up to 45 DAS in the case of the increase of the CO_2 level from 550 to 700 ppm.

Heavy Metals

Heavy metals are significant environmental pollutants, and their toxicity is a problem of increasing significance for ecological, evolutionary, nutritional and environmental reasons. The term "heavy metals" refers to any metallic element that has a relatively high density and is toxic or poisonous even at low concentration (Lenntech Water Treatment and Air Purification, 2004). "Heavy metals" in a general collective term, which applies to the group of metals and metalloids with atomic density greater than 4 g/cm3, or 5 times or more, greater than water (Hawkes, 1997). However, chemical properties of the heavy metals are the most influencing factors compared to their density. Heavy metals include lead (Pb), cadmium (Cd), nickel (Ni), cobalt (Co), iron (Fe), zinc (Zn), chromium (Cr), iron (Fe), arsenic (As), silver (Ag) and the platinum group elements. Environment is defined as totally circumstances surrounding an organism or group of organisms especially, the combination of external physical conditions that affect and influence the growth, development and survival of organisms (Farlex Incorporated 2005). The environment is considered in food, and the less tangible, through no less important, the communities we live in (Gore 1997). A pollutant is any substance in the environment, which causes objectionable effects, impairing the welfare of the environment, reducing the quality of life and may eventually cause death. Such a substance has to be present in the environment beyond a set or tolerance limit. Hence, environmental pollution is the presence of a pollutant in the environment air, water and soil, which may be poisonous or toxic and will cause harm to living things in the polluted environment. Heavy metals are largely found in dispersed form in rock formations. Industrialization and urbanization have increased the anthropogenic contribution of heavy metals in biosphere. Heavy metals have largest availability in soil and aquatic ecosystems and to a relatively smaller proportion in atmosphere as particulate or vapors. Heavy metal toxicity in plants varies with plant species, specific metal, concentration, chemical form and soil composition and pH, as many heavy metals are considered to be essential for plant growth. Some of these heavy metals like Cu and Zn either serve as cofactor and activators of enzyme reactions *e.g.*, informing enzymes/substrate metal complex (Mildvan

1970) or exert a catalytic property such as prosthetic group in metalloproteins. These essential trace metal nutrient stake parts in redox reactions, electron transfer and structural functions in nucleic acid metabolism. Some of the heavy metal such as Cd, Hg and As are strongly poisonous to metal-sensitive enzymes, resulting in growth inhibition and death of organisms. An alternative classification of metals based on their coordination chemistry, categorizes heavy metals as class B metals that come under non-essential trace elements, which are highly toxic elements such as Hg, Ag, Pb, Ni (Nieboer and Richardson 1980). Some of these heavy metals are bioaccumulative, and they neither break down in the environment nor easily metabolized. Such metals accumulate in ecological food chain through uptake at primary producer level and then through consumption at consumer levels. Plants are stationary, and roots of a plant are the primary contact site for heavy metal ions. In aquatic systems, whole plant body is exposed to these ions. Heavy metals are also absorbed directly to the leaves due to particles deposited on the foliar surfaces.

Rising Temperatures and Insects Pest

- ☆ Increased development rates: more generations per season
- ☆ Insect species diversity and feeding intensity tend to increase with increasing temperatures
- ☆ Expanded overwintering ranges
- ☆ Reduced overwintering mortality

In New York a network of pheromone traps in sweet corn fields has been used to monitor corn earworm (*Helicoverpa zea*) throughout the central and western part of the state since 1994 (Petzoldt *et al.*, 2012) (Figure 15.8). Corn earworm is thought not to overwinter in upstate New York and is generally considered to be a late season, migratory pest of sweet corn, so trapping was initiated in mid-July. The graphs compare the trap catches in 1995 with those in 2003 in Eden Valley, NY. During the early years of the trap network, CEW traps remained empty until mid-late August. After an unexpected early season infestation in Eden in 1999 trapping was initiated in early June, and typically, low levels of moths are caught through the early season, increasing when the migratory flight arrives. It is yet to be determined if the earlier arrival of corn earworm indicates it is overwintering in Eden, but since CEW management recommendations are based on trap catches, it is clear that control of this pest is already costing farmers more than it did 9 years ago.

Increased CO_2 and Insect-Pest

- ☆ Generally, effects will be indirect through host plant changes
- ☆ Increased CO_2 may cause chemical changes in host plant tissue that result in increased feeding.

Field studies have found that several insect pests increase their feeding on soybeans in high CO_2 atmospheres. This is thought to be the result of insect feeding stimulation caused by increased simple sugars in the leaves of the soybean plants. Greenhouse and lab studies have shown that plants grown in high CO_2 atmosphere

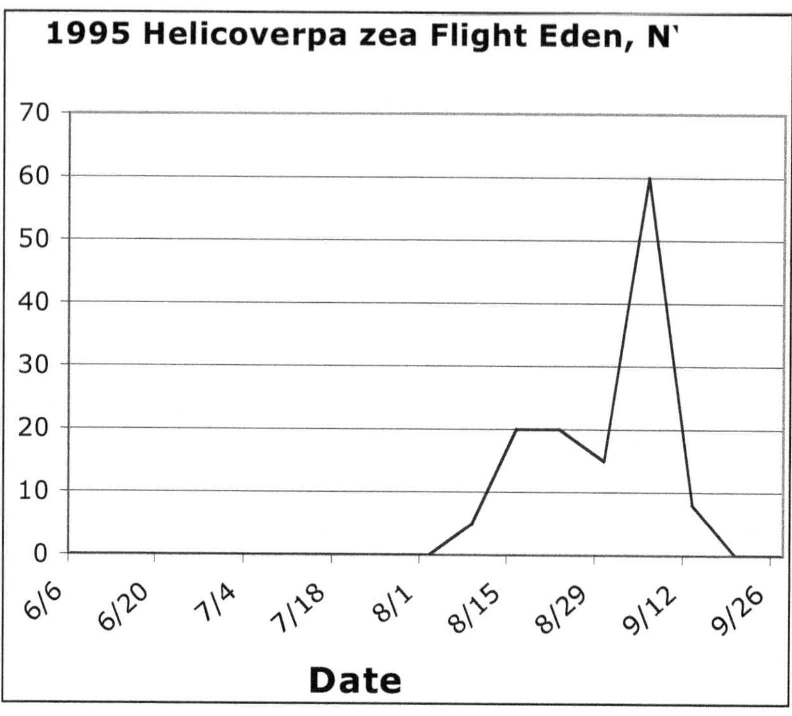

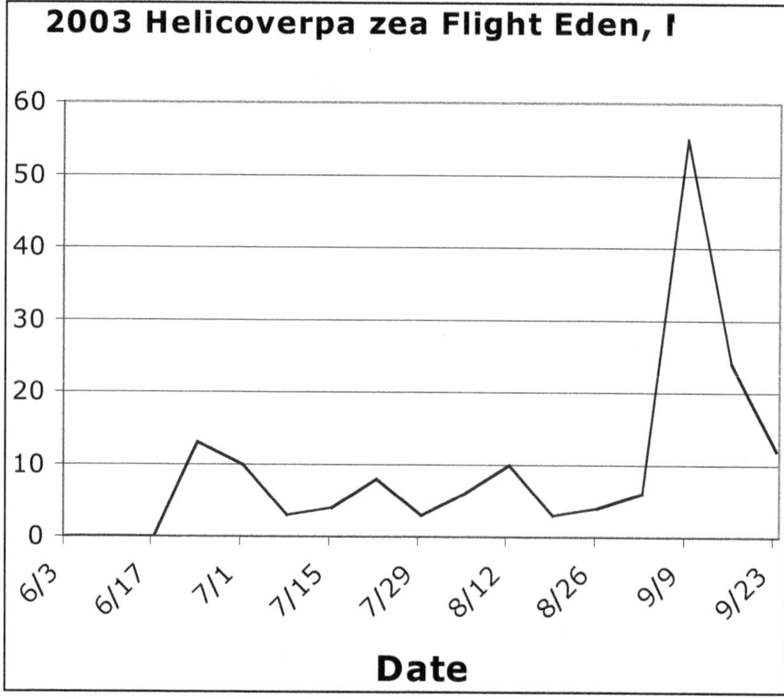

Figure 15.8: Increase in CO_2 Effected Corn Earworm (*Helicoverpa zea*) Infestation in Eden Valley, NY (Petzoldt *et al.*, 2012).

have a lower higher C: N ratio. Insects have been shown to respond to this by increasing their feeding in order to fulfill their metabolic needs for nitrogen.

Increased CO_2 Impacts on Plant Disease

- Increased CO_2 may cause increased sporulation of pathogens.
- Higher atmospheric water vapor concentrations favor fungal spore production, accelerating epidemic development
- Faster plant growth and/or ample moisture results in denser canopies, increasing humidity, favoring pathogens

Cows and Heat Stress

- Higher body temperatures
- Increased respiration rates (>70/hour)
- Less activity
- Increased water intake
- Seek shade and cool areas in the barn
- Energy requirements for maintenance will increase
- Milk yield will decrease
- Reproduction potential will decreases in heat stress situations

Climate Projections

- Average Surface temperature : Increase by 2 - 4 C during 2050s
- Monsoon Rainfall : Marginal changes in monsoon months
- Large changes during non-monsoon months
- No. of rainy days : set to decrease by more than 15 days
- Intensity of rains : to increase by 1-4 mm/day
- Cyclonic storms : Increase in frequency and intensity of cyclonic storms is projected

Adaptation Measures

Adaptation to climate change involves deliberate adjustments in natural or human systems and behaviours to reduce the risks to the people's lives and livelihoods.

a) **Crop Substitution:** Traditional crops/varieties, which are inefficient, utilize of soil moisture, less responsive to production input and potentially low producers should be substituted by more efficient ones.

b) **Changing planting dates**

c) **More use of Intercropping:** *Intercropping has a higher biological efficiency than sole cropping (usually at least 30 per cent, if farmers intercrop), because of:*

1) Better buffering against climatic extremes
2) More efficient use of resources (light, nutrients, water)
3) Less problems with pests and diseases

d) **Tillage and mulching:** Conventional tillage leaves less than 30 per cent crop residue which leads to loss of carbon and nutrients from the soil. If we go for conservation tillage which leaves more than 30 per cent crop residue can help in addition of carbon in the soil and will also help in preventing the nutrient losses through leaching. The residue left on the soil will also act as mulch and reduce the evaporation losses from the field.

Table 15.4: Climate Change Scenarios for India

Year/Scenarios	*Season*	*Temperature Change (ºC)*		*Rainfall Change (per cent)*	
		Lowest	*Highest*	*Lowest*	*Highest*
2020s	Annual	1.0	1.41	2.16	5.97
	Rabi	1.08	1.54	1.95	4.36
	Kharif	0.87	1.17	–1.81	5.1
2050s	Annual	2.23	2.87	5.36	9.34
	Rabi	2.54	3.18	–9.92	3.82
	Kharif	1.81	2.37	7.18	10.52
2080s	Annual	3.53	5.55	7.48	9.9
	Rabi	4.14	6.31	–24.83	–4.5
	Kharif	2.91	4.62	10.1	15.18

(Joshi and Amal Kar, 2009).

That recent global warming is not the result of natural processes but is instead the result of human activity, it must be shown that

- ☆ Global warming is taking place.
- ☆ Natural causes cannot account for the observed warming. (established theory fails).
- ☆ Human-activity can account for the observed warming (new theory succeeds,) plus, the new theory must be consistent with historical data.

e) **Rainwater management:** Efficient rainwater management can increase agricultural production from dry land areas. Application of compost and farmyard manure and raising legumes add the organic matter to the soil and increase the water holding capacity. The water, which is not retained by the soil, flows out as surface runoff. This excess runoff water can be harvested in storing dugout ponds and recycled to donor areas in the server stress during rainy season or for raising crops during winter.

f) Agri-Insurance:

- ☆ Protection against risk of production loss due to drought, floods, etc.
- ☆ State sponsored so far
- ☆ Covers risk for the entire insured area

Mitigation Strategies

These are the actions to reduce green house emission and sequester or store carbon in the short term and developmental choices that will be lead to low emission in the long term.

1) **Reducing emissions of carbon dioxide, methane and nitrous oxide**: Reduction in the emission of green house gases by changing the practice of transplanting rice with the direct seeded rice/aerobic rice which require less water and due to aerobic conditions the emission of CH_4 and CO_2 gases will be less.
2) **Sequestering carbon:** Sequestering atmospheric carbon is very important as without reducing the atmospheric level of carbon mitigation is not possible to mitigate the climate change we can switch over to recommended practices as given below:

Table 15.5: Agricultural Practices for Enhancing Productivity and Increasing the Amount of Carbon in Soils

Conventional Practice	*Recommended Practice*
Plough tilling	Conservation tilling/zero tillage
Residue removal/burning	Residue return as mulch
Summer fallow	Growing cover crops
Low off-farm inputs	Judicious use of fertilizers and INM
Regular fertilizer use	Site-specific soil management
No water control	Water management/conservation, irrigation, water table management
Fence-to-fence utilization	Conversion of marginal lands to nature conservation
Monoculture	Improved farming systems with several crop rotations
Drainage of wetlands	Restoring wetlands

3) **Resource conservation technologies:** Any method, material or tool which enhances the input use efficiency, crop productivity and farm gate income is termed as resource conservation technologies. It includes
   - ☆ Crop establishment system (zero tillage, minimum tillage or reduced tillage *etc.*)
   - ☆ Water management (adoption of laser land leveller technique)
   - ☆ Nutrient management (use of SPAD or SSNM or slow release fertilizers *etc.*)

4) **Enriching soil organic matter**: By applying FYM, compost or by practicing organic farming we can improve the soil organic matter which can help in improvement of soil health.
5) **Bio fuels like jatropa, sweet sorghum**: Use of bio-fuels could be an alternative source as they emit fewer amounts of gases which lead to reduction in the GHG.

Each of these points must be true in order to conclude that human activity is the cause of recent global warming.

Scientists have concluded that most of the observed warming is very likely due to the burning of coal, oil, and gas. This conclusion is based on a detailed understanding of the atmospheric greenhouse effect and how human activities have been tweaking it. At the same time, other reasonable explanations, most notably changes in the Sun, have been ruled out.

The atmospheric greenhouse effect naturally keeps our planet warm enough to be livable. Sunlight passes through the atmosphere. Light-colored surfaces, such as clouds or ice caps, radiate some heat back into space. But most of the incoming heat warms the planet's surface. The Earth then radiates some heat back into the atmosphere. Some of that heat is trapped by greenhouse gases in the atmosphere, including carbon dioxide (CO_2).

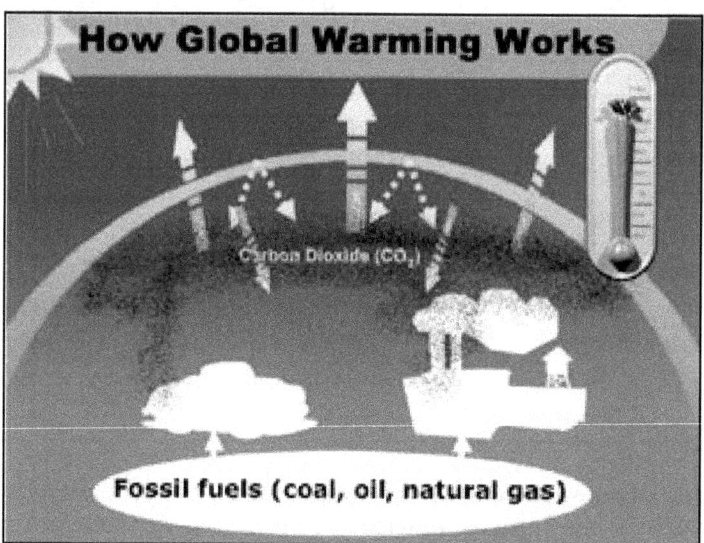

Figure 15.9

Source: http://www.nwf.org/Wildlife/Threats-to-Wildlife/Global-Warming/Global-Warming-is-Human-Caused.aspx

Human activity such as burning fossil fuels causes more greenhouse gases to build up in the atmosphere. As the atmosphere "thickens" with more greenhouse gases, more heat is held in. Fossil fuels such as oil, coal and natural gas are high in carbon and, when burned, produce major amounts of carbon dioxide or CO_2. A single gallon of gasoline, when burned, puts 19 pounds of carbon dioxide into the atmosphere.

The role of atmospheric carbon dioxide (CO_2) in warming the Earth's surface was first demonstrated by Swedish scientist Svante Arrhenius more than 100 years ago. Scientific data have since established that, for hundreds of thousands of years, changes in temperature have closely tracked with atmospheric CO_2 concentrations. Since the Industrial Revolution, the burning of coal, oil and natural gas has emitted roughly 500 billion tons of CO_2, about half of which remains in the atmosphere. This CO_2 is the biggest factor responsible for recent warming trends.

Conclusion

Abiotic and biotic stresses cause alterations in the normal physiological processes of all plant organisms, including the economically important crops. Plant damage and decrease in their productivity take place most often due to naturally occurring unfavourable factors of the environment (natural stress factors) - extreme temperatures; water deficit or abundance; increased soil salinity; high solar irradiance; early autumn or late spring ground frosts; pathogens *etc.* Recently, along with these factors plant organisms are exposed to a large scale of new stressors related to human activity (anthropogenic stress factors) – toxic pollutants such as pesticides, noxious gasses (SO_2, NO, NO_2, NOx, O_3 and photochemical smog); photooxidants; soil acidification and mineral deficit due to acid rains; overdoses of fertilizers; heavy metals; intensified UV-B irradiation etc. All these stresses decrease the biosynthetic capacity of plant organisms, alter their normal functions and cause damages which may lead to plant death (Lichtenthaler 1996, Levitt 1980).

When plants suffer many adverse situations, it causes different types of stresses which increase the production of reactive oxygen species. This damages the cells thus preventing them from reaching maximum development. Anti-oxidative defence systems are present in plants which keep ROS under control or detoxify it. Glutathione cycle is an important detoxifying system in which APX systems present in sub cellular compartments such as chloroplasts, mitochondria, *etc* play a central role in stress tolerance. Various stimulating or opposing effects of multiple stressors on ecosystems can cause technical problems in the evolution of stress indices (Niemi *et al.*, 2004) such as i). Collection and treatment of stress exhibiting data (Smith *et al.*, 2000) and ii). Combining information from various sources into an overall quantitative expression (Locantore *et al.*, 2004).

Independently of the type of stress influence (natural or anthropogenic) the accumulation of reactive oxygen species is an undeniably established fact. It is well known that water deficit (Sell *et al.*, 1992), low temperatures (Badiani *et al.*, 1993; Bridger *et al.*, 1994; Wise and Naylor, 1987), application of pesticides (Babbs *et al.*, 1989; Banas *et al.*, 1993; Kenyon and Duke, 1985; Sergiev *et al.*, 2000; Ivanov, 2003; Ivanov *et al.*, 2003a, b), noxious gasses (Wingsle and Hallgren, 1993), radiation (Dunning *et al.*, 1994), heavy metals (Pahlsson 1989), acid rains (Velikova *et al.*, 1998, 2000), high solar irradiation (Lichtenthaler 1988; 1996) etc., cause an increased production of toxic oxygen species (O_2–, H_2O_2,.OH, $^1O_2$), which are highly detrimentalto all biological systems *i.e.* an oxidative stress occurs (Foyer *et al.*, 1997; Halliwel and Gutteridge 2002).

References

Aeduini, I., Godbold, D. L. and Onnis, A., 1996. Influence of copper on root growth and morphology of *Pinus pinea* L. and *Pinus pinaster Ait.* seedlings. *Tree Physiology*, **15**: 411–415.

Ahakoni, N. and Richmond, A.E., 1978. Endogenous gibberellin and ABA content as related to senescence of detached lettuce leaves. *Plant Physiol.*, **62**: 224-229

Angelone, M. and Bini, C., 1992. Trace elements concentrations in soils and plants of Western Europe. In: Adriano D. C, ed. Biogeochemistry of trace metals. Boca Raton, FL: Lewis Publishers, 19–60.

Ankumah, R.O., W.A. Dick and G. McClung., 1995. Metabolism of carbamothioate herbicide, EPTC, by Rhodococcus strain JE1 isolated from soil. *Soil Science Society of America Journal*, **59**:1071-1077.

Ashraf, M., Mcneilly, T. and Bradshaw, A.D., 1986. The response to NaCl and ionic content of selected salt-tolerant and normal lines of three legume forage species in sand culture. - *New Phytol.*, **104**: 463-471.

Babbs, C., Pham, J. and Coolbaugh, R., 1989. Lethal hydroxyl radical production in paraquat-treated plants. *Plant Physiol.*, **90**: 1267–1270.

Badiani, M., Paolacci, A., D'Annibale, A. and Sermanini, G., 1993. Antioxidant and Photosynthesis in the leaves of *Triticum durum* L. seedlings acclimated to low, non-chilling temperatute. *J. Plant Physiol.*, **142**: 18–24.

Banas, A., Johansson, I., Stenlid, G. and Stymne, S., 1993. Free radical scavengers and inhibitors of lipooxygenases as antagonists against the herbicides haloxyfop and alloxydim. *Swedish J. Agric. Res.*, **23**: 67–75.

Binzel, M.L., Hess, F.D., Bressan, R.A. and Hasegawa, M., 1988. Intracellular compartmentation of ions in salt adapted tobacco cells. *Plant Physiol.*, **86**: 607-614.

Bohnert, H.J. and Jensen, R.G., 1996. Strategies for engineering water-stress tolerance in plants. *Trends Biotechnol.*, **14**:89-97.

Bray, E.A., 1997. Plant responses to water deficit. *Trends Plant Sci.*, **2**:48-54.

Breckle, C.W., 1991. Growth under heavy metals. In: Waisel, Y., Eshel, A., Kafkafi, U. eds. Plant roots: the hidden half. New York, NY: Marcel Dekker, 351–373.

Bridger, G., Yang, W., Falk, D. and McKerstie, B., 1994. Cold acclimation increased tolerance of activated oxygen in winter cereals. *J. Plant Physiol.*, **144**: 235–240.

Cairns John Jr., 2013. Environmental Stress. In: Levin S.A. (ed.) Encyclopedia of Biodiversity, second edition, Waltham, MA: Academic Press, Volume 7, pp. 39-44.

Davenport, T.L., Morgan, P.W. and Jordan, W.R., 1980. Reduction of auxin transport capacity with age and internal water deficit in cotton petioles. *Plant Physiol.*, **65**: 1023-1025.

Di Toppi, L. S., Lambardi, M., Pazzagli, L., Cappugi, G., Durante, M. and Gabbrielli, R., 1999. Response to cadmium in carrot in vitro plants and cell suspension cultures. *Plant Science*, **137**: 119–129.

Dimes, J., Cooper, P. and Rao, K. P. C., 2008. Climate change impact on crop productivity in the semiarid tropics of Zimbabwe in the 21st century. CGIAR Challenge Program on Water and Food, pp 189-198.

Dunning, C., Chalker-Scott, L. and Scott, J., 1994. Exposure to UV-B radiation increased cold hardiness in Phododendron. *Physiol. Plant.*, **92**: 516–520.

El-Shahaby, O. A., 1981. Studies on growth and metabolism of certain plants. Ph. D. Thesis, Mansoura Univ., Mansoura, Egypt.

El-Shahaby, O.A., Abo-Hamed, S.A. and Haroun S.A., 1990. Plant growth, metabolism and adaptation in relation to stress conditions. X. Hormonal control of growth and pigment content of salinized *Pisum sativum* plants. — Submitted to Acta bot. neerl.

El-Shahaby, O.A., Younis, M.E., Abo-Hamed, S.A. and Haroun, S.A., 1990. Plant growth, metabolism and adaptation in relation to stress conditions. VIII. Water stress induced changes in growth, pigment and metabolite contents of *Pisum sativum* plants. - Submitted to Field Crops Research.

Epstein, E., 1961. The essential role of calcium in selective cation transport by plant cells. *Plant Physiol.*, **36**: 437-444.

Farlex Incorporated 2005. Definition: environment, the free dictionary, Farlex Inc. Publishing, USA. Source: http://www.thefreedictionary.com

Foyer, C. H., Lopez-Delgado, H., Dat, J. E. and Scott, I. M., 1997. Hydrogen peroxide and glutathione- associsted mechanisms of acclimatory stress tolerance and signalling. *Physiol. Plantarum*, **100**: 241–254.

Fuhrer, J., 2003. Agroecosystem responses to combinations of elevated CO_2, ozone and global climate change. *Agric Ecosys Environ.*, **97**:1–20.

Galloway, J. N., Thornton, J. D., Norton, S. A., Volcho, H. L. and McLean, R. A., 1982. Trace metals in atmospheric deposition: a review and assessment. *Atmospheric Environment*, **16**: 1677.

Godbold, D. L. and Huttermann, A., 1985. Effect of zinc, cadmium and mercury on root elongation of Picea abies (Karst.) seedlings, and the significance of these metals to forest die-back. *Environmental Pollution*, **38**: 375–381.

Gore, A., 1997. Respect the land, our precious olant. Time Magaz, **150**:8–9.

Goudrian, J.and Zadoks, J.C., 1995. Global climate change: modelling the potential response of agroecosystems with special reference to crop protection. *Environ Pollut.*, **87**:215–224.

Halliwell, B. and Gutteridge, J., 2002. Free radicals in biology and medicine. Third edition, Oxford Univ. Press.

Hasaneen, M.N.A., Younis, M.E. and El-Saht, H.M., 1989. Plant growth, metabolism and adaptation in relation to stress conditions. VI. Differential effects of sodium sulphate on growth and element conditions. VI. Differential effects of sodium sulphate on growth and element content of *Phaseolus vulgaris* and *Zea mays*. - Qatar Univ. Sei. Bull., **9**: 113-123.

Hasaneen, M.N.A., Younis, M.E., El-Shahaby, O.A. and Hussein M.H., 1987. Plant growth, metabolism and adaptation in relation to stress conditions. III. Effects of different levels of salinity on element and acid composition of germinating *Viciafaba* seeds. *Mans. Sei. Bull. Mansoura, Egypt.*, **14**(1): 167-184.

Hawkes, J.S., 1997. Heavy metals. *J Chem Edu.*, **74**:1369–137.

Heagle, A.S., Miller, J.E. and Pusley, W.A., 1998. Influence of ozone stress on soybean response to carbon dioxide enrichment III. Yield and seed quality. *Crop Sci.*, **38**:128–134.

Heagle, A.S., Miller, J.E. and Pusley, W.A., 2000. Effects of ozone and carbon dioxide interactions on growth and yield of winther wheat. *Crop Sci.*, **40**:1656–1664.

Heagle, A.S., Miller, J.E., Booker, F.L. and Pusley, W.A., 1999. Ozone stress, carbon dioxide enrichment and nitrogen fertility interactions in cotton. *Crop Sci.*, **39**:731–741.

Heikal, M.M., Ahmed, A.M. and Shaddad, M.A., 1980. Changes in dry weight and mineral composition of some oil producing plants over a range of salinity stresses. *Biol. Plant*, **22**: 25-33.

Henderson, S., Holman, S. and Mortensen, L., 2002. (Eds.) Human Activity and Climate Change Modified with permission from Global Climates - Past, Present, and Future, EPA Report No. EPA/600/R-93/126. U.S. Environmental Protection Agency, Office of Research and Development, Washington, DC., pp. 91 - 94. Source: http://www.ucar.edu/learn/1_4_2_20t.htm

Ignarajah, K., Jennings, D. and Handley, F., 1975. The effect of salinity on growth of *Phaseolus vulgaris* L. II. Effect of internal solute concentration. *Ann. Bot.*, **39**: 1039-1055.

Imamul-Huq, S.M. and Larher, F., 1983. Osmoregulation in higher plants: Effects of NaCl salinity on non-nodulated *Phaseolus* Osmoregulation in higher plants: Effects of NaCl salinity on non-nodulated *Phaseolus aureus* L. I. Growth and mineral content. *New Phytol.* **93**: 203-208.

IPCC, 2007. In: Solomon, S., Qin, D., Manning, M., Chen, Z., Marquis, M., Averyt, K.B., Tignor, M. and Miller, H.L. (Eds.), Climate Change 2007: The Physical Science Basis. Contribution of I to the Fourth Assessment Report of the Intergovernmental Panel on Climate Change. Cambridge University Press, Cambridge, United Kingdom and New York, NY, USA, pp. 996.

Iriti., M. and Faoro, F., 2007. Review on innate and specific immunity in plants and animals. *Mycopathol*, **164**:57–64.

Ivanov, S. V., Alexieva, V. S. and Karanov, E. N., 2003a. Cumulative effect of low and high atrazine concentrations on *Arabidopsis thaliana* (L.) Heynh plants. *Russian J. Plant Physiol.* (in press).

Ivanov, S., 2003. Effect of sub-herbicide concentrations of atrazine and 2, 4-D and their interaction with high temperature on some physiologo-biochemical parameters in pea plants. PhD thesis.

Ivanov, S., Alexieva, V. and Karanov, E., 2003b. Effect of interaction between sub-herbicide concentrations of 2,4-D and high temperatures on the activities of some stress defence enzymes in pea (*Pisum sativum* L.) plants. *Compt. Rend. Acad. Bulg. Sci.*, **56**(6), 67–72.

Jackson, A. P. and Alloway, B. J., 1991. The transfer of cadmium from sewage sludge amended soils into the edible component of food crops. *Water, Air and Soil Pollution*, **57**: 873–881.

Jain, R.K. and Selvaraj, G., 1997. Molecular genetic improvement of salt tolerance in plants. *Biotechnol Annu Rev.*, **3**:245-267.

Joshi, N. L. and Kar, A., 2009. Contingency crop planning for dryland areas in relation to climate change. *Indian Journal of Agronomy*, **54**(2): 237-243.

Kenyon, W. and Duke, S., 1985. Effect of acifluoren on endogenous antioxidants and protective enzymes in cucumber cotyledons. *Plant Physiol.*, **79**: 862–866.

Langebartels, C., Wohlgemuth, H., Kschieschan, S., Gr¨un, S. and Sandermann, H., 2002. Oxidative burst and cell death in ozone-exposed plants. *Plant Physiol Biochem.*, **40**:567–575.

Lantsy, R.J. and Mackensie, F.T., 1979. Atmospheric trace metals: global cycles and assessment of man's impact. *Geochimica et Cosmochimica Acta* 43, 511.

Lenntech Water Treatment and Air Purification 2004. Water treatment. Lenntech, Rotterdamseweg, Netherlands. Source: http://www. excelwater.com/thp/filters/Water- Purification.htm

Levitt, J., 1980. Responses of Plants to environmental stresses. Vol. 1, Acad. Press, 496.

Lichtenthaler, H. K., 1996. Vegetation stress: an introduction to the stress concept in plants. *J. Plant Physiology*, **148**: 4–14.

Lichtenthaler, H., 1988. In vivo chlorophyll fluorescence as a tool for stress detection in plants. In: Applications of chlorophyll fluorescence, Eds. H. Lichtenthaler, Kluwer Acad. Publ., 129– 42.

Locantore, N.W., Tran, L.T., O'Neill, R.V., McKinnis, P. W., Smith, E.R. and O'Connell, M., 2004. An overview of data integration methods for regional assessment. *Envir Monit Assess*, **94**: 249–261

Lorenzini, G., Medeghini, B., Nali, C., Baroni, F. R. 1994. The protective ffect of rust infection against ozone, sulphur dioxide and paraquat toxicity symptoms in broad bean. *Physiol Mol Plant Pathol.*, **45**:263–279.

Manning, W.G. and Von Tiedemann, A., 1995. Climate change: potential effects of increased atmospheric carbon dioxide (CO_2), ozone (O_3) and ultraviolet-B (UV-B) radiation on plant diseases. *Environ Pollut.*, **88**:219–245.

McKee, I.F., Eiblmeier, M.and Pollc, A., 1997. Enhanced ozone-tolerance in wheat grown at an elevated CO_2 concentration: ozone exclusion and detoxification. *New Phtytol.*, 137:275–284.

McKee, I.F., Farage, P.K. and Long, S.P., 1995. The interactive effects of elevated CO_2 and O_3 concentration on photosynthesis spring wheat. *Photosynth Res.*, **45**:111–119.

Mehrotra, C.L., 1971. Salt tolerance of some agricultural crops during early growth stages. *Indian J. Agric. Sei.*, **41**: 882-890.

Mildvan, A.S., 1970. Metal in enzymes catalysis. In: Boyer DD (ed) The enzymes, vol 11. Academic Press, London, pp 445–536.

NATCOM, 2004. India's initial national commission to the united nation's framework convention on climate change. National communication project Ministry of Environment and Forests, Govt. of India.

Nelson, D.E., Shen, B. and Bohnert, H.J., 1998. Salinity tolerance — mechanisms, models, and the metabolic engineering of complex traits. In *Genetic Engineering*, vol 20. Edited by JK Setlow. New York: Plenum Press.

Nieboer, E. and Richardson, DH.S., 1980. The replacement of the nondescript term 'heavy metal' by a biologically significant and chemically significant classification of metal ions. *Environmental Pollution*, **B1**: 3–26.

Nieboer, E. and Richardson, D.H.S., 1980. The replacement of the nondescript term heavy metals by a biologically and chemistry significant classification of metal ions. *Environ Pollut Series B* 1:3–26.

Niemi, G. J., Wardrop, D., Brooks, R., Anderson, S., Brady, V., Paerl, H., Rakocinski, C., Brouwer, M., Levinson, B. and McDonald, M., 2004. Rationale for a new generation of indicator for coastal waters. *Envir Health Perspect*, **112**, 979–986

Nies, D.H., 1999. Microbial heavy-metal resistance. *Applied Microbiology Biotechnology*, **51**: 730–750.

Pahlsson, A. M., 1989. Toxicity of heavy metals to plants. *Water Air and Soil Pollut.*, **47**: 287– 19.

Petzoldt, C. and Seaman, A., 2012. Climate Change Effects on Insects and Pathogens. *ed.*: Climate Change and Agriculture: Promoting Practical and Profitable Responses. Source: http://umaine.edu/oxford/files/2012/01/III.2Insects.Pathogens1.pdf.

Rivetta, A., Negrini, N. and Cocucci, M., 1997. Involvement of Ca^{2+} calmodulin in Cd^{2+} toxicity during the early phases of radish (*Raphanus sativus* L.) seed germination. *Plant, Cell and Environment*, **20**: 600–608.

Sell, W., Hendry, G. and Lee, J., 1992. Effect of desiccation of some activated oxygen processing enzymes and anti-oxidants in mosses. *J. Exp. Bot.*, **43**(253), 1031–1037.

Sergiev, I., Alexieva, V., Yanev, S. and Karanov, E., 2000. Effect of atrazine and spermine on free proline and some antioxidants in pea (*Pisum sativum* L.) plants. *Comp. Rend. Acad. Bulg. Sci.*, **53**(10), 63–66.

Smith, E.R., O'Neill, R.V., Wickham, J.D., Jones, K.B., Jackson, L., Kilaru, J.V. and Reuter, R., 2000. The U.S. EPA's Regional Vulnerability Assessment Program: a research strategy for 2001–2006. U.S. Environmental Protection Agency, Office of Research and Development, Research Triangle Park, North Carolina.

Solomon, S., Qin, D., Manning, M., Chen, Z., Marquis, M., Averyt, K.B., Tignor, M. and Miller, H.L., 2007. (eds.). IPCC (2007): *Climate Change 2007: The Physical Science Basis. Contribution of Working Group I to the Fourth Assessment Report of the Intergovernmental Panel on Climate Change* Cambridge University Press, Cambridge, United Kingdom and New York, NY, USA

Street, H.E. and Cockburn, W., 1970. Plant metabolism. 2nd edition. Pergamon Press. New York.

Van Assche, F. and Clijsters, H., 1986. Inhibition of photosynthesis in Phaseolus vulgaris by treatment with toxic concentration of zinc: effect on ribulose-1, 5-bisphosphate carboxylaseu oxygenase. *Journal of Plant Physiology*, **125**: 355–360.

Vanaja, M., Reddy, P.R., Lakshmi, N.J., Maheswari, M., Vagheera, P., Ratnakumar, P., Jyothi, M., Yadav, S.K. and Venkateswarulu, B., 2007. Effect of elevated atmospheric CO_2 concentration on growth and yield of balckgram (*Vigna mungo* L. Hepper) a rainfed pulse crop. *Plant Soil Environ*, **53**(2), 81-88.

Velikova, V., Yordanov, I. and Edreva, A., 2000. Oxidative stress and some antioxidant systems in acid rain-treated bean plants. Protective role of exogenous polyamines. *Plant Sci.*, **151**: 59–66.

Velikova, V., Yordanov, I., Georgieva, K., Tsonev, T. and Goltsev, V., 1998. Effect of exogenous polyamines applied separately and in combination with simulated acid rain on functional activity of photosynthetic apparatus. *J. Plant Physiol.*, **153**: 299–307.

Violini, G., 1995. Ozone and plant pathogens: an overview. *Riv Pat Veg.*, **5**:113–130.

Von Tiedemann, A. and Firsching, K.H., 2000. Interactive effects of elevated ozone and carbon dioxide on growth and yield of leaf rust-infected versus non-infected wheat. *Environ Pollut.*, **72**:205–224.

Weast, R.C., 1984. CRC Handbook of Chemistry and Physics, 64th *edn*. Boca Raton, CRC Press.

Wickham, J.D., Jones, K.B., Riitters, K.H., O'Neill, R.V., Tankersly, R.D. and Smith, E.R., 1999. An integrated environmental assessment of the U.S. mid-Atlantic region. *Envir Manage* **24**: 553–560.

Wildner, G.F. and Henkel, J., 1979. The effect of divalent metal ion on the activity of Mg^{2+} depleted ribulose-1,5-bisphosphate oxygenase. *Planta*, **146**: 223–228.

Wilson, J.R., Haydock, K.P. and Robins, M.F., 1970. Response to salinity in *Glycine.* 5-Changes in the chemical composition of three Australian species of G. *wrightii* over a range of salinity stress. *Aust. J. Expt. Agric. Ani. Hus.*, **10**: 156-166.

Wingsle, G. and Hallgren, J.E., 1993. Influence of SO_2 and NO_2 exposure on glutathione, superoxide dismutase and glutathione reductase activities in scots pine needles. J. *Exp. Bot.*, **44**(259), 463–470.

Wise, R. and Naylor, A., 1987. Chilling enhanced photooxidation. *Plant Physiol.*, **83**: 272-277.

Younis, M.E. and El-Tigani, S., 1970. The role of N-dimethylamino-succinamic acid in the control of vegetative development of tomato plants. Z. *Pflanzenphysiol.*, **63**: 225-237.

Zeevaart, J.A.D., 1983. Metabolism of ABA and its regulation in *Xanthium* leaves during and after water stress. *Plant Physiol.*, **71**: 477-481.

Ziska, L.H., Teasdale, J.R., and Bunce, J.A., 1999. Future atmospheric carbon dioxide concentrations may increase tolerance to glyphosate. *Weed Science*, **47**: 608-615.

Ziska, L.H., 2003. Evaluation of yield loss in field sorghum from a C_3 and C_4 weed with increasing CO_2. *Weed Science*, **51**:914-918.

2016, Recent Advances in Plant Stress Physiology *Pages* **329–377**
Editors: ***Praduman Yadav, Sunil Kumar and Veena Jain***
Published by: **DAYA PUBLISHING HOUSE, NEW DELHI**

Chapter 16

Breeding for Abiotic Stress Tolerance in Crop Plants

H.P. Meena[1]*, Naresh Kumar Bainsla[2] and D.K. Yadav[3]

[1]ICAR – Indian Institute of Oilseeds Research, Rajendranagar, Hyderabad – 500 030, India
[2]Division of Field Crops, ICAR – Central Agricultural Research Institute, Port Blair, Andaman and Nicobar Islands, India
[3]Department of Agriculture, Buchawas (Mahendergarh), Haryana, India

ABSTRACT

Human exploitation of earth's resources is leading to new problems day by day and hence newer methods are devised to overcome the same. The long-term goal of crop improvement for abiotic stress tolerance in plants is a traditional objective of breeders. World population is expected to increase by 1.8 billion as of 2030 and by 2.5 billion as of 2050, reaching 9.2 billion. By 2030 food demand is expected to increase by 50 percent because of continued population growth and higher incomes (www.popcouncil.org). If global population reaches 9.1 billion by 2050, the FAO says that world food production will need to rise by 70 per cent, and food production in the developing world will need to double. On the other hand food productivity is decreasing due to the effect of various abiotic stresses; therefore minimizing these losses is a major area of concern for all nations to cope with the increasing food requirements. However, abiotic stresses and climate change are becoming increasingly serious threats to crop production worldwide at a time when food staple supply will need to be significantly higher to meet the demand of the growing human population. Drought is the main abiotic factor followed by soil salinity. Now-a-days, agriculture has new huge challenges due to population growth, the pressure on agriculture liability on the environmental conservation, and climate change. To cope with these new challenges, many plant breeding programs have reoriented their breeding scope to stress tolerance in the last years. One such case is breeding for abiotic stress, utilizing even non arable land, to feed the

**Corresponding author.* E-mail: hari9323@gmail.com

ever increasing population. Research is the key to meeting the challenges of modern agriculture in a sustainable and positive fashion. So, in this chapter, emphasis has been given on importance of abiotic stresses specially on drought and salinity, sources of abiotic stress tolerance, genetics of abiotic stress tolerance, plant breeding presents the conventional (selection and introduction, pedigree method, modified bulk pedigree method, shuttle breeding, mutation breeding, diallele selective mating system supplemented by MAS, back cross method and recurrent selection), and most recent advances and discoveries (somaclonal approach, F_1 anther culture, marker assisted selection and genetically modified crop) applied to abiotic stresses, discussing the breeding methods, and modern molecular biological approaches to develop improved cultivars tolerant to most sorts of abiotic stresses. In this chapter more discussed about successful example of conventional breeding methods and non conventional methods because the breeding procedure of each breeding methods is not possible in one chapter.

Introduction

Abiotic stress is the primary cause of crop loss worldwide, reducing average yields for most major crop plants by more than 50 per cent (Boyer, 1982; Bray *et al.*, 2000). The major abiotic stresses (drought, high salinity, cold and heat) negatively influence the survival, biomass production and yields of staple food crops up to 70 per cent (Vorasoot *et al.*, 2003; Kaur *et al.*, 2008; Thakur *et al.*, 2010; Rodriguez *et al.*, 2005; Acquaah, 2007) hence, threaten the food security worldwide. However, the productivity of crops is not increasing in parallel with the food demand. Drought and salt stress, together with low temperature, are the major problems for agriculture worldwide (Caccarelli and Grando, 1996; Epstein *et al.*, 1980; Abrol *et al.*, 1988) because these adverse environmental factors prevent plants from realizing their full genetic potential. The great concern is that these stresses (drought and salinity) will be increasingly important due to climate change, land degradation and declining water quality (Tester and Langridge, 2010; Witcombe *et al.*, 2008). The scientists working in stress tolerance are becoming important not only because agricultural production need to keep pace with increasing demand for agricultural produce from dwindling resources, but also due to possible changes in climate that may make the environment much more hostile for agricultural production than what it is today. Similarly in India also 67 per cent of the area is rainfed and crops in these areas invariably experience droughts of different magnitudes. Although in the country 33 per cent of the cropped area is under irrigation, but here also crop production is constrained by environmental stresses like thermal. Further due to climate change the levels of stresses may further increase that may adversely influence the crop yields. Curtailing crop losses due to various environmental stressors is a major area of concern to cope with the increasing food requirements (Shanker and Venkateswarlu, 2011).

The task of developing genotypes with a higher yield under stress has been a difficult one, due to the lower heritability of the character under selection (Byerlee and Moya, 1993; Trethowan *et al.*, 2002). Relief from abiotic stresses is possible either by changing/avoiding the environment or changing the genotype of the plant itself. Breeding for abiotic tolerance may be direct (selection pressure under stress) or indirect (selection pressure under stress-free environment) (Lewis and Christiansen, 1981). The recent advent of molecular tools, especially molecular markers, has revolutionized the genetic analysis of crop plants and provided not only geneticists but also physiologists, agronomists, and breeders with valuable new tools to identify traits of importance in improving resistance to abiotic stresses (Quarrie, 1996). In fact these stresses, threaten the sustainability of agricultural industry. It may be pointed out that the relative importance of different abiotic stresses varies greatly depending on the geographical region and the specific location within a given region. The objective of this chapter is to report the importance of conventional breeding methods for abiotic stress tolerance and the recent biotechnological applications in plants.

Why Study Abiotic Stress Tolerance in Plants?

1. Abiotic stresses, such as high or low temperature, drought and salinity are most important factors that limit the geographical distributions of plants on earth.
2. Abiotic stresses usually cause severe loss of crop yields.
3. Abiotic stresses affect global crop production system and endanger food security of human being.

Importance of Abiotic Stress

According to world estimates, an average of 50 per cent yield losses in agricultural crops is caused by abiotic factors (Wang *et al.*, 2007; Oerke *et al.*, 1994; Theilert, 2006). These comprise mostly of high temperature (40 per cent), salinity (20 per cent), drought (17 per cent), low temperature (15 per cent) and other forms of stresses (Ashraf *et al.*, 2008). Only 9 per cent of the world area is conducive for crop production, while 91 per cent is afflicted by various stresses. As per the current estimates (ICAR, 2010), 120.8million ha constituting 36.5 per cent of geographical area in India are degraded due to soil erosion, salinity/alkalinity, soil acidity, water logging, and other edaphic problems. In the context of global warming and climate change, further severe productivity loss due to escalating adverse effects of abiotic stresses can be expected. Among environmental factors, as given in Table 16.1 abiotic stress are the major constraints limiting the productivity (Buchanan *et al.*, 2000).

Drought

Among all abiotic stresses, drought stress is considered to be a major threat to sustaining food security under current and more so in future climates. About 58 per cent (80 M ha) of the net sown area in India continues to be rainfed that contributes 40 per cent of the food grain production and supports two-third of the livestock population. Especially 85 per cent of coarse cereals, 83 per cent of pulses and 70 per cent of oilseeds continue to be rainfed. Worldwide, crop production is limited by drought more than by any other environmental stress (Cattivelli *et al.*, 2008). Drought stress moreover occurs across regions, countries, and continents, for example in South Asia; severe drought years during 1987 and 2002/2003 affected more than 50 per cent of the total cropped area and almost 300 million people in India (Pandey *et al.*, 2007), in south East Asia; drought in 2004 affected 20 per cent of the rice land and more than eight million people in Thailand (Pandey *et al.*, 2007) and among the globally recorded disasters over the past three decades, 20 per cent are accounted

Table 16.1: Record Yield, Average Yields and Yield Losses Due to Biotic and Abiotic Stress of Eight Major Crops

Crop	*Record Yield (kg/ha)*	*Average Yield (kg/ha)*	*Biotic Stresses*	*Abiotic Stresses*	*Yield Loss (Per cent of record yield)*	
					Abiotic Stress	*Biotic Stress*
Corn (*Zea mays* L.)	19300	4600	1952	12700	65.8	10.1
Wheat (*Triticum aestivum* L.)	14500	1880	726	11900	82.1	5.0
Soybean (*Glycine max* L.)	7390	1610	666	5120	69.3	9.0
Sorghum (*Sorghum bicolor* L.)	20100	2830	1051	16200	80.6	5.2
Oat (*Avena sativa* L.)	10600	1720	924	7690	75.1	8.7
Barley (*Hordium vulgare* L.)	11400	2050	765	8590	75.4	6.7
Potato (*Solanum tuberosum* L.)	94100	28300	17775	50900	54.1	18.9
Sugar beet (*Beta vulgaris* L.)	121000	42600	17100	61300	50.7	14.1

Source: Buchanan *et al.*, 2000.

by Africa with nearly half of them caused by extreme weather, particularly drought (Cornford, 2003). India has faced 26 drought years in the last 130 years; with 1987 and 2002 and 2012 being the major drought years in recent times. The drought of 2002 led to reduced acreage of more than 15Mha of the *kharif* crops and resulted in a loss of more than 10 per cent in food production. Only about 15 per cent of the world's agricultural land is irrigated, but irrigated land accounts for almost half of global food production (Fereres and Connor, 2004). The seriousness of drought stress depends on its timing, duration and intensity (Serraj *et al.*, 2005).

Salinity

Agricultural productivity is severely affected by soil salinity because salt levels harmful to plant growth affect large terrestrial areas. Soil salinity is becoming a more acute problem, primarily because of declining irrigation water quality (Rhoades and Loveday, 1990; Ghassemi *et al.*, 1995; Flowers, 2004). Salinity leads to lower agricultural production in arid and semi-arid regions (Bai *et al.*, 2011). As per the report of FAO, 2010, over 800 million ha of worldwide land are severely salt affected and approximately 20 per cent of irrigated areas (about 45 million ha) are estimated to suffer from salinization problems by various degrees. In Asia, 21.5 million hectares of land areas are affected by salinity and estimated to cause the loss of up to 50 per cent fertile land by the 21st midcentury (Nazar *et al.*, 2011). Increased salinization may affect up to 50 per cent of arable land by the year 2050 (Wang *et al.*, 2003). By year 2025, 60-70 per cent more food will be needed from water resources similar to those available today. This condition is alarming and necessitates increasing agricultural productivity per unit of land and water. The development of crop plants able to grow in these adverse conditions assumes importance in the emerging global context.

Sources of Abiotic Stress Resistance

There are several informations on the sources of abiotic tolerance in crop plants. Tolerance may exist in land races, wild relatives, high yielding varieties, initial breeding materials and advance breeding materials. Landraces from dry habitats have been used successfully in breeding toward developing open pollinated varieties or hybrids for water limited environments. Wild species and progenitors of our cultivated crops were always on the agenda as possible donors for abiotic stress. The choice of genetic resource to use as donor for abiotic stress tolerance depends on the probability of discovering the required genes as well as the expected difficulties and projected success in introgression of these genes into the chosen recurrent cultivar. Biotic and abiotic stress may be introduced by genetic engineering.

Cultivated Varieties

Large genetic variation for drought and salinity resistance attributes exists in the breeding materials, and even in some of the improved cultivars of different crop species. The first priority of a breeder should be to identify and utilize such sources of drought resistance since this is least problematic of all sources.

Land Races

Land races (old or desi varieties) developed in and adapted to drought and salinity environments are ordinarily valuable sources of drought and salinity resistance attributes. Utilization of these materials in breeding programmmes may be somewhat problematic due to undesirable linkages. When the variability present in the elite breeding materials and land races is exhausted, only then efforts should be focused on using wild relatives.

Table 16.2: Wild Sources of Resistance to Drought and Salinity in some Crop Plants

Name of Crop	*Name of Wild Species*	*Resistance Available for*
Wheat (*Triticum aestivum* L.)	*Aegliops kotsehyi*	Drought
	Ae. variabills	Drought
	Ae. speltoider	Drought
	Ae. umbellulata	Drought
	Ae. squarrosa	Drought
	Agropyron seirpea	Salinity
	A. pontica	Salinity
	A. junceiformis	Salinity
	A. cliac	Salinity
	Triticum urartu	Drought
	T. boeticum,	Drought
	T. dicoccoide	Drought
	A. geniculata	Drought
	A. taushii	Drought
	T. (Aegilops) tauschii	Salinity
Rice (*Oryza sativa* L.)	*Porteresia coarctata*	Salinity
Sugarcane (*Saccharum officinarum* L.)	*Saccharum spontaneum*	Drought and Salinity
	Lycopersicon cheesmanii	Salinity
Sesame (*Sesamum indicum* L.)	*Sesamum angustifolium*	Drought
	S. malabaricum	Heavy rainfall
	S. laciniatum	Drought
	S. prostratum	Salinity tolerance
	S. occidentale	Drought
	S. radiatum	Drought
Sunflower (*Helainthus annuus* L.)	*H. argophylus*	Drought
	H. debilis	Salinity
	H. paradoxus	Salinity and alkalinity
	H. annuus populations	Salinity

Wild Relatives

Wild sources of drought and salinity resistance have been reported in wheat, sugarcane, tomato, and several other crops (Table 16.2). When the source of drought and salinity resistance is in a wild species, the transfer of resistance poses several problems such as cross incompatibility, hybrid inviability, hybrid sterility and linkage of several undesirable genes with desirable ones.

Genetics of Abiotic Stress Tolerance

To plan efficient breeding programs for developing abiotic stress tolerant varieties, information on the genetic basis of abiotic stress tolerance, mode of inheritance, magnitude of gene effects, heterosis, combining ability and their mode of action are necessary.

Inheritance Studies

The study of genetic mechanism controlling abiotic stress is the foremost important. The genetic control of the drought tolerant characters ranges from oligogenic to polygenic. Generally leaf characters like waxy bloom and glabourous leaves are under monogenic control. Some other traits like ABA accumulation in wheat, constitutive proline accumulation in barley mutant and resistance to flower abscission and ability to support pod formation in rajma (*Phaseolus vulgaris*) seems to be determined by oligogenes (Singh, 2003). Polygenic inheritance of root traits has been reported by Ekanayake *et al.* (1985). Leaf rolling (Singh and MacKill, 1991) and osmotic adjustment have shown monogenic inheritance. The relative water content of rice was noted as quantitative character controlled by polygene (Wu *et al.*, 2004). Genetic control of drought resistance in maize is assigned to 3-5 gene pairs (Ali and Naidu, 1982). The polygenic inheritance for salinity tolerance in rice has been reported by Mishra *et al.* (1996). Mishra *et al.* (1998) inferred that salinity tolerance trait is polygenic in nature and lacks maternal influence. Singh *et al.* (2001) also suggesting that sodicity tolerance is controlled by polygenic trait acting additively in most of the cases along with interactions between the alleles at some loci. Salt tolerance is a complex and multigenic trait (Flowers, 2004). Similarly, in cowpea drought resistance is reported to be controlled by a single dominant gene (Mai-Kodom *et al.*, 1999).

Association Studies

Association studies are important for indirect selection for a desired trait(s) using associated traits if phenotyping of the latter is relatively easy. The association could mainly be due to two reasons: either the traits are closely linked or the gene has pleiotropic effects. Tomar and Prasad (1996) reported a drought resistance gene, *Drt1* in rice, which is linked to genes for plant height, pigmentation, hull colour and awn, and has pleiotropic effect on the root system. The long root and high root numbers are controlled by dominant alleles and thick tip by recessive alleles (Armento-Sato *et al.*, 1983). Under the drought, seed yield per plant of canola showed high significant positive correlation with plant height, number of siliquae per plant, number of primary and secondary branches per plant (Sadaquat *et al.*, 2003). In one association study of 15 F_1 populations of rice, grain yield was highly and positively correlated with filled grains per plant, filled grains per panicle, panicle weight, K content, and 100-grain weight but had a strong negative correlation with percent spikelet sterility and reproductive stage salinity tolerance score, which is highly desirable (Mishra, 1994). Gupta (1999) also studied relationship between eight traits under salt stress environments. The role of Na^+ and K^+ became apparent from the study as grain yield correlated negatively with Na content as well as Na^+/K^+ ratio, whereas it had a positive correlation with K^+ content under both alkali and saline soil conditions. Ali and Naidu (1982) in maize found significant and positive correlation between seed yield under stress and plant height, number of leaves, leaf area index, ear number, length and girth, and 1000 kernel weight.

Gene Action

The mode of gene action is one of the most important parameter for deciding breeding procedures for quantitative characters. Heyne and Bruson (1940) found that drought tolerance in maize was heritable and gene action was intermediate to drought. Williams (1966) reported partial dominance with epistatic variability in sweet corn. Predominance of non-additive gene action was observed for all the traits in sorghum under water deficit conditions (Khidse *et al.,* 1982). Initial diallel studies on salinity tolerance in rice revealed the presence of both additive and non-additive gene action for almost for all the characters associated with salt tolerance (Akbar *et al.,* 1985; Gregoria and Senadhira, 1993; Mishra *et al.,* 1996). Gupta (1999) also reported additive and non-additive gene action for all the studied traits. Narayanan and Rangasamy (1990) reported the non-significance of additive effects for grain yield per plant under normal conditions from a 6 x 6 diallel study. Previous works on wheat showed that salinity tolerance in this crop is controlled by additive and non-additive gene effects (Singh and Singh, 2000; Dehdari *et al.,* 2007; Dashti *et al.,* 2010). In an experiment conducted on inheritance of salt tolerance in barley, generation mean analysis revealed that dominance and epistasis gene action contribute to control of K^+, Na^+ and K^+/Na^+ (Farshadfar *et al.,* 2008). In wheat Farshadfar *et al.* (2011) reported that chlorophyll fluorescence was controlled by additive type of gene action. Qiu and Li (2009) investigated genetics of salt tolerance in *Brassica rapa* and showed that salt tolerance was mainly controlled by dominant genes with an additive effect. Recently, Long *et al.* (2013) reported that salt tolerance was dominated by at least three major genes with additive-dominance effects in *Brassica napus*.

Heritability

High heritability coupled with high genetic advance was recorded by Gomez and Kalamani, (2003) for biological yield per plant, plant height and number of panicles per plant, indicated that those characters may be considered during selection for drought tolerance in rice. Gregoria and Senadhira (1993) reported high environmental effects with low narrow-sense heritability estimates for low Na^+/K^+ ratio in 9 x 9 diallel in rice. However, Mishra *et al.* (1990) and Gupta (1999) reported a high magnitude of narrow sense heritability for Na^+/K^+ ratio for both alkali and saline soils and for grain yield. Recent studies at IRRI have shown moderate to high heritability of grain yield under drought (Bernier *et al.,* 2007; Venuprasad *et al.,* 2007;

Kumar *et al.*, 2007). High broad sense heritability for chlorophyll a and b in wheat was a report by Zhang *et al.* (2009). High narrow-sense heritability estimates were observed for Ca^{2+}, K^+, Na^+, K^+/Na^+, Ca^{2+}/Na^+ and stress tolerance index, indicating the prime importance of additive effects in their genetic control in *Brassica napus* (Rezai and Saeidi, 2005). If the narrow sense heritability is low, early generation population size should be large; replications and locations over years should be increased, if possible, to screen the right genotypes; and selection should be done in the later generations. However, if the trait is associated with salt tolerance has high narrow-sense heritability, then early selection could be practiced.

Combining Ability Analysis

General combining ability for plant height, panicle exsertion, panicle length, grain yield and seed weight was significant among male parents indicating the prevalence of additive gene action in determining these traits in sorghum (Taye *et al.*, 2008). Additive gene action controlling plant height was reported (Toure *et al.*, 1996) for both plant height and inflorescence length (Kenga *et al.*, 2004). Gregoria and Senadhira (1993) showed the significance of both general combining ability (GCA) and specific combining ability (SCA) for salinity tolerance in rice. They also found significant reciprocal effects that indicate the role of maternal inheritance. Mishra (1991) reported that salt tolerance rice varieties CSR-10 and CSR-1 were good combiners for salt tolerance contributing traits. Gupta (1999) also reported the predominance of both *gca* and *sca* effects, but their ratio suggested a higher degree of dominance variance. Abd *et al.* (2009) research showed highly significant positive general combining ability effects for chlorophyll content in rice. Liang (2010) also reported the predominance of both *gca* and *sca* effects for drought tolerance traits during seedling stage in *Brassica napus*. Wang (2011) reported the predominance of GCA and SCA for shoot dry weight, root dry weight, net photosynthetic rate, stomatal conductance, transpiration rate and light compensation point in maize.

Heterosis

Study undertaken by Gupta (1999) at the Central Soil Salinity Research Institute (CSSRI), in India showed significant heterosis over the mid parent and better parent for almost all characters studied in rice. Grain yield is the prime concern of breeders. Out of 15 F_1crosses, Pokkali x IR-28 (79.87 per cent), CSR-10 x IR-28 (67.18 per cent), CSR-13 x IR-28 (54.58 per cent), and CSR-1 x IR-28 (48.56 per cent) had a positive and significant heterotic response over the mid parent, whereas Pokkali x IR-28 (49.31 per cent), exhibited heteriosis over the better parent in alkali soil. On the other hand, the crosses CSR-13 x IR-28 (35.17 per cent) and CSR-10 x IR-28 (26.72 per cent) gave heterotic response over the mid parent and only one cross CSR-10 x CSR-13 (24.53 per cent), performed better than the better parent in saline soil.

Breeding Methodology

When adopting any breeding method for any crops, the type of reproduction of the species should be considered, whether it is self-pollinating, cross-pollinating or asexual. The cultivar type associated to the genetic control of the traits. Because individual plants react so differently to similar abiotic stress factors, it can be difficult to breed a species for more than one resilient trait at a time, but that is precisely what plant breeders are looking to do. Breeding methods for drought resistance are the same as for yield and other economic characters. In general, pedigree and bulk method could be used for self-pollinated crops and recurrent selection for cross-pollinated crops. However, if transfer of few traits relating to drought resistance to a high-yielding genotype is the aim, then back cross is the appropriate methodology. On the other hand, biparental mating (half-sib and full-sib) maintains the broad genetic base as well as provides the scope to evolve the desired genotype of drought resistance (Yunus and Paroda, 1982). Drought tolerant crop cultivars can be developed through introduction, selection, hybridization, and mutations. Salinity tolerant variety were developed through pedigree, modified bulk pedigree and anther culture approach. Complex traits of abiotic stress phenomena in plants make genetic modification for efficient stress tolerance difficult to achieve (Wang *et al.*, 2007). Breeding methodology for drought resistance has been discussed by Hurd (1969, 1971, 1976), Rao *et al.* (1971) and Chang *et al.* (1975). Abiotic stress breeding is essential ways to combat yield reduction. Scientists around the world are putting in their best efforts to produce varieties and hybrids with improved heterosis under stress affected environments. Developing crops varieties with in-built salt, drought and heat tolerance are realized as the most promising, less resource consuming, economically viable and socially acceptable approach.

Conventional Approaches

The choice of suitable breeding programme for the development of tolerant cultivars to a defined abiotic stress depends upon a number of factors: screening techniques, sources and mechanisms of tolerance, modes of gene action and heritability, and their relationship to agronomic traits. Three important elements of drought characterization for successful breeding of stress tolerance are timing, duration and intensity of the stress. Drought tolerant lines of crops such as peanut, common bean, safflower, chickpea, wheat, tall fescue, soybean, wheatgrass, barley and maize have been developed using conventional breeding techniques. One challenge to applying this approach is identifying phenotypes that correlate well with drought tolerance. Strategies for breeding for salt tolerance in cross-pollinating species by cycles of recurrent selection were described long ago (Dewey, 1962): for a self-pollinating species the same process would require the use of male sterile lines to facilitate out-crossing (Ramage, 1980). These approaches depend on adequate heritability of the overall trait, for which there is evidence for wild grasses (Ashraf *et al.*, 1986), sorghum (Maiti *et al.*, 1994), maize (Maiti *et al.*, 1996), and tomato (Saranga *et al.*, 1992, 1993).

Selection and Introduction

Breeding for any trait start with the assesmbly of genetic variation through the collection and evaluation of the available germplasm. If the desirable variability is not available within a locality or species, introduction of the exotic germplasm can be resorted to. This classical approach is still very relevant in all breeding strategies. Few varieties with salinity tolerance have been developed worldwide using selection and introduction approaches. The salt tolerant rice varieties Damodar

(CSR-1), Dasal (CSR-2), and Getu (CSR-3) were pureline selections from the local salt tolerant traditional cultivars originating from the Sunderbans deltaic areas in West Bengal, India. Similarly, Jhona-349, SR-26B, Bhura rata 4-10, Patnai-23, Hamiltn, and Vytilla-1 were also very site specific selections from land races. Besides, the rice varieties Jaladhi-1 (a selection from Kalakhersail), Jaladhi-2 (a selection from Baku), Jalaprabha (a selection from composite), Neeraja (a selection from a landrace), Dinesh [CN-570-652-39-2(Jaladhi-2/Pankaj)], and Hangseswari (a pure line selection) were developed through selection for tolerant to deepwater (Indrani *et al.*, 2013). Mahsuri is one such example: it was introduced to India and other countries from Malaysia and become popular with farmers in eastern India because of its moderate resistance to lodging and good grain quality (Rao Balakrishna and Biswas, 1979). When drought resistance is available in an exotic variety, such variety can be introduced and after through testing, if found suitable, can be released in the new area. From local land races desirable types suited to the environmental conditions are selected and evaluated along with existing varieties if found superior selected strains released as improved varieties in rice. JL-24 (Phule Pragati) is a selection from 94943. Kopergoan 3 is also a selection from local collection. Karad 4-11 is a selection from local collection. ICGS 11 is a selection from Robout 33-1. The initial effort for varietal improvement for the rainfed lowland ecosystem was mainly confined to the purification of landraces through pure line selection (Mallik *et al.*, 2002). Such breeding activities began with selection of superior lines within local landraces. In Bihar, BR-8, BR-9, BR-34, Sugandha, Rajshree, and T-141 were released through this process. These cultivars are tall and photoperiod sensitive, and have varying degrees of submergence tolerance (Mallik *et al.*, 1995).

Pedigree Method

Pedigree selection method can be used to identify superior genotypes for grain yield in a cultivar development program. Pedigree selection for grain yield/plant needs to evaluate selections under a series of environments such as different planting dates (Zakaria, 2004; Ali and Abo-El-Wafa, 2006), different water stress (El-Morshidy *et al.*, 2010). The pedigree method has been described in different text books. Pedigree method consists two lines in crossing, each providing desirable genes which are to be incorporated into the resulting lines. The F_2 segregating generation is grown and selection is made on selfed plants. This is the classical method in which the lineage of the plant selection in the segregating generation is maintained until it is stabilized in the F_7 or F_8 generation. The procedure provide selection opportunities generation after generations. It allow breeder to identify best combination with considerable uniformity. But, due to cumbersome procedures and the involvement of more resources, breeders are modifying this method and not adhering to it strictly. Rice varieties from Bihar, Rajendra dhan, Panidhan-2, Radha, Shakuntala, and Mahsuri have been released, of which Mahsuri is popular among the farmers. The modified pedigree method (single panicle selection) was adopted by breeders at Narendra Deva University of Agriculture and Technology (NDUAT), Faizabad, Uttar Pradesh (Mishra *et al.*, 1996). Pedigree selection method was fund effective to produce new lines tolerant to drought stress (Tammam *et al.*, 2004). CSR0-10, CSR-13, CSR-23, CSR-27, CSR-30, CSR-36, CR Dhan-402, CR

Dhan-403, TRY-1, TRY-2, TRY3, White ponni, CO-43 etc. were developed through recombinant varieties (Reddy *et al.*, 2014).

Modified Bulk Pedigree Method

A combination of pedigree and bulk breeding methods, this is almost effective as the pedigree method, with relatively less use of resources. It has flexibility and is useful for less heritable traits, with the individual F_2 plant harvested in bulk up to the F_4 or F_5 generations, followed by individual panicle selection and handling of the population as in the pedigree method. However, for highly heritable traits, the individual plants are selected in the earlier generations (F_2 or F_3), followed by bulking for a few generations and ultimately single plant or panicle selection in the F_5 or F_6 generation. The wheat variety "Veery" and their progenies, such as the Kauz, Attila, Pastor, and Baviacora groups of lines have demonstrated a superior level of abiotic tolerance to a number of stresses (drought, heat, etc.) and improved nutrient efficiencies (N and P efficiency). The schematic diagrams for the modified bulk pedigree method applied for salt tolerance breeding at IRRI are shown in Figure 16.1. This method is mainly used when the lack of land/laboratory facilities and labor, or when the selection environment is not appropriate to discriminate desirable genotypes from undesirable ones, particularly in breeding for biotic or abiotic stresses, judicious selection of parents tolerant to drought and submergence tolerance. Under such situations, the F_2 generation is raised under optimal conditions and F_3 populations are bulked for up to five generations and the segregating characters are advanced by the bulk method of breeding until the selection of elite breeding lines is made in later generations. The selected genotypes are then evaluated for yield and other component traits in the next season. ICARDA scientist identified wheat genotypes such as '6', '27' and '31' which showed high yield potential combined with drought-tolerance properties. (http://www.icarda.org/bread-wheat-combining-yield-potential-and-drought-tolerance).

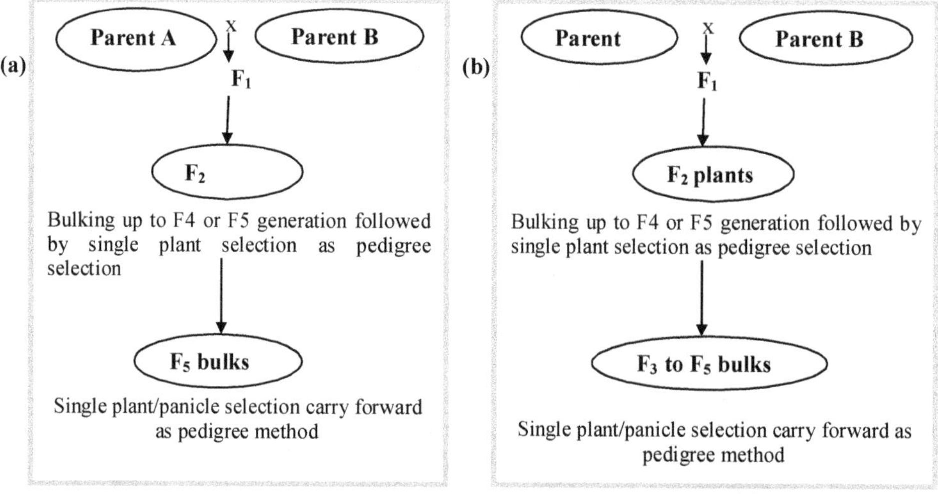

Figure 16.1: Diagrammatic Representation of the Modified Bulk Pedigree Method (a) for low heritability traits and (b) high heritability traits.

Advantages of Modified Bulk Pedigree Method Compare to Simple Pedigree Method

1. The modified bulk-pedigree method of breeding is simple compare to pedigree selection
2. More convenient
3. Requires less labor because bulking up to F_4 or F_5 generation
4. Less expensive to conduct during the early segregating generations.

Shuttle Breeding

To improve the yield potential of the rainfed lowland rice varieties of eastern India (Assam, Bihar, Chhattisgarh, Eastern Uttar Pradesh, Odisha and West Bengal), an ICAR-IRRI Collaborative Shuttle Breeding Program began in the *kharif* season of 1992 (Reddy *et al.*, 2013) for improving rice. The programme was funded by IRRI and jointly coordinated by CRRI, Cuttak, and IRRI, Philippines. Since most abiotic stresses are very location specific, the varieties developed need to fulfill specific plant type requirements. In the shuttle breeding approach, prebreeding or advanced breeding materials are evaluated at different locations for their adaptability and the best adapted materials are again crossed and evaluated at different target site in replicated trials. The letter step functions like the multiplication testing for advanced breeding materials. Through the Eastern Indian Rainfed Lowland (EIRL) shuttle breeding network, new breeding varieties/line like Kishori, Jagabandhu, Upahar, Bhudev, Varshadhan, Prafulla, NDR-8002, Cr-2003-2, CR-2003-3, CR-978-8-2 etc. with submergence tolerance were developed (Mallik *et al.*, 2002; Singh *et al.*, 1998). The salt tolerance rice variety CSR-23 and CSR-27 were developed through a shuttle breeding approach under an ICAR-IRRI collaborative project for salt tolerance (Mishra, 1994). The flood tolerant variety FR13A found to have good degree of submergence tolerance, but it has poor combining ability. Many improved breeding lines such as CR-2003-13, CR-2006-7, Cr-2056-5, CR-2003-24, CR-2033-1, NDR-9830116, NDR-9830125, IR-76522-7-14-4-1, IR-74601-17, TTB-225-2-138, and many more were found to have good submergence tolerance and are being used in breeding programs (Reddy *et al.*, 2013). In addition, numerous improved lines, including CRLC-899 and CR-2003-2, have been identified under the shuttle breeding network as suitable for waterlogged situations in Odisha and other eastern Indian states (Table 16.3).

Mutation Breeding

This method is used in depleted gene pool situation. Chemical mutagenes EMS provide broad spectrum genetic changes with lesser sterility effects, as compared to X ray or particular mutation. Mutation is used when the desired genes are not available in the germplasm. This method was followed for the development of the first salt tolerant rice variety, CSR-10. The female parent of variety M40-431-24-114 was derived from γ ray-irradiated (10kR) F_1 seed of the cross CSR-1/IR-8. This female parent was crossed with Jaya and the progeny was handled as in the pedigree method. Using mutation breeding, two widely accepted rainfed lowland rice cultivars, Biraj and Jagannath, were developed from OC-1393 and T141, respectively. They are suitable for use in 5-30 cm standing water (Rao Balakrishna and Biswas,

Table 16.3: Promising Lines Tested/Identified from Shuttle Breeding Programme in different States

State	*Name of the Variety*	*Designation*	*Pedigree*	*Year of Release*
Assam	Namdang	PSR 1119-13-3-1	Barogar/IR 8	Recommended
	Gadapani	BR 4974-3-6-3-2	BR 113/Swetha Soke	–
	Gitesh	TTB 283-1-26	Kushal/Mugi Sali	2011
	Prafulla	TTB 283-3-38	Kushal/Akisali	2011
Bihar	Kishori	PSR 1119-13-3-1	Barogar/IR 8	1999
	Satyam	SBR 3025-2-1B-2-1	RD 19/Dasari	1999
	Shakuntala	RAU 73-16-1-40	Pankaj/BR-8	1996
	Rajendra Mahsuri-1	RAU 83-500	BR-51-46/Mahsuri	2003
	Rajendra Sweta	RAU 710-99-22	Sita/Pusa Basmati 1//Katarni	2004
	Santosh	RAU 1306-4-3-2-2	Pankaj/BR 34	2000
	Swarna-Sub1	IR82809-237	Swarna*3/IR 49830-7-1-2-3	2010
Chhattisgarh	BKP 232	–	–	–
Odisha	Varshadhan	CRLC 899 (IR 54112-B2-1-6-2-2-2-CR 1-2)	IR 31432-8-3-2/IR 31406-3-3-3-1//IR 26940-3-3-3-1	2006
	Uphar	OR 1234-12-1	Mahslaxmi/IR 62	2006
	Jagabandu	OR 1206-25-1	Savitri/IR 4819 Sel.//IR 27301 Sel	2002
	Swarna-Sub1	CR 2539-1 (IR 82809-237)	Swarna*3/IR 49830-7-1-2-3	2009
	Mrunalini	OR 1898-18	Mahalaxmi/OR 633-7	2009
Uttar Pradesh	NDR 8002	NDR 8002(IR-67493-M-2)	IR60290-CPA 5-1-1-1-1/IR 52533-52-2-1-2-1-B-2-3	2005
	ImprovedSwarna	IR 82809-237	Swarna*3/IR 49830-7-1-2-3	2009
West Bengal	Bhudeb	CN 1035-61(IR 57540)	Pankaj/IR 38699-49-3-1-2//IR 41389-20-1-5	2002

Source: Reddy *et al.* (2013)

1979). Myriana is a winter bread wheat cultivar developed through crossing with a mutant line at the Institute of Agriculture, Karnobat, and released in 1991. Two other cultivars Avangard and Guinness/1322 were registered in 2005 by the Bulgarian Executive Agency for Variety Testing, Field Inspection and Seed Control (IASAS) (Rachovska and Rachovski, 2005). High productivity, ecological adaptability and drought tolerance are its distinctive advantages compared to other cultivars. Maize hybrids such as Kneja HP 556, Kneja 712, Kneja 674, Kneja 570, Kneja 509, Kneja 682, for drought tolerance, Kneja 641 for acid tolerance, released in from Maize Research Institute (MRI), in Kneja through mutation breeding. Mutant 17 MM, released in 1999 from n Bulgaria demonstrates better drought tolerance (Mihov *et al.*, 1999; 2001). A new feed barley cultivar IZ Bori (Kt3026) was released in 2009 in Bulgaria; it was developed by sodium azide mutagensis of a breeding line 280-7 and belongs to *Hordeum sativum* Jess, subsp. *Vulgare,* var. *paralelim.* It is a winter type cultivar, highly tolerant to dip with good tolerance to cold temperatures (Tomlekova, 2010). As an example, the M_2 plants of rice were sown in high saline paddies and the M_1 calli are transplanted on the medium with high salt concentration (Lee *et al.*, 2003). From this experiment, Lee *et al.* (2003) selected two promising salinity tolerant lines from regenerated plants. Huge numbers of potentially useful genotypes have been generated from mutation breeding research and breeding programmes and are available in germplasm collections, *e.g.* tolerance to cold, heat, day length and drought. Diamant variety in barley, created by irradiation of dormant seeds with X-rays, has very high yield, short stem, very good grain, and malting quality as well as lodging resistance. Calrose 76 was the first semi-dwarf table rice variety released in the US produced by irradiation with g-rays. Some outstanding examples of mutant rice cultivars are VND 95-20 and VND 99-3 which have long grains with excellent grain quality and wide adaptation to acid sulphate soil and salinity. Drought tolerant sunflower variety TAS-82 and groundnut variety TG-37A were released from BARC, Mumbai India (http://www.barc.gov.in/rcaindia).

Diallele Selective Mating System Supplemented by MAS

The biggest bottleneck in breeding of self pollinated crops is the narrow genetic background in the resulting progenies as breeders can exercise parental control on only two individuals for a single cross, on three and four way cross, and at the four most for a double cross. To the increase parental control, broaden the genetic base, and break up linkage blocks, IRRI is employing diallele selective mating system (DSMS), which is modified from the DSMS suggested by Jensen (1970). This modified DSMS becomes a permanent breeding scheme for developing multiple abiotic stress tolerance genotypes with wider adaptability. Basically, it is a type of recurrent selection scheme in which only the desired alleles (selected genotypes) are advanced further for selective intermating. The philosophy behind this scheme is the "recurrent selection of the desired genotypes" and "intensive crossing among the selected genotypes" so that the probability of getting the desired recombinant genotypes becomes high. The major goals of this long term breeding strategy are to develop genetic materials with multiple abiotic stress tolerance, break up the very tight linkage blocks common in self-pollinated crops such as rice, maximize the chances that only the desired gene frequencies are forwarded to subsequent generations, tap germplasm donors available at IRRI from different geographical niches, and enable large scale validation and use of marker aided selection (MAS) in combination with conventional breeding. The target areas envisaged for the materials generated are submergence prone costal saline soils, inland salt affected soils, Fe toxic soils and Zn deficient soils (Singh *et al.*, 2006). This method involving marker assisted selection (MAS) for salinity (*SALTOL* locus) and submergence (*SUB 1* locus) tolerance, while phenotypic selection is exercised for other traits (Singh *et al.*, 2008).

Back Cross Method

The backcross method is a form of recurrent hybridization by which a gene for a superior characteristic may be added to desirable variety. To transfer of traits relating to drought tolerance to a high yielding cultivar, backcross method of breeding is appropriate (Edmeades *et al.*, 1992). However, transferring drought tolerance in high yielding genotypes is complicated due to lack of understanding of physiological and genetic basis of adaptation in drought condition. A comprehensive scheme for breeding for drought resistance and high yield potential has been proposed, which takes advantages of recurrent selection procedure (Paroda, 1986). The number of backcrosses to be made depends on the relative performance of the donor compared to the recurrent parent. Dudely, (1984) also suggested that backcrossing is advantageous if one parent has more loci with favourable alleles than the other, if the parents are diverse or if the level of dominance is high. A successful example is the transfer of a major gene for Al tolerance from the Carazinho variety of wheat to the Egret variety (Fisher and Scott, 1987) in Australia. Blum (1979) also suggested the selective incorporation of relevant drought-resistance factors into cultivars with superior yielding ability under optimum conditions, thus making them better adapted to the suboptimal conditions. This method of breeding is presently used by International Rice Research Institute (IRRI), for improving drought tolerance in rice (Lafitte *et al.*, 2006). Large number of introgression lines with enhanced abiotic stress tolerance has been developed in a massive backcross breeding program using three recurrent parents and 203 donor lines with tolerance to several abiotic stresses in rice (Ali *et al.*, 2006).

Recurrent Selection Method

Crosbic *et al.* (1981) suggested that photosynthetic rate could be improved by recurrent selection and thereby drought tolerance. Eight cycle of recurrent selection for improved drought tolerance in a tropical maize population resulted in a yield gain of 500-800 kg/ha (Edmeades *et al.*, 1992). Recurrent selection for drought tolerance for three to eight cycles has increased grain yield under drought at flowering by 30-50 per cent in three low land tropical maize populations (Chapman and Edmeades, 1999). They found that under drought condition, changes per cycle with recurrent selection were as gain (12.6 per cent), fertile ears per plant (8.9 per cent), grain per fertile ear (6.3 per cent), grain number per square meter (12.2 per cent), 1000-grain weight (no change), anthesis-silking interval (22.0 per cent), days from sowing to 50 per cent anthesis (0.7 per cent), plant height (2.0 per cent), primary tassal branch number (5.9 per cent), and senesced leaf area (2.7 per cent).

Grain yield, ear per plant and grains per fertile ear were strongly correlated with anthesis-silking interval across entries under drought, though not when water was plentiful. Half-sib and Full-sib family recurrent selection has been used extensively at CIMMYT to improve populations for drought and low N tolerance. Recurrent selection has been suggested as an alternative method to exploit the additive gene action related to Al tolerance in wheat (Carver *et al.*, 1988), and maize (Granados *et al.*, 1993). Genotypic recurrent selection utilizing S_1 progeny or test cross evaluation have been effectively used in sunflower hybrid breeding to improve drought resistance (Fernandez-Martinez *et al.*, 1990). Edmeades *et al.* (1994) demonstrated that improvements in grain yields under flowering drought stress was achieved through the use of recurrent Full-sib or S_1 selection within the lowland tropical maize population. Four open pollinated populations were improved for tolerance to drought at Chitala (600 masl) using the S_1 recurrent selection scheme. S_1 lines were selected under artificial drought and were recombined to form improved drought populations. Maize varieties ZM621, ZM521, ZM421 and ZM303, have been released. ZM 301 C1 is a result of an S_1 recurrent selection scheme in Botswana of which the parental material is of CIMMYT origin (Lekgari *et al.*, 2002).

Practical Achievements in Drought and Salt Resistant Breeding using Conventional Approaches

Drought Tolerance in India

Breeding for drought resistance is an important objective of plant breeding programmes in many crops. Hence, breeding work on drought resistance is carried out in almost all the International and National Crop Research Institutes and in India. The successful example of conventional breeding for drought tolerance in India is seven salt tolerant varieties of rice (CSR-43 in 2011, CSR-36 in 2005, CSR-30 in 2001, CSR-27 in 1998, CSR-23 in 2004, CSR-13 in 1998, and CSR-10 in 1989) and three salt tolerant varieties of mustard (CS-52, CS-54 and CS-56) and one chickpea (Karnal Chana-1) through hybridization breeding approach have been developed and released (Table 16.4). A new salt tolerant variety of rice (CSR-43) has been released by the U.P. State Variety Release Committee during the year 2011. This variety developed through integration of farmers' participatory variety selection. Four salt tolerant varieties of wheat (KRL-14 in 1990, KRL-19 in 2000, KRL-210 and KRL-213 in 2010) have been developed and released. In sesbania (*Dhaincha*), two genetic stocks CSD-137 and CSD-123 have been registered with NBPGR as salt tolerant genetic stocks (www.cssri.org). The line 'CSS-1', of dill (*Anethum graveolens*) a native collection from farmer's fields in Raibareilly district in Uttar Pradesh gave consistently highest seed yield of about 9 tonnes/ha as against lowest of 5.64 tonnes/ha of 'CSS-9'. It has been registered as unique sodicity tolerant germplasm with NBPGR, New Delhi during 2008 and has been assigned national identity number IC563951 (Gautam *et al.*, 2011).

Drought Tolerance in other Countries

During the last century, conventional breeders at different renowned international research centers have made considerable strides in developing drought tolerant lines/cultivars of some important food crops (Table 16.5). For example,

Table 16.4: Drought Tolerance Genotypes of different Crop Species in India

Crop	*Genotype*	*Reference*
Rice (*Oryza sativa*)	ARC-10327, DJ-129, DV-110, DZ-41, DNJ-60	Khush and Coffman, 1977
	MTU-17, CH-45, PTB-29, B-76, TKM-1, N-22, Sudha	Ram and Singh, 1994
	Sahbhagi Dhan (IR74371-70-1-1)	Verulkar *et al.*, 2010
Wheat (*Triticum aestivum*)	*Triticum sphaerococcum, T. vavilovii, T. eastivum* cv. C-306	–
Maize (*Zea mays*)	ZPBL-1304	Ristic *et al.*, 1998
	HI-209, HI-295, HI-536, HI-1040	Meena Kumari *et al.*, 2004
Cowpea (*Vigna radiata*)	TVu-11979, TVu-14914	Watanbe, 1998
	IT-90K-59-2, Kanannado, Dan IIa	Singh *et al.*, 1999
Chickpea (*Cicer arietinum*)	BG-365, BG-364, Pusa-362, Pusa-3351, Pusa-256 and IPC-94-32	Kumar *et al.*, 2004
Mustard/rape (*Brassica* spp.)	Oscar, range, Tarnab-2	Sadaqat *et al.*, 2003

breeding approach started at the International Maize and Wheat Improvement Center (CIMMYT), Mexico in the 1970s for developing drought tolerant maize is worth mentioning. Based on the selection and breeding program, maize yield improvement of 59 to 233 kg ha–1 cycle–1 of recurrent full-sib or S_1 selection, was recorded (Bänziger *et al.*, 2004). The CIMMYT breeding program was very outcome-oriented and multi-faceted, because it focused on a multitude of imperative problems including drought, low N, and common leaf and ear diseases (Bänziger *et al.*, 2004). A number of maize hybrids developed by the CIMMYT scientists were found superior to all those developed by private enterprises, in terms of growth and grain yield under drought-prone environments (Bänziger *et al.*, 2004). Breeders at IRRI have also used conventional breeding strategies to generate Sahbhagi Dhan, a rice variety that flowers weeks earlier than other types. This variety, which is now cultivated throughout South Asia, "can provide farmers with a yield advantage of one tonne per hectare under drought".

Plant breeders, at the Crops Research Institute (CRI) based at Kumasi, Ghana, have developed a highly drought tolerant maize cultivar 'Obatanpa GH' in 2006 in collaboration with the International Institute of Tropical Agriculture (IITA), Ibadan, the CIMMYT, Mexico, and the Sasakawa Global 2000 (Published online April 25, 2006). Similarly, 16 early maturing maize inbred lines (from TZEI 1 to TZEI 16) resistant to a scrounging weed *Striga hermonthica* (Del.) Benth were produced by the IITA. All these lines were found to be highly resistant to water limited conditions (Published online April 25, 2006). Wheat, which is one of the important staple food crops of the world, is adversely affected by drought. Drought Tolerant Maize for Africa (DTMA) project was has been launched in 2006, to mitigate drought and other constraints to maize production in sub-Saharan Africa. Large number of varieties and open pollinated varieties were developed by this project in many countries. Over 2007-12, 60 drought tolerant hybrids and 57 open pollinated varieties of maize made available to smallholder farmers (www.cimmyt.org).

Limitations in Selection

Although research on drought tolerance achieved many goals, certain constraints still remained unsolved. Drought tolerance has appeared to be difficult research aim in breeding programmes due to several technical limitations (Mitra, 2001).

1. Drought stress is highly heterogeneous in time (over the season and years) and space (between and within sites), and is extremely unpredictable. This makes it very difficult to identify a representative drought stress condition.
2. In any case the environment is not defined in terms of rainfall, soil moisture, temperature and vapour pressure deficit. Consequently, the performance of the same genotypes varies year after year.
3. *Environmental influence*: since the phenotype is the ultimate expression of the interaction between genotype and environment, assessment of the desired genotypes is highly dependent on the proper environmental conditions. The unpredictable and variable forms in which drought stress

Table 16.5: Drought Tolerant Cultivars/Lines of different Crops Developed through Conventional Breeding at different Centers/Institutions

Crop	*Cultivar/Line*	*How Developed*	*Centers/Institutions Involved*	*Reference*
Peanut (*Arachis hypogaea* L.)	ICGV 87354	Derived from a cross between Argentine and PI259747 and developed through nine generations of bulk selection	Plant Materials Identification Committee of the International Crops Research Institute for the Semi-Arid Tropics (ICRISAT), India	Reddy *et al.* (2001)
Common bean (*Phaseolus vulgaris* L.)	SEA 5	Developed from the interracial double cross population BAT 477/San Cristobal 83/ Guanajuato 31/Rio Tibag	International Center for Tropical Agriculture (CIAT), Cali, Colombia	Singh *et al.* (2001)
	SEA 13	Derived from double cross population BAT 477/ San Cristobal 83/BAT 93/Jalo EEP 558	International Center for Tropical Agriculture (CIAT), Cali, Colombia	Singh *et al.* (2001)
	A 195	Developed from the single cross Red Kloud × ICA 10009	Centro Internacional de Agricultura Tropical (CIAT), Palmira, Colombia	Singh *et al.* (2007)
	Line CO46348	The complete pedigree of CO46348 was unknown, however it was derived from a single cross with the pinto cultivar Othello	Colorado Agricultural Experiment Station in cooperation with the University of Idaho and USDA-ARS	Brick *et al.* (2008)
Safflower (*Carthamus tinctorius* L.)	Morlin	Derived from a single plant selection from F11 population	Eastern Agricultural Research Center and Montana Agricultural Experiment Station, Sidney	Bergman *et al.* (2001)
Chickpea (*Cicer arietinum* L.)	FLIP 87-59C	Developed by crossing ILC3843 with FLIP87	International Center for Agricultural Research in the Dry Areas (ICARDA)	Singh *et al.* (1996)
Wheat (*Triticum aestivum* L.)	Willow Creek	Through breeding in single replication observation (SROB) nurseries	Montana Agricultural Experiment Station, Sydney	Cash *et al.* (2009)
	Ripper	Developed by using a modied bulk breeding procedure	Colorado Agricultural Experiment Station, USA	Haley *et al.* (2007)
	NE01643	A bulk breeding procedure was used and approximately 50 per cent of F3 population was visually selected on the basis of agronomic appearance	Nebraska Agricultural Experiment Station and the USDA-ARS	Baenziger *et al.* (2008)
	Prairie Red	Derived from the crosses and backcrosses of CO850034/PI372129/5 TAM 107	Colorado Agricultural Experiment Station, USA	Quick *et al.* (2001)

Contd...

Table 16.5–*Contd...*

Crop	*Cultivar/Line*	*How Developed*	*Centers/Institutions Involved*	*Reference*
	Jinmai 50	Developed from the cross, Pingyang181 × Qingfeng1	Wheat Breeding Innovation Group (WBIG) in the Cotton Research Institute of Shanxi Agri. Sci. Academy, Yuncheng, China and Testing and Appraising Committee of Crop Cultivars of Shanxi Province (TACCCSP)	Xinglai *et al.* (2006)
Soybean [*Glycine max* (L.) Merr.]	R01-416F and R01-581F	Both lines were originated after selection from F9 population. developed from a cross between Jackson and KS4895	Arkansas Agricultural Experiment Station, USA	Chen *et al.* (2007)
Barley (*Hordeum vulgare* L.)	Lenetah	Developed using a pedigree selection procedure with all early generation population and selected from the cross, 94Ab12981 × 91Ab3148	Agricultural Research Service, Aberdeen, ID, in cooperation with the Idaho Agricultural Experimental Station	Obert *et al.* (2008)
	Giza 126	Selected for drought resistance in an F3 population received from ICARDA, initially originating from a single cross Baladi Bahteem/ SD729-Por 12762BC	International Center for Agricultural Research in the Dry Areas (ICARDA)	Noaman *et al.* (1995)
	Giza 2000	The pedigree breeding method was used for development and it was originated from the cross between the Egyptian local cultivar Giza 121 and the line 366/13/1 (Giza 117/Bahteem 52/ Giza 118/FAO 86)	Barley Research Department, Agricultural Research Center at Giza, Egypt	Noaman *et al.* (2007)
	Giza 121, Line 366/13/1	Plant selection within superior F4 populations	Sakha Research Station Northern Delta Region, Egypt	Noaman *et al.* (2007)
	Bentley	Derived from crossing I92125 with TR229	Field Crops Development Centre (FCDC), Lacombe, AB, Canada	Juskiw *et al.* (2009)
	Giza 126	Developed from a single cross, Baladi Bahteem × SD729-Por 12762BC	International Center for Agricultural Research in the Dry Areas (ICARDA) Aleppo, Syria	Noaman *et al.* (1995)

Contd...

Table 16.5–*Contd...*

Crop	*Cultivar/Line*	*How Developed*	*Centers/Institutions Involved*	*Reference*
Maize (*Zea mays* L.)	Oh605 22	Selected full-sib progenies from the AAE population were intermated with 30 full-sib progenies obtained from OhS3267LAN plants Ohio State University (OSU), Ohio	Agricultural Research and Development Center, USA	Pratt and Casey (2006)
	Sixteen tropically adapted in bredlines (TZEI 1 to TZEI 16)	Developed from four diverse germplasm sources with tolerance to drought namely, TZE-W Pop DT STR C0, TZE-Y Pop DT STR C0, TZE Comp 5-Y C6, and TZE-W Pop 3 1368 STR C0	International Institute of Tropical Agriculture (IITA), Ibadan, Nigeria	Badu-Apraku *et al.* (2006)
	ND2005 and ND2006	Developed through an integration of recurrent selection and pedigree breeding methods	North Dakota Agricultural Experiment Station, USA	Carena and Wanner (2009)
	TZE-W Pop DT STRC4 and TZE-Y PopDT STR C4	Developed through backcrossing, inbreeding, hybridization selection	International Institute of Tropical Agriculture (IITA), Ibaban, Nigeria	Badu-Apraku and Yallou, (2009)

Source: Ashraf, 2010.

will manifest itself, makes selection of promising individual plants and hence breeding for drought tolerance become extremely difficult.

4. Lack of repeatable and precise screening techniques. There is hardly any basis of selection in F_2 and F_3 segregating generations, but subsequently some lines may be identified which appear promising.
5. *Negative association of stress tolerance traits with yield*: Some drought tolerance characters are negatively associated with yield. For instance, leaf rolling and stomata closure conserve water in plant, but reduce light interception and entry of carbon dioxide into leaf and in turn, reduce the yield.
6. *Multiple stresses*: Drought tolerance is hardly is separated from other important abiotic stress such as temperature and salinity. Due to these interrelations, no single mechanism exists by which multiple stresses are alleviated. Drought stress reduces nutrient uptake and is associated with temperature stress, and higher elevation with cold. This association makes the breeding programme more complicated.
7. So far breeding for drought tolerance mainly confined in cereal crops. The evaluation of rice genotypes in upland represents for drought tolerance. There is no such directed effort in pulses and oilseed.
8. Similarly, biotechnological approaches of breeding *viz.*, MAS and transgenic development, mainly centred on crops which are easy to handle, like *Arabidopsis, Tobacco, Oryza sativa, Triticum* and some extent in maize.
9. Knowledge is incomplete about reliable attributes as indices of drought tolerance, selection criteria and influence of environment or drought related traits (Paroda, 1986).
10. *Multiple gene involvement*: Drought tolerance has been noted to be a highly complex trait, influenced by many different genes. In fact, drought tolerance should be regarded as a unique heritable trait, but as complex of often fully unrelated plant properties.
11. *Multidisciplinary approach*: There is lack of multidisciplinary approach to understand the integrated plant response to drought and complex genetic control of different mechanisms of drought resistance.

Salinity tolerance in India: Different crop plants have different reaction of salinity. Crops are categorized into different groups (Table 16.3). In India, breeding work for salinity resistance is mainly carried out at Central Soil salinity Research Institute (CSSRI), Karnal (Haryana). Several salt resistant varieties of rice, barley, wheat and sugarcane (CO-453 and CO-62175) have been released. In India, salinity resistant varieties have been developed in sugarcane, okra, rice, onion, and barley. Of 32 salt tolerant rice varieties developed by Central Soil Salinity Research Institute (CSSRI), Karnal, CSR-10 was the first dwarf high yielding salt tolerant early maturing rice variety released in the year of 1989. Seven salt tolerant varieties of rice (CSR-43 in 2011, CSR 36 in 2005, CSR-30 in 2001, CSR-27 in 1998, CSR-23 in 2004, CSR-13 in 1998, and CSR-10 in 1989) and three salt tolerant varieties of mustard (CS-52, CS-

54 and CS-56) and one chickpea (Karnal Chana-1) through hybridization breeding approach have been developed and released from Central Soil Salinity Research Institute (CSSRI), Karnal (http://plantstress.com/files/salt_karnal.htm). A new salt tolerant variety of rice (CSR-43) has been released by the U.P. State Variety Release Committee during the year 2011. This variety developed through integration of farmers' participatory variety selection. Four salt tolerant varieties of wheat (KRL 1-4 in 1990, KRL-19 in 2000, KRL-210 and KRL-213 in 2010) have been developed and released from CSSRI. In sesbania (*Dhaincha*), two genetic stocks CSD-137 and CSD-123 have been registered with NBPGR as salt tolerant genetic stocks (www.cssri.org). The line 'CSS-1', of dill (*Anethum graveolens*) a native collection from farmer's fields in Raibareilly district in Uttar Pradesh gave consistently highest seed yield of about 9 tonnes/ha as against lowest of 5.64 tonnes/ha of 'CSS-9'. It has been registered as unique sodicity tolerant germplasm with NBPGR, New Delhi during 2008 and has been assigned national identity number IC-563951 (Gautam *et al.*, 2011). One salt tolerant cultivar 'CS-56' of *B. juncea* have been developed in India (Zhang *et al.*, 2014).

Table 16.6: Categorized of Salt Tolerance Crop into different Group (Shanon, 1997)

Category	*Crop*
Salt tolerant Crop	Barley, Sugar beet, Cotton, Canola (Brassica spp.), Asparagus, red beet, zucchini squash, date palm, pomegranate, grape, wheat grass, Bermuda grass
Intermediate tolerant crop	Sorghum, sunflower, safflower, sugarcane, potato, alfalfa, faba bean, almond, plum, orange, grapefruit, peanut, chrysanthemum, carnation
Sensitive crop	Rice, corn, wheat, legumes, linseed, cowpea, lentil, chickpea, citrus, avocado, stone fruits, apricot, peach, blackberry, strawberry, aster, poinsettia, gladiolus, azalea, gardenia, amarylis, African violet

Besides, CSSRI, Regional Research Station Canning has also developed salt tolerant rice varieties like CST-1, Sumati, Bhutnath and CSR(S)-7-1-4 suited for the coastal saline areas of the eastern coast. The other Central Variety Release Committee (CVRC) released salt tolerant varieties are Lunishree and Jarava for other agro-climatic regions. Similarly, from time to time State Variety Release Committee (SVRC) have released different salt tolerant varieties for respective states like Haryana (Vikas or CSR-5), Kerala (Vyttila-1 to Vyttila-6), Maharashtra (Panvel-1, Panvel-2 and Panvel-3), Andhra Pradesh (MCM-1), West Bengal (Mohan or CSR-4), Karnataka (CSR-22), Orissa (SR-26B, Sonamani), Tamil Nadu (PVR-1, CO-43, ASD-16, TRY-1 and TRY-2), Gujarat (SLR-51214 or Dandi), Andaman and Nicobar Islands (BTS 24) and Uttar Pradesh (Usardhan-1, Narendra Usardhan-2 and Narendra Ushar dhan-3). The hybrid CORH 2 developed by three line breeding system (A/B/R) has been found suitable for solving Tamil Nadu inland salinity problems.

Salinity Tolerance in other Countries

Salt resistant variety of rice has been released from Indonesia and that of wheat from china. In other countries, BRRIdhan-40, BRRIdhan-41 and BRRI-47 (Bangladesh); OM-576, OM-2717, OM-2517, OM-3242, AS-996 (ietnam), and Giza-

177, Giza-178, Sakha-104, Sakha-111 (Egypt) have been released as salt tolerant rice varieties (Singh *et al.*, 2008). Some successful example of conventional breeding for salt tolerance are mentioned here, for example, some lines/cultivars of alfalfa (*Medicago sativa* L.) such as AZ-Germ Salt II (Dobrenz *et al.*, 1989), AZ-90NDC-ST (Johnson *et al.*, 1991), AZ-97MEC and AZ-97MEC-ST (Al-Doss and Smith, 1998), ZS-9491 and ZS-9592 (Dobrenz, 1999) were tested under natural field conditions. Similarly, two salt-tolerant lines/cultivars of bread wheat (*Triticum aestivum* L.) such as S-24 (Ashraf and O'Leary, 1996) and KRL1-4 (Hollington, 2000) were evaluated on natural salt-affected soils. The salt tolerance varieties of wheat successful releases have been the Indian KRL1-4, KRL213 and KRL 19, released by the Central Soil Salinity Research Institute (CSSRI) at Karnal, the Pakistani LU26S and SARC-1, released by the Saline Agriculture Research Cell (SARC) at Faisalabad, and the Egyptian Sakha-8, released by the Agricultural Research Centre at Giza (Rana Munns *et al.*, 2006). Two salinity tolerant lines HA 429 and HA 430 were released and registered from USDA by an interspecific cross of *H. paradoxus* and *H. annuus* with cultivated sunflower (Miller and Seiler, 2003). In Thailand a number of new varieties was developed and released. However, other varieties except Khao Dawk Mali105 (KDML105) have an inferior in cooking quality. Progress in breeding for salinity tolerance and its related abiotic stresses has been slow owing to: (1) limited knowledge concerning the genetics of tolerance, (2) the involvement of several complex tolerance mechanisms (Yeo and Flowers, 1986), (3) inadequate screening techniques, (4) low selection efficiency, and (5) poor understanding of the stress and environmental interactions.

Abiotic Resistant/Tolerance Genotypes of Oilseeds Crops

Many genotypes of oilseed crops found tolerance/resistance against abiotic stress have been identified by many researchers in India. All the tolerant genotypes are presented in Table 16.6.

Limitations of Conventional Breeding

Traditional approaches to breeding crop plants with improved abiotic stress tolerances have so far met limited success (Richards, 1996). This is due to a number of contributing factors, including: (1) the focus has been on yield rather than on speciûc traits; (2) the difficulties in breeding for tolerance traits, which include complexities introduced by genotype by environment, or G x E, interactions and the relatively infrequent use of simple physiological traits as measures of tolerance, have been potentially less subject to G x E interferences; and (3) desired traits can only be introduced from closely related species. Progress in developing high-yielding, drought-tolerant rice cultivars by conventional breeding has been slow, largely because of difficulties in precisely defining the target environment, complex interactions of drought tolerance with environments, and lack of appropriate screening methodology (Cooper *et al.*, 1999; Wade *et al.*, 1999).

Non-conventional Approaches

Conventional breeding has major limitations, including the need for multiple backcrosses to eliminate undesirable traits, restriction to loci that give a clearly

Table 16.7: Resistant/Tolerant Genotypes of Oilseed Crops against Abiotic Stresses

Crop	*Stress Components*	*Genotypes*	*References*
Groundnut (*Arachis hypogea* L.)	Soil moisture deficit	NRCG-7085, 9773, and 664	Basu, 2003
	Cold (low temp.)	NRCG-1339 and *A. monticola*	
	Soil salinity	NRCG-7453, 3554, 3665, and 168	
	Al-toxicity	ICG-813, 1001, 1021, 1048, 1056, 1064, 1355, 3606, 10964, and 11183	
	Ca deficiency	NRCG-7085-1, 6820, 6919, MOR-161, and ICGHNG-88448	
	P deficiency	NRCG-7085-1, 6155, PBS-13, 18057, 20016, MOR-204, PBS-11037, 20016, 20057, MOR-139	
	Fe deficiency	NRCG-5389, 6450, 7027, 4255, 5513, and 7267	
	Drought (Advanced breeding lines)	PBS-14014, 14021, 14029, 11023, 12118, 20055, code-9, and code11 (all tolerant)	
	Fe deficiency (Advanced breeding lines)	PBS-11010, 11016, 21014, 11015, 11040, and 21018 (all tolerant)	
Rapeseed mustard (*Brassica* spp.)	Drought	Vaibhav, RH-819, RH- 781 and Aravalli, *B. juncea*	Kumar and Premi, 2003
	Frost	Raya variety RH-781	
	Salinity	Varuna, CS-52, RLC-1357, NDR-8604, RH-30, and DIRA-343	
Soybean (*Glycine max*)	Drought	Hardee, Kalitur, Pusa-24, NRC-7, PK-327, PK-564, and JS-71-05	Joshi and Bhatia, 2003
	Light duration	JS-71-05, NRC-7, JS-335, and NRC-2	Bhati *et al.*, 1997
	P deficiency	PK-472 and PK-416	Dravid *et al.*, 2000
Castor (*Ricinus communis*)	Drought	GAUCH-1 and HC-8	Hanumantha and Lavanya, 2003
	Salinity	SA-2, Aruna, 48-1, Kranti, DCS-33, GAUCH-1, DCH-158 and DCH-141	

Contd...

Table 16.7–*Contd...*

Crop	*Stress Components*	*Genotypes*	*References*
Sesame (*Sesamum indicum*)	Drought	*Sesamum angustifoluim,* RT-46, RT-125, RT-54, and RT-127	Duhoon, 2003
	Salinity	TMV-1	
	High temperature	RT-46, RT-125, RT-54, and RT-127	
Sunflower (*Helianthus annuus*)	Drought	CMS-106A, CMS-335A, RCR-119, R-7, CSFI-5075, AKSFI-42-1	Anonymous, 2012-13
Coconut (*Cocos nucifera*)	Drought	WCT, LCT, WCT x COD, LCT x GBGD, LCt x COD	Rajagopal, 2003

observable phenotype and inadequacy if the gene pool lacks sufficient variation in the trait of interest. Therefore, the focus is currently on marker assisted breeding, which allows 'pyramiding' of desirable traits for more rapid crop improvement with less input of resources.

Somaclonal Approach

Pokkali type of rice varieties are highly salt-tolerant and commonly grow in coastal areas of Kerala State, India. It is a traditional, tall, susceptible to lodging, photoperiod-sensitive rice variety with low tillering and long, broad, dark, and droopy leaves. It has red pericarp and poor grain quality. It is highly tolerant of salinity but yielding ability is low. Pokkali was subjected to cell culture for induction of somaclonal variation. Somaclonal variants of Pokkali with improved agronomic traits were identified. The variant isolated (TCCP 266-2-49-B-B-3) had desirable levels of all the characteristics tested. Its level of tolerance for salinity remains as high as the donor. It has vigorous growth and, unlike the parent Pokkali, is semi-dwarf, a trait essential in increasing the yield potential without lodging. The variant had white pericarp which is a distinct advantage. There was also improvement in cooked rice quality, with the semi-dwarf variant having medium consistency. These characteristics make the somaclonal variant TCCP 266-2-49-B-B-3 superior to Pokkali as a donor of salt tolerance in hybridization programs. It has become a popular donor parent and has produced new high-yielding salinity-tolerant lines, some of which were already released as varieties. Using *In vitro* recurrent selection, potato salt tolerant cell lines and plants were developed by Ochatt *et al.* (1999). Salt tolerance rice plants have been obtained through in vitro selection in tissue culture (Zapata and Abrigo, 1986), and lines with tolerance to aluminium toxicity has been identified (Abrigo *et al.*, 1985). *In vitro* screening for salt stress in different national rice hybrids was performed by Shanmuganathan, (2001). It led to the identification of three hybrids CORH 2, DRRH 1 and PSD 1 as salt tolerant and hybrids KRRH 1 and CNRH 3 as moderately salt tolerant.

F_1 Anther Culture Derivatives

Development of a promising salinity-tolerant line through hybridization and selection procedure could take at least 8-10 years where more than one rice crop can be taken in a year, while it takes much more time where only one crop per year is grown. This period could be substantially reduced by applying the F_1 anther culture technique. In 1996, some high-yielding salt-tolerant anther culture-derived lines were generated at IRRI. It only took 3 years for the tolerant lines to be isolated. These AC-derived lines were IR51500-AC11-1, IR51500-AC17, IR51485-AC6534-4, IR72132-AC6-1, IR69997-AC1, IR69997-AC2, IR69997-AC3, and IR69997-AC4 (Senadhira *et al.*, 1994). IR51500-AC11-1 was released as a salt-tolerant variety in the Philippines with the name PSBRc50 or "Bicol" (Senadhira *et al.*, 2002). This is the first time for anther culture-derived lines from *indica-indica* cross to be released as a variety and also the first to be recommended for cultivation in adverse environments. IR51500-AC17 and IR51485-AC6534-4, named CSR21 and CSR28, respectively, were also identified as tolerant rice varieties in India (IRRI, 1997). The other rice varieties AC-6534-1 developed through same method tolerant to salinity and sodicity (Singh and Mishra, 1997; Senadhira *et al.*, 2002). Rahman *et al.* (2010) developed DH lines from the crosses involving salt-tolerant IRRI-derived lines using anther culture and in a field study one line AC-1 was promising for cultivation in saline area of Bangladesh. Anther culture–derived haploid embryos were used as explants for Agrobacterium mediated genetic transformation of bread wheat (*Triticum aestivum* L. cv CPAN1676) using barley *HVA1* gene for drought tolerance (Chauhan and Khurana, 2011). Thus, F_1 anther culture may be utilized in development of metal-toxicity tolerant in crop plants.

Marker Assisted Selection (MAS)

Marker assisted selection will be an effective way to save time in breeding if: (1) the heritability of the trait is high and field evaluation is very costly or simply cannot be done at your location, (2) environmental effects are significant, heritability is low, and classical selection is expensive or slow, or if the conditions for selection are only present occasionally (*e.g.* selection for drought tolerance in the rainy season), and (3) if you want to backcross a known gene into an inbred lines as rapidly as possible. DNA MAS has been seen as a means of improving the speed and efficiency of plant breeding programs because it is growth stage independent, unaffected by environment; no dominance effect and efficient to use in early generations. Most widespread use of MAS to date has done in the marker assisted backcrossing (MAB) of major genes to into already established varieties, mega-varieties (which occupies a large area within the country on across the countries) or elite cultivars. These markers could reduce the linkage around the target gene, and also recover the recurrent parent background within less number of generations in comparison to conventional breeding. The advent of marker added selection opens up real prospects for new strategies in breeding that combine conventional and marker technologies that suit the genetics of both the trait and the plant. Molecular marker linked to Al-toxicity was mapped by Wu *et al.* (2000). In wheat, QTL for salinity tolerance have been mapped by Semikhodskii *et al.* (1997). The mutation (*ari-e*.GP) for salinity tolerant in barley has been mapped on chromosome 5H (Thomas *et al.*, 1984). Some physiological traits associated with salt tolerant in barley were mapped to chromosome 1 (7H), 4 (4H), 5 (1H) and 6 (6H) (Ellis *et al.*, 1997). QTL for water use efficiency (Handley *et al.*, 1994), adaptation to drought environments (Chalmers *et al.*, 1992), and salt tolerance (Ellis *et al.*, 1997) in barley. Major loci and QTLs controlling Al tolerance have been identified in alfalfa (Narasimhamoorthy *et al.*, 2007), Soybean (Bianchi-Hall *et al.*, 2000), rice (Xue *et al.*, 2007), Sorghum (Magalhaes *et al.*, 2007), maize (Ninamango-Cardenas *et al.*, 2003), barley (Wang *et al.*, 2007), wheat (Raman *et al.*, 2005), oat (Wight *et al.*, 2006), and rye (Matos *et al.*, 2005).

Successful Example of MAS for Abiotic Stress

MAS for Drought Tolerance

In 2007, MAS 946-1 became the first drought tolerant aerobic rice variety released in India (www.hindu.com/2007/11/17/stories/2007111752560500.htm). To develop the new variety, scientists at the University of Agricultural Sciences (UAS), Bangalore, crossed a deep rooted upland japonica rice variety from the Philippines with a high yielding *indica* variety (Chandrashekar, 2007). Bred with

MAS, the new variety consumes up to 60 per cent less water than traditional varieties. It saves 3,000 litres of water for every kg of rice. In addition, MAS 946-1 gives yields comparable with conventional varieties (Gandhi, 2007). The new variety is the product of five years of research by a team lead by Shailaja Hittalmani at UAS, with funding from the International Rice Research Institute and the Rockefeller Foundation (Gandhi, 2007). Another aerobic variety, MAS-26 is in advanced stage. By using these MABC products, a variety namely "Birsa Vikas Dhan 111 (PY 84)" was developed and released in Jharkhand State of India (Ashraf, 2010). In sorghum, four major QTL's that control stay green and grain yield (*Stg1-Stg4*) have been identified (Harris *et al.*, 2007) and the favorable allele introgressed via MAS in elite lines. Ribaut and colleagues introgressed five QTL alleles for short anthesis-silking interval (ASI) through MABC from a drought tolerant donor to an elite, drought susceptible line (Ribaut *et al.*, 2004). Major QTLs for seed weight and grain yield at different moisture conditions have also been identified in durum wheat (Maccaferri *et al.*, 2008a). In cotton, extensive work has been carried out by Saranga and colleagues for the identification of drought related QTLs (Levi *et al.*, 2009a and b). Lebreton *et al.* (1995) mapped QTL for physiological traits associated with drought tolerance in maize. The stay-green trait has been associated with drought tolerance in sorghum (Borrell *et al.*, 1999). Yadav *et al.* (1999) reported QTL for drought tolerance and yield components in two mapping populations of pearl millet.

MAS for Salinity Tolerance

Major efforts in MAS to improve salinity tolerance in rice have been undertaken at the International Rice Research Institute for introgressing *Saltol*, a major QTL located at chromosome no. 1. A major QTL for salt tolerance named *Saltol* was mapped on chromosome 1 using F_8 recombinant inbred lines (RILs) of Pokkali/IR29 cross, which is responsible for low Na+, high K+ uptake and maintaining Na+/K+ homeostasis in the rice shoots (Nejad *et al.*, 2008). In Bangladesh, for example, the salt tolerance QTL "*Saltol*" was transferred into two mega rice varieties, BR-11 and BR-28 using marker assisted backcrossing approach (Rahman *et al.*, 2008). Rice variety BT-7 was developed a salt tolerance by using marker-assisted backcross, which was controlled by a major *Saltol* QTL (Linh *et al.*, 2012). Flowers *et al.* (2005), while reporting the putative AFLP markers for ion transport and selectivity for salinity tolerance from a custom made mapping population of rice, also cautioned against any expectation of general applicability of markers for these physiological traits. Subsamples, initially of 23 lines and later increased to 55 lines, of IR59462 were selected to test whether the AFLP markers associated with high Na uptake, high K uptake and low Na : K ratio in IR55178 were also markers for the same traits in IR59462. The presence or absence of markers was tested in two subsamples of the IR59462 population in which one group contained a high quantity of sodium in its shoots of at least 1.8 times that of the other (low) sample. None of the markers associated with salt tolerance in the mapping population (IR55178) was associated either the high or the low subsets of IR59462. Two loci, *Nax1* and *Nax2*, controlling shoot Na^+ accumulation were identified by QTL mapping in durum wheat (genome AABB; Munns *et al.*, 2003; James *et al.*, 2006).

MAS for Submergence Tolerance in Rice

In collaboration with Directorate of Rice Research, IRRI scientists succeeded in converting submergence-susceptible Swarma into submergence-tolerant Swarma-Sub (CR2539-1; IR82809-237; IET20266) in only three years (Neeraja *et al.*, 2007). Swarna-Sub1 shows a twofold or higher yield advantage over Swarna after submergence for 10 days or more during the vegetative stage (Septiningsih *et al.*, 2009). The *Sub 1* introgression lines are the first example of marker assisted breeding approach products in rice introgressed with a major QTL for abiotic stress tolerance. *Sub1* gene also introgressed in rice varieties like, Samba Mashuri-Sub1, IR64-Sub1, TDK1-Sub1, CR1009-Sub1and BR11—Sub1(Septiningsih *et al.*, 2009). Using MABC, *SUB1* was introgressed into Swarna, Samba Mahsuri and Savitri (CR1009) with great success as a high level of submergence tolerance was present in the new "enhanced" varieties and no other undesirable attributes were detected (Singh *et al.*, 2009). MABC was also used to pyramid a submergence-tolerant QTL allele together with three genes for tolerance to different biotic stresses into KDML-105, an *indica* cultivar widely grown in Thailand (Toojinda *et al.*, 2005).

Genetically Modified (GM) Crop

Plant genetic engineering techniques could be effectively utilized in exploiting untapped tolerance mechanisms to improve the harvestable crop yield. Genetic engineering is alternative because of its potential to improve abiotic stress tolerance more rapidly. Scientists in Egypt have produced drought-tolerant wheat by transferring a gene from barley into a local wheat variety. The researchers, at Cairo's Agricultural Genetic Engineering Research Institute (AGERI), showed that by transferring a gene called '*HVAI1*' from barley to wheat, the plants could tolerate low water levels for longer before their leaves wilted (Chauhan and Khurana, 2010). GM related work in wheat for tolerant to drought is going on in China (Xia *et al.*, 2012). Deak *et al.* (1999) developed transgenic rice with *ferritin* gene to enhance high iron storage and they suggested that the enhanced Fe storage ability can reduce reactive oxygen species. *Ferritin* gene has been also transferred into rice (Khalekuzzaman *et al.*, 2006), which enhanced iron level in the endosperm. Sengupta *et al.* (1993) transferred *SOD* (superoxide dismutase) gene from pea to tobacco and the transgenic were found to be drought tolerance. Transgenic soybean plants transferred with antisense *P5CR* (L-D'-pyrroline-5-carboxylate reductase) gene showed increased proline accumulation, leading to higher tolerance to water stress (De Ronde *et al.*, 2000). Transgenic rice carrying barley *HVA1* gene had shown drought tolerance (Xu *et al.*, 1996). Rohila *et al.* (2002) transformed Pusa Basmati 1 with *HVA1* to increase tolerance against abiotic stresses. Some other important examples of plant genetic engineering for tolerance to abiotic stresses were given by Pattanayak and Ananda Kumar, (2000). Over-expression of the gene *Alfin1*in alfalfa increased its salt tolerance and promoted root growth and shoot growth, under normal and saline conditions, producing larger plants than the wild type (Winicov, 2000). *HAL1* alters the salt tolerance of tomato (Gisbert *et al.*, 2000) and increases the K/Na ratio in transgenic plants. A yeast kinase (a functional homologue of the yeast Dbf2 kinase) enhanced tolerance of tobacco cells to salt in tissue culture (Lee *et al.*, 1999). Potato and rice (Yeo *et al.*, 2000 and Garg *et al.*, 2002, respectively) transformed with trehalose synthesis genes displayed tolerance to drought (in case of potato), and salt, drought, and low-temperature stress (in case of rice). Transgenic *Arabidopsis* overexpressing wheat vacuolar Na+/H+ anti porter (*TNHX1*) and H+ pyrophosphatase (*TVP1*) genes simultaneously resulted in tolerance to salt and drought stress (Brini *et al.*, 2007). For, instance, over expression of *Pea DNA Helicase 45* in tobacco increased sat tolerance without affecting the yield (Mishra *et al.*, 2005). Liu *et al.* (2008) have studied the expression of *DEAD box Helicase 1* from halophyte *Apocynum venatum* and found it to be up regulated on exposure to salt or cold stress. Similarly, the over expression of medicago helicase (*MH1*) led to a drastically increased level of drought tolerance in Arabidopsis (Luo *et al.*, 2009). Hoshida *et al.* (2000) have reported on enhanced tolerance to salt stress in transgenic rice that over expresses chloroplast glutamine synthetase gene. Salt tolerance of transgenic rice with *mtlD* gene and *gutD* gene were analysed by Wang *et al.* (2000). Chen and Guo (2008) have reported that tobacco *OPBP1* enhances salt tolerance and disease resistance of transgenic rice.

Conclusion

Majority of crops losses are due to the abiotic stresses that causes more than 50 per cent harvest losses. Based on a plenty of research findings, it is clear that salt and drought stresses have devastating effect on the growth, development, physiology and yield of plants. The conventional plant breeding played a considerable role during the last century not only for improving the quality and yield of crops, but also for improving abiotic stress tolerance including drought and salinity tolerance. But it will take more time to developing abiotic stress tolerant varieties/hybrids. The development of crop varieties/lines/hybrids with increased tolerance to drought, salinity, heat, high temperature, and nutrient deficiency, by both conventional and molecular breeding methods and by genetic engineering is an important strategy to meet global food demands with less water. Combined knowledge of traditional breeding along with marker assisted selection makes it easier and more efficient to induce drought tolerance through the genotypic data in crop plants to enhance and sustain productivity in drought prone environments. To be effective, plant breeders of the future will need a combination of knowledge, skills, and experience in plant breeding, genetics, genomics, statistics, experimental design, genetic diversity, and germplasm management (Repinski *et al.*, 2011). They must be skillful communicators and have the ability to work well in interdisciplinary teams. An understanding of genetic basis of drought and salinity resistance in crop plants is a pre-requisite for a geneticist to evolve superior genotype through either conventional breeding methodology or biotechnological approach. Given rapid changes in crop science and genetic technology, breeders must continually upgrade their knowledge and skills to stay current. Although, scientists are working hard to increase the average yield of various economically important crop plants, a limited success is achieved due to the increase extent of abiotic and biotic stresses. Therefore, there is a urgent need to devise methodologies to enhance crop production particularly in the stressed regions of the world. The present strategy is to develop MAS techniques which will accelerate the breeding process, increase selection efficiency and affordable to National Agricultural Research and Extension Systems (NARES) partners. Conventional breeding methods in developing tolerance to salinity and other related abiotic stresses should not be neglect but be supplemented with studies on soil and water management, and biotechnological techniques. Through recent developments in biotechnology, breeders can now review their wish list of desirable traits to be incorporated through gene pyramiding. The challenge is how breeders adopt and harness these novel techniques. Every day, it becomes more evident that successful breeding for stable high yield under abiotic stress conditions will only be possible when a true integration of traditional breeding with physiology and genomics is achieved. Breeding for abiotic stress tolerance is vital not only in current climates, but also in future projected climates with increased frequency of abiotic stress events. New sources for abiotic stress tolerance need to be exploited. Crop wild relatives and landraces have the potential to contribute abiotic stress tolerance to grain crops, but wild accessions and landraces that may provide the highest number of tolerance genes are under-represented in many germplasm collections. Improvements in molecular technology are contributing to faster identification of traits in wild relatives, and innovative methods have helped develop new and improved varieties. Most of these resources remain untapped and are potentially precious. Changing climates and climate variability highlight the need to discover more rapidly genetic variability and new sources of tolerance and to exploit all available genetic resources, identify new resources and conserve them for the future.

References

Abd Allah, A.A., Mohamed, A.A.A., Gaballah, M.M., 2009. Genetic studies of some physiological and shoot characters in relation to drought tolerance in rice. *Journal Agric. Res.*, 4: 964-990.

Abrigo, W.M., Novero, A.U., Coronal, V.P., Cabuslay, G.S., Blanco, L.C. and Parao, F.T., 1985. Somatic cell culture at IRRI. In: Biotechnology in International Agricultural Research, 149-158. IRRI: Los Banos, Philippines.

Abrol, I.P., Yadav, J.S.P. and Massoud, F.I., 1988. Salt affected soils and their management-United Nations, Rome, FAO Soils Bull. 39.

Acquaah, G., 2007. Principles of plant genetics and breeding. Blackwell, Oxford, UK.

Akbar, M. Khush, G.S. and Hillerislambers, D., 1985. Genetics of salt tolerance. In: Rice genetics. IRRI, Los Bafios, Phillipines, pp. 399-409.

Al-Doss, A.A. and Smith, S.E., 1998. Registration of AZ-97MEC and AZ-97MEC-ST very non-dormant alfalfa germplasm pools with increased shoot weight and differential response to saline irrigation. *Crop Sci*, 38: 568.

Ali, A.J., Xu, J.L., Ismail, A.M., Fu, B.Y., Vijayakumar, C.H.M., Gao, Y.M., Domingo, J., Maghirang, R., Yu, S.B., Gregorio, G., Yanaghihara, S., Cohen, M., Carmen, B., Mackill, D. and Li, Z.K., 2006. Hidden diversity for abiotic and biotic stress tolerances in the primary gene pool of rice revealed by a large backcross breeding program. *Field Crops Res.*, 97: 66–76.

Ali, M.A. and Abo-El-Wafa, A.M., 2006. Inheritance and selection for earliness in spring wheat under heat stress. *Assiut J. Agric. Sci.*, 37: 77-94.

Ali, S.M. and Naidu, A.P., 1982. Screening for drought tolerance in maize. *Ind. J. Genet.*, 42: 381-388.

Armento-Sato, J., Chang, T.T., Loresto, G.C. and O'Toole, J.C., 1983. Genetic analysis of root characters in rice. *SABRAOJ.*, 15: 103-116.

Ashraf, M. and O'Leary, J.W., 1996. Responses of some newly developed salt tolerant genotypes of spring wheat to salt stress. 1. Yield components and ion distribution. *J. Agron. Crop Sci.*, 176: 91–101.

Ashraf, M. and O'Leary, J.W., 1996. Responses of some newly developed salt tolerant genotypes of spring wheat to salt stress. I. Yield components and ion distribution. *J. Agron. Crop Sci.*, 176: 91–101.

Ashraf, M., 2010. Inducing drought tolerance in plants: Recent advances. *Biotech. Advances*, 28: 169–183. Ashraf, M., 2010. Inducing drought tolerance in plants: recent advances. *Biotechnology Advances*, 28: 169-183.

Ashraf, M., Athar, H.R., Harris, P.J.C. and Kwon, T.R., 2008. Some prospective strategies for improving crop salt tolerance. *Adv. Agron.*, 97: 45–110.

Ashraf, M., McNeilly, T. and Bradshaw, A.D., 1986. Tolerance of sodium chloride and its genetic basis in natural populations of four grass species. *New Phytologist*, 103: 725-734.

Badu-Apraku, B. and Yallou, C.G., 2009. Registration of striga-resistant and drought tolerant tropical early maize populations TZE-W Pop DT STR C4 and TZE-Y Pop DT STR C4. *J. Plant Registr.*, 3(1):86–90.

Badu-Apraku, B., Menkir, A., Kling, J.G. and Fakorede, M.A.B., 2006. Registration of 16 striga resistant early maturing tropical maize inbred lines. *Crop Sci.*, 46:1410–1411.

Baenziger, P.S., Beecher, B., Graybosch, R.A., Ibrahim, A.M.H., Baltensperger, D.D., Nelson, L.A., 2008. Registration of'NEO1643' wheat. *J. Plant Registr.*, 2(1):36–42.

Bai, R., Zhang, Z., Hu, Y., Fan, M. and Schmidhalter, U., 2011. Improving the salt tolerance of Chinese spring wheat through an evaluation of genotype genetic variation. *Aust. J. Crop Sci.*, 5(10): 1173-1178.

Bänziger, M., Setimela, P.S., Hodson, D. and Vivek, B., 2004. Breeding for improved drought tolerance in maize adapted to southern Africa. Proceedings of the 4th International Crop Science Congress. Brisbane, Australia: Published on CDROM; 2004. 26 Sep – 1 Oct.

Basu, M.S., 2003. Stress management in groundnut. National Seminar On Stress Management in Oilseeds for Attaining Self-Reliance in Vegetable Oils, January, 28-30, pp. 1-6.

Bergman, J.W., Riveland, N.R., Flynn, C.R., Carlson, G.R. and Wichman, D.M., 2001. Registration of 'Morlin'safower. *Crop Sci.*, 41:1640.

Bernier, J., Kumar, A., Ramaiah, V., Spaner, D. and Atlin, G., 2007. A large-effect QTL for grain yield under reproductive-stage drought stress in upland rice. *Crop Sci.*, 47: 505-516.

Bhati, V.S., Manglik, Priti, Bhatnagar, P.S. and Guruprasad, K.N., 1997. Variation in sensitivity of soybean genotypes to varying photoperiods in India. *Soybean Genetics Newsletter*, 24: 99-103.

Bianchi-Hall, C.M., Carter, T.E., Bailey, M.A., Mian, M.A.R., Rufty, T.W., Ashley, D.A., Boerma, H.R., Arellano, C., Hussey, R.S. and Parrott, W.A., 2000. Aluminum tolerance associated with quantitative trait loci derived from soybean PI 416937 in hydroponics. *Crop Science*, 40: 538–545.

Blum, A., 1979. Genetic improvement of drought resistance in crop plants: A case for sorghum. Pages 429-445 in H. Mussell and R.C. Staples, eds. Stress physiology in crop plants. Wiley Interscience, New York.

Borrell, A.K., Bidinger, F.R. and Sunitha, K., 1999. Stay-green trait associated with yield in recombinant inbred sorghum lines varying in rate of leaf senescence. *Int. Sorghum and Millets Newslett.*, 40: 31-34. Boyer, J.S., 1982. Plant productivity and environment. *Science*, 218: 443-448.

Boyer, J.S., 1982. Plant productivity and environment. *Science*, 218: 443-448.

Bray, E.A., Bailey-Serres, J. and Weretilnyk, E., 2000. Response to abiotic stresses. In: Biochemistry and Molecular Biology of plants. Gruissem W, Buchannan B, Jones R. eds. American Society of Plant Physiologists, Rockville, MD, pp. 1158-1249.

Brick, M.A., Ogg, J.B., Singh, S.P., Schwartz, H.F., Johnson, J.J. and Pastor-Corrales, M.A., 2008. Registration of drought-tolerant, rust-resistant, high-yielding pinto bean germplasm line CO46348. *J. Plant Registr.*, 2(2):120–124.

Brini, F., Hanin, M., Mezghani, I., Berkowitz, G.A. and Masmoudi, K., 2007. Overexpression of wheat Na+/H+ antiporter TNHX1 and H+ pyrophosphatase TVP1 improve salt and drought stress tolerance in Arabidopsis thaliana plants. *J. Exp. Bot.*, 58: 301-308.

Buchanan, B.B., Gruissem, W. and Jones, R.L., 2000. Biochemistry and molecular biology of plants. American Soc. Plant Physiologists. Rockville. Maryland. USA FAO. (1985). Production year book, 39. Roam: FAO.

Byerlee, D. and Moya, P., 1993. Impacts of International Wheat Breeding Research in the Developing World, 1966–1990. CIMMYT, México, D.F, 104p.

Carena, M.J. and Wanner, D.W., 2009. Development of genetically broad-based inbred lines of maize for early-maturing (70-80RM) hybrids. *J. Plant Registr.*, 3:107–111.

Carver, B.F., Inskeep, W.P., Wilson, N.P. and Westerman, R.L., 1988. Seedling tolerance to aluminium toxicity in hardred winter wheat germplasm. *Crop Sci.*, 28: 463-467.

Cash, S.D., Bruckner, P.L., Wichman, D.M., Kephart, K.D. Berg, J.E. and Boyner, R., 2009. Registration of Willow Creek' forage wheat. *J. Plant Registr.*, 3(2):185–190.

Cattivelli, L., Rizza, F., Badeck, F.W., Mazzucotelli, E., Mastrangelo, E.M., Francia, E., Marè, C., Tondelli, A. and Stanca, A.M., 2008. Drought tolerance improvement in crop plants: an integrated view from breeding to genomics. *Field Crops Res.*, 105:1–14.

Ceccarelli, S. and Grando, S., 1996. Drought as a challenge for the plant breeder. *Plant Growth Regulation,* 20: 149-155.

Chalmers, K.G., Waugh, R., Watters, J., Forster, B.P., Nevo, E., Abbott, R.J. and Powell, W., 1992. Grain isozyme and ribosomal DNA variability in Hordeum spontaneum populations from Israel. *Theor. Appl. Genet.,* 84: 313-322.

Chandrashekar, M.V., 2007. UAS develops aerobic rice. Food and Beverage News. www.fnbnews.com/article/detarchive.asp?articleid=22118 and sectionid=34 (verified 20 March 2009).

Chang, T.T., De Datta, S.K. and Coffman, W.R., 1975. Breeding methods for upland rice. In: Major research in upland rice. IRRI, Los Banos, Philippines. 143-157.

Chapman, S.C., and Edmeades, G.O., 1999. Selection improves tolerance to mid/late season drought in tropical maize populations. II. Direct and correlated responses among secondary traits. *Crop Sci.,* 39: 1315–1324.

Chauhan, H. and Khurana, P., 2010. Development of drought tolerant transgenic doubled haploid in wheat through Agrobacterium-mediated transformation. *Plant Biotech. J.,* 9: 408-417.

Chauhan, H. and Khurana, P., 2011. Use of doubled haploid technology for development of stable drought tolerant bread wheat (*Triticum aestivum* L.) transgenics. *Plant Biotech. J.,* 9: 408–417.

Chen, M., Wang, Q.Y., Cheng, X.G., Xu, Z.S., Li, L.C., Ye, X.G., 2007. GmDREB2, a soybean DRE binding transcription factor, conferred drought and high-salt tolerance in transgenic plants. *Biochem. Biophys. Res. Commun.,* 353:299–305.

Chen, X. and Guo, Z., 2008. Tobacco *OPBP1* Enhances salt tolerance and disease resistance of transgenic rice. *Int. J. Mol. Sci.,* 9(12): 2601–2613.

Cooper, M., Fukai, S. and Wade, L.J., 1999. How can breeding contribute to more productive and sustainable rainfed lowland rice systems? *Field Crops Res.,* 64: 199–209.

Cornford, S.G., 2003. The socio-economic impacts of weather events in 2002. *WMO Bull.,* 52:269–290.

Crosbic, T.M., Pearce, R.B. and Mock, J.J., 1981. Recurrent phenotypic selection for high and low photosynthesis in two maize populations. *Crop Sci.,* 21: 736-740.

Dashti, H., Naghavi, M.R. and Tajabadipour, A., 2010. Genetic analysis of salinity tolerance in a bread wheat cross. *J. Agric. Sci. Tech.,* 12: 347-356.

De Ronde, J.A., Spreeth, M.H. and Cress, W.A., 2000. Effects of antisense 1-Pyrroline-5-carbohydrate reductase transgenic soybean plant subjected to osmotic and drought stress. *Plant Growth Reg.,* 32: 13-26.

Deak, M., Horvath, G.V., Davletova, S., Torok, K., Sass, L., Vass, I., Barna, B., Kiraly Z. and Dudits, D., 1999. Plants ectopically expressing the iron-binding protein, ferritin, are tolerant to oxidative damage and pathogens. *Nat. Biotech.,* 17: 192-196.

Dehdari, A., Rezai, A. and Mirmohammadi Maibody, S.A., 2007. Genetic control of salt tolerance in wheat plants using generation means and variances analysis. *J. Sci. Tech. Agric. Natu. Resou.*, 40: 179-192.

Dewey, P.R., 1962. Breeding crested wheatgrass for salt tolerance. *Crop Science,* 2: 403-407. Dobrenz, A.K. Robinson, D.L. Smith, S.E. and Poteet, D.C., 1989. Registration of AZ-germ salt-II nondormant alfalfa germplasm. *Crop Sci.,* 29: 493.

Dobrenz, A.K. Robinson, D.L. Smith, S.E. and Poteet, D.C., 1989. Registration of AZ-germ salt-II nondormant alfalfa germplasm. *Crop Sci.,* 29: 493.

Dobrenz, A.K., 1999. Salt-tolerant alfalfa. US: Agripro Seeds, Inc., Shawnee Mission, KS; x. Dravid, M.S., Joshi, O.P. and Vyas, B.N., 2000. P utilization in soybean as influenced by plant genotypes and growth promoter (Vipul) under field conditions. *J. Nuclear Res. in Agric. And Biol.*, 29: 83-86.

Dravid, M.S., Joshi, O.P. and Vyas, B.N., 2000. P utilization in soybean as influenced by plant genotypes and growth promoter (Vipul) under field conditions. *J. Nuclear Res. in Agric. And Biol.*, 29: 83-86.

Dudely, J.W., 1984. A method of identifying lines for use in improving parents of a single cross. *Crop Sci.,* 24: 355-357. Duhoon, S.S., 2003. Stress management in sesame and niger. National Seminar on Stress Management in Oilseeds for Attaining Self-Reliance in Vegetable Oils, January, 28-30, pp. 42-54.

Duhoon, S.S., 2003. Stress management in sesame and niger. National Seminar On Stress Management in Oilseeds for Attaining Self-Reliance in Vegetable Oils, January, 28-30, pp. 42-54.

Edemeades, G.O., Chapman, S.C., Bolanos, J., Banziger, M. and Lafite, H.R., 1994. Recent evaluations of progress in selection for drought tolerance in tropical maize. In. Jeyvell *et al.* (Eds). Maize Research for stress environments. *Proc. of 4th Eastern and Southern Africa Regional maize Conf. Harare, Zimbabwe.* 94-100.

Edmeades, G.O., Bolanos, J. and Lafitte, H.R., 1992. Progress in breeding for drought tolerance in maize proceedings of the Fourthy-Seventh Annual Corn and Sorghum Industry Research Conference. Pp. 93-111.

Ekanayake, I.J.,'O Toole, J.C., Garrity, D.P. and Masajo, T.M., 1985. Inheritance of root characters and their relations to drought resistance in rice. *Crop Sci.,* 25: 927-933.

Ellis, R.P., Forster, B.P., Waugh, R., Bonar, N., Handley, L.L., Robinson, D., Gordon, D.C. and Powell, W., 1997. Mapping physiological traits in barley. *New Phytologist,* 137: 149-157.

El-Morshidy, M.A., Kheiralla, K.A., Ali, M.A. and Ahmed, A.A.S., 2010. Efficiency of pedigree selection for earliness and grain yield in two what populations under water stress conditions. *Assiut J. Agric. Sci.,* 37: 77-94.

Epstein, E., Norlyn, J.D., Rush, D.W., Kingsbury, R., Kelley, D.B. and Wrana, A.F., 1980. Saline culture of crops: a genetic approach. *Science,* 210: 399-404.

Farshadfar, E., Aghaie Sarbarzah, M., Sharifi, M. and Yaghotipour, A., 2008. Assessment of salt tolerance inheritance in barley via generation mean analysis. *J. Biol. Sci.*, 8: 461-465.

Farshadfar, E., Valiollah, R., Jaime Teixeira, D. and Farshadfar, M., 2011. Inheritance of drought tolerance indicators in bread wheat (*Triticum aestivum* L.) using a diallel technique. *Aust. J. Crop Sci.*, 5(3): 308-317.

Fereres, E. and Connor, D.J., 2004. Sustainable water management in agriculture. In: Cabrera E, Cobacho R (eds.) Challenges of the new water policies for the XXI century. A.A. Balkema, Lisse, pp. 157–170.

Ferna'ndez_Martinez, J.M., Dominguez, J., Jime'nez, C. and Fereres, E., 1990. Registration of three sunflower high oil non restorer germplasm populations. *Crop Sci.*, 30: 965.

Fisher, J.A. and Scott, B.J., 1987. In: P.G.E. Searle, B.G. Davey, eds. Priorities in Plant Soil Relations Research for Plant Production. Sydeny: University of Sydeny. pp. 135-137.

Flowers, T.J., 2004. Improving crop salt tolerance. *J. Exper. Botany*, 55: 307–319.

Flowers, T.J., Koyama, M.L., Flowers, S.A., Sudhakar, C., Singh, K.P., and Yeo, A.R., 2005. QTL: their place in engineering tolerance of rice to salinity. *J. of Exp.Bot.*, 51: 342.

Food and Agriculture Organization, "Report of salt affected agriculture," 2010, http://www.fao.org/ag/agl/agll/spush/.

Gandhi, D., 2007. UAS scientist develops first drought tolerant rice. The Hindu. www.thehindu.com/2007/11/17/stories/2007111752560500.htm (verified 20 March 2009).

Garg, A.K., Kim, J.K., Owens, T.G., Ranwala, A.P., Choi, Y.C., Kochian, L.V. and Wu, R.J., 2002. Trehalose accumulation inrice plants confers high tolerance levels to different abiotic stresses. *Proc. Natl. Acad. Sci., USA*, 99: 15898–15903.

Gautam, R.K., Chauhan, M.S., Panesar, B. Nayak, A.K., Sharma, D.K. Qadar, A. and Gurbachan Singh., 2011. Stress tolerance and heritability in dill (*Anethum graveolens*) lines from central and western India under contrasting soil sodicity environments. *Ind. J. of Agric. Sci.*, 81(4): 353–358.

Ghassemi, F., Jakeman, A.J. and Nix, H.A., 1995. Salinisation of Land and Water Resources Human Causes Extent Management and Case Studies. CABI Publishing, Wallingford, Oxon, p. 526.

Gisbert, C., Rus, A.M., Bolarin, M.C., Lopez-Coronado, J.M., Arrillaga, I., Montesinos, C., Caro, M., Serrano, R. and Moreno, V., 2000. The yeast HAL1 gene improves salt tolerance of transgenic tomato. *Plant Physiology*, 123: 393-402.

Gomez, S.M. and Kalamani, A., 2003. Scope of landraces for future drought tolerance breeding program in rice. *Archives*, 3(1): 77-79.

Granados, G., Pandey, S. and Ceballos, H., 1993. Response to selection for tolerance to acid soils in a tropical maize population. *Crop Sci.*, 33: 936-940.

Gregorio, G.B. and Senadhira, D., 1993. Genetic analysis of salinity tolerance in rice (*Oryza sativa* L.). *Theor. Appl. Genet.,* 86: 333-338.

Gupta, D., 1999. Genetic of salt tolerance and ionic uptake in rice (*Oryza sativa* L.). Ph. D. thesis. BRA University, Agra, India, p 112. Haley, S.D., Johnson, J.J., Peairs, F.B., Quick, J.S., Stromberger, J.A., Clayshulte, S.R., 2007. Registration of 'Ripper' wheat. *J. Plant Registr.,* 1:1–6.

Haley, S.D., Johnson, J.J., Peairs, F.B., Quick, J.S., Stromberger, J.A., Clayshulte, S.R., 2007. Registration of 'Ripper' wheat. *J. Plant Registr.,* 1:1–6.

Handley, L.L., Nevo, E., Raven, J.A., Martinez-Carrasco, R., Scrimgeour, C.M., Pakniyat, H. and Forster, B.P., 1994. Chromosome 4 controls potential water use efficiency in barley. *Journal of Experimental Botany,* 45: 1661-1663.

Hanumantha, C. and Lavanya, C., 2003. Stress management in castorbean. National Seminar On Stress Management in Oilseeds for Attaining Self-Reliance in Vegetable Oils, January, 28-30, pp. 37-41.

Harris, K., Subudhi, P.K., Borrell, A., Jordan, D., Rosenow, D., Nguyen, H., Klein, P., Klein, R. and Mullet, J., 2007. Sorghum stay-green QTL individually reduce post-flowering drought-induced leaf senescence. *J. Exp. Bot.,* 58: 327–338.

Heyne, *E.G.* and Bruson, A.M., 1940. Genetic studies of heat and drought tolerance in maize. *J. Amer. Soc. Agron.,* 32: 803-814.

Hollington, P.A., 2000."Technological breakthroughs in screening/breeding wheat varieties for salt tolerance", Gupta, S.K. and Sharma, S.K. and Tyagi, N.K. (Eds.), Proceedings of the National Conference 'Salinity Management in Agriculture, 2000, p.273-289.

Hoshida, H., Tanaka, Y., Hibino, T., Hayashi, Y., Tanaka, A., Takabe, T. and Takabe, T., 2000. Enhanced tolerance to salt stress in transgenic rice that over expresses chloroplast glutamine synthetase. *Plant Mol. Biol.,* 43(1):103-11.

Hurd, E.A., 1969. A method of breeding for yield of wheat in semi-arid climates. *Euphytica,* 18:217-226.

Hurd, E.A., 1971. Can we breed for drought resistance ? In : Larson, K. L. and Eastin, J. D. (Eds) : Drought injury and resistance in crops. CSSA Special Publication 2. Crop Science Society of America, Madison, Wisconsin. 77-88.

Hurd, E.A., 1976. Plant breeding for drought resistance. In: Kozlowski, T.T. (Ed.): Water deficits and plant growth 4. Academic Press, New York and London. 317-353.

Indrani Dana, Sitesh Chatterjee and Chinmoy Kundu, 2013. Twenty years of achievements of the EIRLSBN at the Rice Research Station, Chinsurah. Collard, B.C.Y., Ismail, I.S. and Hardy, B. (eds.), IRRI. EIRLSBN Twenty years of achievements in rice breeding. Pp. 53-64.

IRRI, 1997. Program report for 1996. International Rice Research Institute, Los Banos, Philippines. James, R.A., Davenport, R.J. and Munns, R., 2006. Physiological characterization of two genes for Na^+ exclusion in durum wheat, *Nax1* and *Nax2*. *Plant Physiol.,* 142: 1537–1547.

James, R.A., Davenport, R.J. and Munns, R., 2006. Physiological characterization of two genes for *Na1* exclusion in durum wheat, *Nax1* and *Nax2*. *Plant Physiol.,* 142: 1537–1547.

Jensen, N.F., 1970. A diallel selective mating system for cereal breeding. *Crop Sci.,* **10:** 629-635. Johnson, D.W. Smith, S.E. and Dobrenz, A.K., 1991. Registration of AZ-90NDC-ST nondormant alfalfa germplasm with improved forage yield in saline environments. *Crop Sci.,* **31:** 1098–1099.

Johnson, D.W. Smith, S.E. and Dobrenz, A.K., 1991. Registration of AZ-90NDC-ST nondormant alfalfa germplasm with improved forage yield in saline environments. *Crop Sci.,* **31:** 1098–1099.

Joshi, O.P. and Bhatia, V.S., 2003. Stress management in soybean. National Seminar On Stress Management in Oilseeds for Attaining Self-Reliance in Vegetable Oils, January, 28-30, pp. 13-25.

Juskiw, P.E., Helm, J.H., Oro, M., Nyachiro, J.M. and Salmon, D.F., 2009. Registration of 'Bentley' barley. *J. Plant Registr.,* 3(2):119–123.

Kaur, G., Kumar, S., Nayyar, H. and Upadhyaya, H.D., 2008. Cold stress injury during the pod- filling phase in chickpea (*Cicer arietinum* L.): effects on quantitative and qualitative components of seeds. *J. Agron. Crop Sci.,* 194:457–464.

Kenga, R., Alabi, S.O. and Gupta, S.C., 2004. Combing ability studies in tropical sorghum (Sorghum bicolor (L.) Moench). *Field Crops Res.,* 88:251–260.

Khalekuzzaman, M., Datta, K., Oliva, N., Alam, M.F., Jorder, A.I. and Datta, S.K., 2006. Stable integration, expression and inheritance of ferritin gene in transgenic elite indica cultivar Br29 with enhanced iron level in the endosperm. *Indian J. Biotech.,* **5**(1): 26-31.

Khidse, S.R., Bhale, N.L. and Borikar, S.T., 1982. Combining ability for leaf water deficit and grain yield in *rabi* sorghum. *Ind. J. Genet.,* 42: 82-86.

Khush, G.S. and Coffman, W.R., 1977. Genetic evaluation and utilization (GEU), the rice improvement program at the International Rice Research Institute. *Theor. Appl. Genet.,* 51: 97-110.

Kumar, A. and Premi, O.P., 2003. Stress management in rapeseed-mustard. National Seminar On Stress Management in Oilseeds for Attaining Self-Reliance in Vegetable Oils, January, 28-30, pp. 7-13.

Kumar, J., Yadav, S.S. and Kumar, S., 2004. Influence of moisture stress on quantitative characters in chickpea (*Cicer arietinum* L.). *Indian J. Gen. Plant Breed.,* 64(2):149-150.

Kumar, R., Venuprasad, R. and Atlin, G.N., 2007. Genetic analysis of rainfed lowland rice drought tolerance under naturally-occurring stress in eastern India: Heritability and QTL effects. *Field Crops Res.,* 103: 42-52.

Lafitte, H.R., Li, Z.K., Vijayakumar, C.H.M., Gao, Y.M., Shi, Y., Xu, J.L., Fu, B.Y., Yu, S.B., Ali, A.J., Domingo, J., Maghirang, R., Torres, R. and Mackill, D., 2006. Improvement of rice drought tolerance through backcross breeding: Evaluation of donors and selection in drought nurseries. *Field Crops Research,* 97: 77–86.

Lebreton, C., Lazic-Jancic, V., Steed, A., Pekic, S. and Quarrie, S.A., 1995. Identification of QTL for drought responses in maize and their use in testing causal relationships between traits. *J. Exp. Bot.*, 46: 853-865.

Lee, J.H., van Montagu, M. and Verbruggen, N., 1999. A highly conserved kinase is an essential component for stress tolerance in yeast and plant cells. *Proceedings of the National Academy of Sciences, USA.* 96: 5873-5877.

Lee, K., Choi, W., Ko, J., Kim, T. and Gregorio, G.B., 2003. Salinity tolerance of japonica and indica rice (*Oryza sativa* L.) at the seedling stage. *Planta,* 216(6): 1043-1046.

Lekgari, A. L. and Setimela, P.S., 2001. Selection of suitable maize genotypes in Botswana. Seventh Eastern and Southern Africa Regional Maize Conference 11th – 15th February, 2001. pp. 213-215.

Lekgari, A. L. and Setimela, P.S., 2001. Selection of suitable maize genotypes in Botswana. Seventh Eastern and Southern Africa Regional Maize Conference 11th – 15th February, 2001. pp. 213-215.

Levi, A., Ovnat, L., Paterson, A.H. and Saranga, Y., 2009b. Photosynthesis of cotton near-isogenic lines introgressed with QTLs for productivity and drought related traits. *Plant Science*, 177(2): 88-96.

Levi, A., Paterson, A.H., Barak, V., Yakir, D., Wang, B., Chee, P.W. and Saranga, Y., 2009a. Field evaluation of cotton near-isogenic lines introgressed with QTLs for productivity and drought related traits. *Molecular Breeding*, 23(2):179-195.

Lewis, C. F. and Christiansen, M.N., 1981. Breeding Plant for Stress Environments. In: "*Plant Breeding II*". (Ed.) Frey, K. J. Proceedings of Plant Breeding Symp. 16 March 1979. Iowa State University Press. Ames, Iowa, pp.151-157.

Liang, L.M., 2010. Combining ability and genetic effect analysis of drought tolerant traits during seedling stage in *Brassica napus* L. M.Sc. Thesis. Southwestern University. Linh, L.H., Linh, T.H., Xuan, T.D., Ham, L.H., Ismail, A.M. and Khanh, T.D., 2012. Molecular Breeding to Improve Salt Tolerance of Rice (Oryza sativa L.) in the Red River Delta of Vietnam. *International Journal of Plant Genomics*, Volume 2012, Article ID 949038, 9 pages doi:10.1155/2012/949038.

Linh, L.H., Linh, T.H., Xuan, T.D., Ham, L.H., Ismail, A.M. and Khanh, T.D., 2012. Molecular Breeding to Improve Salt Tolerance of Rice (*Oryza sativa* L.) in the Red River Delta of Vietnam. *Intern. J. Plant Genomics*, Volume 2012, Article ID 949038, 9 pages doi:10.1155/2012/949038.

Liu, H.H., Liu, J., Fan, S.L., Song, M.Z., Han, H.L., Liu, F. and Shen, F.F., 2008. Molecular cloning and characterization of a salinity stress induced gene encoding DEAD-box Helicase from the halophyte *Apocynum venatum*. *J. Exp. Bot.*, 59(3): 633-644.

Long,W.H., Pu, H.M., Zhang, J.F., Qi, C.K. and Zhang, X.K., 2013. Screening of *Brassica napus* for salinity tolerance at germination stage. *Chin. J. Oil Crop Sci.*, 35: 271–275.

Luo, Y., Liu, Y.B., Drog, Y.X., Gao, X.Q. and Zhang, X.H., 2009. Expression of a putative alfalfa Helicase increases tolerance to abiotic stress in Arabidopsis by enhancing the capacities for ROS scavenging and osmotic adjustment. *J. Plant Physiol.*, 166: 385-394.

Maccaferri, M., Sanguineti, M.C., Corneti, S., Ortega, J.L.A, Ben Salem, M., Bort, J., DeAmbrogio, E., del Moral, L.F.G., Demontis, A., El-Ahmed, A., Maalouf, F., Machlab, H., Martos, V., Moragues, M., Motawaj, J., Nachit, M., Nserallah, N., Ouabbou, H., Royo, C., Slama, A. and Tuberosa, R., 2008. Quantitative trait loci for grain yield and adaptation of durum wheat (*Triticum durum* Desf.) across a wide range of water availability. *Genetics,* 178: 489–511.

Magalhaes, J.V., Liu, J., Guimarães, C.T., Lana, U.G.P., Alves, V.M.C., Wang, Y.H., Schaffert, R.E., Hoekenga, O.A., Piñeros, M.A., Shaff, J.E., Klein, P.E., Carneiro, N.P., Coelho, C.M., Trick, H.N. and Kochian, L.V., 2007. A gene in the multidrug and toxic compound extrusion (MATE) family confers aluminum tolerance in sorghum. *Nat. Genetics*, 39: 1156–1161.

Mai-Kodomi, Y., Singh, B.B., Terao, T., Myers, O., Yopp, J.H.J.R. and Gibson, P.J., 1999. Inheritance of drought tolerance in cowpea. *Indian J. Genet.*, 59(3): 317-323.

Maiti, R.K., Amaya, L.E.D., Cardona, S.I., Dimas, A.M.O., de la RosaIbarra, M. and Castillo, H.D., 1996. Genotypic variability in maize cultivars (*Zea mays* L.) for resistance to drought and salinity at the seedling stage. *Journal of Plant Physiology,* **148**: 741-744.

Maiti, R.K., de la Rosa-Ibarra, M. and Sandoval, N.D., 1994. Genotypic variability in glossy sorghum lines for resistance to drought, salinity and temperature stress at the seedling stage. *Journal of Plant Physiology*, 143: 211-244.

Mallik, S., 1995. Rice germplasm evaluation and improvement for stagnant flooding. KT, editor. Rainfed lowland rice: agricultural research for high-risk environments, Manila (Philip pines): International Rice Research Institute. p. 97-109.

Mallik, S., Mandal, B.K., Sen, S.N. and Sarkarung, S., 2002. Shuttle Breeding: an effective tool for rice varietal improvement in rainfed lowland ecosystem in eastern India. *Curr. Sci.,* 83(9): 1097-1102.

Matos, M., Camacho, M.V., Pérez-Flores, V., Pernaute, B., Pinto-Carnide, O. and Benito, C., 2005. A new aluminum tolerance gene located on rye chromosome arm 7RS. *Theoretical and Applied Genetics*, 111: 360–369.

Meena Kumari, Das, S., Vimala, Y.and Arora, P., 2004. Physiological parameters governing drought tolerance in maize. *Ind. J. of Plant Physiol.*, 9(2): 203-204.

Mihov M., Stoyanova M. and Mehandjiev A., 2001. Some results of application of experimental mutagenesis on lentil (*Lens culinaris* Medic.). Jubilee Scientific Session "80th Anniversary of the Higher Agricultural Education in Bulgaria". Agricultural University-Plovdiv, Bulgaria, Scientific Works, XLVI (2):397-400.

Mihov M., Stoyanova M., Atanasova D. and Mehandjiev A., 1999. Enrichment of lentil's germplasm (*L. culinaris* Medic.). Agrarian Sciences Res. Commun. Of USB, branch Dobrich, 1:57-60.

Miller, J.F. and Seiler. G.J., 2003. Registration of five oilseed maintainer (HA 429-HA 433) sunflower germplasm lines. *Crop Sci.*, 43:2313-2314.

Mishra, B. Akbar, M. Seshu, D.V. and Senadhira, D., 1996. Genetics of salinity tolerance and ionic uptake in rice. *Int. Rice Res. Newsl.*, 21: 38-39.

Mishra, B., 1991. Combining ability and heterosis for yield and yield components related to reproductive stage salinity and sodicity tolerance in rice, Oryza sativa L. IRRI, Rice Genetics II, Manila, Phillipines, p 761.

Mishra, B., 1994. Breeding for salt tolerance in crops. In: Rao *et al.* (eds) Salinity management for Sustainable agriculture: 25 years research at CSSRI. Central Soil Salinity Research Institute, Karnal, India, pp. 226-259.

Mishra, B., Akbar, M. and Seshu, D.V., 1990. Genetics studies on salinity tolerance in rice towards better productivity in salt-affected soils. Rice Research Seminar, International Rice Research Institute, Los Baños, Philippines, 12 July.

Mishra, B., Singh, R.K. and Jetly, Vandna., 1998. Inheritance pattern of salinity tolerance in rice. *Journal of Genetics and Breeding* (Rome) 52: 325-331.

Mishra, N.S., Pham, X.H., Sopory, S.K. and Tuteja, N., 2005. Pea DNA Helicase 45 overexpressioon in tobacco confers high salinity tolerance without affecting yield. PNAS, 102:509-514.

Mitra, J., 2001. Genetics and genetic improvement of drought resistance in crop plants. *Curr. Sci.*, 80: 758-763. Munns, R., Rebetzke, G.J., Husain, S., James, R.A. and Hare, R.A., 2003. Genetic control of sodium exclusion in durum wheat. *Aust. J. Agric. Res.*, 54: 627–635.

Munns, R., 2012. The impact of salinity stress. http//www.plantstress.com/Articles/salinity.htm.Accessed 27 March, 2012.

Narasimhamoorthy, B., Bouton, J.H., Olsen, K.M. and Sledge, M.K., 2007. Quantitative trait loci and candidate gene mapping of aluminum tolerance in diploid alfalfa. *Theor. Appl. Genet.*, 114: 901–913.

Narayanan, K.K. and Rangasamy, S.R.S., 1990. Genetic analyses for salt tolerance in rice. Proceedings of the International Rice Genetic Symposium, May 14-18, Manila, Philippines, pp: 68-68.

Nazar, R., Iqbal, N., Masood, A., Syeed, S. and Khan, N.A., 2011. Understanding the significance of sulfur in improving salinity tolerance in plants," *Environ. and Exper. Bot.*, 70(2-3): 80–87.

Neeraja, C.N., Maghirang-Rodriguez, R., Pamplona, A., Heuer, S., Collard, B.C.Y., Septiningsih, E.M., Vergara, G., Sanchez, D., Xu, K., Ismail, A.M. and MacKill, D.J., 2007. A marker-assisted backcross approach for developing submergence-tolerant rice cultivars. *Theor. Appl. Genet.*, 115: 767– 776.

Nejad, G.M., Arzani, A., Rezai, A.M., Singh, R.K. and Gregorio, G.B., 2008. Assessment of rice genotypes for salt tolerance using microsatellite markers associated with the *saltol* QTL. *Afr. J. of Biotech.*, 7(6): 730-736.

Ninamango-Cárdenas, F.E., Guimarães, C.T., Martins, P.R., Parentoni, S.N., Carneiro, N.P., Lopes, M.A., Moro, J.R. and Paiva, E., 2003. Mapping QTLs for aluminum tolerance in maize. *Euphytica*, 130: 223–232.

Noaman, M.M., Ahmed, I.A., El-Sayed, A.A., Abo-El-Enin, R.A., El-Gamal, A.S. and El-Sherbiny, A.M., 2007. Registration of 'Giza 2000' drought-tolerant six-rowed barley for rainfed and newreclaimed areas in Egypt. *Crop Sci.*, 47:440.

Noaman, M.M., El Sayad, A.A., Asaad, F.A., El Sherbini, A.M., El Bawab, A.O. and El Moselhi, M.A., 1995. Registration of 'Giza 126' barley. *Crop Sci.*, 35(6):1710.

Obert, D.E., Evans, C.P., Wesenberg, D.M., Windes, J.M., Erickson, C.A. and Jackson, E.W., 2008. Registration of 'Lenetah' spring barley. *J. Plant Registr.*, 2(2):85–87.

Ochatt, S.J., Marconi, P.L., Radice, S., Arnozis, P.A. and Caso, O.H., 1999. In vitro recurrent selection of potato: production and characterization of salt tolerant cell lines and plants. *Plant Cell, Tissue and Organ Culture*, 55: 1–8.

Oerke, E.C., Debne, H.W., Schonbeck, F. and Weber, A., 1994. Crop production and crop protection. Elsevier, Amsterdam. Pandey, S., Bhandari, H., Ding, S., Prapertchob, P., Sharan, R., Naik, D., Taunk, S.K. and Sastri, A., 2007. Coping with drought in rice farming in Asia: insights from a cross country comparative study. *Agric. Econ.*, 37:213–224.

Pandey, S., Bhandari, H., Ding, S., Prapertchob, P., Sharan, R., Naik, D., Taunk, S.K. and Sastri, A., 2007. Coping with drought in rice farming in Asia: insights from a cross country comparative study. *Agric. Econ.*, 37:213–224.

Paroda, R.S., 1986. Breeding approaches for drought resistance in crop plants. In: Approaches for Incorporating Drought and Salinity Resistance in Crop Plants, Chopra VL, Paroda RS (eds.). Oxford and IBH Publishing Co. Pvt. Ltd., New Delhi. pp. 87-107.

Pattanayak, D. and Ananda Kumar, P., 2000. Plant biotechnology: current advances and future perspectives. *Proc. Indian Nation. Sci. Acad.* (*PINSA*). 66(6): 265-310.

Pratt, R.C. and Casey, M.A., 2006. Registration of maize germplasm line Oh605. *Crop Sci.*, 46:1004–1005. Qiu,Y. and Li, X.X., 2009. Genetic analysis of salinity tolerance in *Brassica campestris* L. *China Vegetab.*, 1: 21–25.

Quarrie, S.A., 1996. New molecular tools to improve the efficiency of breeding for increased drought resistance. *Plant growth regulators*, 20: 167-178.

Quick, J.S., Stromberger, J.A., Clayshulte, S., Clifford, B., Johnson, J.J. and Peairs, F.B., 2001. Registration of 'Prairie Red' wheat. *Crop Sci.*, 41:1362–1363.

Rachovska, G. and Rachovski, G., 2005. Winter common wheat cultivar Guinneess/1322. *Field Crops Studies*, 2(2):187-191.

Rahman, M.S., Das, K.C., Das, D.K., Biswas, K., Chaudhary, M.B.H., Karim, N.H.,Salam, M.A. and Seraj, J.I., 2010. Breeding and anther derived lines of rice (*Oryza sativa* L.) for saline coastal areas of Bangladesh. *Bangladesh J. Bot.*, 39: 71-78.

Rahman, S., Haque, T., Rahman, M.S. and Seraj, Z.I., 2008. Salt tolerant-BR11 and salt tolerant-BR28 through Marker Assisted Backcrossing (MAB). http://pbtlabdu.net/abs_rahman_et_al.pdf.

Rajagopal, V., 2003. Stress management in coconut. National Seminar On Stress Management in Oilseeds for Attaining Self-Reliance in Vegetable Oils, January, 28-30, pp. 69-78.

Ram, H.H. and Singh, H.G., 1994. Rice. In: Crop Breeding and Genetics. Kalyani Publishers, New Delhi. pp. 58-103.

Ramage, R., 1980. Genetic methods to breed salt tolerance in plants. In: Rains D, Valentine R, Hollaender A, eds. Genetic engineering of osmoregulation impact on plant productivity for food, chemicals and energy. New York: Plenum Press, 311-318.

Raman, H., Zhang, K.R., Cakir, M., Appels, R., Garvin, D.F., Maron, L.G., Kochian, L.V., Moroni, J.S., Raman, R., Imtiaz, M., Drake-Brckman, F., Waters, I., Martin, P., Sasaki, T., Yamamoto, Y., Matsumoto, H., Hebb, D.M., Delhaize, E. and Ryan, P.R., 2005. Molecular characterization and mapping of *ALMT1*, the aluminium-tolerance gene of bread wheat (*Triticum aestivum* L.). *Genome,* 48: 781–791.

Rana Munns, Richard A. James and Andre´ La¨uchli., 2006. Approaches to increasing the salt tolerance of wheat and other cereals. *J. Exper. Bot.,* 57(5): 1025–1043.

Rao Balakrishna, M.J. and Biswas, S., 1979. Rainfed lowland rice in India. In: Rainfed lowland rice: selected papers from the 1978 International Rice Research Conference. Los Banos. International Rice Research Institute, P.O. Box 933, Manila, Philippines, 87-94.

Rao, M.J.B., Murth, K.S., Choudhury, D., Srinivasulu, K. and Gangacharan, G., 1971. Breeding for drought and upland conditions in rice. *Oryza,* 8(2):75-84.

Reddy, J.N. Patnaik, S.S.C., Sakar, R.K., Das, S.R. Singh, V.N. Indrani Dana, Singh, N.K., Sharma, R.N., Ahmed, T., Sharma, K.K., Satish Verulkar, Collard, B.C.Y., Pamplona, A.M., Singh, U.S., Sarkarung, S., Mackill, D.J., and Ismail, A.M., 2013. Overview of the eastern india rainfed lowland shuttle breeding network (eirlsbn). *SABRAO Journal of Breeding and Genetics,* 45 (1) 57-66.

Reddy, L.J., Nigam, S.N., Rao, R.C.N. and Reddy, N.S., 2001. Registration of ICGV 87354 peanut germplasm with drought tolerance and rust resistance. *Crop Sci.,* 41:274–275.

Reddy, M.A., Francies, R.M., Rasool, Sk. N. and Reddy, V.R. P., 2014. Breeding for tolerance to stress triggered by salinity in rice. *Internat. J. Appl. Biol. and Pharmaceutical Tech.,* 5(1): 167-176.

Repinski, S.L., Hayes, K.N., Miller, J.K., Trexler, C.J. and Bliss, F.A., 2011. Plant breeding graduate education: opinions about critical knowledge, experience, and skill requirements from public and private stakeholders worldwide. *Crop Sci.,* 51:2325–2336.

Rezai, A.M. and Saeidi, G., 2005. Genetic analysis of salt tolerance in early growth stages of rapeseed (*Brassica napus* L.) genotypes. *Ind. J. Genet. and Plant Breed.*, 65: 269–273.

Rhoades, J.D. and Loveday., 1990. Salinity in irrigated agriculture. In B. A. Stewart and D. R. Nielsen (eds) Irrigation of Agricultural Crops. ASA Monograph. pp. 1089–1142.

Ribaut, J.M., Hoisington, D., Banziger, M., Setter, T. and Edmeades, G., 2004. Genetic dissection of drought tolerance in maize: a case study. *In* JM Ribaut, ed, Physiology and Biotechnology Integration for Plant Breeding. Marcel Dekker, New York, pp. 571–609.

Richards, R.A., 1996. Dening selection criteria to improve yield under drought. *Plant Growth Regul.*, 20: 57–166.

Ristic, Z., Yang, G.P. Sterzinger, A. and Zhang, L.Q., 1998. Higher chilling tolerance in maize is not always related to the ability for greater and faster abscisic acid accumulation. *J. Plant Physiol.*, 153: 154-162.

Rodriguez, M., Canales, E. and Borras-Hidalgo, O., 2005. Molecular aspects of abiotic stress in plants. *Biotecnol. Aplic.*, 22: 1–10.

Rodriguez, M., Canales, E. and Borras-Hidalgo, O., 2005. Molecular aspects of abiotic stress in plants. *Biotecnol. Aplic.*, 22: 1–10.

Rohila, J.S., Jain, R.K. and Wu, R., 2002. Genetic improvement of Basmati rice for salt and drought tolerance by regulated expression of barley Hva1 cDNA. *Plant Sci.*, 163(3): 525-532.

Sadaquat, H.A., Tahir, M.H.N. and Hussain, M.T., 2003. Physiogenetic aspects of drought tolerance in canola (*Brassica napa*). *Intr. J. Agri. Biol.*, 5(4): 611-614.

Saranga, Y., Cahaner, A., Zamir, D., Marani, A. and Rudich, J., 1992. Breeding tomatoes for salt tolerance-inheritance of salt tolerance and related traits in interspecific populations. *Theor. and Appl. Genet.*, 84: 390-396.

Saranga, Y., Zamir, D., Marani, A. and Rudich, J., 1993. Breeding tomatoes for salt tolerance-variations in ion concentrations associated with response to salinity. *J. of the American Soc. for Hort. Sci.*, 118: 405-408.

Semikhodskii, A.G., Quarrie, S.A. and Snape, J.W., 1997. Mapping quantitative trait loci for salinity responses in wheat. Proceedings of the conference: *Drought and Plant Production*, Lepenski Vir, Serbia, 83-92.

Senadhira, D., Neue, H.U., and Akbar, M., 1994. Development of improved donors for salinity tolerance in rice through somaclonal variation. *SABRAO J.*, 26: 19-25.

Senadhira, O., Zapata-Arias, F.J., Gregorio, G.B., Alejar, M.S., dela Cruz, H.C., Padolina, T.F., and Galvez, A.M., 2002. Development of the first salt-tolerant rice cultivar through indica/indica anther culture. 76: 103-110.

Sengupta, A., Heinen, J.L., Haladay, A.S., Barke, J.J. and Allen, R.D., 1993. Increased resistance to oxidative stress in transgenic plants that overexperss cloroplastic Cu/Zn superoxide dismutase. *Proc. Natl. Acad. Sci. USA*. 90: 1629-1633.

Septiningsih, E.M., Pamplona, A.M., Sanchez, D.L., Neeraja, C.N., Vergara, G.V., Heuer, S., Ismail, A.M. and Mackill, D.J., 2009. Development of submergence-tolerant rice cultivars: the Sub1 locus and beyond. *Annals of Bot.*, 103: 151 – 160.

Serraj, R., Hash, C.T. and Rivzi, S.M.H., 2005. Recent advances in marker assisted selection for drought tolerance in pearl millet. *Plant Prod. Sci.*, 8: 334–337.

Shanker, A.K. and Venkateswarlu, B., 2011. Abiotic stress in plants – mechanisms and adaptations. InTech, Rijeka, p ix.

Shanmuganathan, M., 2001. Stability analysis and *in vitro* screening for salt stress in different national rice hybrids (*Oryza sativa* L.). M.Sc. (Ag.). Thesis, TNAU, Coimbatore.

Shanon, M.C., 1997. Adaptation of plants to salinity. *Adv. Agron.*, 60: 75-120.

Singh, B.B., Mai-Kodomi, Y. and Terao, T., 1999. A simple screening method for drought tolerance in cowpea. *Ind. J. of Genet.*, **59:** 211–220.

Singh, B.D., 2003. Breeding for resistance to abiotic stresses. I. Drought resistance. In: Plant Breeding, Kalyani Publishers, New Delhi. pp. 381-409.

Singh, B.N. and MacKill, D.J., 1991. Genetics of leaf rolling under drought stress. In: Rice Genetics. International Rice Research Institute, Los Banos, Philippines. Pp. 159-160.

Singh, K.B., Omar, M., Saxena, M.C. and Johansen, C., 1996. Registration of FLIP 87-59C, a drought tolerant chickpea germplasm line. *Crop Sci.*, 36(2):1–2.

Singh, R.K. and Mishra, B., 1997. Stable genotypes of rice for sodic soils. *Ind. J. Genet.*, 57(4): 431-438.

Singh, RK., Gregrio, G.B. and Ismail, A.M., 2008. Breeding rice varieties with tolerance to salt stress. *J. Indian Soc. Coastal Agric. Res.*, 26(1): 16-21.

Singh, S. and Singh, M., 2000. Genotypic basis of response to salinity stress in some crosses of spring wheat *Triticum aestivum* L. *Euphytica*, 115: 209-214.

Singh, S., Mackill, D.J., Ismail, A.M., 2009. Responses of *SUB1* rice introgression lines to submergence in the field: Yield and grain quality. *Field Crops Research*, **113**:12-23.

Singh, S., Singh, O.N. and Singh, R.K., 1998. A shuttle breeding approach to rice improvement for rainfed lowland ecosystem in eastern India. In: Sustainable agriculture for food, energy and industry. James and James, Science Publishers, Ltd, UK, pp. 105-115.

Singh, S.P., Teran, H. and Gutierrez, J.A., 2001. Registration of SEA 5 and SEA 13 drought tolerant dry bean germplasm. *Crop Sci.*, 41:276–277.

Singh, S.P., Teran, H., Lema, M., Schwartz, H.F. and Miklas, P.N., 2007. Registration of white mold resistant dry bean germplasm line A 195. *J. Plant Registr.*, 1:62–63.

Steele, K.A., Virk, D.S., Kumar, R., Prasad, S.C. and Witcombe, J.R., 2007. Field evaluation of upland rice lines selected for QTLs controlling root traits. *Field Crops Res.*, 101:180–186.

Tammam, A.M., El-Ashmoony, M.S.F., El-Sherbeny, A.A. and Amin, A.L., 2004. Selection responses for drought tolerance in two bread wheat crosses. *Egypt J. Agric. Res.*, 82: 1213-1226.

Taye, T., Tesfaye T. and Gebisa E., 2008. Combining ability of introduced sorghum parental lines for major morpho-agronomic traits. An Open Access Journal published by ICRISAT, 6: 1-7.

Tester, M. and Langridge, P., 2010. Breeding technologies to increase crop production in a changing world. Science, **327:** 818-822. Thakur, P., Kumar, S., Malik, J.A., Berger, J.D. and Nayyar, H., 2010. Cold stress effects on reproductive development in grain crops: an overview. *Environ. Exp. Bot.*, 67:429–443.

Thakur, P., Kumar, S., Malik, J.A., Berger, J.D. and Nayyar, H., 2010. Cold stress effects on reproductive development in grain crops: an overview. *Environ. Exp. Bot.*, 67:429–443.

Theilert, W., 2006. A unique product: The story of the imidacloprid stress shield. *Pflanzenschutz-Nachrichten Science Forum Bayer*, 59: 73-86.

Thomas, W.T.B., Powell, W. and Wood, W., 1984. The chromosome location of the dwarfing gene in the spring variety Golden Promise. *Heredity*, 53: 177-183.

Tomar, J.B. and Prasad, S.C., 1996. Relationship between inheritance and linkage for drought tolerance in upland rice (*Oryza sativa*). *Ind. J. Agric. Sci.*, 66: 459-465.

Tomlekova, N.B., 2010. Induced mutagenesis for crop improvement in Bulgaria. Plant Mutation Reports, June 2010, 2(2): 6.

Toojinda, T., Tragoonrung, S., Vanavichit, A., Siangliw, J.L., Pa-In, N., Jantaboon, J., Siangliw, M. and Fukai, S., 2005. Molecular breeding for rainfed lowland rice in the Mekong region. *Plant Prod. Sci.*, 8: 330–333.

Toure, A., Miller, F.R. and Rosenow, D.T., 1996. Heterosis and combining ability for grain yield and yield components in guinea sorghums. *African Crop Science Journal*, 4:383–391.

Trethowan, R.M., Ginkel, V.M. and Rajaram, S., 2002. Progress in breeding wheat for yield adaptation in global drought affected environments. *Crop Science*, 42: 1441-1146.

Venuprasad, R., Lafitte, H.R. and Atlin, G.N., 2007. Response to direct selection for grain yield under drought stress in rice. *Crop Science*, **47:** 285-293.

Vorasoot, N., Songsri, P., Akkasaeng, C., Jogloy, S. and Patanothai, A., 2003. Effect of water stress on yield and agronomic characters of peanut. *Songklanakarin J. Sci. Technol.*, 25:283–288.

Wade, L.J., McLaren, C.G., Quintana, L., Harnpichitvitaya, D., Rajatasereekul, S., Sarawgi, A.K., Kumar, A., Ahmed, H.U., Sarwoto, Sing, A.K., Rodriguez, R., Siopongco, J. and Sarkarung, S., 1999. Genotype by environment interactions across diverse rainfed lowland rice environment. *Field Crops Res.*, **64**: 35–50.

Wang, H., Huang, D., Lu, R., Liu, J., Qian, Q. and Peng, X., 2000. Salt tolerance of transgenic rice (*Oryza sativa* L.) with *mtlD* gene and *gutD* gene. *Chinese Science Bulletin*,. 45(18):1685-1690.

Wang, J.P., Raman, H., Zhou, M.X., Ryan, P.R., Delhaize, E., Hebb, D.M., Coombes, N. and Mendham, N., 2007. High-resolution mapping of the Alp locus and identification of a candidate gene *HvMATE* controlling aluminium tolerance in barley (*Hordeum vulgare* L.). *Theoretical and Applied Genetics*, 115: 265–276.

Wang, L.L., 2011. Difference analysis of growth and photosynthetic characteristics in six maize lines and hybrids via a diallel design. Master degree thesis, Nanjing Agricultural College.

Wang, W., Vinocur, B. and Altman, A., 2003. Plant responses to drought, salinity and extreme temperatures: towards genetic engineering for stress tolerance. *Planta*, 218:1–14.

Wang, W., Vinocur, B. and Altman, A., 2007. Plant responses to drought, salinity and extreme temperatures towards genetic engineering for stress tolerance. *Planta*. 218:1-14.

Watanabe, I., 1998. Drought tolerance of cowpea (*Vigna unguiculata* (L.) Walp.). I. Method for evaluation of drought tolerance. *JIRCAS Journal*, 6: 21-28.

Wight, C.P., Kibite, S., Tinker, N.A. and Molnar, S.J., 2006. Identification of molecular markers for aluminium tolerance in diploid oat through comparative mapping and QTL analysis. *Theoretical and Applied Genetics*, 112: 222–231.

Williams, T.V., 1966. The inheritance in sweet corn. Diss Abstr Intr B. 27. Order No. 66-11, 222: pp. 1688-1689. Winicov, I., 2000. *Alfin1* transcription factor overexpression enhances plant root growth under normal and saline conditions and improves salt tolerance in alfalfa. Planta, 210: 416-422.

Winicov, I., 2000. *Alfin1* transcription factor over expression enhances plant root growth under normal and saline conditions and improves salt tolerance in alfalfa. *Planta*, 210: 416-422.

Witcombe, J.R., Hollington, P.A., Howarth, C.J., Reader, S. and Steele, K.A., 2008. Breeding for abiotic stresses for sustainable agriculture. Philosophical Transactions of the Royal Society B: *Biological Sc.*, 363: 703-716.

Wu, M.G., Lin, J.R. and Zhang, G.H., 2004. Genetic analysis of leaf water content in paddy upland hybrid rice. *Chinese J. Rice Sci.*, 18(6): 570-572.

Wu, P., Liao, C.Y., Hu, B., Yi, K.K., Jin, W.Z., Ni, J.J. and He, C., 2000. QTLs and epistasis for aluminium tolerancein rice (*Oryza sativa* L.). *Rice Genet Newsl.*, 100: 1295-1303. www.hindu.com/2007/11/17/stories/2007111752560500.htm.

Xia, L., Ma, Y., He, Y., and Jones, H.D., 2012. GM Wheat Development in China: Current Status and Challenges to Commercialization. *J. Experimental Bot.*, 63(5), 1785–1790.

Xinglai, P., Sangang, X., Qiannying, P. and Yinhong, S. 2006. Registration of'Jinmai 50' wheat. *Crop Sci.*, 46:983–985.

Xu, D., Duan, X., Wang, B., Hong, B., Ho, T.D. and Wu, R.D., 1996. Expressions of a embryogenesis abundant protein gene, HVA1, from barley confers tolerant to water deficit and salt stress in transgenic rice. *Plant Physiol.*, 110: 249-257.

Xue, Y., Jiang, L., Su, N., Wang, J.K., Deng, P., Ma, J.F., Zhai, H.Q. and Wan, J.M., 2007. The genetic basis and fine-mapping of a stable quantitative-trait locus for aluminium tolerance in rice. *Planta*, 227: 255–262.

Yadav, R.S., Hash, C.T., Bidinger, F.R. and Howarth, C.J. 1999., QTL analysis and markerassisted breeding of traits associated with drought tolerance in pearl millet. In: Ito O'Toole J, Hardy B (eds) Genetic improvement of rice for water-limited environments. International Rice Research Institute, Manila, Philippines, pp. 221- 233.

Yeo, A.R. and Flowers, T.J., 1986. Salinity resistance in rice and a pyramiding approach to breeding varieties for saline soils. In: Plant Growth, Drought, and Salinity. CSIRO, Melbourne, Australia, pp. 161-173.

Yeo, E.T., Kwon, H.B., Han, S.E., Lee, J.T., Ryu, J.C. and Byu, M.O., 2000. Genetic engineering of drought resistant potato plants by introduction of the trehalose-6-phosphate synthase (TPS1) gene from *Saccharomyces cerevisiae.* Mol. Cells, 10:263–268.

Yunus, M. and Paroda, R.S., 1982. Impact of biparental mating on correlation coefficients in Bread wheat. *Theor. Appl. Genet.*, 62: 337-343.

Zakaria, M.M., 2004.Selection for earliness and grain yield in bread wheat (Triticm aestivum L.) under different environments. Ph.D Thesis, Assiut University, Egypt.

Zapata, F.J. and Abrigo, E.M., 1986. Plant regeneration and screening for long term. Nacl-stressed rice callus. *IRRN,* 11: 24-25.

Zhang, K., Zhang, Y., Chen, G. and Tian, J., 2009. Genetic analysis of grain yield and leaf chlorophyll content in common wheat. *Cereal Res. Communications,* 37:499–511.

Zhang, X., Lu, G., Long, W., Zou, X., Li, F. and Nishio, T., 2014. Recent progress in drought and salt tolerance studies in *Brassica* crops. *Breeding Science*, 64: 60–73.

2016, Recent Advances in Plant Stress Physiology Pages 379–411
*Editors: **Praduman Yadav, Sunil Kumar and Veena Jain***
Published by: DAYA PUBLISHING HOUSE, NEW DELHI

17

Breeding for Resistance to Biotic Stresses in Plants

Naresh Kumar Bainsla[1]* and H.P. Meena[2]

[1]Division of Field Crops, ICAR – Central Agricultural Research Institute, Port Blair, Andaman and Nicobar Islands

[2]ICAR – Indian Institute of Oilseeds Research, Rajendranagar, Hyderabad – 500 030, India

ABSTRACT

Biotic stresses cause significant loss in crop plants and management of biotic stresses (Diseases and pests) not only increases the cost of production but also has implications on environment and ecology. Increasing use of chemical agents for biotic stress management is concern for growers, exporters and animal and human health. The best method is to use resistant varieties which are economical, healthier and eco-friendly approach. Breeding for disease and pest resistance is major objective in most of the breeding programs across the crop species and world. The resistance breeding requires the resistant source or donor which may be the same species, related species of same genera or family, or altogether an alien species. There is need of recipient or target species and method of transfer of resistance. There are different approaches for breeding for resistance against different kinds of stress which involve both conventional and modern tools. Resistance breeding approach is to be at least one step ahead of the pathogen or pest in question. Achieving the goal of development of a resistant variety also needs attention on the durability of resistance for which the knowledge of resistance on the inheritance, expression and interaction with fellow genes and environment aspects. In this chapter emphasis has been given on importance of biotic stresses, inheritance, resistance sources, breeding approaches, and modern tools.

**Corresponding author.* E-mail: bainslahau@gmail.com

Introduction

Biotic stresses as the name suggests is the stress due to living or biotic factors. In the world of life there are different kinds of species which coexist on the planet earth. This coexistence is not that beautiful as it looks in first impression, in fact living beings do behave in numerous ways with the kinds of their own as well as other kinds of life in the presence of non living factors called environment. Therefore, there are mainly three factors of any life form *i.e.* Individual in question, other individuals of same or different species and environment. These three factors can interact in all sorts of permutations and combinations which decide the survival, longevity, reproduction and future of individuals. The relations can be mutually beneficial, parasitic, competitive and destructive with various degrees of expression. Plants like any other life form are also affected by biotic and abiotic factors. But here we are more concerned about the community or population of plants of same species grown for its economic use to benefit the mankind called crop. Crop plants will be referred to the scope of this chapter in terms of biotic stresses and breeding approaches. The breeding approaches shall cover the points of view on classical methods as well as modern tools. The biotic agents which cause biotic stresses include bacteria, nematodes, fungi, viruses, insect pests (crop as well as stored grain) and weeds. Each of the categories of these biotic agents has its own way of attacking or harming the crops. The stress may be air borne like in case of many rusts, seed borne as in case of smuts and bunts, soil borne like in case of many wilts, blights, nematodes or it may be transmitted through carriers which include insects and weeds. Weeds generally compete for space, water and nutrients with the crop plants and sometimes also act as secondary hosts to facilitate the other biotic agents to complete their life cycle.

Different biotic agencies based upon the available circumstances can cause varying degree of losses to yield or quality of the plants and its produce. The minor reaction can be loss below economic threshold level wherein control of the disease or pest is not desirable. The losses can be higher causing significant loss in the yield or quality of the crop in a locality or sometimes so devastating that it can cause epidemic by spreading to larger areas even continent forcing to famine like situations. Therefore, degree of loss defines the economic importance of a disease or pest and accordingly called a minor for very low or major for significantly higher incidence of disease or pest.

The importance of biotic stresses can be imagined from the facts epidemic spread of *Phytopthora* could result into Irish famine during nineteenth century. During the famine approximately 1 million people died and a million more emigrated from Ireland. Single outbreak of locusts could devastate the vegetation across the continents. The desert locust, the most notorious of about a dozen locust species for its ability to rapidly multiply and travel long distances, could threaten an area of 32 million square kilometres, stretching across 50 countries from west Africa to India. Similarly the rusts of wheat, mildew of maize and millets have caused serious losses to crops and affected a large number of human populations. The outbreak of Ug99 race of has shaken the world with its possible implications and many projects involving huge money were started across the world in order to identify the source of resistance.

It is generally thought that molecular biology is altogether different from biochemistry and similarly molecular genetics or so called biotechnology is some kind of super science which is different from genetics. In fact as the science grows the different terminologies also come in to knowledge but the relevance of new terms is actually stated by the basics behind it. For many beginners it sometimes becomes difficult to come to basics and understand the fundamentals behind each terminology which creates lot of confusion. Therefore efforts have been made to show the co linearity between the modern and classical terminologies and fundamentals. Further the chapter includes more of the general and common findings based upon the understanding like a common man rather than making it difficult to understand by including large number of references and complex definitions.

What is Stress and Biotic Stress?

Whenever any combination of an individual with the fellow individuals of same species or other species and environment becomes detrimental to its survival or reproduction or for future the individual is said to be in stress. If the major cause of this stress is from individuals of same species or any other species but definitely with the help of environment it will be known as biotic stress. Therefore biotic stress per se requires three factors to play role in its existence. Biotic factors affecting plants can be bacteria, fungi, virus, mycoplasma, insects, nematodes, rodents and even mammalian pests. But to cause a stress interaction of a susceptible host, aggressive or virulent pathogen or pest and favorable environment is must.

Why Biotic Stress Resistant Varieties Required?

Crop loss caused by plant pathogens have been reviewed extensively in a number of review articles over the past 50 years (Gaunt, 1995; James, 1974), beginning with Chester in 1950. However, throughout the world, approximately 800 million people are malnourished and the demand for food will increase as the population increases. It is predicted that by 2050 the world population will increase from approximately 6.7 billion to over 9 billion and that the current trend for more resource-intensive diets, which include more dairy and meat products, will continue. It is estimated that current global production of wheat must increase annually by about 2 per cent (Singh and Trethowan, 2007). At the same time, the demand for land from uses other than agriculture is increasing. Due to biotic stress up to 35 per cent of the total food production is lost. The estimated crop loss was of the value of Rs 90,000 crores in 2004 and in 2009, Rs. 1,40,000 crores (Suresh and Malathi, 2013). Overall, pests accounted for pre-harvest losses of 42 per cent of the potential value of output, with 15 per cent attributable to insects and 13 per cent each to weeds and pathogens. An additional 10 per cent of the potential value was lost postharvest (Orke *et al.*, 1994). Out of a US$ 1.3 trillion annual food production capacity worldwide, the biotic stresses caused by insects, diseases and weeds cause 31-42 percent loss (US$ 500 billion), with an additional 6-20 percent (US$ 120 billion) lost post harvest to insects and to fungal and bacterial rots. Crop losses due to pathogens are often more severe in developing countries (*e.g.* cereals,

22 percent) when compared to crop losses in developing countries (*e.g.* cereals, 6 percent) (Oerke *et al.*, 1994). Since more than 42 per cent of the potential world crop yield is lost owing to biotic stresses (15 per cent attributable to insects, 13 per cent to weeds, and 13 per cent to other pathogens), a reduction in this incidence will be one of the more important possibilities for improving plant production (Pimentel, 1997). Humans and insects have always competed for food and fibre, so they have been constantly at was with each other. Insects cause millions of dollars worth of losses annually to food crops and other plants all over the world. Crop losses due to insects and nematodes, estimated at 10–20 per cent for major crops, are signicant factors in limiting crop yields. The development of insect- and nematode resistant plants is therefore an important objective of plant breeding strategies with relevant implications for both farmers and the seed and agrochemical industries. In the United States in 1987, crop losses caused by diseases and insects in specific vegetables were, respectively: cole crops 9 and 13 per cent, lettuce 12 and 7 per cent, potato 20 and 6 per cent, tomato 21 and 7 per cent, sweet corn 8 and19 per cent, onion 21 and 4 per cent, cucumber 15 and 21 per cent, pea 23 and 4 per cent, and pepper 14 and 7 per cent. World-wide losses from diseases range from 9 per cent to 16 per cent in rice, wheat, barley, maize, potato, cotton, soybean (Chakraborty *et al.*, 2000). If pest related losses are taken into consideration, it contributes to 14 to 25 per cent on average of the total global agricultural production (see DeVilliers and Hoisington, 2011). Estimated losses due to pests among some major crops were estimated to be 26 per cent for soybean, 28 per cent for wheat, 31 per cent for maize, 37 per cent for rice and 40 per cent for potatoes (Oerke and Dehne, 2004). The global loss potential caused by pests is particularly high in crops grown under high productivity environments and also in the tropics and sub-tropics where climatic conditions favour the damaging function of pests (Oerke, 2006). Furthermore, losses due to pathogens, animal pests and weeds were estimated to be 16, 18, and 34 per cent, respectively.

Table 17.1: Overview of Potential and Actual Losses Attributable to Fungal and Bacterial Pathogens, Viruses, Animals Pests and Weeds as well as the Efficacy of the Applied Pest Control Operations in Maize, Wheat, Rice, Barley, Potatoes, Soybean, Sugar Beet, and Cotton Maize, Wheat, Rice, Barley, Potatoes, Soybean, Sugar Beet, and Cotton

	Pests and Pathogens				
	Fungi and Bacteria	*Viruses*	*Animal Pests*	*Weeds*	*Total*
Loss potential (per cent)[a]	14.9	3.1	17.6	31.8	67.4
Actual losses (per cent)[a]	9.9	2.7	10.1	9.4	32.0
Efficacy (per cent)[b]	33.8	12.9	42.4	70.6	52.5

Source: Modied from Oerke and Dehne (2004)

a: As percentage of attainable yields; b: As percentage of loss potential prevented.

Past Research Achievements

In upland cotton, variety MCU-5 is tolerant to *verticilium* wilt. In wheat, variety Rescue with solid stem is resistant to stem sawfly. It was released in Canada in 1946. In okra, variety Prabhani Kranti is resistant to yellow mosaic virus. In USA barley variety 'Will' is resistant to green bugs was released. In alfalfa, varieties 'Cody, Moapa and Zia are resistant to alfalfa spotted aphid. These varieties were released in USA. In oat, variety Greta is resistant to stem eelworm. It was developed in Belgium. In upland cotton, varieties B-1007, Khandwa-2, SRT-1, DHY-286 and PKV-081 are resistant to jassids. Other cotton varieties Kanchana, Supriya and LK-861 were resistant to whitefly. Intra-hirsutum cotton hybrid LHH-144 is resistant to leaf curl virus. It was developed from Punjab Agricultural University, Ludhiana.

Susceptibility, Tolerance, Resistance, Immunity, Escape?

Great Darwin's theory of "survival of fittest" seems to be quite relevant to biotic stress but here comes the factor of the degree of stress. An individual may surrender and die due to stress (Susceptible) or may survive in stress and can bear the parasite for whole life (Tolerant) or can eliminate the parasite or pathogen or so called pest without affecting its performance (Resistant), the individual may show a hypersensitive or non acceptance of entry of pathogen or pest into its system (Immune) or it may acquire a mechanism to complete its cycle or go into dormant stage or protective state before the parasite causes the loss to life (Escape). Therefore all these situations are quite relative in terms of their degree but at the same time quite different in terms of mechanisms devised for healthy life. Plants have their own naturally devised system of defense such as pre-formed structures and compounds that contribute to resistance prior to immune response these may include plant cuticle/surface, plant cell walls, antimicrobial chemicals (for example: glucosides, saponins), antimicrobial proteins, Enzyme inhibitors, Detoxifying enzymes that break down pathogen-derived toxins, receptors that perceive pathogen presence and activate inducible plant defences. Plants do have inducible plant defences that are generated after infection these mechanisms involve cell wall reinforcement (callose, lignin, suberin, cell wall proteins), antimicrobial chemicals (including reactive oxygen species such as hydrogen peroxide, or peroxynitrite, or more complex phytoalexins such as genistein or camalexin), antimicrobial proteins such as defensins, thionins, or PR-1, antimicrobial enzymes such as chitinases, beta-glucanases, or peroxidases, hypersensitive response - a rapid host cell death response associated with defence mediated by "Resistance genes."Endophyte assistance: Plant's roots release chemicals that attract beneficial bacteria to fight off infections.

Plant immune systems show some mechanistic similarities with the immune systems of insects and mammals, but also exhibit many plant-specific characteristics. Plants can sense the presence of pathogens and the effects of infection via different mechanisms than animals. As in most cellular responses to the environment, defences are activated when receptor proteins directly or indirectly detect pathogen presence and trigger ion channel gating, oxidative burst, cellular redox changes, protein kinase cascades, and many other responses that either directly activate cellular changes (such as cell wall reinforcement or the production of antimicrobial compounds), or activate changes in gene expression that then elevate plant defence responses. Two major types of pathogen detection systems are observed in plant immune systems: PAMP-Triggered Immunity (PTI; also known as MAMP-triggered immunity or MTI), and Effector-Triggered Immunity (ETI). The two systems detect different types of pathogen molecules, and tend to utilize different classes of plant receptor proteins to activate antimicrobial defences. Although many specific examples of plant-pathogen detection mechanisms are now known that defy clear classification as PTI or ETI, the larger trend across many well-studied plant-pathogen interactions supports continued use of PTI/ETI concepts.

Treatment or Prevention?

It is said that prevention is better than cure. But can we always depend upon the prevention or cure alone? We may have to opt for the suitable strategy to tackle the ailment through preventive as well as curative methods and at times combination of both the techniques seems suitable for minimizing the losses. The curative approaches wherever successful find the quickest control on the situation but not permanent in nature and are costly to the pocket of farmer as well to environment. The preventive methods however can be of mechanical, management and resistance and are time consuming but sustainable and eco-friendly and economic and the only solution in many cases.

Plant Breeding and Biotic Stresses

Plant breeding is the science which is a complex compound of art of culturing plants with skills of selection, standing on the platform of science of genetics at meta as well as micro/nano/pico or atomic level using statistics as its main tool may be through simple chi sqaures, random tables, calculators to high throughput computational software. In common man's words we can say plant breeding is the system of getting educated young ones by virtue of education done to their parents. Plant breeding plays very important role in preventive, immunized and curative designer crop plants against biotic stresses. Plant breeding has been effective tool in controlling many diseases and insect pests wherein the chemical methods were too much costly to be affordable by a common farmer especially from poor countries.

Guiding Principle for Development of Biotic Stress Resistant or Acclaimed Crop Plants

The plants do have the inherent capacity to response against any kind of unwanted pressure due to environment or its biotic components. The genetic makeup of the plant makes it either resistant or susceptible. In a group of large number of plants of a species having sufficient chance of random segregation of alleles some plants are ought to be resistant as compared to other plants for a given stress. Similarly the pathogens do have complementary genes for virulence. Thus the guiding principle for the resistance in plants lies in genome of host as well as pathogen.The **gene-for-gene relationship** was discovered by Harold Henry Flor in 1942 who was working with rust (*Melampsoralini*) of flax (*Linum usitatissimum*). Flor was the first scientist to study the genetics of both the host and parasite and to integrate them into one genetic system (Flor 1942, 1947 and 1955). Gene-for-gene relationships are a widespread and very important aspect of plant disease resistance. The combination of suitable genes makes it resistant while that of deleterious genes makes it susceptible. The inheritance of the resistance trait under study and its heritability is quite important in the development of a resistant variety or line. Based upon the inheritance and heritability of the resistance trait the breeding methodology can be prioritized for the given species which is further dependent on the pollination behaviour of species in question. The simply inherited genes are easier to transfer through directional selection in bi-parental crossing programmes.There are several different classes of R Genes. The major classes are the NBS-LRR genes and the cell surface pattern recognition receptors (PRR). The protein products of the NBS-LRR R genes contain a nucleotide binding site (NBS) and a leucine rich repeat (LRR). The protein products of the PRRs contain extracellular membrane, trans membrane and intracellular non-RD kinase domains. Within the NBS-LRR class of R genes are two subclasses the first class is Toll/Interleukin 1 receptor homology region (TIR) and has an amino-terminal. This includes the *N* resistance gene of tobacco against tobacco mosaic virus (TMV).The other subclass does not contain a TIR and instead has a leucine zipper region at its amino terminal.

Plant Breeding for Disease Resistance Typically includes

- ☆ Identification of resistant breeding sources (plants that may be less desirable in other ways, but which carry a useful disease resistance trait). Ancient plant varieties and wild relatives are very important to preserve because they are the most common sources of enhanced plant disease resistance.
- ☆ Crossing of a desirable but disease-susceptible plant variety to another variety that is a source of resistance, to generate plant populations that mix and segregate for the traits of the parents.
- ☆ Growth of the breeding populations in a disease-conducive setting. This may require artificial inoculation of pathogen onto the plant population. Careful attention must be paid to the types of pathogen isolates that are present, as there can be significant variation the effectiveness of resistance against different isolates of the same pathogen species.
- ☆ Selection of disease-resistant individuals. Breeders try to sustain or improve numerous other plant traits related to plant yield and quality, including other disease resistance traits, while they are breed for improved resistance to any particular pathogen.

Factors Affecting Breeding Methodology

Breeding approach may vary from the point of view of different workers based upon the economic importance of stress, the stress in question, the genetics of resistance, availability of resistance sources, expertise and the facilities available etc.

The most important from the point of view of a plant breeder is the genetics of resistance of the stress. There are numerous successful examples wherein the conventional breeding approaches have done wonders some of these include the resistant varieties against the rusts and smuts in wheat, downy mildew in maize and pearl millets. In fact the fruits of green revolution in India could be harvested amply due to conventionally bred rust resistant varieties in wheat. The change of cytoplasm in case of maize and pearl millet restored the hybrids in these crops. The breeding for monogenic or oligogenic traits with high heritability gives faster results as compared to the complex traits. The availability of resistance source is another factor which keeps its prime importance. Lack of proper resistance source against many insect pests has been the source of business to many pesticide industries with multimillion dollar turnovers. Later on the resistance source identified in totally unrelated organisms like *Bacillus thuringiensis* has again resulted to non-conventional answers to dreaded pests and weeds. Similarly the concept of non host resistance is also gaining importance. The approach may seem to be modern and non-conventional but the basics of genetics of resistance and the selection methods breeding tools still remain conventional. The first and foremost thing is to dissect the inheritance of resistance followed by identification of resistance source, selection of method to introgress the resistance in target germplasm and lastly to evaluate the final product.

Sources of Disease Resistance

- ☆ **A known variety:** Resistant plants were isolated from commercial varieties in the cases of cabbage yellows in cabbage (*Brassica oleracea* var. *capitata*), currely top in resistant in sugar beet (*Beta vulagaris* L.) etc.
- ☆ **Germplasm collection:** Resistant to net bloch in barley (*Hordeum vulgare* L.), resistant to wilt in watermelon (*Citrullus lanatus* L.) etc.
- ☆ **Related species:** Prabhani Kranti a YVM resistant variety has been developed in which resistant is transferred from *Abelmoschas album*. Resistant to grassy stunt virus in rice (*Oryza sativa* L.) has been transferred from *Oryza nivara*.
- ☆ **Mutation:** Resistant to some disease may be obtained through mutations arising spontaneously or induced. Resistant to victoria blight in Oats (*Avena sativa* L.) was introduced by irradiation with X-ray or thermal neutrons.
- ☆ **Somaclonal variation:** Ono variety (a somaclonal from variety Pindar) of sugarcane (*Saccharum officinarum* L.) is resistant to Fiji disease.
- ☆ **Unrelated organism:** *e.g.* coat protein genes of a pathogenic virus, genes for novel phytoalexins.

Breeding Methods for Biotic Stress Resistance

Methods of breeding for disease resistance are, (1) selection, (2) introduction, (3) mutation, (4) hybridization (pedigree and backcross method), (5) somaclonal variation, and (6) genetic engineering.

Selection

Selection is the most ancient and basic procedure in plant breeding. Selection of biotic resistance plants from a commercial variety is the cheapest and the quickest method of developing a resistant variety. This method has been useful in many cases in the past, but it has only a limited usefulness at the present level of crop improvement. For example, Kufri Red potato is a disease resistant selection from Darjeeling Red Round. Other such examples include resistance to curly-top in sugarbeets, to mildew and leaf spot in alfalfa, to cabbage yellows in cabbage and to Periconia root ro in sorghum. Pusa Sawani bhindi is a selection from a collection from Bihar; it was comparatively resistant to yellow mosaic under field conditions. Cotton (*Gossypium hirsutum* L.) variety MCU-1 (Madras, Combodia, Uganda-1) was selected from the variety Coimbatore-4 (CO-4) for resistance to black-arm; it had an acceptable level of resistance under field conditions. For example, selection for resistance to potato leafhopper and spotted alfalfa aphid in two broad based germplasms of alfalfa was highly effective in both the populations. A relatively recently released pigeonpea sterility virus (PSV) and wilt resistant variety of pigeon pea, Maviya Vikalp (MA3), is a pureline selection from material collected from a farmers field in Mirzapur (U.P.). Pusa Sawani okra variety is a YVM resistant selection from Bihar.

Introduction

Introductions have served as a useful method of disease control. Resistance varieties may be introduced for cultivation in a new area. This offers a relatively simple and quick means of obtaining resistant varieties. For example, Ridley wheat introduced from Australia had been useful as a rust resistant variety. Early varieties of groundnut introduced from USA were resistant to leaf spot or tikka disease. Kalyan Sona and Sonalika wheat varieties originated from the segregating materials introduced form CIMMYT, Mexico, and were rust resistant. Introductions also serve as sources of resistance in breeding programmes. For example, African pearl millet introductions have been used for developing Downey mildew resistant male sterile lines (Tift23 cytoplasm) for use in hybrid seed production. There are many examples of introduction of insect resistant varieties, the most striking of them being the introduction of D. Vitifoliae resistant grape root-stocks from USA into France; this saved the grape industry of France from virtual extinction. Ridley wheat variety introduced from Australia has been useful as a rust resistant variety.

Mutation

A mutagenic treatment may convert a susceptible genotype into a resistant one. If the mutation is a point mutation, the resistant mutant will be identical to the original cultivar, except for its resistance. Usually, however, there are undesirable side-effects of the mutagenic treatment. Several other genes may also have undergone changes, or the mutation for resistance has undesirable pleiotropic effects. As a consequence, the selection of a resistant mutant should be followed by further breeding efforts (*i.e.* backcrossing) to produce a commercially acceptable cultivar. The value of induced mutations is usually highlighted using the example of peppermint (*Mentha piperita*) cultivation in USA with annual value of 20 million U.S. dollars. This monoclonal crop became susceptible to *Verticillium* wilt, and its commercial cultivation was in jeopardy. A large scale mutation breeding project permitted the isolation of *Verticillium* wilt resistant mutant that had the same specific quality of peppermint oil as the parent variety; this saved the peppermint crop from being wiped out. Another impressive example of the economic success of mutant varieties is provided by the over 200 million Dutch florius chrysanthemum industry of the Netherlands; in 1981, mutants accounted for 35 per cent of the total sale of 600 million cuttings. Using γ-rays, amber grained mutants of Sonora-64 and Lerma Rojo were produced and released as Sharbati Sonora and Pusa Lerma respectively. In North America, one variety of common bean resistance to Anthracnose namely 'Samilac' was released by Down and Anderson in Michigan in 1956. Some other examples of disease resistance induced by mutagenic agents include resistance to stripe rust in wheat, crown rust in oats, mildew in barley, flax rust and leaf spot and stem rot resistance in pea nuts.

Hybridization

The most frequently employed plant breeding technique is hybridization. The aim of hybridization is to bring together desired traits found in different plant lines into one plant line via cross pollination. Hybridization is the most common method of breeding for disease resistance. It serves the following two chief purposes: (1) transfer of disease resistance from an agronomically undesirable variety to a susceptible but otherwise desirable variety (by back cross method), and (2) combining disease resistance and some other desirable characters of one variety with the superior characteristics of another variety (by pedigree method). In both these cases, one parent is selected for disease resistance; it should have a high degree of resistance to as many races of the pathogen as possible, and the resistance should be governed by few oligogenes. When the resistant variety is unadapted and agronomically undesirable, back cross method is the obvious choice. But when the resistant variety is well adapted and has some other desirable features as well, the pedigree method of breeding is preferred. Kufri Jyoti a prominent variety of potato developed by hybridization. It is resistant to late blight of potato. Kufri Khyati produces white oval tubers with shallow eyes, is moderately resistant to late blight, and is suitable for cultivation in plains of India.

Pedigree Method

Pedigree method consists of selecting individual F_2 plants for desirable features, including resistance to diseases. Progenies of these selections are reselected in each succeeding generation until homozygosity is obtained. This is very common method used to develop improved cultivars to diseases. This method is quite suited for breeding for horizontal or polygenic resistance, for which backcross method is of limited value. Pedigree method for breeding of disease resistance is not materially different from the method used for other quantitative traits, *e.g.*, yield. However, in breeding for disease resistance, artificial disease epidemics are generally created to enable an effective selection. A vast majority of disease resistant commercial varieties have been developed through this method *e.g.*, Kalyan Sona, Sonalika, Malviya-12, Malviya-37, Malviya-206, Malviya-234, and many other wheat varieties. *G. hirsutum* variety Laxmi resistant to red leaf blight was developed using pedigree method from the cross between susceptible parent Gadag-1 and the resistant parent Coimbatore Combodia-2. Resistance to diseases such as anthracnose and bean common mosaic necrosis virus (BCMNV) have been successful, and in recent years pedigree breeding has also used marker assisted selection to identify specific genes for disease resistance (Miklas *et al.*, 2001). But an important limitation to pedigree selection is the amount of time needed (Fehr, 1987).

Backcross Method

This method is useful in transferring genes for resistance from a variety that is undesirable in agronomic characteristics to a susceptible variety, which is widely adapted and is agronomically highly desirable. The backcross programme would differ depending upon whether the allele for resistance is recessive or dominant to that for susceptibility. A rigid selection is done for disease resistance, and the plant type of recurrent parent is also selected for. In general, 4-6 backcrosses are made, but with an effective use of marker assisted selection three backcrosses may be adequate. At the end of backcross programme, the progeny are selfed and resistant plants are selected. Progenies derived from different homozygous resistant plants that are identical in agronomic characteristics are usually bulked to produce the new disease resistant variety. The new variety would be almost identical to the recurrent parent, except for the disease resistance; therefore, extensive yield trials are usually not required for its release. There are many examples of interspecific transfer of genes; "Transfer" was the first commercial wheat variety in which rust resistance was transferred from a related species. Rust resistance has been transferred to Kalyan Sona from several diverse sources, *e.g.*, Robin, K-1, Bluebird, Tobari, Frecor, HS-19, etc., using backcross method. Three multiline varieties, *viz.*, KSML-3, KSML-11 and KSML-7406, have been released for cultivation as a result of the above programmes using Kalyan Sona as recurrent parent and exotic lines resistance to leaf rust as non-recurrent parent. The Advantages of this method: 1) when resistant variety is unadapted and agronomically undesirable, backcross method is an obvious choice. 2) Useful to transfer one or few major genes (vertical resistance). 3) Extensive yield trials are usually not required before its release for commercial cultivation. 4) Multiple resistance breeding is possible. Disadvantage: No advancement in yield potential is possible. Incorporation of genes for rust resistance in Indian wheat varieties through backcross method led to the development of several near isogenic lines like HW 2004 (backcross line of C 306 carrying Lr24), HW 2044 (backcross line of PBW 226) for commercial cultivation.

Somaclonal Variation

Disease resistant somaclonal variants can be obtained in the following two ways. Firstly, plants regenerated from cultured cells or progeny of such plants are subjected to disease test and resistant plants are isolated (screening). Secondly, cultured cells are selected for resistance to the toxin or culture filtrate produced by the pathogen and plants are regenerated from the selected cells (cell selection). In most cases, these plants are also resistant to the disease in question. Cell selection strategy is most likely to be successful in cases where the toxin is involved in disease development.

Genetic Engineering

In this approach genes expected to confer disease resistance are isolated, cloned and transferred into the crop in question. In case of viral pathogens, several transgenes have been evaluated, *viz.*, virus coat protein gene, DNA copy of viral satellite RNA, defective viral genome, antisense constructs of critical viral genes, and ribozymes. Viral coat protein gene approach seems to be the most successful (Singh, 2011), and a virus resistant transgenic variety of squash is in commercial cultivation in USA. Genes conferring insect resistance in plants have been transferred from *B. thuringiensis* (*cry* gene) and from other plants through genetic engineering. The *cry* gene transfers are the most successful, and insect resistant transgenic varieties of maize, soybean, cotton, etc. Expressing this gene is being cultivated in USA and other countries. In India, insect resistant *Bt*-cotton hybrids are in cultivation since 2002.

Generations for Dissection of Inheritance Behaviour

Generally the inheritance studies are done in early filial generation in case of biparental crosses. The half of the F_1 seed can be used for knowing dominance of the trait, if all the F_1 survive or resist than inheritance will be said to be governed by dominant alleles and if reverse is true then recessive alleles are the player. The F_2 population is considered best for the study of inheritance but it should be large enough to represent all possible combinations of alleles. Some workers prefer to go for Back cross population but the practical limitation lies in case of crops where low seed set is there a large number of back crosses need to be attempted and then verified to represent a larger back cross population set. Use of F_2 population also becomes tricky when we are dealing with destructive stress and there is no recovery of seed from the diseased or affected plant. The size of F_2 or back cross population can be decided based upon the pollination behaviour of crops. For self pollinated crops where survival nature of species is balanced through high homozygosity level and resistance being highly heritable the population size can be just above the perfect F_2 population size for number of genes in question. For example for single gene where three classes are supposed to be there and at least four plants are required but practically a population size of 30- 40 plants can be ideal in order to signify the results and sufficient statistical exercise. For two genes at least 16 plants are required but a population size of 60 to 70 is required for proper scoring. For three genes 64 plants are required but the population should have at least 200 individuals. Similarly for often cross pollinated crops where the survival of species lies in some degree of heterozygocity a fairly higher number (10 - 30 per cent) of plants is required for population. For cross pollinated crops the population size should always be higher side at least 1.5 times that of self pollinated crops. Odeny *et al.* (2009) conducted a study to determine the inheritance of resistance to Fusarium wilt in pigeonpea, which remains unknown, and to assess whether its genetic control would differ between African and Indian germplasm. Two resistant lines; one from African germplasm (ICEAP00040) and another of Indian origin (ICP8863) were used to make three different crosses (NPP670 x ICEAP00040, ICPL87091 x ICP8863, and KAT60/8 x ICP8863). Tests of F_1, F_2 and backcross generations under controlled conditions indicated involvement of one recessive gene in ICEAP0040 and 2 recessive genes in ICP8863. F_1, F_2 and backcross populations were developed by crossing resistant accessions (ICEAP 00554, ICEAP 00557) with susceptible accessions (KAT 60/8, ICP 7035). The Parents, F_1, F_2, backcrosses (BC_1F_1 and BC_2F_1) populations were evaluated for Fusarium wilt resistance. F_2 populations derived from ICEAP 00554 KAT 60/8, ICEAP 00557 KAT 60/8, ICEAP 00554 ICP 7035, ICEAP 00557 ICP 7035 crosses exhibited a 3:1 ratio which indicated that resistance to Fusarium wilt was under the control of major gene, however, a recessive gene was detected from ICP 7035 KAT 60/8 cross. The genes detected could be valuable for wilt resistance breeding (Karimi *et al.*, 2010).

Most of the workers generally prefer the F_1, F_2 and back cross population for dissection of genetic behaviour and inheritance. However, use of F_3 can be practiced wherein the stress is destructive and chances of losing transgressive segregants are more such as wilts. The representative progeny families comprising of 30- 50 plants from each F_2 plant can be tested for phenotyping against the stress which by virtue of Hardy Weinberg Law depict the same results in a more decisive manner.

Quantitative Inheritance and Resistance against Biotic Stresses

Although the resistance traits are generally thought to be of qualitative nature suggesting in majority of cases the resistance is generally governed by monogenic to oligogenic inheritance. Yet the proper dissection of a monogenic trait also reveals the presence of subclasses when the stress scored in larger categories. Generally the biotic stresses or any kind of stress cannot be governed by a single gene if we go at biochemical level simply because the pathogen or pest also evolves in many forms. Resistance mechanism involves the pathway of several products to express the final reaction of resistance. Any breakdown at any step of pathway results in compromise in the degree of resistance. The monogenic and simply inherited trait of resistance is one way simpler to transfer but possess the chances of busting the resistance in the presence of altered pathotype and can cause epidemic due to vertical resistance reaction. A classical example of bacterial blight of rice can be taken, the resistance is pathotype specific and there are more than 30 gene (*Xa*/xa) having been discovered so far. The single gene concept is although valid for its transfer but resistance is not stable. Therefore efforts are being made to accumulate multiple genes to confer horizontal resistance. A detailed analysis of the genetic basis of resistance to powdery mildew in wheat revealed the presence of more than 15 QTLs. Moreover the qualitative inheritances do behave like quantitative when scored on a normal scale over the time. Resistance against Fusarium wilt in pigeonpea is thought to be of monogenic and digenic by many workers however when scored on percent scale the resistance behaved liked a quantitative trait (Kumar, 2010).

Phenotyping of Biotic Stresses

The phenotyping of any disease or any biotic stress is quite important aspect. Generally the grades of effect of the stress are decided by the experts who by and large are uniform for most of the diseases and insect pests. The diseases are scored on 0-5 or 1-9 scales or percent disease index. The stresses scored in different scales helps in making the categories or classes of immune, resistant, moderately resistant, moderately susceptible, susceptible etc. But sometimes the scores are given arbitrarily in nature and differ for person to person. In such cases importance should be given to more scientific and quantitative indices rather than qualitative scores. It is always better to score the stress on a normal curve rather than in discrete classes. The screening of germplasm or varieties against any biotic stress is also a very crucial step in most of the crop improvement programmes. The screening can be done in more than one ways depending upon the nature of stress and pathogen and available resources.

Natural Screening

The screening under natural conditions has been practiced in past in majority of the cases. But the major drawback of the natural screening is that there are high chances of error due to escape of otherwise susceptible host due to environment. This method is however quite useful in hot spots of stress where the chances of escape of are minimized due to continued presence of virulent pathogen or pest and congenial environment. This method requires least efforts and resources. This method is the only way in cases where the pathogen or pest cannot be cultured.

Sick Plots

Another method of screening of the stress is to use sick plot technique. This technique is very simple and effective provided the sick plots are maintained regularly. But this method is quite specific to race of pathogen as the same sick plot cannot be used for different pathovars and variants. This method requires initial efforts in development of sick plot but later can be used with ease. The test entries of different genotypes are planted in the sick plot in a requisite randomized design or augmented design where a large number of entries are there. The beauty of the system is that crop can be raised in similar way to normal crop except the ideal situation for the stress. The sick plot technique is quite effective in case of soil borne pathogens and pests. Sick plot techniques has been used in case of fungal, bacterial, nematodes, stem borers etc.

Artificial Epiphytic Conditions

Artificial inoculation is the best method as it minimizes the time, highly specific and efficient. This method is useful in case of pathogen and pests which can be cultured on specific media without losing the virulence. This method requires both skills and science and therefore resource demanding. The culture of the pathogen is maintained in laboratory and used for inoculating the test entries as and when required. The artificial inoculation technique examples are *Xanthomonas oryzae* pv *oryzae* isolates through leaf cutting methods in rice, leaf stappling technique in case of inducing viral infections in groundnut, *Fusarium udum* inoculation in pigeon pea. The cultures of pathogens can be maintained locally or requested from a laboratory where it is being maintained. Institute of Microbial Technology (IMTech), Chandigarh, India maintains the cultures of a variety of pathogens.

Methods to Minimize the Time of Phenotyping

All the methods mentioned above are very stage specific and require lot of time as one full generation of the plant is required. However, with the advances in diagnostic tools the phenotyping for stress can be highly specific and faster. Biochemistry has provided the solutions for faster assays of stresses by utilizing the basic property of metabolism of plants or pathogen or pest. Such assays provide an early answer to screening a large number of samples. ELISA test for viral infections and presence of aflatoxins in seeds, Protein and Carbon analysis are few examples of daily diagnostic tools. Use of DNA or Protein based markers can significantly reduce the time and efforts in growing all the plants in field. DNA fingerprints and linked markers are now the tools for the selection of desired plant types.

Molecular Approaches

Molecular markers are useful in disease resistance breeding as they can substitute phenotypic screening in the early phase of breeding program and to identify resistant lines at juvenile stage to save time and cost of screening. It helps in easy identification and transfer of recessive genes and to monitor alien gene introgression, reduces the linkage drag and aids in eliminating undesirable traits in much shorter time frame than those expected through conventional breeding programs. It facilitates map-based cloning of disease resistance genes and pyramiding of genes for multiple disease resistance in a single cultivar, faster recovery of the recurrent parent genome in the backcross breeding programme (Tanksley *et al.*, 1989). It could also reduce the need for phenotypic selection that may be inappropriate in identifying genotypic differences and in selection of rare recombinants between tightly linked resistance genes. Molecular markers offer great scope for improving the efficiency of conventional plant breeding. The essential requirements for developing MAS system are (i) availability of germplasm with substantially contrasting phenotypes for the traits of interest, (ii) highly accurate and precise screening techniques for phenotyping mapping population for the trait of interest, (iii) identification of flanking markers closely associated with the loci of interest and the flanking region on either side and (iv) simple robust DNA marker technology to facilitate rapid and cost-effective screening of large population (Paterson *et al.*, 2004).

The use of DNA marker systems, such as random amplified polymorphic DNAs (RAPDs) (Williams *et al.*, 1990), amplified fragment length polymorphisms (AFLPs) (Vos *et al.*, 1995), and simple sequence repeats (SSRs) (Akkaya *et al.*, 1992), has contributed greatly to the development of genetic linkage maps for many important crop species including cowpea (Fatokun *et al.*, 1993; Waugh *et al.*, 1997). In combination with the bulked segregant analysis (BSA) method, (Michelmore and Meyers, 1998) the use of RAPDs, AFLPs, and SSRs has made it possible to rapidly identify molecular markers linked to genes of agronomic importance (Lee, 1995; Young, 1999). The development and use of molecular marker technologies has also facilitated the subsequent cloning and characterization of disease, insect, and pest resistance genes from a variety of plant species (Meyers *et al.*, 1999). Tiwari *et al.* (1998) identified coupling and repulsion phase RAPD markers linked to powdery mildew resistant gene er-1 in pea using bulk segregant analysis of F3 individuals. Marker OPO-18 was found to be linked in coupling phase while, the markers OPE 16 and OPL 6 were in repulsion phase to resistant gene *ER* -1. Kotresh *et al.* (2006) identified RAPD markers associated with pigeonpea wilt using F2 population derived from contrasting parents GS 1 (susceptible), ICPL 87119 (resistant) and ICP 8863 (resistant). PCR testing revealed presence of two amplicons at 704 bp and 500 bp linked with susceptibility. Analysis of individual F2 plants showed a segregation ratio of 3:1 for the presence: absence of amplicons in the crosses. Selvi *et al.* (2006) identified three RAPD markers *viz.*, OPT16, OPS7 and OPAK 19 specific to MYMV resistant parent and resistant bulk but absent in MYMV susceptible parent and susceptible bulk. From linkage analysis, one RAPD marker OPS 7900 was identified to be associated with mungbean yellow mosaic virus resistance. Ganapathy*et al.* (2009) used two AFLP primer pairs generating 4 markers (E-CAA/M-GTG150, E-CAA/M-GTG60, E-CAG/M-GCC120 and E-CAG/M-GCC150) which were polymorphic between the resistant and susceptible bulks indicating these markers are linked to SMD and located at a map distance of 5.7, 4.8, 5.2 and 20.7 cM respectively. The markers E-CAA/M-GTG150, E-CAA/M-GTG60 were linked in coupling phase to the susceptible dominant allele amplifying only in susceptible individuals which, can be effectively used for marker assisted selection.

Trait Mapping

With advancement in technology driven tools in modern Plant Breeding the use of map based approach are getting momentum. The resistance genes have been mapped in several crops and with the current advancements in high throughput and super throughput next generation sequencing facilities the genome wide information is available for consensus studies. The presence of the gene can be diagnosed using flanking DNA markers without waiting for the gene effect to be present in the phenotype (Paterson *et al.*, 1991). Bulked segregant analysis (BSA), first described by Michelmore *et al.* (1991), is a method used to identify molecular markers linked to phenotypic traits controlled by single major genes. This method relies on the availability of bulked DNA samples collected from individuals that segregate for the two extreme divergent phenotypes within a single population. One bulk contains the DNA of the trait being targeted, while the other contains DNA from individuals lacking the trait. DNA Polymorphisms between the bulks are therefore, likely to be linked to genes that govern the trait. In lentil, this method has been used to identify markers that are tightly linked to genes for resistance to Fusarium, vascular wilt and Ascochyta blight (Chowdhury *et al.*, 2001; Eujayl *et al.*, 1998b; Ford *et al.*, 1999). Eujayl *et al.* (1998b) used an RIL mapping population to identify molecular markers linked to the single dominant gene conditioning Fusarium vascular wilt resistance (*Fw*). They also identified a RAPD marker (OPS16750) that was 9.1 cM from the radiation-frost tolerance locus (*Frt*) (Eujayl *et al.*, 1999). However, most probably due to insufficient genome map coverage, the *Frt*locus and the linked RAPD marker were unable to be placed on the existing linkage map developed by Eujayl *et al.* (1998a). Ford *et al.* (1999) identified RAPD markers, RV01 and RB18, approximately 6 and 14 cM, respectively, from and flanking the foliar Ascochyta blight resistance locus *Ral*1 (*AbR*1) in ILL5588. These were subsequently converted to locus-specific sequence characterized amplified region (SCAR) markers and screened for applicability across parental lines in the Australian breeding program. Two RAPD markers, UBC2271290 and OPD-10870, were identified that flanked and were linked in repulsion phase to the resistance gene *ral*2 in the cultivar Indianhead at 12 and 16 cM, respectively (Chowdhury *et al.*, 2001). Kotresh *et al.* (2006) identified RAPD markers associated with pigeonpea wilt using F_2 population derived from contrasting parents GS 1 (susceptible), ICPL 87119 (resistant) and ICP 8863 (resistant). PCR testing revealed presence of two amplicons at 704 bp and 500 bp linked with susceptibility. Analysis of individual F_2 plants showed a segregation ratio of 3:1 for the presence: absence of amplicons in the crosses. *Fusarium equiseti* causes a discoloration on ginseng roots that significantly affects their marketability. The cellular and biochemical changes in affected roots that lead to this symptom, as well as differential gene expression following pathogen inoculation were studied.

Marker-Assisted Selection and Trait Pyramiding

Marker assisted selection (MAS) is the ability to select for and breed for a desirable trait with a marker, or suite of markers, from within a plant genotype without the need to express the associated phenotype. Marker assisted selection can be applied in such cases where linked markers are available. Therefore, MAS offers great opportunity for improved efficiency and effectiveness in the selection of plant genotypes with a desired combination of traits. This approach relies upon the establishment of a tight linkage between a molecular marker and the chromosomal location of the gene(s) governing the trait to be selected in a particular environment. Once this has been achieved, selection can be conducted in the laboratory and does not require the expression of the associated phenotype. For example, using MAS, disease resistance can be evaluated in the absence of the disease and in early stages of plant development. For example, Abenes *et al.* (1993) used MAS for selection of brown planthopper resistance (Bph3) and bacterial blight resistance (Xa21) using PCR-based markers in rice during F_2 generation. Similarly, PCR-based MAS for Xa21 gene was employed by Reddy *et al.* (1997) in rice improvement program. Sequence tagged sites (STS) are ideal markers for MAS.

STS markers are mapped loci for which all or part of the corresponding DNA sequence has been determined. The sequence information is used to design PCR primers for amplification of all or part of the original sequence. They are more robust and reproducible than the arbitrary sequences they are designed from, such as RAPD markers, as they are developed from the known sequences and produce an amplicon from longer primers. Differences in the lengths of amplified fragments serve as genetic markers for the locus. If no length polymorphism is detected, the amplified fragments can be cleaved with restriction enzymes to observe subsequent length differences. This technique is often referred to as cleaved amplified polymorphic sequences or CAPS (Jarvis *et al.*, 1994). The use of converted locus-specific PCR markers is also referred to as a specific polymorphic locus amplification test (SPLAT), as well as sequence characterized amplified region (SCAR) markers and allele specific associated primer (ASAP) markers. SPLAT markers are designed from sequencing the insert of a polymorphic RFLP marker (Gale and Witcombe, 1992), whereas SCAR and ASAP markers are developed from sequencing specific RAPD markers (Gu *et al.*, 1995; Ford *et al.*, 1999; Paran and Michelmore, 1993). The conversion of more technically-demanding RFLP markers into PCR based markers (*e.g.* SPLAT) may provide a more rapid, cost-effective and efficient tool in lentil breeding. Pyramiding of multiple resistance genes to foliar fungal pathogens should provide a broader and more durable resistance, as similarly shown in rice against bacterial blight (Singh *et al.*, 2001).

Table 17.2: Selected Examples of Improved Lines/Cultivars/Varieties Developed with Resistance to Pest and Diseases through Molecular Breeding Approaches in India and Abroad

Crop	*Cultivar/Breeding Line/Parental Line*	*Resistant to Trait*	*Gene(s)/QTL (Q)*	*Reference*
Rice (*Oryza sativa* L.)	PR-106	Bacterial blight	xa5 + xa13 + Xa21	Singh *et al.* (2001)
	Improved Pusa Basmati-1	Bacterial blight	xa13 + Xa21	Gopalakrishnan *et al.*, 2008
	Samba Mahsuri	Bacterial blight	xa5 + xa13 + Xa21	Sundaram *et al.* (2008)
	RPBio-226 (IET 19046) and RPBio- 210 (IET 19045)	Bacterial blight	xa5 + xa13 + Xa21	Sundaram *et al.* (2010)
	Mahsuri	Bacterial blight	Xa4 + xa13+ xa5+ Xa21	Shanti *et al.* (2010)
	Restorer line R-8012	Bacterial blight and Blast	Pi25/Xa21/xa13/xa5	Zhan *et al.* (2012)
	ADT-43	Bacterial blight	xa5 + xa13 + Xa21	Perumalsamy *et al.* (2010)
	Pusa Basmati-1	Bacterial blight, blast and sheath blight	xa13 + Xa21 + Pi54 + qSBR11-1	Singh *et al.* (2012)
	Pusa-6B and PRR-78	Bacterial blight	xa13 + Xa21	Basavaraj *et al.* (2010)
	CO-39	Bacterial blight, blast	Pi1 + Piz5 + Xa21	Narayanan *et al.* (2004)
	NSIC Rc142 and NSIC Rc154	Bacterial blight	Xa4+xa5+Xa21	Toenniessen *et al.* (2003)
	PGMS 3418s	Bacterial blight	Xa21	Luo *et al.* (2003)
	Restorer lines R-8006 and -1176	Bacterial blight	Xa21	Cao *et al.* (2003)
	Kang-4183	Bacterial blight	Xa21	Luo *et al.* (2005)
	Pusa-1602	Blast	Piz-5	Singh *et al.* (2012)
	Pusa-1603	Blast	Pi55	Singh *et al.* (2012)
	Pusa-6B and PRR-78	Blast	Pi-K^h and Piz5	Prabhu *et al.* (2009)
	IR-50	Blast	Piz	Narayanan *et al.* (2002)
	Duokang #1, Phalguna	Asian rice gall midge	Gm-2, Gm-6(t)	Katiyar *et al.* (2001)
	Improved Samba-Mahsuri (ISM)	Asian rice gall midge	Gm8	Sama *et al.* (2012)

Contd...

Table 17.2–*Contd...*

Crop	*Cultivar/Breeding Line/Parental Line*	*Resistant to Trait*	*Gene(s)/QTL (Q)*	*Reference*
	IIYou 8006	Bacterial blight	Xa21	Wu *et al.* (2008)
	Guodao 3	Bacterial blight	Xa21	Cao *et al.* (2006)
	RGD-7S/RGD-8S	Blast	Pi1, Pi2	Liu *et al.* (2008)
	Yueza 746/763	Blast	Pi1, Pi2	Jin *et al.* (2007)
	Huahui 1	Insects	Cry1Ac/Cry1Ab	Liu *et al.* (2012)
	Bph68S/Luohong4A	BPH	Bph14, Bph15	Zhu *et al.* (2013)
	Ning 9108	Stripe blight	Stv-bi, Wx-mq	Yao *et al.* (2010)
	T16S	Insects	Bt	Wu *et al.* (2010)
	TP309	Blast	Pi9	Qu *et al.* (2006)
Wheat	Zak	Stripe rust	Yr15	Randhawa *et al.* (2009)
(*Triticum aestivum* L.)	HD-2329	Leaf rust	Lr24 + Lr28	Kumar *et al.* (2010)
	Biointa-2004	Leaf rust	Lr47	Bainotti *et al.* (2009)
	HD-2329	Leaf rust	Lr9 + Lr24 + Lr28	Prabhu *et al.* (2009)
	PBW343	Leaf rust	Lr24 + Lr48Lr28 + Lr48	Prabhu *et al.* (2009)
	Patwin	Stripe and leaf rust	Yr17 and Lr37	Helguera *et al.* (2005)
	UC-1113 (PI-638741)	Stem rust (Ug99)	Lr19 + Sr25	Yu *et al.* (2009)
	CM-137, CM-138, CM-139, CM-150and CM-151	Turcicum leaf blight and Polysora rust	RppQ	Prasanna *et al.* (2010a); Prasanna *et al.* (2010b)
	Mace Wheat	streak mosaic virus	Wsm-1	Graybosch *et al.* (2009)
	D3-8-3, D3-8-5	Cyst nematode	CreX + CreY	Barloy *et al.* (2007)
	Yang 158	Powdery mildew	–	Liu *et al.* (2000)
Maize (*Zea mays* L.)	NA QTLs on	Corn borer resistance	chrom. 7, 9 and 10	Willcox *et al.* (2002)
	Steptoe	Stripe rust	Bmy1	Toojinda *et al.* (1998)

Contd...

Table 17.2–*Contd...*

Crop	*Cultivar/Breeding Line/Parental Line*	*Resistant to Trait*	*Gene(s)/QTL (Q)*	*Reference*
Barley	NA	leaf rust	Rphq6	van Berloo *et al.* (2001)
(*Hordeum vulgare* L.)	NA	Barley yellow dwarf virus	Yd2	Jefferies *et al.* (2003)
	DH-lines	Barley yellow dwarf virus	Ryd2 + Ryd3	Riedel *et al.* (2011)
	NA	Yellow mosaic virus	rym4 + rym5 + rym9+ rym11	Werner *et al.* (2005)
	HHB 67-improved	Downy mildew	–	Hash *et al.* (2006)
	SloopSA	Cyst nematode	–	Barr *et al.* (2000)
	Tango	Stripe rust	–	Hayes *et al.* (2003)
	Igri	BYDV	–	Scholz *et al.* (2009)
Pearl Millet	Nema TAM	Nematode resistance	Rma	Simpson *et al.* (2003)
Groundnut (*Arachis hypogea* L.)	CR41-10	Cassava mosaic disease	CMD2	Okogbenin *et al.* (2007)
Cassava	JTN-5503	Soybean cyst nematode	rhg1 + Rhg4 + Rhg5	Arelli *et al.* (2006, 2007)
Soybean	JTN-5109	Soybean cyst nematode	rhg1 + Rhg4 + Rhg5	Arelli and Young (2009)
(*Glycine max* L.)	Essex	Soybean mosaic virus	Rsv1 + Rsv3 + Rsv4	Maroof *et al.* (2008)
	S99-2281	frogeye leaf spot	–	Shannon *et al.* (2009)
White Bean	Verano	Bean golden yellow mosaic virus	–	Beaver *et al.* (2008)

Molecular breeding done for biotic stress resistance in India from 2005 to till 2009 (ICAR) (Table 17.3) and molecular breeding for biotic stress resistance in India from 2009 to till 2014 (DBT-GCP/ACIP) (Table 17.4). Some of the successful example mentioned below incorporating bacterial leaf blight resistance genes using marker aided selection in Indian rice cultivars *viz.*, IR-24, IR-64, Samba Mashuri, PR-106, Tapaswini, Pusa Basmati-1, Lalat and Swarna. Using marker assisted selection two cereal cyst nematode genes (*CreX* and *CreY*) have been pyramided from *Aegilops variabilis* in a wheat background (Barloy *et al.*, 2007).

Table 17.3: ICAR-Molecular Breeding for Biotic Stress Resistance in India from 2005-2009

Centre	*Cultivar*	*Genes for Resistance to*		
		Bacterial Blight	*Blast*	*Gall Midge*
DRR, Hyderabad	BPT-5204	xa13 + Xa21	Pi2 + Pi-kh	Gm1 + Gm4
CRRI, Cuttack	Tapaswani, Lalat, IR-64, Swarna	xa13 + Xa21	Pi2 + Pi9	Gm1 + Gm4
IARI, New Delhi	Pusa Basmati-1, Pusa6A/B, PRR-78	xa13 + Xa21	Pi-kh + Piz-5	Not required

Table 17.4: DBT-GCP/ACIP-Molecular Breeding for Biotic Stress Resistance in India from 2009-2014

Name of the Centres	*Target Varieties*	*Bacterial Blight*	*Blast*	*Gall Midge*	*Brown Plant Hopper*
DRR, Hyderabad	Sampada, Akshyadhan, DRR-17B and RPHR-1005 (Hybrid rice parental lines)	xa13 + Xa21	Pi-kh + Pi9	GM1 + GM4	Bph13 + Bph18
IARI, New Delhi	Pusa-1121 and Pusa-1401	xa13 + Xa21	Piz-5 + Pi-kh	–	Bph-18 + Bph-20 + Bph-21
PAU, Ludhiana	PAU-201, PAU-3075-3-38 and PAU-3105-45	xa13 + Xa21 + Xa30	–	–	Bph13 + Bph18

Quantitative Trait Loci Mapping and Identification of Genes

When a trait is governed by multiple and quantitative trait loci (QTL) and/or co-dominantly inherited genes, a more holistic genome mapping approach may be undertaken to identify genomic locations, interaction and subsequent molecular markers for accurate trait selection. QTLs have been detected in many crops but success in transfer of QTLs in desired background still remains a challenge. More than 80 independent QTL and 51 resistance genes from 62 different mapping populations were projected onto the consensus map of wheat for powdery mildew resistance (Marone *et al.*, 2013). In fact QTL per se is not faulty rather the associated facts such as the biased populations to identify the QTLs. Wherein the QTL effects are overestimated due to lesser population size, ignoring outliers, less number of markers, ignoring small effect QTLs, QTL interactions with other QTLs and residual factors. In majority of the recent cases the information on QTLs is published based upon the catchy interpretations and smart work of software tools. A large number of workers interpret QTLs as additive effect of variance however, in biology two plus two does not necessarily mean four. A small effect QTL may be likely to be excluded when we think of QTL transfer but for a given population the effect of a large QTL may be induced by the presence of small effect QTL. The small effect QTLs may be different trans factors which say a lot about QTL X QTL interactions. Moreover linkage and QTL mapping is not any different science from classical biometrical plant breeding. As it was difficult for classical breeders to transfer the interaction component of variance it is so difficult to quantify and use interaction component of QTLs.

Association Mapping

Association or linkage disequilibrium (LD) mapping was used successfully to discover genetic determinants to traits initially in humans is now being used in plants (Flint-Garcia *et al.*, 2003; Thornsbury *et al.*, 2001). Using association mapping, entire genomes can be scanned for markers associated with qualitative and quantitative traits. The association mapping approach may allow plant breeders to break out of restrictive F_1-derived mapping populations and employ any plant population including those from breeding programs or germplasm collections to conduct marker-trait association studies (Flint-Garcia *et al.*, 2003). Gebhardt *et al.* (2004) clearly summarized the four potential benefits of the association mapping approach: (1) it allows assessment of the genetic potential of specific genotypes before phenotypic evaluation; (2) it allows the identification of superior trait alleles in germplasm; (3) it can assist in high resolution QTL mapping; and (4) it can be used to validate candidate genes responsible for individual traits (Gebhardt *et al.*, 2004). The important issues to consider in designing and implementing any association mapping studies in plants are: (1) determination of the population structure (Pritchard *et al.*, 2000); (2) estimation of nucleotide diversity (Zhu *et al.*, 2003); (3) estimates of haplotype frequencies and LD (non random association of alleles at different loci) (Flint-Garcia *et al.*, 2003); and (4) precise evaluation of phenotypes (Neale and Savolainen, 2004).

Why genetic engineering for biotic stress: Availability of genetic variation in most of the crop species is one of the problems encountered by conventional breeders. The conventional approach as a whole is time consuming and labour intensive; undesirable genes are often transferred in combination with desirable ones; and reproductive barriers limit transfer of favourable alleles from inter specific and inter generic sources. Due to these reasons genetic engineering is being employed as a potential option worldwide for improving biotic stress tolerance.

Breeder versus Molecular Breeder and Genetic Engineer

It is a generally conception that breeder who goes to field and spends lot of time in hard work for doing crossing, selection, take data and then bring out the desired product in the form of a variety or breeding line and Molecular breeder on the contrary does his part of smart work while sitting on computer and doing experimentation in air conditioned laboratory. This is a matter of debate and concern that whether expressed genome which talks about the phenotype of a plant and a breeder relies on his skills for selection for desired plant out of a population of hundreds to millions of plants. And in laboratory a molecular breeder who relies on single base pair (SNP) or few hundred base pairs (SSRs and STS) or on few thousand base pairs (RFLP, AFLP) based markers can do wonders in selection of plants. The matter is clear that if any marker system is closely linked to gene of interest with great heritability is as good as selection of desired plant in field through experience and acquired skills. As far as chances of mistake are not there both the system serve the purpose well. However, a great degree of confusion and efforts are there before getting the selection skills and developing a sure shot genomic tool or marker system for a given crop and trait. As far as science behind the two ways is concerned nothing is different and both systems religiously rely on genetic principles, statistical analysis and interpretation of data. The fact that laboratory based very good genotyping requires very good phenotyping as compulsory step to exclude false positives and proof reading and if one is good in phenotyping then there is no need of genotyping if the purpose is to just bring out the product. But to understand the basis of final product derived in field one has to go for genotyping or phenotyping at biochemical level.

Genetic engineering mainly refers to the non conventional methods of designing a plant with one or more genes from quaternary gene pool through alien gene transfer methods using targeted approach. This science has shown its potential in terms of biotic stress resistance. Use of glyphosate resistance gene in different crops such as soybean and maize and use of *Bt* genes for resistance against lepidopteran pests has revolutionized the modern day Agriculture. Engineering a single or set of genes with modified cassette of promoter and controlling regions for target specific expression is the future of Plant Breeding. This technique has drastically reduced the dependence on harmful chemicals and pesticides. This technique although very effective yet has some reservations on its broad utilization mainly due to business behind it. The end user has to pay handsome royalty for the single gene working in the background of some half a million genes that is agronomically superior genotype which also involves the due consideration. Moreover certain issues pertaining to bio-safety and its regulation make hindrance in the use of this technology. War for economic gains between genetic engineering and chemical industry tycoons has misguided policymakers, farmers and scientific community and created propaganda for slow acceptance of technology. The next line of approach for a win- win situation will be targeting genes from plants which are wild relatives and possess resistance genes against biotic stresses. The use of plants from extreme climates such as mangroves, desserts and marshy lands can feed the bellies of ever-growing population.

Multidisciplinary Approach

The chances of success of getting a desired variety or plant type are multifold increased with the collective effort of a Plant Breeder, Molecular tools, Biochemist, Physiologist, Pathologist or Entomologist and a crop manager. Without the minimum company of these fellows desired product with full scientific dissection is not likely to come. The members of ideal team may opt to work together for faster results or they will have to work separately for considerably longer to time to reveal the portion of knowledge of their respective domain. A Plant Breeder looks at the phenotype of a plant with certain traits say disease resistant, high yielding, timely maturing and highly competitive plant. The pathologist has to ensure the type and degree of stress and its mechanism, Physiologist and biochemist has to go in to system of source to sink pathways, enzymes and system of plant as well as pathogen. A crop manager has to ensure a good crop with proper package of practices to express its fullest potential. Molecular Breeder with the help of markers and genomic tools can drastically reduce the generation time and can establish the linkages between genome and trait with mechanisms and functional properties. The genetic engineering approach can answer the problems of crossing incompatibilities and species borders.

Utilization of Resistance Source

The use of resistance source is the key factor wherein the breeding objective is to develop a resistant variety. Generally the resistance source for a particular biotic stress taken from the gene pool of biodiversity. Nature has provided the antidote of every problem and it is up to the Breeder where from he tries to extracts the resistant gene or genes. If the resistance is available in same species and varieties can be crossed easily it means the resistance source is primary gene pool. When the resistance source is available in another species of same genera and crossing can be done easily or with some limited difficulties which can be overcome by simpler methods the resistance source is said to be from secondary gene pool for example, *Solanum torvum* (a wild brinjal) is resistant to all kinds of soil borne diseases and is an excellent source of resistance for *S. melongena* (cultivated brinjal). Tertiary gene pool is the distantly related species with different genus but same family such gene pools cannot be used with classical methods of gene transfer and pose a great degree of difficulty. Such gene pool can be used with the help of bridging species which can cross with both resistant source as well as susceptible variety. When the resistance source is altogether different family or even kingdom and there is no relation between the donor and recipient species the source is said to be from quaternary gene pool. The classical example is the Bt gene from bacteria transferred in a number of crop species through genetic engineering. This area of science is getting momentum where in all the boundaries at genomic level are taken care through genetic engineering. Some workers also prefer to use mutation and somaclonal variation approaches for creating variability in the available germplasm these techniques are technically dependent on the chance of getting useful mutant and the frequency of such mutant is generally very low ($1x\ 10^{-6}$ to $1x\ 10^{-9}$). But the science behind these techniques is also same that is change in the genome through deletions, repetitions, translocations or substitutions in the base pairs of a gene or its trans regulating factors which in turn modifies the expression profile of proteins or enzymes which can bring about the desired change.

Challenges and Way Ahead

The development of biotic stress resistant varieties particularly in NARS faces the basic challenges of lack of focus on utilization of secondary and tertiary gene pools from the great biodiversity vastly available in country. The monotony of public research efforts on utilization of resistance sources from patented technology has virtually hijacked the efforts for mining and developing our own resistance genes.

There are several reports on microbes to higher plants possessing resistance and genes. India being one of the largest hub of biotechnologists however hardly looking into the immense possibilities of utilization of resistance sources from related species, wild relatives and genes from extreme climatic plants. Shortage of technical people on taxonomy and systemics is another missing link to further utilization of our rich biodiversity. With the advancement in science of cross species gene transfer India can be leader with a record number of gene patents from indigenous flora and fauna. The concerted team efforts on silencing bad genes and knock out of faulty pathways and introgression of useful genes is the way ahead to fill the bellies of billions.

References

Akkaya, M.S., Bagwhatt, A.A. and Cregan, P.B. (1992). Length polymorphisms of simple sequence repeat DNA in soybean. *Genetics*, 132: 1131–1139.

Arelli, P.R., Pantalone, V.R., Allen, F.L. and Mengistu, A. (2007). Registration of soybean germplasm JTN-5303. *Journal of Plant Registrations*, 1: 69-70.

Arelli, P.R., Young, L.D. and Concibido, V.C. (2009). Inheritance of resistance in soybean PI 567516C to LY1 nematode population infecting cv. Hartwig. *Euphytica*, 165: 1-4.

Arelli, P.R., Young, L.D. and Mengistu, A. (2006). Registration of high yielding and multiple disease resistant soybean germplasm JTN-5503. *Crop Science*, 46: 2723-2424.

Bainotti, C.T., Fraschina, J., Salines, J., Nisi, J., Dubcovsky, J., Lewis, S., *et al.* (2009). Registration of 'BIOINTA 2004' wheat. *Journal of Plant Registrations* 3: 165-169.

Barloy, D., Lemoine, J., Abelard, P., Tanguy, A.M., Rivoal, R. and Jahier, J. (2007). Marker-assisted pyramiding of two cereal cyst nematode resistance genes from Aegilops variabilis in wheat. *Molecular Breeding*, 20: 31-40.

Barr, A.R., Jefferies, S.P., Warner, P., Moody, D.B., Chalmers, K.J. and Langridge, P. (2000). Marker-Assisted Selection in Theory and Practice. *Prodeedings of the 8th International Barley Genetics Symposium, Vol I.*, Adelaide, Australia, 167-178.

Basavaraj, S.H., Singh, V.K., Singh, A., Singh, A., Singh, A., Yadav, S., Ellur, R.K., Singh, D., Gopala Krishnan, S., Nagarajan, M., Mohapatra, T., Prabhu, K.V. and Singh, A.K. (2010). Marker-assisted improvement of bacterial blight resistance in parental lines of Pusa RH-10, a superfine grain aromatic rice hybrid. *Molecular Breeding*, DOI 10.1007/s11032-010-9407-3.

Beaver, J.S., Porch, T.G. and Zapata, M. (2008). Registration of 'Verano' White Bean. Journal of Plant Registrations, 2: 187-189.

Cao, L., He, L., Zhan, X., Zhuang, J. and Cheng, S. (2006). Guodao 3, a new hybrid rice combination with good quality, high yield and disease resistance. *Hybrid Rice*, 21:83–84.

Cao, L.Y., Zhuang, J.Y., Zhan, X.D., Zheng, K.L. and Cheng, S.H. (2003). Hybrid rice resistant to acterial blight developed by marker assisted selection. Zhongguo Shuidao Kexue (*Chinese Journal of Rice Science*), 17(2): 184-186.

Chakraborty, S., Tiedemann, A.V. and Teng, P.S. (2000). Climate change: potential impact on plant diseases. *Environmental Pollution*, 108: 317-326.

Chester, K.S. (1950). Plant disease losses: their appraisal and interpretation. *Plant Disease Report* (Suppl.) 193: 189-362.

Chowdhury, M.A., Andrahennadi, C.P., Slinkard, A.E. and Vandenberg, A. (2001). RAPD and SCAR markers for resistance to ascochyta blight in lentil. *Euphytica*, 118: 331–337.

Daniela, M., Maria, A.R., Giovanni, L., Pasquale, D.V., Roberto, P., Antonio, B., Agata, G., Diego, R. and Mastrangelo, A.M. (2013). Genetic basis of qualitative and quantitative resistance to powdery mildew in wheat: from consensus regions to candidate genes. *BMC Genomics*, 14: 562.

Down, E.E. and Anderson, A.L. (1956). Agronomic use of an X-ray induced mutant. *Science*, 124: 223-224.

Eujayl, I., Baum, M., Erskine, W., Pehu, E. and Muehlbauer, F.J. (1997). The use of RAPD markers for lentil genetic mapping and the evaluation of distorted F_2 segregation. *Euphytica*, 96: 405–412.

Eujayl, I., Baum, M., Powell, W., Erskine, W. and Pehu, E. (1998a). A genetic linkage map of lentil (*Lens* sp.) based onRAPDand AFLP markers using recombinant inbred lines. *Theoretical and Applied Genetics*, 97: 83–89.

Eujayl, I., Erskine, W., Baum, M. and Pehu, E. (1999). Inheritance and linkage analysis of frost injury in lentil. *Crop Science*, 39: 639–642.

Eujayl, I., Erskine, W., Bayaa, B., Baum, M. and Pehu, E. (1998b). Fusarium vascular wilt in lentil: Inheritance and identification of DNA markers for resistance. *Plant Breeding*, 117: 497–499.

Fatokun, C.A., Danesh, D., Menancio-Hautea, D.I. and Young, N.D. (1993). A linkage map for cowpea [*Vignaunguiculata*(L) Walp] based on DNA markers (2*n* = 22). *In Genetic maps 1992: A compilation of linkage and restriction maps of genetically studied organisms.* Edited by J.S. O'Brien. Cold Spring Harbour Laboratory Press, New York. pp. 6256–6258.

Flint-Garcia, S.A., Thornsberry, J.M. and Buckler, E.S. (2003). Structure of linkage disequilibrium in plants. *Annual Review of Plant Biology*, 54: 357– 374.

Flor, H.H. (1942). Inheritance of pathogenicity in *Melampsoralini. Phytopathology*, 32: 653–669.

Flor, H.H. (1947). Inheritance of reaction to rust in flax. *Journal of Agricultural Research*, 74: 241–262.

Flor, H.H. (1955). Host-parasite interaction in flax rust - its genetics and other implications. *Phytopathology*, 45: 680–685.

Ford, R., Pang, E.C.K. and Taylor, P.W.J. (1999). Genetics of resistance to ascochyta blight (*A. lentis*) of lentil and the identification of closely linked RAPD markers. *Theoretical and Applied Genetics*, 98: 93–98.

Frederick, J., Muehlbauer, S.C., Sarker Ashutosh, McPhee Kevin, E., Coyne Clarice, J., Rajesh, P.N. and Ford, R. (2006). Application of biotechnology in breeding lentil for resistance to biotic and abiotic stress. *Euphytica*, 147: 149–165.

Gale, M.D. and Witcombe, J.R. (1992). DNA markers and marker mediated applications in plant breeding, with particular reference to pearl millet breeding. In: J.P. Moss (Ed.), Biotechnology and Crop Improvement in Asia, pp. 323–332. ICRISAT, Patancheru.

Ganapathy, K.N., Gowda Byre, M., Venkatesha, S.C., Rama Chandra, R., Gnanesh, B.N. and Girish, G. (2009). Identification of AFLP markers linked to sterility mosaic disease in pigeonpea *Cajanus cajan*(L.) Millsp. *International Journal of Integrative Biology*, 7**:** 145-149.

Gaunt, R.E. (1995). The relationship between plant disease severity and yield. *Annual Review of Phytopathology*, **33:** 119.

Gebhardt, C., Ballvora, A., Walkemeir, B., Oberhagemann, P. and Schuler, K. (2004). Assessing genetic potential in germplasm collections of crop plants by marker-trait association: A case study for potatoes with quantitative variation of resistance to late blight and maturity type. *Molecular Breeding*, 13: 93–102.

Gopalakrishnan, S., Sharma, R.K., Rajkumar, K.A., Joseph, M., Singh, V.P., Singh, A.K., *et al.* (2008). Integrating marker assisted background analysis with foreground selection for identification of superior bacterial blight resistant recombinants in Basmati rice. *Plant Breeding*, 127: 131-139.

Graybosch, R.A., Peterson, C.J., Baenziger, P.S., Baltensperger, D.D., Nelson, L.A., Jin, Y, *et al.* (2009). Registration of 'Mace' hard red winter wheat. *Journal of Plant Registrations*, 3: 51-56.

Gu, W.K., Weeden, N.F., Yu, J. and Wallace, D.H. (1995). Large-scale, cost-effective screening of PCR products in marker-assisted selection applications. *Theoretical and Applied Genetics*, 91: 465–470.

Hash, C.T., Sharma, A., Kolesnikova-Allen, M.A., Singh, S.D. Thakur, R.P. Raj, A.G.B., *et al.* (2006). Teamwork delivers biotechnology products to Indian small-holder crop-livestock producers: Pearl millet hybrid "HHB 67 Improved" enters seed delivery pipeline. *Journal of SAT Agricultural Research*, 2: 1-3.

Hayes, P.M., Corey, A.E., Mundt, C., Toojinda, T. and Vivar, H. (2003). Registration of 'Tango' Barley. *Crop Science*, 43: 729-731.

Helguera, M., Vanzetti, L., Soria, M., Khan, I. A., Kolmer, J. and Dubcovsky, J. (2005). PCR Markers for Triticum speltoides Leaf Rust Resistance Gene Lr51 and Their Use to Develop Isogenic Hard Red Spring Wheat Lines. *Crop Science*, 45: 728-734.

James, W.C. (1974). Assessment of plant diseases and losses. *Annual Review of Phytopathology*, 12: 27.

Jarvis, P., Lister, C., Szab´O, V. and Dean, C. (1994). Integration of CAPS markers into the RFLP map generated using recombinant inbred lines of *Arabidopsis thaliana*. *Plant Molecular Biology*, 24: 685– 687.

Jefferies, S.P, King, B.J., Barr, A.R., Warner, P, Logue, S.J. and Langridge, P. (2003). Marker-assisted backcross introgression of the Yd2 gene conferring resistance to barley yellow dwarf virus in barley. *Plant Breeding*, 122: 52-56.

Jin, S., Liu, W., Zhu, X., Wang, F., Li, J., Liu, Z., Liao, Y., Zhu, M., Huang, H. and Liu, Y. (2007). Improving blast resistance of a thermo-sensitive genic male sterile line GD-8S by molecular marker- assisted selection. *Chinese Journal of Rice Science*, 21:599–604.

Karimi, R., Owuoche James, O. and Silim, S.N. (2010). Inheritance of fusarium wilt resistance in pigeonpea [*Cajanus cajan* (L.) Millspaugh]. *Indian Journal of Genetics and Plant Breeding*, 70(3): 271-276.

Katiyar, S,K, Tan, Y., Huang, B., Chandel, G., Xu, Y., Zhang, Y. *et al.* (2001). Molecular mapping of gene Gm-6(t) which confers resistance against four biotypes of Asian rice gall midge in China. *Theoretical and Applied Genetics*, 103: 953-961.

Kotresh, H., Fakrudin, B., Punnuri, S.M., Rajkumar, B.K., Thudi, M., Paramesh, H., Lohithswa, H. and Kuruvinashetti, M.S. (2006). Identification of two RAPD markers genetically linked to a recessive allele of a fusarium wilt resistance gene in pigeonpea (*Cajanus cajan* L. Millsp.). *Euphytica*, 149: 113 – 120.

Kumar, J., Mir, R.R., Kumar, N., Kumar, A., Mohan, A., Prabhu, K.V., Balyan, H. S. and Gupta, P. K. (2010). Marker assisted selection for pre-harvest sprouting tolerance and leaf rust resistance in bread wheat. *Plant Breeding*, 129: 617-621.

Kumar, N. (2010). Molecular mapping of *Fusarium wilt* in pigeonpea *Cajanus cajan* L. Millsp. PhD thesis, CCSHAU, Hisar.

Lee, M. (1995). DNA markers and plant breeding programs. *Advance Agronomy*, 55: 265–344.

Liu, J., Liu, D., Tao, W., Li, W., Wang, C.P., Cheng, S. and Gao, D. (2000). Molecular marker-facilitated pyramiding of different genes for powdery mildew resistance in wheat. *Plant Breeding*, 119: 21–24.

Liu, N., Jiang, L., Zhang, W., Liu, L., Zhai, H. and Wan, J. (2008). Role of embryo LOX3 Gene under adversity stress in rice (*Oryza sativa* L.). *Chinese Journal of Rice Science*, 22:8–14.

Liu, Z., Zhao, J., Li, Y., Zhang, W., Jian, G. and Peng, Y Qi, F. (2012). Non uniform distribution pattern for differentially expressed genes of transgenic rice huahui 1 at different developmental stages and environments. PLoS One, 7:e37078.

Luo, Y.C., Wang, S.H., Li, C.Q., Wang, D.Z., Wu, S. and Du, S.Y. (2003). Breeding of the photo-period-sensitive genetic male sterile line '3418S' resistant to bacterial bright in Rice by molecular marker assisted selection rice. Zuowu Xuebao (*Acta Agronomica Sinica*), 29(3): 402-407.

Luo, Y.C., Wang, S.H., Li, C.Q., Wang, D.Z., Wu, S. and Du, S.Y. (2005). Improvement of bacterial blight resistance by molecular marker-assistedselection in a wide compatibility restorer line of hybrid rice. Zhongguo Shuidao Kexue (*Chinese Journal of Rice Science*), 19(1): 36-40.

Marone D., Maria, A. R., Giovanni L., Pasquale D.V., Roberto P., Antonio B., Agata G., Diego R. and Mastrangelo, A.M. (2013). Genetic basis of qualitative and quantitative resistance to powdery mildew in wheat: from consensus regions to candidate genes. *BMC Genomics*, 14:562.

Maroof, S., Jeong, S.C., Gunduz, I., Tucker, D.M., Buss, G.R. and Tolin, S.A. (2008). Pyramiding of soybean mosaic virus resistance genes by marker-assisted selection. *Crop Science*, 48: 517-526.

Meyers, B.C., Dickerman, A.W., Michelmore, R.W., Sivaramakrishnam, S., Sobral, B.W. and Young, N.D. (1999). Plant disease resistance genes encode members of an ancient and diverse protein family within the nucleotide-binding super family. *Plant Journal*, 20: 317–332.

Michelmore, R.W. and Meyers, B.C. (1998). Clusters of resistance genes evolve by divergent selection and a birth and death process. Genome Research, 8: 1113–1130.

Michelmore, R.W., Paran, I. and Kesseli, R.B. (1991). Identification of markers linked to disease-resistance genes by bulked segregant analysis: A rapid method to detect markers in specific genomic regions by using segregating populations. *PNAS* (USA) 88: 9828– 9832.

Narayanan, N.N., Baisakh, N., Cruz, V., Gnanamanickam, S.S., Datta, K. and Datta, S.K. (2002). Molecular breeding for the development of blast and bacterial blight resistance in rice cv. IR50. *Crop Science*, 42: 2072-2079.

Narayanan, N.N., Baisakh, N., Oliva, N.P., Vera Cruz, C.M., Gnanamanickam, S.S. and Datta, K. (2004). Molecular breeding: marker-assisted selection combined with biolistic transformation for blast and bacterial blight resistance in indica rice (cv. CO39). *Molecular Breeding*, 14: 61-71.

Neale, D.B. and Savolainen, O. (2004).Association genetics of complex traits in conifers. *Trends in Plant Science*, 325–330.

Odeny, D.A., Jayashree, B., Gebhardt, C. and Crouch, J. (2009). New microsatellite markers for pigeonpea (*Cajanus cajan*(L.) millsp.). *BMC Research Notes* **2:** 35.

Oerke, E.C. (2006). Crop losses to pests. *Indian Journal of Agricultural Sciences*, 144: 31-43.

Okogbenin, E., Porto, M.C.M., Egesi, C., Mba, C., Espinosa, E., Santos, L.G., Ospina, C., Marín, J., Barrera, E., Gutiérrez, J., Ekanayake, I., Iglesias, C. and Fregene, M.A. (2007). Marker assisted introgression of resistance to cassava mosaic disease into Latin American germplasm for the genetic improvement of cassava in Africa. *Crop Science*, 47: 1895-1904.

Paran, I. and Michelmore, R.W. (1993). Development of reliable PCR based markers linked to downy mildewresistance genes in lettuce. *Theoretical and Applied Genetics*, 85: 985–993.

Paterson, A.H., Stalker, H.T., Gallo-Meagher, M., Burrow, M.D., Dwivedi, S.L., Crouch, J. H. and Mace, E.S. (2004). Genomics and Genetic Enhancement of Peanut. In: *Legume Crop Genomics*. (Eds.) Wilson, R. F., Stalker, H. T. and Brummer, E. C., Champaign, III: AOCS Press. pp. 97-109.

Paterson, A.H., Tanksley, S.D. and Sorrells, M.E. (1991). DNA markers in plant improvement. *Advance Agronomy*, 46: 39–90.

Perumalsamy, S., Bharani, M., Sudha, M., Nagarajan, P., Arul, L., Saraswathi, R., Balasubramanian, P. and Ramalingam, J. (2010). Functional marker-assisted selection for bacterial leaf blight resistance genes in rice (*Oryza sativa* L.). *Plant Breeding*, 129: 400-406.

Prabhu, K.V., Singh, A.K., Basavaraj, S.H., Cherukuri, D. P., Charpe, A., Gopala Krishnan, S., Gupta, S.K., Joseph, M., Koul, S., Mohapatra, T., Pallavi, J.K., Samsampour, D., Singh, A., Vikas K. Singh, Singh, A. and Singh, V.P. (2009). Marker assisted selection for biotic stress resistance in wheat and rice. *Indian Journal of Genetics*, 69(4) (Spl. issue): 305-314.

Prasanna, B.M., Hettiarachchi, K., Mahatman, K., Rajan, A., Singh, O.N., Kaur, B., Kumar, B., Gowda, K.T.P., Pant, S.K. and Kumar, S. (2010). Molecular marker-assisted pyramiding of genes conferring resistance to Turcicum leaf blight and Polysora rust in maize inbred lines in India. In: Zaidi PH, Azrai M, Pixley KN, eds. Maize for Asia: Emerging trends and technologies. Proceedings of 10th Asian Regional Maize Workshop, Makassar, Indonesia 20–23 October 2008. Mexico: CIMMYT.

Prasanna, B.M., Pixley, K., Warburton, M.L. and Xie, C. (2010). Molecular marker-assisted breeding options for maize improvement in Asia. *Molecular Breeding*, 26: 339-356.

Pritchard, J.K., Stephens, M. and Donnelly, P. (2000). Association mapping in structured populations. *American Journal of Human Genetics*, 67: 170–181.

Randhawa, H.S., Mutti, J.S., Kidwell, K., Morris, C.F., Chen, X. and Gill, K.S. (2009). Rapid and targeted introgression of genes into popular wheat cultivars using marker-assisted background selection. PLoS ONE 4: e5752.

Rao, P.S. and Kauffman, H.E. (1971). A new Indian host of Xanthamonas oryzae, incitant of bacterial leaf blight of rice. *Curr. Sci.*, 40: 271-272.

Riedel, C., Habekuß, A., Schliephake, E., Niks, R., Broer, I., Ordon, F. (2011). Pyramiding of Ryd2 and Ryd3 conferring tolerance to a German isolate of barley yellow dwarf virus-PAV (BYDV-PAV-ASL-1) leads to quantitative resistance against this isolate. *Theoretical and Applied Genetics*, 123: 69-76.

Sama, V.S.A.K., Himabindu, K., Naik, S.B., Sundaram, R.M., Viraktamath, B.C. and Bentur, J.S. (2012). Mapping and marker-assisted breeding of a gene allelic to the major Asian rice gall midge resistance gene Gm8. *Euphytica*, 187: 393-400.

Scholz, M., Ruge-Wehling, B., Habekub, A., Schrader, O., Pendinen, G., Fischer, K. and Wehling, P. (2009). Ryd4Hb: a novel resistance gene introgressed from *Hordeum bulbosum* into barley and conferring complete and dominant resistance to the barley yellow dwarf virus. *Theoretical and Applied Genetics*, 119: 837-849.

Selvi, R., Muthaih, A.R. and Manivannan, N. (2006). Tagging of RAPD marker for MYMV resistance in mungbean [*Vigna radiata*(L.) wilczek]. *Asian Journal of Plant Sciences*, 5: 277- 280.

Shannon, J.G., Lee, J.D., Wrather, J.A., Sleper, D.A., Mian, M.A.R., Bond, J.P. and Robbins, R.T. (2009). Registration of S99-2281 Soybean Germplasm Line with Resistance to Frogeye Leaf Spot and Three Nematode Species. *Journal of Plant Registrations*, 3: 94-98.

Shanti, M.L., Shenoy, V.V., Lalitha Devi, G., Mohan Kumar, V., Premalatha, P., Naveen Kumar, G., Shashidhar, H.E., Zehr, U.B. and Freeman, W.H. (2010). Marker-assisted breeding for resistance to bacterial leaf blight in popular cultivar and parental lines of hybrid rice. *Journal of Plant Pathology*, 92 (2): 495-501.

Simpson, C.E., Starr, J.L., Church, G.T., Burrow, M.D. and Paterson, A.H. (2003). Registration of Nema TAM peanut. *Crop Science*, 43: 1561.

Singh, B.D. (2011). Plant Biotechnology, 2nd edn. Kalyani Publishers, New Delhi.

Singh, S., Sidhu, J.S., Huang, N., Vikal, Y., Li, Z., Brar, D.S., Dhaliwal, H.S. and (2001). Pyramiding three bacterial blight resistance genes (xa5, xa13 and Xa21) using marker-assisted selection into indica rice cultivar PR106. *Theoretical and Applied Genetics*, 102: 1011-1015.

Singh, V.K., Singh, A., Singh, S.P., Ellur, R.K., Choudhary, V., Sarkhel, S., Singh, D., Gopala Krishnan, S., Nagarajan, M., Vinod, K.K., Singh, U.D., Rathore, R., Prashanthi, S.K., Agrawal, P.K., Bhatt, J.C., Mohapatra, T., Prabhu, K.V. and Singh, A.K. (2012). Incorporation of blast resistance into 'PRR78', an elite Basmati rice restorer line, through marker assisted backcross breeding. *Field Crops Research*, 128: 8-16.

Srivastava, D.N., Rao, Y.P. and Durgapal, J.C. (1966). Can Taichung Native-1 stand upto Bacterial blight ? *Indian Farming*, 16(2): 15.

Sundaram, R.M., Vishnupriya, M.R., Biradar, S.K., Laha, G.S., Reddy, G.A., Rani, N.S., Sarma, N.P. and Sonti, R.V. (2008). Marker assisted introgression of bacterial blight resistance in Samba Mahsuri, an elite indica rice variety. *Euphytica*, 160: 411-422.

Sundaram, R.M., Vishnupriya, M.R., Shobha Rani, N., Laha, G.S., Viraktamath, B.C., Balachandran, S.M., Sarma, N.P., Mishra, B., Ashok Reddy, G. and Sonti, R.V. (2010). RPBio-189 (IET19045) (IC569676; INGR09070), a paddy (Oryza sativa) germplasm with high bacterial blight resistance, yield and fine-grain type. *Indian Journal of Plant Genetic Resources*, 23: 327-328.

Suresh, S. and Malathi, D. (2014). Gene pyramiding for biotic stress tolerance in crop plants. *Weekly Science Research Journal*, 1(23, 26): 1-14.

Tanksley, S.D., Young, N.D., Paterson, A.H. and Bonierbal, M.W. (1989). RFLP mapping in plant breeding- new tools for an old science. *Bio-Technology*, 7: 257-264.

Thornsbury, J.M., Goodman, M.M., Doebley, J., Kresovich, S., Nielsen, D. and Buckler, E.S. (2001). Dwarf8 polymorphisms associate with variation in flowering time. *Nature Genetics*, 28: 286–289.

Tiwari, K.R., Penner, G.A. and Warkentin, T.D. (1998). Identification of coupling and repulsion phase RAPD markers for powdery mildew resistance gene *er-1* in pea. *Genome*, 41: 440-444.

Toenniessen, G.H., O'Toole, J.C. and DeVries. J. (2003). Advances in plant biotechnology and its adoption in developing countries. *Current Opinion in Plant Biology*, 6: 191–198.

Toojinda, T., Baird, E., Booth, A., Broers, L., Hayes, P., Powell, W., Thomas, W., Vivar, H. and Young, G. (1998). Introgression of quantitative trait loci (QTLs) determining stripe rust resistance in barley: an example of marker assisted line development. *Theoretical and Applied Genetics*, 96: 123-131.

Van Berloo, R., Aalbers, H., Werkman, A. and Niks, R.E. (2001). Resistance QTL confirmed through development of QTL-NILs for barley leaf rust resistance. *Molecular Breeding*, 8: 187-195.

Vos, P., Hogers, R., Bleekers, M., Rejians, M., Van de Lee, T., Hornes, M., Frijters, A., Pot, J., Peleman, J., Kiuper, M. and Zabeau, M. (1995). AFLP: A new technique for DNA fingerprinting. *Nucleic Acids Research*, 23: 4407-4414.

Waugh, R., Bonar, N., Baird, E., Thomas, B., Graner, A., Hayes, P. and Powell, W. (1997). Homology of AFLP products in three mapping populations of barley. *Molecular and General Genetics*, 255: 311–321.

Werner, K., Friedt, W. and Ordon, F. (2005). Strategies for pyramiding resistance genes against the barley yellow mosaic virus complex BaMMV, aYMV, BaYMV-2). *Molecular Breeding*, 16: 45-55.

Willcox, M.C., Khairallah, M.M., Bergvinson, D., Crossa, J., Deutsch, J.A., Edmeades, G.O., González-de-León, D., Jiang, C., Jewell, D.C., Mihm, J.A., Williams, W. P. and Hoisington, D. (2002). Selection for resistance to southwestern corn borer using marker-assisted and conventional backcrossing. *Crop Science*, 42: 1516-1528.

Williams, J.G.K., Kubelik, A.R., Livak, K.J., Rafalski, J.A. and Tingey, S.V. (1990). DNA polymorphisms amplified by arbitrary primers are useful as genetic markers. *Nucleic Acids Research*, 18: 6531–6535.

Wu, W., Cao, L., Zhan, X., Shen, X., Chen, S. and Cheng, S. (2008). The selection and breeding of hybrid rice IIyou-8006 and its cultivation techniques. *Food Crop* 4:77–78.

Wu, X., Fei, Z., Zhang, J., Wei, L., Dong, H. and Zhou,.P. (2010). Breeding and Characteristic Analysis of a Photoperiod-sensitive Genetic Male Sterile Rice T16S with Pest-resistant *Bt Gene. Hubei Agriculture Science*, 49:2974–2977.

Xu, J.R., Zhao, X. And Dean, R.A. (2007). From genes to genome: a new paradigm for studying fungal pathogenesis in Magnaporthe oryzae. *Adv. Conet.*, 57: 175-218.

Yao, S., Chen, T., Zhang, Y., Zhu, Z., Zhao, L., Zhao, Q., Zhou, L. and Wang, C. (2010). Pyramiding of translucent endosperm mutant gene *Wx- mq* and rice stripe disease resistance gene *Stv-bi* by marker assisted selection in rice (Oryza sativa). *Chinese Journal of Rice Science*, 24:341–347.

Young, N.D. (1999). A cautiously optimistic vision for marker-assisted breeding. *Molecular Breeding*, 5: 505–51.

Yu, L.X., Abate, Z., Anderson, J.A., Bansal, U.K., Bariana, H.S., Bhavani, S., Dubcovsky, J., Lagudah, E.S., Liu, S.X., Sambasivam, P.K., Singh, R.P. and Sorrells, M.E. (2009). Developing and optimizing markers for stem rust resistance in wheat. Proceedings, oral papers and posters, 2009 Technical Workshop, Borlaug Global Rust Initiative, Cd. Obreg_on, Sonora, Mexico, 17-20 March, 2009. *Borlaug Global Rust Initiative.*

Zhan Xiao-deng., Zhou Hai-peng., Chai Rong-yao., Zhuang Jie-yun., Cheng Shi-hua. and Cao Li-yong. (2012). Breeding of R8012, a rice restorer line resistant to blast and bacterial blight through marker-assisted selection. *Rice Science*, 19 (1): 29-35.

Zhu, Y.L., Song, Q.J., Hyten, D.L., Van Tassell, C.P., Matukumalli, L.K., Grimm, D.R., Hyatt, S.M., Fickus, E.W., Young, N.D. and Cregan, P.B. (2003). Single-nucleotide polymorphisms in soybeam. *Genetics*, 163: 1123–1134.

Zhu, R., Huang, W., Hu, J., Liu, W. and Zhu, Y. (2013). Breeding and utilization of hybrid rice Lianyou 234 and NMS Line Bph68 resistance to brown plant hopper. *Natural Sciences Education*, 59: 24–28.

Qu, S., Liu, G., Zhou, B., Bellizzi, M., Zeng, L., Dai, L., Han, B. and Wang, G.L. (2006). The broad-spectrum blast resistance gene Pi9 encodes a nucleotide binding site-leucine-rich repeat protein and is a member of a multigene family in rice. *Genetics*, 172: 1901–1914.

2016, Recent Advances in Plant Stress Physiology *Pages* **413–444**
Editors: ***Praduman Yadav, Sunil Kumar and Veena Jain***
Published by: DAYA PUBLISHING HOUSE, NEW DELHI

Chapter 18

Transgenic Approaches for Abiotic Stress Tolerance in Plants

B. Usha Kiran* and M. Sujatha

Crop Improvement Section, Indian Institute of Oilseeds Research, Rajendranagar, Hyderabad – 500 030, India

ABSTRACT

Abiotic stresses such as drought, ooding or submergence, heat and cold spells, freezing and salinity are harmful factors which reduce growth and productivity of crops and are a serious threat to agriculture and environment. Understanding the mechanism(s) involved in the response of plants to abiotic stresses is the first step in development of stress tolerant crops. Transgenic plants that express/over-express genes regulating osmolytes, ion homeostasis, reactive oxygen species (ROS) scavenging, late embryogenesis abundant (LEA) and heat shock proteins, transcription factors and signal transduction showed improved tolerance to many abiotic stresses. Although, many genes associated with plant response(s) to abiotic stresses have been identified and used to generate stress tolerant plants, the success in producing commercially stress-tolerant crops is rather limited. Accurate and systematic phenotyping under conditions that mimic the natural field situation for a prolonged period is a major challenge limiting the success of abiotic stress tolerant transgenics. This chapter highlights the most significant achievements of the transgenic approach to improve tolerance to drought, salinity, chilling and heat stress and the limitations that hinder the commercialization of abiotic stress tolerant transgenic crops.

Keywords: *Transgenic, Abiotic stress, Transformation, Stress tolerance, Genetic engineering.*

Introduction

Abiotic stress is best defined as the stress exerted by any non-living factor in the environment on the optimal functioning of a plant. It is one of the most harmful

**Corresponding author.* E-mail: ushakirandor@gmail.com

factors concerning the growth, productivity of crops and worldwide agricultural loss, posing a substantial challenge in the face of an ever increasing world population (Gao *et al.*, 2007). The most commonly encountered abiotic stress factors are drought, ooding or submergence, temperature extremes (heat stress, cold spells and freezing) and soil ion content, with the latter typically in the form of increased salinity (sodium content) (Bohnert and Jensen, 1996). Often crops are exposed to multiple stresses, and the manner in which a plant senses and responds to different environmental factors appears to be overlapping. It was reported that abiotic stressors are at their most harmful stage when they occur together in various combinations (Palta *et al.*, 2006). In order to meet the increasing global agricultural demands under the worsening weather conditions due to climate change there is an immediate need to develop plants which can tolerate adverse conditions. Understanding the mechanisms involved in the response of plants to adverse environmental conditions is, without a doubt, the first step in the generation of crops with higher tolerance to stress. In the last few years, there has been a significant progress in understanding the mechanisms and processes underlying abiotic stress adaptation and defense in different plant species (Fraire-Velazquez, 2011; Hirayama and Shinozaki, 2010).

The conventional breeding or molecular breeding techniques in developing stress tolerant plants met with limited success. This is because of the lack of precise knowledge of the key genes underlying the QTLs (Quantitative Trait Loci) and the associated genetic drag from the donor parent. Therefore, the development of genetically engineered plants through the introduction and/or over expression of selected genes involved in the biosynthetic pathways associated with stress response seems to be a viable option. Also, it would be the only option when genes of interest originate from cross barrier species, distant relatives, or from non-plant sources. In recent years, many genes have been successfully introduced in plants to improve stress tolerance through transgenic approaches (Blumwald, 2000; Hong and Vierling, 2000; Bhatnagar *et al.*, 2008) despite the problems associated with them (Flowers, 2004). With the advent of functional genomics approaches, identification of candidate genes related to abiotic stress response has increased, thus fueling this approach for genetic engineering. In future, the combination of transcriptomic, proteomic and metabolomic analyses will also be useful for gene discovery in the genetic engineering of abiotic stress tolerance. In this chapter, we are focusing on the strategies and genes where the function of a gene is analyzed and validated using transgenic approach especially for abiotic stresses like drought, salinity, chilling and heat stress.

Single Genes

Osmoprotectants

The biosynthesis and accumulation of low molecular weight water-soluble compounds known as "compatible solutes" or "osmolytes", is an important adaptive mechanism to combat abiotic stresses and enables protection of cell turgor, restores water status of cells by maintaining cellular water potential, stabilizes membranes and/or scavenges ROS (McNeil *et al.*, 1999; Diamant *et al.*, 2001). These compatible solutes include amines (polyamines and glycine betaine), amino acids (proline), sugars (trehalose, fructan), and sugar alcohols (trehalose, mannitol and galactinol) (Rontein *et al.*, 2002). It is believed that osmoregulation would be the best strategy for abiotic stress tolerance, especially if osmoregulatory genes could be triggered in response to drought, salinity and high temperature (Bhatnagar-Mathur *et al.*, 2008). Over expression of these endogenous or ectopic genes in plants has therefore, been successfully used to synthesize compatible solutes in several target crops to improve tolerance to abiotic stresses (Taji *et al.*, 2002; Abebe *et al.*, 2003; Umezawa *et al.*, 2006; Kawakami *et al.*, 2008). This has resulted in a profusion of reports involving osmoprotectants such as glycine-betaine (Holmstrom, 2000; McNeil *et al.*, 2000) and proline (Yamada *et al.*, 2005). In addition, a number of "sugar alcohols" (mannitol, trehalose, myo-inositol and sorbitol) have also been targeted for the engineering of compatible-solute overproduction, thereby protecting the membrane and protein complexes during stress (Abebe *et al.*, 2003; Zhao *et al.*, 2000; Garg *et al.*, 2002; Cortina and Culianez, 2005; Gao *et al.*, 2000). Similarly, transgenics engineered for the over expression of polyamines have also been developed (Roy and Wu, 2002; Waie and Rajam, 2003; Capell *et al.*, 2004). Another noteworthy point is that increased levels of solutes often have enhanced tolerance for drought, salt and cold stress at the same time (Jang *et al.*, 2003; Park *et al.*, 2007), implying that genetic engineering by altering osmolytes is a fruitful approach for obtaining combined tolerance to different abiotic stresses.

Polyamines and Glycine Betaine

Polyamines are polycationic compounds of low molecular weight that are ubiquitously present in all living organisms were shown to be involved in the response to abiotic stress (Alcazar *et al.*, 2010). Several changes in concentrations of polyamines in plant cells take place while responding to the stressful conditions (Alcazar *et al.*, 2006b; 2010; Groppa and Benavides, 2008). Though the exact mechanism of involvement of polyamines during stressful conditions is not fully understood, studies are underway to examine the molecular mechanisms (Liu *et al.*, 2007; Alcazar *et al.*, 2010). In plant cells, the diamine putrescine (Put), triamine spermidine (Spd) and tetramine spermine (Spm) constitute the major polyamines. The genes which have been reported to be involved in the polyamine metabolism are arginine decarboxylase (ADC), ornithine decarboxylase (ODC), S-adenosylmethionine decarboxylase (SAMDC), or Spd synthase (SPDS). There are many reports on genetic engineering of these genes for polyamine metabolism in rice (Capell *et al.*, 2004; Prabhavathi and Rajam, 2007), potato (Kasukabe *et al.*, 2006), arabidopsis (Alcázar *et al.*, 2006a), tobacco (Wi *et al.*, 2006), cucurbits (Kasukabe *et al.*, 2004), apple (He *et al.*, 2008) to enhance the tolerance to different abiotic stresses.

Glycine betaine (GB), a fully N-methyl-substituted derivative of glycine, accumulates in the chloroplasts and other plastids of many species in response to abiotic stress and is considered to be the major osmolyte involved in cell membrane protection (Munns and Tester, 2008). It is one of the most studied osmoprotectant, the accumulation of which has been studied with respect to modifications of several metabolic steps. Plants over expressing the genes responsible for GB synthesis, choline monooxygenase (CMO), betaine aldehyde dehydrogenase (BADH), an NAD-dependent dehydrogenase have an enhanced tolerance to a wide range of abiotic stresses, including salt, heat, drought, and chilling, by the increased accumulation of GB in plant cells (Park, 2004; Quan, 2004; Yang *et al.*, 2005a) The crops which does not have the capacity to accumulate glycine betaine such as rice, tomato are the potential candidates for genetic engineering of betaine biosynthesis. BADH is a key enzyme in the biosynthesis of glycine betaine, and a BADH gene, originated from *Atriplex hortensis*, transformed into alfalfa was reported to increase the salt tolerance of the transgenic plants (Liu *et al.*, 2011). Transgenic bread wheat plants overexpressing a BADH gene, showed improved osmotic adjustment and antioxidative defense capacity which resulted in increased tolerance to drought and heat (Wang *et al.*, 2010). Transgenic plants expressing bacterial choline-oxidizing enzymes displayed increased tolerance to various stresses such as high salt concentrations and extreme temperatures (Sakamoto and Murata, 2000). Introducing the betA (encoding choline dehydrogenase) gene to maize (Zea mays) (Quan *et al.*, 2004) and wheat (He *et al.*, 2010), choline monooxygenase (CMO) in cotton resulted in improved yield under stressful conditions in the field. Genetic engineering of elite indica rice *codA* gene, which encodes choline oxidase improved salinity tolerance (Mohanty *et al.*, 2002). Similarly, tomato plants transformed with a bacterial *codA* gene targeted to the chloroplast were highly tolerant to chilling and oxidative stress (Park *et al.*, 2004).

Proline

Proline, known to occur widely in higher plants, acts as a store of carbon and nitrogen and functions as a molecular chaperone stabilizing the structure of proteins, and plays a vital role in plants, particularly in abiotic stress conditions (Ashraf and Foolad, 2007). Many genes are involved in the synthesis and degradation of proline under different stress situations. Two enzymes, pyrroline-5-carboxylate synthetase (P5CS) and pyrroline-5-carboxylate reductase (P5CR), are known to play a major role in the proline biosynthetic pathway (Ashraf and Foolad, 2007). Among these, P5CS has been mostly used for genetic engineering of crop plants for abiotic stress tolerance. Transgenic Arabidopsis with proline dehydrogenase antisense (Mani, 2002) or a knockout of this enzyme (Nanjo *et al.*, 2003) resulted in increased proline accumulation and better growth performance under osmotic stress (Simon-Sarkadi *et al.*, 2005). It has also been reported that in tobacco and petunia, over expression of *P5CS* gene from Arabidopsis and rice increased proline accumulation and enhanced drought and salt tolerance (Hong *et al.*, 2000; Yamada *et al.*, 2005). In tobacco, *P5CS* gene also showed freezing tolerance (Konstantinova *et al.*, 2002). Similarly, *P5CS* gene showed drought tolerance in common bean (Chen *et al.*, 2009), soybean (DeRonde *et al.*, 2004), wheat (Vendruscolo *et al.*, 2007) and osmotic stress in potato (Hmida-Sayari *et al.*, 2005), alfalfa (Armengaud *et al.*, 2004). On the other hand, transgenic chickpea (*Cicer arietinum*) expressing *P5CSF129A* constitutively showed a modest increase in transpiration efficiency, suggesting that enhanced proline had little bearing on yield in chickpea (Bhatnagar-Mathur, 2009). There are few reports in which mutated version of *P5CS* gene was used for stress tolerance (Hong *et al.*, 2000; Bhatnagar-Mathur *et al.*, 2009; Kumar *et al.*, 2010). Transgenic tobacco carrying mutagenized *P5CSA129A* gene showed drought tolerance by further elevating proline content under drought stress (Gubis *et al.*, 2007). Similarly, over-expression of the mutated form of *Vigna aconitifolia P5CS* in transgenic tobacco accumulated about 2-fold more proline than plants expressing wild-type and showed tolerance to osmotic stress. (Hong *et al.*, 2000). Over-expression of Arabidopsis d-OAT has been shown to enhance proline levels and increased the stress tolerance of rice and tobacco (Roosens *et al.*, 2002; Qu *et al.*, 2005).

Sugars and Sugar Alcohols

Sugars and "sugar alcohols" (mannitol, trehalose, myo-inositol and sorbitol) which play an essential role in osmotic adjustment have been targeted for the engineering of compatible-solute overproduction for abiotic stress tolerance in plants. Mannitol is a sugar alcohol that is accumulated during salt and water stress and thus alleviates abiotic stress. Transgenic tobacco plants carrying bacterial gene coding for mannitol-1-phosphate dehydrogenase (*mt1D*), resulted in mannitol accumulation in the cytoplasm and enhanced tolerance to salinity (Tarczynski, 1993). Overproducing mannitol in wheat, which does not synthesize mannitol normally, by constitutively expressing the mannitol-1- phosphate dehydrogenase (*mtl*D) gene resulted in improved tolerance to drought and salinity in terms of growth. However, under control conditions, growth was associated with sterility (Abebe, 2003). Sickler *et al.* (2007) showed that Arabidopsis plants transformed with mannose- 6-phosphate reductase (*M6PR*) gene from celery produced mannitol and grew normally under salt stress. When engineering these sugar polyols in plants, sometimes pleiotropic effects were observed. Plants with low levels of sorbitol developed normally, but necrotic lesions, reduced shoot and root growth and low fertility, were associated with an excessive sorbitol accumulation. Abnormal phenotypes with high levels of sorbitol accumulation were found in transgenic tobacco transformed with *stpd1*, a cDNA encoding sorbitol-6-phosphate dehydrogenase from apple (Sheveleva *et al.*, 1998). Trehalose (a rare, non-reducing sugar) acts as a compatible solute, protecting membranes and proteins and conferring desiccation tolerance on cells in the absence of water (Crowe *et al.*, 1984). It is specially accumulated in desiccation tolerant "resurrection plants" (Adams, 1990), and was genetically engineered in crop plants for abiotic stress tolerance. Transgenic rice (Garg *et al.*, 2002; Jang *et al.*, 2003) over expressing two *E. coli* trehalose biosynthetic genes (*otsA* and *otsB*) was shown to have improved tolerance to drought, freezing, salt and heat stress condition. The modest increase in trehalose levels in the transgenic plants resulted in a higher photosynthetic rate and a decrease in photo-oxidative damage during stress. Similarly, different groups have generated stress-tolerant plants by transferring microbial trehalose biosynthetic genes in tobacco (Lee *et al.*, 2003; Han *et al.*, 2005; Karim *et al.*, 2007), alfalfa (Suarez, 2009), Arabidopsis (Karim *et al.*, 2007; Miranda *et al.*, 2007) tomato (Cortina and Culianez-Macia, 2005), and potato (Goddijn *et al.*, 1997).

Transporter Genes

During abiotic stress, maintenance and re-establishment of water and ion homeostasis is extremely important for plant survival and growth to avoid water deficit and/or ion toxicity. Aquaporins are membrane proteins which are either tonoplast- (TIP) or plasma membrane- (PIP) localized and facilitate water, glycerol, small molecule, and gas transfer through membranes and, therefore, have a role in water and ion homeostasis upon stress (Bartels *et al.*, 2007). Ion transporters selectively transport ions and maintain them at physiologically relevant concentrations while Na^+/H^+ antiporters play a crucial role in maintaining cellular ion homeostasis under saline conditions (Serrano *et al.*, 1999). This has been and continues to be a major approach to improve salt tolerance in plants through genetic engineering, where the target is to achieve Na^+ excretion out of the root, or their storage in the vacuole (Bhatnagar-Mathur *et al.*, 2008). The Na^+/H^+ exchanger plays a key role in the regulation of cytosolic pH, cell volume and Na^+ homeostasis in plant cells (Wiebe *et al.*, 2001). Sodium transporters in plant cells have been extensively studied. Plant Na^+/H^+ antiporters have been isolated from Arabidopsis (*AtNHX1*, SOS1), (Gaxiola *et al.*, 1999; Shi *et al.*, 2000), rice plants (Fukuda *et al.*, 1999), halophytic plant *Atriplex gmelini* (Hamada *et al.*, 2001), forage plants (Li *et al.*, 2009; Tang *et al.*, 2010) and crystalline ice plant (Chauhan *et al.*, 2000). Several attempts have been made to produce transgenic plants tolerant to abiotic stress by increasing the cellular levels of proteins (such as vacuolar antiporter proteins) that control the transport functions. The Na^+/H^+ antiporter salt overly sensitive 1 (SOS1), is the only Na^+ efflux protein on the plasma membrane characterized so far in plants involved in Na^+ extrusion and long distance Na^+ transport (Shi *et al.*, 2003). Similarly, transgenic rice plants constitutively expressing the yeast Na^+/H^+ antiporter sodium2 (*SOD2*) gene, showed higher accumulation of cations in the shoots as compared to wild type plants (Zhao *et al.*, 2006).

Transgenic Arabidopsis that over-express *AtNHX1* accumulated abundant quantities of the transporter in the tonoplast and exhibited substantially enhanced salt tolerance (Quintero *et al.*, 2000). The seed/fruit yield and quality were not altered by the high salt levels in the transgenic brassica and tomato plants over-expressing *AtNHX1* (Zhang *et al.*, 2001; Zhang and Blumwald, 2001). It also improved salt tolerance in cotton (He *et al.*, 2005), wheat (Xue *et al.*, 2004), beet (*Beta vulgaris*) (Yang *et al.*, 2005[b]) and tall fescue (*Festuca arundinacea*) (Zhao *et al.*, 2007). The expression of *MsNHX1* isolated from *Medicago sativa* in transgenic Arabidopsis (An *et al.*, 2008) and *SsNHX1*, isolated from the halophyte *Salsola soda* in transgenic alfalfa (Li *et al.*, 2011[c]) under the stress-inducible *rd29A* promoter significantly enhanced salt-tolerance. Likewise, expression of the rice ortholog, *OsNHX1*, in rice (Apse *et al.*, 2003) and maize (Chen *et al.*, 2007) showed improved salt stress tolerance. Recently, over-expression of the Arabidopsis intracellular Na^+/H^+ antiporter *AtNHX5* resulted in enhanced salt and drought tolerance in rice seedlings and paper mulberry (Li *et al.*, 2011a; 2011b) Transgenic melon (Bordas *et al.*, 1997) and tomato (Gisbert *et al.*, 2000) plants expressing the *HAL1* gene involved in the regulation of K^+ showed a certain level of salt tolerance. Such transgenic lines also demonstrated better fruit yield under salt stress (Rus *et al.*, 2001). When the *AVP1* gene, a vacuolar H^+-pyrophosphatase gene from *A. thaliana* was over-expressed in alfalfa there was an improvement in salt and drought tolerance (Gaxiola *et al.*, 2001; Bao, *et al.*, 2009). Recently, heterologous expression of the potassium-dependent *TsVP* gene from the halophyte *T. halophyta* showed improved drought tolerance in maize (Li *et al.*, 2008).

Attempts have also been made to improve drought tolerance of plants by altering the expression of aquaporins (Jang *et al.*, 2007; Cui *et al.*, 2008; Miyazawa *et al.*, 2008; Zhang *et al.*, 2008c), although contrasting results were observed in transformed plants. Transgenic rice plants over-expressing a barley *HvPIP2;1*, showed more sensitivity (reduction in growth rate) to salinity stress (Katsuhara *et al.*, 2003). Similarly, Arabidopsis plants expressing the wild soybean (*Glycine soja*) tonoplast intrinsic protein (TIP), *GsTIP2;1*, showed more sensitivity to salt and dehydration presumably due to enhanced water loss of the transgenic plants (Wang *et al.*, 2011). In contrast, heterologous overexpression of rice *OsPIP-1* and *OsPIP2-2* in Arabidopsis resulted in improved salinity and dehydration tolerance (Guo *et al.*, 2006) and over-expression of wheat nodulin 26-like intrinsic proteins (*NIP*) gene, *TaNIP*, in Arabidopsis enhanced plants tolerance to abiotic stresses. Additionally, some aquaporins genes, such as *AtPIP1b* have been shown to diminish the drought tolerance capability of some plants, while others, such as the *Vicia faba PIP1*, *Panax ginseng PgTIP1*, *Brassica napus BnPIP1*, and *Brassica juncea BjPIP1*, were shown to improve drought tolerance (Aharon *et al.*, 2003; Yu *et al.*, 2005; Peng *et al.*, 2006; Cui *et al.*, 2008; Zhang *et al.*, 2008[b]).

Detoxifying Genes

Abiotic stress induced production of reactive oxygen species (ROS) such as O_2, O^{-2}, H_2O_2, and HO^- is another aspect of environmental stress in plants (Mittler, 2002) which causes oxidative damage to proteins, DNA and lipids (Miller *et al.*, 2010). To protect themselves against these toxic oxygen intermediates, plants have developed several antioxidant defense systems to scavenge these toxic compounds. Antioxidants involved in plant strategies to degrade ROS include: (1) enzymes such as catalase, glutathione peroxidase superoxide dismutase (SOD), ascorbate peroxidase (APX), and glutathione reductase; and (2) non-enzymes such as ascorbate, glutathione, carotenoids, and anthocyanins (Wang *et al.*, 2003[b]). Additional compounds, such as osmolytes, proteins (*e.g.* peroxiredoxin) and amphiphilic molecules (*e.g.* tocopherol), can also function as ROS scavengers (Noctor and Foyer, 1998). Enhancement of antioxidant defense in plants through genetic engineering can increase tolerance to different abiotic stresses.

A number of transgenic improvements for abiotic stress tolerance have been achieved through detoxification strategy. Catalase enzymes have potential to directly dismutate H_2O_2 into H_2O and O_2 and is indispensable for ROS detoxification during stress condition. Transgenic rice plants overexpressing *OsMT1a* showed increase in CAT activity and thus enhanced tolerance to drought (Yang *et al.*, 2009). As an antioxidant enzyme, glutathione peroxidase (GPX) reduces hydroperoxides in the presence of glutathione to protect cells from oxidative damage, including lipid peroxidation (Maiorino *et al.*, 1995). Gaber *et al.* (2006) generated transgenic Arabidopsis plants overexpressing 2GPX genes in cytosol (*AcGPX2*) and chloroplasts (*ApGPX2*) which showed enhanced tolerance to chilling with high light intensity, high salinity or drought. SOD is the first enzyme in the enzymatic antioxidative pathway which plays a major role in defending against severe abiotic stresses. Transgenic plants over expressing Mn-superoxide dismutase that mediates the conversion of O^{2-} to H_2O_2 in alfalfa (McKersie *et al.*, 1996), wheat (Gusta *et al.*, 2009) and potato (Waterer *et al.*, 2010) showed higher tolerance to different abiotic stresses and improved yields under field conditions. The over expression of Mn-SOD also showed enhanced tolerance against salt stress in tomato (Wang, 2007), arabidopsis (Wang *et al.*, 2004) and water stress in alfalfa (McKersie *et al.*, 1996). Overexpression of chloroplast Cu/Zn SOD showed increased resistance to oxidative stress caused by high light and low temperatures in transgenic tobacco (Gupta *et al.*, 1993a; 1993b) and potato plants (Perl *et al.*, 1993). Prashanth *et al.* (2008) reported that transformed rice plants with the cytosolic Cu/Zn SOD gene from the mangrove plant *Avicennia marina* showed better tolerance to salinity stress. In another study, transgenic alfalfa plants expressing either Mn-SOD or Fe-SOD had increased winter survival rates and yields (McKersie *et al.*, 1999; 2000). In plant cells, APXs are directly involved in catalyzing the reduction of H_2O_2 to water. Xu *et al.* (2008) transformed Arabidopsis plants with a pAPX gene from barley (*HvAPX1*) which resulted in enhanced tolerance to salt stress. It has also been noted that over-expression of APX in tobacco chloroplasts enhanced plant tolerance to salt and water deficit (Badawi *et al.*, 2004). Yang *et al.* (2009) correlated enhanced tolerance of *OsMT1a* over-expressing transgenic rice plants to drought stress with the increase in APX activity. Transgenic Arabidopsis plants over-expressing *OsAPXa* or *OsAPXb* exhibited increased salt tolerance. Lee and Jo (2004) introduced *BcGR1*, a Chinese cabbage gene that encodes cytosolic GR into tobacco plants which resulted in increased tolerance to oxidative stress. Transgenic tobacco plants over-expressing glutathione peroxidase showed increased tolerance to oxidative stress (Yoshimura, 2004) and chilling and salt stress (Roxas *et al.*, 1997). In tobacco, co-expression of three antioxidant enzymes, copper zinc superoxide dismutase (CuZnSOD), APX, and dehydroascorbate reductase (DHAR) resulted in a higher tolerance to salt stress (Lee, 2007). Overexpression of the aldehyde dehydrogenase *AtALDH3* gene in Arabidopsis conferred tolerance to drought and salt stress (Sunkar *et al.*, 2003). In Arabidopsis, overexpression of the *chyB* gene that encodes b-carotene hydroxylase (an enzyme active in the zeaxanthin biosynthetic pathway) resulted in a 2-fold increase in the pool of the xanthophyll cycle (Davison *et al.*, 2002).

LEA Proteins

LEA proteins are proteins in animals and plants that are associated with desiccation tolerance. They are called "Late embryogenesis abundant" because they are abundant during late embryogenesis than during mid-embryogenesis (Galau *et al.*, 1986). LEA proteins were initially discovered accumulating late in embryogenesis of cotton seeds (Dure *et al.*, 1981). These proteins accumulate not only in seeds but also in vegetative tissues after exogenous ABA treatment or during environmental stresses such as chilling, freezing, drought, and salinity stimulation (Mingdershih, 2008). Although the causes of LEA protein induction have not yet been determined, conformational changes in transcription factors or integral membrane proteins due to water loss have been suggested (Caramel *et al.*, 2009) and particularly mitochondrial membrane protection against dehydration damage (Tolleter *et al.*, 2010). According to the appearance of different sequence motifs/patterns or biased amino acid composition, plant LEA proteins have been separated into different groups (Battaglia *et al.*, 2008). These proteins are linked to cellular water deficit and are proposed to function in common mechanisms to combat the effects of dehydration stress in both plant and non-plant organisms (Kikawada *et al.*, 2006; Tyson *et al.*, 2007). LEA proteins are found in different tissue and cell types. They have been found to accumulate in cytoplasm and plastids and the very nature of their wide cellular distribution infers a protective function (Schneider *et al.*, 1993; Bartels and Salamini, 2001).

A single LEA gene introduced into model systems may provide systematic tool to measure the degree of stress tolerance. Barley *HVA1* gene (*LEA III*) introduced into rice was the first successful attempt to confer resistance to drought or salinity stresses (Xu *et al.*, 1996). Transgenic rice overexpressing *hva1* has been shown to confer dehydration tolerance via maintaining cell membrane stability and a higher Relative Water Content (RWC) and lesser reduction in plant growth under drought conditions (Rohila *et al.*, 2002; Babu *et al.*, 2004). Similarly, transgenic wheat harboring *HVA1* showed increased desiccation tolerance, biomass productivity, and water use efficiency under high salt, osmotic, or drought conditions via membrane protection (Patnaik and Khurana 2003; Sivamani *et al.*, 2000). In others plants like oat (Maqbool *et al.*, 2002; Oraby *et al.*, 2005), mulberry (Shalini Lal *et al.*, 2008) and bent grass (Fu *et al.*, 2007) also, correlation between HVA1 over expression and stress tolerance has been established.

Transgenic tobacco that ectopically expressed LEA IV genes, *BhLEA1* and *BhLEA2* from the resurrection plant *Boea hygrometrica* showed higher RWC of leaves, photosystem II activity and lower membrane permeability which confer dehydration tolerance through the protection of membrane proteins during drought stress (Xia *et al.*, 2009). Rapeseed (*Brassica napus*) *LEA III* gene, *ME-leaN4* introduced into Chinese cabbage (*Brassica campestris*) or lettuce (*Lactuca sativa* L.) resulted in improved drought tolerance (Park *et al.*, 2005a; b). The accumulation of a hot pepper hydrophobic LEA V protein, *CaLEA6*, in transgenic tobacco or chinese cabbage enhanced tolerance to dehydration and salt stresses (Kim *et al.*, 2005; Park *et al.*, 2003). The cold tolerance of transgenic tobacco was increased by the expression of a citrus gene encoding a LEA protein, CuCOR19 (Hara *et al.*, 2003). Likewise, the freezing tolerance of Arabidopsis was increased by the ectopic expression of the wheat gene *WCS19* (NDong *et al.*, 2002), the arabidopsis gene *COR15A* (Artus *et al.*, 1996) and the co-expression of the genes *RAB18* and *COR47*, and *XERO2* and *ERD10* (Puhakainen *et al.*, 2004). The freezing tolerance of strawberry leaves was enhanced by expression of the wheat dehydrin gene *WCOR410* (Houde *et al.*, 2004). On the other hand, the expression of two cold-induced LEA proteins from spinach and three desiccation-induced LEA proteins from *C. plantagineum* (Iturriaga *et al.*, 1992) in tobacco did not induce any significant changes in the freezing or drought tolerance of the respective transgenic plants. This may indicate either that not all LEA proteins make a significant contribution to plant stress tolerance, or that they need a particular background to function in, as suggested for transgenic strawberry plants (Houde *et al.*, 2004).

Heat Shock Proteins

Maintaining proteins in their functional conformations and preventing the aggregation of non-native proteins are particularly important for cell survival under stress. Moreover, many molecular chaperones are stress proteins and majority of them were originally identified as heat-shock proteins (Hsps) (Lindquist and Craig, 1988). Five major families of Hsps/chaperones are conservatively recognized, the Hsp70 (DnaK) family; the chaperonins (GroEL and Hsp60); the Hsp90 family; the Hsp100 (Clp) family; and the small Hsp (sHsp) family. The two best-studied families are the chaperonins and the Hsp70 family chaperones. Each major HSP family has a unique mechanism of action. Some promote the degradation of misfolded proteins, others bind to different types of folding intermediates and prevent them from aggregating (Hsp70 and Hsp60) and still another (Hsp100) promotes the reactivation of proteins that have already aggregated (Parsell and Lindquist, 1993; 1994). However, except for the sHsp family, relatively little focus has been given to the role of the many other H_sp_s/chaperones in plant response to abiotic stress and direct support for Hsp/chaperone function in plant abiotic stress tolerance is rather limited (Hong *et al.*, 2003; Burke, 2001; Wang *et al.*, 2003[a]). This is despite the fact that Hsps/chaperones are known to be expressed in plants not only when they experience high temperature stress but also in response to a wide range of other environmental severities such as water stress, salinity and osmotic, cold and oxidative stress (Boston *et al.*, 1996; Waters *et al.*, 1996).

The only direct evidence for the function of an individual HSP in stress tolerance in plants comes from transgenic carrot cell cultures and plants. Increased thermo tolerance by constitutive expression of HSP17.7 in transgenic carrot cell lines and other plants of carrot was observed due to modest changes in growth rates of tissue culture cells and electrolyte leakage of leaves (Malik *et al.*, 1999). Mitochondrial small HSP (*MT-sHSP*) gene from tomato is used for the production of thermotolerant transformed tobacco (Sanmiya, 2004). In addition, overexpression of transcription factors (*AtHSF1, AtHSF3*) responsible for HSPs constitutively enhanced thermo tolerance without requiring prior heat treatment in transgenic Arabidopsis (Lee *et al.*, 1995; Prändl *et al.*, 1998). The successful cases of incorporation of *HSP* genes to improve heat tolerance were reported in rice. Transgenic rice overexpressing an Arabidopsis *HSP101* gene showed enhanced thermo tolerance (Katiyar-Agarwal, 2003) and sHSP17.7 conferred both heat tolerance and UV-B resistance to rice plants (Murakami *et al.*, 2004).

Regulatory Genes

Transcription Factors

The transgenic approach using single gene encoding metabolic pathway had very limited success due to involvement of multiple pathways and the complexity of interactions between them for abiotic stress tolerance. On the other hand, targeting regulatory genes which affect multiple pathways could help to restore plant metabolic homeostasis during stress (Reguera *et al.*, 2012). Transcription factor (TF) genes play important role in the regulation of gene expression in response to abiotic stresses. Different families of TF such as ERF/AP2, HSF, bZIP, MYB, MYC, NFY, NAC, WRKY, Cys2His2, MADS-box and zinc-finger have been shown to regulate the expression of stress-responsive genes (Yamaguchi-Shinozaki and Shinozaki, 2006). TFs are powerful targets for genetic engineering of stress tolerance because overexpression of a single TF can lead to the up-regulation of a wide array of stress response genes that are controlled by the TF (Nakashima *et al.*, 2009). Individual members of the same family often respond differently to various stress stimuli. On the other hand, some stress responsive genes may share the same transcription factors, as indicated by the significant overlap of the gene expression profiles that are induced in response to different stresses (Seki *et al.*, 2002; Chen and Murata, 2002). Introduction of transcription factors in the ABA signaling pathway can also be a mechanism of genetic improvement of plant stress tolerance. In the Arabidopsis genome, 145 DREB/ERF related proteins are classified into five groups- the AP-2 subfamily, RAV subfamily, DREB subfamily, ERF subfamily, and others (Sakuma *et al.*, 2002).

Significant improvement of stress tolerance was found upon over expression of a single TF in engineered *Arabidopsis thaliana* plants. There have been numerous efforts in enhancing tolerance towards multiple stresses such as cold, drought and salt stress in crops other than the model plants like Arabidopsis, rice, tobacco and alfalfa. Over expression of CBF4 resulted in the expression of cold- and drought-induced genes under non-stress conditions, and the transgenic Arabidopsis plants showed more tolerance to freezing and drought conditions (Haake *et al.*, 2002). Another study showed that CBF4 expression was induced by salt, but not by drought, cold, or ABA (Sakuma *et al.*, 2002). Gilmour *et al.* (2000) reported not only freezing tolerance when Arabidopsis was transformed with CBF3 but also the elevated levels of proline and total soluble sugars. Similar observations of higher levels of soluble sugars and proline have been recorded during many CBF overexpression studies, which suggest that the use of constitutively expressed CBF/DREB genes may not be applicable to the development of crops with improved drought tolerance. It is thought that the use of stress-inducible promoters that have low expression levels under non-stress conditions could be used in conjunction with CBF genes to alleviate the retarded growth observed in CBF over-expression studies (Zhang *et al.*, 2004). Strong tolerance to freezing stress was observed in transgenic Arabidopsis plants that over express CBF1 (DREB1B) under the control of the CaMV 35S promoter without any negative effect on the development and growth characteristics (Jaglo-Ottosen *et al.*, 1998). DREB1A gene in transformed Arabidopsis showed increased tolerance to freezing, water and salinity stress with both constitutive promoter (CaMV 35S) and DRE containing promoter from the dehydration-induced gene (*rd29A*) (Kasuga *et al.*, 1999). Similarly, a dehydration responsive element (DRE) identified in *A. thaliana*, is also involved in gene expression of drought, low temperature and high salt stress conditions in many dehydration responsive genes like *rd29A* (Yamaguchi-Shinozaki and Shinozaki, 1994; Iwasaki *et al.*, 1997; Nordin *et al.*, 1993; Liu *et al.*, 1998). Over-expression of CBF1/DREB1B genes resulted in improved tolerance to the drought, salinity and temperature stress in crop plants such as rice, wheat, canola, tobacco, tomato, wheat and groundnut (Dubouzet *et al.*, 2003; Jaglo *et al.*, 2001; Kasuga *et al.*, 2004; Pellegrineschi *et al.*, 2004; Hsieh *et al.*, 2002; Behnam *et al.*, 2006; Bhatnagar-Mathur *et al.*, 2004; 2006). At the same time, the transgenic plants showed negative phenological abnormalities such as severe growth retardation under control conditions (Ito *et al.*, 2006). This problem was reduced when using more specific promoter, such as the ABA-inducible (*rd29a*) promoter (Yi *et al.*, 2010). Furthermore, when driven by three copies of an ABA-responsive complex (ABRC1) from the barley *HAV22* gene, transgenic tomato plants expressing CBF1 exhibited enhanced tolerance to chilling, water deficit, and salt stress, and maintained normal growth and yield under normal growing conditions when compared with control plants (Lee *et al.*, 2003). Transgenic rice over-expressing SNAC1 (STRESS-RESPONSIVE NAC 1) was found to significantly improve plant resistance to severe drought stress during reproductive and vegetative growth and was not associated with any negative phenotypic effects or yield penalty (Hu *et al.*, 2006). Recently, expression of sNAC10 under control of a root-specific promoter (RCc3) yielded more grain in the field under drought conditions (Jeong *et al.*, 2010). Other TFs involved in mediation of ABA-dependent and ABA-independent signal transduction and gene expression include NAC, WRKY, RING finger, and zinc-finger TFs (Seki *et al.*, 2003; Zhang *et al.*, 2004; Chen *et al.*, 2006). Transgenic rice seedlings, expressing OsWRKY11 under the control of a rice heat shock protein (HSP) promoter, HSP101, were shown to survive longer and lost less water under a short, severe drought treatment, than WT plants (Wu *et al.*, 2008[a]). A salt- and drought-induced RING-finger protein, SDIR1, was found to confer enhanced drought tolerance to tobacco and rice (Zhang *et al.*, 2008[b]). In tobacco, recently drought-tolerant gene, HDG11 which encodes a protein from the homeodomain (HD)-START TF family was found to confer drought resistance via enhanced root growth and decreased stomatal density when constitutively over-expressed in transgenic tobacco (Yu *et al.*, 2008). Transgenic maize constitutively expressing ZmNF-YB2 showed enhanced tolerance to severe drought stress in field trials. Under water-limiting conditions, transgenic plants displayed improved grain yield, as well as reduced wilting and lower leaf temperature (Nelson *et al.*, 2007). Jiang *et al.* (2008) recently characterized SAZ, an Arabidopsis gene from the SUPERMAN (SUP) family of plant-specific zinc-finger genes which was found to be rapidly down-regulated in response to drought and other abiotic stresses and *SAZ* gene knockouts resulted in elevated expression of the ABA-responsive genes *rd29B* and *rab18* under stressed and non-stressed conditions. This shows that gene knockouts and gene silencing may also be applicable to the development of crops with improved drought resistance.

Signal Transduction Genes

Signalling pathways are induced in response to environmental stresses, and molecular and genetic studies have revealed that these pathways involve a host of diverse responses (Chinnusamy *et al.*, 2004). It includes external stimulus to end response in plants, *i.e.*, receptor activation, then the generation of second messengers translating the primary external signal to intracellular signals. These messengers will be further stimulated by other signals in the cells resulting in the inspiration of downstream pathways which leads to physiological changes, such as guard cell closure, or to the expression of genes and resultant modification of molecular and cellular processes (Knight and Knight, 2001). Simply saying, the signal transduction pathway is a delicate cooperating process conducted by each single participator including receptor-coupled phospho-relay, phosphoionositol- induced Ca^{2+} changes, mitogen activated protein kinase (MAPK) cascade, and transcriptional activation of stress responsive genes (Xiong and Zhu, 2001) in which few are ABA dependent. In addition, protein posttranslational modifications and adapter or scaffold-mediated protein-protein interactions are also important in abiotic stress signal transduction (Xiong and Zhu, 2001). A number of signaling components are associated with the plant response to high temperature, freezing, drought and anaerobic stresses (Grover *et al.*, 2001). One of the advantages for manipulation of signaling factors through transgenic approach is that they can control a broad range of downstream events that can result in superior tolerance for multiple aspects (Umezawa *et al.*, 2006).

Receptor molecules are the initial stress signal perceivers and they convey the signal to the other molecules. Phospholipases involved in ABA-mediated signal transduction, catalyzes the release of lipid and lipid derived secondary messengers (Sang *et al.*, 2001). Transgenic maize plants transformed with *ZmPLC1* gene had increased photosynthetic activity, better recovery, and higher grain yield when subjected to drought stress (Wang *et al.*, 2008). MAPKs function *via* cascades which involve sequential phosphorylation of a kinase by its upstream kinase, which is important for transferring signals. Calcium-dependent protein kinases (CDPKs) and serine-threonine-type protein kinases are other protein kinases involved in stress signaling (Xiong and Ishitani 2006; Bartels *et al.*, 2007). The constitutive expression of a heterologous tobacco MAPKKK (Nicotiana PK1) gene in maize activates an oxidative signal cascade resulting in cold, heat, salinity and drought tolerance in transgenic plants (Kovtun *et al.*, 2000; Shou *et al.*, 2004). Similarly in alfalfa, P44MMK4 kinase was specifically activated under drought and cold treatment conditions, but not induced by high salt concentrations or heat shock. In contrast, an alfalfa mitogen-activated protein kinase (MAPK), *Msapk1* gene, harbouring a unique ankyrin repeat, was found to be highly regulated by salt stress (Baln Ka *et al.*, 2000; Chinchilla *et al.*, 2003). Over-expression of an osmotic-stress-activated protein kinase, SRK2C resulted in a higher drought tolerance in *A. thaliana*, which coincided with the upregulation of stress-responsive genes (Umezawa *et al.*, 2004). However, suppression of signaling factors could also effectively enhance tolerance to abiotic stress (Wang *et al.*, 2005). This hypothesis was based on previous reports indicating that a and b subunits of farnesyl transferase ERA1 functions as a negative regulator of ABA signaling (Cutler *et al.*, 1996). Ca^{2+} sensors are important for coupling extracellular signaling to intercellular responses and comprise calmodulin (CaM), CaM-related proteins and calcineurin B-like proteins (CBL) (Sneddon and Fromm, 2001). The calcineurin B–like (CBL) protein family represents a unique group of calcium sensors in plants. Pardo *et al.* (1998) developed salt stress-tolerant transgenic plants by over-expressing calcineurin (a Ca^{2+}/Calmodulin dependent protein phosphatase). *CBL1*-overexpression in transgenic Arabidopsis showed enhanced tolerance to salt and drought but reduced tolerance to freezing. (Yong *et al.*, 2003). Transgenic plants over-expressing the transgenes calcineurin B-like protein-interacting protein kinases (CIPKs) *OsCIPK genes* in rice showed significant improvement in the tolerance to cold, drought, and salt stress (Xiang, 2007). Transgenic tobacco plants produced by altering stress signaling through functional reconstitution of activated yeast calcineurin not only opened-up new routes for study of stress signaling, but also for engineering transgenic crops with enhanced stress tolerance (Grover *et al.*, 1999).

Limitations of Transgenic for Abiotic Stress Tolerance

Employing transgenic engineering to instill abiotic stress tolerances in plants stands out as a promising approach to develop stress resistant plants. However, the extreme challenge involved in translating these methods from research labs to farmers proves to be a major limitation (Mittler and Bluwald, 2010). Besides, several major hurdles which includes lack of in-depth scientific knowledge and understanding on the abiotic stress tolerances, availability of sequence information and transformation protocols, incomplete knowledge on genetic diversity between model species and crops, multigenic characters, and systematic phenotyping limits the potential of transgenic engineering in developing a completely foolproof abiotic stress resistant plants (Vinocur, 2005; Yamaguchi, 2005). The stress tolerance achieved by engineering single genes in plants is limited. Highly efficient and sustainable stress tolerances can only be attained by compounding action of multiple genes involved in various cellular processes; however, these processes are not fully understood. With the advent of genomics, transcriptomics and proteomics, it became clear that the response to abiotic stresses is regulated at transcriptional level. Posttranscriptional control, translation, posttranslational, and epigenetic effects play an important role in the stress regulation. Another important limitation for abiotic stress tolerance is the method of evaluation of stress itself. Most of the transgenic engineered plants are studied and evaluated for the stress under controlled laboratory conditions; in reality the laboratory conditions are not ideal or akin to environmental conditions in which the crops grow. In addition, the time frame during which the abiotic stresses are evaluated is also important and is recommended to study the effects of abiotic stress for a longer period on plant growth and yield. Improved and systematic phenotyping of plants will greatly contribute to overcome the limitations associated with the methods to evaluate stress tolerance (Bhatnagar -Mathur, 2009).

Conclusions

Abiotic stress tolerance is a complex trait involving genes regulating complex networks which controls morphological, physiological, biochemical, and molecular changes. Although research on transgenics for stress tolerance using single genes, transcription factors, combination of genes in the regulatory pathways is present, the number of genes that were successfully engineered for stress tolerance to improve crop performance is low and certain areas are still unexplored. Most of the basic research findings on abiotic stresses originate from model species which have to be proven in other crop species. The role of miRNA, small peptides and epigenetics during regulation and abiotic stress signaling is yet to be unexplored. Stress response by plants is complex in nature, modifying a functional process or even multiple interconnected functional processes instead of one single function, appears to be the way forward. However, this approach is often limited by the lack of knowledge on the network of molecular mechanisms underpinning stress tolerance. Efforts are now being made to develop computational models to study abiotic stress tolerance in plants. By including modeling in biological studies, our understanding of the plants' response to abiotic stress will increase which will further accelerate the transgenic approach towards abiotic stress tolerance in all the major crops.

References

Abebe, T., Guenzi, A.C., Martin, B and Cushman, J.C., 2003. Tolerance of mannitol-accumulating transgenic wheat to water stress and salinity. *Plant Physiol.*, 131: 1748–1755.

Adams, R.P., Kendall, E and Kartha, K.K., 1990. Comparison of free sugars in growing and desiccated plants of *Selaginella lepidophylla. Biochem. Syst. Ecol.*, 18: 107–110.

Aharon, R., Shahak, Y., Wininger, S., Bendov, R., Kapulnik, Y and Galili, G., 2003. Overexpression of a plasma membrane aquaporin in transgenic tobacco improves plant vigor under favorable growth conditions but not under drought or salt stress. *Plant Cell*, 15: 439–447.

Alcazar, R., Altabella, T., Marco, F., Bortolotti, C., Reymond, M., Koncz, C., Carrasco, P and Tiburcio, A. F., 2010. Polyamines: molecules with regulatory functions in plant abiotic stress tolerance. *Planta.*, 231: 1237–1249.

Alcázar, R., Cuevas, J.C., Patrón, M., Altabella, T and Tiburcio, A.F., 2006[a]. Abscisic acid modulates polyamine metabolism under water stress in *Arabidopsis thaliana. Plant Physiol.*, 128: 448–455.

Alcazar, R., Marco, F., Cuevas, J.C., Patrón, M., Ferrando, A., Carrasco, P., Tiburcio, A.F and Altabella, T., 2006[b]. Involvement of polyamines in plant response to abiotic stress. *Biotechnol. Lett.*, 28: 1867–1876.

An, B.Y., Luo, Y., Li, J.R., Qiao, W.H., Zhang X.S., and Gao. X.Q., 2008. Expression of a vacuolar Na^+/H^+ antiporter gene of alfalfa enhances salinity tolerance in transgenic Arabidopsis. *Acta Agronomica Sinica.*, 34(4): 557-564.

Apse, M.P., Sottosanto, J.B and Blumwald, E., 2003. Vacuolar cation/H^+ exchange, ion homeostasis, and leaf development are altered in a T-DNA insertional mutant of AtNHX1, the Arabidopsis vacuolar Na^+/H^+ antiporter. *Plant J.*, 36: 229–239.

Armengaud, P., Thiery, L., Buhot, N., Grenierde March, G and Savouré, A., 2004. Transcriptional regulation of proline biosynthesis in *Medicago truncatula* reveals developmental and environmental specific features. *Physiologia Plantarum.*, 120: 442-450.

Artus, N. N., Uemura, M., Steponkus, P. L., Gilmour, S. J., Lin, C and Thomashow, M. F., 1996. Constitutive expression of the coldregulated Arabidopsis thaliana COR15a gene affects both chloroplast and protoplast freezing tolerance. *Proc. Natl. Acad. Sci. USA.*, 93: 13404–13409.

Ashraf, M and Foolad, M., 2007. Roles of glycine betaine and proline in improving plant abiotic stress resistance. *Env. Exptl. Bot.*, 59: 206-216.

Babu, R.C., Zhang, J., Blum, A., Ho, T.H.D., Wu, R and Nguyen, H.T., 2004. *HVA1*, a LEA gene from barley confers dehydration tolerance in transgenic rice (*Oryza sativa* L.) via cell membrane protection. *Plant Sci.*, 166: 855–862.

Badawi, G.H., Yamauchi, Y., Shimada, E., Sasaki, R., Kawano, N and Tanaka, K., 2004. Enhanced tolerance to salt stress and water deficit by over expressing superoxide dismutase in tobacco (*Nicotiana tabacum*) chloroplasts. *Plant Sci.*, 166: 919-928.

Baln Ka, F., Ovecka M and Hirt, H., 2000. Salt stress induces changes in amounts and localization of the mitogen-activated protein kinase SIMK in alfalfa roots. *Protoplasma.*, 212: 262-267.

Bao, A.K., Wang, S.M., Wu, G.Q., Xi, J.J., Zhang J.L and Wang, C.M., 2009. Overexpression of the Arabidopsis H^{+}-PPase enhanced resistance to salt and drought stress in transgenic alfalfa (*Medicago sativa* L.). *Plant Sci.*, 176: 232-240.

Bartels, D., Phillips, J and Chandler, J., 2007. Desiccation tolerance: gene expression, pathways, and regulation of gene expression. In: *Plant Desiccation Tolerance*, (Ed.) M.A. Jenks, A.J. Wood, Blackwell, Ames, IA, 115–137.

Battaglia, M., Olvera-Carrillo, Y., Garciarrubio, A., Campos, F and Covarrubias, A. A., 2008. The enigmatic LEA proteins and other hydrophilins. *Plant Physiol.*, 148: 6–24.

Behnam, B., Kikuchi, A., Celebi-Toprak, F., Yamanaka, S., Kasuga, M., Yamaguchi-Shinozaki, K. and Watanabe, K.N., 2006. The Arabidopsis DREB1A gene driven by the stress-inducible rd29A promoter increases salt-stress tolerance in proportion to its copy number in tetrasomic tetraploid potato (*Solanum tuberosum*). *Plant Biotechnol.*, 23: 169–177.

Bhatnagar-Mathur, P, Devi, M. J., Reddy, D. S., Vadez, V., Yamaguchi-Shinozaki, K., Sharma, K. K., 2006. Overexpression of *Arabidopsis thaliana* DREB1A in transgenic peanut (*Arachis hypogaea* L.) for improving tolerance to drought stress (poster presentation). In: Arthur M. Sackler Colloquia on "From Functional Genomics of Model Organisms to Crop Plants for Global Health", *Natl. Acad Sci, Washington, DC.*

Bhatnagar-Mathur, P., Sharma, K.K., Devi, M.J., Serraj, R., Yamaguchi-Shinozaki, K and Vadez, V., 2004. *Evaluation of transgenic groundnut lines under water limited conditions. Intl. Arachis Newslett.*, 24: 33-35.

Bhatnagar-Mathur, P., Vadez V and Sharma, K.K., 2008. Transgenic approaches for abiotic stress tolerance in plants: retrospect and prospects. *Plant Cell Rep.*, 27: 411-424.

Bhatnagar-Mathur, P., Devi, M.J., Vadez, V and Sharma, K. K., 2009. *Differential antioxidative responses in transgenic peanut bear no relationship to their superior transpiration efficiency under drought stress. J. Plant Physiol.*, 166: 1207-1217.

Blumwald, E., Aharon, G.S and Apse, M.P., 2000. Sodium transport in plant cells. *Biochimica et Biophysica Acta (BBA)-Biomembranes.*, 1465: 140-151.

Bohnert, H.J and Jensen, R.G., 1996. Strategies for engineering water-stress tolerance in plants. *Trends Biotechnol.*, 14: 89–97.

Bordas, M., Montesinos, C., Debauza, M., Salvador, A., Roig, L.A., Serrano, R and Moreno, V., 1997. Transfer of the yeast salt tolerance gene HAL1 to *Cucumis melo* L. cultivars and *in vitro* evaluation of salt tolerance. *Transgenic Res.*, 6: 41–50.

Boston, R.S., Vittanen, P.V and Vierling, E., 1996. Molecular chaperones and protein folding in plants. *Plant Mol. Biol.*, 32: 191–222.

Burke, J.J., 2001. Identification of genetic diversity and mutations in higher plant acquired thermotolerance. *Physiol. Plant.*, 112: 167–170.

Capell, T., Bassie, L and Christou, P., 2004. Modulation of the polyamine biosynthetic pathway in transgenic rice confers tolerance to drought stress. *Proc. Natl. Acad. Sci. USA.*, 101: 9909–9914.

Caramelo, J.J and Iusem, N.D., 2009. When cells lose water: Lessons from biophysics and molecular biology. *Progress Biophysics Mol. Biol.*, 99(1): 1–6.

Chauhan, S., Forsthoefel, N., Ran, Y., Quigley, F., Nelson, D.E and Bohnert, H.J., 2000. Na^+/myo-inositol symporters and Na^+/H^+-antiport in *Mesembryanthemum crystallinum*. *Plant J.*, 24: 511–522.

Chen, J.R., Lü, J.J., Wang, T.X., Chen, S.Y and Wang, H.F., 2009. Activation of a DREbinding transcription factor from *Medicago truncatula* by deleting a Ser/Thr-rich region. *In Vitro Cell Dev. Biol. Plant.*, 45: 1-11.

Chen, M., Chen, Q.J., Niu, X.G., Zhang, R., Lin, H.Q., Xu, C.Y., Wang, X.C., Wang, G.Y and Chen, J., 2007. Expression of OsNHX1 gene in maize confers salt tolerance and promotes plant growth in the field, Plant. *Soil Environ.*, 53: 490–498.

Chen, T.H.H. and Murata, N., 2002. Enhancement of tolerance of abiotic stress by metabolic engineering of betaines and other compatible solutes. *Curr. Opin. Plant Biol.*, 5: 250–257.

Chen, Y., Yang, X., He, K., Liu, M., Li, J., Gao, Z., Lin, Z., Zhang, Y., Wang, X., Qiu, X., Shen, Y., Li, Z., Deng, X., Luo, J., Deng, X.W., Chen, Z., Gu, H. and Qu, L.J., 2006. The MYB transcription factor superfamily of *Arabidopsis*: Expression analysis and phylogenetic comparison with the rice MYB family. *Plant Mol Biol.*, 60: 107-124.

Chinchilla, D., Merchan, F., Megias, M., Kondorosi, A., Sousa, C and Crespi, M., 2003. Ankyrin protein kinases: a novel type of plant kinase gene whose expression is induced by osmotic stress in alfalfa. *Plant Mol. Biol.*, 51: 555-566.

Chinnusamy, V., Schumaker, K and Zhu, J., 2004. Molecular genetic perspectives on cross-talk and specificity in abiotic stress signaling in plants. *J. Exptl. Bot.*, 55: 225-236.

Cortina, C and Culianez-Macia, F., 2005. Tomato abiotic stress enhanced tolerance by trehalose biosynthesis. *Plant Sci.*, 169: 75–82.

Crowe, J.H., Crowe, L.M and Chapman, D., 1984. Preservation of membranes in anhydrobiotic organisms: the role of trehalose. *Science.*, 223: 701–703.

Cui, X.H., Hao, F.S., Chen, H., Chen, J and Wang, X.C., 2008. Expression of the *Vicia faba VfPIP1*gene in *Arabidopsis thaliana* plants improves their drought resistance. *J. Plant Res.*, 121: 207–214.

Cutler, S., Ghassemian, M., Bonetta, D., Cooney, S and McCourt, P., 1996. A protein farnesyl transferase involved in abscisic acid signal transduction in *Arabidopsis*. *Science.*, 273: 1239-1241.

Davison, P.A., Hunter, C.N and Horton, P., 2002. Overexpression of β-carotene hydroxylase enhances stress tolerance in Arabidopsis. *Nature.*, 418: 203–206.

DeRonde, J.A., Cressc, W.A., Krügerd, G.H.J., Strasserd, R.J and Stadenb, J.V., 2004. Photosynthetic response of transgenic soybean plants, containing an *Arabidopsis P5CR* gene, during heat and drought stress. *J. Plant Physiol.*, 161: 1211–1224.

Diamant, S., Eliahu, N., Rosenthal, D and Goloubinoff, P., 2001. Chemical chaperones regulate molecular chaperones *in vitro* and in cells under combined salt and heat stresses. *J. Biol. Chem.*, 276: 39586–39591.

Dubouzet, J.G., Sakuma, Y., Ito, Y., Kasuga, M., Dubouzet, *E.G.*, Miura, S., Seki, M., Shinozaki, K. and Yamaguchi-Shinozaki, K., 2003. *OsDREB* genes in rice, *Oryza sativa* L., encode transcription activators that function in drought-, high-salt- and cold-responsive gene expression. *Plant J.*, 33: 751–763.

Dure III, L., 1993. A repeating 11-mer amino acid motif and plant desiccation. *Plant J.*, 3: 363–369.

Flowers, T.J., 2004. Improving crop salt tolerance. *J. Exptl. Bot.*, 55: 307–319.

Fraire-Velázquez, S., Rodríguez-Guerra, R and Sánchez-Calderón, L., 2011. Abiotic and biotic stress response cross talk plants. In: *Abiotic Stress Response in Plants—Physiological, Biochemical and Genetic Perspectives*, (Ed.) A. Shanker, InTech, Rijeka, Croatia, 3–26.

Fu, D., Huang, B., Xiao, Y., Muthukrishnan, S and Liang, G. H., 2007. Overexpression of barley *HVA1* gene in creeping bent grass for improving drought tolerance. *Plant Cell Rep.*, 26: 467–477.

Fukuda, A., Nakamura, A and Tanaka, Y., 1999. Molecular cloning and expression of the Na^+/H^+ exchanger gene in *Oryza sativa*. *Biochem. Biophys. Acta.*, 1446: 149–155.

Gaber, A., Yoshimura, K., Yamamoto, T., Yabuta, Y., Takeda, T., Miyasaka, H., Nakano, Y and Shigeoka, S., 2006. Glutathione peroxidase-like protein of Synechocystis PCC 6803 confers tolerance to oxidative and environmental stresses in transgenic Arabidopsis. *Physiol. Plant.*, 128: 251–262.

Galau G.A., Hughes D.W and Dure L III., 1986. Abscisic acid induction of cloned cotton late embryogenesis-abundant (Leu) mRNAs. *Plant Mo1. Biol.*, 7: 155-170

Gao, Ji-Ping., Dai-Yin Chao and Hong-Xuan Lin., 2007. Understanding abiotic stress tolerance mechanisms: Recent studies on stress response in rice. *J. Intgr. Plant Biol.*, 49: 742–750.

Gao, M., Sakamoto, A., Miura, K., Murata, N., Sugiura, A and Tao, R., 2000. Transformation of Japanese persimmon (*Diospyros kaki Thunb.*) with a bacterial gene for choline oxidase. *Mol. Breed.*, 6: 501–510.

Garg, A.K., Kim, J.K., Owens, T.G., Ranwala, A.P., Choi, Y.D., Kochian, L.V and Wu, R.J., 2002. Trehalose accumulation in rice plants confers high tolerance levels to different abiotic stresses. *Proc Natl Acad Sci USA.*, 99: 15898-15903.

Garg. A.K., Ju-Kon, K., Owens, G.T., Anil, P.R., Yang, Do Choi., Leon, V.K and Ray, J.W., 2002. Trehalose accumulation in rice plants confers high tolerance levels to different abiotic stress. *Proc. Natl. Acad. Sci. USA.*, 99(25): 15898-15903.

Gaxiola, R.A., Rao, R., Sherman, A., Grisafi, P., Alper, S.L and Fink, G.R., 1999. The *Arabidopsis thaliana* proton transporters, AtNhx1 and Avp1, can function in cation detoxification in yeast. *Proc. Natl. Acad. Sci. USA.*, 96: 1480–1485.

Gilmour, S.J., Sebolt, A.M., Salazar, M.P., Everard, J.D. and Thomashow, M.F., 2000. Overexpression of the Arabidopsis CBF3 transcriptional activator mimics multiple biochemical changes associated with cold acclimation. *Plant Physiol.*, 124: 1854–1865.

Gisbert, C., Rus, A.M., Bolar n, M.C., Lopez-Coronado, J.M., Arrillaga, I., Montesinos, C., Caro, M., Serrano, R and Moreno, V., 2000. The yeast HAL1 gene improves salt tolerance of transgenic tomato. *Plant Physiol.*, 123: 393–402.

Goddijn, O.J.M., Verwoerd, T.C., Voogd, E., Krutwagen, R.W.H.H., de Graaf, P.T.H.M., Poels, J., VanDun, K., Ponstein, A.S., Damm, B and Pen, J., 1997. Inhibition of trehalase activity enhances trehalose accumulation in transgenic plants. *Plant Physiol.*, 113: 181–190.

Goyal, K., Pinelli, C., Maslen, S.L., Rastogi, R.K., Stephens, E and Tunnacliffe, A., 2005. Dehydration-regulated processing of late embryogenesis abundant protein in a desiccation-tolerant nematode. *FEBS Lett.*, 579: 4093–4098.

Groppa, M.D and Benavides, M.P., 2008. Polyamines and abiotic stress: recent advances. *Amino Acids.*, 34: 35–45.

Grover, A., Kapoor, A., Satya Lakshmi, O., Agrawal, S., Sahi, C., Katiyar- Agarwal, S., Agarwal M and Dubey, H., 2001. Understanding molecular alphabets of the plant abiotic stress responses. *Curr. Sci.*, 80: 206–216.

Grover, A., Sahi, C., Sanan, N and Grover, A., 1999. Taming abiotic stresses in plants through genetic engineering: current strategies and perspective. *Plant Sci.*, 143: 101–111.

Gubis, J., Vankova, R., Cervena´, V., Dragunova, M. M., Hudcovicova, M., Lichtnerova, H. T., Dokupil, T and Jurekova, Z., 2007. Transformed tobacco plants with increased tolerance to drought. *South Afr J Bot.*, 73: 505–511.

Guo, L., Wang, Z.Y., Lin, H., Cui, W.E., Chen, J., Liu, M., Chen, Z.L., Qu, L.J and Gu, H., 2006. Expression and functional analysis of the rice plasma-membrane intrinsic protein gene family. *Cell Res.*, 16: 277–286.

Gupta, A.S., Heinen, J.L., Holaday, A.S., Burkem J.J and Allenm R.D., 1993[a]. Increased resistance to oxidative stress in transgenic plants that overexpress chloroplastic Cu/Zn superoxide dismutase. *Proc. Natl. Acad. Sci. USA.*, 90: 1629–1633.

Gupta, A.S., Webb, R.P., Holaday, A.S and Allen, R.D., 1993[b]. Overexpression of superoxide dismutase protects plants from oxidative stress. *Plant Physiol.*, 103: 1067–1073.

Gusta, L., Benning, N., Wu, G., Luo, X., Liu, X., Gusta, M and McHughen, A., 2009. Superoxide dismutase: an all-purpose gene for agri-biotechnology. *Mol. Breed.*, 24: 103–115.

Haake, V., Cook, D., Riechmann, J.L., Pineda, O., Thomashow, M.F. and Zhang, J.Z., 2002. Transcription factor CBF4 is a regulator of drought adaptation in *Arabidopsis*. *Plant Physiol.*, 30: 639-648.

Hamada, A., Shono, M., Xia, T., Ohta, M., Hayashi, Y., Tanaka, A and Hayakawa, T., 2001. Isolation and characterization of a Na^+/H^+ antiporter gene from the halophyte *Atriplex gmelini*. *Plant Mol. Biol.*, 46: 35–42.

Han, S.E., Park, S.R., Kwon, H.B., Yi, B.Y., Lee, G.B and Byun, M.O., 2005. Genetic engineering of drought-resistant tobacco plants by introducing the trehalose phosphorylase (TP) gene from *Pleurotus sajorcaju*. *Plant Cell Tissue Organ Cult.*, 82: 151–158.

Hara, M., Terashima, S., Fukaya, T and Kuboi, T., 2003. Enhancement of cold tolerance and inhibition of lipid peroxidation by citrus dehydrin in transgenic tobacco. *Planta.*, 217: 290–298.

He, C., Yang, A., Zhang, W., Gao, Q and Zhang, J., 2010. Improved salt tolerance of transgenic wheat by introducing betA gene for glycine betaine synthesis. *Plant Cell Tissue Org Cult.*, 101: 65–78.

He, L., Ban, Y., Inoue, H., Matsuda, N., Liu, J and Moriguchi, T., 2008. Enhancement of spermidine content and antioxidant capacity in transgenic pear shoots overexpressing apple spermidine synthase in response to salinity and hyperosmosis. *Phytochem.*, 69: 2133–2141.

Hirayama, T and Shinozaki, K., 2010. Research on plant abiotic stress responses in the postgenome era: past, present and future. *Plant J.*, 61: 1041–1052.

Hmida-Sayari, A., Gargouri-Bouzid, R., Bidani, A., Jaoua, L., Savoure, A and Jaoua, S., 2005. Overexpression of D1-pyrroline-5-carboxylate synthetase increases proline production and confers salt tolerance in transgenic potato plants. *Plant Sci.*, 169: 746–752.

Hong, S.W and Vierling, E., 2000. Mutants of *Arabidopsis thaliana* defective in the acquisition of tolerance to high temperature stress. *Proc. Natl. Acad. Sci. USA.*, 97: 4392–4397.

Hong, S.W., Lee U and Vierling E., 2003. Arabidopsis hot mutants define multiple functions required for acclimation to high temperatures. *Plant Physiol.*, 132: 757–767.

Hong, Z., Lakkineni, K., Zhang, K and Verma, D.P.S., 2000. Removal of feedback inhibition of D1-pyrroline-5-carboxylate synthetase results in increased proline accumulation and protection of plants from osmotic stress. *Plant Physiol.*, 122: 1129–1136.

Houde, M., Sylvain, D., NDong, D and Sarhan, F., 2004. Overexpression of the acidic dehydrin WCOR410 improves freezing tolerance in transgenic strawberry leaves. *Plant Biotechnol. J.*, 2: 381–387

Hsieh, T.H., Lee, J.T., Yang, P.T., Chiu, L.H., Charng, Y.Y., Wang, Y.C. and Chan, M.T., 2002. Heterology expression of the Arabidopsis C-repeat/dehydration response element binding factor 1 gene confers elevated tolerance to chilling and oxidative stresses in transgenic tomato. *Plant Physiol.*, 129: 1086–1094.

Hu, H., Dai, M., Yao, J., Xiao, B., Li, Xianghua., Zhang, Q. and Xiong, L., 2006. Over expressing a NAM, ATAF, and CUC (NAC) transcription factor enhances drought resistance and salt tolerance in rice. *Proc. Natl. Acad.Sci. USA.*, 103: 12987–12992.

Ito, Y., Katsura, K., Maruyama, K., Taji, T., Kobayashi, M., Seki, M., Shinozaki, K. and Yamaguchi-Shinozaki, K., 2006. Functional analysis of rice DREB1/CBF-type transcription factors involved in cold-responsive gene expression in transgenic rice. *Plant Cell Physiol.*, 47: 141–153.

Iturriaga, G., Schneider, K., Salamini, F and Bartels, D., 1992. Expression of desiccation-related proteins from the resurrection plant in transgenic tobacco. *Plant Mol Biol.*, 20: 555–558.

Iwasaki, T., Kiyosue, T., Yamaguchi-Shinozaki, K. and Shinozaki, K., 1997. The dehydration-inducible *rd17* (*cor47*) gene and its promoter region in *Arabidopsis thaliana*. *Plant Physiol.*, 115: 1287.

Jaglo, K.R., Kleff, S., Amundsen, K.L., Zhang, X., Haake, V., Zhang, J.Z., Deits, T. and Thomashow, M.F., 2001. Components of the Arabidopsis C-Repeat/ dehydration-responsive element binding factor cold-response pathway are conserved in *Brassica napus* and other plant species. *Plant Physiol.*, 127: 910–917.

Jaglo-Ottosen, K.R., Gilmour, S. J., Zarka, D.G., Schabenberger, O. and Thomashow, M.F., 1998. Arabidopsis *CBF1* over expression induces COR genes and enhances freezing tolerance. *Science.*, 280: 104–106.

Jang, I C., Oh, S.J., Seo, J.S., Choi, W.B., Song, S.I., Kim, C.H., Kim, Y.S., Seo, H.S., Choi, Y.D., Nahm, B.H and Kim, J.K., 2003. Expression of a bifunctional fusion of the *Escherichia coli* genes for trehalose-6-phosphate synthase and trehalose-6-phosphate phosphate in transgenic rice plants increases trehalose accumulation and abiotic stress tolerance without stunting growth. *Plant Physiol.*, 131: 516–524.

Jang, J.Y., Lee, S.H., Rhee, J.Y., Chung, G.C., Ahn, S.J and Kang, H., 2007. Transgenic Arabidopsis and tobacco plants overexpressing an aquaporin respond differently to various abiotic stresses. *Mol. Biotechnol.*, 40: 280–292.

Jeong, J.S., Kim, Y.S., Baek, K.H., Jung, H., Ha, S.H., Do Choi, Y., Kim, M., Reuzeau, C and Kim, J.K., 2010. Root-specic expression of OsNAC10 improves drought tolerance and grain yield in rice under eld drought conditions. *Plant Physiol.*, 153: 185–197.

Jiang, C.J., Aono, M., Tamaoki, M., Maeda, S., Sugano, S., Mori, M. and Takatsuji, H., 2008. SAZ, a new SUPERMAN-like protein, negatively regulates a subset of ABA-responsive genes in *Arabidopsis*. *Mol. Genet. Genomics.*, 279: 183–192.

Karim, S., Aronsson, H., Ericson, H., Pirhonen, M., Leyman, B., Welin, B., Mantyla, E., Palva, E.T., van Dijck, P and Holstrom, K.O., 2007. Improved drought tolerance without undesired side effects in transgenic plants producing trehalose. *Plant Mol. Biol.*, 64: 371–386.

Kasuga, M., Liu, Q., Miura, S., Yamaguchi-Shinozaki, K. and Shinozaki, K., 1999. Improving plant drought, salt, and freezing tolerance by gene transfer of a single stress inducible transcription factor. *Nat. Biotechnol.*, 17: 287–291.

Kasuga, M., Miura, S., Shinozaki, K. and Yamaguchi-Shinozaki, K., 2004. A combination of the *Arabidopsis* DREB1A gene and stress-inducible *rd29A* promoter improved drought and low-temperature stress tolerance in tobacco by gene transfer. *Plant Cell Physiol.*, 45: 346–350.

Kasukabe, Y., He,L., Watakabe, Y., Otani, M., Shimada, T and Tachibana, S., 2006. Improvement of environmental stress tolerance of sweet potato by introduction of genes for spermidine synthase. *Plant Biotechnol.*, 23: 75–83.

Kasukabe, Y., He, L., Nada, K., Misawa, S., Ihara, I and Tachibana, S., 2004. Overexpression of spermidine synthase enhances tolerance to multiple environmental stresses and up-regulates the expression of various stress-regulated genes in transgenic *Arabidopsis thaliana*. *Plant Cell Physiol.*, 45: 712–722.

Katiyar-Agarwal, S., Agarwal, M and Grover, A., 2003. Heat-tolerant basmati rice engineered by over-expression of hsp101. *Plant Mol Biol.*, 51: 677–686.

Katsuhara, M., Koshio, K., Shibasaka, M., Hayashi, Y., Hayakawa, T and Kasamo, K., 2003. Overexpression of a barley aquaporin increased the shoot/root ratio and raised salt sensitivity in transgenic rice plants. *Plant Cell Physiol.*, 44: 1378–1383.

Kawakami, A., Sato, Y and Yoshida, M., 2008. Genetic engineering of rice capable of synthesizing fructans and enhancing chilling tolerance. *J Exptl Bot.*, 59: 793–802.

Kikawada, T., Nakahara, Y., Kanamori, Y., Iwata, K., Watanabe, M., McGee, B., Tunnacliffe, A and Okuda, T., 2006. Dehydration-induced expression of LEA proteins in an anhydrobiotic chironomid. BBRC., 348: 56–61.

Kim, H. S., Lee, J. H., Kim, J. J., Kim, C. H., Jun, S. S and Hong, Y. N., 2005. Molecular and functional characterization of CaLEA6, the gene for a hydrophobic LEA protein from *Capsicum annuum*. *Gene.*, 344: 115–123.

Knight, H and Knight, M.R., 2001. Abiotic stress signalling pathways: specificity and cross-talk. *Trends Plant Sci.*, 6: 262–267

Konstantinova, T., Parvanova, D., Atanassov, A and Djilianov, D., 2002. Freezing tolerant tobacco, transformed to accumulate osmoprotectants. *Plant Sci.*, 163: 157–164.

Kovtun, Y., Chiu, W. L., Tena, G and Sheen, J., 2000. Functional analysis of oxidative stress-activated mitogen-activated protein kinase cascade in plants. *Proc. Natl. Acad. Sci. USA.*, 97: 2940–2945.

Kumar, V., Shriram, V., Kavi Kishor, P.B., Jawali, N and Shitole, M.G., 2010. Enhanced proline accumulation and salt stress tolerance of transgenic indica rice by over expressing *P5CSF129A* gene. *Plant Biotechnol. Rep.*, 4: 37–48.

Lee, H and Jo, J., 2004. Increased tolerance to methyl viologen by transgenic tobacco plants that overexpress the cytosolic glutathione reductase gene from *Brassica campestris*. *J Plant Biol.*, 47: 111–116.

Lee, J.H., Hubel, A and Schoffl, F., 1995. Derepression of the activity of genetically engineered heat shock factor causes constitutive synthesis of heat shock proteins and increased thermotolerance in transgenic *Arabidopsis*. *Plant J.*, 8: 603–612.

Lee, J.T., Prasad, V., Yang, P.T., Wu, J.F., David, Ho, T.H., Charng, Y.Y and Chan, M.T., 2003. Expression of Arabidopsis *CBF1* regulated by an ABA/stress inducible promoter in transgenic tomato confers stress tolerance without affecting yield. *Plant Cell Environ.*, 26: 1181–1190.

Lee, Y.P., Kim, S.H., Bang, J.W., Lee, H.S., Kwak, S.S and Kwon, S.Y., 2007. Enhanced tolerance to oxidative stress in transgenic tobacco plants expressing three antioxidant enzymes in chloroplasts. *Plant Cell Rep.*, 26: 591–598.

Li, B., Wei, A., Song, C., Li, N and Zhang, J., 2008. Heterologous expression of the TsVP gene improves the drought resistance of maize. *Plant Biotechnol.*, J 6: 146–159.

Li, M., Li, Y., Li, H and Wu, G., 2011[a]. Overexpression of AtNHX5 improves tolerance to both salt and drought stress in *Broussonetia papyrifera* (L.) Vent. *Tree Physiol.*, 31: 349–357.

Li, M., Lin, X., Li, H., Pan, X and Wu, G., 2011[b]. Overexpression of AtNHX5 improves tolerance to both salt and water stress in rice (*Oryza sativa* L.). *Plant Cell Tissue Org. Cult.*, 13: 283-293.

Li, W., Wang, D., Jin, T., Chang, Q., Yin, D., Xu, S., Liu, B and Liu, L., 2011[c]. The vacuolar Na^+/H^+ antiporter gene *SsNHX1* from the halophyte salsola soda confers salt tolerance in transgenic alfalfa (*Medicago sativa* L.). *Plant Mol. Biol. Rep.*, 29: 278-290.

Li, X., Wang, Z., Wang, X.M., Gao, H.W., Chen, X.F., Dong, J and Xu, B., 2009. Cloning and analysis of a Vacuolar Na^+/H^+ Antiporter Gene in *Galega orientalis*. *Plant physiol. Commun.*, 45(5): 1-5.

Lindquist, S and Craig, E.A., 1988. The heat-shock proteins. *Annu. Rev. Genet.*, 22: 631–677.

Liu, J.H., Kitashiba, H., Wang, J., Ban, Y and Moriguchi, T., 2007. Polyamines and their ability to provide environmental stress tolerance to plants. *Plant Biotechnol.*, 24: 117–126.

Liu, Q., Kasuga, M., Sakuma, Y., Abe, H., Miura, S., Yamaguchi-Shinozaki, K. and Shinozaki, K., 1998. Two transcription factors, DREB1 and DREB2, with an EREBP/AP2 DNA binding domain separate two cellular signal transduction pathways in drought and low-temperature-responsive gene expression, respectively, in Arabidopsis. *Plant Cell.*, 10: 1391–1406.

Liu, Z.H., Zhang, H.M., Li, G.L., Guo, X.L., Chen, S.Y., Liu, G.B and Zhang, Y.M., 2011. Enhancement of salt tolerance in alfalfa transformed with the gene encoding for betaine aldehyde dehydrogenase. *Euphytica.*, 178: 363-372.

Maiorino, F.M., Brigelius-Flohe, R., Aumann, K.D., Roveri, A., Schomburg, D and Flohe, L., 1995. Diversity of glutathione peroxidases. *Methods Enzymol.*, 252: 38–48.

Malik, M.K., Slovin, J.P., Hwang, C.H and Zimmerman, J.L., 1999. Modified expression of a carrot small heat shock protein gene, *Hsp17.7*, results in increased or decreased thermo tolerance. *Plant J.*, 20: 89–99.

Maqbool, B., Zhong, H., El-Maghraby, Y., Ahmad, A., Chai, B., Wang, W., Sabzikar, R and Sticklen, B., 2002. Competence of oat (*Avena sativa* L.) shoot apical meristems for integrative transformation, inherited expression, and osmotic tolerance of transgenic lines containing *hva1*. *Theor. Appl. Genet.*, 105: 201–208.

McKersie, B.D., Bowley, S.R and Jones,K.S., 1999. Winter survival of transgenic alfalfa overexpressing superoxide dismutase. *Plant Physiol.*, 119: 839-847.

McKersie, B.D., Bowley, S.R., Harjanto, E and Leprince, O., 1996. Water-deficit tolerance and field performance of transgenic alfalfa overexpressing superoxide dismutase. *Plant Physiol.*, 111: 1177-1181.

McKersie, B.D., Murnaghan, J., Jones, K.S and Bowley, S.R., 2000. Iron-superoxide dismutase expression in transgenic alfalfa increases winter survival without a detectable increase in photosynthetic oxidative stress tolerance. *Plant Physiol.*, 122: 1427-1437.

McNeil, S.D., Nuccio, M.L and Hanson, A.D., 1999. Betaines and related osmoprotectants. Targets for metabolic engineering of stress resistance. *Plant Physiol.*, 120: 945–949.

McNeil, S.D., Nuccio, M.L., Rhodes, D., Shachar-Hill, Y and Hanson, A.D., 2000. Radiotracer and computer modeling evidence that phosphobase methylation is the main route of choline synthesis in tobacco. *Plant Physiol.*, 123: 371–380.

Miller, G., Suzuki, N., Ciftci-Yilmaz, S and Mittler, R., 2010. Reactive oxygen species homeostasis and signalling during drought and salinity stresses, *Funct. Plant Biol.*, 33: 453–467.

Mingder Shih., Folkert, A. Hoekstra., and Yu-Ie C. Hsing., 2008. Late embryogenesis abundant proteins In. Advances in Botanical Research, Vol. 48. Incorporating Advances in Plant Pathology DOI: 10.1016/S0065-2296(08)00404-7.

Mittler, R and Blumwald, E., 2010. Genetic engineering for modern agriculture: challenges and perspectives. *Ann. Rev. Plant Biol.*, 61: 443–462.

Mittler, R., 2002. Oxidative stress, antioxidants and stress tolerance. *Trends Plant Sci.*, 7: 405-410.

Miyazawa, S-I., Yoshimura, S., Shinzaki, Y., Maeshima, M and Miyake, C., 2008. Deactivation of aquaporins decreases internal conductance to CO_2 diffusion in tobacco leaves grown under long-term drought. *Funct. Plant Biol.*, 35: 556–564.

Mohanty, A., Kathuria, H., Ferjani, A., Sakamoto, A., Mohanty, P., Murata, N and Tyagi, A. K., 2002. Transgenics of an elite indica rice variety Pusa Basmati 1 harbouring the *codA* gene are highly tolerant to salt stress. *Theor. Appl. Genet.*, 106: 51–57.

Munns, R and Tester, M., 2008. Mechanisms of salinity tolerance. *Ann Rev Plant Biol.*, 59: 651–681.

Murakami, T., Matsuba, S., Funatsuki, H., Kawaguchi, K., Saruyama, H., Tanida, M and Sato, Y.,2004. Overexpression of a small heat shock protein, sHSP17.7, confers both heat tolerance and UV-B resistance to rice plants. *Mol. Breed., 13*: 165–175.

Nakashima, K., Ito, Y. and Yamaguchi-Shinozaki, K., 2009. Transcriptional regulatory networks in response to abiotic stresses in arabidopsis and grasses. *Plant Physiol.*, 149: 88–95.

Nanjo, T., Fujita, M., Seki, M., Kato, T., Tabata, S and Shinozaki, K., 2003. Toxicity of free proline revealed in an Arabidopsis T-DNA-tagged mutant decient in proline dehydrogenase. *Plant Cell Physiol.*, 44: 541–548.

Ndong, C., Danyluk, J., Wilson, K.E., Pocock, T., Huner, N.P. A and Sarhan, F., 2002. Cold-regulated cereal chloroplast late embryogenesis abundant-like proteins. Molecular characterization and functional analyses. *Plant Physiol.*, 129: 1368–1381.

Nelson, D.E., Repetti, P.P., Adams, T.R., Creelman, R.A., Wu, J., Warner, D.C., Anstrom, D.C., Bensen, R.J., Castiglioni, P.P., Donnarummo, M.G., Hinchey, B.S., Kumimoto, R.W., Maszle, D.R., Canales, R.D., Krolikowski, K.A., Dotson, S.B., Gutterson, N., Ratcliffe, O.J. and Heard, J.E., 2007. Plant nuclear factor Y (NF-Y) B subunits confer drought tolerance and lead to improved corn yields on water-limited acres. *Proc. Natl. Acad. Sci. U. S. A.*, 104: 16450–16455.

Noctor, G., Foyer and C.H., 1998. Ascorbate and glutathione: keeping active oxygen under control. *Annu. Rev Plant Biol.*, 49: 249–279.

Nordin, K., Vahala, T. and Palva, E.T., 1993. Differential expression of two related low-temperature-induced genes in *Arabidopsis thaliana* (L.) Heynh. *Plant Mol. Biol.*, 21: 641-653.

Oraby, H.F., Ransom, C.B., Kravchenko, A.N and Sticklen, M.B., 2005. Barley HVA1 gene confers salt tolerance in r3 transgenic oat. *Crop Sci.*, 45: 2218–2227.

Palta, Jiwan, P and Farag, Karim., September, 2006. Methods for enhancing plant health, protecting plants from biotic and abiotic stress related injuries and enhancing the recovery of plants injured as a result of such stresses., *United States Patent 7101828.*

Pardo, J.M., Reddy, M.P and Yang, S., 1998. Stress signaling through Ca^{+2}/ Calmodulin dependent protein phosphates calcinerin mediates salt adaptation in plants. *Proc. Natl. Acad. Sci. USA.*, 95: 9681-9683.

Park, B.J., Liu. Z., Kanno. A and Kameya, T., 2005[a]. Genetic improvement of Chinese cabbage for salt and drought tolerance by constitutive expression of a *Brassica napus* LEA gene. *Plant Sci.*, 169: 553–558.

Park, B.J., Liu, Z., Kanno, A and Kameya, T., 2005[b]. Increased tolerance to salt- and water-deficit stress in transgenic lettuce (*Lactuca sativa* L.) by constitutive expression of LEA. *Plant Growth Regul.*, 45: 165–171.

Park, E.J., Jeknic, Z., Sakamoto, A., DeNoma, J and Yuwansiri, R., Murata, N and Chen T. H., 2004. Genetic engineering of glycine betaine synthesis in tomato protects seeds, plants, and flowers from chilling damage. *Plant J.*, 40: 474–487.

Park, J.A., Cho, S.K., Kim, J.E., Chung, H.S., Hong, J.P., Hwang, B., Hong, C.B and Kim, W. T., 2003. Isolation of cDNAs differentially expressed in response to drought stress and characterization of the Ca-LEAL1 gene encoding a new family of atypical LEA-like protein homologue in hot pepper (*Capsicum annum* L. cv. Pukang). *Plant Sci.*, 165: 471–481.

Park, M.R., Baek, S.H., de los Reyes, B.G and Yun, S.J., 2007. Overexpression of a high-affinity phosphate transporter gene from tobacco (*NtPT1*) enhances phosphate uptake and accumulation in transgenic rice plants. *Plant Soil.*, 292: 259–269.

Parsell, D.A and Lindquist, S., 1993. The function of heat-shock proteins in stress tolerance: Degradation and reactivation of damaged proteins. *Annu. Rev. Genet.*, 27: 437–496.

Parsell, D. A and Lindquist, S., 1994. Heat shock proteins and stress tolerance. In: *The Biology of Heat Shock Proteins and Molecular Chaperones*, (Ed.) R.I. Morimoto, A. Tissieres, and C. Georgopoulos, Cold Spring Harbor Laboratory Press, 457–494.

Patnaik, D and Khurana, P., 2003. Genetic transformation of Indian bread (*T. aestivum*) and pasta (*T. durum*) wheat by particle bombardment of mature embryo-derived calli. *BMC Plant Biol.*, 3: 3–5.

Pellegrineschi, A., Reynolds, M., Pacheco, M., Brito, R.M., Almeraya, R., Yamaguchi-Shinozaki, K. and Hoisington, D., 2004. Stress-induced expression in wheat of the *Arabidopsis thaliana DREB1A* gene delays water stress symptoms under greenhouse conditions. *Genome.*, 47: 493–500.

Peng, L. X, Gu, L. K., Zheng, C. C., Li, D. Q and Shu, H. R., 2006. Expression of MaMAPK gene in seedlings of *Malus* L. under water stress. *Acta Biochim Biophys Sin.*, 38: 281–286.

Perl, A., Perl-Treves, R., Galili, S., Aviv, D., Shalgi, E., Malkin, S and Galun, E., 1993. Enhanced oxidative-stress defence in transgenic potato expressing tomato Cu, Zn superoxide dismutases. *Theor. Appl. Genet.*, 85: 568-576.

Prabhavathi, V.R and Rajam, M.V., 2007. Polyamine accumulation in transgenic eggplant enhances tolerance to multiple abiotic stresses and fungal resistance. *Plant Biotechnol.*, 24: 273–282.

Prandl, R., Hinderhofer, K., Eggers-Schumacher, G and Schoffl, F., 1998. HSF3, a new heat shock factor from *Arabidopsis thaliana*, derepresses the heat shock response and confers thermotolerance when overexpressed in transgenic plants. *Mol. Gen. Genet.*, 258: 269–278.

Prashanth, S.R., Sadhasivam, V and Parida, A., 2008. Over expression of cytosolic copper/zinc superoxide dismutase from a mangrove plant *Avicennia marina* in indica Rice var Pusa Basmati-1confers abiotic stress tolerance. *Transgen Res.*, 17: 281–291.

Puhakainen, T., Hess, M.W., Makela, P., Svensson, J., Heino, P and Palva, E.T., 2004. Overexpression of multiple dehydrin genes enhances tolerance to freezing stress in *Arabidopsis*. *Plant Mol. Biol.*, 54: 743–753.

Qu, L.J., Wu, L.Q., Fan, Z.M., Guo, L., Li, Y.Q and Chen, Z.L., 2005. Over-expression of the bacterial *nhaA* gene in rice enhances salt and drought tolerance. *Plant Sci.*, 168: 297-302.

Quan, R.D., Shang, M., Zhang, H., Zhao, Y.X and Zhang, J.R., 2004. Engineering of enhanced glycine betaine synthesis improves drought tolerance in maize. *Plant Biotechnol J.*, 2: 477–486.

Quintero, F.J., Blatt, M.R and Pardo, J.M., 2000. Functional conservation between yeast and plant endosomal Na^+/H^+ antiporters. *FEBS Lett.*, 471: 224–228.

Reguera, M. Peleg, Z and Blumwald, E., 2012. Targeting metabolic pathways for genetic engineering abiotic stress-tolerance in crops. *Biochem. Biophys. Acta.*, 19(2): 186-194.

Rohila, J.S., Jain, R.K and Wu, R., 2002. Genetic improvement of basmati rice for salt and drought tolerance by regulated expression of a barley Hva1 cDNA. *Plant Sci.*, 163: 525–532.

Rontein, D., Basset, G and Hanson, A.D., 2002. Metabolic engineering of osmoprotectant accumulation in plants. *Metab. Eng.*, 4: 49–56.

Roxas, V.P., Smith, R.K., Allen, E.R and Allen, R.D., 1997. Overexpression of glutathione S-transferase/glutathione peroxidase enhances the growth of transgenic tobacco seedlings during stress. *Nature Biotechnol.*, 15: 988–991.

Roy, M and Wu, R., 2002. Overexpression of S-adenosylmethionine decarboxylase gene in rice increases polyamine level and enhances sodium chloride-stress tolerance. *Plant Sci.*, 163: 987-992.

Sakamoto, A and Murata, N., 2000. Genetic engineering of glycine betaine synthesis in plants: current status and implications for enhancement of stress tolerance. *J. Exptl. Bot.*, 51: 81–88.

Sakuma, Y., Liu, Q., Dubouzet, J.G., Abe, H., Shinozaki, K. and Yamaguchi-Shinozaki, K., 2002. DNA binding specificity of the ERF/AP2 domain of

Arabidopsis DREBs, transcription factors involved in dehydration- and cold-inducible gene expression. *Biochem. Biophys. Res. Commun.*, 290: 998–1009.

Sang, Y., Zheng, S., Li, W., Huang, B and Wang, X., 2001. Regulation of plant water loss by manipulating the expression of phospholipase D. *Plant J.*, 28: 135–144.

Sanmiya, K., Katsumi Suzuki., Yoshinobu Egawa and Mariko Shono., 2004. Mitochondrial small heat-shock protein enhances thermo tolerance in tobacco plants. *FEBS Lett.*, 557: 265-268.

Schneider, K., Wells, B., Schmelzer, E., Salamini, F and Bartels, D., 1993. Desiccation leads to the rapid accumulation of both cytosolic and chloroplastic proteins in the resurrection plant *Craterostigma plantagineum* Hochst. *Planta.*, 189: 120–131.

Seki, M., Kamei, A., Satou, M., Sakurai, T., Fujita, M., Oono, Y., Yamaguch-Shinozaki, K. and Shinozaki, K., 2003. Transcriptome analysis in abiotic stress conditions in higher plants. *Topics Curr Genet.*, 4: 271–295.

Seki, M., Narusaka, M., Ishida, J., Nanjo, T., Fujita, M., Oono, Y., Kamiya, A., Nakajima, M., Enju, A., Sakurai, T., Satou, M., Akiyama, K., Taji, T., Yamaguchi-Shinozaki, K., Carninci, P., Kawai, J., Hayashizaki, Y. and Shinozaki, K., 2002. Monitoring the expression profiles of 7000 *Arabidopsis* genes under drought, cold and high salinity stresses using a full-length cDNA microarray. *The Plant J.*, 31: 279–292.

Serrano, R., Mulet, J.M., Rios, G., Marquez, J.A., De Larrinoa, I.F., Leube, M.P., Mendizabal, I, Pascual-Ahuir, A., Proft, M., Ros, R and Montesinos, C., 1999. A glimpse of the mechanisms of ion homeostasis during salt stress. *J. Exptl. Bot.*, 50: 1023–1036.

Shalni Lal, S., Gulyani, V and Khurana, P., 2008. Overexpression of *HVA1* gene from barley generates tolerance to salinity and water stress in transgenic mulberry (*Morus indica*). *Transgenic Res.*, 17: 651–663.

Sheveleva, E.V., Marquez, S., Chmara, W., Zegeer, A., Jensen, R.G and Bohnert, H.J., 1998. Sorbitol-6-phosphate dehydrogenase expression in transgenic tobacco. *Plant Physiol.*, 117: 831–839.

Shi, H., Ishitani, M., Kim, C and Zhu, J. K., 2000. The *Arabidopsis thaliana* salt tolerance gene SOS1 encodes a putative Na^+/H^+ antiporter. *Proc. Natl. Acad. Sci. USA.*, 97: 6896–6901.

Shi, H., Lee B.H., Wu, S.J and Zhu, J.K., 2003. Over expression of a plasma membrane Na^+/H^+ antiporter gene improves salt tolerance in *Arabidopsis thaliana. Nat. Biotechnol.*, 21: 81–85.

Shou, H.X., Bordallo, P and Wang, K., 2004. Expression of the Nicotiana protein kinase (NPK1) enhanced drought tolerance in transgenic maize. *J. Exptl Bot.*, 55: 1013–1019.

Sickler, C.M., Edwards, G.E., Kiirats, O., Gao Z and Loescher, W., 2007. Response of mannitol-producing *Arabidopsis thaliana* to abiotic stress. *Funct Plant Biol.*, 34: 382–391.

Simon-Sarkadi, L., Kocsy, G., Varhegyi, A, Galiba G and de Ronde, J.A., 2005. Genetic manipulation of proline accumulation inuences the concentrations of other amino acids in soybean subjected to simultaneous drought and heat stress. *J Agric Food Chem.*, 53: 7512–7517.

Sivamani, E., Bahieldin, A., Wraith, J.M., AlNiemi, T., Dyer, W. E., Ho T-HD QuR., 2000. Improved biomass productivity and water use efficiency under water deficit conditions in transgenic wheat constitutively expressing the barley *HVA1* gene. *Plant Sci.*, 155: 1–9.

Sneddon, W.A and Fromm, H., 2001. Calmodulin as a versatile calcium signal transducer in plants. *New Phytol.*, 151: 35–66.

Suarez, R., Calderon, C and Itturiaga G., 2009. Enhanced tolerance to multiple abiotic stresses in transgenic alfalfa accumulating trehalose. *Crop Sci.*, 49: 1791–1799.

Sunkar, R., Bartels, D and Kirch, H.H., 2003. Overexpression of a stress inducible aldehyde dehydrogenase gene from *Arabidopsis thaliana* in transgenic plants improves stress tolerance. *Plant J.*, 35: 452–464.

Taji, T., Ohsumi, C., Luchi, S., Seki, M., Kasuga, M., Kobayashi, M., Yamaguchi-Shinozaki, K and Shinozaki, K., 2002. Important role of drought and cold-inducible genes for galactinol synthase in stress tolerance in *Arabidopsis thaliana. Plant J.*, 29: 417–426.

Tarczynski, M. C., Jensen R. G and Bohnert H. J. 1993. Stress protection in transgenic tobacco producing a putative osmoprotectant, mannitol. *Science.*, 259: 508–510.

Tolleter, D., Hincha, D.K and Macherel, D., 2010. A mitochondrial late embryogenesis abundant protein stabilizes model membranes in the dry state. *Biochemica Biophys Acta.*, 1798 (10): 1926–1933.

Tyson, T., Reardon, W., Browne, J.A and Burnell, A.M., 2007. Gene induction by desiccation stress in the entomopathogenic nematode *Steinernema carpocapsae* reveals parallels with drought tolerance mechanisms in plants Int. *J. Parasitol.*, 37: 763–776.

Umezawa, T., Fujita, M., Fujita, Y., Yamaguchi-Shinozaki, K and Shinozaki, K., 2006. Engineering drought tolerance in plants: discovering and tailoring genes to unlock the future. *Curr. Opin. Plant Biotech.*, 17: 113–122.

Umezawa, T., Yoshida, R., Maruyama, K., Yamaguchi-Shinozaki, K and Shinozaki, K., 2004. SRK2C, a SNF1-related protein kinase 2, improves drought tolerance by controlling stress-responsive gene expression in *Arabidopsis thaliana. Proc. Natl. Acad. Sci. USA.*, 101: 17306–17311.

Vendruscolo, E.C.G., Schuster, I., Pileggi, M., Scapim, C.A, Molinari, H.B.C, Marur, C.J and Vieira, L.G.E., 2007. Stress-induced synthesis of proline confers tolerance to water decit in transgenic wheat. *J. Plant Physiol.* 164: 1367–1376.

Vinocur, B and Altman, A., 2005. Recent advances in engineering plant tolerance to abiotic stress: achievements and limitations. *Curr. Opin. Biotechnol.*, 16: 123–132.

Waie, B and Rajam, M.V., 2003. Effect of increased polyamine biosynthesis on stress responses in transgenic tobacco by introduction of human S-adenosylmethionine gene. *Plant Sci.*, 164: 727–734.

Wang, C.R., Yang, A.F., Yue, G.D., Gao, Q., Yin, H.Y and Zhang, J.R., 2008. Enhanced expression of phospholipase C 1 (ZmPLC1) improves drought tolerance in transgenic maize. *Planta.*, 227: 1127–1140.

Wang, G.P., Hui, Z., Li, F., Zhao, M.R., Zhang, J and Wang, W., 2010. Improvement of heat and drought photosynthetic tolerance in wheat by over accumulation of glycine betaine. *Plant Biotechnol.*, 4: 213-222.

Wang, M., Gu, D., Liu, T., Wang, Z., Guo, X., Hou, W., Bai, Y., Chen, X and Wang, G., 2007 Over expression of a putative maize calcineurin B-like protein in *Arabidopsis* confers salt tolerance. *Plant Mol Biol.*, 65: 733–746.

Wang, R.G., Chen, S.L., Liu, L.Y., Hao, Z.Y., Weng, H.J., Li, H., Yang, S and Duan, S., 2005. Genotypic differences in antioxidative ability and salt tolerance of three poplars under salt stress. *J. Beijing For. Univ.*, 27: 46–52.

Wang, W. X., Barak, T., Vinocur, B., Shoseyov, O and Altman, A., 2003[a]. Abiotic resistance and chaperones: possible physiological role of SP1, a stable and stabilizing protein from Populus. In: Vasil IK (ed) Plant biotechnology 2000 and beyond. *Kluwer, Dordrecht.*, pp. 439–443.

Wang, W., Vinocur, B and Altman, A., 2003[b]. Plant responses to drought, salinity and extreme temperatures: towards genetic engineering for stress tolerance. *Planta.*, 218: 1-14.

Wang, X., Li Y., Ji, W., Bai, X., Cai, H., Zhu, D., Sun, X.L., Chen, L.J and Zhu Y. M., 2011. A novel *Glycine soja* tonoplast intrinsic protein gene responds to abiotic stress and depresses salt and dehydration tolerance in transgenic *Arabidopsis thaliana*, *J. Plant Physiol.*, 168: 1241–1248.

Wang, Y., Ying, Y., Chen, J and Wang, X.C., 2004. Transgenic Arabidopsis overexpressing Mn-SOD enhanced salt-tolerance. *Plant Sci.*, 167: 671-677.

Waterer, D., Benning, N., Wu, G., Luo, X., Liu, X., Gusta, M., McHughen, A and Gusta, L., 2010. Evaluation of abiotic stress tolerance of genetically modified potatoes (*Solanum tuberosum* cv. Desiree). *Mol Breed.*, 25: 527–540.

Wi, S. J., Kim, W. T and Park, K.Y., 2006. Over expression of carnation s-adensylmethionine decarboxylase gene generates a broad spectrum tolerance to abiotic atresses in transgenic tobacco plants. *Plant Cell Rep.*, 25: 1111-1121.

Wiebe, C.A., DiBattista, E.R and Fliegel, L., 2001. Functional role of polar amino acid residues in Na^+/H^+ exchangers. *Biochem J.*, 357: 1-10.

Xia L., Zhi Wanga., Lili Wanga., Renhua Wua., Jonathan Phillips., Xin Deng., 2009. LEA 4 group genes from the resurrection plant *Boea hygrometrica* confer dehydration tolerance in transgenic tobacco. *Plant Sci.*, 176: 90–98.

Xiang, Y., Huang, Y and Xiong, L., 2007. Characterization of stress-responsive CIPK genes in rice for stress tolerance improvement. *Plant Physiol.*, 144: 1416-1428.

Xiong, L and Ishitani, M., 2006. Stress signal transduction: components, pathways, and network integration. In: *Abiotic stress tolerance in plants: toward the improvement of global environment and food*, (Ed.). A.K. Rai and T. Takabe Springer-Dordercht, The Netherlands, 3–29.

Xiong, L and Zhu, J.K., 2001. Abiotic stress signal transduction in plants: molecular and genetic perspectives. *Physiol Plant.*, 112: 152-166.

Xu, D., Duan, X., Wang, B., Hong, B and Ho H.D WuR., 1996. Expression of a late embryogenesis abundant protein gene, HVA1, from barley confers tolerance to water deficit and salt stress in transgenic rice. *Plant Physiol.*, 110: 249–257.

Xu, W.F., Shi, W.M., Ueda, A and Takabe, T., 2008. Mechanisms of salt tolerance in transgenic *Arabidopsis thaliana* carrying a peroxisomal ascorbate peroxidase gene from Barley. *Planta.*, 218: 1–14.

Xue, Z.Y., Zhi, D.Y., Xue, G.P, Zhang, H., Zhao, Y.X and Xia, G.M., 2004. Enhanced salt tolerance of transgenic wheat (*Triticum aestivum* L.) expressing a vacuolar Na^+/H^+ antiporter gene with improved grain yields in saline soils in the field and a reduced level of leaf Na^+. *Plant Sci.*, 167: 849–859.

Yamada, M., Morishita, H., Urano, K., Shiozaki, N., Yamaguchi-Shinozaki, K., Shinozaki, K and Yoshiba, Y., 2005. Effects of free proline accumulation in petunias under drought stress. *J Exptl Bot.*, 56: 1975–1981.

Yamaguchi, T and Blumwald, E., 2005. Developing salt-tolerant crop plants: challenges and opportunities. *Trends Plant Sci.*, 10: 615–620.

Yamaguchi-Shinozaki, K and Shinozaki, K., 1994. A novel *cis*-acting element in an Arabidopsis gene is involved in responsiveness to drought, low-temperature, or high-salt stress. *Plant Cell.*, 6: 251–264.

Yang, A.F., Duan, X.G., Gu, X.F., Gao, F and Zhang, J.R., 2005[b]. Efficient transformation of beet (*Beta vulgaris*) and production of plants with improved salt-tolerance. *Plant Cell Tissue Org Cult.*, 83: 259–270.

Yang, O., Popova, O.V., Süthoff, U., Lüking, I., Dietz,K.J and Golldack., 2009. The Arabidopsis basic leucine zipper transcription factor AtbZIP24 regulates complex tran- scriptional networks involved in abiotic stress resistance. *Gene.*, 436: 45–55.

Yang, X.H., Liang, Z and Lu, C.M., 2005[a]. Genetic engineering of the biosynthesis of glycine betaine enhances photosynthesis against high temperature stress in transgenic tobacco plants. *Plant Physiol.*, 138: 2299–2309.

Yi, N., Youn Shic Kim., Min Ho Jeong., Se Jun Oh., Jin Seo Jeong., Su-Hyun Park., Harin Jung., Yang Do Choi and Ju-Kon Kim., 2010. Functional analysis of six drought-inducible promoters in transgenic rice plants throughout all stages of plant growth. *Planta.*, 232: 743-754.

Yong H.C., Kyung Nam Kim., Girdhar K. Pandey., Rajeev Gupta., John J. Grant and Sheng Luan., Aug 2003. CBL1, a calcium sensor that differentially regulates salt, drought, and cold responses in *Arabidopsis*. *Plant Cell.*,15: 1833–1845.

Yoshimura, K., Miyao, K., Gaber, A., Takeda, T., Kanaboshi, H., Miyasaka, H and Shigeoka, S., 2004. Enhancement of stress tolerance in transgenic tobacco plants overexpressing Chlamydomonas glutathione peroxidase in chloroplasts or cytosol. *Plant J.*, 37: 21-33.

Yu, H., Chen, X., Hong, Y.Y., Wang, Y., Xu, P., Ke, S.D., Liu, H.Y., Zhu, J.K., Oliver, D.J. and Xiang, C. B., 2008. Activated expression of an *Arabidopsis* HD-START protein confers drought tolerance with improved root system and reduced stomatal density. *Plant Cell.*, 20: 1134–1151.

Yu, Q., Hu, Y., Li, J., Wu, Q and Lin, Z., 2005. Sense and antisense expression of plasma membrane aquaporin BnPIP1 from *Brassica napus* in tobacco and its effects on plant drought resistance. *Plant Sci.*, 169: 647–656.

Zhang, H.X and Blumwald, E., 2001. Transgenic salt-tolerant tomato plants accumulate salt in foliage but not in fruit. *Nature Biotech.*, 19: 765–768.

Zhang, H.X., Hodson, J.N., Williams, J.P and Blumwald, E., 2001. Engineering salt-tolerant Brassica plants: characterization of yield and seed oil quality in transgenic plants with increased vacuolar sodium accumulation. *Proc Natl Acad Sci USA.*, 98: 12832– 12836.

Zhang, J.Z., Creelman, R.A and Zhu, J.K., 2004. From laboratory to field. Using information from Arabidopsis to engineer salt, cold, and drought tolerance in crops. *Plant Physiol.*, 135: 615–621.

Zhang, J., Tan, W., Yang, X.H and Zhang, H.X., 2008[a]. Plastid-expressed choline monooxygenase gene improves salt and drought tolerance through accumulation of glycine betaine in tobacco. *Plant Cell Rep.*, 27: 1113–1124.

Zhang, Y., Wang, Z., Chai, T., Wen, Z and Zhang, H., 2008[b]. Indian mustard aquaporin improves drought and heavy-metal resistance in tobacco. *Mol Biotechnol.*, 40: 1–13.

Zhao, F., Guo, S., Zhang, H and Zhao, Y., 2006. Expression of yeast SOD2 in transgenic rice results in increased salt tolerance. *Plant Sci.*, 170: 216–224.

Zhao, H.W., Chen, Y.J., Hu, Y.L., Gao, Y and Lin, Z.P., 2000. Construction of a trehalose-6-phosphate synthase gene driven by drought responsive promoter and expression of drought-resistance in transgenic tobacco. *Acta Bot Sin.*, 42: 616–619.

Zhao, J., Zhi, D., Xue, Z., Liu, H and Xia, G., 2007. Enhanced salt tolerance of transgenic progeny of tall fescue (*Festuca arundinacea*) expressing a vacuolar Na^+/H^+ antiporter gene from Arabidopsis. *J. Plant Physiol.*, 164: 1377–1383.

2016, Recent Advances in Plant Stress Physiology *Pages* **445–459**
Editors: ***Praduman Yadav, Sunil Kumar and Veena Jain***
Published by: **DAYA PUBLISHING HOUSE, NEW DELHI**

Chapter 19

Transgenic Approaches for Biotic Stress Tolerance in Plants

K. Lakshmi[1]*, C. Appunu[1], Ravinder Kumar[2], M.R. Meena[2] and Deepak Kumar Yadav[3]**

[1]Sugarcane Breeding Institute (ICAR), Coimbatore, Tamil Nadu, India
[2]Sugarcane Breeding Institute, Regional Centre-Karnal, Haryana, India
[3]Department of Agriculture, Buchawas, Mahendergarh, Haryana, India

ABSTRACT

Biotic stresses such as viruses, fungi, bacteria, weeds, insects and other pests and pathogens are major constraints to agricultural productivity. These stresses not only widen the gap between the mean yield and the potential yield, but also cause yield instability in agriculturally important crops. As sessile organisms, plants are inevitably exposed to one or more combinations of these stress factors every now and then throughout their life cycle. Plants are categorized according to their stress responses viz., stress-susceptible and stress-tolerant. Crop improvement has been a traditional issue to increase yield and enhance stress tolerance; however, genetic improvement through conventional breeding for biotic stress tolerance in crop is constrained due to the complex nature of stress tolerance (timing, duration, intensity, frequency) and genome composition of the crop species. Transgenic approach is one of the important tools available for modern plant improvement programs. Transgenic plants are majorly developed by particle bombardment and Agrobacterium mediated transformation methods. Gene discovery and functional genomics projects have revealed multitudinous mechanisms and gene families, which confer improved productivity and adaptation to biotic stresses. Further, pyramiding of desirable genes with similar effects can also be achieved by transgenic approaches. This chapter outlines the major yield-limiting biotic stresses and the plant stress resistance mechanisms and transgenic approaches to overcome biotic stresses.

Keywords: *Biotic stress, Bt Toxin, Viral resistance, Genetic engineering, Insect pest.*

**Corresponding author*: E-mail: lakshmimbb@gmail.com or cappunu@gmail.com

Introduction

Agriculture production in Asia, particularly in India, has increased considerably during the last five decades. This has happened largely due to the development and large-scale cultivation of new higher-yielding varieties, increase in area under such varieties and greater application of irrigation and nutrients. This increase in food production has made the Asian region self-sufficient and contributed tremendously to food security. Despite surplus buffer stocks currently available in many parts of South-Asia, it is projected that food security of this region might be at risk shortly due to increasing population and pressure for alternate land uses. Indian sub-continent is now home for almost one quarter of the world population and it is projected that by 2050, India's population is expected to grow to 1.6 billion, making it the most populous country of the world. This rapid and continuing increase in population implies a greater demand for food. It is projected that by 2020 our food grain demand will be 294 million tons, against our current production level of 208 million tons (Kumar 1998).

High yield varieties are confronted by various biotic stresses (fungi, bacteria, viruses, insects, nematodes) which causes huge loss in production and productivity. Plants have developed defense mechanisms manifested through altered physiology to endure combat challenges arising from biotic stresses. India is classified by FAO as a low income, food deficit country and nearly 26 per cent of India's populations are considered food insecure, consuming less than 80 per cent of minimum energy requirements. Therefore, the pace of genetic progress must now increase to meet the projected food grains demand of 294 million tonnes by the year 2020 to feed the growing population and demand for food grains certainly increases with respect to that of population growth. There is an urgent need for identifying a novel solution to overcome losses due to biotic stresses. Great challenge of food security has directed plant scientists towards gene revolution which involves direct modification of qualitative and quantitative traits in an organism by transferring desired genes using 'transgenic approach' or 'genetic transformation'. In contrast to classical breeding, genetic engineering offers an excellent tool for asexually inserting gene(s) of unrelated organisms into plant cells. This process takes less time thus accelerating the process of genetic improvement of crop plants. In addition, this exciting technology allows access to an unlimited gene pool without the constraint of sexual compatibility. Crop plants engineered to suit the environment better through incorporation of genes for tolerance to biotic and abiotic stresses have been suggested to represent an improvement in crop production (Ortiz 1998). Over the past few decades, breeding possibilities have been broadened by genetic engineering and gene transfer technologies, as well as by gene mapping and identification of the genome sequences of model plants and crops which resulted in efficient transformation and generation of transgenic lines in a number of crop species (Gosal *et al.*, 2009). Further, pyramiding of desirable genes with similar effects can also be achieved by using these approaches. Transgenic plants are majorly developed by particle bombardment and *Agrobacterium* mediated transformation methods. Genetic transformation by *Agrobacterium* mediated approaches has been successfully demonstrated for economically important traits like biotic resistance

(Wani and Sanghera 2010) and abiotic stress tolerance in different crop plants. Use of agrochemicals for control of crop pests and diseases is however inefficient and environmentally unsound (Robinson 1999). Thus, Genetic engineering offers a remedy through more precise targeting of pest and viral diseases (Wani and Sanghera 2010). Tremendous loss in yield of several economically important crops occurs due to biotic stresses (insect infestation and diseases). Significant researches have been made to produce plants with high resistance to insect pest and diseases following transgenic technology.

I. **Pest resistance**: Through the use of *Bacillus thuringieneses* (*Bt*) cry protein toxins, protease, lectins, etc.

II. **Virus resistance**: Achieved by means of incorporating virus coat proteins, antisense RNA technology, ribosome, etc.

III. **Fungal and bacterial disease resistance:** By incorporating pathogenesis related proteins, phytoalexins.

Engineering for Insect Pests

Humans have searched for crop plants that can survive and produce in spite of insect pests. Knowingly or unknowingly, ancient farmers selected for pest resistance genes in their crops, sometimes by actions as simple as collecting seed from only the highest yielding plants in their fields. With the advent of genetic engineering, genes for insect resistance now can be moved into plants more quickly and deliberately. Insect-resistant crops have been one of the major successes of applying plant genetic engineering technology to agriculture; cotton (*Gossypium hirsutum*) resistant to lepidopteran larvae (caterpillars) and maize (*Zea mays*) resistant to both lepidopteran and coleopteran larvae (rootworms) have become widely used in global agriculture and have led to reductions in pesticide usage and lower production costs (Toenniessen *et al.*, 2003; Brookes and Barfoot 2005). *Bacillus thuringiensis*, commonly known as Bt, is a bacterium that occurs naturally in the soil. For years, bacteriologists have known that some strains of Bt produce proteins that kill certain insects with alkaline digestive tracts. When these insects ingest the protein produced by Bt, the function of their digestive systems is disrupted, producing slow growth and, ultimately, death. Strains of Bt are effective against European corn borers and cotton bollworms (Lepidoptera), Colorado potato beetles (Coleoptera), and certain flies and mosquitos (Diptera). Bt is not harmful to humans, other mammals, birds, fish, or beneficial insects. Not only Bt toxins used in plants work against biting insect pests, there are large groups of insect pests, including sucking insects (Homoptera, such as leaf hoppers and aphids) against which no natural *Bt* toxins are known. Synthetic, chimeric toxins have been developed which extend the range of these toxins to other insect groups (Mehlo *et al.*, 2005). Furthermore, as for fungi, there are examples of transgenic strategies using genes encoding other insecticidal proteins, for example lectins, which have led to resistance against these types of insects in the laboratory (Saha *et al.*, 2006).

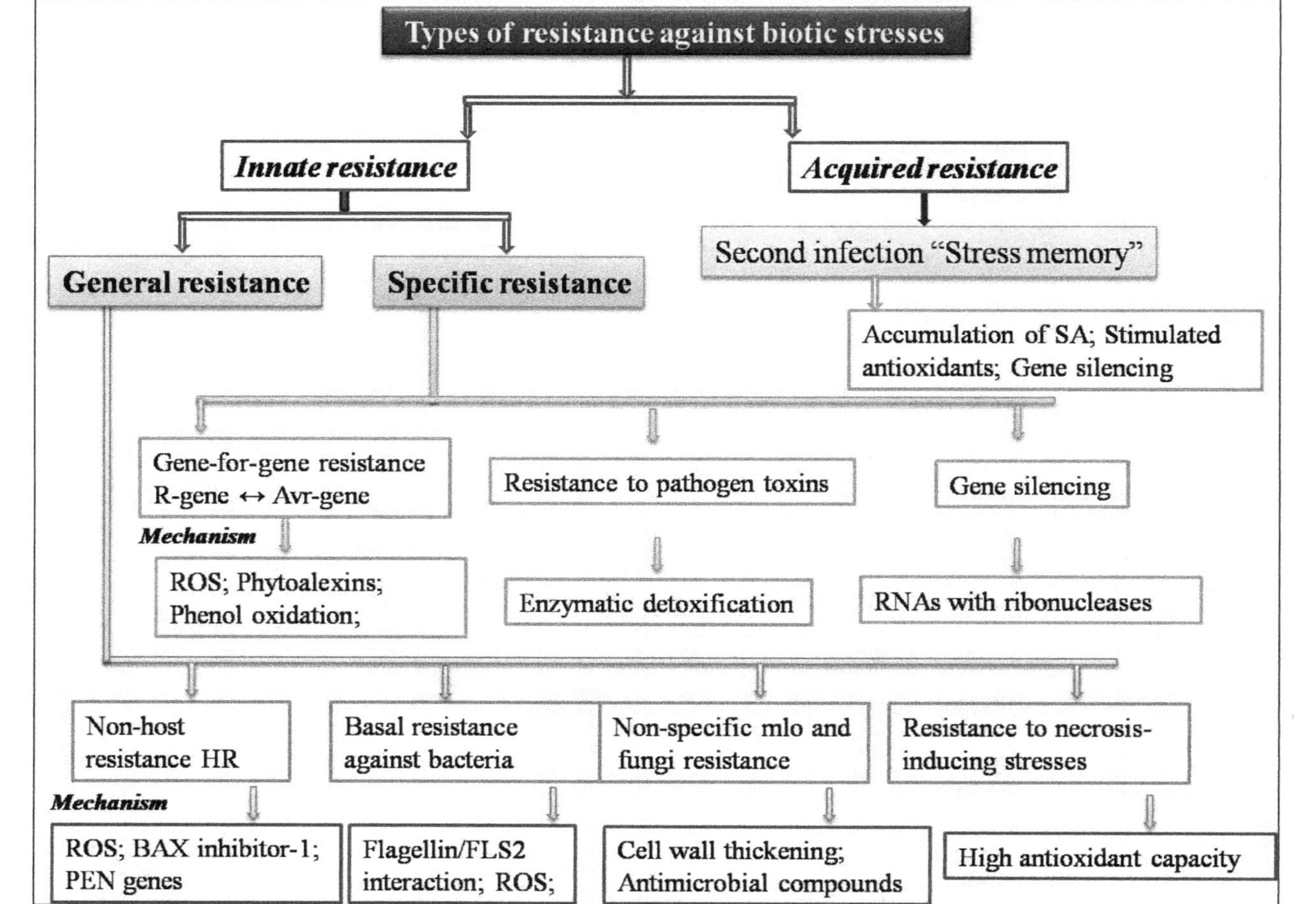

Figure 19.1: Types of Biotic Stresses Resistance and Mechanisms by which Plants Protected from Biotic Stresses.

Expression Level of *Bt* Toxins

The Expression level of Bt toxins in transgenic plants needs to be sufficient to confer adequate protection against target pests. Transformation of the nuclear genome with genes encoding Bt toxins gives very low levels of expression unless extensive modifications, which include removal of AT-rich regions from the coding sequence and use of modified constitutive or tissue-specific promoters, are carried out. But this level of expression can be achieved by introducing the unmodified Bt genes into the chloroplast genome which results into high levels of toxin accumulation (3 per cent - 5 per cent of total leaf protein; McBride *et al.*, 1995), as the plastid genome is bacterial in origin. This method has not been widely adopted, due to significant technical problems in achieving stable transformation of the plastid genome and in transforming plastids in species other than tobacco (*Nicotiana tabacum*). Nevertheless, these difficulties overcome and carious *Cry* proteins have been expressed in plastids of tobacco (Kota *et al.*, 1999; De Cosa *et al.*, 2001), and

Table 19.1: Transgenic Crops Engineered for Enhanced Resistance against different Insect Pests

Transgenics	*Gene*	*Targeted Insect*
Rice	*cry1Ac, 2A*	*Scirpophaga incertulas* and *Scirpophaga medinalis*
Rice	*cry1Ac*	*Scirpophaga incertulas* and *Nilaparvata lugens*
Rice	*cry1Ab*	*Chilo suppressalis*
Rice	*cry1Ab*	*Scirpophaga incertulus*
Rice	*cry1Ac*	*Scirpophaga incertulus*
Rice	*PI-II*	*Chilo suppressalis*
Rice	*Gna*	*Nilaparvata lugens*
Rice	*SKTI*	*Nilaparvata lugens*
Rice	*pin2*	*Scirpophaga incertulas*
Rice	*ASAL*	*Nilaparvata virescens* and *Nilaparvata lugens*
Rice	*CpTi*	*Scirpophaga incertulas*
Sorghum	*cry1Ac*	*Chilo partellus*
Tobacco	*Magi 6*	*Spodoptera frugiperda*
Tobacco	*CpTi*	*Heliothis virescens*
Tobacco	*PI-II*	*Menduca sexta*
Tobacco	*PI-II*	*Spodoptera exigua*
Tobacco	*M. sexta PI*	*Bemisia tabaci*
Potato	*cry3a*	*Leptinotarsa decemlineata*
Potato	*CpTi*	*Lacanobia oleracea*
Pea	*α-amylase*	*Callosobrunchus* spp.
Chickpea	ASAL	*Aphis craccivora*
Corn	*cry1Ab*	*Ostrinia nubilalis*
Cotton	*cry1Ab/1Ac*	*Pectinophora gossypiella*

Sanghera *et al.*, 2011 and references therein.

*Cry*1Ab has been expressed in soybean (*Glycine max*) plastids (Dufourmantel *et al.*, 2005). Overexpression of the *Cry*2Aa2 operon in plastids is effective in giving broad spectrum protection against a range of pests. This technique has the potential advantage that plastid-encoded characteristics are predominantly maternally inherited in most plants, so that pollen from transgenic crops is less likely to disperse the transgene to non transgenic plant a stock which is easy to overcome the biosafety issues. Protein Engineering is also used as a tool to improve the toxicity level in plants to confer tolerance to insect pests. The three domain *Cry* proteins have been extensively studied; their mechanism of action involves a proteolytic activation step, which occurs in the insect gut after ingestion, followed by interaction of one or both of domains II and III with "receptors" on the surface of cells of the insect gut epithelium. This interaction leads to oligomerization of the protein, and domain I is responsible for the formation of an open channel through the cell membrane (Bravo *et al.*, 2007). The resulting ionic leakage destroys the cell, leading to breakdown of the gut, bacterial proliferation, and insect death.

Engineering for Virus Resistance

Genetic resistance to plant viruses has been used for at least 80 years to control agricultural losses to viral diseases. To date, hundreds of naturally occurring genes for resistance to plant viruses have been reported from studies of both monocot and dicot crops.

RNA-mediated gene silencing is a predominant approach (viral RNA is degraded and viral DNA is inactivated by methylation) for viral resistance integration in plant system. RNA-mediated resistance can be obtained without transgenic expression of a protein and this strategy thereby minimizes toxicological and allergenic risks. In the 1990s, the Papaya industry on Hawaii suffered a 50 per cent decline in production due to an outbreak of the potyvirus - Papaya ring spot virus, PRSV (Gonsalves 1998). Virus resistance was obtained in a high-yielding papaya hybrid using the viral coat protein sequence as the transgene. Following distribution of transgenic seeds to farmers, a 50 per cent rebound of total papaya production on Hawaii was achieved within 4 years (Gonsalves 2004). A similar approach has been successfully applied in US cucurbit production, although the situation has been more complicated due to the presence of several different viruses (Fuchs *et al.*, 1997; Fuchs and Gonsalves, 2007). For DNA viruses, strategies for transgenic resistance involves both RNA-mediated resistance and several approaches using mutated viral proteins exerting trans dominant negative effects on viral replication (Vanderschuren *et al.*, 2007 and references therein).

Engineering for Enhanced Resistance against Fungal and Bacterial Pathogens

Transgenic plants have been produced with genes involved in different pathways to enhance disease resistance against fungal pathogens. The first report on developing fungus resistant transgenic plant expressing bean chitinase in tobacco and *Brassica napus* to enhance resistance towards *Rhizoctonia solani* (Broglie *et al.*, 1991). Similarly, antifungal genes have been engineered in various crop plants

Table 19.2: Transgenic Crops Engineered for Enhanced Resistance against Viral Pathogens

Plants	*Genes*	*Against Viral Diseases*
RNA interference		
Common bean	Replication initiator protein (*rep*; *AC1*), Transactivator protein (*TrAP*; *AC2*), Replication enhancer protein (*REn*; *AC3*) and movement protein (*BC1*)	*Bean golden mosaic virus* (*BGMV*)
Coat protein mediated resistance		
Tobacco	coat protein (CP)	*Cucumber mosaic virus*
Tomato	coat protein (CP)	*Cucumber mosaic virus*
Tomato	coat protein (CP)	*Cucumber mosaic virus*
Tomato	*N* gene	*Tomato spotted wilt virus*
Papaya	coat protein (CP)	*Papaya ring spot virus*
Melon	coat protein (CP)	*Cucumber mosaic virus*
Cucumber	coat protein (CP)	*Cucumber mosaic virus*
Tobacco	coat protein (CP)	*Cucumber mosaic virus*
Tomato	coat protein (CP)	*Cucumber mosaic virus*
Tobacco	coat protein (CP)	*Cowpea aphid-borne mosaic virus* (*CABMV*)
Tobacco	coat protein (CP)	*Cucumber mosaic virus*
Tobacco	coat protein (CP)	*Tobacco mosaic viruses*
Arabidopsis	P69 (TYMV) and HC-Pro (TuMV)	Turnip yellow mosaic virus (*TYMV*) and turnip mosaic virus (*TuMV*)
Tobacco	coat protein (CP)	*Tobacco mosaic virus*
Tobacco	coat protein (CP	*Cucumber mosaic virus*
Tobacco	coat protein (CP)	*Cucumber mosaic virus*
Pepper	coat protein (CP)	*Cucumber mosaic virus*
Tobacco	coat protein (CP)	*Cucumber mosaic virus*
Tobacco	coat protein (CP)	*Cucumber mosaic virus sub group IB*
Tomato	coat protein (CP)	*Cucumber mosaic virus*
Squash	coat protein (CP)	*Cucumber mosaic virus*
Tobacco	Coat Protein (CP)	*Alfalfa mosaic virus*
Tobacco	*N* gene *TSWV*	*Impatiens necrotic spot virus, Groundnut ring spot virus*
Tomato	coat protein (CP)	*Cucumber mosaic virus*

Sanghera *et al.*, 2011; David *et al.*, 2008 and references therein.

to render resistance against fungal pathogens (Sahrawat *et al.*, 2003). Genetically transformed plants showed spontaneous activation of different defense mechanisms. This defense mode greatly enhances the plant's ability to quickly react to a pathogen invasion and to overcome the infection. Antimicrobial proteins, peptides, and lysozymes that naturally occur in insects (Jaynes *et al.*, 1987), plants (Broekaert *et al.*, 1997), animals (Vunnam *et al.*, 1997), and human beings (Nakajima *et al.*, 1993) are

Table 19.3: Transgenic Crops Engineered for Enhanced Resistance against Fungal and Bacterial Pathogens

Plants	*Gene*	*Against the Fungal and Bacterial Pathogen*
Cotton	N*PR1*	*Verticillium dahliae*, *Fusarium oxysporum*f. sp. *vasinfectum*, *Rhizoctonia solani*, and *Alternaria alternata*
Carrot	Acidic wheat class IV chitinase + acidic wheat β 1,3-glucanase + rice cationic peroxidase (*POC1*)	*Sclerotinia sclerotiorum*
Carrot	Microbial factor 3 (*MF3*)	*Pseudomonas fluorescence, Alternaria dauci, Alternaria radicina* and *Botrytis cinerea*
Tobacco	*hrp N*	*Erwinia amylovora, Botrytis cinerea*
Rice	PRm5	Enhanced resistance to multiple fungal pathogen
Carrot	Lipid transfer protein gene and chitinase	Wheat, Barley Foliar fungal pathogen
Tobacco	WRKY 1	Multiple fungal pathogen
Rice	RCH10, RAC22, β-glucanase, β- RIP	*Magnaporthe grisea*
Tobacco	PR1	*W. japonica B. cinerea*
Arabidopsis	SAR 8.2 gene (CASAR82A)	*Fusarium* and *Botrytis*
Rice	Allene oxide synthase	*Magnaporthe grisea*
Wheat	Ace-AMP1	Enhanced antifungal activity
Tobacco	*GAFP* (Gastrodia antifungal)	*Rhizoctonia* spp., *Phytophthora* spp.
Gastrodia (Orchid)	PR3	*T. viride R. solani*
Tobacco	PR3	*R. solani*
Rice	*Cercosporin A*	*Magnaporthe grisea*
Tobacco	*hrp N*	*Botrytis cinerea, Erwinia amylovora*
Pearl millet	*Afp*	*Aspergillus giganteus*
Wheat	*NPR 1*	*Fusarium graminaereum*
Rice	*AFP*	*Aspergillus giganteus, Magnaporthe grisea*
Wheat	*Stilbene synthase*	*Puccinia recondite*

Contd...

Table 19.3–*Contd...*

Plants	*Gene*	*Against the Fungal and Bacterial Pathogen*
Tobacco	Cercopin-A-melittin peptide	*Fusarium solani*
Tomato	*NPR1*	Resistance to fungal and bacterial disease
Rice	*AFP*	*Aspergillus giganteus, Magnaporthe grisea*
Arabidopsis	*Fusarium* specific antibody linked to antifungal peptides	Multiple fungal pathogens
Rice	*ech*42, *nag* 70, *gluc*	*Trichoderma atroviridae, Magnaporthe grisea*
Peanut	PR2	*S. minor*
Rice	*Chitinase C* (Chi C)	*Streptomyces griseus, Magnaporthe grisea*
Wheat	*Thaumatin like protein,chitinase, glucanase*	*Fusarium graminearum*
Tomato	*Pn-AMPs* (hevein like protein)	*Pharbitis nil Phytophthora* spp., *Fusarium* spp.
Tobacco and Banana	MSI-99 peptide	*Alternaria, Botrytis, Mycosphaerella musicola*
Tomato	PR2	*F. oxysporum*
Tobacco	Mannitol dehydrogenase	*Alternaria alternata*
Tobacco	Spi-2 (peroxidase gene)	*Phytophthora* spp.
Rice	Puroindolines (antimicrobial peptide gene)	*Magnporthe grisea, Rhizoctonia solani*
Wheat	Chitinase	*Blumeria graminis, Puccinia recondita*
Poplar	Oxalate oxidase	*Septoria musiva*
Rice	Chitinase	*Rhizoctonia solani*
Grape	Endochitinase	*Trichoderma harzianum, Botrytis cinerea*
Tobacco	Chitinase	*Baculovirus Alternaria alternate*
Potato	Defensins (alfAFP)	*Verticillium dalhiae*
Potato	Cercosporin-melittin cationic peptide Synthetic gene	Multiple pathogens

Contd...

Table 19.3–*Contd...*

Plants	*Gene*	*Against the Fungal and Bacterial Pathogen*
Tomato	Gene1-2	*Fusarium* spp.
Tobacco	Sarcotoxin peptideGene, Sarcophaga peregrine	*Rhizoctonia solani, Pythium aphanidermatum, Phytophthora nicotianae*
Tobacco	Chloroperoxidase	*Pseudomonas Colletotrichum destructivum*
Wheat	RIP	*Blumeria graminis*
Alfalfa	Resveratrol synthase	*Phoma medicaginis*
Tomato	Defensin	*Alternaria solani*
Carrot	Human lysozyme	*Erysiphe heraclei, Alternaria dauci*
Grape	Chitinase	*Uncinulla necatar, Elsinoe ampelina*
Grape	Polygalacturoase inhibiting protein	*Botrytis cinerea*
Tobacco	Salicylic acid synthase	Bacterial origin *Oidium lycopersicon*
Carrot	Chitinase	*Alternaria dauci, A.radicina,Colletotrichum corotae*
Tobacco	Antimicrobial Synthetic peptide	*Colletotrichum destructivum*
Tomato	Oxalate decarboxylase	*Sclerotinia sclerotiorum*
Wheat	TL protein	*Fusarium graminearum*
Rice	TL protein	*Rhizoctonia solani*
Rice	RIP	*R. solani, M. grisea*
Potato	Osmotin gene	*Phytophthora infestans*
Chrysanthemum	Chitinase	*Botrytis cinerea*
Geranium	Antimicrobial protein	*Botrytis cinerea*
Wheat	PR5	*E. graminis*
Tobacco	β-cryptogein elicitor	*Phytophthora cryptogea, Phytophthora parasitica*
Tobacco	PAPII	*Phytolacca americana*
Potato	Endochitinase	*Trichoderma harzianum*

Contd...

Table 19.3–*Contd...*

Plants	*Gene*	*Against the Fungal and Bacterial Pathogen*
Apple	*attacin E gene* (*attE*)	*Sarcophaga peregrine, E. amylovora*
Potato	*sarco* gene coding for sarcotoxin IA	*Sarcophaga peregrine, E. carotovora, P. syringae pv. lachrymans* and *R. solanacearum*
Tobacco	*argK*/ROCT ornithine carbamoyl transferase	*Pseudomonas syringae, Pseudomonas syringae* pv. phaseolicola
Potato	chimaeric peptide MsrA1+ melittin	*E. carotovora* ssp. *atroseptica*
Tobacco	*bO*/Bacterio-opsin (BO)	*Halobacterium halobium, Pseudomonas syringae* pv. tabaci
Tobacco	*expI*/N-oxoacyl-homoserine lactone	*Erwinia carotovora*
Tobacco	*aiiA*/Acyl-homoserinelactonase	*Erwinia carotovor, Bacillus* sp.

Sanghera *et al.*, 2011 and references therein.

the potential source of plant resistance. The role of chitinases in fungal protection has been documented in rice (Datta 2004; Itoh *et al.*, 2003). Some proteins, called defensins, are small cysteine-rich peptides with antimicrobial activity. Expression of genes for suitable pathogenesis related proteins and defensins offers a suitable approach for controlling diseases. Transgenic expression of plant defensins has been reported to enhance protection in vegetative tissues against pathogen attack. The constitutive expression of an alfalfa defensin in potato provided robust resistance against the agronomically important fungus *Verticillium dahlia* under field conditions (Gao *et al.*, 2000). Overexpression of *BSD1* (stamen specific defensin) in transgenic tobacco plants enhanced their tolerance against the pathogen *Phytophthora parasitica* (Park *et al.*, 2002). Another alfalfa defensin was shown to inhibit the growth of the FHB pathogen *Fusarium graminearum* in wheat (Spelbrink *et al.*, 2004).

Genetically transformed plants show spontaneous activation of different defense mechanisms. This defense mode greatly enhances the plant's ability to quickly react to a pathogen invasion and to overcome the infection. Antimicrobial proteins, peptides, and lysozymes that naturally occur in insects (Jaynes *et al.*, 1987), plants (Broekaert *et al.*, 1997), animals (Vunnam *et al.*, 1997), and human beings (Mitra and Zhang 1994; Nakajima *et al.*, 1993) are the potential source of plant resistance. Antibacterial lytic peptides like cecropins are found in the hemolymph of the giant silk moth (*Hyalophora cecropia*) (Durell *et al.*, 1992, Tripathi *et al.*, 2004), attacins is another group of antibacterial proteins produced by *Hyalophora cecropia* pupae (Hultmark *et al.*, 1983). Transgenic plants expressing cecropins have increased resistance to *P. syringae* pv. *tabaci*, (Huang *et al.*, 1997) and Attacin expressed in transgenic potato enhanced its resistance to bacterial infection by *E. carotovora* subsp. *atrospetica* (Arce *et al.*, 1999). Synthetic lytic peptide analogs, *Shiva-1* and *SB-37*, produced from transgenes in potato plants reduce bacterial infection caused by *Erwinia carotovora* subsp. *atroseptica* in transgenic potato plants (Arce *et al.*, 1999).

Similarly, transgenic rice plants overexpressing *cecropin B* gene showed a significant reduction in development of lesions caused by *X. oryzae* pv. *oryzae* (Coca *et al.*, 2004). Beside this, a new approach to protect plants against bacterial diseases is based on interference with the communication system, quorum-sensing, used by several phytopathogenic bacteria to regulate expression of virulence genes according to population density (Cui and Harling 2005). The enzyme, *AiiA*, isolated from bacterial strain, *Bacillus* sp.240B1, was found to degrade the quorum-sensing signalling molecule of the soft rot pathogen, *Erwinia carotovora*, and thereby rendering the bacteria incapable of infecting the host (Dong *et al.*, 2000). Transgenic expression of *AiiA in planta* was subsequently demonstrated to provide significant enhancement of resistance against soft rot in potato (Dong *et al.*, 2001).

References

Arce P, Moreno M, Gutierrez M, Gebauer M, Dell'Orto P, Torres H, Acuna I, Oliger P, Venegas A, Jordana X, Kalazich J and Holuigue L (1999). Enhanced resistance to bacterial infection by Erwinia carotovora subsp. atroseptica in transgenic potato plants expressing the attacin or the cecropin SB-37 genes. *Amer J Potato Res* **76**: 169- 177.

Bravo A, Gill SS and Sobero´n M (2007). Mode of action of *Bacillus thuringiensis* Cry and Cyt toxins and their potential for insect control. *Toxicon* **49**: 423–435.

Broekaert WF, Cammue BPA, De Bolle M FC, Thevissen K, De Samblanx G W and Osborn RW (1997). Antimicrobial peptides from plants. *Critical Rev Pl Sci* **16**: 297-323.

Brogue K, Chet I, Holliday M, Cressman R, Biddle P, Knowlton S, Mauvais CJ and Broglie R (1991). Transgenic plants with enhanced resistance to the fungal pathogen *Rhizoctonia solani*. *Science* **254**: 1194-1197.

Brookes G and Barfoot P (2005). GM crops: the global economic and environmental impact: the first nine years 1996-2004. *AgBioForum* **8**: 187-196.

Coca M, Bortolotti C, Rufat M, Peñas G, Eritja R, Tharreau D, del Pozo AM, Messeguer J and San Segundo B (2004). Transgenic rice plants expressing the antifungal AFP protein from *Aspergillus giganteus* show enhanced resistance to the rice blast fungus *Magnaporthe grisea*. *Plant Mol Biol* **54**: 245-259.

Cui X and Harling R (2005). N-acyl-homoserine lactone-mediated quorum sensing blockage, a novel strategy for attenuating pathogenicity of Gram-negative bacteria plant pathogens. *Euro J Pl Path* **111**: 327–339.

Datta, S. K. (2004). Rice biotechnology: A need for developing countries. *AgBioForum*, 7, 31–35.

David B. Collinge and Ole Søgaard Lund and Hans Thordal-Christensen (2008). what are the prospects for genetically engineered, disease resistant plants? *Eur J Plant Pathol* **121**:217–231.

De Cosa B, MoarW, Lee SB, Miller M and Daniell H (2001). Overexpression of the Bt cry2Aa2 operon in chloroplasts leads to formation of insecticidal crystals. *Nat Biotechnol* **19**: 71–74.

Dong YH, Wang LH, Xu JL, Zhang HB, Zhang XF, and Zhang LH (2001). Quenching quorum-sensing-dependent bacterial infection by an N-acyl homoserine lactonase. *Nature* **411**: 813–817.

Dong YH, Xu JL, Li XZ and Zhang LH (2000). AiiA, an enzyme that inactivates the acylhomoserine lactone quorum- sensing signal and attenuates the virulence of *Erwinia carotovora*. *Proc Natl Acad Sci* USA, **97**: 3526–3531.

Dufourmantel N, Tissot G, Goutorbe F, Garcon F, Muhr C, Jansens S, Pelissier B, Peltier G and Dubald M (2005). Generation and analysis of soybean plastid transformants expressing *Bacillus thuringiensis* Cry1Ab protoxin. *Plant Mol Biol* 58: 659–668.

Durell SR, Raghunathan G and Guy HR (1992). Modeling the ion channel structure of cecropin. *Biophys J* **63**: 1623-1631.

Fuchs M, Ferreira S and Gonsalves D (1997). Management of virus diseases by classical and engineered protection. Molecular Plant Pathology On-Linc http://www.bspp.org.uk/mppol/] 1997/0116fuchs.

Fuchs M and Gonsalves D (2007). Safety of virus-resistant transgenic plants two decades after their introduction: Lessons from realistic field risk assessment studies. *Annual Review of Phytopathology* **45:** 173–202.

Gao AG, Hakimi SM, Mittanck CA, Wu Y, Woerner BM, Stark DM, Shah DM, Liang J and Rommens CM (2000). Fungal pathogen protection in potato by expression of a plant defensin peptide. *Nature Biotech* **18**: 1307-1310.

Gonsalves D (1998). Control of papaya ringspot virus in papaya: a case study. *Ann Rev Phytopathology* **36**:415–37.

Gonsalves D (2004). Transgenic papaya in Hawaii and beyond. *AgBioForum*, **7:** 36–40.

Gosal SS, Wani SH, Kang MS, (2009). Biotechnology and drought tolerance. *Journal of Crop Improvement* **23**: 19-54.

Huang Y, Nordeen RO, Di M, Owens LD, McBeth JH (1997). Expression of an engineered cecropin gene cassette in transgenic tobacco plants confers disease resistance to *Pseudomonas syringae* pv. tabaci. *Phytopathol* **87:** 494–499.

Hultmark D, Engström Å, Andersson K, Steiner H, Bennich H, Boman HG (1983). Insect immunity. Attacins, a family of antibacterial proteins from Hyalophora cecropia. *Euro Mol Biol Organi* J **2:** 571-576.

Itoh Y, Takashahi K, Takizawa H, Nikaidou N, Tanaka H, Nishihashi H, Watanabe T and Nishizawa Y (2003). Family 19 chitinase of Streptomyces griseus HUT6037 increases plant resistance to the fungal disease. *Bioscience Biotechnology and Biochemistry* **67:** 847-855.

Jaynes JM, Xanthopoulos KG, Destéfano-Beltrán L and Dodds JH (1987). Increasing bacterial disease resistance in plants utilizing antibacterial genes from insects. *Bioassays* **6:** 263-270.

Kota M, Daniell H, Varma S, Garczynski SF, Gould F and Moar WJ (1999). Overexpression of the *Bacillus thuringiensis* (*Bt*) Cry2Aa2 protein in chloroplasts confers resistance to plants against susceptible and Bt-resistant insects. *Proc Natl Acad Sci* USA **96**: 1840–1845.

Kumar, P. (1998). Food Demand and Supply Projections for India. Paper No. 98-01, Indian Agricultural Research Institute, New Delhi.

McBride KE, Svab Z, Schael DJ, Hogan PS, Stalker KM and Maliga P (1995). Amplification of a chimeric Bacillus gene in chloroplasts leads to an extraordinary level of an insecticidal protein in tobacco. *Bio/Technology* **13**: 362–365.

Mehlo L, Gahakwa D, Nghia PT, Loc NT, Capell T, Gatehouse JA, Gatehouse AMR and Christou P (2005). An alternative strategy for sustainable pest resistance in genetically enhanced crops. *Proc Natl Acad Sci* USA **102**: 7812–7816.

Mitra A and Zhang Z (1994). Expression of a human lactoferrin cDNA in tobacco cells produces antibacterial protein(s). *Plant Physiol* **106:** 977- 981.

Nakajima M, Hayakawa T, Nakamura I and Suzuki M (1993). Protection against Cucumber mosaic virus (CMV) strains O and Y and chrysanthemum mild mottle virus in transgenic tobacco plants expressing CMV-O coat protein. *J Gen Virol* **74:** 319-322.

Ortiz, R. (1998). Critical role of plant biotechnology for the genetic improvement of food crops: Perspectives for the next millennium. *Electronic J Biotech* [on line] **1(3)**: Issue of August 15.

Park HC, Kang YH, Chun HJ, Koo JC, Cheong YH, Kim CY, Kim MC, Chung WS, Kim JC, Yoo JH, Koo YD, Koo SC, Lim CO, Lee SY and Cho MJ (2002). Characterization of a stamen-specific cDNA encoding a novel plant defensin in Chinese cabbage. *Plant Molecular Biology* **50:**59–69.

Robinson J (1999). Ethics and transgenic crops: a review. *Electronic J Biotech* **2**:71-81.

Saha P, Majumder P, Dutta I, Ray T, Roy SC and Das S (2006). Transgenic rice expressing *Allium sativum* leaf lectin with enhanced resistance against sap-sucking insect pests. *Planta* **223**: 1329–1343.

Sahrawat AK, Becker D, Lütticke S and Lörz H (2003). Genetic improvement of wheat via alien gene transfer, an assessment. *Plant Sci* **165:**1147–1168.

Sanghera, GS. Wani, SH Singh, G Kashyap PL and. Singh NB (2011). Designing crop plants for biotic stresses using transgenic approach. *Vegetos* Vol. **24** (1) : 1-25

Spelbrink RG, Dilmac N, Allen A, Smith TJ, Shah DM and Hockerman GH (2004). Differential antifungal and calcium channel blocking activity among structurally related plant defensins. *Plant Physiology* **135:**2055–2067.

Toenniessen GH, O'Toole JC, DeVries J (2003). Advances in plant biotechnology and its adoption in developing countries. *Curr Opin Plant Biol* **6**: 191–198.

Tripathi L, Tripathi JN and Tushemereirwe WK (2004). Strategies for resistance to bacterial wilt disease of bananas through genetic engineering. *African J Biotech* **3:** 688-692.

Vanderschuren H, Stupak M, Futterer J, Gruissem W. and Zhang P (2007). Engineering resistance to geminiviruses – Review and perspectives. *Plant biotechnology J.* **5**: 207-220.

Vunnam S, Juvvadi P and Merrifield RB (1997). Synthesis and antibacterial action of cecropin and proline-arginine-rich peptides from pig intestine. *J Peptide Res* **49:** 59-66.

Wani S H and Sanghera G S (2010). Genetic engineering for viral disease management in plants. *Notulae Scientia Biologicae* **2**: 20-28.

2016, Recent Advances in Plant Stress Physiology *Pages* **461–472**
Editors: ***Praduman Yadav, Sunil Kumar and Veena Jain***
Published by: **DAYA PUBLISHING HOUSE, NEW DELHI**

Chapter 20

Advances in Biotechnology for Development of Submergence Tolerance in Crops

Nikhil Khurana[1*] and Jitender Kumar[2]

[1]School of Biotechnology, Gautam Buddha University, Greater Noida – 201 310, U.P., India
[2]Department of Molecular Biology and Biotechnology, CCS Haryana Agricultural University, Hisar – 125 004, Haryana, India

ABSTRACT

More than 16 million ha of rice lands of the world in lowland and deep-water rice areas are unfavorably affected by flooding due to complete submergence. In eastern India, about 13 m ha of rice lands are negatively affected by excess water and periodically suffer from flash-floods and complete submergence. Barley, maize and soybean crops are also affected by waterlogging which causes chlorosis, necrosis, stunting, degradation of proteins that ultimately lead to 20-30 per cent yield loss. Submergence has been identified as the third most important constraint for higher rice productivity in eastern India, because it sometimes resulted in near total yield loss. Improvement of germplasm is likely the best option to withstand submergence and stabilize productivity in these environments. The trait for submergence tolerance can be improved if the genetic bases of submergence are well understood. Research outcomes suggested both simple and quantitative inheritances are responsible for submergence tolerance. This chapter focuses on physiological understanding of tolerance to submergence in crops and to explore the potential recent technologies like genetic engineering and molecular breeding to enhance the tolerance for submergence in water logged crops.

Keywords: *Breeding, Flooding, Genetic engineering, Submergence.*

**Corresponding author*: E-mail: nickhil90@gmail.com

Introduction

Approximately 30 per cent of the world's rice (*Oryza sativa* L.) farmlands are at a low elevation and irrigated byrain (Bailey-Serres *et al.*, 2010). In South Asia, about 12 m ha of rice is prone to flash flooding. In India, about 17.4 m ha of rainfed lowland rice are grown each year; of which 5.2 m ha are submergence-prone (Singh *et al.*, 2013). Flash flooding and the resulting submergence of rice adversely affects at least 16 per cent of the rice lands of the world. Rice has adapted to flood-prone environments, *i.e.* where excessive flooding limits growth and yields, by either submergence tolerance or elongation ability (Mohanty and Chaudhary,1986). In India, food grain production has increased from 50.8 to 255 m tones from the year 1950-1951 to 2012-2013. Rice and wheat production has reached to 103 and 92.46 m tones respectively in the year 2012-2013. Rice is grown in a wide range of ecologies ranging from irrigated to uplands, rainfed lowland, deep water and tidal wetlands. About 29 per cent of India's total rice area, *i.e.* ~13 m ha is rainfed lowland, which contributes only 19 per cent to national rice production. In a normal year, about 4 m ha is drought-prone, while 3 m ha is favorable, another 3 m ha is medium deep waterlogged with water standing for up to 50 cm. The remaining 3 m ha is submergence or flood-prone, where plants are completely submerged for 1–2 weeks or so, resulting in partial or even complete crop failure. Rainfed lowlands constitute highly fragile ecosystems, always prone to flash-floods (submergence) with an average productivity of only 1.2 t ha in normal years and hardly 0.5 t ha (Singh *et al.*, 2002) in case of submergence. Among the 42 biotic and abiotic stresses affecting rice production, submergence has been identified as the third most important constraint for higher rice productivity in eastern India (Hossain *et al.*, 2002) because it sometimes resulted in near total yield loss. Besides India, flooding is widespread in other South and Southeast Asian countries such as Bangladesh, Thailand, Vietnam, Myanmar and Indonesia. Suitable germplasm as well as management technologies are therefore needed to enhance and stabilize rice productivity in these areas.

Plant breeding has produced elite lines with up to four-fold greater yields and submergence tolerance equal to the world's most tolerant cultivars, but successful introduction of these elite lines in the field is elusive. Development in molecular marker assisted breeding and functional genomic tools for identification of QTLs and candidate genes for a particular trait can provide solutions to the abiotic stresses. New technologies are essential to genetically dissect complex quantitative traits and use them for abiotic stress tolerance in high yielding varieties. The present article focuses on the complicated problems to which rice crop is exposed during submergence. It also emphasizes various tolerant traits or mechanisms that are necessary for high and stable productivity in submergence prone areas.

Physiology of Submergence Tolerance in Plants

Waterlogging is a serious problem in low-laying rain fed areas. Lack of oxygen supply for the plant is the main reason of damage in waterlogging condition, which hampers nutrient and water uptake, as a reason the plant shows wilting. Characterization of flood water environments in most rice growing areas pointed to gas diffusion as the most important limiting factor under complete submergence.

This is because gas diffusion is much slower in water than in air. Oxygen levels in flood water vary with location and time of day, usually below air saturation during the night but may become super saturated during the day. Anoxia for 24 h can kill sensitive rice varieties presumably because of the need for O_2 for respiration to maintain survival and elongation growth processes. Plants which are resistant to submergence have mechanism such as increased availability of soluble sugars (sucrose, glucose and fructose), aerenchyma formation, greater activity of glycolytic pathway and fermentation enzyme and use of antioxidant defense mechanism to cope with oxidative stress induced by waterlogging (Ahmed *et al.*, 2013). Under flash flooding, few characters were identified as playing a key role in submergence tolerance in rice, the most critical are: maintenance of high carbohydrate concentration, optimum rates of alcoholic fermentation and energy conservation by maintaining low elongation growth rates during submergence. Under submerged conditions, mainly ethylene, gibberllin and ABA hormones are synthesized in plants which are involved in regulation of shoot elongation (Voesenek *et al.*, 2003). Ethylene plays a key role in submergence tolerance partly through regulating gibberllin and ABA which are positive and negative regulators of shoot elongation respectively. After the ethylene concentration rises, there is first decrease in the level of ABA and then an increase in the level of endogenous gibberllin. The rapid decrease in ABA level thought to be prerequisite for submergence-induced gibberllin biosynthesis thereby promoting internode or petiole elongation (Kende *et al.*, 1998).

Alcoholic fermentation is the primary means by which plants produce energy under anaerobiosis (Kennedy *et al.*, 1992; Drew *et al.*, 1997) and much importance is given to the enzymes produced during alcoholic fermentation and endogenous substrate carbohydrate (Sarkar *et al.*, 1996; Settler *et al.*, 1997; Dubey *et al.*, 2003; Kato-Noguchi *et al.*, 2004). It is considered as a source of energy and has been recognized as an important factor in submergence tolerance of rice. It has been found that submergence tolerant rice cultivars have around 30-50 per cent more non-structural carbohydrates (Chauturvedi *et al.*, 1995, Settler *et al.*, 1997; Sarkar *et al.*, 1996). Non-structural carbohydrates are metabolized to generate energy for growth and maintenance. (Sarkar *et al.*, 1996; Huang *et al.*, 2005). Due to prolong flooding, old leaves die and further initiation of new leaves and their subsequent growth requires availability of non- structural carbohydrates and those cultivars which maintain more than 6 per cent of their initial non- structural carbohydrates at the time of re-aeration were found to be capable of developing new leaves quickly (Das *et al.*, 2001; Das *et al.*, 2005). Alcoholic fermentation is recognized as the principle catalytic pathway for recycling NAD to maintain glycolysis and substrate level phosphorylation in absence of O_2 and is known for energy production during anoxic conditions (limited conditions).

The submerged plants are promptly exposed to an environment with high light and oxygen tension, leading to formation of reactive oxygen species (ROS). Anoxia and low light are found to be the important factors for reducing the life of submerged plants and is responsible for severe leaf damage. Tolerant rice plants have appeared to develop two mechanisms against oxidative damage.

1. **Natural antioxidants:** Many natural compounds such as carotenoids, ascorbate, tocopherol, phenols have found to be involved in oxidative stress response mechanisms and are known for their antioxidaive properties. Most of the work has been done on ascorbate and the latter plays an important role in protecting the plants against damage by ROS (Kawano *et al.*, 2002; Das *et al.*, 2004; Hancock *et al.*, 2005).
2. **Antioxidant enzyme system:** Under submergence conditions, the activity of superoxide dismutase (SOD) decreases and after exposure to air, the activity of this enzyme remains below the level of controlled plants (Sarkar *et al.*, 2001) and the pathway involving this enzyme is not so efficient in removing reactive oxygen species (ROS) in rice. Upon de-submergence, the ascorbate-glutathione cycle (Asada-Halliwel pathway) may play an important role in scavenging ROS. Ascorbate-glutathione cycle regenerates the reduced forms of ascorbate and glutathione that are primary scavengers. Maintenance of higher levels of reduced forms of ascorbate is essential to mitigate the oxidative damage both under submergence as well as salt-stress situations. Ascorbate works in cooperation with glutathione and also helps maintain the regeneration of α-tocopherol, providing a synergetic role in the protection of cellular membranes (Thomas *et al.*, 1992; Garnczarska, 2005). Under submerged conditions crop plants adopt various strategies to survive under the anoxic environment (Figure 20.1).

The average rice productivity of submergence-prone areas in eastern India is 0.5 - 0.8 tonnes (t) ha^{-1}, whereas it is about 2.0 t ha^{-1} for favourable rainfed lowlands, being much lower than the input-intensive irrigated system (5.0 tha^{-1}) (Bhowmick *et al.*, 2014). Submergence also deteriorates the grain quality in rice. The kernels in flood-affected samples became soft and developed fissures which contributed to low head rice recoveries and the milled rice had lower kernel weight and protein content but showed higher amylose and ash content (Singh *et al.*, 1990).

In other crops like barley, maize, soybean and wheat also submergence and waterlogging causes severe loss to yield and quality of seeds. In barley and maize, waterlogging causes chlorophyll, protein, and degradation of RNA, and also reduces the concentration of nutrients like nitrogen, phosphorus, metal ions, and minerals in shoot (Ahmed *et al.*, 2013; Setter *et al.*, 1999; Rathore and Warsi, 1998). Soybean is also affected by submergence and because of waterlogging chlorosis, necrosis, stunting, defoliation, reduced nitrogen fixation, and plant death occurred which causes yield loss (Reyna *et al.*, 2003; Bacanamwo and Purcell, 1999). In case of wheat about 39-40 per cent yield loss is recorded under waterlogging condition (Collaku and Harrison, 2002; Musgrave and Ding, 1998). Under waterlogging conditions a combining effect of reduced kernel and tiller numbers are responsible for reducing yield of wheat (Collaku and Harrison, 2002).

The coastal regions are prone to flooding, therefore, the plants present here face salinity and submergence stresses. Flooding causes waterlogging and sometimes complete submergence, as well as salinity and flooded soil typically become anoxic where CO_2, ethylene, Mn^{2+}, Fe^{2+}, S^{2-} and carboxylic acids can accumulate,

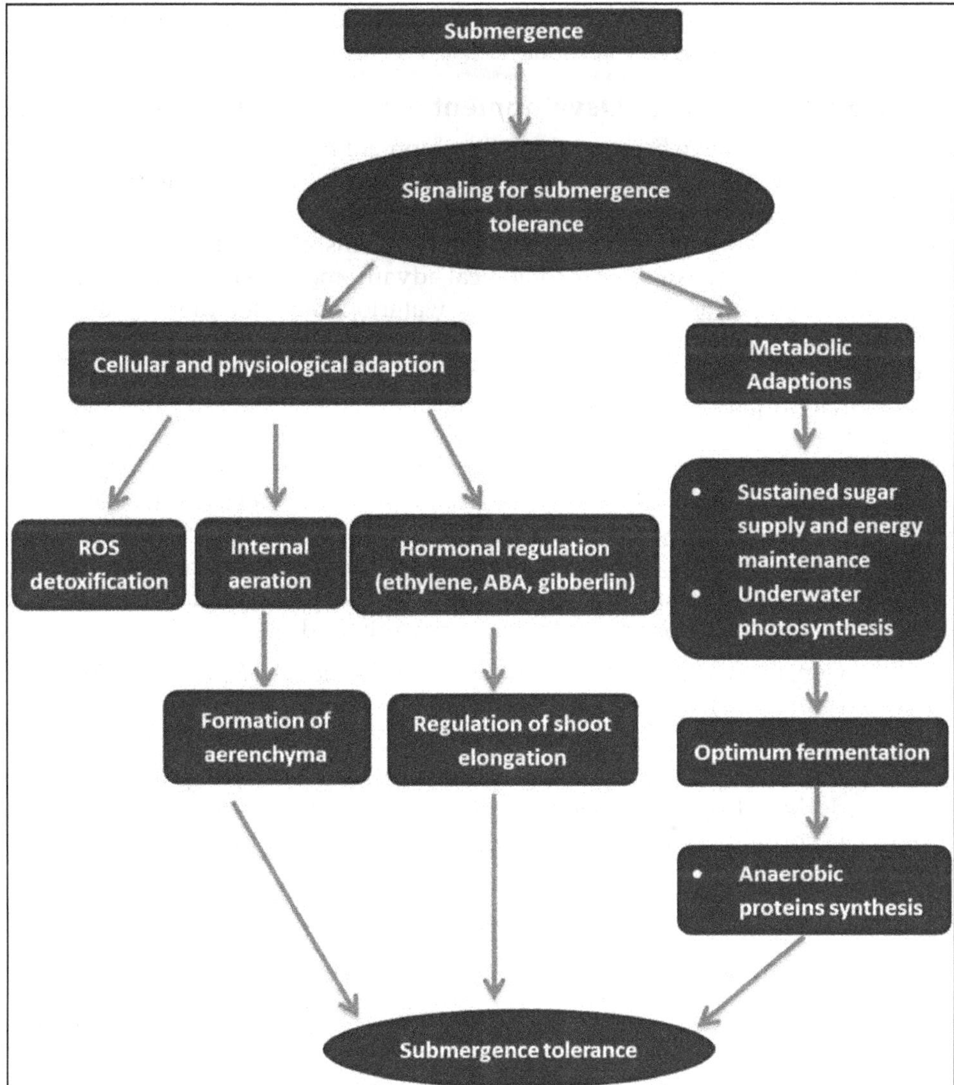

Figure 20.1: Strategies Adopted by Plants to Survive under Submerged Conditions.

impacting negatively on roots of poorly adapted plants (Greenway *et al.*, 2006). Colmer *et al.*, 2009 studied tolerance of combined submergence and salinity in the halophytic *Tecticorniaper granulata* and concluded that this plant tolerates complete submergence, even in waters of high salinity. A 'quiescence response', *i.e.* no shoot growth, would conserve carbohydrates, but tissue sugars still declined with time. A low K^+: Na^+ ratio, typical for tissues of succulent halophytes, was tolerated even during prolonged submergence, as evidenced by maintenance of underwater net photosynthesis (P_N) at up to 400 mMNaCl. Underwater P_N provides O_2 and sugars, and thus should enhance survival of submerged plants. More research is required to completely understand the various metabolic changes occurring during submergence under saline conditions.

Transgenic Strategy for Development of Submergence Tolerant Rice

Various genes have been used for development of submergence tolerance in rice. Development in various methods and use of specific promoters for production of transgenic plants has led to advancement in production of transgenic rice with specific attribute like tolerance to virus, insects, insects, pests and salt can be produced. The lacunae of this technological advancement lie in the identification, isolation, cloning of associated genes with waterlogging. Due to involvement of carbohydrates in abiotic stress tolerance mechanism researchers are focusing on carbohydrate metabolism. It has been reported that there is switch from respiration to fermentation pathway. Ethanol fermentation involves two enzymes pyruvate decarboxylase (PDC) and alcohol dehydrogenase (ADH). Research group at CISRO (Division of Plant Industry), Australia are trying to alter the levels of two genes *adh* and *pdc* in rice. The two *adh* genes present in rice have been cloned (Xie and Wu, 1989). Three different *pdc*genes in rice have been cloned and characterized (Ahmed *et al.*, 2013; Hossain *et al.*, 1994; Huq *et al.*, 1995). The coding regions of *pdc* or *adh* was inserted into the sequences which allow their expression after introduction into the rice genome. Two types of promoters were used:

1. Constitutive promoter (*CaMV35S* and the rice *Act1* promoter; McElroy *et al.*, 1991) which promotes the expression of downstream sequences irrespective of environment conditions.
2. Inducible promoter (*6XARE* promoter; Last *et al.*, 1991) and originally derived from maize *adh* gene and was engineered in such a way that it contains six copies of the promoter motif required for anaerobic inducibility (Olive *et al.*, 1990).

Thus the levels of ADH and PDC specifically modified under conditions of the anaerobic stress.

Molecular Breeding for Development of Submergence Tolerance in Rice

During submergence in the field, plants are exposed to various biophysical stresses including reduced gas exchange between plant surfaces and outside atmosphere. Gases are known to diffuse 10,000 times slower in water than air (Armstrong, 1979). Reduced air exchange adversely affects plant growth and metabolic processes. Reduced O_2 supply limits respiration, reduced CO_2 supply limits photosynthesis and reduced ethylene diffusion away from the plant triggers chlorosis and excessive elongation of leaves of intolerant cultivars (Ella *et al.*, 2003). Rice cultivars vary in their capacity to tolerate complete submergence; quantitative trait loci analyses have revealed that a large portion of this variation in submergence tolerance can be explained by one locus (*Sub1*) on chromosome 9.

Analysis of the completed rice genome sequence provided the identification of thousands of new targets for DNA markers, especially SSRs. Using publically available BAC and PAC clones, more than 200 validated SSRs were released in 2002 (McCouch *et al.*, 2002). This was soon followed by 18828 Class I (di-,tri-, tetra-repeats) SSRs that were released after the completion of the Nipponbare genome sequence in 2005 (Matsumoto *et al.*, 2005). The extremely high density of SSRs (~51 SSRs per Mb) has provided a considerable "tool kit" for map construction and MAS for numerous applications.

Research on the development of flood tolerant rice varieties started as early as 1987 at the International Rice Research Institute (IRRI). *SUB1A*, *SUB1B* and *SUB1C* were the 3 genes identified within the *Sub1* quantitative trait locus (QTL), encoding putative ethylene response factors. *SUB1*A was subsequently identified as the major determinant of submergence tolerance (Xu *et al.*, 2006; Xu *et al.*, 2000; Xu and Mackill, 1996; Septiningsih *et al.*, 2009; Bailey-Serres *et al.*, 2010; Singh *et al.*, 2010). The *SUB1A* gene, which confers tolerance to flooding, was characterized and fine-mapped from one of the landraces of Odisha (FR13A or Dhalaputia) by David Mackill and his team (Xu *et al.*, 2006). The discovery and cloning of the *SUB1A* gene facilitated the introgression of the gene through marker assisted back crossing (MABC) to several high yielding varieties within a short span of 2-3 years and now released in South and South-East Asian countries (Bailey *et al.*, 2010; Collard *et al.*, 2013; Mackill *et al.*, 2012) These varieties with *SUB1A* gene showed a yield advantage of 1 to 1.3 t/ha over the original varieties following submergence for a few days up to 18 days (Gregorio *et al.*, 2013). Submergence tolerant varieties were released by Bangladesh, India, Indonesia, Nepal and the Philippines within last five years by introgression of *Sub1* into mega varieties (Table 20.1).

Table 20.1: Submergence Tolerant Rice Varieties Released

Country	*Name of the Variety*	*Year*
Bangladesh	BRRI dhan51	2010
	BRRI dhan52	
India	Swarna-Sub1	2009
Indonesia	INPARA-3	2008
	INPARA-4	
	INPARA-5	
Phillipines	Submarino	2009
Nepal	Swarna-Sub1	2011
	S. Mahsuri-Sub1	

Source: Gregorio *et al.*, 2013.

In case of rice identification of *Sub1* QTL has played a significant role in development of submergence tolerant rice varieties. In some other crops also some QTLs (*Qwt4-1* in Barley; *Rps* in soybean and QTLs on chromosome 4 and 9 in maize) have been identified that can play important role in development of submergence tolerant crops (Ahmed *et al.*, 2013).

Future Prospects

Rice is the only cereal crop that is well adapted to the conditions of waterlogging or partial flooding. Advances in research tools and availability of rice genome sequence has revealed the unknown mechanisms of abiotic stress tolerance in rice. There is a strong need of development of rice cultivars with improved tolerance to abiotic stresses. Rice genome sequence has provided a lot of information and development of new molecular markers that can play important roles in identification of novel genes/QTLs associated with abiotic stress tolerance mechanisms in rice and other cereal crops. Knowledge of new effective QTLs providing tolerance to submergence can be used for designing next generation rice cultivars tolerant to abiotic stresses. Marker assisted selection and marker assisted backcrossing has opened new way for pyramiding beneficial genes/QTLs in high yielding varieties. Combination of marker assisted breeding and conventional breeding can be used to improve abiotic stress tolerance in plants. Performance of high-yielding varieties can further be augmented if their submergence tolerance is enhanced, which is now becoming more feasible. Current understanding of the physiological and biochemical bases of submergence tolerance has progressed well in recent years, making it possible to design efficient phenotyping protocols and has laid the infrastructure for further genetic and molecular studies, to discover genes underlying component traits associated with tolerance. Still there is need to develop more efficient biotechnological tools that can be used by researchers and plant breeders for increasing the agricultural productivity under abiotic stress conditions.

References

Ahmed, F., Rafii, M.Y., Ismail, M.R., Juraimi, A.S., Rahim, H.A., Asfaliza, R. and Latif, M.A., 2013.Waterlogging Tolerance of Crops: Breeding, Mechanism of Tolerance, Molecular Approaches, and Future Prospects. *Bio. Med. Res. International*,2013:1-10.http://dx.doi.org/10.1155/2013/963525

Armstrong, W., 1979.Aeration in higher plants. *Adv. Bot. Res.*, 7:225–331.

Bacanamwo, M. and Purcell, L.C., 1999.Soybean root morphological and anatomical traits associated with acclimation to flooding. *Crop Sc.*, 39(1):143–149.

Bailey-Serres, J., Fukao, F., Ronald, P., Ismail, A., Heuer, S. and Mackill, D., 2010. Submergence tolerant rice: SUB1's journey fromlandrace to modern cultivar. *Rice*, 3:138-147.

Bhowmick, M.K., Dhara, M.C., Singh, S., Dar, M.H. and Singh, U.S., 2014. Improved Management Options for Submergence-Tolerant (Sub1) Rice Genotype in Flood-Prone Rainfed Lowlands of West Bengal. *Am. J. Plant Sc.*, 5:14-23.

Chaturvedi, G.S., Misra, C.H., Singh, O.N., Pandey, C.B., Yadav, V.P., Singh, A.K., Dwivedi, J.L., Singh, B.B. and Singh, R.K., 1995. Physiological basis and screening for tolerance for flash flooding. In Rainfed Lowland Rice, Agricultural Research for High-Risk Environments (ed. Ingram, K. T.), IRRI, The Philippines, 1995, pp. 79–96.

Collaku, A. and Harrison, S.A., 2002.Losses in wheat due to waterlogging. *Crop Sc.*, 42(2):444–450.

Collard, B.C.Y., Septiningsih, E.M., Das, S.R., Carandang, J.J., Pamplona, A.M., Sanchez, D.L., Kato, Y., Ye, G., Reddy, J.N., Singh, U.S., Iftekharuddaula, K.M., Venuprasad, R., Vera-Cruz, C.N., Mackill, D.J. and Ismail, A.M., 2013. Developing new flood-tolerant varieties at the International Rice Research Institute (IRRI). *SABRAO J. Breed. Genet.*,45:42-56.

Colmer, T.D., Vos, H. and Pedersen, O., 2009. Tolerance of combined submergence and salinity in the halophytic stem-succulent *Tecticornia pergranulata. Annals Bot.*, 103:303–312.

Das, K.K. and Sarkar, R.K., 2001. Post flood changes on the status of chlorophyll, carbohydrate and nitrogen content and its association with submergence tolerance in rice. *Plant Arch.*, 1:15–19.

Das, K.K., Panda, D., Nagaraju, M., Sharma, S.G. and Sarkar, R.K., 2004. Antioxidant enzymes and aldehyde releasing capacity of rice cultivars (*Oryza sativa* L.) as determinants of anaerobic seedling establishment capacity. *Bulg. J. Plant Physiol.*, 30: 34–44.

Das, K.K., Sarkar, R.K. and Ismail, A.M., 2005. Elongation ability and non-structural carbohydrate levels in relation to submergence tolerance in rice. *Plant Sci.*, 168:131–136.

Drew, M.C., 1997. Oxygen deficiency and root metabolism: Injury and acclimation under hypoxia and anoxia. *Annu. Rev. Plant Physiol. Mol. Biol.*, 48:223–250.

Dubey, H., Bhatia, G., Pasha, S. and Grover, A., 2003. Proteome maps of flood-tolerant FR 13A and flood-sensitive IR 54 rice types depicting proteins associated with O_2 deprivation stress and recovery regimes. *Curr. Sci.*, 84:83–89.

Ella, E.S., Kawano, N., Yamauchi, Y., Tanaka, K. and Ismail, A.M., 2003. Blocking ethylene perception enhances flooding tolerance inrice seedlings. *Funct. Plant Biol.*, 30: 813–819.

Garnczarska, M., 2005.Response of ascorbate–glutathione cycle to reaeration following hypoxia in lupine roots. *Plant Physiol. Biochem.*, 43:583–590.

Greenway, H., Armstrong, W. and Colmer, T.D., 2006. Conditions leading to high CO_2 (5 kPa) in waterlogged-flooded soils and possible effects on root growth and metabolism.*Annals Bot.*, 98:9–32.

Gregorio, G.B., Islam, M.R., Vergera, G.V. and Thirumeni, S., 2013. Recent advances in rice science to design salinity and other abiotic stress tolerant rice varieties. *SABRAO J. Breed. Genet.*, 45:31-41.

Hancock, R.D. and Viola, R., 2005. Biosynthesis and catabolism of Lascorbic acid in plants. *Crit. Rev. Plant Sci.*, 24:167–188.

Hossain, M.A., Huq, E. and Hodges, T.K., 1994. Sequence of cDNA from *Oryza sativa* L. encoding pyruvate decarboxylase I gene. *Plant Physiol.*, 106:799-800.

Hossain, M.A., McGee, J.D., Grover, A., Dennis, E., Peacock, W.J. and Hodges, T.K., 1994. Nucleotide sequence of a rice genomic pyruvate decarboxylase gene that lacks introns: a pseudo-gene? *Plant Physiol.*, 106:1697-1698.

Hossain, M. and Laborte, A., 2002. Differential growth in rice production in eastern India: agroecological and socio-economic constraints, 2002. In Physiology of Stress Tolerance of Rice, NDUAT and IRRI, Los Banos, The Philippines, pp. 221–239.

Huang, C., He, W., Guo, J., Chang, X., Su, P. and Zhang, L., 2005.Increased sensitivity to salt stress in an ascorbate-deficient Arabidopsis mutant. *J. Exp. Bot.*, 56:3041–3049.

Huq, M.E., Hossain, M.A., Hodges, T.K., 1995. Cloning and sequencing of a cDNA encoding pyruvate decarboxylase 2 gene (Accession No. U27350) from rice (PGR95-072). *Plant Physiol.*, 109:722.

Kato-Noguchi, H., Sasaki, R. and Ichii, M., 2004.Anoxia tolerance in a root hair defective mutant of rice. *Russ. J. Plant Physiol.*, 51:116–119.

Kawano, N., Ella, E., Ito, O., Yamauchi, Y. and Tanaka, K., 2002. Metabolic changes in rice seedlings with different submergence tolerance after desubmergence. *Environ. Exp. Bot.*, 47:95–203.

Kende, H., Vander, K.E. and Cho, H.T., 1998. Deepwater rice: a model plant to study stem elongation. *Plant Physiol.*, 118:1105–1110.

Kennedy, R.A., Rumpho, M.E. and Fox, T.C., 1992.Anaerobic metabolism in plants. *Plant Physiol.*, 100:1–6.

Last, D.I., Brettell, R.I.S., Chamberlain, D.A., Chaudhury, A.M., Larkin, P.J., Marsh, E.L., Peacock, W.J. and Dennis, E.S., 1991. PEMU: an improved promoter for gene expression in cereal cells. *Theor. Appl. Genet.*, 81:581-588.

Mackill, D.J., Ismail, A.M., Singh, U.S., Labios, R.V. and Paris, T.R., 2012. Development and rapid adoption of submergence-tolerant (Sub1) rice varieties. *Adv. Agron.*,115:303-356.

Matsumoto, T., Wu, J. and Kanamori, H., 2005.The map-based sequence of the rice genome. *Nature*, 436 (7052):793-800.

McCouch, S.R., Teytelman, L., Xu, Y.B., Lobos, K.B., Clare, K., Walton, M., Fu, B.Y., Maghirang, R., Li, Z.K., Xing, Y.Z., Zhang, Q.F., Kono, I., Yano, M., Fjellstrom, R., DeClerck, G., Schneider, D., Cartinhour, S., Ware, D. and Stein, L., 2002. Development and mapping of 2,240 new SSR markers for rice (*Oryza sativa* L.). *DNA Res.*, 6:199-207.

McElroy, D., Blowers, A.D., Jenes, B. and Wu, R., 1991. Construction of expression vectors based on the rice acting (Actl) 5' region for use in monocot transformation. *Mol.General Genet.*, 231:150-160.

Mohanty, H.K. and Chaudhary, R.C., 1986.Breeding for submergence tolerance in rice in India. In: Progress in rainfed lowland rice. Manila: International Rice Research Institute, 191-200.

Musgrave, M.E. and Ding,N., 1998.Evaluating wheat cultivars for waterlogging tolerance. *Crop Sc.*, 38(1):90–97.

Olive, M.R, Walker, J.C., Singh, K. and Dennis, E.S., 1990. Functional properties of the anaerobic responsive element of the maize *Adh I* gene. *Plant Mol. Biol.*, 15: 593-604.

Rathore, T.R. and Warsi, M.Z.K., 1998. Production of maize under excess soil moisture (waterlogging) conditions in Proceedings of the 2nd Asian Regional Maize Workshop PACARD, Laos Banos, Phillipines, February 1998.

Reyna, N., Cornelious, B., Shannon, J.G. and Sneller, C.H., 2003.Evaluation of a QTL for waterlogging tolerance in Southern Soybean Germplasm. *Crop Sc.*, 43(6):2077–2082.

Sarkar, R.K., Das, S. and Ravi, I., 2001. Changes in certain antioxidative enzymes and parameters as a result of complete submergence and subsequent re-aeration of rice cultivars differing in submergence tolerance. *J. Agron. Crop Sci.*, 187:69–74.

Sarkar, R.K., De, R.N., Reddy, J.N. and Ramakrishnayya, G., 1996.Studies on submergence tolerance mechanism in relation to carbohydrate, chlorophyll and specific leaf weight in rice.*J. Plant Physiol.*, 149:623–625.

Septiningsih, E.M., Pamplona, A.M., Sanchez, D.L., Neeraja, C.N., Vergara, GV.,Heuer, S., Ismail, A.M. and Mackill, D.J., 2009. Development of submergence-tolerant rice cultivars: the Sub1 locus and beyond. *Annals Bot.*, 103(2): 151-160. doi: 10.1093/aob/mcn206

Setter, T.L., Burgess, P., Water, I. and Kuo, J., 1999. "Genetic diversity of barley and wheat for waterlogging tolerance in Western Australia," in Proceeding of the 9th Australian Barley Technical Symposium, Melbourne, Australila.

Setter, T.L., Ellis, M., Laureles.E.V., Ella, E.S., Senadhira, D., Mishra, S.B., Sarkarung, S. and Datta, S. 1997. Physiology and genetics of submergence tolerance in rice. *Annals Bot.* 79:67–77.

Singh, B.N., 2002. High yielding rice varieties in India. *Rice,* 12:5–6.

Singh, N., Dang, T.T. M.,Vergara, G.V., Pandey, D.V., Sanchez, D., Neeraja, C.N., Septiningsih, E.M., Mendioro, M., Tecson-Mendoza, E.M., Ismail, A.M., Mackill, D.J. and Heuer, S., 2010.Molecular marker survey and expression analyses of the rice submergence tolerance gene *SUB1A*. *Theor. Appl. Genet.*, 121:1441-1453.

Singh, N., Sekhon, K.S. and Kaur, A., 1990. Effect of pre-harvest flooding of paddy on the milling and cooking quality of rice. *J. Sc. Food and Agri.*, 52: 23-34.

Singh, U.S., Dar, M.H., Singh, S., Zaidi, N.W., Bari, M.A., Mackill, D.J., Collard, B.C.Y., Singh, V.N., Singh, J.P., Reddy, J.N., Singh, R.K. and Ismail, A.M., 2013. Field performance, dissemination, impact and tracking of submergence tolerant (sub1) rice varieties in south asia. *SABRAO J. Breed. Genet.*,45(1):112-131.

Thomas, C.E., McLean, L.R., Parker, R.A. and Ohlweiler, D.F., 1992.Ascorbate and phenolic antioxidant interactions in prevention of liposomal oxidation.*Lipids*, 27:543–550.

Voesenek, L.A.C.J., Benschop, J.J., Bou, J., Cox, M.C.H., Groeneveld, H.W., Millenaar, F.F., Vreeburg, R.A.M. andPeeters, A.J.M., 2003. Interactions between plant hormones regulate submergence-induced shoot elongation in the flooding-tolerant dicot Rumexpalustris. *Annals Bot.*, 91:205–211.

Xie, Y. and Wu, R., 1989. Rice alcohol dehydrogenase genes: anaerobic induction, organ specific expression and characterization of cDNA clones. *Plant Mol. Biol.*, 13: 53-58.

Xu, K. and Mackill, D.J., 1996. A major locus for submergence tolerance mapped on rice chromosome 9. *Mol. Breed.*, 2:219-224.

Xu, K., Xia, X., Fukao, T., Canlas, P., Maghirang-Rodriguez, R., Heuer, S., Ismail, A.M., Bailey-Serres, J., Ronald, P.C. and Mackill, D.J., 2006. Sub1A is an ethylene response factor-like gene that confers submergence tolerance to rice. *Nature*, 442:705-708.

Xu, K., Xu, X., Ronald, P.C. and Mackill, D.J., 2000. A high-resolution linkage map in the vicinity of the rice submergence tolerance locus Sub1.*Mol. Gen. Genet.*, 263:681-689.

2016, Recent Advances in Plant Stress Physiology *Pages* **473–505**
Editors: ***Praduman Yadav, Sunil Kumar and Veena Jain***
Published by: **DAYA PUBLISHING HOUSE, NEW DELHI**

Chapter 21

Proteomic Perspectives on Understanding Plant Stress Response

Prathap Reddy Kallamadi[1*], Kamakshi Dandu[1], Sudheendra Hebbar[2], Pardeep Kumar[3] and Ramesh Bellamkonda[1]

[1]Sri Venkateswara University, Tirupati – 517 502, India
[2]University of Hertfordshire, Hatfield, Hertfordshire, United Kingdom
[3]National Bureau of Plant Genetic Resources, New Delhi, India

ABSTRACT

Research on the plant-stress has become one of most interesting and rapidly moving fields in the plant biology. The advances in this area have lead in development of new strategies of plant protection. Proteomics, the systematic evaluation of changes in the protein constituencies of cell, is more than just the generations of lists of proteins that increase or decrease in expression as a cause or consequence of abiotic and biotic stresses. Developments in new proteomic techniques have contributed in identifying several proteins, which are expressed in plants during abiotic and biotic stresses. The proteomic approaches to study stress mechanism helped in understanding and deciphering the mechanism of plant protection against the abiotic and biotic stresses and to arrive new solution to address the evolving pathogen against plant. This chapter would emphasis on the proteomic techniques used in studying proteins expressed during abiotic and biotic stresses. It also provides information on databases available for proteomic studies.

Keywords: *Abiotic stress, Biotic stress, Proteomics, Proteins.*

**Corresponding author.* E-mail: prathapreddykallamadi@gmail.com

Introduction

The increasing human population and parallel decrease in agriculture land due to deforestation, urbanization, industrialization, and climatic changes impose serious threat to agriculture. Today India enjoys the second position all over the world in agricultural production. In economic point of view, agriculture is the broadest sector which plays an important role in socio economic status of the country. Furthermore, world food production needs to be doubled by the year 2050 to meet the ever-growing demands of the population (Tilman *et al.*, 2002). Plant growth and productivity is adversely affected by biotic stresses like bacteria, fungi, viruses, insects, nematodes and abiotic stresses like low temperature, salt, drought, heat, heavy metal toxicity. These stresses are major threat to plant growth and halt them from reaching their full genetic potential and eventually limiting their productivity (Mahajan *et al.*, 2005). Acclimation to stress is mediated via profound change in gene expression which results in change in composition of plant transcriptome, proteome and metabolome. Several studies (Gygi *et al.*,1999; Ideker *et al.*, 2001; Bogeat-Triboulot *et al.*, 2007) have already proven that change in gene expression at transcript level do not often correspond with change at protein level. So, study of change in plant proteome is highly important since protein act directly on biochemical processes involved in plant stress response.

Classical genomics studies were unable to clear up confusion related to the functional aspects of the metabolic processes taking place during stress conditions because, many proteins are modified by post-translational modifications such as phosphorylation, glucosylation, ubiquitinylation, sumoylation etc (Mann and Jensen, 2003; Schweppe *et al.*, 2003; Canovas *et al.*, 2004) which significantly influence protein functions. Knowledge of the full complement of proteins expressed by the genome of a cell, tissue or organism at a specific time point is necessary to understand the biology of a cell or an organism. Modern techniques of proteomics allow us to study the protein content of cells which are differentially synthesized under diverse physiological and stress related conditions. Thus, proteomic techniques that allow the identification and quantification of stress-related proteins, mapping protein expression and post-translational modifications could contribute to better understanding of physiological mechanism underlying plant stress response.

Techniques Used in Proteomics

Plant acclimation to stress involves changes in proteome composition. Current proteomic tools allow large scale, high throughput analyses for the detection, identification and functional investigation of proteome. The following section provides a brief overview of currently available proteomic techniques.

Two Dimensional 'Gel' Electrophoresis

Two-dimensional gel electrophoresis separates proteins based on two separate parameters, isoelectric point (*pI*) in the first dimension and molecular mass (Mr) in the second dimension by combining isoelectric focusing and sodium dodecyl sulphate polyacrylamide gel electrophoresis. Methodology involves two electrophoresis in two dimensions the first dimension is preferably performed in individual IPG strips laid side by side on a cooling platform, with the sample often adsorbed into the entire strip during rehydration. At the end of the IEF step, the strips have to be interfaced with the second dimension, almost exclusively performed by mass discrimination via saturation with the anionic surfactant SDS. After the equilibration step, the strip is embedded on top of an SDS-PAGE slab, where the 2D run is carried out perpendicular to the 1D migration. The 2D map displayed at the end of these steps is the stained SDS-PAGE slab, where polypeptides are seen, after staining, as spots, each characterized by an individual set of p*I*/*M*r coordinates. The strengths of 2DE include, relative quantification and PTM information.The major limitation of this method is poor separtion of acidic, basis, hydrophobic and low bundant proteins.

Fluorescence 2 D difference Gel Electrophoresis (2D-DIGE)

A recent technical advance in 2DGE has been the development of the multiplexing fluorescent 2D-DIGE method. The technique of two-dimensional (2D) gel electrophoresis is a powerful tool for separating complex mixtures of proteins. In 2D-DIGE, the protein samples are labelled with fluorescent dyes (Cy2, Cy3, or Cy5) prior to 2DE. CyDyes are cyanine dyes carrying an N-hydroxysuccinimidyl ester reactive group that covalently binds the e-amino group of lysine residues in proteins. During DIGE, proteins in each of up to 3 samples can be characterized with one of these fluorescent dyes, and the differentially characterized samples can be mixed and loaded together on one single gel, allowing the quantitative comparative analysis of three samples using a single gel. The gels are scanned by laser scanners and the image of 2D-PAGE for multiple samples is obtained from single gels. The different samples as many as the number of fluorescent dyes can be studied in single gels. In 2D-DIGE, the gel staining is replaced by laser scanning. In contrast, 2D-DIGE images are obtained within one hour by simple laser scanning. The DIGE technique has exhibited higher sensitivity as well as linearity, eliminated post electrophoresis processing (fixing and distaining) steps and enhanced reproducibility by directly comparing samples under similar electrophoresis conditions. The resulting images are then analyzed by software. The major advantages of 2D-DIGE are the high sensitivity and linearity of its dyes, its straight forward protocol, as well as its important reduction of inter gel variability, growing the possibility to definitely identify biological variability, and reducing bias from experimental variation. The DIGE technique has dramatically improved the reproducibility, sensitivity, and accuracy of quantisation; however, its labelling chemistry has some limitations; proteins without lysine cannot be labelled, and they need particular equipment for visualization, and fluorophores are very costly.

Isotope-Coded Affinity Taq (ICAT)

An Isotope-coded affinity tag (ICAT) is a method in which a protein was labelled with isotopes and is used in quantitative proteomics by mass spectrometry that uses chemical labelling reagents. ICAT is relatively new technique that is used for the quantititative determination of protein content. With the help of this method differences in the cellular protein expression can be detected. The principle behind this method is based on measurement of protein expression in two different cell lines, which are grown in different states and comparing the expression between the cell lines. In this method the tags that were used are of biotinylated origin. These can function as molecular handles in determining the protein expression levels. These tags are very highly selective in nature and are reversible. These tags are separated from the complex biological mixture very easily.

ICAT mainly involves three functional moieties; a cysteine reactive moiety, a linker moiety and a biotin moiety. A linker moiety is generally contains either eight hydrogens (light form) or eight deuteriums (heavy form). In this technique cysteine side chains in complex mixture of proteins are reduced and alkylated with the help of light form of reagent in one cell and heavy form of reagent in another cell. The mixtures are then mixed and are subjected to the proteolytic digestion. After the digestion peptides that containing generated cysteine are affinity purified using an avidin column which results in a peptide mixture that contains 10 fold fewer peptides that the original mixture. These peptides that are obtained are analyzed using mass spectroscopy and quantified based on the information from the relative abundance of the d0 and d8 isotopes. The proteins were identified based on the peptide molecular weight and MS/MS derived amino acid sequence that is obtained.

Isobaric Taq for Relative and Absolute Quantification (iTRAQ)

Isobaric tags for relative and absolute quantisation (iTRAQ) is an isobaric labelling method that is widely used in quantitative proteomics by tandem mass spectrometry in order to determine the amount of proteins from different sources in a single experiment. It uses stable isotope labelled molecules that can be covalent bonded to the N-terminus and side chain amines of proteins. The main advantages of iTRAQ are; the chemical labelling method can be implemented after the cell or tissue lysis which indicates that iTRAQ can be used for the quantification of proteins in any biological system and is not limited to that system which can accommodate incorporation of stable isotopes during cell culture. In this method many number of samples (four at maximum) can be compared in one mass spectrometry-based experiment. The results cause concomitant increase in precursor ion intensity in sample difficulty relative to multiplexed, precursor-based quantitative methods.

The principle behind this method is utilization of isobaric reagents to label the primary amines that are present in peptides and proteins. For example iTRAQ reagents for quantification include a reporter group with different masses, a balance group with different masses and a reactive group that can react with primary amines of peptides. The balance group that is present in each of the iTRAQ reagents function in such a way that labelled peptides is made from each sample isobaric with same mass. Quantification is performed through the analysis of reporter groups that are generated after fragmentation in the mass spectrometer.

The steps that are involved are; the samples that need to be quantified are prepared by the lyses of cell to extract the proteins. After the protein concentration of each sample are estimated proteins are digested with an enzyme in order to generate the proteolytic peptides. After the digestion each peptide is labelled with a different iTRAQ reagent and then the labelled digests are combined into one sample mixture. The collective peptide mixture is analyzed by LC-MS/MS for both recognition and quantification. The sequence of a peptide is determined from the product ions that are generated from cleavage about peptide inter residue bonds. The relative quantification of a peptide among the treated samples is determined by comparing the intensities of reporter ion signals also present in the MS/MS scan (Ross *et al.*, 2004).

Stable Isotope Labelling with Amino Acids in Cell culture (SILAC)

General procedure in Stable Isotope Labelling by Amino Acids in Cell Culture experiment, cells are grown in such a medium that contains specific essential amino acids usually Arginine and Lysine that contain naturally occurring atoms which are light or a their isotope counterparts that are heavy. Once the labelled amino acids are fully incorporated the light and heavy chains are subjected to different perturbations the harvested and then combined. Peptides that are obtained from the combined cell lysate after the enzymatic digestion are detected by mass spectrometry in form of ion pairs (doublets) and their ratios indicate the relative changes in protein abundance between the light and heavy chain (Schutz *et al.*, 2011).

Protein Microarray Technology

Protein microarray technology evolved for the study of protein interactions and modifications. High sensitivity, specificity and high-throughput are the advantages of Protein microarray technology. Many limitation of the technology are still unsolved and prevent protein microarray technology from reaching its full potential. These limitations include the generation of content and the conservation of protein functionality during immobilization as well as the provision of the required absolute and relative sensitivity (Philipp, 2005). The wide application of this technology include in quantitative specific proteins used in diagnostic (biomarkers or antibody) and discovery research.

Mass Spectrometry

Mass spectrometry has been widely used to analyze biological samples and has evolved into an important tool for proteomics research. Among the various proteomic techniques available to investigate proteins, Mass Spectroscopy (MS) has gained popularity because of its ability to handle the complexities associated with proteome (Han *et al.*, 2006). The basic principle of MS measures the mass-to-charge ratio (m/z) of gas phase ions. Mass Spectroscopy consist of an ion source that convert analyte molecules into gas-phase ions, a mass analyzer that separate ionized analytes on the basis of m/z ratio and a detector that records the number of ions at each m/z value. Recent developments of electrospray ionization (ESI) and matrix assisted laser desorption/ionization (MALDI) had revolutionized protein analysis using MS. The mass analyzer is the central to MS technology. Four types of mass analyzer are widely used in MS are; Quadrupole ion trap (QIR), Linear ion trap (LIT or LTQ), Time-of-flight (TOF) and Fourin-transform ion cyclotron resonance (FTILR) mass analyse. They vary in their physical principles and analytical performance.

MS is not only used to demonstrate the presence of a protein or PTM within a sample but it can be used to measure dynamic changes in protein and PTM abundances. The quantification may involve use of stable isotopes for sample labelling or it may use label-free methods. Irrespective of the proteomic parting technique, gel based or gel free, a mass spectrometer is always the primary tool for protein identification. In gel-free methods, samples are directly analysed by MS, whereas in gel-based separation technique, the protein spots are first excised from the gel and then digested with trypsin. The resulting peptides are then separated by LC and directly analysed by MS (Chandramouli and Qian, 2009). High sensivity, specificity and high-throughput analysis are the major strengths of MS.

Plant Abiotic Stress-Proteomics

Now-a-day, with the help of proteomics study of proteins at large scale has become a very important tool in the research biology. Proteomics is a very important tool because the proteins are the key players in majority of cellular events. Proteomics can also be used to detect translational and post-translational regulations and provide new insights into complex biological phenomena like abiotic stresses in plants. The use of proteomics in deduction of proteins that are involved in water stress has been researched extensively. For instance, with the help of proteomics different expression of proteins and phosphoproteins that induce water stress are identified in *Oryza. sativa*. This approach identified 18 different proteins that have an impact on the water stress. For example, out of 10 phosphoproteins that are identified in response to drought 7 have not been reported under conditions of water stress (Ke *et al.*, 2009). Several studies have been performed by comparative analysis of proteome of plants subjected to salt stress. For example studies performed in wheat revealed five different proteins in the chloroplast and these proteins are involved in chloroplast function in plants under salt stress in leaves and roots (Abbasi and Komatsu, 2004). Based on the proteomic studies on the plants that are grown under flooding stress revealed that many number of proteins are involved in the flood-induced response mechanism. For example proteins such as UDP-glucose dehydrogenase, UGP, β-glucosidase G4 and rhamnose synthase), amino acids (aspartate aminotransferase), and lipids (lipoxygenase) are induced as early flood responsive proteins (Jackson and Ram, 2003; Nanjo *et al.*, 2010). Studies performed on cold tolerant and sensitive plants helped to understand the overall response as well as recovery mechanism against cold stress. For example, activation of metabolic processes was observed in rice roots upon 24–72 h of chilling stress as indicated by the enhanced levels of several metabolism-associated proteins (Lee *et al.*, 2009).

Water Stress

Drought is one of the major problems of great importance in world agriculture, especially in arid and semiarid regions, which represent a one third of the world. The response of plants to deficit of water is through series of processes at the physiological, cellular and molecular levels that lead to the stress tolerance. The effect of water stress on the plant responses include stomata closure, reduction in cell growth, decreased photosynthetic activity and increased respiration. Most common responses include the chlorophyll damage, decreased antioxidant system; increase in H_2O_2 production etc.Various proteomic studies and their significant finding in understanding drought mechanism in different plants were provided in Table 21.1.

Table 21.1: Differential Proteins Expressed in Plants during Various Abiotic Stress

Abiotic Stress	*Plant*	*Method*	*Important Findings*	*References*
Cold	*Arabidopsis thaliana*	2-DE MALDI-TOF	Proteomic analyses led to the identification of 184 spots corresponding to 158 different proteins implicated in a variety of cellular functions.	Bae *et al.*, 2003
	Arabidopsis thaliana	2-DE MALDI-TOF	Identified 38 proteins that changed in quantity during cold acclimation	Kawamura *et al.*, 2003
	Arabidopsis thaliana	2D-DIGE, MALDI-TOF, ESI-MS/MS	When plants were exposed to 10 degrees C, 18 of these 22 spots still showed a 2-fold change; however, the alterations were, in general, more moderate than observed under 6 degrees C.	Amme *et al.*, 2006
	Pea	2-DE LC-MS/MS	Together these data suggest that although many of the molecular events identified by chemical stresses of mitochondria from a range of model eukaryotes are also apparent during environmental stress of plants, their extent and significance can vary substantially.	Taylor *et al.*, 2005
	Pea	2-DE HPLC ESI-MS/MS	Out of 112 statistically variable spots (44 in leaves, 38 in stems, and 30 in roots), 32 proteins were identified. These proteins were related to frost response or frost resistance. It seems that Champagne could resist to frost with the reorientation of the energy metabolism	Dumont *et al.*, 2011
	Rice	2-DE MALDI-TOF	Seventy protein spots were differentially displayed after four days of cold treatment at the young microspore stage. Of these, 12 protein spots were newly-induced, 47 were up-regulated, and 11 were down-regulated by cold treatment at the early microspore stage	Imin *et al.*, 2004
	Rice	2-DE, MALDI-TOF, ESI-MS/MS	The functional proteomes illuminate the facts, at least in plant cell, that protein quality control mediated by chaperones and proteases and enhancement of cell wall components play important roles in tolerance to cold stress	Cui *et al.*, 2005
	Rice	2-DE MALDI-TOF	Mass spectrometry analysis allowed the identification of 85 differentially expressed proteins, including well known and novel cold-responsive proteins	Yan *et al.*, 2006
	Rice	2D-DIGE, MALDI-TOF, ESI-MS/MS	Of the 250-400 protein spots from each organ, 39 proteins changed in abundance after cold stress, with 19 proteins increasing, and 20 proteins decreasing.	Hashimoto *et al.*, 2007

Contd...

Table 21.1–*Contd...*

Abiotic Stress	*Plant*	*Method*	*Important Findings*	*References*
	Rice	2D-DIGE, MALDI-TOF, ESI-MS/MS	A total of 27 up-regulated proteins were identified by matrix-assisted laser desorption ionization-time of flight (MALDI-TOF) mass spectrometry or electro spray ionization-tandem mass spectrometry (ESI-MS/MS) analysis.	Lee *et al.,* 2009
	Rice	2-DE MALDI-TOF	A total of 34 and 39 protein spots were found in the PEG supernatant and pellet fractions, respectively. Using MALDI-TOF MS, a total of 48 proteins were identified.	Lee *et al.,* 2007
	Soybean	2-DE MALDI-TOF	Total 40 protein spots were differentially expressed in response to low temperature, in which, 25 protein spots were up-regulated and fifteen protein spots were down-regulated.	Cheng *et al.,* 2010
	Wheat	2-DE, nano LC-MS (Q-TOF)	The level of the accumulation of WCS120, WCS66 and WCS40 distinguished our two frost-tolerant winter wheat cultivars.	Vítámvás *et al.,* 2007
Heat	Wheat	2-DE MALDI-TOF	Proteome analysis allowed the identification of 48 differentially expressed proteins.	Skylas *et al.,* 2002
	Wheat	2-DE MALDI-TOF	Of the 43 heat-changed proteins, 24 were found to be up-regulated whereas 19 spot proteins were down-regulated.	Majoul *et al.,* 2004
Drought	Chickpea	2-DE LC-MS/MS	Approximately 205 protein spots were found to be differentially regulated under dehydration. Mass spectrometry analysis allowed the identification of 147 differentially expressed proteins, presumably involved in a variety of functions including gene transcription and replication, molecular chaperones, cell signaling, and chromatin remodeling	Pandey *et al.,* 2008
	Maize	2-DE LC-ESI-MS TOF	The results revealed major and predominantly region-specific changes in protein profiles between well-watered and water-stressed roots. In total, 152 water deficit-responsive proteins were identified and categorized into five groups based on their potential function in the cell wall: reactive oxygen species (ROS) metabolism, defense and detoxification, hydrolases, carbohydrate metabolism, and other/unknown.	Zhu *et al.,* 2007

Contd...

Table 21.1–*Contd...*

Abiotic Stress	*Plant*	*Method*	*Important Findings*	*References*
	Pea	2-DE LC-MS/MS	Together these data suggest that although many of the molecular events identified by chemical stresses of mitochondria from a range of model eukaryotes are also apparent during environmental stress of plants, their extent and significance can vary substantially.	Taylor *et al.*, 2005
	Soyabean	2-DE, nano-liquid chromatography (LC)-MS/MS	In the leaves of PEG-treated and drought-stressed seedlings, metabolism-related proteins increased and energy production- and protein synthesis-related proteins decreased.	Mohammadi *et al.*, 2012
	Soybean	2-DE MALDI-TOF	28 proteins were identified by mass spectrometry; the levels of 5 protein spots were increased, 21 were decreased and 2 spots were newly detected under drought condition.	Alam *et al.*, 2010
	Sugar beet	2-DE LC-MS/MS	79 spots showed significant changes under drought. Some proteins showed genotype-specific patterns of up or down regulation in response to drought.	Hajheidari *et al.*, 2005
Water logging		2-DE	Comparative analysis of 2-D gels of control and 3 day flooding-experienced soybean root samples revealed 70 differentially expressed protein spots, from which 80 proteins were identified.	Salavati *et al.*, 2012
	Soybean	2-DE MALDI-TOF	14 proteins were upregulated; 5 proteins were down regulated; and 5 were newly induced in waterlogged roots. The identified proteins include well-known classical an aerobically induced proteins as well as novel water logging-responsive proteins that were not known previously as being water logging responsive.	Alam *et al.*, 2010
	Tomato	2-DE MALDI-TOF, ESI-MS/MS	The identified proteins are involved in several processes, *i.e.* photo-synthesis, disease resistance, stress and defense mechanisms, energy and metabolism and protein biosynthesis. Results from 2-DE analysis, combined with immunoblotting clearly showed that the fragments of Rubisco large subunit were significantly degraded.	Ahsan *et al.*, 2007

Contd...

Table 21.1–*Contd...*

Abiotic Stress	*Plant*	*Method*	*Important Findings*	*References*
Salinity	*Arabidopsis thaliana*	2DE-DIGE MALDI-TOF	Identified 75 salt and sorbitol responsive spots. These fall into 10 functional categories that include H(+) transporting ATPases, signal transduction related proteins, transcription/translation related proteins, detoxifying enzymes, amino acid, purine biosynthesis related proteins, proteolytic enzymes, heat-shock proteins, carbohydrate metabolism-associated proteins and proteins with no known biological functions.	Ndimba *et al.*, 2005
	Barley	MALDI-TOF-TOF MS	Identified 44 cellular proteins, which represented 18 different proteins and were classified into seven categories and a group with unknown biological function. These proteins were involved in various many cellular functions.	Fatehi *et al.*, 2012
	Barley	2-DE, MALDI-TOF/TOF	Mass spectrometry analysis using MALDI-TOF/TOF led to the identification some proteins involved in several salt responsive mechanisms which may increase plant adaptation to salt stress including higher constitutive expression level and up-regulation of antioxidant, up-regulation of protein involved in signal transduction, protein biosynthesis, ATP generation and photosynthesis	Rasoulnia *et al.*, 2011
	Rice	2-DE, Edman sequencing	The expression of SOD was a common response to cold, drought, salt and abscisic acid (ABA) stresses while the expression of LSY081, LSY363 and OEE2 was enhanced by salt and ABA stresses.	Abbasi *et al.*, 2004
	Rice	2-DE nano ESI-LC-MS/MS	Most of these identified proteins belonged to major metabolic processes like photosynthetic carbon dioxide assimilation and photorespiration, suggesting a good correlation between salt stress-responsive proteins and leaf morphology.	Kim *et al.*, 2005
	Rice	2-DE, MALDI-TOF/TOF	Mass spectrometry analysis and database searching helped us to identify 12 spots representing 10 different proteins. Three spots were identified as the same protein, enolase. While four of them were previously confirmed as salt stress-responsive proteins, six are novel ones, *i.e.* UDP-glucose pyrophosphorylase, cytochrome c oxidase subunit 6b-1, glutamine synthetase root isozyme, putative nascent polypeptide associated complex alpha chain, putative splicing factor-like protein and putative actin-binding protein.	Yan *et al.*, 2005

Contd...

Table 21.1–*Contd...*

Abiotic Stress	*Plant*	*Method*	*Important Findings*	*References*
	Rice	2-DE MALDI-TOF/TOF	Nine proteins were up-regulated and nine were down-regulated. MS analysis indicated that most of these membrane-associated proteins are involved in important physiological processes such as membrane stabilization, ion homeostasis, and signal transduction. In addition, a new leucine-rich-repeat type receptor-like protein kinase, OsRPK1, was identified as a salt-responding protein.	Cheng *et al.*, 2009
	Rice	MALDI TOF/TOF	Identified 18 protein spots that were involved in several processes that might increase plant adaptation to salt stress, such as carbohydrate/energy metabolism, anther wall remodeling and metabolism, and protein synthesis and assembly	Sarhadi *et al.*, 2012
	Tobacco	2-DE LC-MS/MS	Quantitative evaluation and statistical analyses of the resolved spots in treated and untreated samples revealed 20 polypeptides whose abundance changed in response to salt stress.	Dani *et al.*, 2005
	Wheat	2-DE MALDI-TOF	Quantitative evaluation, statistical analyses and MALDI-TOF MS characterization of the resolved spots in treated and untreated samples enabled us to identify 38 proteins whose levels were altered in response to salt stress.	Caruso *et al.*, 2008
	Wheat	MALDI-TOF-TOF MS	Comparative analysis showed that 41 DEPs were salt responsive with significant expression changes in both varieties under salt stress, and 99 (52 in Jing-411 and 47 in Chinese Spring) were variety specific. Only 15 and 9 DEPs in Jing-411 and Chinese Spring, respectively, were up-regulated in abundance under all three salt concentrations.	Guo *et al.*, 2012
Ozone	Maize	2-DE MALDI-TOF	Out of the 48 protein spots analyzed, 46 were identified by matching to the protein database. We describe the expression of proteins involved in a wide range of cellular functions, including previously reported anaerobic proteins, and discuss their possible roles in adaptation of plants to low-oxygen stress.	Chang *et al.*, 2000

Contd...

Table 21.1–*Contd...*

Abiotic Stress	*Plant*	*Method*	*Important Findings*	*References*
	Rice	2-DE, Edman sequencing	Most prominent change in leaves, within 24 h post-treatment with ozone, was the induced accumulation of a pathogenesis related (PR) class 5 protein, three PR 10 class proteins, ascorbate peroxidase(s), superoxide dismutase, calcium-binding protein, calreticulin, a novel ATP-dependent CLP protease, and an unknown protein.	Agrawal *et al.*, 2002
Heavy metals				
Arsenic	Rice	2-DE MALDI-TOF	These results suggest that SAMS, CS, GSTs, and GR presumably work synchronously wherein GSH plays a central role in protecting cells against - stress.	Ahsan *et al.*, 2008
Cadmium	*Arabidopsis thaliana*	2-DE MALDI-TOF	Most of the identified proteins belong to four different classes: metabolic enzymes such as ATP sulphurylase, glycine hydroxymethyltransferase, and trehalose-6-phosphate phosphatase; glutathione S-transferases; latex allergen-like proteins; and unknown proteins.	Roth *et al.*, 2006
	Arabidopsis thaliana	2-DE MALDI-TOF, nano ESI-LC-MS/MS	This study revealed that the main variation at the metabolite level came from the presence of six different families of phytochelatins, in *A.thaliana* cells treated with Cd, whose accumulation increases with Cd concentrations.	Sarry *et al.*, 2006
	Arabidopsis thaliana	2-DE, LC-ESI-MS/MS	Of the 730 reproducible proteins among all gels, 21 were statistically un-regulated in response to Cd. These proteins can be functionally grouped into 5 classes: proteins involved in (1) oxidative stress response, (2) photosynthesis and energy production, (3) protein metabolism, (4) gene expression and finally, (5) proteins with various or unknown function.	Semane *et al.*, 2010
	Rice	2-DE MALDI-TOF	Plant responses were analysed by measuring the levels of gluthatione (GSH) and phytochelatins (PCs), the thiol-peptides involved in heavy metal tolerance mechanisms. Results showed a progressive increase of GSH up to 10μM of Cd treatment, whereas a significant induction only of PC_3 was detected in roots of plants exposed to 100μM of Cd.	Aina *et al.*, 2007

Contd...

Table 21.1–*Contd...*

Abiotic Stress	*Plant*	*Method*	*Important Findings*	*References*
	Rice	2-DE MALDI-TOF	A total of 36 proteins were up- or down-regulated following Cd treatment. As expected, total glutathione levels were significantly decreased in Cd-treated roots, and approximately half of the up-regulated proteins in roots were involved in responses to oxidative stress.	Lee *et al.*, 2010
Multiple heavy metals	Rice	2-DE-Edman sequencing	Most of the proteins showed homology to RuBisCO protein, and some to defense/stress-related proteins, like the pathogenesis related class 5 protein (OsPR5), the probenazole-inducible protein (referred to as the OsPR10), superoxide dismutase, and the oxygen evolving protein.	Hajduch *et al.*, 2001

Salinity Stress

Salts that are present in the soil have a great impact on the plant growth and this is because of the two reasons. Firstly, osmotic effect reduces the capacity of plant water uptake and causes a slow plant growth and secondly, the toxic effect due to the entry of salts, which causes the tissue damage to the leaves. This is because of the influx of ions and salts exceeds the normal level and entry into the cytoplasm causes the inhibition of enzyme activity and because of this cell wall is altered and the cell becomes dehydrated and die (Khan *et al.*, 2007; Munns, 2005). Another important impact of large amount of salts on plants causes an osmotic imbalance that causes a nutritional imbalance. This causes the changes in the ion concentration in the cytoplasm that have toxic effect on the membrane structure and the enzyme systems (Ashraf and Harris, 2004). This osmotic imbalance also causes the oxidative stress that leads to the production of toxic reactive oxygen that has an impact on the lipid peroxidation production (Qureshi *et al.*, 2005). Various proteomic studies and their important discoveries in understanding salt stress in different crops were provided in Table 21.1.

Flooding Stress

Whenever there is heavy or continuous rainfall causes severe environmental stress that affects plants especially at their early growth stages. Some of the plants those are sensitive to this kind of stress are soybean, wheat, barley and maize. The stress is caused by the hypoxic environment that is due to the submerged state during flooding stress and affects aerobic respiration that leads to the boosted production of ATP and regeneration of NAD^+ through anaerobic respiration. Because of this protein synthesis is altered (Bailey-Serres *et al.*, 2008; Gibbs *et al.*, 2003; Komatsu *et al.*, 2012). Studies that were done for understanding water logging cope up mechanism in different crops were provided in Table 21.1.

Cold Stress

Cold or low-temperature stress is one of the major abiotic stresses that severely affect plant growth and survival. Chilling (<20 C) or freezing (<0 C) temperatures can induce ice formation in plant tissues, leading to cellular dehydration. To cope with this adverse condition, plants adopt several strategies such as produce more energy by activation of primary metabolisms, raising the level of antioxidants and chaperones, and maintain osmotic balance by altering membrane structure (Chinnusamy *et al.*, 2007; Prasad, 1996). A variety of proteomic studies and their significant finding in understanding cold stress in different crops were provided in Table 21.1.

Ozone and Heavy Metal Stress

Studies on plant stress due to pollutants like ozone and heavy metals were given in the Table 21.1. At proteome level, ozone causes a severe oxidative stress which results in accumulation of several ROS scavenging enzymes and also proteins involved in regulation of protein redox status and folding. Heavy metals such as cadmium, lead, mercury and arsenic reveals inhibitory effects on many vital processes in plant cells since they act as inhibitors of many enzymes with metal cofactors.

Plant Biotic Stress-Proteomics

Biotic stress caused by other living organisms like bacteria, fungi, viruses, nematodes, insects, is one of the major stresses limiting crop productivity of many important crops worldwide. An understanding of how plants and pathogens recognize each other and differentiate to establish either a successful or an unsuccessful relationship is crucial in biotic stress studies. Plants responding to biotic stresses produce several protective compounds and proteins such as pathogenesis related (PR) proteins, directly disease related proteins and other proteins. The first step in the understanding of disease resistance is to identify the proteins expressed during plant-pathogen interactions. The next step will be to determine which proteins confer pathogenicity and disease resistance, and the mechanisms by which they do so (Navi and Toorchi, 2013). Hence, the proteomic analysis is a very useful tool for providing complex information about differences in the plant proteome during biotic stresses.

Plant-Bacterium Interactions

Plant pathogenic bacteria cause diseases in important crops which seriously and negatively impact agricultural production. So, an understanding of the mechanisms by which plants resist bacterial infection at the stage of the basal immune response or mount a successful specific R-dependent defense response is crucial. Since a better understanding of the biochemical and cellular mechanisms underlying these interactions will enable molecular and transgenic approaches to crops with increased biotic resistance. Proteomics has been used to gain in-depth understanding of many aspects of the host defense against pathogens and has allowed monitoring differences in abundance of proteins in different cells as well as post-transcriptional and post-translational processes, protein activation/inactivation, and turnover. It also offers a window to study protein trafficking and routes of communication between organelles through different pathways.

Proteomic analyses are particularly useful for understanding the mechanisms of host plant against the pathogen attack and their interaction. Recent advances in the field of proteome analyses have initiated a new research area, *i.e.*, the analysis of more complex microbial communities and their interaction with plant (Afroz *et al.*, 2013). Such areas hold great potential to elucidate, not only the interactions between bacteria and their host plants, but also of bacteria-bacteria interactions between different bacterial taxa, symbiotic, pathogenic bacteria, and commensal bacteria. During biotic stress, plant hormonal signaling pathways prioritizes defense over other cellular functions. Some plant pathogen stake advantage of hormone dependent regulatory system by mimicking hormones that interfere with host immune responses to promote virulence.

Rice bacterial leaf streak (BLS), caused by the pathogen *Xanthomonas campestris* pv. Oryzicola, has become a serious threat to rice production in tropical and subtropical regions of Asia. Chen *et al.*, 2012 identified the proteins that are up-regulated in rice leaves after infection. Approximately 1,500 protein spots were detected on each 2D gel after silver-staining; those with increased protein levels were selected for MALDI-TOF-MS analysis and identified 32 up-regulated proteins that might be involved in disease resistance signal transduction, pathogenesis, and regulation of cell metabolism was determined. By using publicly available microarray data, mRNA transcripts of 23 proteins were expressed in the leaves. Some of the upregulated protein was involved in cell metabolism like repair mechanism of photosystem II, RuBisCO large and small chains, tRNA processing, cyclophilins which facilitate protein-folding by catalyzing the cis-trans isomerization of the peptide bonds that precede proline residues, retrograde golgi transport, enzymes in the metabolism of nucleic acids, proteins, or carbohydrates.

Protein phosphorylation is the major mechanism by which eukaryotic cells respond to extracellular signals, the directed proteomics approach is a powerful addition to the growing arsenal of techniques used for functional genomics. Peck *et al.*, 2001 found that AtPhos43, and related proteins in tomato and rice, are phosphorylated within minutes after treatment with flagellin or chitin fragments. By measuring ^{32}P incorporation into AtPhos43 immunoprecipitated from extracts of elicitor-treated hormone and defense-response mutants, it was found that phosphorylation of AtPhos43 after flagellin treatment but not chitin treatment is dependent on FLS2, a receptor-like kinase involved in flagellin perception. Induction by both elicitors is not dependent on salicylic acid or EDS1, a putative lipase involved in defense signaling.

Plant-bacteria interactions are highly complex due to multiple bacterial factors and plant-signaling events which take place, and ultimately define the susceptibility or resistance of the plant exposed to the pathogen. Several recent and current investigations are directed towards gaining a better understanding of the molecular mechanisms implicated in basal and specific plant defense responses against plant bacterial pathogens and their interactions. The details of some more proteomic studies related to plant-bacterial interaction were provided in Table 21.2.

Plant-Virus Interactions

Plant viruses represent a major threat for wide range of host species causing severe losses in agricultural practices. The full comprehension of mechanism underlying event of virus-host plant interaction is crucial to devise novel plant resistance strategies. Plant-virus interaction reveals a really complex patho-physiological context which triggers expression response of hundreds of genes. The development of proteomic methods has revolutionized our ability to assess protein interactions and cellular changes on a global scale, allowing the discovery of previously unknown connections.

Proteomics approaches were used by Delalande *et al.* (2005) to study the compatible interaction between *Oryza sativa* and rice yellow mottle sobemovirus (RYMV). This analysis led to the identification of a phenylalanine ammonia-lyase, a mitochondrial chaperonin-60 and an aldolase C, but the role of these proteins during RYMV infection of rice remains to be determined. Brizard *et al.*, 2006 investigated RYMV–rice (susceptible *O. sativa* indica IR64) protein complexes (formed *in vivo* or

Table 21.2: Proteins Expressed in Plant–Pathogen Interactions and Identified in Plants using Proteomic Approaches

Biotic Stresses Reference		*Pathogen*	*Studied organism*	*Methods*	*Protein expressed*
Viruses	RYMV	*Oryza sativa*	Nano LC-MS/MS	Three proteins: Phenylalanine ammonia-lyase, Mitochondrial chaperonin-60, Aldolase C-1were identified, but the role of these proteins during RYMV infection of rice remains to be determined.	Delalande *et al.*, 2005
	TMV	*Capsicum annuum*	2DE, MALDI-TOF MS	A hypothetical protein and five annotated nuclear proteins were identified, including four defence-related proteins.	Lee *et al.*, 2006
	PPV	*Prunus persica*	2DE, MALDI-TOF MS	Four proteins were identified: one thaumatin-like and three mandelonitrile lyases (MDLs). Thaumatins are proteins involved in the plant response against fungal infection, and MDLs in the catabolism of (R)-amygdaline.	Diaz-Vivancos *et al.*, 2006
	CMV	*Lycopersicon esculentum*	DIGE- MS/MS	Detected down-regulation of proteins involved in photosynthesis, photorespiration, carbon metabolism, and defense.	Carli *et al.*, 2010
	SCMV	*Zea mays*	2D DIGE, MALDI-TOF MS/MS	Eighty seven proteins were detected related to stress and defence response; energy and metabolism; photosynthesis, and protein synthesis and folding.	Wu *et al.*, 2013
Bacteria	*Clavibacter michiganensis* ssp. *michiganensis*	*Lycopersicon hirsutum*	2DE, ESI-MS DMS	Twenty six differentially regulated proteins were identified, 12 of which were directly related to defence and stress responses.	Coaker *et al.*, 2004
	Xanthomonas oryzae pv. oryzae	*Oryza sativa*	2DE	Twenty proteins were differentially expressed in response to bacterial inoculation. Out of these, four were defence-related proteins.	Mahmood *et al.*, 2006
	Pseudomonas syringae	*Arabidopsis thaliana*	2DE, LC-MS/MS	Seventy three differential spots representing 52 unique proteins were successfully identified, related to defence-related antioxidants and metabolic enzymes.	Jones *et al.*, 2006
	Xanthomonas oryzae pv. oryzae	*Oryza sativa*	2DE, MS/MS	Out of 20 proteins differentially regulated by pathogen, eight putative plasma membrane-associated and two non-plasma membrane-associated proteins were identified with potential functions in rice defence.	Chen *et al.*, 2007

Contd...

Table 21.2–*Contd...*

Abiotic Stress	*Plant*	*Method*		*Important Findings*	*References*
	Pseudomonas putida	*Brassica napus*	2DE, MS	Identify the increased level of proteins involved in nutrient utilization, root colonization and decreased in bacterial communication (chemotaxis, quorum sensing) proteins.	Cheng *et al.*, 2009
	Bradyrhizobium japonicum	*Glycine max*	LC-MS/MS, LTQ-Orbitrap MS	Proteins involved in translation, post transcriptional regulation, nitrogenase complex, aspartate amino transferase, carbon metabolism, translation, and nucleic acid metabolism were identified.	Delmotte *et al.*, 2010
Nematodes	*Heterodera schachtii*	Host plants	2DE, LCQ-MS/MS	Identification of 4 nematode-secreted pharyngeal gland proteins.	De Meutter *et al.*, 2001
	Meloidogyne incognita	Host plants	2DE, internal micro-sequencing	Identification of 7 stylet-secreted proteins.	Jaubert *et al.*, 2002
	Meloidogyne spp.	Host plants	nano-LC coupled to ion-trap MS	Identified the six proteins: actin, enolase, CG3752-PA protein similar to aldehyde dehydrogenase, HSP-60 and tzhangranslation initiation factor eIF-4A.	Calvo *et al.*, 2005
	Melodogyne hapla	Host plants	LC/MSE	A total of 516 non-redundant proteins were identified with an average of 9 unique peptides detected per protein.	Mbeunkui *et al.*, 2010
Fungus	*Aphanomyces euteiches*	*Medicago truncatula*	2DE	The majority of the indentified proteins belonged to the PR-10 family, whereas others corresponded to putative cell wall proteins and enzymes of the phenylpropanoid–isoflavonoid pathway.	Colditz *et al.*, 2004
	Fusarium verticillioides	*Zea mays*	2DE, MS	The proteins identified included PR proteins, antioxidant enzymes and proteins involved in protein synthesis, folding and stabilization.	Campo *et al.*, 2004
	Fusarium graminearum	*Triticum aestivum*	2DE, LC-MS/MS	Identify 33 plant proteins which were divided into two groups, each related to defence response or metabolism.	Zhou *et al.*, 2006
	Fusarium oxysporum	*Lycopersicon esculentum*	2DE, MS	Identified 33 different tomato proteins. Of these, 16 proteins were found in the xylem sap for the first time.	Houterman *et al.*, 2007

Contd...

Table 21.2–*Contd...*

Abiotic Stress	*Plant*	*Method*		*Important Findings*	*References*
	Thielaviopsis basicola	*Gossypium hirsutum*	LC-MS/MS, MASCOT MS/MS	Identified pathogen-induced cotton proteins implicated in post-invasion defence responses (PR-proteins related to oxidative burst), nitrogen metabolism, amino acid synthesis and isoprenoid synthesis.	Coumans *et al.*, 2009
Insect	*Myzus persicae*	Brassicaceae and Solanaceae	2D, ESI MS/MS	Fourteen aphid proteins were found to vary according to host plant switch; ten of them were down regulated (proteins involved in glycolysis, TCA cycle, protein and lipid synthesis) while four others were over expressed (mainly related to the cytoskeleton).	Francis *et al.*, 2006
	Spodoptera exigua	*Arabidopsis thaliana*	2DE, MALDI-TOF MS, MALDI-TOF/TOF MS	Nineteen insect-feeding-responsive proteins were identified, all of which were involved in metabolic regulation, binding functions or cofactor requirement of protein, cell rescue, defense and virulence.	Zhang *et al.*, 2010
	Leptinotarsa decemlineata, Macrosiphum euphorbiae	*Solanum tuberosum*	2DE	Thirty one proteins were up- or down regulated. Of these proteins, 29 were regulated by beetle chewing, 8 by wounding and 8 by aphid feeding.	Duceppe *et al.*, 2012
	Macrosyphum euphorbiae	*Solanum lycopersicum*	2DE, MALDI TOF-MS	Fifty seven proteins were identified, most of the proteins involved in the detoxification of oxygen radicals.	Coppola *et al.*, 2013

in vitro) to identify plant proteins putatively involved in the virus–host interactions. They revealed the presence of 223 different proteins and classified them into three functional categories using SDS-PAGE and nano-LC-MS/MS. In the metabolism category, a number of enzymes involved in glycolysis, citrate and malate cycles were found, probably for the production of energy to support viral replication. In the protein synthesis category, proteins involved in translation, elongation factors, chaperones, protein-disulfide isomerases and proteins involved in protein turnover with the 20S proteasome were observed. In the defence category, proteins involved in the generation and detoxification of reactive oxygen species were identified, probably to maintain an oxidation-reduction environment compatible with viral replication.

Also, rice stripe disease is one of the most devastating and widespread rice viral diseases caused by rice stripe virus (RSV). Yang *et al.*, 2013 employed proteomic tools including two-dimensional electrophoresis and mass spectrometry to globally identify differential proteins between a RSV-resistant rice cultivar Xudao 3 and a RSV -susceptible rice cultivar Wuyujing 3. They showed that out of 27 proteins, 15 were up-regulated and the other 12 proteins were down-regulated in Wuyujing 3. Some proteins such as heat shock protein, protein disulfide isomerase, glyoxalase and Os04g0435700 related to stress and/or defense responses were down-regulated in Wuyujing 3 which may be implicated in its susceptibility to RSV.

The proteomics studies of Mungbean Yellow Mosaic India Virus (MYMIV)-susceptible and resistant *Vigna mungo* genotypes helped to get an insight in the molecular events during compatible and incompatible plant-virus interactions (Kundu *et al.*, 2013). Hence, proteomic approach of viral pathogen not only provides candidate proteins and genes for subsequent studies, but also useful to identify novel proteins or genes in plants. The details of some more proteomic studies related to plant-virus interaction were provided in Table 21.2.

Plant-Fungus Interactions

Fungi are characterized by their lack of photosynthetic pigment and their chitinous cell wall. The majority of fungal species are saprophytes, a number of them are parasitic and around 15,000 of them causing disease in plants, the majority belonging to the ascomycetes and basidiomycetes (Hawksworth *et al.*, 2001). Studies of fungal pathogens and their interactions with plants have been performed using several approaches, from classical genetical approaches to the modern, holistic, and high-throughput omic techniques. In recent years, functional genomics analysis, including transcriptomics, proteomics, and metabolomics are unravelling molecular host-pathogen crosstalk, the complex mechanisms involving pathogenesis and host avoidance (Walter *et al.*, 2010).

The ascomycete, *Botrytis cinerea* is a necrotrophic plant pathogen and the causal agent of the disease commonly known as "gray mold", causing significant yield losses on more than 200 crop species worldwide (Staats *et al.*, 2005), such as flowers (rose and gerbera flowers), fruits and legumes (grapes, strawberries, tomatoes, french beans, chickpeas) and oil crops (sunflower). Fernandez-Acero *et al.*, 2006 developed a partial proteome map of *B. cinerea*. They have identified twenty-two proteins by MALDI-TOF or ESI IT MS/MS, with some of them corresponding to forms of malate dehydrogenase, glyceraldehyde-3-phosphate dehydrogenase, and a cyclophilin, proteins that have been related to virulence. In 2009, Fern´ andez-Acero *et al.*, also tried to establish a proteomic map of *B. cinerea* during cellulose degradation using 2-DE and MALDI-TOF/TOF MS/MS. They were able to identified 306 proteins, mostly representing unannoted proteins.

Cagas *et al.*, 2011 has quantitatively evaluated the proteomic response of *Aspergillus fumigatus* to caspofungin (an antifungal compound) by subcellular fractionation (for localization) and MALDI TOF/TOF MS identification. In the cell wall/plasma membrane fraction, an altered expression of 56 proteins was evident (26 upregulated and 30 downregulated), 81 per cent of the upregulated proteins were ribosomal proteins, the most highly upregulated protein was a UFP and chitinase was the most significantly downregulated protein. In the secreted fraction, the expression of six proteins (including ATP citrate lyase, Asp f4, hsp60, and transaldolase) was elevated, while eight proteins were significantly downregulated, including the ribotoxin Asp f1 (12-fold). The intracellular or extracellular protein expression studies of fungicide should improve our understanding of fungal drug resistance and facilitate the development of strategies to circumvent drug resistance.

A comparative proteomic analysis was performed by Gao *et al.* (2013) in resistant cotton (*Gossypium barbadense* cv. '7124') upon infection with *Verticillium dahliae* to identify differentially expressed proteins by mass spectrometry (MALDI-TOF/TOF). A total of 188 differentially expressed proteins were identified and classified into 17 biological processes based on gene ontology. Virus-induced gene silencing (VIGS) analyses reveal that gossypol, brassinosteroids and jasmonic acid contribute to the resistance of cotton to *V. dahliae.* The details of some more proteomic studies related to plant-fungus interaction were provided in table.

Plant-Insect Interactions

Insect pests remain a major reason for crop loss worldwide despite extensive use of chemical insecticides. More than 50 per cent of all insecticides are organophosphates, followed by synthetic pyrethroids, organochlorines, carbamates, and biopesticides, and their continued use may have many environmental, agricultural, medical, and socioeconomic issues. Dawkar *et al.*, 2013 focused on adaptations of Lepidopteran insects to phytochemicals and synthetic pesticides in native and modern agricultural systems. Because of heavy use of chemical insecticides, a strong selection pressure is imposed on insect populations, resulting in the emergence of resistance against candidate compounds. Current knowledge suggests that insects generally implement a three-tier system to overcome the effect of toxic compounds at physiological, biochemical, and genetic levels. Using the proteomics approach, insect proteins interacting with insecticides can be identified, and their modification in resistant insects can be characterized.

Jurat-Fuentes and Adang, (2007) used a proteomic approach to identify Cry1Ac binding proteins from intestinal brush border membrane (BBM) prepared from *Heliothis virescens* larvae. Cry1Ac binding BBM proteins were detected in 2D blots and identified using peptide mass fingerprinting (PMF) or de novo sequencing.

Among other proteins, the membrane bound alkaline phosphatase (HvALP), and a novel phosphatase, were identified as Cry1Ac binding proteins. Reduction of HvALP expression levels correlated directly with resistance to Cry1Ac in the YHD2-B strain of *H. virescens*.

The proteomic study was conducted by Collin *et al.* (2010) to investigate physiological factors affecting feeding behaviour by larvae of the insect, *Plutella xylostella*, on herbivore-susceptible and herbivore-resistant *Arabidopsis thaliana*. The leaves of 162 recombinant inbred lines (RILs) were screened to detect genotypes upon which *Plutella* larvae fed and 2D-PAGE revealed significant differences in the proteomes between the identified resistant and susceptible RILs. The proteomic results showed that there was increased production of reactive oxygen species (ROS), in particular H_2O_2 in *Plutella*-resistant RILs. Many of the proteins such a carbonic anhydrases, malate dehydrogenases, glutathione S-transferases, isocitrate dehydrogenase-like protein, and lipoamide dehydrogenase were detected abundant in the *Plutella*-resistant RILs are known to be involved in limiting ROS damage in other biological systems. Thus, the potential applications of proteomics include both novel compound development and production of genetically modified plants resistant to insect pests. The details of some more proteomic studies related to plant-insect interaction were provided in Table 21.2.

Plant-Nematode Interactions

Nematodes are one of the most ancient and diverse groups of animals on earth. Most nematodes are microscopic, free-living and feed on bacteria, fungi, protozoans and other nematodes, but about 15 percent of the known nematode species are parasites of plants. Various nematodes feed on all plant parts: roots, stems, leaves, flowers and seeds. Most plant parasitic nematodes are root feeders and live in the soil. The main diagnostic signs and symptoms of nematode infestations are root cysts or root galls, and "nematode wool" on bulbs and corms. The most studied plant-parasitic nematodes are the sedentary endoparasitic root knot and cyst nematodes. These nematodes have evolved complex interactive relationships with host cells to form highly specialized nematode feeding sites called giant-cells, in infected plant roots from which they withdraw nutrients to sustain a sedentary parasitic lifestyle (Curtis, 2007).

Cell death and resistance (*cdr*) mutants are good materials to analyze defense-related proteins in rice because these mutants induce a series of defense responses, including expression of defense-related genes and high accumulation of phytoalexins. Tsunezuka *et al.*, 2006 carried out a proteome analysis of the *cdr2* mutant in rice. Out of 37 proteins that were differentially expressed between *cdr2* and wild type, 28 spots were up-regulated and nine were down-regulated in the *cdr2* mutant. The changes of these protein levels were quantitatively analyzed and classified into three classes; defense-related proteins (*e.g.* probenazole-induced protein, caffeic acid 3-O-methyltransferase etc), molecular chaperone (*e.g.* Heat shock protein 70 and DnaK-type molecular chaperone etc.) and metabolic enzymes (*e.g.* S-adenosylmethionine synthase, Glyceraldehydes-3-phosphate dehydrogenase etc.). This study shows that proteome analysis is a useful approach to study programmed cell death and defense signaling in plants.

Zhang *et al.*, 2011 identified 46 differentially expressed proteins in soybean hypocotyls infected with *Phytophthora sojae*, using two-dimensional electrophoresis and matrix-assisted laser desorption/ionization tandem time of flight (MALDI-TOF/TOF). Out of 26 proteins studied, 12 were up-regulated and 14 down-regulated at various time points in the tolerant soybean line, Yudou25. In contrast, only 20 proteins were significantly affected (11 up-regulated and 9 down-regulated in the sensitive soybean line, NG6255. Among these proteins, 26 per cent were related to energy regulation, 15 per cent to protein destination and storage, 11 per cent to defense against disease, 11 per cent to metabolism, 9 per cent to protein synthesis, 4 per cent to secondary metabolism, and 24 per cent were of unknown function. The details of some more proteomic studies related to plant-nematode interaction were provided in Table 21.2.

Bioinformatics Tools in Proteomics

Bioinformatics has emerged into a discipline due to the availability of huge amount of data. The collection of biological information is very vast and as a result there are number of databases which are available online. Protein databases are more specialized than primary sequence databases. They contain information derived from the primary sequence databases. Some contain protein translations of the nucleic acid sequences and some contains sets of patterns and motifs derived from sequence homologs. Table 21.3 provides various protein databases available and their uses.

Future Prospective

In the past decades breakthrough in genetic crop improvement has been done by green revolution. To meet the current challenges of food insecurity, proteins that control stress resistance in a wide range of environments need to be identified to facilitate the biological improvement of crop productivity. Most likely, the current decade will rely on functional genomics to identify genes that enhance crop yield with minimizing the losses due to different stress. With the rapid progress in protein extraction, separation, and identification techniques, proteomics technology is playing a key role in identifying proteins responsive to abiotic and biotic stresses. However, certain disadvantages are limiting the use of proteomics. Metabolite identification is still a major hurdle and annotation of more metabolites can hasten progress in this area.

Protein get denaturized on exposure to high temperature, extremes of pH, oxidation, specific chemicals etc and causes loss in secondary and tertiary structure important during their analysis. There are some classes of proteins which are difficult to analyze due to their poor solubility. Furthermore, proteomic analysis suffers from the inability to directly analyze proteins because of their large size and unavailability of precise libraries. A systematic and efficient analysis of vast genomic and proteomic data sets is also a major challenge for researchers today. Currently, there are few experts available for functional interpretation of data obtained from integration of proteomics with genomics and metabolomics. The other challenge remains to saturate the proteome responsive to various biotic stresses. This saturation will

Table 21.3: Plant Proteome Databases and their Uses

Name	*Web Address*	*Uses*
PIR	http://pir.georgetown.edu/	The Protein Information Resource (PIR) is an integrated public bioinformatics resource to support genomic, proteomic and systems biology research and scientific studies.
UniProtKB/Swiss-Prot	http://web.expasy.org/docs/swiss-prot_guideline.html	UniProtKB/Swiss-Prot is the manually annotated and reviewed section of the UniProt Knowledgebase (UniProtKB). It is a high quality annotated and non-redundant protein sequence database, which brings together experimental results, computed features and scientific conclusions. Since 2002, it is maintained by the UniProt consortium and is accessible via the UniProt website.
ALIGN	http://www.bioinf.man.ac.uk/dbbrowser/ALIGN/	ALIGN is a compendium of sequence alignments: it is a companion resource to PRINTS.
InterPro: protein sequence analysis and classification	http://www.ebi.ac.uk/interpro/	InterPro provides functional analysis of proteins by classifying them into families and predicting domains and important sites. We combine protein signatures from a number of member databases into a single searchable resource, capitalising on their individual strengths to produce a powerful integrated database and diagnostic tool.
PRINTS	http://bioinf.man.ac.uk/dbbrowser/PRINTS/index.php	PRINTS are a compendium of protein fingerprints. A fingerprint is a group of conserved motifs used to characterize a protein family; its diagnostic power is refined by iterative scanning of a SWISS-PROT/TrEMBL composite.
Pfam	http://www.sanger.ac.uk/resources/databases/pfam.html	The Pfam database is one the most important collections of information in the world for classifying proteins. The database categorises 75 per cent of known proteins to form a library of protein families - a 'periodic table' of biology.
PROSITE	http://prosite.expasy.org/	PROSITE consists of documentation entries describing protein domains, families and functional sites as well as associated patterns and profiles to identify them.
PDBeFold	http://www.ebi.ac.uk/msd-srv/ssm/	Pair wise comparison and 3D alignment of protein structures, multiple comparison and 3D alignment of protein structures, examination of a protein structure for similarity with the whole PDB archive or SCOP archive.
RCSB PDB	http://www.rcsb.org/pdb/home/home	Single international repository for the processing and distribution of 3-D macromolecular structure data primarily determined experimentally.
CATH/Gene3D	http://www.cathdb.info/	CATH is a classification of protein structures downloaded from the Protein Data Bank.
Enzyme nomenclature database	http://enzyme.expasy.org/	ENZYME is a repository of information relative to the nomenclature of enzymes. It is primarily based on the recommendations of the Nomenclature Committee of the International Union of Biochemistry and Molecular Biology (IUBMB) and it describes each type of characterized enzyme for which an EC (Enzyme Commission) number has been provided.

Contd...

Table 21.3–*Contd...*

Name	*Web Address*	*Uses*
Biological magnetic Resonance Data Bank	http://www.bmrb.wisc.edu/	A repository for data from NMR spectroscopy on proteins, peptides, nucleic acids and other biomolecules.
Enzyme Structures Database	http://www.biochem.ucl.ac.uk/bsm/enzymes/index.html	This database contains the known enzyme structure that have been deposited in the Brookhaven Protein Data Bank.
KEGG LIGAND Database	http://www.genome.jp/dbget/ligand.html	KEGG LIGAND contains our knowledge on the universe of chemical substances and reactions that are relevant to life. It is a composite database consisting of COMPOUND, GLYCAN, REACTION, RPAIR, RCLASS, and ENZYME databases, whose entries are identified by C, G, R, RP, RC, and EC numbers, respectively.
RADAR	http://www.ebi.ac.uk/Tools/pfa/radar/	RADAR identifies gapped approximate repeats and complex repeat architectures involving many different types of repeats.
ProteinProspector	http://prospector.ucsf.edu/prospector/mshome.htm	Proteomics tools for mining sequence databases in conjunction with Mass Spectrometry experiments.
Standard Protein BLAST	http://blast.ncbi.nlm.nih.gov/Blast.cgi?PROGRAM=blastp and PAGE_TYPE=BlastSearch and LINK_LOC=blasthome	Search protein database using a protein query.
SignalP 4.1 server	http://www.cbs.dtu.dk/services/SignalP/	SignalP 4.1 server predicts the presence and location of signal peptide cleavage sites in amino acid sequence from different organism.
FASTA	http://www.ebi.ac.uk/Tools/sss/fasta/	This tool provides sequence similarity searching against protein databases using the FASTA suite of programs.
Sequences Annotated by Structure	http://www.biochem.ucl.ac.uk/bsm/sas/sasquery.html	The following option will perform a FASTA search of the given sequence against the proteins of known structure in the PDB and returns a multiple alignment of all hits, each annotated by structural features.
PRINTS Peptide Scan Tool	http://biotools.umassmed.edu/cgi-bin/biobin/pscan	Pscan locates multiple motif features (fingerprints) in a protein sequence.
P-val finger PRINT scan	http://www.bioinf.manchester.ac.uk/dbbrowser/fingerPRINTScan/	Provides the closest matching PRINTS fingerprints to your protein sequence. Tabular results provide different levels of diagnostic details, and direct links to GRAPHScan.

Contd...

Table 21.3–*Contd...*

Name	*Web Address*	*Uses*
Conserved Domain Database (CDD)	http://www.ncbi.nlm.nih.gov/Structure/cdd/cdd.shtml	CDD is a protein annotation resource that consists of a collection of well-annotated multiple sequence alignment models for ancient domains and full-length proteins. These are available as position-specific score matrices (PSSMs) for fast identification of conserved domains in protein sequences via RPS-BLAST. CDD content includes NCBI-curated domains, which use 3D-structure information to explicitly define domain boundaries and provide insights into sequence/structure/function relationships, as well as domain models imported from a number of external source databases (Pfam, SMART, COG, PRK,TIGRFAM).
ScanProsite	http://prosite.expasy.org/scanprosite/	This form allows you to scan proteins for matches against the PROSITE collection of motifs as well as against your own patterns.
Mobyle	http://mobyle.pasteur.fr/cgi-bin/portal.py?#forms::pepstats	Calculates statistics of protein properties.
BTPRED	http://www.biochem.ucl.ac.uk/bsm/btpred/index.html	BTPRED is a method for predicting the location and type of beta-turns in protein sequences. Predictions are made using a combination of artificial neural networks and simple filtering rules.
Yino Yang 1.2	http://www.cbs.dtu.dk/services/YinOYang/	The Yino Yang server produces neural network predictions.
ProtParam	http://web.expasy.org/protparam/	ProtParam is a tool which allows the computation of various physical and chemical parameters for a given protein stored in Swiss-Prot or TrEMBL or for a user entered sequence.
COILS	http://www.ch.embnet.org/software/COILS_form.html	Prediction of coiled coil regions in proteins.
CPH models 3.2	http://www.cbs.dtu.dk/services/CPHmodels/index.php	CPH models 3.2 are a protein homology modeling server. The template recognition is based on profile-profile alignment guided by secondary structure and exposure prediction.
Garnier peptide structure tool	http://biotools.umassmed.edu/cgi-bin/biobin/garnier	Garnier predicts protein secondary structure.
TMpred	http://www.ch.embnet.org/software/TMPRED_form.html	Prediction of transmembrane regions and orientation.
PDBeFold. Structure Similarity	http://www.ebi.ac.uk/msd-srv/ssm/	Pair wise comparison and 3D alignment of protein structures, multiple comparison and 3D alignment of protein structures, examination of a protein structure for similarity with the whole PDB archive or SCOP archive best Cα-alignment of compared structures.
TMHMM server v 2.0	http://www.cbs.dtu.dk/services/TMHMM-2.0/	Prediction of transmembrane helices in proteins.

reveal the potential biomarkers unique to each stress, and such biomarkers could be useful for plant biologists and breeders in marker-assisted breeding for biotic stress.

In conclusion, proteomics play a key role in the development of functional genomics. The combination of genomics and proteomics will play a major role in understanding molecular mechanisms in plant stress biology. It will accelerate the development of high yield varieties, with better resistance towards various abiotic and biotic stresses. So, there is the need for more sensitive analytical systems and effective methods for large scale data comparison in order to advance our knowledge on protein expression.

References

Abbasi, F. and Komatsu, S., 2004. A proteomic approach to analyze salt responsive proteins in rice leaf sheath. *Proteomics* 4, 2072-2081.

Afroz, A., Zahur, M., Zeeshan, N. and Komatsu, S., 2013. Plant-bacterium interactions analyzed by proteomics. *Frontier in Plant Science,* doi: 10.3389/fpls.2013.00021.

Agarwal, G.K. Rakwal, R. Yonekura, M. Saji, H., 2002. Proteome analysis of differentially displayed proteins as a tool for investigating ozone stress in rice (*Oryza sativa* L.) seedlings. *Proteomics* 2:947–59.

Ahsan, N. Lee. D.G. Lee, S.H. Kang, K.Y. Bahk, J.D. Choi, M.S., 2007. A comparative proteomic analysis of tomato leaves in response to waterlogging stress. *Physiol Plant.* 131:555–70.

Aina, R. Labra, M. Fumagalli, P. Vannini, C. Marsoni, M., Cucchi, U., 2007. Thiol-peptide level and proteomic changes in response to cadmium toxicity in *Oryza sativa* L. roots. *Env Exp Bot* 59:381–92.

Amme, S. Matros, A. Schlesier, B. Mock, H.P., 2006. Proteome analysis of cold stress response in Arabidopsis thaliana using DIGE-technology. *J Exp Bot.* 57:1537–46.

Ashraf, M. and P.J.C. Harris. 2004. Potential biochemical induction of salinity tolerance in plants. *Plant Sci.* 166: 3-16.

Bae, M.S Cho, E.J. Choi, E.Y. Park, O.K., 2003. Analysis of the Arabidopsis nuclear proteome and its response to cold stress. *Plant J.* 36:652–63.

Bailey-Serres, J. Voesenek, L. A. C. J., 2008. Flooding stress: acclimations and genetic diversity. *Annu. Rev. Plant Biol.* 59:313–339.

Bogeat-Triboulot, M.B., Brosché, M., Renaut, J., Jouve, L., Le Thiec, D., Fayyaz, P., Vinocur, B., Witters, E., Laukens, K., Teichmann, T., Altman, A., Hausman, J.F., Polle, A., Kangasjärvi, J. and Dreyer, E., 2007. Gradual soil water depletion results in reversible changes of gene expression, protein profiles, ecophysiology, and growth performance in *Populus euphratica*, a poplar growing in arid regions. *Plant Physiol.,* 143: 876–92.

Brizard, J.P., Carapito, C., Delalande, F., Van Dorsselaer, A. and Brugidou, C., 2006. Proteome analysis of plant–virus interactome: comprehensive data for virus multiplication inside their hosts. *Mol. Cell. Proteomics,* 5: 2279–2297.

Cagas, S.E., Jain, M.R., Li, H. and Perlin, D.S., 2011. Profiling the *Aspergillus fumigatus* proteome in response to caspofungin. *Antimicrob Agents Ch.*, 55: 146–154.

Calvo, E., Flores-Romero, P., Lopez, J.A., Navas, A., 2005. Identification of proteins expressing differences among isolates of *Meloidogyne* spp. (Nematoda: Meloidogynidae) by nano-liquid chromatography coupled to ion-trap mass spectrometry. *J. Proteome Res.* 4:1017–1021.

Campo, S., Carrascal, M., Coca, M., Abi´an, J., Segundo, BS., 2004.The defense response of germinating maize embryos against fungal infection: a proteomics approach. *Proteomics.* 4(2): 383–396.

Canovas, F.M., Dumas-Gaudot, E., Recorbet, G., Jorrin, J., Mock, H.P. and Rossignol, M., 2004. Plant proteome analysis. *Proteomics,* 4: 285–298.

Carli, M.D., Villani, M.E., Bianco, L., Lombardi, R., Perrotta, G., Benvenuto, E., Donini M., 2010. Proteomic analysis of the plant-virus interaction in cucumber mosaic virus (CMV) resistant transgenic tomato. *J. Proteome Res.* 9(11): 5684-97.

Chang, W.W.P. Huang, L. Shen, M. Webster, C. Burlingame, A.L. Roberts, J.K.M., 2000. Patterns of protein synthesis and tolerance to anoxia in root tips of maize seedlings acclimated to a low-oxygen environment, and identification of proteins by mass spectrometry. *Plant Physiol.* 122:295–317.

Chen, F., Huang, Q., Zhang, H., Lin, T., Guo, Y., Lin, W. and Chen, L., 2007. Proteomic analysis of rice cultivar Jiafuzhan in the responses to *Xanthomonas campestris* pv. Oryzicola infection. *Acta. Agron. Sin.,* 33: 1051–1058.

Chen, L., Li, D., Wang, L., Teng, S., Zhang, G., Guo, L., Mao, Q., Wang, W. and Li, M., 2012. Proteomics analysis of rice proteins up-regulated in response to bacterial leaf streak disease. *J. Plant Biol.,* 55: 316-324.

Cheng, Y. Qi, Y. Zhu, Q. Chen, X. Wang, N. Zhao, X., 2009. New changes in the plasma-membrane-associated proteome of rice roots under salt stress. *Proteomics* 9:3100–14.

Cheng, Z., Duan, J., Hao, Y., McConkey, B.J., Glick, B.R., 2009. Identification of bacterial proteins mediating the interactions between *Pseudomonas putida* UW4 and *Brassica napus* (Canola). *Mol. Plant Microbe Interact.* 22: 686e694.

Chinnusamy, V. Zhu, J. Zhu, J. K., 2007. Cold stress regulation of gene expression in plants. *Trends Plant Sci.* 12 : 444–451.

Coaker, G.L.,Willard, B., Kinter, M., Stockinger, E.J., Francis, D.M., 2004. Proteomic analysis of resistance mediated by Rcm 2.0 and Rcm 5.1, two loci controlling resistance to bacterial canker of tomato. *Mol. Plant Microbe Interact.* 17: 1019–1028.

Colditz, F., Nyamsuren, O., Niehaus, K., Eubel, H., Braun, H.P., Krajinski, F., 2004. Proteomic approach: identification of *Medicago truncatula* proteins induced in roots after infection with the pathogenic oomycete *Aphanomyces euteiches*. *Plant Mol. Biol.* 55(1): 109–120.

Collins, R.M., Afzal, M., Ward, D.A., Prescott, M.C.,Sait, S.M., Rees, H.H. and Tomsett, A.B., 2010. Differential proteomic analysis of *Arabidopsis thaliana* genotypes exhibiting resistance or susceptibility to the insect herbivore, *Plutella xylostella. Plos one*, 5(4): e10103.

Coppola, V., Coppola, M., Rocco, M., Digilio, M.C., Ambrosio, C.D., Renzone, G., Martinelli, R., Scaloni, A., Pennacchio, F., Rao, R., Corrado, G., 2013. Transcriptomic and proteomic analysis of a compatible tomato-aphid interaction reveals a predominant salicylic acid-dependent plant response. *BMC Genomics*. 14: 515.

Coumans, J.V.F., Poljak, A., Raftery, M.J., Backhouse, D., Pereg-Gerk, L., 2009. Analysis of cotton (*Gossypium hirsutum*) root proteomes during a compatible interaction with the black root rot fungus *Thielaviopsis basicola*. *Proteomics*. 9(2): 335–349.

Cui, S. Huang, F. Wang, J. Ma, X. Cheng, Y. Liu, J., 2005. A proteomic analysis of cold stress responses in rice seedlings. *Proteomics*. 5:3162–72.

Curtis, R.H.C. 2007. Plant parasitic nematode proteins and the host–parasite interaction. *Brief Funct. Genomic Proteomic*, 6: 50–58.

Danchenko, M. Skultety, L. Rashydov, N.M. Berezhna, V.V. Mátel Å, Salaj T., 2009. Proteomic analysis of mature soybean seeds from the Chernobyl area suggests plant adaptation to the contaminated environment. *J Proteome Res*. 8:2915–22.

Dani, V. Simon, W.J. Duranti, M. Croy, R.R.D., 2005. Changes in the tobacco leaf apoplast proteome in response to salt stress. *Proteomics*. 5:737–45.

Dawkar, V.V., Chikate, Y.R., Lomate, P.R., Dholakia, B.B., Gupta, V.S. and Giri, A.P., 2013. Molecular insights into resistance mechanisms of Lepidopteran insect pests against toxicants. *J. Proteome Res.*, **DOI:**10.1021/pr400642p.

De Meutter, J., Vanholme, B., Bauw, G., Tytgat, T., Gheysen, G., Gheysen, G., 2001. Preparation and sequencing of secreted proteins from the pharyngeal glands of the plant parasitic nematode *Heterodera schachtii*. *Mol. Plant Pathol*. 2: 297–301.

Delalande, F., Carapito, C., Brizard, J. P., Brugidou, C., Van Dorsselaer, A., 2005. Multigenic families and proteomics: extended protein characterization as a tool for paralog gene identification. *Proteomics*, 5(2): 450-60.

Delmotte, N., Ahrens, C.H., Knief, C., Qeli, E., Koch, M., Fischer, H., Vorholt, J.A., Hennecke H., Pessi, G., 2010. An integrated proteomics and transcriptomics reference data set provides new insights into the *Bradyrhizobium japonicum* bacteroid metabolism in soybean root nodules. *Proteomics*. 10(7): 1391-1400.

Diaz-Vivancos, P., Rubio, M., Mesonero, V., Periago, P.M., Barcelo, A.R., Martinez-Gomez, P., Hernandez, J.A., 2006. The apoplastic antioxidant system in *Prunus*: response to long-term plum pox virus infection. *J. Exp. Bot*. 57: 3813–3824.

Duceppe, M., Cloutier, C., Michaud, D., 2012. Wounding, insect chewing and phloem sap feeding differentially alter the leaf proteome of potato, *Solanum tuberosum* L. *Proteome Sci*. 10: 73.

Dumont, E. Bahrman, N. Goulas, E. Valot, B. Sellier, H. Hilbert, J.L., 2011. A proteomic approach to decipher chilling response from cold acclimation in pea (*Pisum sativum* L.). *Plant Sci.* 180:86–98.

Fernández-Acero, F.J., Carbú, M., Garrido, C., Collado, I.G., Cantoral, J.M. and Vallejo, I., 2006. Screening study of potential lead compounds for natural product-based fungicides against *Phytophthora* Species. *J. Phytopathol.,* 154: 616-21.

Fernández-Acero, F.J., Colby, T., Harzen, A., Cantoral, J.M. and Schmidt, J., 2009. Proteomic analysis of the phytopathogenic fungus *Botrytis cinerea* during cellulose degradation. *Proteomics,* 9: 2892-902.

Francis, F., Gerkens, P., Harmel, N., Mazzucchelli, G., Pauw, E.D., Haubruge, E., 2006. Proteomics in *Myzus persicae*: Effect of aphid host plant switch. *Insect Biochem. Mol. Biol.* 36: 219–227.

Gao, W., Long, L., Zhu, L., Xu, L., Gao, W., Sun, L., Liu, L. and Zhang, X., 2013. Proteomic and virus-induced gene silencing (VIGS) analyses reveal that Gossypol, Brassinosteroids and Jasmonic acid contribute to the resistance of cotton to *Verticillium dahlia.* MCP Papers in Press. M113.031013.

Gibbs, J. and Greenway H. 2003. Mechanisms of anoxia tolerance in plants. I. Growth, survival and anaerobic catabolism. *Funct. Plant Biol.* 30, 1–47 10.1071/PP98095.

Gygi, S.P., Rist, B., Gerber, S.A., Turecek, F., Gelb, M.H. and Aebersold, R., 1999. Quantitative analysis of complex protein mixtures using isotope-coded affinity tags. *Nature Biotechnology,* 17: 994–999.

Hajduch, M. Rakwal, R. Agrawal, G.K. Yonekura, M. Pretova, A., 2001. High-resolution two-dimensional electrophoresis separation of proteins from metal-stressed rice (Oryza sativa L.) leaves: drastic reductions/fragmentation of ribulose-1,5-bisphosphate carboxylase/oxygenase andinduction of stress-relatedproteins. *Electrophoresis.* 22:2824–31.

Han, X., Aslanian, A., Yates, J.R.,2008. Mass spectrometry for proteomics. Current opinion in chemical biology, 12:483-490.

Hawksworth, D.L., Kirk, P.M., Sutton, B.C. and Pegler, D.N., 2001. Ainsworth and Bisby's Dictionary of the Fungi, International Mycological Institute, Surrey, UK.

Houterman, P.M., Speijer, D., Dekker, H.L., de Koster, C.G., Cornelissen, B.J.C., Rep, M., 2007. The mixed xylem sap proteome of *Fusarium oxysporum*-infected tomato plants. *Mol. Plant Pathol.* 8(2): 215–221.

Ideker, T., Thorsson, V., Ranish, J.A., Christmas, R., Buhler, J., Eng, J.K., Bumgarner, R., Goodlett, D.R., Aebersold, R. and Hood, L., 2001. Integrated genomic and proteomic analyses of a systematically perturbed metabolic network. *Science,* 292: 929–934.

Imin, N, Kerim, T. Rolfe, B.G., 2004. Weinman JJ. Effect of early cold stress on the maturation of rice anthers. *Proteomics.* 4:1873–82.

Jackson, M. B. Ram, P.C., 2003. Physiological and molecular basis of susceptibility and tolerance of rice plants to complete submergence. *Ann. Bot.* 91, 227–241 10.1093/aob/mcf242.

Jaubert, S., Ledger, T.N., Laffaire, J.B., Piotte, C., Abad, P., Rosso, M.N., 2002. Direct identification of stylet secreted proteins from root-knot nematodes by a proteomic approach. *Mol. Biochem. Parasitol.* 121: 205–211.

Jones, A.M.E., Thomas, V., Bennett, M.H., Mansfield, J., Grant, M., 2006. Modifications to the Arabidopsis defence proteome occur prior to significant transcriptional change in response to inoculation with *Pseudomonas syringae. Plant Physiol.* 142: 1603–1620.

Jurat-Fuentes, J.L. and Adang, M.J., 2007. A proteomic approach to study Cry1Ac binding proteins and their alterations in resistant *Heliothis virescens* larvae. *Journal of Invertebrate Pathology*, 95: 187–191.

Kawamura, Y. Uemura, M., 2003. Mass spectrometric approach for identifying putative plasma membrane proteins of Arabidopsis leaves associated with cold acclimation. *Plant J* 36:141–54.

Ke, Y., G. Han, H. He, and J. Li. 2009. Differential regulation of proteins and phosphoproteins in rice under drought stress. *Biochem. Biophys. Res. Comm.* 379, 133-138.

Khan, S.V., L. Hoffmann, J. Renauty, and J.F. Hausman. 2007. Current initiatives in proteomics for the analysis of plant salt tolerance. Curr. Sci. 93(6), 807-817.

Komatsu, S. Hiraga, S. Yanagawa, Y. 2012. Proteomics techniques for the development of flood tolerant crops. *J. Proteome Res.* 11: 68–78 10.1021/pr2008863.

Kundu, S., Chakraborty, D., Kundu, A. and Pal, A., 2013. Proteomics approach combined with biochemical attributes to elucidate compatible and incompatible plant-virus interactions between *Vigna mungo* and Mungbean Yellow Mosaic India Virus. *Proteome Science,* 11:15.

Lee, B.J., Kwon, S.J., Kim, S.K., Kim, K.J., Park, C.J., Kim, Y.J., Park, O.K., Paek, K.H., 2006. Functional study of hot pepper 26S proteasome subunit RPN7 induced by tobacco mosaic virus from nuclear proteome analysis. *Biochem. Biophys. Res. Commun.* 351: 405–411.

Lee, D.G. Ahsan, N. Lee, S.H. Kang, K.Y. Bahk, J.D. Lee, I.J., 2007. A proteomic approach in analyzing heat-responsive proteins in rice leaves. *Proteomics.* 7:3369–83.

Lee, D.G. Ahsan, N. Lee, S.H. Lee, J.J. Bahk, J.D. Kang, K.Y., 2009. Chilling stress-induced proteomic changes in rice roots. *J Plant Physiol.* 166:1–11.

Lee, K. Bae, D.W. Kim S.H. Han, H.J. Liu, X. Park, H.C., 2010. Comparative proteomic analysis of the short-term responses of rice roots and leaves to cadmium. *J Plant Physiol.* 167:161–8.

Mahajan, S. Tuteja, N., 2005. Cold, salinity and drought stresses: an overview. *Arch Biochem Biophys.* 15:444(2):139-58

Mahmood, T., Jan, A., Kakishima, M., Komatsu, S., 2006. Proteomic analysis of bacterial-blight defence responsive proteins in rice leaf blades. *Proteomics.* 6: 6053–6065.

Mann, M. and Jensen, O.N., 2003. Proteomic analysis of post-translational modifications. *Natural Biotechnology,* 21: 255–261.

Mbeunkui, F., Scholl, E.H., Opperman, C.H., Goshe M.B., Bird, D.M., 2010. Proteomic and bioinformatic analysis of root-knot nematode *Meloidgyne hapla:* The basis for plant parasitism. *J. Proteome Res.* 9(10): 5370-81.

Munns, R. 2005. Genes and salt tolerance: Bringing them together. New Phytol. 167:645-663.

Nanjo Y., Skultety L., Ashraf Y., Komatsu S. (2010). Comparative proteomic analysis of early-stage soybean seedlings responses to flooding by using gel and gel-free techniques.

Navi, M.F.T. and Toorchi, M., 2013. Importance of proteomics approach on identifying defense protein in response to biotic stress in rice (*Oryza sativa* L.). *Int. J. Biosci.* 3(10): 221-232.

Peck, C. S., Nuhse, S. T., Hess, D., Lglesias, A., Meins, F. and Boller, T., 2001. Directed proteomics identifies a plant-specific protein rapidly phosphorylated in response to bacterial and fungal elicitors. *The Plant Cell,* 13: 1467–1475.

Prasad T. K. 1996. Mechanisms of chilling-induced oxidative stress injury and tolerance in developing maize seedlings: changes in antioxidant system, oxidation of proteins and lipids, and protease activities. Plant J. 10, 1017–1026 10.1046/j.1365-313X.1996.10061017.x.

Qureshi, M.I., M. Israr, M.Z. Abdin, and M. Iqbal. 2005. Responses of *Artemisia annual* L. to lead and salt-induced oxidative stress. Environ. Exp. Bot. 53, 185-195.

Ross, P.L. and Huang, Y.N., 2004. "Multiplexed protein quantitation in Saccharomyces cerevisiae using amine-reactive isobaric tagging reagents". *Mol. Cell. Proteomics* 3: 1154–69.

Roth, U. von, Roepenack-Lahaye, E. Clemens, S., 2006. Proteome changes in Arabidopsis thaliana roots upon exposure to Cd2+.*J Exp Bot* 57:4003–13.

Sarry, J.E. Kuhn, L. Ducruix, C. Lafaye, A. Junot, C. Hugouvieux, V., 2006. The early responses of Arabidopsis thaliana cells to cadmium exposure explored by protein and metabolite profiling analyses. *Proteomics.* 6:2180–98.

Schutz, Wolfgang, Niklas, Hausmann, Karsten Krug, Rudiger Hampp, Boris Macek. 2011. Extending SILAC to Proteomics of Plant Cell Lines. *The Plant Cell.* 23: 1701–1705.

Schweppe, R.E., Haydon, C.E., Lewis, T.S., Resing, K.A. and Ahn, N.G., 2003. The characterization of protein post-translational modifications by mass spectrometry. *Accounts of Chemical Research,* 36: 453–461.

Semane, B. Dupae, J. Cuypers, A. Noben, J.P. Tuomainen, M. Tervahauta, A., 2010. Leaf proteome responses of Arabidopsis thaliana exposed to mild cadmium stress. *J Plant Physiol.* 167:247–54.

Staats, M., van Baarlen, P. and van Kan, J.A., 2005. Molecular phylogeny of the plant pathogenic genus *Botrytis* and the evolution of host specificity. *Mol. Biol. Evol.*, 22: 333-346.

Staats, M., van Baarlen, P. and van Kan, J.A., 2005. Molecular phylogeny of the plant pathogenic genus *Botrytis* and the evolution of host specificity. *Mol. Biol. Evol.*, 22: 333-346.

Taylor, N.L. Heazlewood, J.L. Day, D.A. Millar, A.H., 2005. Differential impact of environmental stresses on the pea mitochondrial proteome. *Mol Cell Proteomics.* 4:1122–33.

Tsunezuka, H., Fujiwara, M., Kawasaki, T. and Shimamoto, K., 2006. Proteome analysis of programmed cell death and defense signaling using the rice lesion mimic mutant *cdr2. Molecular Plant-Microbe Interactions,* 18(1): 52-59.

Walter, S., Nicholson, P. and Doohan, F.M., 2010. Action and reaction of host and pathogen during Fusarium head blight disease. *New Phytologist,* 185(1): 54–66.

Wu, L., Han, Z., Wang, S., Wang, X., Sun, A., Zu, X., Chen, Y., 2013. Comparative proteomic analysis of the plant–virus interaction in resistant and susceptible ecotypes of maize infected with sugarcane mosaic virus. *J. Proteomics.* 89: 124 – 140.

Yan, S. Tang, Z. Su, W. Sun, W., 2005. Proteomic analysis of salt stress-responsive proteins in rice root. *Proteomics.* 5: 235–44.

Yang, Y., Dai, L., Xia, H., Zhu, K., Liu, X. and Chen, K., 2013. Comparative proteomic analysis of rice stripe virus (RSV)-resistant and -susceptible rice cultivars. *Australian journal of crop science,* 7(5): 588-593.

Zhang, J.H., Sun, L.W., Liu, L.L., Lian, J., An, S.L., Wang, X., Zhang, J., Jin, J.L., Li, S.Y., Xi, J.H., 2010. Proteomic analysis of interactions between the generalist herbivore *Spodoptera exigua* (Lepidoptera: Noctuidae) and *Arabidopsis thaliana. Plant Mol. Biol. Rep.* 28: 324–333.

Zhang, Y., Zhao, J., Xiang, Y., Bian, X., Zuo, Q., Shen, Q., Gai, J. and Xing, H., 2011. Proteomics study of changes in soybean lines resistant and sensitive to *Phytophthora sojae. Proteome Science,* 9: 52.

Zhou, W., Eudes, F., Laroche, A., 2006. Identification of differentially regulated proteins in response to a compatible interaction between the pathogen *Fusarium graminearum* and its host, *Triticum aestivum. Proteomics.* 6(16): 4599–4609.

2016, Recent Advances in Plant Stress Physiology *Pages* ***507–537***
Editors: ***Praduman Yadav, Sunil Kumar and Veena Jain***
Published by: **DAYA PUBLISHING HOUSE, NEW DELHI**

Chapter 22

Role of Molecular Markers for Improvement of Stress Tolerance in Crops

P. Kadirvel[1]*, S. Senthilvel[1] and S. Geethanjali[2]

[1]ICAR-Indian Institute of Oilseeds Research, Rajendranagar, Hyderabad – 500 030, India
[2]Department of Millets, Tamil Nadu Agricultural University, Coimbatore - 641 003, Tamil Nadu, India

ABSTRACT

Molecular marker technology is a modern tool for crop improvement. DNA markers are used to understand genetic basis of complex traits through quantitative trait loci (QTL) mapping approach. A substantial progress has been achieved in mapping genes/QTLs associated with resistance/tolerance to major biotic (diseases, nematodes, insect-pests) and abiotic (drought, salinity, heat, cold, flooding and nutrient deficiency) stresses. Marker-assisted selection (MAS) of target QTLs/genes enables plant breeders to select desirable progenies more effectively and rapidly in breeding programmes. In this chapter, developments in DNA marker technology and their contributions towards genetics and breeding of stress tolerance in crops are discussed.

Keywords: *DNA markers, Biotic stresses, Abiotic stresses, Major genes, QTL mapping, Association mapping, Marker-assisted selection, Gene pyramiding, Cultivars.*

Introduction

Genetic markers are important tools for characterization and utilization of genetic diversity for crop improvement. They are referred to morphological

**Corresponding author*: E-mail: kadirvel.palchamy@icar.gov.in characters, gene products (biochemical) and pieces of DNA sequences, which represent genetic differences (polymorphism) among plants. Morphological markers are phenotypic polymorphisms that can be observed and scored easily. Biochemical markers are structural polymorphisms of proteins such as isozymes. Morphological and biochemical markers have long been used in genetic studies but with inherent limitations: limited in number and influenced by plant growth stages and environmental conditions. DNA markers or molecular markers are sequence polymorphisms in the genome of individuals. They are highly advantageous over morphological or biochemical markers because they fulfill most of the ideal properties of a genetic marker system: abundance, neutrality, reliability and amenability for automation.

Molecular markers are widely used for characterizing germplasm, mapping of genes/quantitative trait loci (QTLs) associated with important traits, positional cloning of genes/QTLs, mining of alleles at candidate gene loci and marker-assisted selection (MAS) for pyramiding of desirable genes/QTLs for cultivar improvement purposes. Host plant resistance to biotic stresses in crops has been well studied. Numerous genes conferring resistance to biotic stresses have been genetically mapped and are being used in breeding programmes through MAS. In contrast, tolerance to abiotic stresses is genetically complex - exhibits quantitative inheritance controlled by multiple genes and influenced by the environment – that make genetic enhancement through plant breeding more challenging. Molecular markers have become powerful tools to detect genes controlling abiotic stress tolerance in crops.

DNA Markers: An Overview

Numerous kinds of DNA markers have been described in the literature, which basically differ in their ability to detect DNA polymorphisms among genotypes. Reiter *et al.* (2001) may be referred for detailed description of marker systems. The most commonly used DNA markers in genetic mapping and marker-assisted selection are briefly described below.

Restriction Fragment Length Polymorphism (RFLP)

The RFLP markers are the oldest kind of DNA markers (Botstein *et al.*, 1980). The RFLP technique exploits the differences in restriction sites of endonuclease enzymes in the genome. The RFLP markers can be generated by digesting the genomic DNA with restriction enzyme, separating the DNA fragments of varied length in agarose gel and autoradiography. The RFLPs are co-dominant and bi-allelic markers. They are highly reliable markers; however, low level of polymorphism and tedious assay procedure make them unattractive for genotyping applications now-a-days.

Randomly Amplified Polymorphic DNA (RAPD)

The RAPD technique exploits PCR principle (Williams *et al.*, 1990). Briefly, the RAPDs are produced by the following steps: PCR amplification using a single synthetic oligonucleotide primer (~10 base length) with low annealing temperatures (35-40 C), separation of amplicons in agaraose gel electrophoresis and visualization of DNA bands after staining with ethidium bromide. The RAPDs are the simplest kind of markers, which can be performed without the need for any DNA sequence information. The poor reproducibility of RAPD markers is a serious concern; hence, they are not preferred for genetic analysis. However, they may still be useful for genetic diversity analysis, hybrid confirmation etc. but may be less informative because they are dominant marker type which cannot distinguish heterozygotes from homozygotes.

Amplified Fragment Length Polymorphism (AFLP)

The AFLP technique beautifully combines the procedure of RFLP and RAPD markers (Vos *et al.*, 1995). Briefly, the AFLP assay is carried out as follows: (i) genomic DNA is digested by a pair of restriction enzymes and a large number of DNA fragments of various sizes are generated, (ii) an adapter sequence is ligated to those DNA fragments, (iii) pre-amplification is carried out using the primers that are complementary to the sequences of adapters and the adjoining restriction sites, (iv) selective amplification of a sub set of pre-amplified DNA fragments is carried out using the primers that contain sequences complementary to those of adapters and restriction sites, along with one to three selective bases added at their 3′ ends, and (v) polymorphisms are visualized after polyacrylamide gel electrophoresis (PAGE) and silver staining. The AFLP technique is an excellent choice for generating a large number of markers quickly and cheaply for any given genotype without the need for any DNA sequence information. A single assay can generate about 20-200 bands. The AFLPs are supposed to be more reliable than RAPDs. However, AFLPs are dominant markers and the assay procedure is tedious which make them less attractive for genetic analysis. They may still be useful for genetic diversity analysis, hybrid confirmation etc. but may be less informative.

Sequence Characterized Amplified Polymorphic Regions (SCAR)

The SCAR markers can be generated by PCR amplification of genomic DNA with pairs of specific oligonucleotide primers (Paran and Michelmore, 1993). The SCARs are just an improved version of RAPDs wherein a pair of primers provides more specificity and reliability. The SCAR markers could either be dominant or codominant types. The RAPD or AFLP markers can be converted to SCARs for convenience in application.

Sequence Tagged Sites (STS)

The STS markers are similar to SCARs but amplify short unique sequences by pairs of primers designed from specific genomic sequences such as RFLP probes.

Cleaved Amplified Polymorphic Sequence (CAPS)

The CAPS markers can be developed by amplifying a gene or expressed sequence tag (EST) sequence in PCR using specific primer pair and digesting the PCR product with restriction enzymes to produce polymorphic fragments. The assay is simple and cheap as the polymorphisms can be detected in agarose gels.

Microsatellites or Simple Sequence Repeats (SSR)

The SSRs are short DNA sequences with tandem repeat (mono, di, tri, tetra or penta) motifs in the genome Akkaya *et al.* (1992). They are hyper variable regions in the genome; the number of repeats differs between individuals resulting in polymorphism. DNA sequence information must be available for a species in order to develop SSR markers. Genomic sequences and expressed sequence tags (ESTs) can be searched for repeat motifs using the data mining tools and primers could be designed from the flanking sequences and tested for their amplification. The assay procedure involves PCR amplification of a repeat motif using a forward and reverse primers, which flanking the repeat motif followed by PAGE and silver staining to visualize polymorphisms. The SSR markers are locus specific, multi-allelic, abundant, randomly and widely distributed throughout the genome, co-dominant, simple to assay, highly reliable, reproducible and can be easily automated. Therefore, SSRs are considered ideal marker system for genetic diversity and genome mapping studies in plants.

Inter-Simple Sequence Repeats (ISSR)

The ISSR markers are generated by amplifying the DNA fragments lying between two identical microsatellite repeat regions using only one of the primers complementary to a target microsatellite (Zietkiewicz *et al.*, 1994). The ISSRs are multi-locus markers, detect more polymorphism, easy to assay and reproducible.

Diversity Arrays Technology (DArT)

The DArT was developed as a low-cost high throughput marker system to overcome some of the limitations associated with RFLP, SSR and AFLP (Jaccoud *et al.*, 2001). DArT is a generic, hybridization-based, cost-effective whole-genome fingerprinting method. DArT markers are scored for polymorphism in parallel rather than serial analysis as done in gel based marker systems. A DArT assay simultaneously detects hundreds to thousands of sequence variations resulting from Single Nucleotide Polymorphisms (SNPs)/insertion-deletion polymorphisms (InDels)/methylation, across the genome. It does not require any prior sequence information, hence very attractive for the crops not having enough genomic resources.

Single Nucleotide Polymorphism (SNP)

The SNPs refer to DNA polymorphisms at the level of individual base level (Jordan and Humphries, 1994). SNP markers can be discovered by sequencing/ resequencing of candidate genes or PCR products or whole genomes in more than one genotype. A variety of genotyping methods are available for SNP markers. The SNPs are locus specific, bi-allelic, extraordinarily abundant, uniformly distributed throughout a genome, easily amenable for automation, highly reliable, and therefore considered to be the most attractive marker system for molecular breeding in crop plants.

Gene Targeted and Functional Markers

The markers described above are random markers as they are derived from the entire genomic regions randomly. Gene targeted markers or functional markers refer to markers derived from polymorphic sites in the genic regions or within fully characterized candidate genes. These markers have great potential for marker-assisted selection purposes because there may be no recombination, which results in perfect co-segregation of marker with the target gene or trait. The prospects of gene targeted and functional markers are reviewed by Poczai *et al.* (2013).

Next Generation Sequencing Technologies and Marker Development

The advent of next generation sequencing (NGS) technologies have provided opportunities for development of sequence based markers for any organism with less cost and time (Glenn, 2011). The complete genome or the transcriptome of genotypes can be sequenced and genome wide markers such as SSR, insertion-deletion (In-del) and SNPs can be rapidly identified and converted into genetic markers. However, the assay development for SNPs for routine genotyping is hindered due to the low conversion rate and high cost. Recently, genotyping by sequencing (GBS) approaches have emerged wherein sequences are directly used for marker-trait association thereby avoiding SNP identification and assay development (Elshire *et al.*, 2011). This has become possible due to reduction in genome sequencing costs and developments in bioinformatics tools.

Choice of Markers

The choice of markers for molecular breeding applications depends on the attributes of markers such as abundance, genomic distribution, level of polymorphism, basis of polymorphism, technical requirements, cost of genotypic assay development, etc. Multi-locus markers such as RAPD, ISSR and AFLP are cheaper but less preferred because they are dominant markers, less reproducible and cannot be automated for high throughput application. The single locus markers such as SSR and SNPs are highly preferred because they are co-dominant markers, highly reproducible and can easily be automated but the assay is costlier. GBS is highly cost prohibitive at the moment.

Molecular Markers in Mapping Genes Associated with Stress Tolerance Traits in Crop Plants

Genetic analyses of stress tolerance traits require information on genotypic, phenotypic and environmental components. Molecular markers and genomics technologies have adequately strengthened the genotypic component by which a large amount of genotypic data of segregating populations can be quickly generated. Accurate phenotyping is the most fundamental requirement for gene mapping and molecular breeding efforts. In most cases, phenotyping for resistance to biotic stresses has been well established but phenotyping for tolerance to abiotic stresses is still a challenge due to trait complexity. However, substantial efforts have been made to understand the phenotypic traits that contribute for abiotic stress tolerance in plants. For instance, drought tolerance is manifested by a combination of primary traits (phenological - roots, waxes, leaf area etc.), secondary traits (physiological - water use efficiency, water potential, relative water content, canopy temperature, leaf senescence, stay green, leaf rolling, seedling vigour, flowering time etc.) and integrated traits (yield components). Genetic variability for abiotic stress tolerance traits in crop germplasm collections is extensively investigated (Ali *et al.*, 2006, Rao *et al.*, 2013). Though abiotic stress tolerance is generally viewed as a complex trait, Blum (2011) argues that it may be complex at molecular level but it is simpler when viewed from physiological and agronomic aspects at whole plant or crop levels. Recently, the advent of plant phenomics provides opportunities for high resolution phenotyping for abiotic stress tolerance (Furbank and Tester, 2011). The phenomics technologies are based on thermal and spectral imaging technologies, which help to profile the physiological responses of plants under abiotic stress conditions in a precise and non-destructive manner. The heritability of the stress tolerance traits is expected to improve under the controlled phenomics facility; therefore, the chances of detecting minor genes would be high.

Methodologies for genetic analysis differ for simple (qualitative) and complex (quantitative) traits. Traditionally, Mendelian and quantitative genetics principles are applied to study qualitative and quantitative traits, respectively. In case of Mendelian genetics approach, segregating populations (F_2 or backcross) are created by crossing tolerant and susceptible genotypes and number of genes controlling the trait is inferred by grouping the segregants into two discrete phenotypes (resistant and susceptible) followed by analyzing the segregation ratios. This approach has been quite successfully used for resistance to biotic stresses, which are mostly qualitative in nature and controlled by major genes. Classical quantitative genetics approaches have limitations in dealing with complex traits because of the involvement of polygenes and genotype (G) x environment (E) interactions. Molecular marker technology has been recognized as a potential tool to map polygenes (quantitative trait loci-QTLs), associated with complex traits and further use in breeding programmes. QTLs are genomic region(s) associated with quantitative traits, which can be located on the genome precisely using molecular markers; the approach is called genome mapping (Paterson, 1996).

Genome mapping has evolved into two major approaches: linkage mapping and association mapping. Linkage mapping involves locating major genes and QTLs on chromosomes using mapping populations produced from two diverse parents (bi-parental crosses) for the trait of interest. Association mapping involves use of natural populations for mapping genes of interest. A brief description on genome mapping principles and methods is provided here.

Linkage Mapping

Broadly, linkage mapping involves the following steps: selection of two genotypes that are highly diverse for the target trait, development of mapping population (F_2, recombinant inbred lines-RILs, backcross or backcross inbred lines – BILs and doubled haploids – DH), genotyping the mapping populations using DNA markers and construction of genetic linkage map, phenotyping the mapping population for the target traits accurately and establishing marker-trait association through appropriate QTL detection methods such as single marker analysis, interval mapping, composite interval mapping and multiple-interval mapping. Several software have been developed for use in construction of linkage maps: Mapmaker, MapManager and JoinMap as well as for QTL analysis: MapMaker/QTL, MQTL, QTL Cartographer, Qgene, PLABQTL etc. Though QTL analysis is a powerful tool, adequate caution is required on several issues. For instance, the accuracy of QTL mapping depends on the type of segregating population, its size, the heritability of the trait, the number and contribution of each QTL to the total genotypic variance, their interactions, their distribution over the genome, the number and distance between consecutive markers, the reliability of the order of the markers in the linkage map, the reliability of the phenotyping method and the ability of the statistical method to determine the location and to estimate the genetic effect of the QTL (Asíns, 2002). Linkage mapping has been extensively applied in mapping of major genes and QTLs associated with stress tolerance in crops.

Mapping of Major Genes Associated with Resistance to Biotic Stresses in Crop Plants

Host plant resistance to biotic stresses is classified into two categories: qualitative and quantitative. Qualitative resistance is mostly controlled by single genes or oligogenes (major genes) whereas quantitative resistance is controlled by polygenes (QTLs). Traditionally, identification of major genes conferring resistance to biotic stresses was based on differential reaction of a resistant plant to different pathotypes (of pathogen) or biotypes (of insect-pest). When a resistant plant survives or succumbs to a particular pathotype or biotype that signals the involvement of different genes for resistance. Based on this, a set of resistant genotypes that show differential reaction to pathotypes or biotypes (selection differentials) is identified for differentiating genes. This approach was highly successful; but has limitations. It would be impossible to differentiate genes if no pathotype or biotype information is available. Molecular markers help to localize genes on chromosomes, which can easily be recognized without the need for screening against pathotypes or biotypes. Numerous genes conferring resistance to biotic stresses have been genetically mapped across crop plants using linkage mapping approach. The list is highly exhaustive; however, a few classical examples of mapped major genes for biotic stresses in major crop plants are provided in Table 22.1.

QTL Mapping for Stress Tolerance Traits in Crop Plants

Quantitative resistance to biotic stresses has been well documented in crop plants. Extensive literature is available on mapping of QTLs associated with resistance to diseases, nematodes and insect-pests (Soundararajan *et al.*, 2004; Geethanjali *et al.*, 2008; Kadirvel *et al.*, 2013; Rosewarne *et al.*, 2013). A substantial progress has also been made in mapping QTLs associated with tolerance to major abiotic stresses. Table 22.2 summarizes a few examples of mapping QTLs associated with tolerance to major abiotic stresses: drought, salinity, heat, flooding/water-logging/submergence, low temperature/freezing and nutrient deficiency in crop plants. Linkage mapping has been a useful tool to trace genes/QTLs associated with stress tolerance traits in a single donor genotype. It may take a long time to map QTLs from multiple sources through linkage mapping using bi-parental mapping populations. Furthermore, only small proportion of genetic variation can be analyzed at a time.

Association Mapping

Due to the advances in high throughput marker discovery through NGS and bioinformatics tools, a large number of DNA markers covering the entire genome can

Table 22.1: A Few Classical Examples of Mapped Major Genes Conferring Resistance to Biotic Stresses in Crops

Crop	*Biotic Stress*	*Genes Mapped on Linkage Maps*	*Source*
Rice	Bacterial blight (*Xanthomonas oryzae* pv. *oryzae*)	*Xa1, Xa2, Xa3/Xa26, Xa4, xa5, Xa6, Xa7, xa8, xa9, Xa10, Xa11, Xa12, xa13, Xa14, xa15, Xa16, Xa17, Xa18, xa19, xa20, Xa21, Xa22(t), Xa23, xa24(t), xa25/Xa25(t), Xa25, xa26(t), Xa27, xa28(t), Xa29(t), Xa30(t), xa31(t), Xa32(t), xa33(t), xa34(t), Xa35(t), Xa36(t)*	Khan *et al.* (2014)
	Blast (*Magnaporthe grisea*)	*Pid1(t), Pitq-5, Pig(t), Piy(t), Pi39(t), Pi10(t), Pi23(t), Pi2-1, Pi2-2, Pi8(t), Pi13(t), Pi13(t), Pi22(t), Pi26(t), Pi40(t), Pi50(t), Pigm(t), Piz, Pitq-1, Pi17(t), Pi42(t), Pi33(t), Pi55(t), PiGD-1(t), Pi3(t), Pi15(t), Pi56(t), PiGD-2(t), Pi18(t), Pi38(t), Pi44(t), PiCO39(t), Pilm-2, Pi7(t), Pi47(t), Pi43(t), Piks, Pikg(t), Piy(t), Pizy(t), Pi19(t), Pita-2, Pi6(t), Pi62(t), Pi24(t), Pi12(t), Pi20(t), PiGD-3, Pi51(t), Pi39(t), Pi41(t), Pi157(t), Pi48(t), Pitq-6, Pih1(t)*	Wang *et al.* (2014)
	Brown planthopper (*Nilaparavate lugens*)	*Bph1, bph2, Bph3, bph4, Bph9, Bph10, bph11, Bph12, Bph13*(t), *Bph14, Bph15, Bph16, Bph17, Bph18, bph19, Bph20, Bph21, Bph22, Bph25, Bph26*	Fujita *et al.* (2013)
	Gall midge (*Orseolia oryzae*)	*Gm1, Gm2, gm3, Gm4, Gm5, Gm6, Gm7, Gm8, Gm11*	Sama *et al.* (2014)
Wheat	Leaf rust (*Puccinia triticina*)	*Lr1, Lr3, Lr9, Lr10, Lr13, Lr14a, Lr16, Lr19, Lr20, Lr21, Lr22, Lr24, Lr25, Lr26, Lr28, Lr29, Lr32, Lr34, Lr35, Lr37, Lr39, Lr46, Lr47, Lr50, Lr51, Lr52, Lr57, Lr58*	Ellis *et al.* (2014)
	Stem rust (*Puccinia graminis*)	*Sr2, Sr1RAmigo, Sr6, Sr9a, Sr24, Sr25, Sr26, Sr31, Sr35, Sr36, Sr38, Sr39, Sr40, SrCad, SrWeb, Sr51, Sr52, Sr53*	Ejaz *et al.* (2012)
	Aphid (*Diuraphis noxia*)	*Dn1, Dn2, Dn4, Dn6, Dn7, Dn8, Dn9, Dnx*	Liu *et al.* (2001)
Chickpea	Fusarium wilt (*Fusarium oxysporum* f.sp. *ciceris*)	*foc1, foc2, foc3*	Kumar *et al.* (2011)
Soybean	Aphid (*Aphis glycines*)	*Rag1, Rag2*	Kim *et al.* (2014)
	Cyst nematode (*Heterodera glycines*)	*Rhg1, Rhg2, Rhg3, Rhg4, Rhg5*	Liu *et al.* (2012)
Sunflower	Downy mildew (*Plasmopara halstedii*)	*Pl1-16, Plv, Plw, Plx-z, Mw, Mx, PlArg*	Wieckhorst *et al.* (2010)
Brassica napus	Blackleg (*Leptosphaeria maculans*)	*LepR1, LepR2, LepR3*	Larkan *et al.* (2013)
Tomato	Root-knot nematode (*Meloidogyne* spp.)	*Mi*	Kaloshian *et al.* (1998)
	Tomato yellow leaf curl virus disease	*Ty-1, Ty-2, Ty-3, Ty-4, Ty-5*	De la Peña *et al.* (2010)
Pepper	Root-knot nematode (*Meloidogyne* spp.)	*Me1, Mech2, Me3, Me4, Me7*	Djian-Caporalino *et al.* (2007)

Table 22.2: A Few Examples of QTLs Mapped for Tolerance to Major Abiotic Stresses in Crop Plants

Crop	*Stress*	*Traits Measured under Stress*	*Populations*	*Marker Number/Type*	*No. of QTLs*	*Phenotypic Variance (%)*	*Reference*
Rice	Drought	Grain yield and components	154 DH lines derived from cross: CT9993 (T) x IR62266 (S)	145 RFLPs, 37 SSRs, and 153 AFLPs	77	7.5 - 55.7	Lanceras *et al.* (2004)
		Fitness, yield, and the root system related traits	180 RILs derived from cross: Zhenshan 97 (S) x IRAT109 (T)	245 SSRs	38	5.81-39.17	Yue *et al.* (2006)
		Grain yield	490 RILs ($F_{4:5}$) derived from cross: Apo (T) x Swarna (S)	293 SSRs	DTY2.1, DTY3.1	16-31	Venuprasad *et al.* (2009)
		Grain yield	436 F_3 lines from the cross Vandana (T) x Wayrarem (S)	126 SSRs	qtl12.1	51	Bernier *et al.* (2007)
		Grain yield	$F_{3:4}$ population derived from crosses: N22 (T) x Swarna, IR64, and MTU1010 (S)	125 SSRs	qDTY1.1	16.1 – 29.3	Vikram *et al.* (2011)
		Root growth angle	117 RILs derived from cross: IR64 (shallow rooting) x Kinandang Patong (KP) (deep rooting)	77 SSRs, 24 STSs, and 30 InDels	Deeper Rooting 1 (DRO1)	66.6	Uga *et al.* (2011;2013)
	Salinity	Seedling damage, Na^+ and K^+ concentrations in leaves and roots, seedling height and chlorophyll content	RILs of the cross: IR29 x Pokkali	RFLPs, SSRs	Saltol	43	Bonilla *et al.* (2002), Thomson *et al.* (2010)
	Flooding, water-logging, submergence	Elongation ability (plant height, internode elongation, leaf elongation) and submergence tolerance	165 F_8 derived RILs from the cross IR74 x Jalmagna	AFLPs	7	8.6-36.7	Sripongpangkul *et al.* (2000)

Contd...

Table 22.2–*Contd...*

Crop	*Stress*	*Traits Measured under Stress*	*Populations*	*Marker Number/Type*	*No. of QTLs*	*Phenotypic Variance (%)*	*Reference*
		Submergence tolerance	169 $F_{2:3}$ families of a cross: IR40931-26 x PI543851. IR40931-26 inherited submergence tolerance FR13A.	RFLPs and RAPDs	Sub1A	69	Xu and Mackill (1996); Xu *et al.* (2006)
	Low temperature, cold, freezing	Low temperature-Germination	RILs derived from a cross: USSR5 and N22		qLTG-7, qLTG-9 and qLTG-12	7.08 - 12.12	Li *et al.* (2013)
		Seedling response under low temperature (4°C-5°C)	$F_{2:3}$ population produced from the cross: Guichao2 x IL112	176 SSRs	7	8.0-20.0	Liu *et al.* (2013)
		Cold induced necrosis and wilting	191 RILs from the cross: M-202 x IR50	181 SSRs	qCTS12a	41	Andaya and Mackill (2003), Andaya and Tai (2006)
		Germination under low temperature (15°C)	122 BILs derived from a cross: Italica Livorno and Hayamasari	–	qLTG-3-1	35	Fujino *et al.* (2004; 2008)
	Nutrient deficiency	Plant biomass, grain yield related traits, and grain nitrogen content under three nitrogen levels	82 DH lines derived from cross: IR64 x Azucena	175 RFLPs	16	15.7–50.3	Senthilvel *et al.* (2008)
		Nitrogen deficiency and use efficiency traits: relative grain yield, relative biomass yield, relative grain nitrogen, relative biomass nitrogen, nitrogen response, grain yield response, physiological nitrogen-use efficiency	127 RILs derived from the cross: Zhenshan97 x Minghui63	168 RFLPs and 52 SSRs	26	4.0-16.6 per cent	Wei *et al.* (2012)
		Phosphorus deficiency tolerance - P uptake, internal P-use efficiency, dry weight, and tiller number	98 BILs from the cross: Nipponbare x Kasalath	245 RFLPs	7; a major QTL, Pup1	5.8-54.5	Wissua *et al.* (1998; 2002)

Contd...

Table 22.2–*Contd...*

Crop	*Stress*	*Traits Measured under Stress*	*Populations*	*Marker Number/Type*	*No. of QTLs*	*Phenotypic Variance (%)*	*Reference*
	Heavy metal toxicity	Seedling root growth in aluminium solution	146 DH lines	RFLPs, AFLPs, SSRs	10	10.3-28.7	Nguyen *et al.* (2002)
		Seedling root growth in aluminium solution (40 ppm)	171 F_6 RILs	RFLPs and SSRs	5	9-24.9	Nguyen *et al.* (2003)
		Aluminium content in root apex	183 backcross lines	RFLPs	3	7.3-11.1	Ma *et al.* (2002)
		Relative root growth in alumimium solution (160 μM concentration)	383 accessions, 134 RILs from the cross: Azucena x IR64, 78 back-cross introgression lines of the cross: Nipponbare x Kasalath	SNPs	7 in linkage mapping and 48 in GWA	5.9-16.5	Famoso *et al.* (2011)
		Tolerance to mercury: root growth under Hg^{2+} stress (1.5 umol/litre)	120 RILs	210 SSRs	3	10.8-13.9	Chong-qing *et al.* (2013)
		Cadmium accumulation	F_2 plants		1	85.6	Ueno *et al.* (2010)
Wheat	Drought	Coleoptile length, seedling height, longest root length, root number, seedling fresh weight, stem and leaves fresh weight, root fresh weight, seedling dry weight, stem and leaves dry weight, root dry weight, root to shoot fresh weight ratio, root-to-shoot dry weight ratio	229 and 302 RILs from the crosses: Weimai 8 x Luohan 2, Weimai 8 x Yannong 19, respectively	338 (SSRs, STS, ISSRs, SRAPs, RAPDs)	22	3.33-77.01	Zhang *et al.* (2013)

Contd...

Table 22.2–*Contd...*

Crop	*Stress*	*Traits Measured under Stress*	*Populations*	*Marker Number/Type*	*No. of QTLs*	*Phenotypic Variance (%)*	*Reference*
	Salinity	Leaf symptoms, tiller number, seedling biomass, chlorophyll content, and shoot Na(+) and K(+) concentrations	DH population of the cross: Berkut x Krichauff	527 SSRs, DArT- and gene-based markers	40	–	Genc *et al.* (2010)
	Heat	Terminal heat stress – heat susceptibility index (HSI) of thousand grain weight, grain fill duration, grain yield, canopy temperature depression	148 RILs from the cross: NW1014 x HUW468	160 SSRs	3	7.42 - 20.34	Paliwal *et al.* (2012)
	Heavy metal toxicity	Tolerance to copper – shoot dry weight of seedlings grown in pots with Cu-enriched soil (1000 mg of Cu/kg of dry soil mass) for two weeks	114 F_8 RILs from the cross: W7984 x Opata 85	332 RFLPs	1	28.9	Bálint *et al.* (2007)
Maize	Drought	L-ABA concentration	80 $F_{3:4}$ families derived from cross: Os420 x IABO78	133 RFLPs	16; Root-ABA1	32	Tuberosa *et al.* (1998), Giuliani *et al.* (2005)
		Sugar concentration, root density, root dry weight, total biomass, relative water content, and leaf abscisic acid content, osmotic potential, leaf surface area and grain yield	105 $F_{2:3}$ families	RFLPs	22; one major QTL for sugar concentration	52	Rahman *et al.* (2011)
	Flooding, water-logging, submergence	Water-logging tolerance during seedling stage - root length, rootdry weight, plant height, shoot dry weight, total dry weight and waterlogging tolerance coefficient	288 $F_{2:3}$ lines derived from cross: HZ32 and K12	177 SSRs	>25	4.1-37.3	Qiu *et al.* (2007)

Contd...

Table 22.2–*Contd...*

Crop	*Stress*	*Traits Measured under Stress*	*Populations*	*Marker Number/Type*	*No. of QTLs*	*Phenotypic Variance (%)*	*Reference*
	Nitrogen use efficiency	Ear-leaf area, plant height, grain yield per plant, number of ears per plant and number of kernels per ear at two nitrogen levels	214 $F_{2:3}$ families derived from the cross: B73 x G79	99 RFLPs	Many	10.3-18.3	Agrama *et al.*, 1999
	Heavy metal toxicity-aluminium	Seedling root length	168 $F_{3:4}$ families derived from L53 x L1327	RFLPs and SSRs	5	5.3-15.6	Ninamango-Cardenas *et al.* (2003)
	Arsenic accumulation	Arsenic concentration in different tissues (leaf, stem, bract and kernel)	203 RILs produced from the cross Huang x X4178	217 SSRs	6	5.62-15.23	Ding *et al.* (2011)
Sorghum	Drought	Stay green and chlorophyll content	RILs	145 RFLPs	7	20-30	Xu *et al.* (2000)
Pearl millet	Drought	Grain yield, stover yield, biomass yield, harvest index, 100-grain mass, panicle number, panicle grain number	92 $F_{2:3}$ progenies of the cross: H 77/833–2 x PRLT 2/89–33	50 RFLPs	1	23	Yadav *et al.* (2002; 2011)
Barley	Salinity	Leaf chlorosis and plant survival	188 DH lines from a cross: TX9425 x Naso Nijo	326 SSRs	1	50	Xu *et al.* (2012)
	Heavy metal toxicity-aluminium	Tolerance scare by a modified Al pulse recovery-Solutions in culture methods	Wheat-barley chromosome addition lines	AFLPs and SSRs	1		Raman *et al.* (2003)
		Seedling root length in aluminium solution	145 F_6 RILs produced from M3G and 1-6 x M-7A-1	AFLP	1		Miftahudin *et al.*(2002)
Groundnut	Drought	Transpiration efficiency, SCMR, specific leaf area, shoot plus pod dry weight, total dry matter and carbon discrimination ratio	318 RILs produced from the cross: TAG 24 x ICGV 86031	191 SSRs	>105	3.28-33.36	Ravi *et al.* (2011)

Contd...

Table 22.2–*Contd...*

Crop	*Stress*	*Traits Measured under Stress*	*Populations*	*Marker Number/Type*	*No. of QTLs*	*Phenotypic Variance (%)*	*Reference*
		Transpiration efficiency, SCMR, specific leaf area, shoot plus pod dry weight, total dry matter and carbon discrimination ratio	177 RILs produced from the crosses: ICGS 76 × CSMG 84-1 and ICGS 44 × ICGS 76	293 SSRs	178	4.83-40.1	Gautami *et al.* (2012)
Soybean	Drought	Leaf water status and grain yield traits	184 RILs developed from a cross: Kefeng1 x Nannong1138-2	–	40	–	Du *et al.* (2009)
	Water-logging	Tolerance evaluation under flooded field	RILs from crosses: A5403 x Archer and P9641 x Archer	912 SSRs	2	10-16	Cornelious *et al.* (2005)
Alfalfa	Salinity	Parameters associated with Na^+ and K^+ content in leaves, stems and roots	RILs derived from the cross Jemalong A17 (JA17) and F83005.5 (F83)	70 SSRs	2	9.8-9.9	Arraouadi *et al.* (2012)
	Heavy metal toxicity-aluminium	–	F_2 and backcross lines	SSRs and AFLPs			Sledge *et al.* (2002)
Medicago truncatula	Freezing tolerance	Number of leaves, leaf area, chlorophyll content index, shoot and root dry weights at the end of the acclimation period; visual freezing damage during the freezing treatment; foliar electrolyte leakage after two weeks of regrowth.	RILs	–	3	40	Avia *et al.* (2013)
Field Pea	Salinity	Seedling response	134 RILs produced from Kaspa x Parafield	705 SNPs	4	12-19	Leonforte *et al.* (2013)
Lettuce	Heat	Tip burn and heat stress induced physiological disorders	152 F_7 RILs derived from Emperor x El Dorado	449 SNPs	36; a major QTL, qTPB5.2	38–70	Jenni *et al.* (2013)

easily be developed and used for mining target genes or alleles for a trait of interest in a large germplasm collection. This approach is called 'association mapping' wherein genes/alleles are identified directly in a natural population consisting of unrelated individuals from the germplasm collection, based on the principle of linkage disequilibrium (LD) (Rafalsti, 2000). Association mapping bypasses the requirement of bi-parental mapping populations and allows the germplasm collection to be used directly for mapping genes/QTLs; therefore, a large proportion of genetic variation for a single trait can be effectively analyzed.

Association mapping can be done in two ways: (i) Genome-wide association (GWA) analysis and (ii) candidate gene association mapping (CGAM). Genome-wide association analysis uses anonymous or neutral DNA markers from across the genome. The markers are chosen simply without any assumption regarding their association with a particular gene or trait. This approach requires a large number of markers for genotyping the samples so that the whole genome can be scanned adequately and genes/QTLs can be mapped with linked markers that are in perfect LD. Whereas, CGAM uses markers that are derived from a set of genes presumed to be involved in the trait. CGAM identifies alleles at these candidate genes that contribute to trait variation. More details on association mapping in plants can be found in Oraguzie *et al.* (2007).

Broadly, association mapping involves the following steps: choice of natural population that consists of germplasm accessions and represent high level of genetic diversity for the target trait (association mapping panel-AM panel), determination of LD in AM panel, determination of population structure, determining kinship of the accessions in the AM panel and establishing marker-trait association using these parameters. The software, structures is used to determine population structure in the natural population; SPAGedi (Spatial Pattern Analysis of Genetic Diversity) is used to determine kinship co-efficient and TASSEL (trait analysis by association, evolution and linkage) (http://www.maizegenetics.net) is used for marker-trait association.

Alternative to natural populations, two kinds of artificially created populations *viz.*, nested association mapping (NAM) (Yu *et al.*, 2008) and multi-parent advanced generation intercross (MAGIC) (Cavanagh *et al.*, 2008) populations are proposed to improve the efficiency of QTLs associated with complex traits. Development of a NAM population requires a set of diverse founder lines (~25) and a common parent (elite line). The founder lines are crossed with the common parent and F_2 populations are produced from each cross. A subset of F_2 plants (~200) from each cross are advanced to RILs through single seed descent method by selfing. The resulting population of a large number of RILs would provide higher statistical power and mapping resolution than the bi-parental mapping populations used in linkage mapping.

The MAGIC populations differ from NAM populations wherein a large number of RILs are created by crossing a small set of founder lines (usually eight lines with high potential for breeding) followed by intercrosses among progenies. The main advantage of MAGIC populations is that it offers high recombination and mapping resolution due to intercrossing and shuffling of genome. In brief, the founder lines are crossed in a half-diallel fashion to produce single crosses followed by two-way crosses. A subset of two-way cross recombinants is intercrossed to produce four-way recombinants. A subset of four-way recombinants is intercrossed to produce eight-way recombinants, which are subsequently advanced through single seed descent method by selfing. Recently, Bandillo *et al.* (2013) described development of MAGIC populations in rice using a set of 16 founder lines.

Association Mapping of QTLs and Alleles for Stress Tolerance Traits in Crops

Application of association mapping for detecting QTLs associated with stress tolerance has been limited; but there is an excellent scope for its application in crops. A few examples of association mapping studies in crops addressing stress tolerance is presented here.

Jia *et al.* (2012) detected 10 SSR marker loci on seven linkage groups showing significant association with quantitative resistance to sheath blight disease caused by the soil-borne pathogen *Rhizoctonia solani* in a rice core collection. Zegeye *et al.* (2014) used GWA to map seedling and adult plant resistance to stripe rust disease in wheat. A total of 27 SNPs in nine genomic regions and 38 SNPs in 18 genomic regions were associated with seedling and adult plant resistance, respectively suggesting the genetic complexity of the stripe rust disease in synthetic hexaploid wheat. Wen *et al.* (2014) detected 20 loci associated with quantitative resistance to sudden death syndrome in soybean caused by the soil-borne fungal pathogen *Fusarium virguliforme* through GWA.

Varshney *et al.* (2012) made an effort to use GWA analyses for drought tolerance related traits in barley. A panel of 243 accessions was genotyped with 816 markers including DarT, SNPs and SSRs. Association analysis did not detect major QTLs but revealed only minor QTLs. Cai *et al.* (2013) mapped two novel loci (12.9 per cent and 9.7 per cent of the phenotypic variation) for aluminium tolerance in barley through GWA analysis using 166 genotypes including wild accessions and cultivars with 469 DArT markers. These studies demonstrated that GWA has been useful in mapping minor QTLs associated with complex traits.

Candidate gene based association mapping has been very effective for abiotic stress tolerance traits. An association mapping study in maize using a panel of 282 diverse inbred lines showed that four candidate gene loci: *Zea mays* AltSB like (*ZmASL*), *Zea mays* aluminum-activated malate transporter-2 (*ALMT2*), S-adenosyl-L-homocysteinase (*SAHH*), and Malic Enzyme (*ME*) are associated with aluminum tolerance. These genes were further confirmed by linkage analysis using F_2 populations (Krill *et al.*, 2010). In wheat, Edae *et al.* (2013) characterized allelic diversity of five drought tolerance candidate genes: dehydration responsive element binding 1A (*DREB1A*), enhanced response to abscisic acid (*ERA1-B* and *ERA1-D*), and fructan 1-exohydrolase (*1-FEH-A* and *1-FEH-B*) for their association with drought tolerance phenotypes. *DREB1A* was associated with vegetation index, heading date, biomass, and spikelet number; *ERA1-A* and *ERA1-B* with harvest index, flag leaf width, and leaf senescence; *1-FEH-A* with grain yield, and *1-FEH-B* with 1000 kernel weight. Yu *et al.* (2013) conducted candidate gene association mapping of drought tolerance traits in a germplasm panel of 192 diverse perennial ryegrass (*Lolium perenne* L.) accessions. The panel was first investigated for genetic variation for important drought tolerance traits: leaf wilting, leaf water content, canopy and air temperature difference, and chlorophyll fluorescence under well-watered and drought conditions across six environments. Candidate genes involved in antioxidant metabolism, dehydration, water movement across membranes, and signal transduction were analyzed by obtaining expression-based sequence readings that revealed 346 SNPs. Association analysis revealed that a putative LpLEA3 encoding late embryogenesis abundant group 3 protein and a putative LpFeSOD encoding iron superoxide dismutase associated with leaf water content under drought stress. Similarly, a putative LpCyt Cu-ZnSOD encoding cytosolic copper-zinc superoxide dismutase showed association with chlorophyll fluorescence. Rao *et al.* (2015) reported that allelic variations in the candidate genes, DREB IA and VP1.1 (Vacuolae-type H^+-Pyrophasphatox) associated with salinity tolerance in wild tomato (*Solanum pinpinellifolium*).

Identification of Genes Controlling Stress Tolerance Traits in Crop Plants through Gene/QTL Mapping and Map-Based Cloning Approaches

Map based cloning of genes and major QTLs associated with stress tolerance have been highly successful. Some classical examples of major genes cloned for important biotic stresses in crops include rice resistance to bacterial blight (*Xa1, xa5, xa13, Xa21, Xa3/Xa26 and Xa27*) (Khan *et al.*, 2014), blast (*Pit, Pi37, Pish, Pib, pi21, Pid2, Pi9, Pi2, Piz-t, Pid3, Pi25, Pi36, Pi5, Pi1, Pik, Pikm, Pikp, Pikh, Pi54rh, Pia, Pb1, Pita*) (Wang *et al.*, 2014), brown planthopper (*Bph14, Bph18)* (Fujita *et al.*, 2013) and gall midge (*gm3*) (Sama *et al.*, 2014); wheat resistance to leaf rust (*Lr1, Lr10, Lr21*) (Ellis *et al.*, 2014) and stem rust (*Sr33* and *Sr35*) (Ejaz *et al.*, 2012); soybean resistance to cyst nematode (*Rhg4*) (Liu *et al.*, 2012), and tomato resistance to root knot nematode (*Mi*) (Kaloshian *et al.*,1998). Recently, the number of genes being cloned is on the rise due to advances in NGS technologies.

A substantial progress has also been made in cloning of major QTLs associated with abiotic stress tolerance in crops. Xu and Mackill (1996) first mapped a QTL on chromosome 9, Sub1 for submergence tolerance from FR13A, an unadapted rice variety from North East India. Xu *et al.* (2006) further fine mapped the Sub1 QTL region identified a cluster of genes: *Sub1A*, *Sub1B* and *Sub1C*; of which, *Sub1A* encodes putative ethylene response factors that confers tolerance to submergence in tolerant genotypes. Over-expression of *Sub1A* in a submergence-sensitive genotype resulted in up-regulation of Alcohol dehydrogenase 1 (*Adh1*). Gamuyao *et al.* (2012) identified a gene, phosphorus-starvation tolerance 1 (*PSTOL1*), from the Pup1 QTL region that was previously identified (Wissua *et al.*, 2002). *PSTOL1* acts as an enhancer of early root growth, thereby enabling plants to acquire more phosphorus and other nutrients. Over-expression of *PSTOL1* in phosphorus deficiency sensitive varieties significantly enhanced grain yield in phosphorus-deficient soil. Uga *et al.* (2013) identified a gene, deeper rooting 1 (*DRO1*), controlling root growth angle in rice. The authors previously mapped a QTL, DRO1 on chromosome 9 of rice. Subsequently, they developed a near isogenic line (NIL) with tolerant DRO1 allele. Further high resolution mapping and candidate gene analysis lead to identification of *DRO1* gene that controls root growth angle in rice. *DRO1* is involved in cell elongation in the root tip that leads to deep rooting. Fujino *et al.* (2008) cloned a QTL, qLTG3–1, controlling low temperature germinability in rice. The qLTG3–1 encoded by protein of unknown function and strongly expressed in the embryo during seed germination. Liu *et al.* (2013) detected seven QTLs for cold tolerance in rice using a $F_{2:3}$ population derived from the progeny of tolerant and sensitive parents. Later, by combining microarray analysis with QTL results, they selected four genes as cold QTL-related candidates, of which, one candidate gene, *LOC_Os07g22494*, showed stronger association with cold tolerance in a number of rice varieties. Over-expression of this candidate gene in transgenic rice plants displayed strong tolerance to low temperature at early-seedling stage. Ueno *et al.* (2010) identified a gene (*OsHMA3*) responsible for low cadium (Cd) accumulation in rice, which encodes a transporter belonging to the P1B-type ATPase family. This gene was isolated by map based cloning of a QTL which was detected earlier in a mapping population derived from an indica cultivar, Anjana Dhan and a japonica cultivar, Nipponbare.

Marker-Assisted Breeding for Stress Tolerance in Crop Plants

Marker-assisted selection (MAS) refers to selection of plants carrying desired QTL/gene based on presence or absence of the associated molecular marker with the trait. MAS can be used in various breeding methods: pedigree selection, backcrossing and recurrent selection depending upon the purpose. Collard and Mackill (2008) may be referred for detailed procedures of MAS strategies. Marker-assisted backcrossing (MABC) and marker-assisted recurrent selection (MARS) may be used to transfer single or a few major QTLs/genes and multiple QTLs, respectively for cultivar improvement purposes. Genomic selection is a new method of MAS for complex traits, which is still at evolving stage. MAS has advantages over the conventional method for selection of stress tolerance traits in the following ways: (1) marker assays can be carried out on young plants to predict stress tolerance trait at the very early stage of the plant growth, (2) MAS is non-destructive; can help the breeder to confirm the F_1 and subsequently select desirable F_2 plants where single plants cannot be screened by conventional methods, (3) MAS can save generations - conventionally, selection for stress tolerance can be made only in advanced generations (F_5 and later) whereas, by MAS, selection for stress tolerance trait can be made in earlier generation (F_2) generation itself, (4) MAS can help the breeder to take up timely decisions to reject quickly the unwanted progenies and to follow up only a small number of desirable progenies for subsequent phenotypic confirmation in the laboratory or greenhouse, and (5) MAS help to save time and cost by reducing significantly the amount of breeding material carried through a breeding programme. Marker-assisted breeding requires multi-disciplinary efforts. A schematic diagram on the Marker-assisted breeding work flow is presented in Figure 22.1.

Cultivar Development through MAS

Though numerous genes and QTLs have been mapped for stress tolerance across crops, successful application in plant breeding is still limited. A few successful examples of MABC for improving stress tolerance in breeding lines and cultivars in the public plant breeding programmes are listed in Table 22.3.

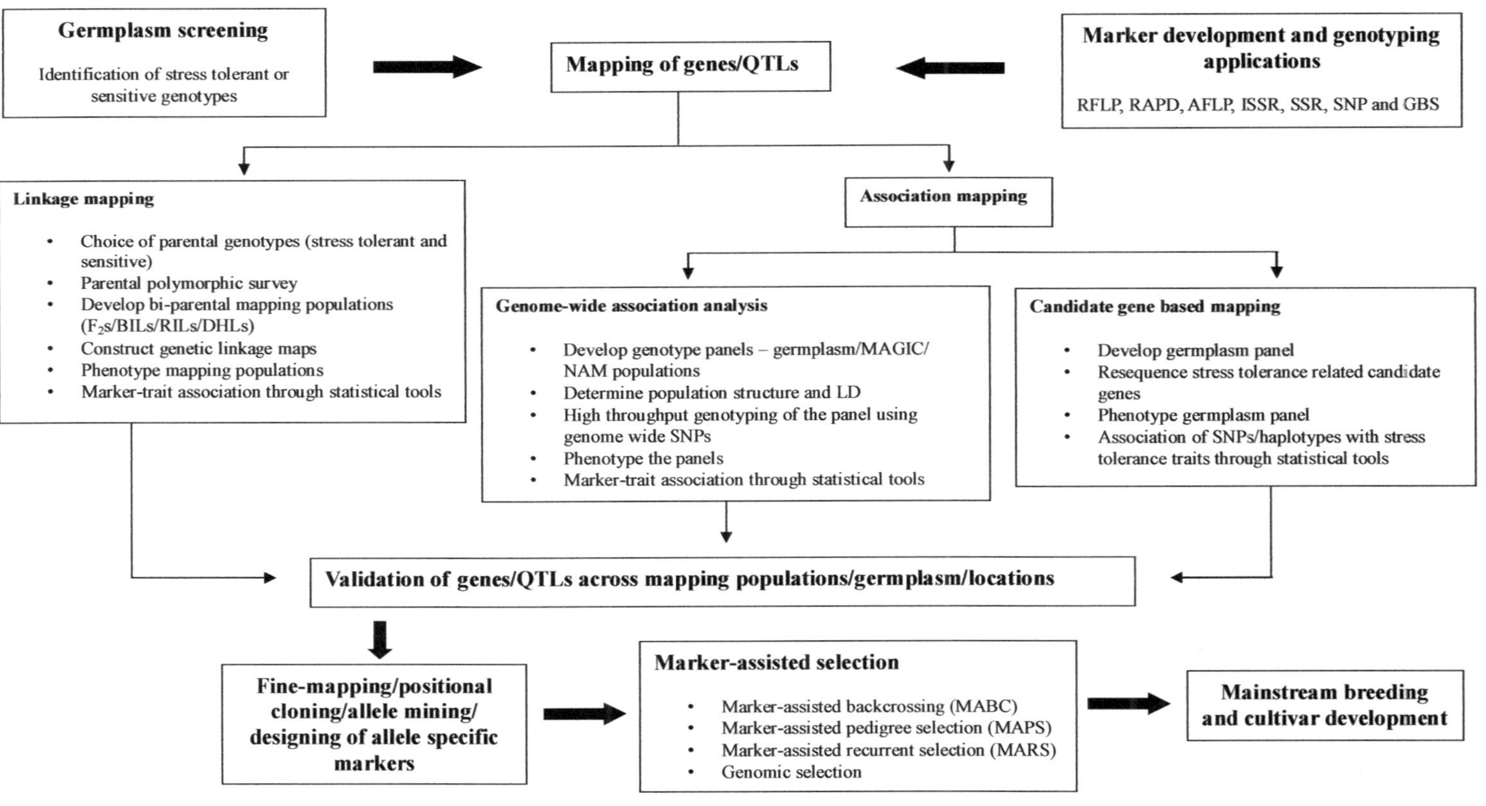

Figure 22.1: Schematic Diagram Showing Marker-Assisted Breeding Work Flow.

Table 22.3: List of Varieties/Breeding Lines Developed/Improved by MAS

Crop	*Trait*	*Genes*	*Variety/Genotype*	*Country*	*Year of Release/Source*
Rice	Bacterial Blight	*Xa4* + *xa5*	Angke	Indonesia	2002 (Khan *et al.*, 2014)
		Xa4 + *Xa7*	Conde	Indonesia	2002 (Khan *et al.*, 2014)
		Xa4 + *Xa21*	NSIC Rc154 (Tubigan 11)	Philippines	2006 (Khan *et al.*, 2014)
		Xa4 + *Xa21*	Guodao 1	China	2004 (Khan *et al.*, 2014)
		xa5 + *xa13* + *Xa21*	Improved Pusa Basmati-1	India	2007 (Khan *et al.*, 2014)
		xa5 + *xa13* + *Xa21*	PR106	India	Singh *et al.* (2001)
		xa5 + *xa13* + *Xa21*	Samba Mahsuri	India	Sundaram *et al.* (2008)
		xa5 + Blast R	RD6	Thailand	Pinta *et al.* (2013)
		Xa21	Zhongyou 6 and Zhongyou 1176 (Hybrids)	China	Cao *et al.* (2003)
	Submergence tolerance	*Sub1* QTL	Swarna, Samba Mahsuri, IR64, TDK1 and CR1009	India	Neeraja *et al.* (2007); Septiningsih *et al.* (2009)
		Sub1 QTL	ASS996	Vietnam	Cuc *et al.* (2012)
	Phosphorus use efficiency	*Pup1* QTL	Dodokan, Batur, and Situ Bagendit	Indonesia	Chin *et al.* (2011)
	Salt tolerance	*Saltol* QTL	ASS996	Vietnam	Huyen *et al.* (2012)
Wheat	Leaf rust resistance	*Lr-28* + *Lr-48*	PBW343	India	Chuneja *et al.* (2011)
		Lr-9 + *Lr-24* + *Lr-48*	HD2329	India	Charpe *et al.* (2012)
Tomato	Tomato yellow leaf curl virus	*Ty1* + *Ty2* + *Ty3*	Breeding lines	Taiwan	De la Peña *et al.* (2012)

Critical Considerations for an Effective MAS Programme

The following critical issues may be kept in mind for a successful integration of MAS in the mainstream breeding programme to develop stress tolerant cultivars.

- ☆ A suitable strategy of MAS must be determined based on the goals of a breeding programme, heritability of a trait, the number of genes governing a trait, strength of linkage between target locus and marker, population size and availability of resources.
- ☆ Simple, cost effective and breeder friendly genotyping platforms must be available depending upon the type of markers, the scale of a MAS programme and the resources. Gel based genotyping platform would be enough and effective when markers such as STS, SCAR, SSR and CAPS are used and only a few genes or a few hundred samples are handled in a MAS programme. SNP markers are ideal for achieving high throughput and automation. A variety of SNP genotyping assays *viz.*, high resolution melt (HRM) (Terracciano *et al.*, 2013), KBiosciences Competitive Allele-Specific PCR (KASPar) (Luciana Rosso *et al.*, 2011), Golden Gate (Thomson *et al.*, 2012), etc. are available.
- ☆ Effects of gene/QTL may be influenced by the genetic backgrounds and the environments (QTL x environment interaction) where it needs to be introduced. For example, Vadez *et al.* (2011) reported that effects of stay green QTLs on water extraction, transpiration efficiency and seed yield varied in different genetic backgrounds. There are numerous examples to suggest that QTL effects vary across environments. Hence, prioritization of genes/QTLs for MAS work must be made after a thorough validation across genetic backgrounds and target environments to avoid false positives. This would demand an effective phenotyping network across target environments.

Conclusion

Molecular marker technology has shown promise for better understanding on genetic basis of stress tolerance traits in major crops like rice. Information in the horticultural and under-utilized, orphan crops are limited. As evident from the literature, stress tolerance traits are controlled by both major and predominantly minor QTLs. Major QTLs have been followed up for fine-mapping or gene discovery and further utilization in marker-assisted breeding for cultivar improvement. More research is needed on minor QTLs, which remains elusive for application in breeding. Substantial efforts have been made by plant physiologists in identifying stress tolerance traits in crop germplasm. Recently, phenomics technologies are emerging as powerful tools to provide precise high throughput phenotyping options which can be exploited for QTL discovery and validation in stress tolerance experiments (Tuberosa, 2012). Genomics approaches in combination with QTL approach will play a greater role in systematic dissection of QTLs and discovering candidate genes associated with stress tolerance in crop plants (Collins *et al.*, 2008). Successful application of MAS for improvement of cultivars with tolerance to submergence, salt etc. demonstrates the power of molecular breeding for stress tolerance in crop plants.

References

Agrama, H.A.S., Zakaria, A.G., Said, F.B. and Tuinstra, M., 1999. Identification of quantitative trait loci for nitrogen use efficiency in maize. *Mol. Breed.*, 5: 187–195.

Akkaya, M.S., Bhagwat, A.A. and Cregan, P.B., 1992. Length polymorphisms of simple sequence repeat DNA in soybean. *Genetics* 132: 1131-1139.

Ali, A.J., Xua, J.L., Ismail, A.M., Fua, B.Y., Vijaykumar, C.H.M., Gaoa, Y.M., Domingoa, J., Maghiranga, R., Yua, S.B., Gregorioa, G., Yanaghihara, S., Cohen, M., Carmen, B., Mackill, D., and Li, Z.K., 2006. Hidden diversity for abiotic and biotic stress tolerances in the primary gene pool of rice revealed by a large backcross breeding program. *Field Crops Res.*, 97: 66–76.

Andaya, V.C. and Mackill, D.J., 2003. Mapping of QTLs associated with cold tolerance during the vegetative stage in rice. *J. Exp. Bot.*, 54: 2579-2585.

Andaya, V.C. and Tai, T.H., 2006. Fine mapping of the qCTS12 locus, a major QTL for seedling cold tolerance in rice. *Theor. Appl. Genet.*, 113: 467–475.

Arraouadi, S., Badri, M., Abdelly, C. and Aouani, M.E., 2012. QTL mapping of physiological traits associated with salt tolerance in *Medicago truncatula* recombinant inbred lines. *Genomics*, 99: 118-125.

Asíns, M.J., 2002. Present and future of quantitative trait locus analysis in plant breeding. *Plant Breed.*, 121: 281–368.

Avia, K., Pilet-Nayel, M.L., Bahrman, N., Baranger, A., Delbreil, B., Fontaine, V., Hamon, C., Hanocq, E., Niarquin, M., Sellier, H., Vuylsteker, C., Prosperi, J.M. and Lejeune-Hénaut, I., 2013. Genetic variability and QTL mapping of freezing tolerance and related traits in *Medicago truncatula*. *Theor. Appl. Genet.*, 126: 2353-2366.

Bálint, A.F., Röder, M.S., Hell, R., Galiba, G. and Börner, A., 2007. Mapping of QTLs affecting copper tolerance and the Cu, Fe, Mn and Zn contents in the shoots of wheat seedlings. *Biologia Plantarum* 51(1): 129-134.

Bandillo, N., Raghavan, C., Muyco, P.A., Sevilla, M.L., Lobina I.T., Dilla-Ermita, C.J., Tung, C-W., McCouch, S., Thomson, M., Mauleon, R., Singh, R., Gregorio, G., Redoña, E. and Leung, H. 2013. Multi-parent advanced generation inter-cross (MAGIC) populations in rice: progress and potential for genetics research and breeding. *Rice*, 6: 11.

Bernier, J., Kumar, A., Venuprasad, R., Spaner, D. and Atlin, G., 2007. A large-effect QTL for grain yield under reproductive stage drought stress in upland rice. *Crop Sci.*, 47: 507–518.

Blum, A., 2011. Drought resistance – is it really a complex trait? *Funct. Plant Biol.*, 38: 753–757.

Bonilla, P., Dvorak, J., Mackill, D., Deal, K. and Gregorio, G., 2002. RFLP and SSLP mapping of salinity tolerance genes in chromosome 1 of rice (*Oryza sativa* L.) using recombinant inbred lines. *Philipp. Agric. Sci.*, 85: 68–76.

Botstein, D., White, R.L., Skolnick, M. and Davis, R.W., 1980. Construction of a genetic linkage map in man using restriction fragment length polymorphisms. *Amer. J. Human Genet.*, 32: 314-331.

Cai, S., Wu, D., Jabeen, Z., Huang, Y., Huang, Y. and Zhang, G., 2013. Genome-wide association analysis of aluminum tolerance in cultivated and Tibetan wild barley. *PLoS ONE*, 8(7): e69776. doi: 10.1371/journal.pone.0069776

Cao, L-y., Zhuang, J-y., Zhan, X-d., Zheng, K-l. and Cheng, S-h., 2003. Hybrid rice resistant to bacterial blight developed by marker assisted selection. *Chinese J. Rice Sci.*, 17: 184-186.

Cavanagh, C., Morell, M., Mackay, I. and Powell, W., 2008. From mutations to MAGIC: resources for gene discovery, validation and delivery in crop plants. *Curr. Opin. Plant Biol.*, 11: 215–221

Charpe, A., Koul, S Gupta, S.K., Singh, A., Pallavi, J.K. and Prabhu, KV., 2012. Marker assisted gene pyramiding of leaf rust resistance genes *Lr-9*, *Lr-24* and *Lr-28* in a bread wheat cultivar HD 2329. *J. Wheat Res.*, 4 (1): 20-28.

Chin, J.H., Gamuyao, R., Dalid, C., Bustamam, M., Prasetiyono, J., Moeljopawiro, S., Wissuwa, M. and Heuer, S., 2011. Developing rice with high yield under phosphorus deficiency: Pup1 sequence to application. *Plant Physiol.*, 156: 1202–1216.

Chong-qing, W., Tao, W., Ping, M., Zi-chao, L. and Ling, Y., 2013. Quantitative trait loci for mercury tolerance in rice seedlings. *Rice Sci.*, 20(3): 238–242.

Chuneja P., Vikal, Y., Kaur, S., Singh, R., Juneja, S., Bain, N.S., Berry, O., Sharma, A., Gupta, S.K., Charpe, A., Prabhu, K.V. and Dhalival, H.S., 2011. Marker-assisted pyramiding of leaf rust resistance genes Lr24 and Lr28 in wheat (*Triticum aestivum*). *Indian J. Agric. Sci.*, 81 (3): 214–218.

Collard, B.C.Y. and Mackill D.J., 2008. Marker-assisted selection: an approach for precision plant breeding in the twenty-first century. *Phil. Trans. R. Soc. B*, 363: 557–572.

Collins, N.C., Tardieu, F. and Tuberosa, R., 2008. QTL approaches for improving crop performance under abiotic stress conditions: Where do we stand? *Plant Physiol.*, 147: 469-486.

Cornelious, B., Chen, P., Chen, Y., de Leon, N., Shannon, J.G. and Wang, D., 2005. Identication of QTLs underlying water-logging tolerance in soybean. *Mol. Breed.*, 16: 103–112.

Cuc, L.M., Huyen, L.T.N., Hien, P.T.M., Hang, V.T.T., Dam, N.Q., Mui, P.T., Quang, V.D., Ismail, A.M. and Ham, L.H., 2012. Application of marker assisted backcrossing to introgress the submergence tolerance QTL *SUB*1 into the Vietnam elite rice variety-AS996. *Amer. J. Plant Sci.*, 3: 528-536.

De la Peña, R.C., Kadirvel, P., Venkatesan, S., Kenyon, L. and Hughes, J., 2010. Integrated approaches to manage tomato yellow leaf curl viruses. In: Biocatalysis and Biomolecular Engineering. Hou, C.T. and Shaw J-F. John (Eds). John Wiley and Sons Inc. pp. 105-132.

Ding, D., Li, W., Song, G., Qi, H., Liu, J. and Tang, J., 2011. Identification of QTLs for arsenic accumulation in maize (*Zea mays* L.) using a RIL population. *PLoS One*, 6(10): e25646. doi: 10.1371/journal.pone.0025646

Djian-Caporalino, C., Fazari, A., Arguel, M.J., Vernie, T., VandeCasteele, C., Faure, I., Brunoud, G., Pijarowski, L., Palloix, A., Lefebvre, V. and Abad, P., 2007. Root-knot nematode (*Meloidogyne* spp.) *Me* resistance genes in pepper (*Capsicum annuum* L.) are clustered on the P9 chromosome. *Theor. Appl. Genet.*, 114: 473-486.

Du, W., Yu, D. and Fu, S., 2009. Detection of quantitative trait loci for yield and drought tolerance traits in soybean using a recombinant inbred line population. *J. Integr. Plant Biol.*, 51: 868-878.

Edae, E.A., Byrne, P.F., Manmathan, H., Haley, S.D., Moragues, M., Lopes, M.S. and Reynolds, M.P., 2013. Association mapping and nucleotide sequence variation in five drought tolerance candidate genes in spring wheat. *The Plant Genome*, 6(2): 1-13.

Ejaz, M., Iqbal, M., Shahzad, A., Rehman, A., Ahmed, I. and Ali, GM., 2012. Genetic variation for markers linked to stem rust resistance genes in Pakistani wheat varieties. *Crop Sci.*, 52: 2638-2648.

Ellis, J.G., Lagudah, E.S., Spielmeyer, W. and Dodds, P.N., 2014. The past, present and future of breeding rust resistant wheat. *Front. Plant Sci.*, Available at doi: 10.3389/fpls.2014.00641

Elshire, R.J., Glaubitz, J.C., Sun, Q., Poland, J.A., Kawamoto, K., Buckler, E.S. and Mitchell, S.E., 2011. A robust, simple genotyping-by-sequencing (GBS) approach for high diversity species. *PLoS ONE*, 6(5): e19379.

Famoso, A.N., Zhao, K., Clark, R.T., Tung, C-W., Wright, M.H., Bustamante, C., Kochian, L.V. and McCouch, S.R., 2011. Genetic architecture of aluminum tolerance in rice (*Oryza sativa*) determined through genome-wide association analysis and QTL mapping. *PLoS Genet.*, 7(8): e1002221. doi: 10.1371/journal.pgen.1002221.

Fujino, K., Sekiguchi, H., Sato, T., Kiuchi, H., Nonoue, Y., Takeuchi, Y., Ando, T., Lin, S.Y. and Yano, M. 2004. Mapping of quantitative trait loci controlling low-temperature germinability in rice (*Oryza sativa* L.). *Theor. Appl. Genet.*, 108: 794–799.

Fujino, K., Sekiguchi, H., Matsuda, Y., Sugimoto, K., Ono, K. and Yano, M., 2008. Molecular identification of a major quantitative trait locus, *qLTG3-1*, controlling low-temperature germinability in rice. *Proc. Natl Acad. Sci. USA* 105: 12623–12628.

Fujita, D., Kohli, A. and Horgan, F.G., 2013. Rice resistance to planthoppers and leafhoppers. *Crit. Rev. Plant Sci.*, 32: 162-191.

Furbank, R.T. and Tester, M., 2011. Phenomics-technologies to relieve the phenotyping bottleneck. *Trends Plant Sci.*, 16: 635-644.

Gamuyao, R., Chin, J.H., Pariasca-Tanaka, J., Pesaresi, P., Catausan, S., Dalid, C., Slamet-Loedin, I., Tecson-Mendoza, E.M., Wissuwa, M. and Heuer, S., 2012.

The protein kinase *Pstol1* from traditional rice confers tolerance of phosphorus deficiency. *Nature*, 488: 535–539.

Gautami, B., Pandey, M.K., Vadez, V., Nigam, S.N., Ratnakumar, P., Krishnamurthy, L., Radhakrishnan, T., Gowda, M.V.C., Narasu, M.L., Hoisington, D.A., Knapp, S.J. and Varshney. R.K., 2012. Quantitative trait locus analysis and construction of consensus genetic map for drought tolerance traits based on three recombinant inbred line populations in cultivated groundnut (*Arachis hypogaea* L.). *Mol. Breed.*, 30: 757–772.

Geethanjali, S., Kadirvel, P., Gunathilagaraj, K. and Maheswaran, M., 2009. Detection of QTLs associated with resistance to whitebacked planthopper (*Sogatella furcifera* Horvath) in rice. *Plant Breed.*, 128: 130-136.

Genc, Y., Oldach, K., Verbyla, A.P., Lott, G., Hassan, M., Tester, M., Wallwork, H. and McDonald, G.K., 2010. Sodium exclusion QTL associated with improved seedling growth in bread wheat under salinity stress. *Theor. Appl. Genet.*, 121: 877-894.

Giuliani, S., Sanguineti, M.C., Tuberosa, R., Bellotti, M., Salvi, S. and Landi, P. 2005. Root-ABA1, a major constitutive QTL, affects maize root architecture and leaf ABA concentration at different water regimes. *J. Exp. Bot.*, 56: 3061–3070.

Glenn, T.C., 2011. Field guide to next-generation DNA sequencers. *Mol. Ecol. Resour.*, 11: 759–769.

Huyen, L.T.N., Cuc, L.M., Ismail, A.M. and Ham, L.H., 2012. Introgression the salinity tolerance QTLs *Saltol* into AS996, the elite rice variety of Vietnam. *Amer. J. Plant Sci.*, 3: 981-987.

Jaccoud, D., Peng, K., Feinstein, D. and Kilian, A. 2001. Diversity arrays: a solid state technology for sequence information independent genotyping. *Nucleic Acids Res.*, 29(4): e25.

Jenni, S., Truco, M. and Michelmore, R.W., 2013. Quantitative trait loci associated with tipburn, heat stress-induced physiological disorders, and maturity traits in crisphead leuttuce. *Theor. Appl. Genet.*, 126: 3065-3079.

Jia, L., Yan, W., Zhu, C., Agrama, H.A., Jackson, A., Yeater, K., Li, X., Huang, B., Hu, B., McClung, A. and Wu, D., 2012. Allelic analysis of sheath blight resistance with association mapping in rice. *PLoS ONE* 7(3): e32703. doi: 10.1371/journal. pone.0032703.

Jordan, S.A. and Humphries, P., 1994. Single nucleotide polymorphism in exon 2 of the BCP gene on 7q31-q35. *Human Mol. Genet.*, 3: 1915.

Kadirvel, P., de la Peña, R., Schafleitner, R., Huang, S., Geethanjali, S., Kenyon, L., Tsai, W. and Hanson, P., 2013. Mapping of QTLs in tomato line FLA456 associated with resistance to a virus causing Tomato yellow leaf curl disease. *Euphytica* 190: 297–308.

Kaloshian, I., Yaghoobi, J., Liharska, T., Hontelez, J., Hanson, D., Hogan, P., Jesse, T., Wijbrandi, J., Simons, G., Vos, P., Zabel, P. and Williamson, V.M., 1998.

Genetic and physical localization of the root-knot nematode resistance locus *mi* in tomato. *Mol. Gen. Genet.*, 257: 376-385.

Khan, M.A., Naeem, M. and Iqbal, M. 2014. Breeding approaches for bacterial leaf blight resistance in rice (*Oryza sativa* L.), current status and future directions. *Eur. J. Plant Pathol.*, 139: 27-37.

Kim, K.S., Chirumamilla, A., Hill, C.B., Hartman, G.L. and Diers, B.W., 2014. Identification and molecular mapping of two soybean aphid resistance genes in soybean PI 587732. *Theor. Appl. Genet.*, 127: 1251–1259.

Krill, A.M., Kirst, M., Kochian, L.V., Buckler, E.S. and Hoekenga, O.A., 2010. Association and linkage analysis of aluminum tolerance genes in maize. *PLoS ONE*, 5(4): e9958. doi: 10.1371/journal.pone.0009958.

Kumar, J., Choudhary, A.K., Solanki, R.K. and Pratap, A., 2011. Towards marker-assisted selection in pulses: a review. *Plant Breed.*, 130: 297-313.

Lanceras, J.C., Pantuwan, G., Jongdee, B. and Toojinda, T., 2004. Quantitative trait loci associated with drought tolerance at reproductive stage in rice. *Plant Physiol.*, 135: 384–399.

Larkan, N.J., Lydiate, D.J., Parkin, I.A.P., Nelson, M.N., Epp, D.J., Cowling, W.A., Rimmer S.R., and Borhan, M.H., 2013. The *Brassica napus* blackleg resistance gene *LepR3* encodes a receptor-like protein triggered by the *Leptosphaeria maculans* effector AVRLM1. *New Phytologist*, 197: 595–605.

Leonforte, A., Sudheesh, S., Cogan, N.O., Salisburry, P.A., Nicolas, M.E., Materne, M., Forster, J.W. and Kaur, S., 2013. SNP marker discovery, linkage map construction and identification of QTLs for enhanced salinity tolerance in field pea (*Pisum sativum* L.). *BMC Plant Biol.*, 13: 161.

Li, L., Liu, X., Xie, K., Wang, Y., Liu, F., Lin, Q., Wang, W., Yang, C., Lu, B., Liu, S., Chen, L., Jiang, L. and Wan, J., 2013. qLTG-9, a stable quantitative trait locus for low-temperature germination in rice (*Oryza sativa* L.). *Theor. Appl. Genet.*, 126: 2313–2322.

Liu, X.M., Smith, C.M., Gill, B.S. and Tolmay, V., 2001. Microsatellite markers linked to six Russian wheat aphid resistance genes in wheat. *Theor. Appl. Genet.*, 102: 504–510.

Liu, S., Kandoth, P.K., Warren, S.D., Yeckel, G., Heinz, R., Alden, J., Yang, C., Jamai, A., El-Mellouki, T., Juvale, P.S., Hill, J., Baum, T.J., Cianzio, S., Whitham, S.A., Korkin, D., Mitchum, M.G. and Meksem, K., 2012. A soybean cyst nematode resistance gene points to a new mechanism of plant resistance to pathogens. *Nature* 492: 256–260

Liu, F., Xu, W., Song, Q., Tan, L., Liu, J., Zhu, Z., Fu, Y., Su, Z. and Sun, C. 2013. Microarray-assisted fine-mapping of quantitative trait loci for cold tolerance in rice. *Mol. Plant*, 6: 757–767.

Luciana Rosso, M., Burleson, S.A., Maupin, L.M. and Rainey, K.M., 2011. Development of breeder-friendly markers for selection of MIPS1 mutations in soybean. *Mol. Breed.*, 28: 127–132.

Ma, J. F., Shen, R., Zhao, Z., Wissuwa, M., Takeuchi, Y., Ebitani, T. and Yano, M., 2002. Response of rice to aluminium stress and identification of QTL for aluminium tolerance. *Plant Cell Physiol.*, 43: 652–659.

Miftahudin, G., Scoles, J. and Gustafson, J. P. (2002) AFLP markers tightly linked to aluminium tolerance gene *Alt3* in rye (*Secale cereale* L.). *Theor. Appl. Genet.*, 104: 626–631.

Neeraja, C.N., Maghirang-Rodriguez, R., Pamplona, A., Heuer, S., Collard, B.C.Y., Septiningsih, E.M., Vergara, G., Sanchez, D., Xu, K., Ismail, A.M. and Mackill, D.J., 2007. A marker-assisted backcross approach for developing submergence-tolerant rice cultivars. *Theor. Appl. Genet.*, 115: 767–776.

Nguyen, V. T., Nguyen, B. D., Sarkarung, S., Matinez, C., Paterson, A. H. and Nguyen, H. T., 2002. Mapping of genes controlling aluminium tolerance in rice: comparison of different genetic backgrounds. *Mol. Genet. Genom.*, 267: 722–780.

Nguyen, B., Brar, D. S., Bui, B. C., Nguyen, T. V., Pham, L. N. and Nguyen, H. T., 2003. Identification and mapping of the QTL for aluminium tolerance introgressed from the new source, *Oryza rufipogon* Griff., into indica rice (*Oryza sativa*). *Theor. Appl. Genet.*, 106: 583–593.

Ninamango-Cardenas, F. E., Guimaraes, C.T., Martins, P.R., Parentoni, S.N., Carneiro, N.P., Lopes, M.A., Moro, J.R. and Paiva, E., 2003. Mapping QTLs for aluminium tolerance in maize. *Euphytica*, 130: 223–232.

Oraguzie, N.C., Rikkerink, E.H.A., Gardiner, S.E. and De Silva, H.N., 2007. Association mapping in plants. Springer Science+Business Media, LLC.

Paliwal, R., Röder, M.S., Kumar. U., Srivastava. J.P. and Joshi, A.K. 2012. QTL mapping of terminal heat tolerance in hexaploid wheat (*T. aestivum* L.). *Theor. Appl. Genet.*, 125: 561-575.

Paran, I. and Michelmore, R.W., 1993. Development of reliable PCR-based markers linked to downy mildew resistance genes in lettuce. *Theor. Appl. Genet.*, 85: 985-993.

Paterson, A.H. 1996. Genome mapping in plants. Austin, TX: Landes, 330p.

Pinta, W., Toojinda, T., Thummabenjapone, P. and Sanitchon, J., 2013. Pyramiding of blast and bacterial leaf blight resistance genes into rice cultivar RD6 using marker assisted selection. *African J. Biotech.*, 12: 442-448.

Poczai, P., Varga, I., Laos, M., Cseh, A., Bell, N., Valkonen, J.P.T. and Hyvönen, J. 2013. Advances in plant gene-targeted and functional markers: a review. *Plant Methods* 9: 6

Qiu, F., Zheng, Y., Zhang, Z. and Xu, S. 2007. Mapping of QTL associated with waterlogging tolerance during the seedling stage in maize. *Ann. Bot.*, 99: 1067–1081.

Rafalski, J.A., 2010 Association genetics in crop improvement. *Curr. Opin. Plant Biol.*, 13: 174–180.

Rahman, H., Pekic, S., Lazic-Jancic, V., Quarrie, S.A., Shah, S.M., Pervez, A. and Shah, M.M., 2011. Molecular mapping of quantitative trait loci for drought tolerance in maize plants. *Genet. Mol. Res.*, 10: 889-901.

Raman, H., Karakousis, A., Moroni, J.S., Raman, R., Read, B.J., Garvin, D.F., Kochian, L.V. and Sorrells, M.E., 2003. Development and allele diversity of microsatellite markers linked with an aluminium tolerance gene *Alp* in barley. *Aust. J. Agric. Res.*, 54: 1315–1321.

Rao, E.S., Kadirvel, P., Symonds, R.C. and Ebert, A.W., 2013. Relationship between survival and yield related traits in *Solanum pimpinellifolium* under salt stress. *Euphytica*, 190: 215–228.

Ravi, K., Vadez, V., Isobe, S., Mir, R.R., Guo, Y., Nigam, S.N., Gowda, M.V., Radhakrishnan, T., Bertioli, D.J., Knapp, S.J. and Varshney, R.K., 2011. Identification of several small main-effect QTLs and a large number of epistatic QTLs for drought tolerance related traits in groundnut (*Arachis hypogaea* L.). *Theor. Appl. Genet.*, 122: 1119-1132.

Reiter, R., Phillips, R.L. and Vasil, I.K. (*Eds*.). 2001. DNA-based markers in plants. Kluwer Academic, Dordrecht. pp. 9–29.

Rosewarne, G.M., Herrera-Foessel, S.A., Singh, R.P., Huerta-Espino, J., Lan, C.X. and He, Z.H., 2013. Quantitative trait loci of stripe rust resistance in wheat. *Theor. Appl. Genet.*, 126: 2427–2449.

Sama, V.S.A.K., Rawat, N., Sundaram, R.M., Himabindu, K., Naik, B.S., Viraktamath, B.C. and Bentur, J.S. 2014. A putative candidate for the recessive gall midge resistance gene *gm3* in rice identified and validated. *Theor. Appl. Genet.*, 127: 113–124

Senthilvel, S., Vinod, K.K., Malarvizhi, P. and Maheswaran, M. (2008)QTL and QTL x environment effects on agronomic and nitrogen acquisition traits in rice. *J. Integr. Plant Biol.*, 50: 1108-1117.

Septiningsih, E.M., Pamplona, A.M., Sanchez, D.L., Neeraja, C.N., Vergara, G.V., Heuer, S., Ismail A.M. and Mackill, D.J., 2009. Development of submergence tolerant rice cultivars: the *SUB*1 locus and beyond. *Ann. Bot.*, 103: 151-160.

Singh, S., Sidhu, J.S., Huang, N., Vikal, Y., Li, Z., Brar, D.S., Dhaliwal, H.S. and Khush, G.S., 2001. Pyramiding three bacterial blight resistance genes (*xa5, xa13* and *Xa21*) using marker-assisted selection into indica rice cultivar PR106. *Theor. Appl. Genet.*, 102: 1011-1015.

Sledge, M.K., Bouton, J.H., Agnoll, M.D., Parrot, W.A. and Kochert, G., 2002. Identification and confirmation of aluminium tolerance QTL in diploid *Medicago sativa* subsp. *coerulea*. *Crop Sci.*, 42: 1121-1128.

Soundararajan, R.P., Kadirvel, P., Gunathilagaraj, K. and Maheswaran, M., 2004. Mapping of QTLs associated with resistance to brown planthopper (*Nilaparvata lugens* Stal) by means of a doubled-haploid population in rice. *Crop Sci.*, 44: 2214-2220.

Sripongpangkul, K., Posa, G.B.T., Senadhira, D.W., Brar, D., Huang, N., Khush, G.S. and Li, Z.K., 2000. Genes/QTLs affecting flood tolerance in rice. *Theor. Appl. Genet.*, 101: 1074–1081.

Sundaram, R.M., Vishnupriya, M.R., Biradar, S.K., Laha, G.S., Reddy, G.A., Shobha Rani, N., Sarma, N.P., Sonti, R.V., 2008. Marker assisted introgression of bacterial blight resistance in Samba Mahsuri, an elite indica rice variety. *Euphytica*, 160: 411-422.

Terracciano, I., Maccaferri, M., Bassi, F., Mantovani, P., Sanguineti, M.C., Salvi, S., Simkova, H., Dolezel, J., Massi, A., Ammar, K., Kolmer, J. and Tuberosa, R. 2013. Development of COS-SNP and HRM markers for high throughput and reliable haplotype-based detection of *Lr14a* in durum wheat (*Triticum durum* Desf.). *Theor. Appl. Genet.*, 126: 1077–1101.

Thomson, M.J., Zhao, K., Wright, M., McNally, K.L, Rey, J., Tung, C.W., Reynolds, A., Schefer, B., Eizenga, G., McClung, A., Kim, H., Ismail, A.M., de Ocampo, M., Mojica, C., Reveche, M.Y., Dilla-Ermita, C.J., Mauleon, R., Leung, H., Bustamante, C. and McCouch, S.R. 2012. High-throughput single nucleotide polymorphism genotyping for breeding applications in rice using the BeadXpress platform. *Mol. Breed.*, 29: 875–886.

Thomson, M.J., de Ocampo, M., Egdane, J., Rahman, M.A., Sajise, A.G., Adorada, D.L., Tumimbang-Raiz, E., Blumwald, E., Seraj, Z.I., Singh, R.K., Gregorio, G.B. and Ismail, A.M., 2010. Characterizing the *Saltol* quantitative trait locus for salinity tolerance in rice. *Rice,* DOI 10.1007/s12284-010-9053-8.

Tuberosa, R., 2012. Phenotyping for drought tolerance of crops in the genomics era. *Front. Physiol.*, 3: 347.

Tuberosa, R., Sanguineti, M.C., Landi, P., Salvi, S., Casarini, E. and Conti. S. 1998. RFLP mapping of quantitative trait loci controlling abscisic acid concentration in leaves of drought-stressed maize (*Zea mays* L.). *Theor. Appl. Genet.*, 97: 744-755.

Ueno, D., Yamaji, N., Kono, I., Huang, C.F., Ando, T., Yano, M. and Ma, J.F., 2010. Gene limiting cadmium accumulation in rice. *PNAS* 107: 16500-16505.

Uga, Y., Sugimoto, K., Ogawa, S., Rane, J., Ishitani, M., Hara, N., Kitomi, Y., Inukai, Y., Ono, K., Kanno, N., Inoue, H., Takehisa, H., Motoyama, R., Nagamura, Y., Wu, J., Matsumoto, T., Takai, T., Okuno, K. and Yano, M., 2013. Control of root system architecture by *DEEPER ROOTING 1* increases rice yield under drought conditions. *Nat. Genet.*, 45: 1097–1102.

Uga, Y., Okuno, K. and Yano, M., 2011. *Dro1*, a major QTL involved in deep rooting of rice under upland field conditions. *J. Exp. Bot.*, 62: 2485–2494.

Vadez, V., Deshpande, S.P., Kholova, J., Hammer, G.L., Borrell, A.K., Talwar, H.S. and Hash, C.T., 2011. Stay-green quantitative trait loci's effects on water extraction, transpiration efficiency and seed yield depend on recipient parent background. *Funct. Plant Biol.*, 38: 553–566.

Varshney, R.K., Paulo, M.J., Grando, S., van Eeuwijk, F.A., Keizer, L.C.P., Guo, P., Ceccarelli, S., Kilian, A., Baum, M. and Graner, A. 2012. Genome wide association analyses for drought tolerance related traits in barley (*Hordeum vulgare* L.). *Field Crops Res.*, 126: 171–180.

Venuprasad, R., Dalid, C.O., Del Valle, M., Zhao, D., Espiritu, M., Sta Cruz, M.T., Amante, M., Kumar, A. and Atlin, G.N., 2009. Identification and characterization of large-effect quantitative trait loci for grain yield under lowland drought stress in rice using bulk-segregant analysis. *Theor. Appl. Genet.*, 120: 177–190.

Vikram, P., Mallikarjuna Swamy, B.P., Dixit, S., Uddin Ahmed, H., Sta Cruz, M.T., Singh, A.K. and Kumar, A., 2011. *qDTY1.1*, a major QTL for rice grain yield under reproductive-stage drought stress with a consistent effect in multiple elite genetic backgrounds. *BMC Genetics*, 12: 89. doi: 10.1186/1471-2156-12-89.

Vos, P., Hogers, R., Bleeker, M., Reijans, M., Van De Lee, T., Hornes, M., Frijters, A., Pot, J., Peleman, J., Kuiper, M. and Zabeau, M., 1995. AFLP: a new technique for DNA fingerprinting. *Nucleic Acids Res.*, 23: 4407-4414.

Wang, X., Lee, S., Wang, J., Ma, J., Bianco, T. and Jia, Y. 2014. Current advances on genetic resistance to rice blast disease. In: Rice - Germplasm, Genetics and Improvement. INTECH. Pp 195-217. Available at http: //dx.doi.org/10.5772/56824

Wei, D., Cui, K., Ye, G., Pan, J., Xiang, J., Huang, J. and Nie, L., 2012. QTL mapping for nitrogen-use efficiency and nitrogen-deficiency tolerance traits in rice. *Plant Soil*, 359: 281–295.

Wen, Z., Tan, R., Yuan, J., Bales, C., Du, W., Zhang, S., Chilvers, M.I., Schmidt, C., Song, Q., Cregan, P.B. and Wang, D., 2014. Genome-wide association mapping of quantitative resistance to sudden death syndrome in soybean. *BMC Genomics* 15: 809.

Wieckhorst, S., Bachlava, E., Dußle, C.M., Tang, S., Gao, W., Saski, C., Knapp, S.J., Schön, C.C., Hahn, V. and Bauer, E., 2010. Fine mapping of the sunflower resistance locus Pl_{ARG} introduced from the wild species *Helianthus argophyllus*. *Theor. Appl. Genet.*, 121: 1633–1644.

Williams, J.G.K., Kubelik, A.R., Livak, K.J., Rafalsky, J.A. and Tingey, S.V., 1990. DNA polymorphisms amplified by arbitrary primers are useful as genetic markers. *Nucleic Acids Res.*, 18: 6531-6535.

Wissuwa, M., Yano, M. and Ae, N., 1998. Mapping of QTLs for phosphorus-deficiency tolerance in rice (*Oryza sativa* L.). *Theor. Appl. Genet.*, 97: 777-783.

Wissuwa, M., Wegner, J., Ae, N. and Yano, M., 2002. Substitution mapping of Pup1: a major QTL increasing phosphorus uptake of rice from a phosphorus-deficient soil. *Theor. Appl. Genet.*, 105: 890–897.

Xu, K. and Mackill, D.J., 1996. A major locus for submergence tolerance mapped on rice chromosome 9. *Mol. Breed.*, 2: 219-224.

Xu, W., Subudhi, P.K., Crasta, O.R., Rosenow, D.T., Mullet, J.E. and Nguyen, H.T., 2000. Molecular mapping of QTLs conferring stay-green in grain sorghum (*Sorghum bicolor* L. Moench). *Genome*, 43: 461-469.

Xu, K., Xu, X., Fukao, T., Canlas, P., Maghirang-Rodriguez, R., Heuer, S., Ismail, A.M., Bailey-Serres, J., Ronald, P.C. and Mackill, D.J., 2006. *Sub1A* is an ethylene-response-factor-like gene that confers submergence tolerance to rice. *Nature*, 442: 705-708.

Xu, R., Wang, J., Li, C., Johnson, P., Lu, C. and Zhou, M. 2012. A single locus is responsible for salinity tolerance in a Chinese landrace barley (*Hordeum vulgare* L.). *PLoS ONE* 7(8): e43079. doi: 10.1371/journal.pone.0043079.

Yadav, R.S., Hash, C.T., Bidinger, F.R., Cavan, G.P. and Howarth, C.J., 2002. Quantitative trait loci associated with traits determining grain and stover yield in pearl millet under terminal drought-stress conditions. *Theor. Appl. Genet.*, 104: 67-83.

Yadav, R.S., Sehgal, D., and Vadez, V., 2011. Using genetic mapping and genomics approaches in understanding and improving drought tolerance in pearl millet. *J. Exp. Bot.*, 62: 397-408.

Yu, J, Holland, J.B., McMullen, M.D. and Buckler, E.S., 2008. Genetic design and statistical power of nested association mapping in maize. *Genetics*, 178: 539–551.

Yu, X., Bai, G., Liu, S., Luo, N., Wang, Y., Richmond, D.S., Pijut, P.M., Jackson, S.A., Yu, J. and Jiang. Y., 2013. Association of candidate genes with drought tolerance traits in diverse perennial ryegrass accessions. *J. Exp. Bot.*, doi: 10.1093/jxb/ert018.

Yue, B., Xue, W., Xiong, L., Yu, X., Luo, L., Cui, K., Jin, D., Xing, Y. and Zhang. Q., 2006. Genetic basis of drought resistance at reproductive stage in rice: separation of drought tolerance from drought avoidance. *Genetics*, 172: 1213–1228.

Zegeye, H., Rasheed, A., Makdis, F., Badebo, A. and Ogbonnaya, F.C. 2014. Genome-wide association mapping for seedling and adult plant resistance to stripe rust in synthetic hexaploid wheat. *PLoS ONE* 9(8): e105593. doi: 10.1371/journal.pone.0105593.

Zhang, H., Cui, F., Wang, L., Li, J., Ding, A., Zhao, C., Bao, Y., Yang, Q. and Wang, H., 2013. Conditional and unconditional QTL mapping of drought-tolerance-related traits of wheat seedling using two related RIL populations. *J. Genet.*, 92: 213-231.

Zietkiewicz, E., Rafalski, A. and Labuda, D., 1994. Genome fingerprinting by simple sequence repeats (SSR)-anchored PCR amplification. *Genomics*, 20: 176-183.

Index

Q

R

S

www.ingramcontent.com/pod-product-compliance
Lightning Source LLC
Chambersburg PA
CBHW050519190326
41458CB00005B/1599

9789351309482